Advanced Structured Materials

Volume 100

Series Editors

Common engineering materials reach in many applications their limits and new developments are required to fulfil increasing demands on engineering materials. The performance of materials can be increased by combining different materials to achieve better properties than a single constituent or by shaping the material or constituents in a specific structure. The interaction between material and structure may arise on different length scales, such as micro-, meso- or macroscale, and offers possible applications in quite diverse fields.

This book series addresses the fundamental relationship between materials and their structure on the overall properties (e.g. mechanical, thermal, chemical or magnetic etc) and applications.

The topics of *Advanced Structured Materials* include but are not limited to

- classical fibre-reinforced composites (e.g. glass, carbon or Aramid reinforced plastics)
- metal matrix composites (MMCs)
- micro porous composites
- micro channel materials
- multilayered materials
- cellular materials (e.g., metallic or polymer foams, sponges, hollow sphere structures)
- porous materials
- truss structures
- nanocomposite materials
- biomaterials
- nanoporous metals
- concrete
- coated materials
- smart materials

Advanced Structured Materials is indexed in Google Scholar and Scopus.

More information about this series at http://www.springer.com/series/8611

Holm Altenbach · Andreas Öchsner
Editors

State of the Art and Future Trends in Material Modeling

Editors
Holm Altenbach
Faculty of Mechanical Engineering
Otto von Guericke University Magdeburg
Magdeburg, Sachsen-Anhalt, Germany

Andreas Öchsner
Faculty of Mechanical Engineering
Esslingen University of Applied Sciences
Esslingen, Germany

ISSN 1869-8433 ISSN 1869-8441 (electronic)
Advanced Structured Materials
ISBN 978-3-030-30357-0 ISBN 978-3-030-30355-6 (eBook)
https://doi.org/10.1007/978-3-030-30355-6

This Springer imprint is published by the registered company Springer Nature Switzerland AG
The registered company address is: Gewerbestrasse 11, 6330 Cham, Switzerland

Preface

The Springer series on Advanced Structured Materials (ASM) was established in 2010 to provide a new forum to present recent results from research on the fundamental relationships between materials and their structure, and the consequences for their overall properties (e.g. mechanical, thermal, chemical or magnetic, among others) and applications. At the beginning it was planed that four volumes should be published every year within this series.

The constant need and requirement for research in this area is based on the fact that common engineering materials reach their limits for many applications, and that new developments are required to fulfil increasing demands on performance and properties, as well as further considerations such as recyclability and environmental compatibility. The performance of materials can be increased by combining different materials to achieve properties better than single constituent materials, or by shaping the materials or constituents in a specific structure. The interaction between materials and structures may arise on different length scales, such as at the micro-, meso- or macroscale, and offers possible applications in quite diverse fields. On the other hand, new or more accurate material laws allow a more efficient and safer exploitation of existing and novel materials. Many times, these goals can be only achieved in a multi-disciplinary approach.

At the beginning it was not clear what will be the format of this series: proceedings, collections of papers, monographs, among others. Now it is clear that any format was accepted by the scientific community. The main indicator (downloads) shows that there is no preferred format. Within the series some books attract a huge number of scientists world-wide, in other cases the books were published for some specialists. Today this is not a problem since the electronic copies are used by the majority of the readers.

On the occasion of the 100th volume in the Advanced Structured Materials series, we intended to compile a special anniversary monograph to celebrate the success and acceptance of this Springer book series and to give international experts a forum to showcase the actual state-of-the-art and future trends in materials modelling. Finally, 59 authors from 14 countries have submitted 20 papers.

We would like to express our sincere appreciation to all the authors and the representatives of Springer, who made this volume and the book series possible. We delightedly acknowledge Dr. Christoph Baumann (Editorial Director, Springer Publisher) for initiating the Advanced Structured Materials Series and this book project. In addition, we have to thank Dr. Mayra Castro (Senior Editor Research Publishing – Books Interdisciplinary Applied Science) and Mr. Ashok Arumairaj (Production Administrator) giving us the final support. Last but not least, the first editor has to acknowledge the Fundacja na rzecz Nauki Polskiej (Fundation for Polish Science) allowing to finalize this book at the Politechnika Lubelska (host: Prof.dr.hab.inż. Tomasz Sadowski, dr.h.c.) with the help of the Alexander von Humboldt Polish Honorary Research Felloship.

Magdeburg, Esslingen, *Holm Altenbach*
July 2019 *Andreas Öchsner*

Contents

List of Contributors

Alireza Akhavan-Safar
Instituto de Ciência e Inovação em Engenharia Mecânica e Engenharia Industrial (INEGI), Rua Dr. Roberto Frias, 4200-465 Porto, Portugal,
e-mail: akhavan101@gmail.com

Leandro Daniel Lau Alfonso
Instituto de Cibernética, Matemática y Física, ICIMAF, Calle 15 No. 551, entre C y D, Vedado, Habana 4, CP–10400, Cuba,
e-mail: leandro@icimaf.cu

Holm Altenbach
Lehrstuhl für Technische Mechanik, Institut für Mechanik, Fakultät für Maschinenbau, Otto-von-Guericke-Universität Magdeburg, Universitätsplatz 2, 39106 Magdeburg, Germany,
e-mail: holm.altenbach@ovgu.de

Marcus Aßmus
Lehrstuhl für Technische Mechanik, Institut für Mechanik, Fakultät für Maschinenbau, Otto-von-Guericke-Universität Magdeburg, Universitätsplatz 2, 39106 Magdeburg, Germany,
e-mail: marcus.assmus@ovgu.de

Emilio Barchiesi
International Research Center for the Mathematics and Mechanics of Complex Systems, University of L'Aquila, Italy,
e-mail: barchiesiemilio@gmail.com

James T. Boyle
Department of Mechanical & Aerospace Engineering, University of Strathclyde, Glasgow, Scotland, UK,
e-mail: jim.boyle@strath.ac.uk

Julián Bravo-Castillero
Instituto de Investigaciones en Matemáticas Aplicadas y en Sistemas, Universidad Nacional Autónoma de México, Alcaldía Álvaro Obregón, 01000 CDMF, México,
e-mail: julian@mym.iimas.unam.mx

Michael Brünig
Institute for Mechanics and Structural Analysis, Faculty for Civil Engineering and Environmental Science, Bundeswehr University Munich, 85577 Neubiberg, Germany,
e-mail: michael.bruenig@unibw.de

Otto T. Bruhns
Institute of Mechanics, Fakultät für Bau- und Umweltingenieurwissenschaften, Ruhr-University Bochum, 44780 Bochum, Germany,
e-mail: otto.bruhns@rub.de

Raul Duarte Salgueiral Gomes Campilho
Departamento de Engenharia Mecânica, Instituto Superior de Engenharia do Porto, Instituto Politécnico do Porto, Rua Dr. António Bernardino de Almeida, 431 4249-015 Porto, Portugal,
e-mail: rds@isep.ipp.pt

Ricardo João Camilo Carbas
Instituto de Ciência e Inovação em Engenharia Mecânica e Engenharia Industrial (INEGI), Rua Dr. Roberto Frias, 4200-465 Porto, Portugal,
e-mail: carbas@fe.up.pt

Johanna Eisenträger
Lehrstuhl für Technische Mechanik, Institut für Mechanik, Fakultät für Maschinenbau, Otto-von-Guericke-Universität Magdeburg, Universitätsplatz 2, 39106 Magdeburg, Germany,
e-mail: johanna.eisentraeger@ovgu.de

Francesco dell'Isola
International Research Center for the Mathematics and Mechanics of Complex Systems, University of L'Aquila & Dipartimento di Ingegneria Civile, Edile-Architettura e Ambientale, Università degli Studi dell'Aquila, L'Aquila, Italy,
e-mail: francesco.dellisola.me@gmail.com

Victor A. Eremeyev
Faculty of Civil and Environmental Engineering, Gdańsk University of Technology, ul. Gabriela Narutowicza 11/12, 80-233 Gdańsk, Poland,
e-mail: victor.eremeev@pg.edu.pl
Southern Federal University, Milchakova str. 8a, 344090 Rostov on Don & Southern Scientific Center of RASci, Chekhova str. 41, 344006 Rostov on Don, Russia,
e-mail: eremeyev.victor@gmail.com

Artur Ganczarski
Institute of Applied Mechanics, Faculty of Mechanical Engineering, Cracow University of Technology, al. Jana Pawła II 37, 31-864 Kraków, Poland,
e-mail: artur.ganczarski@pk.edu.pl

Steffen Gerke
Institute for Mechanics and Structural Analysis, Faculty for Civil Engineering and Environmental Science, Bundeswehr University Munich, 85577 Neubiberg, Germany,
e-mail: steffen.gerke@unibw.de

Raul Guinovart-Díaz
Facultad de Matemática y Computación, Universidad de la Habana, San Lázaro y L, Vedado, Habana 4, CP–10400, Cuba,
e-mail: guino@matcom.uh.cu

François Hild
Laboratoire de Mécanique et Technologie (LMT), ENS Paris-Saclay, CNRS, Université Paris-Saclay, 94235 Cachan Cedex, France,
e-mail: francois.hild@ens-paris-saclay.fr

Leonhard Hitzler
Institute of Materials Science and Mechanics of Materials, Department of Mechanical Engineering, Technical University of Munich, Boltzmannstrasse 15, 85748 Garching, Germany,
e-mail: hitzler@wkm.mw.tum.de

Elena A. Ivanova
Peter the Great St. Petersburg Polytechnic University, Department of Theoretical Mechanics, Institute of Applied Mathematics and Mechanics, Politechnicheskaya 29, 195251 St. Petersburg & Laboratory of Mechatronics, Institute for Problems in Mechanical Engineering of Russian Academy of Sciences, Bolshoy pr. V.O. 61, 199178 St. Petersburg, Russia,
e-mail: elenaivanova239@gmail.com

Takeshi Iwamoto
Academy of Science and Technology, Hiroshima University, 1-4-1 Kagamiyama, Higashi-Hiroshima, Hiroshima, 739 - 8527 Japan,
e-mail: iwamoto@mec.hiroshima-u.ac.jp

Michael Johlitz
Institute of Mechanics, Faculty for Aerospace Engineering, Bundeswehr University Munich, Germany,
e-mail: michael.johlitz@unibw.de

Vladimir A. Kolupaev
Fraunhofer Institute for Structural Durability and System Reliability (LBF), Schloßgartenstr. 6, 64289 Darmstadt, Germany,
e-mail: Vladimir.Kolupaev@lbf.fraunhofer.de

Anton M. Krivtsov
Peter the Great St. Petersburg Polytechnic University, Department of Theoretical Mechanics, Institute of Applied Mathematics and Mechanics, Politechnicheskaya 29, 195251 St. Petersburg & Laboratoty "Discrete models in mechanics", Institute for Problems in Mechanical Engineering of Russian Academy of Sciences, Bolshoy pr. V.O. 61, 199178 St. Petersburg, Russia,
e-mail: akrivtsov@bk.ru

Vitaly A. Kuzkin
Peter the Great St. Petersburg Polytechnic University, Department of Theoretical Mechanics, Institute of Applied Mathematics and Mechanics, Politechnicheskaya 29, 195251 St. Petersburg & Laboratoty "Discrete models in mechanics", Institute for Problems in Mechanical Engineering of Russian Academy of Sciences, Bolshoy pr. V.O. 61, 199178 St. Petersburg, Russia,
e-mail: kuzkinva@gmail.com

Ruslan L. Lapin
Peter the Great St. Petersburg Polytechnic University, Department of Theoretical Mechanics, Institute of Applied Mathematics and Mechanics, Politechnicheskaya 29, 195251 St. Petersburg, Russia,
e-mail: lapruslan@gmail.com

Frédéric Lebon
Aix-Marseille University, CNRS, Centrale Marseille, LMA, 4 Impasse Nikola Tesla, CS 40006, 13453 Marseille Cedex 13, France,
e-mail: lebon@lma.cnrs-mrs.fr

Alexander Lion
Institute of Mechanics, Faculty for Aerospace Engineering, Bundeswehr University Munich, Germany,
e-mail: alexander.lion@unibw.de

Eduardo André de Sousa Marques
Instituto de Ciência e Inovação em Engenharia Mecânica e Engenharia Industrial (INEGI), Rua Dr. Roberto Frias, 4200-465 Porto, Portugal,
e-mail: emarques@fe.up.pt

Dmitry V. Matias
Peter the Great St. Petersburg Polytechnic University, Department of Theoretical Mechanics, Institute of Applied Mathematics and Mechanics, Politechnicheskaya, 29, 195251, St. Petersburg, Russia,
e-mail: dvmatyas@gmail.com

Markus Merkel
Institute for Virtual Product Development, Aalen University of Applied Sciences, 73430 Aalen, Germany,
e-mail: markus.merkel@hs-aalen.de

Wolfgang H. Müller
Institut für Mechanik, Kontinuumsmechanik und Materialtheorie, Technische Universität Berlin, Sek. MS. 2, Einsteinufer 5, 10587 Berlin, Germany,
e-mail: wolfgang.h.mueller@tu-berlin.de

Nikita D. Muschak
Peter the Great St. Petersburg Polytechnic University, Department of Theoretical Mechanics, Institute of Applied Mathematics and Mechanics, Politechnicheskaya 29, 195251 St. Petersburg, Russia,
e-mail: niky-m@yandex.ru

Andreas Öchsner
Faculty of Mechanical Engineering, Esslingen University of Applied Sciences, Kanalstrasse 33, 73728 Esslingen, Germany,
e-mail: andreas.oechsner@hs-esslingen.de

Jose A. Otero
Instituto Tecnológico de Estudios Superiores de Monterrey, CEM, E.M. CP 52926, México,
e-mail: j.a.otero@tec.mx

Mariana Paulino
School of Engineering, Faculty of Science, Engineering and Built Environment, Deakin University, Geelong, Australia,
e-mail: Mariana.Paulino@deakin.edu.au

Wilhelm Rickert
Institut für Mechanik, Kontinuumsmechanik und Materialtheorie, Technische Universität Berlin, Sek. MS. 2, Einsteinufer 5, 10587 Berlin, Germany,
e-mail: rickert@tu-berlin.de

Reinaldo Rodríguez-Ramos
Facultad de Matemática y Computación, Universidad de la Habana, San Lázaro y L, Vedado, Habana 4, CP–10400, Cuba & Visiting Professor from February 15 to August 15, 2019 at Instituto de Investigaciones en Matemáticas Aplicadas y en Sistemas, Universidad Nacional Autónoma de México, Apartado Postal 20-126, Delegación de Álvaro Obregón, 01000 CDMF, México,
e-mail: reinaldo@matcom.uh.cu

Philipp L. Rosendahl
Institute of Structural Mechanics, Department of Mechanical Engineering, Technische Universität Darmstadt, Darmstadt, Germany,
e-mail: rosendahl@fsm.tu-darmstadt.de

Federico J. Sabina
Instituto de Investigaciones en Matemáticas Aplicadas y en Sistemas, Universidad Nacional Autónoma de México, Alcaldía Álvaro Obregón, 01000 CDMF, México,
e-mail: fjs@mym.iimas.unam.mx

Jonas Schröder
Institute of Mechanics, Faculty for Aerospace Engineering, Bundeswehr University Munich, Germany,
e-mail: jonas.schroeder@unibw.de

Enes Sert
Faculty of Mechanical Engineering, Esslingen University of Applied Sciences, Kanalstrasse 33, 73728 Esslingen, Germany,
e-mail: enes.sert@hs-esslingen.de

Lucas Filipe Martins da Silva
Departamento de Engenharia Mecânica, Faculdade de Engenharia (FEUP), Universidade do Porto, Rua Dr. Roberto Frias, 4200-465 Porto, Portugal,
e-mail: lucas@fe.up.pt

Alexey V. Shutov
Lavrentyev Institute of Hydrodynamics, pr. Lavrentyeva 15 & Novosibirsk State University, ul. Pirogova 1, 630090, Novosibirsk, Russia,
e-mail: alexey.v.shutov@gmail.com

Mario Spagnuolo
International Research Center for the Mathematics and Mechanics of Complex Systems, University of L'Aquila, Italy,
e-mail: mario.spagnuolo.memocs@gmail.com

Igor I. Tagiltsev
Lavrentyev Institute of Hydrodynamics, pr. Lavrentyeva 15 & Novosibirsk State University, ul. Pirogova 1, 630090, Novosibirsk, Russia,
e-mail: i.i.tagiltsev@gmail.com

Filipe Teixeira-Dias
School of Engineering, The University of Edinburgh, Edinburgh, United Kingdom,
e-mail: F.Teixeira-Dias@ed.ac.uk

Samuel Thompson
School of Engineering, The University of Edinburgh, Edinburgh, United Kingdom,
e-mail: Samuel.Thompson@ed.ac.uk

Chuong Anthony Tran
International Research Center for the Mathematics and Mechanics of Complex Systems, University of L'Aquila, Italy,
e-mail: tcanth@outlook.com

Truong Duc Trinh
Graduate School of Engineering, Hiroshima University, 1-4-1 Kagamiyama, Higashi-Hiroshima, Hiroshima, 739 - 8527 Japan,
e-mail: d184860@hiroshima-u.ac.jp

Vadim A. Tsaplin
Peter the Great St. Petersburg Polytechnic University, Department of Theoretical Mechanics, Institute of Applied Mathematics and Mechanics, Politechnicheskaya 29, 195251 St. Petersburg & Laboratoty "Discrete models in mechanics", Institute for Problems in Mechanical Engineering of Russian Academy of Sciences, Bolshoy pr. V.O. 61, 199178 St. Petersburg, Russia,
e-mail: vtsaplin@yandex.ru

Elena Vilchevskaya
Peter the Great St. Petersburg Polytechnic University, Department of Theoretical Mechanics, Institute of Applied Mathematics and Mechanics, Politechnicheskaya 29, 195251 St. Petersburg & Laboratory of Mathematical Methods in Mechanics of Materials, Institute for Problems in Mechanical Engineering of Russian Academy of Sciences, Bolshoy pr. V.O. 61, 199178 St. Petersburg, Russia,
e-mail: vilchevska@gmail.com

Si-Yu Wang
School of Mechanics and Construction Engineering and MOE Key Lab of Disaster Forecast and Control in Engineering, Jinan University, 510632 Guangzhou, China,
e-mail: siyu0904@foxmail.com

Ewald Werner
Institute of Materials Science and Mechanics of Materials, Institute of Materials Science and Mechanics of Materials, Technical University of Munich Technical University of Munich, Boltzmannstrasse 15, 85748 Garching, Germany,
e-mail: hitzler@wkm.mw.tum.de, werner@wkm.mw.tum.de

Hui-Feng Xi
School of Mechanics and Construction Engineering and MOE Key Lab of Disaster Forecast and Control in Engineering, Jinan University, 510632 Guangzhou, China,
e-mail: xihuifeng@jnu.edu.cn

Heng Xiao
School of Mechanics and Construction Engineering and MOE Key Lab of Disaster Forecast and Control in Engineering, Jinan University, 510632 Guangzhou & Shanghai Institute of Applied Mathematics and Mechanics & School of Mechanics and Engineering Science, Shanghai University, 200072 Shanghai, China,
e-mail: xiaoheng@shu.edu.cn

Mustafa Erden Yildizdag
Department of Naval Architecture and Ocean Engineering, Istanbul Technical University, 34469, Maslak, Istanbul, Turkey & International Research Center for the Mathematics and Mechanics of Complex Systems, University of L'Aquila, Italy,
e-mail: yildizdag@itu.edu.tr

Lin Zhan
School of Mechanics and Construction Engineering and MOE Key Lab of Disaster Forecast and Control in Engineering, Jinan University, 510632 Guangzhou, China,
e-mail: 14110290008@fudan.edu.cn

Moritz Zistl
Institute for Mechanics and Structural Analysis, Faculty for Civil Engineering and Environmental Science, Bundeswehr University Munich, 85577 Neubiberg, Germany,
e-mail: moritz.zistl@unibw.de

Chapter 1
On Viscoelasticity in the Theory of Geometrically Linear Plates

Marcus Aßmus and Holm Altenbach

Abstract A phenomenological theory for viscoelastic plates is developed in a geometrically linear framework whereby present work is based on the direct approach for homogeneous plates. We confine our research to isotropic viscoelastic materials, assume stiffness laws by means of rheology, and generalize them in order to describe the behavior of shear-deformable thin-walled structures. The restriction to isotropy enables to utilize eigenspace projectors since stiffness tensors are coaxial in this special case. It is thus possible to formulate the system of tensor-valued differential equations in orthogonal subspaces and to simplify the calculation rules like those for scalar-valued exprcssions. The resulting behavior is illustrated exemplary by means of uniaxial tests. We furthermore provide information on material parameter determination.

Keywords: Plate theory · Viscelasticity · Rheology · Isotropy · Eigenspace projections

1.1 Introduction

1.1.1 Motivation

The application of elasticity theory for problems at thin-walled structural elements that are made of polymeric materials seems to be detrimental. In general, pronounced viscoelastic phenomena can be observed when deforming rubber, caoutchouc, and liquid or solid plastics (Ferry, 1980). The associated inelastic material behavior is called rate-dependent or rheonomous. To an increasing extent, these materials are

---

Marcus Aßmus · Holm Altenbach
Institut für Mechanik, Fakultät für Maschinenbau, Otto-von-Guericke-Universität Magdeburg, Universitätsplatz 2, 39106 Magdeburg, Germany,
e-mail: marcus.assmus@ovgu.de, holm.altenbach@ovgu.de

H. Altenbach and A. Öchsner (eds.), *State of the Art and Future Trends in Material Modeling*, Advanced Structured Materials 100,
https://doi.org/10.1007/978-3-030-30355-6_1

also being used in thin-walled structural elements like plates, shells and composite structures. The theories are logically based on classical continuum theory and derived for the special structural element (Naghdi, 1972). Viscoelastic problems, however, can be solved by two conceptually different approaches.

- one-dimensional viscoelastic equations based on kinematics
- tensor function representations

The latter is phenomenological but simpler in the context of computational handling and has a wider range of applications additionally. In the literature, different models are available in the context of the methodology (Gurtin and Sternberg, 1962; Mālmeisters et al, 1977):

- viscoelastic models of differential-type
- viscoelastic models of integral-type

However, we are not going to have such a strict separation here and leave such classifications to historians in this field.

In the context of phenomenological rheology, a huge number of circuitries have been established at least for one-dimensional representations. Three-dimensional generalizations are widespread in engineering sciences, with isotropic viscoelastic behavior in the foreground. Contrary to classical procedures based on a three-dimensional parent-continuum and a degeneration and condensation of the constitutive equations, it is also possible to introduce viscoelasticity *ab initio* at dimensionally reduced continua, at least in the context of plates.

An interesting concept for plates is the direct approach for homogeneous and inhomogeneous in the thickness direction plates (Altenbach, 1988). It is possible there to introduce the topic of viscoelasticity directly, also. To date, however, a direct introduction of viscoelasticity in the most natural way is absent. Thereby, it is much clearer to distinguish between elastic and viscous deformations while simultaneously considering dilatoric and deviatoric ones with respect to all deformation states admissible. The present work can be classified in this area while we enter the topic for the most simple plate, i.e. isotropic and decoupled. However, we don not want to lapse into details concerning the rheological circuits. Rather, we would like to emphasize the clear physical interpretation and the straightforward algebraic handling. Hereby, we preserve the mathematical structure of the resulting equations, as they are known from a one-dimensional theory. This guarantees a clear and elegant presentation. Generally, this paper offers an outline of the linear theory of viscoelasticity for plates. At the same time, however, it offers an initial introduction to the procedure presented here.

1.1.2 Organisation of the Paper

In context of a slender three-dimensional body $\mathfrak{B}$ in the three-dimensional Euclidean space $\mathbb{E}^3$, we operate on the two-dimensional mid-surface $\mathfrak{S}$, solely. Referring to an orthonormal basis $\{\boldsymbol{e}_i\}\ \forall i = \{1, 2, 3\}$ one can describe this manifold within the volume V occupied by $\mathfrak{B}$ as follows (Aßmus et al, 2020).

$$V = \left\{ X_i \in \mathfrak{B} \subset \mathbb{E}^3 : X_\alpha \in \mathfrak{S} \subset \mathbb{E}^2, X_3 \in \left[-\frac{h}{2}, +\frac{h}{2} \right] \right\} \tag{1.1}$$

Herein $\mathbb{E}^2$ is the two-dimensional subspace which is sufficient for the geometric description of $\mathfrak{S}$. However, in the context of a plane two-dimensional manifold, one can introduce the orthonormal vector-basis $\{\boldsymbol{e}_\alpha, \boldsymbol{n}\}\ \forall \alpha = \{1, 2\}$ which is more convenient in plate theory. This resulting two-dimensional mechanical problem will be introduced *ab initio* in present treatise, i.e. without any derivation.

At first, we recall some basic relationships which are necessary for the subsequent considerations. Initially, we introduce the degrees of freedom, the geometry of deformation, conjugate kinetic quantities, and constitutive relations in the case of isotropic linear elasticity. The latter are postulated in a special form.

Thereafter, we directly enter the topic of viscoelasticity. Our starting point are uniaxial rheological models. After introducing their idiosyncrasies at two-dimensional body manifolds, we present applications of classical circuitries. We limit our attention initially to continuous histories and assume (without essential loss of generality) that the medium is in its undeformed state for all times $\tau \in (-\infty, 0)$. The tensorial generalization is thereby derived by a projection methodology. Two briefly described problems are formulated, and both are more or less solved. Thereby we present the mechanical behavior for all deformation states considered. Finally, we give hints for the identification of material parameters.

1.1.3 Preliminaries and Notation

In present work we make use of the direct tensor calculus. We tend to use this mathematical tool in the sense of a language for everyday use rather than as a body of theorems and proofs. Nevertheless, a few arrangements have to be made.

Tensors of zeroth-order (or scalars) are symbolised by italic letters (e.g. a or α), italic lowercase bold letters denote first-order tensors (or monads) (e.g. $\boldsymbol{a} = a_i\, \boldsymbol{e}_i$), second-order tensors (or dyads) are designated by italic uppercase bold letters (e.g. $\boldsymbol{A} = A_{lm}\, \boldsymbol{e}_l \otimes \boldsymbol{e}_m$), and fourth-order tensors (or tetrads) are symbolised by italic uppercase bold calligraphic letters (e.g. $\boldsymbol{\mathcal{A}} = A_{stuv}\, \boldsymbol{e}_s \otimes \boldsymbol{e}_t \otimes \boldsymbol{e}_u \otimes \boldsymbol{e}_v$), whereas Einstein sum convention is applied. Latin indices run through the values 1, 2, and 3, while Greek indices run through the values 1 and 2.

Essential operations especially for polyadics used in present manuscript are the scalar product $\boldsymbol{a} \cdot \boldsymbol{b} = \alpha$ (with $\alpha \in \mathbb{R}$), the cross product $\boldsymbol{a} \times \boldsymbol{b} = \boldsymbol{c}$, the dyadic product $\boldsymbol{a} \otimes \boldsymbol{b} = \boldsymbol{C}$, the composition of a second and a first-order tensor $\boldsymbol{A} \cdot \boldsymbol{a} = \boldsymbol{d}$, the composition of two second-order tensors $\boldsymbol{A} \cdot \boldsymbol{B} = \boldsymbol{D}$, the cross product between a second and a first-order tensor $\boldsymbol{A} \times \boldsymbol{b} = \boldsymbol{G}$, the double scalar product between two second-order tensors $\boldsymbol{A} : \boldsymbol{B} = \gamma$, the double scalar product between a fourth and a second-order tensor $\mathcal{A} : \boldsymbol{B} = \boldsymbol{F}$, and the fourfold scalar product between two fourth-order tensors $\mathcal{A} :: \mathcal{B} = \omega$. In context of the cross product introduced, the permutation symbol ϵ_{ijk} is needed. In three dimensions it is determined via the triple product of three orthogonal unit vectors.

$$\epsilon_{ijk} = \boldsymbol{e}_i \cdot (\boldsymbol{e}_j \times \boldsymbol{e}_k) \quad \begin{cases} +1 & \text{if } (i,j,k) \text{ is an even permutation of } (1,2,3) \\ -1 & \text{if } (i,j,k) \text{ is an odd permutation of } (1,2,3) \\ 0 & \text{if } (i,j,k) \text{ is not a permutation of } (1,2,3) \end{cases} \tag{1.2}$$

Above introduction is based on a Cartesian coordinate system and orthonormal bases, e.g. $\{\boldsymbol{e}_i\}$. We furthermore make use of the two-dimensional Hamiltonian (Nabla operator). This vector valued operator is defined as.

$$\boldsymbol{\nabla} = \boldsymbol{e}_\alpha \frac{\partial \ldots}{\partial X_\alpha}$$

It is used to determine the divergence $(\boldsymbol{\nabla} \cdot \square)$ and the gradient $(\boldsymbol{\nabla} \square)$ of a tensor $\square$ of any order. Furthermore,

$$\boldsymbol{\nabla}^{\text{sym}} \square = \frac{1}{2} [\boldsymbol{\nabla} \square + \boldsymbol{\nabla}^\top \square]$$

is the symmetric part of the gradient $\boldsymbol{\nabla} \square$. The transposed gradient is defined as $\boldsymbol{\nabla}^\top \square = [\boldsymbol{\nabla} \square]^\top$. The transposition of a tensor is given by $\boldsymbol{a} \cdot \boldsymbol{A}^\top \cdot \boldsymbol{b} = \boldsymbol{b} \cdot \boldsymbol{A} \cdot \boldsymbol{a}$. For detailed penetrations of these operations we refer to e.g. Šilhavý (1997).

The Frobenius norm of fourth- and second-order tensors is determined as follows.

$$\|\mathcal{A}\| = [\mathcal{A} :: \mathcal{A}]^{\frac{1}{2}} \qquad \|\boldsymbol{A}\| = [\boldsymbol{A} : \boldsymbol{A}]^{\frac{1}{2}} \tag{1.3}$$

In present context we also need invariants of second-order tensors. There are principal and main invariants. However, since we are working on two-dimensional body-manifolds, the number of invariants differs slightly in contrast to the classical standpoint. In view that we are interested in symmetric tensors only, the number of independent main invariants is identical to the number of principle invariants. The two invariants of a tensor $\boldsymbol{A} \in \mathbb{E}^2$ are defined by

$$\mathit{In}_I^{\boldsymbol{A}} = \operatorname{tr} \boldsymbol{A} \tag{1.4a}$$

$$\mathit{In}_{II}^{\boldsymbol{A}} = \det \boldsymbol{A} \tag{1.4b}$$

while $\operatorname{tr} \boldsymbol{A} = \boldsymbol{A} \cdot \boldsymbol{P}$ is the trace and $\det \boldsymbol{A} = \prod \lambda_\alpha^{\boldsymbol{A}}$ is the determinant of a second-order tensor $\boldsymbol{A}$. Herein, $\lambda_\alpha^{\boldsymbol{A}}$ are the eigenvalues of $\boldsymbol{A}$, determined by the eigenvalue problem $\boldsymbol{A} \cdot \boldsymbol{a} = \lambda_\alpha \boldsymbol{a}$. Furthermore, $\boldsymbol{P} = \boldsymbol{e}_\alpha \otimes \boldsymbol{e}_\alpha$ is the first metric tensor of a planar two-dimensional body manifold. A tensor of first order $\boldsymbol{a}$ features only one invariant, namely its square ($\mathit{In}^{\boldsymbol{a}} = \boldsymbol{a} \cdot \boldsymbol{a}$) (Schade and Neemann, 2009).

1.2 Linear Elastic Background

First of all, it is sensible to gather a few classical formula of linear isotropic elastic plates in a clear geometric form. Thereby, we follow the ideas of the basic works from Kirchhoff (1850); Reissner (1945), and Mindlin (1951) while we here introduce the subject in a direct manner (Zhilin, 1976).

We consider a plane material surface $\mathfrak{S}$ which is endowed with five kinematic degrees of freedom, cf. Fig. 1.1. These are three translations $\boldsymbol{a}$ and two rotations $\boldsymbol{\varphi}$.

$$\boldsymbol{a} = \boldsymbol{v} + w\boldsymbol{n} \qquad \text{with } \boldsymbol{v} = v_\alpha \boldsymbol{e}_\alpha \tag{1.5a}$$

$$\boldsymbol{\varphi} = \varphi_\alpha \boldsymbol{e}_\alpha \tag{1.5b}$$

Therein, the translations were splitted additively into an in-plane $\boldsymbol{v}$ and an out-of-plane part w. The rotations $\boldsymbol{\varphi}$ arise from the more physical Mindlinean definition

$$\boldsymbol{\psi} = \varphi_1 \boldsymbol{e}_2 - \varphi_2 \boldsymbol{e}_1$$

via

$$\boldsymbol{\varphi} = \boldsymbol{\psi} \times \boldsymbol{n}$$

We are working with the following set of reduced deformation measures.

$$\boldsymbol{G} = \boldsymbol{\nabla}^{\mathrm{sym}} \boldsymbol{v} \tag{1.6a}$$

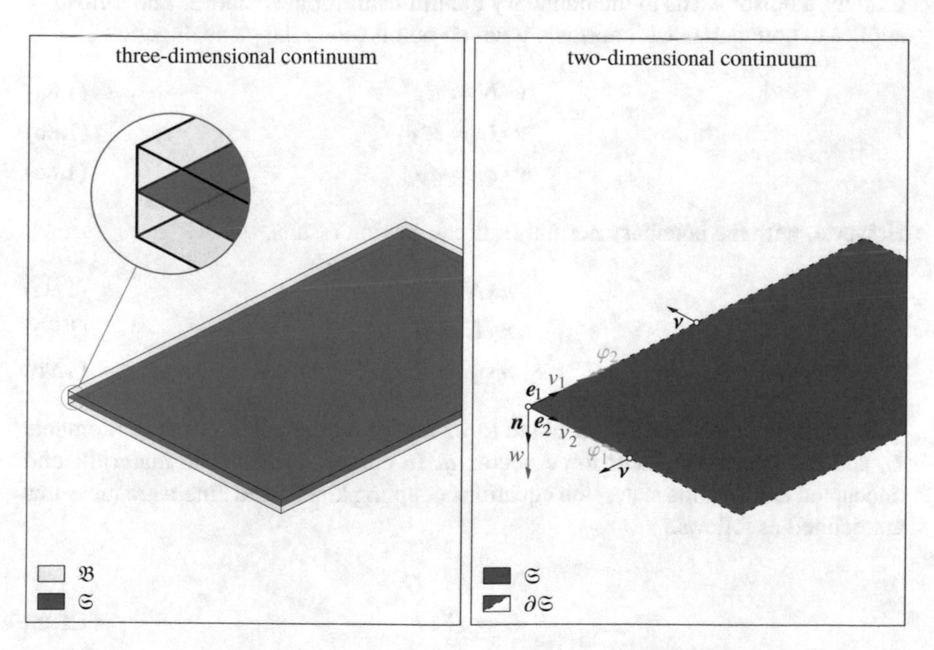

Fig. 1.1 Plane surface continuum in context of a three-dimensional slender body

$$\boldsymbol{K} = \nabla^{\mathrm{sym}} \boldsymbol{\varphi} \tag{1.6b}$$

$$\boldsymbol{g} = \nabla w + \boldsymbol{\varphi} \tag{1.6c}$$

Herein, $\boldsymbol{G}$ denotes plane normal and plane shear strains, $\boldsymbol{K}$ normal curvature changes due to bending and torsion, and $\boldsymbol{g}$ transverse shear strains. Obviously, all three deformation measures are introduced in analogy to deformation states usual in engineering. To be exact, these are the in-plane state, the out-of-plane state, and the transverse shear state.

We introduce dual kinetic measures resulting from principle of cuts. Boundary measures are defined by forces and moments acting at the surface. Thereby, we make use of tangential forces $\boldsymbol{s}_{\mathfrak{S}}$, orthogonal forces $p_{\mathfrak{S}}$, and out-of-plane moments $\boldsymbol{m}_{\mathfrak{S}}$.

$$\boldsymbol{n}_{\boldsymbol{\nu}} = \lim_{\Delta L \to 0} \frac{\Delta \boldsymbol{s}_{\mathfrak{S}}}{\Delta L} \tag{1.7a}$$

$$\boldsymbol{m}_{\boldsymbol{\nu}} = \lim_{\Delta L \to 0} \frac{\Delta (\boldsymbol{m}_{\mathfrak{S}} \times \boldsymbol{n})}{\Delta L} \tag{1.7b}$$

$$q_{\boldsymbol{\nu}} = \lim_{\Delta L \to 0} \frac{\Delta p_{\mathfrak{S}}}{\Delta L} \tag{1.7c}$$

Here, L is a plane measure. The vectors and the scalar of the left-hand sides indicate the boundary resultants of the in-plane state $\boldsymbol{n}_{\boldsymbol{\nu}}$, the out-of-plane state $\boldsymbol{m}_{\boldsymbol{\nu}}$ and the transverse shear state $q_{\boldsymbol{\nu}}$. The orientation of a cut is determined by the corresponding normal. Here, we use the boundary normals $\boldsymbol{n}$ and $\boldsymbol{\nu}$, while $\boldsymbol{n} \cdot \boldsymbol{\nu} = 0$ holds. Following Cauchy, a tensor exists to the boundary quantities introduced above. The following applies to boundaries with normals $\boldsymbol{\nu}$ which points along the plane directions.

$$\boldsymbol{\nu} \cdot \boldsymbol{N} = \boldsymbol{n}_{\boldsymbol{\nu}} \tag{1.8a}$$

$$\boldsymbol{\nu} \cdot \boldsymbol{L} = \boldsymbol{m}_{\boldsymbol{\nu}} \tag{1.8b}$$

$$\boldsymbol{\nu} \cdot \boldsymbol{q} = q_{\boldsymbol{\nu}} \tag{1.8c}$$

However, with the boundary normal $\boldsymbol{n}$, the following results.

$$\boldsymbol{n} \cdot \boldsymbol{N} = \boldsymbol{0} \tag{1.9a}$$

$$\boldsymbol{n} \cdot \boldsymbol{L} = \boldsymbol{0} \tag{1.9b}$$

$$\boldsymbol{n} \cdot \boldsymbol{q} = 0 \tag{1.9c}$$

The resulting tensors are the in-plane force tensor $\boldsymbol{N}$, the polar tensor of moments $\boldsymbol{L}$, and the transverse shear force vector $\boldsymbol{q}$. In context of isotropic materials and uncoupled deformation states, the equations coupling kinetic and kinematic measures are defined as follows.

$$\boldsymbol{N} = \boldsymbol{\mathcal{A}} : \boldsymbol{G} \tag{1.10a}$$

$$\boldsymbol{L} = \boldsymbol{\mathcal{D}} : \boldsymbol{K} \tag{1.10b}$$

$$\boldsymbol{q} = \boldsymbol{Z} \cdot \boldsymbol{g} \tag{1.10c}$$

Here, the constitutive measures can be given by the aid of the projector representation (Aßmus et al, 2017) whereby this idea can be traced back to Lord Kelvin (Thomson, 1878). With respect to the deformation states we get the in-plane stiffness tensor $\mathcal{A}$, the plate stiffness tensor $\mathcal{D}$, and the transverse shear stiffness tensor $\boldsymbol{Z}$.

$$\mathcal{A} = \lambda_J^{\mathcal{A}} \mathcal{P}_J \tag{1.11a}$$

$$\mathcal{D} = \lambda_J^{\mathcal{D}} \mathcal{P}_J \tag{1.11b}$$

$$\boldsymbol{Z} = \lambda^{\boldsymbol{Z}} \boldsymbol{P} \tag{1.11c}$$

Generally, for the the subscript index $J \in \{1, \ldots, N\}$ holds. In the case of isotropy we can restrict our presentation to $J \in \{1,2\}$. The fourth-order isotropic projectors are given as follows (Rychlewski, 1995; Aßmus et al, 2017).

$$\mathcal{P}_1 = \frac{1}{2} \boldsymbol{P} \otimes \boldsymbol{P} \qquad \mathcal{P}_2 = \mathcal{P}^{\text{sym}} - \mathcal{P}_1 \qquad \boldsymbol{P} = \boldsymbol{e}_\alpha \otimes \boldsymbol{e}_\alpha \tag{1.12}$$

Herein,

$$\mathcal{P}^{\text{sym}} = \frac{1}{2} \left(\boldsymbol{e}_\alpha \otimes \boldsymbol{e}_\beta \otimes \boldsymbol{e}_\alpha \otimes \boldsymbol{e}_\beta + \boldsymbol{e}_\alpha \otimes \boldsymbol{e}_\beta \otimes \boldsymbol{e}_\beta \otimes \boldsymbol{e}_\alpha \right)$$

is the symmetric part of the fourth-order identity of the surface. The fourth-order eigenprojectors are used to determine the dilatoric and deviatoric portions of a second-order tensor, cf. Aßmus et al (2017).

$$\mathcal{P}_1 : \boldsymbol{A} = \boldsymbol{A}^{\text{dil}} \qquad \mathcal{P}_2 : \boldsymbol{A} = \boldsymbol{A}^{\text{dev}} \qquad \boldsymbol{A} = \boldsymbol{A}^{\text{dil}} + \boldsymbol{A}^{\text{dev}} \tag{1.13}$$

Herein $\boldsymbol{A} = A_{\alpha\beta} \boldsymbol{e}_\alpha \otimes \boldsymbol{e}_\beta$ is chosen arbitrary. The eigenvalues

$$\lambda_J^{\square} \quad \forall J \in \{1,2\} \quad \wedge \quad \square \in \{\mathcal{A}, \mathcal{D}\}$$

and $\lambda^{\boldsymbol{Z}}$ are determined by following operations.

$$\lambda_J^{\mathcal{A}} = \mathcal{A} :: \frac{\mathcal{P}_J}{||\mathcal{P}_J||^2} \tag{1.14a}$$

$$\lambda_J^{\mathcal{D}} = \mathcal{D} :: \frac{\mathcal{P}_J}{||\mathcal{P}_J||^2} \tag{1.14b}$$

$$\lambda^{\boldsymbol{Z}} = \boldsymbol{Z} : \frac{\boldsymbol{P}}{||\boldsymbol{P}||^2} = \boldsymbol{Z} : \boldsymbol{P} \tag{1.14c}$$

This results in a set of five isotropic eigenvalues for $\mathcal{A}$, $\mathcal{D}$, and $\boldsymbol{Z}$.

$$\lambda_1^{\mathcal{A}} = \frac{Yh}{1-\nu} = 2Bh \qquad \lambda_2^{\mathcal{A}} = \frac{Yh}{1+\nu} = 2Gh \tag{1.15a}$$

$$\lambda_1^{\mathcal{D}} = \frac{Yh^3}{12(1-\nu)} = 2B\frac{h^3}{12} \qquad \lambda_2^{\mathcal{D}} = \frac{Yh^3}{12(1+\nu)} = 2G\frac{h^3}{12} \tag{1.15b}$$

$$\lambda^{\mathbf{Z}} = \frac{\kappa Y h}{2(1+\nu)} = \kappa G h \tag{1.15c}$$

For every fourth-order stiffness tensor, two distinct eigenvalues result, while the second-order stiffness tensor exhibits only one eigenvalue. The eigenvalues are not independent since

$$\lambda_J^{\mathcal{D}} = \frac{h^2}{12} \lambda_J^{\mathcal{A}}$$

holds (Aßmus et al, 2017). To be exact, the engineering parameters Young's modulus Y and Poisson's ratio ν as well as shear G and bulk modulus B are correlated as follows.

$$\begin{aligned}
Y &= \frac{2}{h} \frac{\lambda_1^{\mathcal{A}} \lambda_2^{\mathcal{A}}}{\lambda_1^{\mathcal{A}} + \lambda_2^{\mathcal{A}}} = \frac{24}{h^3} \frac{\lambda_1^{\mathcal{D}} \lambda_2^{\mathcal{D}}}{\lambda_1^{\mathcal{D}} + \lambda_2^{\mathcal{D}}} \\
&= \frac{2\lambda^{\mathbf{Z}}}{\kappa h} \left[1 + \frac{\lambda_1^{\square} - \lambda_2^{\square}}{\lambda_1^{\square} + \lambda_2^{\square}} \right] \qquad \forall \square \in \{\mathcal{A}, \mathcal{D}\} && (1.16a) \\
\nu &= \frac{\lambda_1^{\square} - \lambda_2^{\square}}{\lambda_1^{\square} + \lambda_2^{\square}} \qquad \forall \square \in \{\mathcal{A}, \mathcal{D}\} && (1.16b) \\
B &= \frac{1}{2h} \lambda_1^{\mathcal{A}} = \frac{12}{2h^3} \lambda_1^{\mathcal{D}} && (1.16c) \\
G &= \frac{1}{2h} \lambda_2^{\mathcal{A}} = \frac{12}{2h^3} \lambda_2^{\mathcal{D}} = \frac{1}{\kappa h} \lambda^{\mathbf{Z}} && (1.16d)
\end{aligned}$$

Herein we of course make use of the bulk modulus of the surface

$$B = \frac{Y}{2(1-\nu)},$$

cf. Aßmus et al (2017). Furthermore, $0 < \kappa \leq 1$ is a correction factor accounting for transverse shear, cf. Reissner (1945); Mindlin (1951). To correlate the material parameters introduced by eigenvalues of stiffness tensors to classical stiffnesses of plate theories, we can give following relations.

$$D_{\mathrm{M}} = \frac{1}{2} \left[\lambda_1^{\mathcal{A}} + \lambda_2^{\mathcal{A}} \right] \qquad D_{\mathrm{B}} = \frac{1}{2} \left[\lambda_1^{\mathcal{D}} + \lambda_2^{\mathcal{D}} \right] \qquad D_{\mathrm{S}} = \lambda^{\mathbf{Z}} \tag{1.17}$$

Here, D_{M} is the in-plane stiffness, D_{B} is the bending stiffness, and D_{S} is the transverse shear stiffness. The relation

$$D_{\mathrm{B}} = \frac{h^2}{12} D_{\mathrm{M}}$$

holds true. It is also worth mentioning that

$$\mathcal{D} = \frac{h^2}{12} \mathcal{A}$$

applies.

By the aid of the projectors presented in Eq. (1.12) we can also write our constitutive equations Eqs. (1.10a)–(1.10c) in terms of the eigenvalues of corresponding stiffnesses, cf. Eqs. (1.11a)–(1.11c). Furthermore, we can use these projectors to map the kinetic and deformations measures into their eigenspaces.

$$\boldsymbol{N}_J = \boldsymbol{\mathcal{P}}_J : \boldsymbol{N} \qquad \boldsymbol{G}_J = \boldsymbol{\mathcal{P}}_J : \boldsymbol{G} \tag{1.18a}$$

$$\boldsymbol{L}_J = \boldsymbol{\mathcal{P}}_J : \boldsymbol{L} \qquad \boldsymbol{K}_J = \boldsymbol{\mathcal{P}}_J : \boldsymbol{K} \tag{1.18b}$$

$$\boldsymbol{q}_J = \boldsymbol{P} \quad \cdot \boldsymbol{q} = \boldsymbol{q} \qquad \boldsymbol{g}_J = \boldsymbol{P} \quad \cdot \boldsymbol{g} = \boldsymbol{g} \tag{1.18c}$$

As introduced in Eq. (1.13), if $J \triangleq 1$, the dilatoric portion, and if $J \triangleq 2$, the deviatoric portion results. Obviously, $\boldsymbol{q}$ and $\boldsymbol{g}$ are idempotent. Both measures are localised in the deviatoric eigenspace solely. However, in the course of this procedure, we can also write the constitutive laws in the most natural way.

$$\boldsymbol{N} = \lambda_J^{\mathcal{A}} \boldsymbol{G}_J \tag{1.19a}$$

$$\boldsymbol{L} = \lambda_J^{\mathcal{D}} \boldsymbol{K}_J \tag{1.19b}$$

$$\boldsymbol{q} = \lambda^{\boldsymbol{Z}} \boldsymbol{g} \tag{1.19c}$$

The power of the projector representation introduced here is that we can reduce tensorial stiffness measures to scalar ones. This fact will help us to achieve a clear representation for what follows.

1.3 Constitutive Models for Linear Viscoelasticity

1.3.1 Basic Elements of Rheological Circuits

To describe viscoelastic material behavior in context of rheology, two different basic elements are required which will be introduced in the sequel.

The pure elastic material is symbolized by a spring. In present context we reduce ourselves to linear elastic behavior, so that it is sufficient to utilize a linear spring. This basic element is visualized in Fig. 1.2 on the left-hand side and is called the Hooke element (Reiner, 1960; Giesekus, 1994; Palmov, 1998). Since the Hooke element (Hookean spring) refers to pure elastic behavior we make use of the superscript index ^{E} (for elastic) for all further executions. As already introduced in the previous section, the constitutive relations can be given as follows.

$$\boldsymbol{N}^{\mathrm{E}} = \boldsymbol{\mathcal{A}} : \boldsymbol{G}^{\mathrm{E}} = \lambda_J^{\mathcal{A}} \boldsymbol{G}_J^{\mathrm{E}} \tag{1.20a}$$

$$\boldsymbol{L}^{\mathrm{E}} = \boldsymbol{\mathcal{D}} : \boldsymbol{K}^{\mathrm{E}} = \lambda_J^{\mathcal{D}} \boldsymbol{K}_J^{\mathrm{E}} \tag{1.20b}$$

$$\boldsymbol{q}^{\mathrm{E}} = \boldsymbol{Z} \quad \cdot \boldsymbol{g}^{\mathrm{E}} = \lambda^{\boldsymbol{Z}} \boldsymbol{g}^{\mathrm{E}} \tag{1.20c}$$

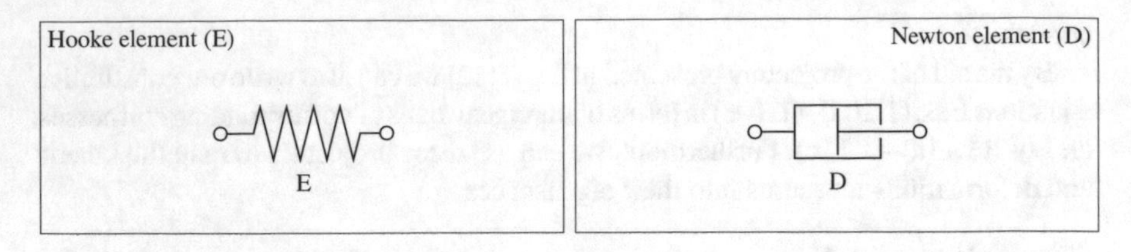

Fig. 1.2 Basic elements of rheological models used to reproduce viscoelastic material behavior

To be exact and for the purpose of comparison with subsequent executions, the elastic material parameters are given again.

$$\lambda_1^{\mathcal{A}} = 2Bh \qquad \lambda_2^{\mathcal{A}} = 2Gh \tag{1.21a}$$

$$\lambda_1^{\mathcal{D}} = 2B\frac{h^3}{12} \qquad \lambda_2^{\mathcal{D}} = 2G\frac{h^3}{12} \tag{1.21b}$$

$$\lambda^{\boldsymbol{Z}} = \kappa G h \tag{1.21c}$$

The second basic element that is needed represents rate-dependency. The simplest approach to introduce viscous behavior is to apply a viscous law. Usually, this is represented by a linear damper. Such an element is is visualized in Fig. 1.2 on the right-hand side and is called the Newton element (or Newtonian dashpot) (Reiner, 1960; Giesekus, 1994; Palmov, 1998).This is representing linear viscous behavior. Since damping is a dissipative process, we make use of the superscript index ^{D} (for dissipative). Since we assume isotropic viscous behavior, we can split the viscous relations in the same fashion as already demonstrated for isotropic stiffnesses. Following constitutive relations result.

$$\boldsymbol{N}^{\mathrm{D}} = \mathcal{B} : \dot{\boldsymbol{G}}^{\mathrm{D}} = \lambda_J^{\mathcal{B}} \dot{\boldsymbol{G}}_J^{\mathrm{D}} \tag{1.22a}$$

$$\boldsymbol{L}^{\mathrm{D}} = \mathcal{E} : \dot{\boldsymbol{K}}^{\mathrm{D}} = \lambda_J^{\mathcal{E}} \dot{\boldsymbol{K}}_J^{\mathrm{D}} \tag{1.22b}$$

$$\boldsymbol{q}^{\mathrm{D}} = \boldsymbol{X} \cdot \dot{\boldsymbol{g}}^{\mathrm{D}} = \lambda^{\boldsymbol{X}} \dot{\boldsymbol{g}}^{\mathrm{D}} \tag{1.22c}$$

Analogously, $\mathcal{B}$ is the in-plane viscosity tensor, $\mathcal{E}$ is the plate viscosity tensor, and $\boldsymbol{X}$ is the transverse shear viscosity tensor. We can specify the isotropic eigenvalues of these viscosities as follows.

$$\lambda_1^{\mathcal{B}} = 2\mu h \qquad \lambda_2^{\mathcal{B}} = 2\eta h \tag{1.23a}$$

$$\lambda_1^{\mathcal{E}} = 2\mu\frac{h^3}{12} \qquad \lambda_2^{\mathcal{E}} = 2\eta\frac{h^3}{12} \tag{1.23b}$$

$$\lambda^{\boldsymbol{X}} = \zeta \eta h \tag{1.23c}$$

Here μ is the surface (dilatational) and η is the shear (deviatoric) viscosity. Furthermore, $0 < \zeta \leq 1$ is a correction factor accounting for transverse shear in pure viscous state. For the sake of simplicity one can consider the coincidence $\zeta = \kappa$ which does not necessarily retain its validity for arbitrary deformation states. These eigenvalues obviously have the same structure as the eigenvalues of the stiffness tensors. In analogy to established stiffnesses, cf. Eq. (1.17), we can introduce subsequent scalar-valued viscosities.

$$V_{\mathrm{M}} = \frac{1}{2}\left[\lambda_1^{\mathcal{B}} + \lambda_2^{\mathcal{B}}\right] \qquad V_{\mathrm{B}} = \frac{1}{2}\left[\lambda_1^{\mathcal{E}} + \lambda_2^{\mathcal{E}}\right] \qquad V_{\mathrm{S}} = \lambda^{X} \tag{1.24}$$

Here, V_{M} is the in-plane viscosity, V_{B} is the bending viscosity, and V_{S} is the transverse shear viscosity. The relation

$$V_{\mathrm{B}} = \frac{h^2}{12} V_{\mathrm{M}}$$

applies analogously since

$$\lambda_J^{\mathcal{E}} = \frac{h^2}{12} \lambda_J^{\mathcal{B}}$$

holds. Consequently,

$$\mathcal{E} = \frac{h^2}{12} \mathcal{B}$$

holds as well.

We reduce our concern to linear-viscoelastic models. Such models can be constructed via parallel or series arrangements and arbitrary combinations by using the rheological elements R^i introduced in present section. For the sake of simplicity we restrict ourselves to two of the most simple models where the basic rules are presented in the following two sections. For explanations of subsequent arrangements we use numerals for different elements so that $i \in \{(1), \ldots, (N)\}$ holds instead of explicitly stating the element type with superscript indices E and D. We are content with two elements to illustrate the rules. Both arrangements are visualized in Fig. 1.3.

1.3.2 Series Arrangement of Elements

A series circuitry of two elements is illustrated in Fig. 1.3 (left-hand side). The series arrangement utilizes the so-called 'iso-stress concept'. To be exact, this entails that in both elements all kinetic measures are equal independently, while the total of the individual dual kinematic measure results from the sum of both elements. Applied to

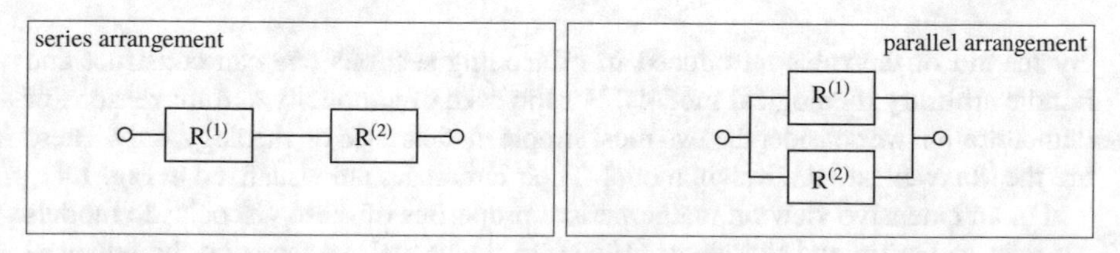

Fig. 1.3 Basic arrangements of elements R^i in rheological models

the kinetic quantities introduced this results in following relations.

$$\boldsymbol{N} = \boldsymbol{N}^{(1)} = \boldsymbol{N}^{(2)} \tag{1.25a}$$

$$\boldsymbol{L} = \boldsymbol{K}^{(1)} = \boldsymbol{L}^{(2)} \tag{1.25b}$$

$$\boldsymbol{q} = \boldsymbol{q}^{(1)} = \boldsymbol{q}^{(2)} \tag{1.25c}$$

Vice versa, the following applies to the deformation tensors.

$$\boldsymbol{G} = \boldsymbol{G}^{(1)} + \boldsymbol{G}^{(2)} \tag{1.26a}$$

$$\boldsymbol{K} = \boldsymbol{K}^{(1)} + \boldsymbol{K}^{(2)} \tag{1.26b}$$

$$\boldsymbol{g} = \boldsymbol{g}^{(1)} + \boldsymbol{g}^{(2)} \tag{1.26c}$$

1.3.3 Parallel Arrangement of Elements

A parallel circuitry of two elements is illustrated in Fig. 1.3 (right-hand side). The parallel arrangement utilizes the so-called 'iso-strain concept'. This means that in both elements all kinematic measures are equal independently, while the total of the individual dual kinetic measure results from the sum of both elements. For the kinematic quantities introduced the following applies.

$$\boldsymbol{G} = \boldsymbol{G}^{(1)} = \boldsymbol{G}^{(2)} \tag{1.27a}$$

$$\boldsymbol{K} = \boldsymbol{K}^{(1)} = \boldsymbol{K}^{(2)} \tag{1.27b}$$

$$\boldsymbol{g} = \boldsymbol{g}^{(1)} = \boldsymbol{g}^{(2)} \tag{1.27c}$$

Contrary, the following applies to the kinetic measures.

$$\boldsymbol{N} = \boldsymbol{N}^{(1)} + \boldsymbol{N}^{(2)} \tag{1.28a}$$

$$\boldsymbol{L} = \boldsymbol{K}^{(1)} + \boldsymbol{L}^{(2)} \tag{1.28b}$$

$$\boldsymbol{q} = \boldsymbol{q}^{(1)} + \boldsymbol{q}^{(2)} \tag{1.28c}$$

1.4 Application to Viscoelastic Plates

By the aid of the rules introduced in protruding sections one can construct and handle arbitrary rheological models. For the sake of simplicity and for reasons of demonstration we consider the two most simple models used on rheology, only. These are the Maxwell and the Kelvin model. Their circuitries are visualized in Fig. 1.4.

For an extensive view on mathematical properties of these viscoelastic models we refer to Gurtin and Sternberg (1962). In the sequel we focus on the tensorial generalization with application to plates.

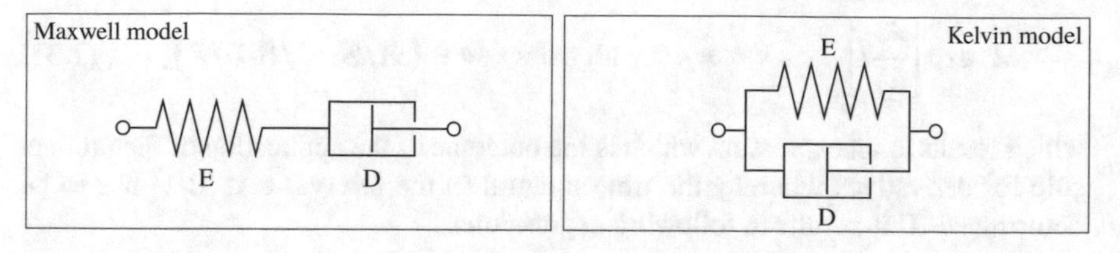

Fig. 1.4 Circuits of most simple rheological models in linear viscoelasticity

For all subsequent operations we consider finite ∘-processes and the dual •-processes between the instants 0 and t, where initial values for ∘(0) and •(0) $\forall\, \circ/\bullet \in \{\mathcal{A}/\mathcal{B}, \mathcal{D}/\mathcal{E}, \boldsymbol{Z}/\boldsymbol{X}\}$ are needed. If the •-process is prescribed, one wants to determine the ∘-process, or vice versa.

1.4.1 Maxwell Model

The Maxwell model consists of a series arrangement of spring and damper. Due to the rules of a series arrangement presented in Subsect. 1.3.2 and the constitutive relations (1.20) and (1.22), we can formulate the following five tensor-valued equations for the one-dimensional analogue visualized in Fig. 1.4 (left-hand side).

$$\dot{\boldsymbol{G}} = \frac{1}{\lambda_J^{\mathcal{A}}}\dot{\boldsymbol{N}}_J + \frac{1}{\lambda_J^{\mathcal{B}}}\boldsymbol{N}_J \tag{1.29a}$$

$$\dot{\boldsymbol{K}} = \frac{1}{\lambda_J^{\mathcal{D}}}\dot{\boldsymbol{L}}_J + \frac{1}{\lambda_J^{\mathcal{E}}}\boldsymbol{L}_J \tag{1.29b}$$

$$\dot{\boldsymbol{g}} = \frac{1}{\lambda^{\boldsymbol{Z}}}\dot{\boldsymbol{q}} + \frac{1}{\lambda^{\boldsymbol{X}}}\boldsymbol{q} \tag{1.29c}$$

By integration of these equations we obtain a form explicit in the deformation measures.

$$\boldsymbol{G}(t) = \boldsymbol{G}(0) + \frac{1}{\lambda_J^{\mathcal{A}}}\left[\boldsymbol{N}_J(t) - \boldsymbol{N}_J(0)\right] + \frac{1}{\lambda_J^{\mathcal{B}}}\int_0^t \boldsymbol{N}_J(\tau)\,\mathrm{d}\tau \tag{1.30a}$$

$$\boldsymbol{K}(t) = \boldsymbol{K}(0) + \frac{1}{\lambda_J^{\mathcal{D}}}\left[\boldsymbol{L}_J(t) - \boldsymbol{L}_J(0)\right] + \frac{1}{\lambda_J^{\mathcal{E}}}\int_0^t \boldsymbol{L}_J(\tau)\,\mathrm{d}\tau \tag{1.30b}$$

$$\boldsymbol{g}(t) = \boldsymbol{g}(0) + \frac{1}{\lambda^{\boldsymbol{Z}}}\left[\boldsymbol{q}(t) - \boldsymbol{q}(0)\right] + \frac{1}{\lambda^{\boldsymbol{X}}}\int_0^t \boldsymbol{q}(\tau)\,\mathrm{d}\tau \tag{1.30c}$$

To gain the explicit forms of kinetics, the initial equations are multiplied by

$$\lambda_J^\circ \exp\left[\frac{\lambda_J^\circ}{\lambda_J^\bullet} t\right] \qquad \forall \circ \neq \bullet \quad \text{with pairs } \circ/\bullet \in \{\mathcal{A}/\mathcal{B}, \mathcal{D}/\mathcal{E}, \boldsymbol{Z}/\boldsymbol{X}\} \tag{1.31}$$

which results in an expression which is the outcome of the application of the product rule for derivatives whereby the time integral in the interval $\tau \in [0, 1]$ has to be determined. This results in following expressions.

$$\boldsymbol{N}(t) = \boldsymbol{N}_J(0) \exp\left[-\frac{\lambda_J^{\mathcal{A}}}{\lambda_J^{\mathcal{B}}} t\right] + \lambda_J^{\mathcal{A}} \int\limits_0^t \dot{\boldsymbol{G}}_J(\tau) \exp\left[\frac{\lambda_J^{\mathcal{A}}}{\lambda_J^{\mathcal{B}}}(\tau - t)\right] \mathrm{d}\tau \tag{1.32a}$$

$$\boldsymbol{L}(t) = \boldsymbol{L}_J(0) \exp\left[-\frac{\lambda_J^{\mathcal{D}}}{\lambda_J^{\mathcal{E}}} t\right] + \lambda_J^{\mathcal{D}} \int\limits_0^t \dot{\boldsymbol{K}}_J(\tau) \exp\left[\frac{\lambda_J^{\mathcal{D}}}{\lambda_J^{\mathcal{E}}}(\tau - t)\right] \mathrm{d}\tau \tag{1.32b}$$

$$\boldsymbol{q}(t) = \boldsymbol{q}(0) \quad \exp\left[-\frac{\lambda^{\boldsymbol{Z}}}{\lambda^{\boldsymbol{X}}} t\right] + \lambda^{\boldsymbol{Z}} \int\limits_0^t \dot{\boldsymbol{g}}(\tau) \quad \exp\left[\frac{\lambda^{\boldsymbol{Z}}}{\lambda^{\boldsymbol{X}}}(\tau - t)\right] \mathrm{d}\tau \tag{1.32c}$$

Handling the integrals by partial integration gives the final set of equations.

$$\begin{aligned}\boldsymbol{N}(t) = \lambda_J^{\mathcal{A}} \boldsymbol{G}_J(t) + \left[\boldsymbol{N}_J(0) - \lambda_J^{\mathcal{A}} \boldsymbol{G}_J(0)\right] \exp\left[-\frac{\lambda_J^{\mathcal{A}}}{\lambda_J^{\mathcal{B}}} t\right] \\ - \frac{\left(\lambda_J^{\mathcal{A}}\right)^2}{\lambda_J^{\mathcal{B}}} \int\limits_0^t \boldsymbol{G}_J(\tau) \exp\left[\frac{\lambda_J^{\mathcal{A}}}{\lambda_J^{\mathcal{B}}}(\tau - t)\right] \mathrm{d}\tau\end{aligned} \tag{1.33a}$$

$$\begin{aligned}\boldsymbol{L}(t) = \lambda_J^{\mathcal{D}} \boldsymbol{K}_J(t) + \left[\boldsymbol{L}_J(0) - \lambda_J^{\mathcal{D}} \boldsymbol{K}_J(0)\right] \exp\left[-\frac{\lambda_J^{\mathcal{D}}}{\lambda_J^{\mathcal{E}}} t\right] \\ - \frac{\left(\lambda_J^{\mathcal{D}}\right)^2}{\lambda_J^{\mathcal{E}}} \int\limits_0^t \boldsymbol{K}_J(\tau) \exp\left[\frac{\lambda_J^{\mathcal{D}}}{\lambda_J^{\mathcal{E}}}(\tau - t)\right] \mathrm{d}\tau\end{aligned} \tag{1.33b}$$

$$\begin{aligned}\boldsymbol{q}(t) = \lambda^{\boldsymbol{Z}} \boldsymbol{g}(t) + \left[\boldsymbol{q}(0) - \lambda^{\boldsymbol{Z}} \boldsymbol{g}(0)\right] \exp\left[-\frac{\lambda^{\boldsymbol{Z}}}{\lambda^{\boldsymbol{X}}} t\right] \\ - \frac{\left(\lambda^{\boldsymbol{Z}}\right)^2}{\lambda^{\boldsymbol{X}}} \int\limits_0^t \boldsymbol{g}(\tau) \quad \exp\left[\frac{\lambda^{\boldsymbol{Z}}}{\lambda^{\boldsymbol{X}}}(\tau - t)\right] \mathrm{d}\tau\end{aligned} \tag{1.33c}$$

Using Eqs. (1.30) and (1.33), the viscoelastic behavior of plates can be completely described by the aid of the Maxwell model.

1.4.2 Kelvin Model

The Kelvin model consists of a parallel arrangement of spring and damper. Due to the rules of a parallel arrangement given in Subsect. 1.3.2 and the constitutive relations (1.20) and (1.22), we can formulate the following tensor-valued equations for the Kelvin model.

$$\boldsymbol{N} = \lambda_J^{\mathcal{A}} \boldsymbol{G}_J + \lambda_J^{\mathcal{B}} \dot{\boldsymbol{G}}_J \tag{1.34a}$$
$$\boldsymbol{L} = \lambda_J^{\mathcal{D}} \boldsymbol{K}_J + \lambda_J^{\mathcal{E}} \dot{\boldsymbol{K}}_J \tag{1.34b}$$
$$\boldsymbol{q} = \lambda^{\boldsymbol{Z}} \boldsymbol{g} + \lambda^{\boldsymbol{X}} \dot{\boldsymbol{g}} \tag{1.34c}$$

Analogously to the Maxwell model, we determine the explicit forms of kinematics. Hereby, we multiply our set of equations by the ansatz

$$\frac{1}{\lambda_J^{\bullet}} \exp\left[\frac{\lambda_J^{\circ}}{\lambda_J^{\bullet}} t\right] \qquad \forall \circ \neq \bullet \quad \text{with pairs } \circ/\bullet \in \{\mathcal{A}/\mathcal{B}, \mathcal{D}/\mathcal{E}, \boldsymbol{Z}/\boldsymbol{X}\} \tag{1.35}$$

which, after minor conversions, results in the following forms.

$$\boldsymbol{G}(t) = \boldsymbol{G}_J(0) \exp\left[-\frac{\lambda_J^{\mathcal{A}}}{\lambda_J^{\mathcal{B}}} t\right] + \frac{1}{\lambda_J^{\mathcal{B}}} \int_0^t \boldsymbol{N}_J(\tau) \exp\left[\frac{\lambda_J^{\mathcal{A}}}{\lambda_J^{\mathcal{B}}} (\tau - t)\right] \mathrm{d}\tau \tag{1.36a}$$

$$\boldsymbol{K}(t) = \boldsymbol{K}_J(0) \exp\left[-\frac{\lambda_J^{\mathcal{D}}}{\lambda_J^{\mathcal{E}}} t\right] + \frac{1}{\lambda_J^{\mathcal{E}}} \int_0^t \boldsymbol{L}_J(\tau) \exp\left[\frac{\lambda_J^{\mathcal{D}}}{\lambda_J^{\mathcal{E}}} (\tau - t)\right] \mathrm{d}\tau \tag{1.36b}$$

$$\boldsymbol{g}(t) = \boldsymbol{g}(0) \exp\left[-\frac{\lambda^{\boldsymbol{Z}}}{\lambda^{\boldsymbol{X}}} t\right] + \frac{1}{\lambda^{\boldsymbol{Z}}} \int_0^t \boldsymbol{q}(\tau) \exp\left[\frac{\lambda^{\boldsymbol{Z}}}{\lambda^{\boldsymbol{X}}} (\tau - t)\right] \mathrm{d}\tau \tag{1.36c}$$

The opposite case, deriving the explicit form in the kinetic measures what corresponds with a relaxation test, is not feasible for this model since finite kinetics cannot introduce discontinuous kinematics. Therefore, using Eqs. (1.36), the viscoelastic behavior of plates is completely described by the aid of the Kelvin model.

1.4.3 Visualization of Model Behavior

When it comes to visualizations of simple deformation tests, we make use of one-dimensional measures for reasons of clarity. These are defined as follows.

$$N = \boldsymbol{e}_1 \cdot \boldsymbol{N} \cdot \boldsymbol{e}_1 \qquad G = \boldsymbol{e}_1 \cdot \boldsymbol{G} \cdot \boldsymbol{e}_1 \tag{1.37a}$$
$$L = \boldsymbol{e}_1 \cdot \boldsymbol{L} \cdot \boldsymbol{e}_1 \qquad K = \boldsymbol{e}_1 \cdot \boldsymbol{K} \cdot \boldsymbol{e}_1 \tag{1.37b}$$

$$q = \boldsymbol{q} \cdot \boldsymbol{e}_1 \qquad g = \boldsymbol{g} \cdot \boldsymbol{e}_1 \tag{1.37c}$$

In this spirit, we can e.g. reduce our constitutive relations to the following expressions

$$N = Yh \quad G \tag{1.38a}$$

$$L = Y\frac{h^3}{12}K \tag{1.38b}$$

$$q = \kappa Yh \quad g \tag{1.38c}$$

at least in the context of elasticity. For the sake of simplicity we consider a constant load $\Delta\Box$ applied at time t_i for a kinetic process

$$\Box_i(t) = \Xi(t - t_i)\,\Delta\Box(t_i) \qquad \forall\,\Box \in \{N_{\text{init}}, L_{\text{init}}, q_{\text{init}}\}, \tag{1.39}$$

whereby Ξ is the Heaviside function

$$\Xi = \begin{cases} 0 & \text{if } \tau < 0 \\ 1 & \text{if } \tau \geq 0 \end{cases}, \tag{1.40}$$

and $\Delta\Box\ \forall\,\Box \in \{N, L, q\}$ is the unit load. Therefore, the unit load is applied for all times $0 < \tau < \infty$ in present investigations (no unloading). Such scenarios are known as retardation tests. Resulting behavior is visualized in Fig. 1.5. The upper row shows the impact while the lower row shows the response. Dual quantities are directly juxtaposed, while results of both models applied here are compared qualitatively.

Vice versa, we can also apply a kinematic process while we have to change $\Delta\Box$ in Eq. (1.39) to $\Box \in \{G_{\text{init}}, K_{\text{init}}, g_{\text{init}}\}$. This results in so called relaxation tests, visualized in Fig. 1.6.

The two figures show the mechanical responses of both models introduced comparatively. However, considering these one-dimensional analogues, the behaviors are well known in the literature, e.g. Krawietz (1986), so we are not going to discuss them in detail, here. Rather, we would like to point out that all three deformation states are completely decoupled, i.e. in-plane, plate and transverse shear state have no influence on each other.

1.4.4 Ansatz for Viscous Parameters

The presented models are limited to their linearity. Nonlinearity can be taken into account by the dependence of the viscosities on the dynamic quantities.

$$\lambda_J^{\mathcal{B}} = \lambda_J^{\mathcal{B}0} \exp(-H_J^{\boldsymbol{N}}) \tag{1.41a}$$

$$\lambda_J^{\mathcal{E}} = \lambda_J^{\mathcal{E}0} \exp(-H_J^{\boldsymbol{L}}) \tag{1.41b}$$

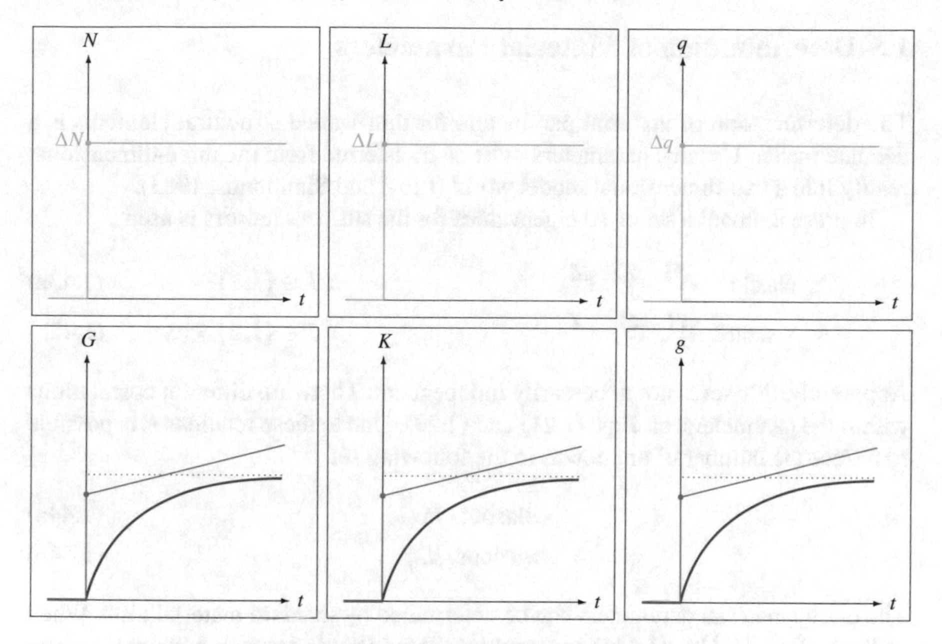

Fig. 1.5 Retardation test with Maxwell model (blue) and Kelvin model (red)

$$\lambda^{\boldsymbol{X}} = \lambda^{\boldsymbol{X}_0} \exp(-H^{\boldsymbol{q}}) \tag{1.41c}$$

Herein, $\lambda_J^{\mathcal{B}_0}$, $\lambda_J^{\mathcal{E}_0}$, and $\lambda^{\boldsymbol{X}_0}$ are non-negative stiffness parameters. The relation

$$\lambda_J^{\mathcal{E}_0} = \frac{h^2}{12} \lambda_J^{\mathcal{B}_0}$$

holds. Furthermore, the exponents $H_J^{\square}$ are linear forms of the i invariants In.

$$H_J^{\square} = \sum_{i=1}^{2} H_{Ji}^{\square} \mathit{In}_i^{\square} \qquad \forall \square \in \{\boldsymbol{N}, \boldsymbol{L}\} \tag{1.42a}$$

$$H^{\boldsymbol{q}} = \sum_{i=1}^{1} H_i^{\boldsymbol{q}} \mathit{In}_i^{\boldsymbol{q}} \tag{1.42b}$$

In context of our note in Sect. 1.2, we can build isotropic invariants In_i by the aid of a reduced set of operations. Since our dynamic measures are symmetric $\forall \boldsymbol{H} \in \{\boldsymbol{G}, \boldsymbol{K}\}$ as well and $\boldsymbol{q}$ is of first order only, these operations result in a whole set of 5 invariants. It is noteworthy that $\lambda_J^{\mathcal{B}}$, $\lambda_J^{\mathcal{E}}$, and $\lambda^{\boldsymbol{X}}$ are constant during monotonous retardation tests.

This ansatz is only one possibility which has established itself in practical application. When restricting to isotropy, the use of von Mises equivalents of present kinetic measures is another option.

1.5 Determination of Material Parameters

The determination of material parameters for thin-walled structural elements is a delicate matter. Usually, parameters must be transferred from the three-dimensional reality into a two-dimensional model world (Libai and Simmonds, 1983).

In present model a set of 10 eigenvalues for the stiffness tensors is used.

$$\text{elastic: } \lambda_J^{\mathcal{A}}, \lambda_J^{\mathcal{D}}, \lambda^{\mathbf{Z}} \qquad \forall J \in \{1,2\} \tag{1.43a}$$

$$\text{viscous: } \lambda_J^{\mathcal{B}}, \lambda_J^{\mathcal{E}}, \lambda^{\mathbf{X}} \qquad \forall J \in \{1,2\} \tag{1.43b}$$

Apparently, this set is not necessarily independent. There are different correlations within the parameters, cf. Eqs. (1.21) and (1.23). Due to these relations it is possible to reduce the number of unknowns to the following set.

$$\text{elastic: } B, G \tag{1.44a}$$

$$\text{viscous: } \mu, \eta \tag{1.44b}$$

The elastic material parameters can be determined by standard material tests. When utilizing Eqs. (1.41a)-(1.41c) to consider the nonlinear viscous behavior we can specify the set of viscous unknowns to the following.

$$\text{viscous: } \lambda_J^{\mathcal{B}0}, \lambda_J^{\mathcal{E}0}, \lambda^{\mathbf{X}0} \qquad \forall J \in \{1,2\} \tag{1.45a}$$

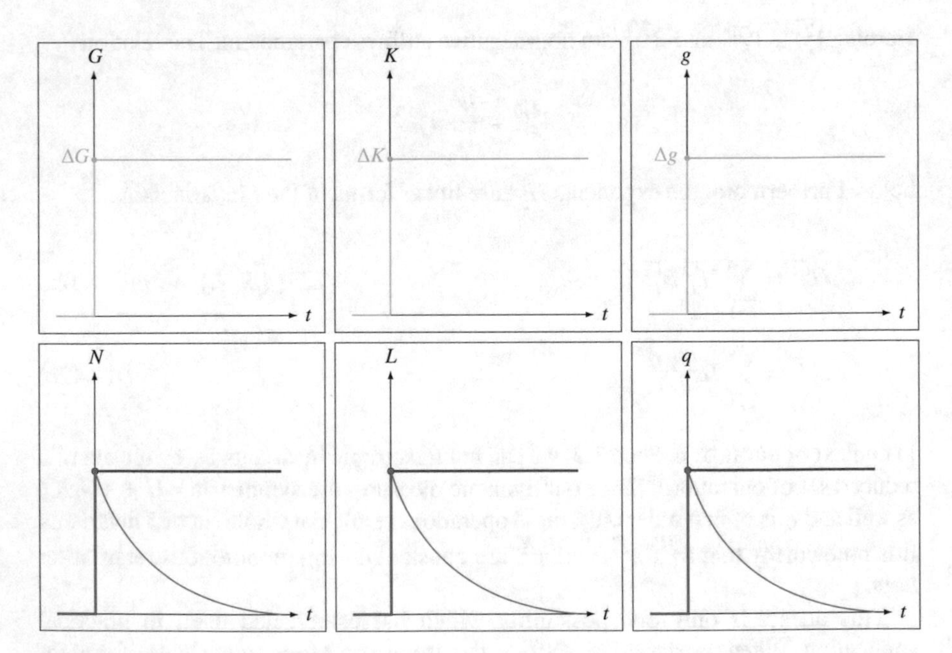

Fig. 1.6 Relaxation test with Maxwell model (blue) and Kelvin model (red)

Since we assume that all viscous deformations are isochoric, μ and thus parameters with $J \triangleq 1$ are irrelevant

$$\frac{1}{\lambda_1^{\mathcal{B}_0}} = \frac{1}{\lambda_1^{\mathcal{E}_0}} = 0 \tag{1.46}$$

and can be set to 1. The remaining three parameters ($\lambda_2^{\mathcal{B}_0}$, $\lambda_2^{\mathcal{E}_0}$, $\lambda^{\mathcal{X}_0}$) can be determined by fitting model responses to material tests, i.e. minimizing the discrepancy between both procedures. For this purpose, data of relaxation and retardation tests is required, preferably in the relevant load and temperature range. Clearly, the reliability of predictions computed with present model depends on the variety of experimental data used for calibration of the material parameters. Since material tests are obviously carried out on three-dimensional specimens, the correlation to the three-dimensional bulk modulus B_{3D} seems useful.

$$B_{3D} = \frac{2(1-\nu)}{3(1-2\nu)} B \tag{1.47}$$

Nevertheless, we can also find such analogy in the case of the volume shear viscosity μ_{3D}.

$$\mu_{3D} = \frac{2(1-\nu)}{3(1-2\nu)} \mu \tag{1.48}$$

Since we restrict our concern to isotropy, it is sufficient to determine the set of unknown material parameters for a single direction which can be chosen arbitrary.

In context of the shear correction factors κ and ζ no directional investigations are required either. Due to their sensitivity to boundary conditions, loading scenarios, etc., their determination is part of special research efforts (Altenbach, 2000a,b; Vlachoutsis, 1992).

1.6 Conclusion

The present framework is a generalization of uniaxial rheological models to describe small viscoelastic deformations for isotropic shear-deformable plates. We have shown that it is possible to formulate a viscoelastic material model for plates *ab initio*. The emphasis was on the discussion of the formal structure of such an approach. Due to the smart choice of the mid surface position within the original body at $X_3 = 0$ within the transverse limits $X_3^{\max} = | \pm \frac{h}{2}|$, the deformation states considered are uncoupled but superposed eventually for both, elastic and viscous material behavior – at least in a geometrically linear framework. This enables to utilize decoupled constitutive relations which leads to substantial simplification of the mathematical executions. Additionally, the projector representation aided transparency. Finally, this mathematical modeling results in a set of ordinary differential equations which can be solved by standard methods. The preceding calculations have been performed on a rather general level. This facilitates the problem description in the most general form, contrary to usual procedures in phenomenological material modeling. However, the whole procedure protrudes by its conceptual clearness. The special case of a viscoelastic shear-rigid plate can be deduced in analogy to the procedure presented in Aßmus et al (2020).

In present work we make use of two most simple circuits. Extensions of present models can be realized by considering extended rheological circuits (i.e. Zener model (Zener, 1948) as well as Poynting (Poynting and Thomson, 1902) or Burgers model (Burgers, 1939)) or generalized models (generalized Maxwell model or generalized Kelvin chain) to include the primary and tertiary phase of what is often referred to as *creep* (The terminology is not used here due to its vague definition.). However, in this course, a generalized system equation can be introduced which is valid for arbitrary rheological models (with respect to the restriction to isotropy and geometrical symmetry). Nevertheless, we would like to point out that the two models used here – Maxwell and Kelvin – represent extremes. All extended models work within these bounds. In this context its also possible to include further rheological basic elements (Coulomb or St. Venant element) for advanced circuits. Hereby, the introduction of a directly formulated theory of plasticity for two-dimensional body manifolds by the aid of eigenspace projectors is an open task. Apparently, however, we are leaving the topic of pure viscoelasticity theory with such circuitry.

Considering coupled deformation states (due to anisotropies, a deviating reference surface or an initially curved configuration) will blow up the present format since we have to increase the number of stiffness and dissipation tensors which must be taken into account. Furthermore, this format will loose its clarity due to the coupling of the stiffness and dissipation tensor underneath each other. However, when considering anisotropies (Weyl, 1997), both, in elastic and viscoelastic behavior, a harmonic decomposition of the stiffness tensors may be used in dealing with theoretical problems which also arise in the discussion of geometrical asymmetry.

Basically, it must be noted that the world of thin-walled structural elements (in-plane loaded plates, out-of-plane loaded plates, shells, and folded structures) lives from special cases. In this context, it is unfeasible to generalize the present concept to the whole group. The reason for this is the non-trivial coupling of different deformation states, at least when transverse shear is to be coupled.

However, the common limit of all presented here rheological models is their linearity. The viscous behavior of natural materials is often strongly non-linear, e.g. of polymers. Therefore a large number of extensions can be found in the literature, e.g. frequency or kinetics dependent damping.

To conclude, the ansatz introduced here is a possible way to describe the viscoelastic behavior of plates with transverse shear sensitivity. The power of the proposed method lies in its extensibility. It can also be applied in a geometrically non-linear framework although the relationships are becoming more complex.

Acknowledgements The work described in this paper was carried out under a grant (RTG 1554, grant no. 83477795) of the German Research Foundation (DFG). We like to thank Jan Kalisch for helpful hints with invariants on two-dimensional vector spaces.

References

Altenbach H (1988) Eine direkt fomulierte lineare Theorie für viskoelastische Platten und Schalen. Ingenieur-Archiv 58(3):215–228

Altenbach H (2000a) An alternative determination of transverse shear stiffnesses for sandwich and laminated plates. International Journal of Solids and Structures 37(25):3503–3520

Altenbach H (2000b) On the determination of transverse shear stiffnesses of orthotropic plates. Zeitschrift für angewandte Mathematik und Physik ZAMP 51(4):629–649

Aßmus M, Eisenträger J, Altenbach H (2017) Projector Representation of Isotropic Linear Elastic Material Laws for Directed Surfaces. ZAMM - Zeitschrift für Angewandte Mathematik und Mechanik 97(12):1625–1634

Aßmus M, Naumenko K, Altenbach H (2020) Subclasses of mechanical problems arising from the direct approach for homogeneous plates. In: Altenbach H, Chróścielewski J, Eremeyev VA, Wiśniewski K (eds) Recent Developments in the Theory of Shells, Advanced Structured Materials, vol 110, Springer, Singapore, pp 1–21

Burgers JM (1939) Mechanical considerations–model systems–phenomenological theories of relaxation and of viscosity. First report on viscosity and plasticity, Prepared by the committee for the study of viscosity of the academy of sciences at Amsterdam, Nordemann Publ, 2nd Edition, New York

Ferry JD (1980) Viscoelastic Properties of Polymers, 3rd edn. John Wiley & Sons, Inc., New York · Chichester · Brisbane · Toronto · Singapore

Giesekus H (1994) Phänomenologische Rheologie: Eine Einführung. Springer, Berlin Heidelberg

Gurtin ME, Sternberg E (1962) On the linear theory of viscoelasticity. Archive for Rational Mechanics and Analysis 11(1):291–356

Kirchhoff GR (1850) Über das Gleichgewicht und die Bewegung einer elastischen Scheibe. Journal für die reine und angewandte Mathematik 40:51–88

Krawietz A (1986) Materialtheorie - Mathematische Beschreibung des phänomenologischen thermomechanischen Verhaltens. Springer-Verlag, Berlin · Heidelberg · New York · Tokyo

Libai A, Simmonds JG (1983) Nonlinear elastic shell theory. Advances in Applied Mechanics 23:271–371

Mālmeisters A, Tamužs V, Teters G (1977) Mechanik der Polymerwerkstoffe. Akademie-Verlag, Berlin

Mindlin RD (1951) Influence of rotatory inertia and shear on flexural motions of isotropic, elastic plates. Trans ASME Journal of Applied Mechanics 18:31–38

Naghdi PM (1972) The Theory of Shells and Plates. In: W Flügge (ed) Encyclopedia of Physics, vol VIa/2: Linear Theories of Elasticity and Thermoelasticity (ed. C. Truesdell), Springer, Berlin · New York, pp 425–640

Palmov V (1998) Vibrations of Elasto-Plastic Bodies. Foundations of Engineering Mechanics, Springer, Berlin, Heidelberg

Poynting JH, Thomson JJ (1902) A Text-Book of Physics: Properties of Matter, vol 1. Charles Griffin and Company, London

Reiner M (1960) Deformation, Strain and Flow: an Elementary Introduction to Rheology. H. K. Lewis, London

Reissner E (1945) The effect of transverse shear deformation on the bending of elastic plates. Trans ASME Journal of Applied Mechanics 12:69–77

Rychlewski J (1995) Unconventional approach to linear elasticity. Archives of Mechanics 47(2):149–171

Schade H, Neemann K (2009) Tensoranalysis. De Gruyter, Berlin
Šilhavý M (1997) The Mechanics and Thermodynamics of Continuous Media. Springer, Berlin · Heidelberg
Thomson W (1878) Elasticity. In: Black A, Black C (eds) Encyclopedia Britannica, Edinburgh, vol 7
Vlachoutsis S (1992) Shear correction factors for plates and shells. International Journal for Numerical Methods in Engineering 33(7):1537–1552
Weyl H (1997) The Classical Groups: Their Invariants and Representations, 2nd edn. Princeton University Press, Princeton, 15. print and 1. paperback print
Zener CM (1948) Elasticity and Anelasticity of Metals. University of Chicago Press, Chicago
Zhilin PA (1976) Mechanics of deformable directed surfaces. International Journal of Solids and Structures 12(9):635 – 648

Chapter 2
Teaching Mechanics

James T. Boyle

Abstract In this Chapter issues associated with teaching mechanics in its broadest sense are discussed. A historical approach is adopted throughout in the belief that this is important for novice students to help them comprehend the conceptual difficulties inherent in this subject. It also highlights the historical problems which its creators faced and the teaching methods used at the time. The Chapter describes a conceptual teaching approach, enhanced through technology, which was initially developed for physics education in large classes, although the pedagogy is based on Socratic Dialogue. For reasons which will become clear the focus is on introductory courses: it is considered crucial that novice students are provided with an opportunity to experience a different way of thinking about mechanics. An emphasis is placed on conceptual understanding before mathematical technique.

Keywords: Teaching mechanics · Teaching physics · Conceptual mechanics · Introductory mechanics · History of technology · History of science · History of higher education · Socratic dialogue · Introductory statics · Continuum mechanics · Royal College of Science & Technology · University of Strathclyde · Mechanics institutes · Technology in teaching · Peer instruction · Just in time teaching · Simulation in teaching · Large lecture format

2.1 Background

In a series of lectures delivered in 1948 to the History of Science Committee in Cambridge, later published as *The Origins of Modern Science* (Butterfield, 1959), the Master of Peterhouse and Professor of Modern History, Herbert Butterfield observed

James T. Boyle
Department of Mechanical & Aerospace Engineering, University of Strathclyde, Glasgow, Scotland, UK,
e-mail: jim.boyle@strath.ac.uk

H. Altenbach and A. Öchsner (eds.), *State of the Art and Future Trends in Material Modeling*, Advanced Structured Materials 100,
https://doi.org/10.1007/978-3-030-30355-6_2

that "... of all the intellectual hurdles which the human mind has confronted and has overcome in the last fifteen hundred years, the one which seems to have been the most amazing in character and the most stupendous in the scope of its consequences is the one relating to the problem of motion ... ". Butterfield's hope was that the lectures would stimulate in the historian some interest in science, and in the scientist some interest in history. It is notable that formal studies of the history of science and technology only began fairly recently with the publication of the magazine *ISIS* in 1913. Academic studies of the pedagogical issues associated with the teaching of science and technology are even more recent. A fundamental feature of the latter has been an awareness that the historical development of the subject is important (Brush, 1989; Allchin, 1992), primarily for the reason Butterfield identified – the central intellectual hurdle. In the context of teaching, if the finest scientific minds in history struggled with the basic concepts of mechanics until at least the early 1600s (and perhaps to this day) how can the young novice readily comprehend the nature of these 'hurdles'? The development of the teaching of mechanics in its broadest sense will be discussed: one aim will be to draw attention to conceptual issues which, in the writer's experience, many teaching practitioners may be unfamiliar. Students in secondary and tertiary education are typically introduced to the minefield of mechanics through mathematical problem solving which presents the fundamental concepts and laws as predetermined. It has now well recognised that a mathematical approach, used exclusively, is insufficient. An understanding of many of the basic concepts remain undeveloped in the novice student's mind. A further aim will be to highlight the strategic importance of using history as a tool (Allchin, 1992). In relation to this aim, the writer must confess a certain bias:

The University of Strathclyde has played a significant role in the history of engineering education and in so doing the teaching of mechanics. With direct origins from Anderson's University, established in Glasgow in 1796 as "... the place of useful learning ... " through the subsequent rapid worldwide expansion of parallel Mechanics' Institutes (Walker, 2016) then to the Royal College of Science & Technology in 1903 (together with the Imperial College in London), Strathclyde has been at the heart of the development of engineering & technology education. The Glasgow University Professor William John Macquorn Rankine (1820-1872) helped develop thermodynamics with his colleague Lord Kelvin and Rudolf Clausius (Truesdell, 1980) in Germany. Rankine also progressed understanding of the phenomenon of fatigue and developed methods to calculate forces and 'stresses' in framed structures. It is less well known that he was one of the main originators of organised programs for the training of young engineers – a common interest and pursuit of professors and college lecturers in Glasgow at the time due to the initiatives of John Anderson and the Mechanics' Institutes. Rankine developed 'Manuals' for engineering science and practice to support these programs. Indeed, his *Manual of Applied Mechanics* (Rankine, 1858) eventually became widely adopted in engineering education at university and college level. When he took up his post at the University of Glasgow engineering was taught in the Faculty of Arts and did not lead to the award of a degree. However, engineering qualifications could be earned at the Technical College; Rankine's *Manual of Applied Mechanics* was used extensively in the College when it was published. Today the *Manual* looks like quite familiar compared to modern calculus-based physics textbooks.

With these comments in mind this Chapter begins with an outline of the historical development of pedagogical views on the teaching of mechanics, followed by an overview of the main issues.

2.2 The Development of Mechanics Teaching

In the days of Rankine and his European peers the whole idea of teaching mechanics, indeed engineering in general, to students at university and college was innovative. As innovators some consideration would have to be given to the manner in which the subject should be presented, even if the delivery mechanism was tied to the formal, didactic lecture format of the day. Rankine would surely have given considerable thought to the order in which kinematics and then dynamics should be presented to the novice. In fact, his original 1858 textbook began with an overview of statics before introducing kinematics as a precursor to dynamics (which was the accepted convention for a logical introduction to mechanics). Later editions, published after his death, reversed this order 'in harmony with modern practice' (Rankine and Bamber, 1873). The history of mechanics appears to be well established, certainly as presented in most engineering textbooks. As we will see, this assumption can be misleading, and certainly conceals related issues which are important for teaching the subject.

Mechanics has a history extending over more than two thousand years (Renn et al, 2003). Many engineers would identify the origin of mechanics with Archimedes' (c. 287-212 BC) *Law of the Lever* (Archimedes, 1952) although the problem had been treated about a century before in Aristotle's (c. 384-382 BC) *Mechanical Problems* (Aristotle, 1936). Common experience in ancient times equated the weight of an object with the force required to lift it up, yet engineers and artisans must have known at the time that this was not a restriction on either concept. Aristotle (1936) demonstrated that levers could be used to move great weights with small forces. This modified the concept of force so that force no longer equates to weight but depends on the position of the weight. The *Mechanical Problems* is one of the earliest links between the practical knowledge of engineers and artisans and the mathematical philosophers' theoretical debates on levers and weight. This treatise consists of some thirty-five problems – simple machines such as the balance, lever, pulley, wedge, oars and rudders, together with a variety of other topics, including the strength of beams and projectile motion. Yet this link was subsequently all but forgotten.

As these concepts became more widely used during Medieval times, the study of balance and weight was considered part of mathematics rather than natural philosophy, which dealt with the general principles of motion. Natural philosophy did not consider its usefulness to society, only in its description of natural phenomenon. Early mechanics was associated with the ideas of natural philosophers so did not include the science of weight, Archimedean statics or Aristotelian dynamics. While in Antiquity, treatises on each of these were available, the practical knowledge of machines, part of the mechanical arts, was customarily disseminated through oral tradition with apprentices. This disconnection of fundamental mechanical concepts (weight, force, motion, machines etc.) possibly impeded the development of mechanics as a science until the Renaissance. Medieval natural philosophers established the mathematical analysis of 'local motion' (Wallace, 1971). Thomas Bradwardine's (Lamar Crosby Jr., 1955) *Treatise on Ratios* provided a new approach to the study of motion including the concepts of instantaneous velocity, 'uniformly accelerated' motion and an early, but undeveloped, perception of the need for the calculus and analytical geometry. Over the following two centuries, Bradwardine's ideas were refined. Natural philosophers at Merton College, Oxford, emphasised the role of mathematical analysis in the description of motion and its causes. The ratio of distance moved to the time elapsed – with velocity as a cause of motion – was considered to be important. These ideas moved to France, Italy and Spain. Indeed, the Spaniard, Domingo do Soto, explicitly postulated, probably without experiment, that falling bodies accelerate uniformly over the time of their fall some eighty years before Galileo (Laird, 1986). Bradwardine's mathematical foundations laid the groundwork for the Renaissance accomplishments in mechanics in Northern Italy:

During the sixteenth century, the *Mechanical Problems* was re-introduced and led to the elevation of mechanics to the rank of a theoretical science (Laird, 1986). It was introduced into the university curriculum at Padua in the 1560s based on a translation and commentary by Niccolo Leonico Tomeo (1456-1531), published in Venice in 1525. Leonico asserted in his commentary that mechanics, according to the Greeks, was that part of the art of construction that used machines. A later commentary in Latin by Alessandro Piccolomini (1508-1579) was published in Rome in 1547 and translated into Italian by the metallurgist Vannoccio Biringuccio in 1582. He emphasised that mechanics was reflective rather than practical, and mathematical in nature, somewhat equivalent in rank to astronomy. This mathematical emphasis, when applied to the problem of motion would eventually lead to Galileo's new mathematical science of motion, published as the *Third Day of his Two New Sciences* in 1638 (Lamar Crosby Jr., 2003). Galileo also considered mechanics as a science which described the principles of machines. It is evident he understood how to transform the static balance into the dynamic lever, relating statics and dynamics for possibly the first time. During the sixteenth century mechanics emerged as a distinctive science on its own. The chief innovative characteristic of mechanics from the Renaissance onwards was that it could be used for societal needs.

Early teaching in mechanics was based on the 'great' treatises, and any accompanying commentaries, which were then debated in small groups directed by the teacher. In Antiquity the influence of Plato (c. 424-348 BC), a student of Socrates and a teacher of Aristotle, is well known. Plato's Dialogues advocated the method of Socratic Dialogue - a pedagogical technique which we will return to later. One Dialogue concerns a mathematician, Theaetetus, with whom the narrator discusses the nature of knowledge. This discussion has some unintended relevance to knowledge and understanding in mechanics, as it developed over the centuries, since all definitions proposed on the nature of knowledge were deemed unsatisfactory.

Plato was known as an accomplished mathematics teacher who differentiated pure and applied mathematics as separate activities.

More is known about university teaching in the period from the Middle Ages until the Renaissance from 1100 to 1700 when the *Scholastic Method* dominated (Makdisi, 1974). Scholasticism was a method of learning which in the classroom was characterised by disputation. A topic was proposed in the form of a question followed by responses, a counterproposal could be argued and any opponents' arguments refuted. It is known these techniques were applied to subjects other than law and theology and in particular to the natural sciences (Hannam, 2011). Interestingly, medieval students had, for the most part, a highly practical view of the university as an institution of direct relevance to society, and were known to voice their views about their education (Cobban, 1971) if this was not the case.

Barely fifty years after Galileo's Two New Sciences, Isaac Newton published his Principia in 1687 (Newton, 1729). Undeniably the *Principia* is considered one of the most significant works in the history of science; certainly, all elementary mechanics textbooks begin with Newton's *Laws of Motion* and it is every student's first introduction to the topic.

Ernst Mach (1960) wrote a well-regarded historical account of the development of mechanics, including critical and, in later editions, philosophical commentaries. In his account Newton's Laws were placed at the centre of mechanics, and in Mach's view did not need to be enhanced. Clifford Truesdell (1968) points out that many of Mach's views originate in Lagrange's equally distinguished *Méchanique Analytique* (Lagrange, 1997), published almost a century after the *Principia*. Lagrange included short historical accounts of statics and dynamics among others. Following Duhem's arguments (Duhem, 1996), Truesdell points out that mechanics was not created solely by Galileo and Newton: indeed, Mach knew that Book I of the *Principia* 'was not entirely original'. At the time of its publication readers were more interested in the novelty of Newton's mathematical and deductive approach to problem solving than the Laws themselves being already known in some form. Newton did not use his own approach later in the *Principia* and does not include any differential equations of motion for systems of more than two point masses. In Duhem and Truesdell's view, Newton did not 'complete the formal enunciation of mechanics', rather he initiated it. What is significant in the present context of teaching, is that the Principia, and much of what followed, did not conclusively explain some central elementary concepts in mechanics. Mach conceded that the concept of force was weak – is the Second Law a definition of force, how do we measure it and where does it fit in nature? Newton was able to discriminate between mass and weight, but was imprecise on what he meant by a *body* – sometimes it was a point mass, whereas in other parts of the *Principia* he studies 'bodies' which have finite volume and distributed mass. Mach (1960) discusses the relation between mass and weight at length in a form now adopted in many student textbooks. The point here is that the Laws of Motion, to this day, are based on *concepts* which were unclear and perhaps have remained so to many. The associated quantities are easy to manipulate as part of a mathematical problem, which is usually emphasised to the novice student, but less easy to conceptualise as real natural entities.

As we know, mechanics developed rapidly over the next two hundred years after Newton and Lagrange primarily due to the efforts of European and Prussian 'mathematicians' (in the absence of a better description at the time). The rise, and eventual nature, of the university during the 19th Century is also well known and descriptions of the importance of the lecture, tutorial and laboratory system which emerged are readily available. Inherent in this system is the role of discussion, usually in small classes, although what this involved is not well documented. William Clark's notion of 'academic charisma' includes the ethos of 'lecturing with applause' which placed the lecturer at the centre of this teaching system (Clark, 2008).

At the beginning of the Industrial Revolution Glasgow was a favourite gateway to the Atlantic and was thus a natural centre for industrial development. The scene was then set to provide education for the workers in industry. Public (rather than military) engineering education started to appear through open lectures in the engineering sciences and mechanics in particular. An accompanying literature in English arose to meet the demand (Emmerson, 1973). Public lecturers were usually itinerant and brought their demonstration equipment into 'courses of lectures' which were often held in clubs and coffee houses. The Scottish scientist James Ferguson (1710-1776) was well known, eventually moving to London where his lectures on mechanics were both very popular and fashionable, with the future King George III occasionally in attendance. Professor John Anderson (1726-1796) of Glasgow University gave popular lectures for the townspeople of Glasgow. His evening lectures increased in number from 30 to 200 'people of every rank, age and employment'. One of Anderson's abler students, John Robison, succeeded Joseph Black (1728-1799). Black introduced the concepts of latent and specific heat while at Glasgow University. He was then appointed as professor of mathematics at Kronstadt Naval School in 1772; where it is known he interacted with Euler who was nearby in St Petersburg. Robison returned to Edinburgh in 1774 as professor of natural philosophy and was regarded as one of the best and most enlightened educators in mechanics, contributing much to the new *Encyclopaedia Britannica* (1797). Robison's talents as a teacher are well illustrated in these articles. Robison was inspired by Black's gift for teaching. One of Black's students, the eminent lawyer Henry Brougham, wrote (Brougham, 1871): '. . . his style of lecturing was nearly perfect . . . it had all the simplicity which is so entirely suited to scientific discourse . . . there was no effort, but it was an easy and graceful conversation . . . in one department of his lectures he exceeded any I have ever known . . . the manipulation of experiments . . . '. In his will Anderson left funds (which were in fact insufficient) to create a new university as a 'place of useful learning' for technical education in 1796 to be named Anderson's University. The University title had to be removed later and the institution became the *Glasgow Mechanics' Institute* (Walker, 2016) in 1824. Similar Institutes were established throughout the country and spread to Paris and the USA (as the Franklin Institute – Benjamin Franklin visited John Anderson in Glasgow). One of the Glasgow Mechanics' Institutes lecturers, George Birkbeck (1776-1841), founded the London Mechanics' Institute (later Birckbeck College now part of the University of London). Birkbeck's lectures were described (Emmerson, 1973) by one prominent member of his audience: '. . . when he at length completed his illustrations (on the Laws of Attraction) . . . a unanimous plaudit burst forth from the delighted audience . . . I was struck that within a century of the death of Newton . . . his most sublime discoveries could be rendered intelligible to eight hundred working mechanics . . . '. A potent example of Clark's 'academic charisma' (Clark, 2008). This was in stark contrast to the practice at the universities such as Oxford and Cambridge, which emphasised the classics and mathematics at the time. Professors did little teaching, being the responsibility of 'fellows' of the colleges - although the real teachers were the many private coaches used by the students. Engineering education was reviewed in Cambridge in 1890. As part of this review it was noted that the technical colleges and institutes '. . . studied a large number of technical problems which were in fact merely problems in Physics or Mechanics . . . '. It was believed that the universities were far better suited to teach these subjects along with the supporting mathematics. The move to teach engineering (and mechanics) in universities in the UK subsequently followed.

It is evident that mechanics teaching until the Twentieth Century focussed on gifted lecturers using original, modern textbooks in a new and exciting subject. The lectures would seem to have been well attended and well received. Skills in mathematical problem solving in mechanics were desirable, and practised in tutorials with small classes. The various sub-topics could be illustrated by laboratory work and a 'hands-on' approach. George Emmerson (1973) provides an excellent overview of the engineering mechanics textbooks in the 19th Century. In the early part of the century the French textbooks were well established: Claude-Louis Navier's *Leçons sur l'Application de la Méchanique* (Navier, 1838) was a standard work for French and German engineers. However, in the later part of the century, '. . . transcending all the engineering textbooks of the century were William J. McQ. Rankine's *A Manual of Applied Mechanics* . . . ' (Rankine, 1858). '. . . in these manuals there was a unique blend of science and technology . . . '. It was used to educate several generations of engineers and was translated into German, French and Italian.

In concluding this section, it should be noted that commentaries on the mode of teaching are scarce. It is known that the lecture, tutorial and laboratory system flourished, but there are few observations on their worth. A recent survey (Fox and Guagnini, 1993) on the development of science and technology education in Europe and the USA between the mid nineteenth century and the 1930s hardly mentions teaching methods at all. The survey only remarks on the difficulty in attracting students away from the colleges to the universities in the UK. This was different in other countries who introduced higher technical education in different ways, for example through specialist schools for new subjects in France and the detailed use of mathematics and calculation in the German *Hochschulen*. The latter were reformed, starting in the 1890s, to include more laboratory work. Indeed, the rise of practical, hands-on instruction in the laboratory was a common feature throughout Europe. Yet the colleges, then universities, in the USA never seriously committed to experimental work, which was considered part of an apprenticeship in a specific industry. The historian Eugene Ferguson (1992) has argued that the loss of the practical apprenticeship in the American system has not been beneficial '. . . engineering education since 1945 has been skewed toward analytical techniques which are easiest to teach and evaluate . . . '. This has been exacerbated by the rise of mass higher education, making the teaching of engineering, and mechanics in general with its inherent conceptual hurdles, more difficult.

2.3 A New Conceptual Approach

Educators at the beginning of the twentieth century started to recognise that the conceptual difficulties associated with introductory mechanics could be problematic in teaching. Commentaries on the development of *dynamics*, for example (Barbour, 2001), highlight these conceptual problems throughout each era but there is little evidence that this was recognised openly in teaching until the twentieth century.

A 1917 discussion by Huntington (1917) proposing a logical structure to teach elementary dynamics identified the issues with the concept of force: '... for example, if no force is acting on a body then there is no change in velocity seems to contradict our commonest experience of motion which appears to die down of themselves ...' and further '... what a force means should not be complicated by any discussion of what a force will do ... force is an active agent by which matter is buffeted about according to our will ... '. Huntington further argued that '... textbooks impose a student's attention on an unpractical group of systems in which the derived units are based on mass, length and time instead of force, length and time ... '. Fawdry (1920) believed that calculus should not be used for beginners in mechanics since the work would more likely to be an exercise in the manipulation of symbols. This was echoed by Filon (1926) '... the investigation of principles is frequently replaced by a few dogmatic statements, and, with an ill-concealed impatience of difficulties which he is unwilling to face, the teacher hurries on to a mass of numerical or algebraic examples, often of a highly artificial nature ... but ... how are we to meet the difficulty about mass? ... '. There was continuing debate, particularly in the *Mathematical Gazette*, around the emphasis on mathematical problem solving; Milne (1954) argued that in school '... one does not get a grasp of the principles of mechanics by thinking about them, however intensely ... one gets a grasp .. by carrying out the solutions of mathematical problems ... '. Yet '... for the university course, some analysis of the origin of the concepts of mass and force must be traced, and in this regard it is difficult to do better than follow Mach. In Mach's presentation, the equality of action and reaction, Newton's Third Law, is not adopted as an empirical law, but adopted as an axiomatic definition leading to a satisfactory definition of equal forces and hence in due course of mass and force ... '.

Perhaps one of the most persuasive discussions on mechanics teaching has been due to the German/Dutch mathematician Hans Freudenthal in 1993 (Freudenthal, 1993). Freudenthal argued that teaching mechanics to students should begin with their own (body) experience and be guided to transform their ideas into mechanical science. Unless this is done at the beginning '... rather than mechanics, the learners are taught mechanising. Forces, unless experienced personally, are doomed to remain phantoms ... '. Further, '... what matters didactically is to get across that force expresses itself by changing the state of motion ... '. Freudenthal then argues that

Descartes original idea of force, which we recognise as work, should form a basis for instruction: '. . . energy, among all mechanical notions the trouble physicists had with it in the past . . . seems to be the most concrete notion . . . '. Indeed, seventy years before, Fawdry (1920) thought that '. . . it is desirable to reduce to a minimum the time spent on the second and third Laws of Motion, so that we can devote our savings to the acquisition of facility in dealing with problems by the application of the Energy principle. If our pupils get into the habit of calculating the acceleration of a moving body, it is difficult to get them out of it . . . '. The emphasis on Newton's Laws in early mechanics teaching is criticised every so often: we can leave a final comment to Freudenthal – '. . . Action and Reaction: never have you heard me pronounce here this pair of words. However interesting it might be from a historical standpoint, I simply do not know how to use it in instruction . . . '.

During the 1980s a vision arose that elementary mechanics teaching needed to be revised and that a new approach should be considered; if this was not addressed then problems with student understanding could continue into later years. A focus of this observation was the work of Sheila Tobias (1990) who examined the 'second tier' of university students - those who had to study science but whose major was not physics. These students dismissed the standard of teaching in physics/mechanics as compared to other classes. From the 'Eric Experiment' Tobias quotes this student as testifying '. . . the class consisted basically of problem solving and not of any interesting or inspiring exchange of ideas. The professor spent the first 15 minutes defining terms and apparently that was all the new information we were going to get on kinematics. Then he spent 50 minutes doing problems from chapter 1. He was not particularly good at explaining why he did what he did to solve the problems, nor did he have any real patience for people who wanted explanations . . . '. Further Tobias pointed out that simple mechanics as taught in the 1980s was not too different from the 1880s, but delivered to a larger and more diverse audience without revision. In fact, concerns with introductory courses in mechanics had existed even in the 1800s. James Clerk Maxwell (1831-1879) was considered a poor teacher: although his classes consisted of less than ten students they were noisy, undisciplined and uninterested. Yet he is reported as having been forward thinking in his pedagogical ideas, which find resonance with today's outlook (Golin, 2013). He believed in the notion of illustrating the mathematical approach with practical experience, either in the laboratory or by demonstration and strongly valued the history of science in teaching. A significant catalyst for change were the studies of David Hestenes and his colleagues (Halloun and Hestenes, 1985; Hestenes et al, 1992). The *Force Concept Inventory* [38] was a revelation: a simple quiz of thirty questions aimed to test a student's understanding of the concept of force, in several different situations, and basic Newtonian mechanics. This was eventually given to over six thousand freshman students, as reported in 1998 by Hake (1998). As a diagnostic pre-test it could show the difference between students' initial knowledge of force as compared to after some mode of instruction towards the end of a first year class (a post-test). The pre-test was consistently discouraging even in the best universities and colleges. There then grew a belief, initially among groups of physics educators in the USA, that a different approach, emphasising the understanding of concepts, was needed.

Early work had already been done by the physicist Paul G. Hewitt who wrote a popular series of textbooks, *Conceptual Physics* (Hewitt, 1987) and co-authored, with Lewis Carroll Epstein, the innovative *Thinking Physics* (Epstein, 1981) which consists of a large number of conceptual mechanics questions which could be used in a lecture or tutorial to provoke discussion. Hewitt's book was the first to attempt to teach introductory mechanics with minimal mathematics using his own drawings and cartoons – an approach he later popularised (Hewitt, 2002). A formal teaching approach, based on the works of Hewitt, Epstein, Tobias, Hestenes et al was later developed by the Harvard physicist Eric Mazur as *Peer Instruction* (Mazur, 1997) and by The University of Massachusetts Physics Education Research Group (UMPERG) as *Minds on Physics* (Gerace et al, 2000; Leonard et al, 1990).

A re-evaluation of basic physics teaching, and mechanics in particular, began to evolve. In addition to the issues already highlighted with conceptual understanding, other problematic areas emerged: should the subject be introduced in the early stages using algebra rather than calculus to aid conceptual understanding, should there be less focus on Newton's laws perhaps treating energy principles equally and so on. How should the large lecture format be dealt with and the ubiquity of online resources? Should practical experience be hands-on in the laboratory or through demonstration or video or simulation? This pedagogical approach, which was based on presenting concepts and was based a form of Socratic Dialogue assisted by technology in large classes, also spread to other subject areas.

2.4 Teaching Introductory Mechanics

The writer adopted concept-based teaching in mechanics over two decades ago and during that time has experienced students' conceptual difficulties working with over three thousand first year students. Numerous seminars on these experiences, and demonstrations, have been given in schools, colleges and universities across the globe. Before this journey, like many other educators or academics in higher education, the writer had seen that a proportion of students had a poor understanding of mechanics but believed that the abler students were comfortable with the specifics of the subject. With hindsight it is very likely that the latter were more successful at preparation for examination questions. A seminar with a large number of secondary school physics teachers presented by the writer in 1998 was intended to demonstrate conceptual difficulties in mechanics. But it was disastrous since it was clear the teachers had similar difficulties. A similar experience has been mentioned by Richard Hake (1991). Hake's solution was to follow the teachings of the University of Washington science educator Arnold Arons (Arons, 1974) who wrote '. . . If a teacher disciplines himself to conduct such Socratic Dialogues . . . he begins to see the difficulties, misconceptions, verbal pitfalls, and conceptual problems encountered by the learner . . . in my own case, I have acquired in this empirical fashion virtually everything I know about the learning difficulties encountered by students. I have never been bright enough or clairvoyant enough to see the more subtle and significant learning hurdles a priori . . . '. Hake (2019) later recommended what he called the *Arons-Advocated Method* for introductory science teaching. This method emphasised conceptual understanding, interactive engagement, Socratic Dialogue, attention to cognitive development, attention to preconceptions of beginning students among several others. All of these are discussed in considerable depth in Aron's monumental work *Teaching Introductory Physics* (Arons, 1997). These ideas eventually led to the pedagogical tools of *Peer Instruction* and *Minds-On-Physics*.

In summary, the importance of paying close attention to conceptual misunderstandings in mechanics in the early years of university, and in particular at the transition from school, cannot be underestimated. For this reason, these aspects are dealt with first here:

2.4.1 Conceptual Misunderstandings in Newtonian Mechanics

One basis for the need to re-think mechanics teaching in the early years at university, has been the very nature of the subject itself: mechanics is counter-intuitive and often goes against 'common sense'. Of course this has been a feature of the subject throughout its history, even with its originators. The central role of adopting Socratic Dialogue – teaching by questioning – in order to support and reveal conceptual misunderstandings has been highlighted above. In this section an assortment of such misunderstandings will be discussed:

The following discussion illustrates some typical *multiple-choice questions* used to initiate Socratic Dialogue in a first year mechanics class as part of revision of school work. The students answer these questions initially using a *classroom voting system* (which will be discussed in the following section) before any discussion. Almost all of these students have come highly qualified from the school system. They have already studied physics and Newton's Laws and their applications; some have also studied connected bodies and circular motion at constant speed. On arrival they naturally view mechanics as mathematical problem solving. Now they are now faced with a class where the focus will be on understanding and discussing concepts.

2.4.2 Kinematics & The Law of Falling Bodies

Most, if not all, introductory mechanics courses start with Kinematics and the Law of Falling Bodies before continuing to Newton's Laws. However, two simple conceptual questions usually confuse many students based on their initial responses (or 'votes') without discussion:

Question 1: If you drop an object in the absence of air resistance, it accelerates downward at 9.8 m/s^2. If instead you throw it downward, its downward acceleration after release is: (1) less than 9.8 m/s^2, (2) 9.8 m/s^2 or (3) more than 9.8 m/s^2.

Question 2: A person standing at the edge of a cliff throws one ball straight up and another ball straight down at the same initial speed. Neglecting air resistance, the ball to hit the ground below the cliff with the greater speed is the one initially thrown: (1) upward, (2) downward or (3) neither, they both hit at the same speed.

Surprisingly the class will usually be split between choices (2) or (3) in Question 1. On initial discussion it becomes apparent that those who answered (3) argue that 'it is going faster since it was thrown', even though the question is about the object's acceleration under gravity. Responses to Question 2 are equally surprising with only about half the class choosing the correct answer. Usually the majority of the rest choose answer (1) with the belief that after being thrown upwards, once it reaches the top it has a greater distance to fall, so will be going faster when it hits the ground. In fact, using other questions and further discussion it becomes apparent that not all students have grasped the difference between velocity and acceleration and so have not really thought too much about the nature of kinematics.

Instead, students tend to concentrate on the simple equations of motion for linear kinematics with constant acceleration - relating initial and final velocities, time elapsed and distance moved - and attempt to interpret situations using these equations and by 'plugging in numbers'. In fact, even in situations where acceleration is not constant most students will initially try to use these equations.

2.4.3 Basic Forces

Before introducing Newton's Laws themselves, and discussing their implications, it can be useful to introduce real situations and see if students can interpret what is happening in terms of forces acting. Three questions are instructive:

Question 3: You are in a cabin on a Ferris observation wheel as it slowly turns at constant speed. Is there a net force on the cabin? (1) No, its speed is constant, (2) Yes or, (3) It depends on the speed of the cabin.

Question 4: You're driving a car up a gentle slope. Put the car into neutral and coast. At the instant of zero velocity, abruptly put on the brakes. What do you feel? (1) Nothing, (2) A jerk or, (3) Not enough information

Responses to Question 3 are often divided between choices (1) and (2). Choice (1) is selected since responders see the phrase 'constant speed' and forget that the cabin is not moving in a straight line so must be accelerating under some force. And what is the force? Question 4 is even more interesting since nearly all students will select answer (1). The notion that acceleration can change seems to come as a revelation to students. When asked to identify how the force on the car changes students are equally puzzled. Then a familiar situation is proposed and discussed:

Question 5: When you climb up a rope, the first thing you do is pull down on the rope. How do you manage to go up the rope by doing that?

Students are generally unable to resolve this situation in terms of the forces acting (in practice they are asked to discuss in groups and volunteer solutions – a few groups come close to the correct analysis). Questions like those given above are useful in getting students to realise that their familiar notions of simple kinematics and the analysis of forces in realistic situations need to be 'upgraded'. The problem (which is well-known amongst science educators but not necessarily by all academics teaching mechanics) is that 'force' is viewed as just a number which is input to Newton's Second Law. The 'force' may even be given a practical-sounding name but most students will not have considered what this force actually represents, where it comes from, or what it does. This is openly debated with the class and the reasons for devoting so much class time in in the discussion of concepts in mechanics can be stressed. (At this point the writer habitually has a diversion to the career of the mathematician Richard E. Bellman (1920-1984). Most students will not have heard of Bellman but his accomplishments are outlined: dynamic programming, the Bellman Equation and its role in economic theory. However, they can now appreciate the relevance of his classic statement '. . . the trick one learns over time, a basic part of mathematical methodology, is to *sidestep the equation and focus instead on the structure of the underlying physical process* . . . " (Bellman, 1984)).

Before examining problem solving methods the writer has found it best to spend substantial class time discussing several *basic forces*: gravity, contact, friction and tension. Newton's Laws of Motion can be introduced formally before or after this discussion, but the writer prefers the latter in order to underline the significance of the various forces as real quantities, rather than as a just number to be input to the Second Law as a mathematical exercise.

The *force of gravity* is understood by most students but they are happier dealing with the acceleration it produces in falling objects:

Question 6: What can you say about the force of gravity acting on a stone and a feather as they fall in a vacuum? (1) The force is greater on the feather, (2) The force is greater on the stone, or (3) The force is equal on both always

An instant response from the majority would favour answer (3), confusing force and acceleration. A follow-up question, without comment on the outcome of Question 6, replacing 'force of gravity' with 'acceleration due to gravity' elicits the correct response and allows for discussion. This leads to the relationship between mass and weight (as a force) and the use of the Second Law as proposed by Mach (1960). The realisation of the distinction between the force of gravity and acceleration due to gravity starts to become apparent to students only at this stage and it is worthwhile outlining the difficulties found in the Principia (Newton, 1729) and those highlighted by Truesdell (1968) to show the students that it's not only themselves who get confused.

While the force of gravity can be perplexing, the force of contact is a complete mystery (and a novelty) to most incoming students. For example, students can be asked to support a heavy object in the palm of their hand and discuss the forces acting on the object. First there is recognition that the force of gravity must be (always) present and that the object is not moving, and the net force is zero – so what force is balancing the force of gravity on the object to stop it moving? Eventually, and this may take time, the existence of a contact force emerges. Further discussion brings out that this force must equal the weight, but is it always equal to the weight? (A notion that was repeatedly discussed in antiquity, Aristotle, 1936; Wallace, 1971; Lamar Crosby Jr., 1955; Laird, 1986; Truesdell, 1968; Freudenthal, 1993). This can now be discussed further in another context:

Question 7: Consider a student pulling or pushing a loaded sled on snow with a force which is applied at an angle. What can we say about the contact force between the sled and the snow?

After considerable discussion the majority of students will finally realise that the contact force is greater when pushed than when pulled, and be able to roughly explain why. The contact force is not equal to the weight in either case. If this realisation takes some time a further situation can be put to the class:

Question 8: Consider two identical blocks, one resting on a flat surface and the other resting on an incline. What can we say about the contact force?

Finally, a classic force problem can be introduced:

Question 9: Consider a person standing in an elevator that is accelerating upward. The upward normal force exerted by the elevator floor on the person is: (1) Larger than, (2) identical to, or (3) smaller than the weight of the person?

This also brings much discussion, since initial responses commonly prefer answer (2). At this point Freudenthal's notion of 'body experience' is useful (Freudenthal, 1993): "what happens when you're in an elevator in a tall building? Do you feel lighter or heavier at any point during the motion, and why?"

Of course, the correct resolution of Questions 7-9 requires a free body diagram: the students are aware of this little picture, but have rarely used in problem solving. This will be discussed later in this section.

Most engineers are familiar with the difficulties associated with the friction force since it is basically treated empirically in most university courses. Once the concepts of kinematic and static friction are summarised (the latter traditionally leading to much confusion) some common misconceptions can be discussed through questions:

Question 10: As we know from everyday experience, the force of kinetic friction tends to oppose motion. Do you agree with this?

The majority will agree, yet faced with the next question (which does need a free body diagram and the students are encouraged to draw this):

Question 11: Consider the following situation: a block sitting on top of another block and the system moving to the right under the action of an applied force. For the upper block, how does the direction of the friction force relate to its direction of motion? (Imagine what happens if we lubricated the surfaces between the blocks).

There is usually a sense of disbelief when confronted with this situation.

The tension force has been used many times in school problems, but is usually given in a diagram as a symbol or number and an arrow. When introducing tension some time has to be taken to talk about ideal ropes/cables etc. in order to reach the concept that the tension will be the same throughout the rope. Similarly, the notion of an ideal pulley needs to be discussed. Once this has been done a pair of questions can be tested:

Question 12: A 10 kg mass is hangs from an ideal cable which passes over an ideal pulley fixed to a ceiling and is then attached to a wall so that the cable is horizontal. What is the tension in the cable?

Question 13: Two 10 kg masses are each attached to the ends of an ideal cable. The cable passes over two ideal pulleys each fixed to the ceiling at the same height. Between the pulleys the cable is horizontal. The masses hang from each end after the cable passes over each pulley. What is the tension in the cable?

The majority of students answer Question 12 correctly but surprisingly in Question 13 work out the tension from the sum of the masses and obtain double the correct value. Experience has shown that the only way to convince them of the correct value for the tension is to draw a free body diagram, and even then some dispute this.

2.4.4 Connected Bodies, Free Body Diagrams and Problem Solving

A significant amount of time is spent defining what constitutes a *valid free body diagram* and working through (again using multiple choice questions and discussion) many examples. Towards the end of this discussion the idea of connected bodies is introduced, with multiple masses in contact, or connected by cables etc. The *spring force* (ideal linear spring) is also introduced, so that connecting objects by springs can be examined. Yet again experience has shown that many students have some difficulty correctly interpreting the forces in (even simple) connected bodies. For example, the following question causes problems:

Question 14: Two boxes, one large and heavy, the other small and light rest on a frictionless level floor. You push with a horizontal force on either the small box or the heavy box. Is the contact force between the two boxes: (1) The same in each case, (2) larger when you push on the large box, or (3) larger when you push on the small box?

Without fail the majority of students will answer (1) without thinking. Prompted to actually draw free body diagrams for each box usually (but not always) convinces most students. During discussion it usually emerges that their first thoughts still relate to applying the Second Law, that is thinking in terms of solving a mathematical problem to the whole system.

The presentation and in-class questioning and discussion of the preceding topics typically takes twenty hours of class time (ten two hour lectures). This may seem like an unwarranted amount of time, but after twenty years it has been found to be essential. The incoming students, while well qualified from school, do not really understand the basic concepts. By the end of experiencing this novel conceptual approach they become acutely aware of this, but at the same time also aware that most their peers have the same difficulty. Up to this stage only point masses have been studied – later in the course distributed mass, statics and rigid body rotational dynamics are also introduced using the same Socratic Dialogue methodology. Before moving on to more advanced topics, mathematical problem solving is finally reintroduced, but applied to connected bodies with multiple masses, not single masses. In this course mechanics is taught only using algebra – the course presentation is intentionally not calculus based, to distinguish the subject from an application of the calculus. The students are presented with a *structured problem solving framework* (Moore, 2009): a diagram of the problem should be sketched and labelled then all variables must be identified, given appropriate symbols and defined in words from the outset. Known or unknown quantities should be identified for each mass and then a valid free body diagram should be drawn. After this the relevant mechanical principle(s) should be stated and appropriate equations written down and the unknown variables solved symbolically – only at the end numbers are used and 'the solution' given. This structured problem solving approach uses *multiple representations* of a problem: there are three representations – visual (or pictorial, drawing a diagram of the problem), *conceptual* (identifying variables and drawing free body diagrams) and *mathematical*. The students are told that good problem solving involves looking at the problem in different ways (multiple representations).

This problem solving methodology is then applied to circular motion with constant speed and then moves on to energy methods for point masses and conservation of momentum (again all with many conceptual questions for in-class discussion). A different structured problem solving framework is introduced for conservation of energy and momentum. Finally, this approach is used in when introducing statics and distributed masses.

2.4.5 Using Socratic Dialogue with Technology

Little has been said so far about the methods and systems used to enable Socratic Dialogue in class. This is easy with a small class, say less than thirty students at most, but much less so in a large class (the writer routinely had first year mechanics classes with around four hundred students). Twenty years ago using Socratic Dialogue in such large classes would be almost impossible, if not chaotic. Affordable computer based 'classroom communication' systems started to emerge in the mid-1990s. The first, *ClassTalk*, was a proprietary system of networked Texas Instruments graphing calculators released in 1994 and described by Dufresne et al (1996) who used it in freshman physics classes at the University of Massachusetts, Amherst (and was later adopted by the writer). This system allowed multiple choice questions to be asked using an overhead projector (or projected on the screen from the accompanying software) with each student selecting their answer on a shared calculator. The results were collected by a personal computer linking all calculators in the first 'wired classroom'. The combined responses could then be shown graphically on screen as histograms to initiate a discussion. Soon after ClassTalk, cheaper handheld systems began to appear, using firstly infrared then RF technology to collect responses. Each student could use their own handset which was small and portable. The intention here is not to describe this type of technology in any detail – it is continually developing and has already merged with mobile phone technology and Wi-Fi. The current market leader is Turning Technologies *TurningPoint* and the interested reader is referred to their website (Turning Technologies, 2019). These technologies are now used in thousands of university courses in many disciplines: these are described in more detail in several surveys (Banks, 2006; Duncan, 2004; Goldstein and Wallis, 2015).

Given the ready availability of this type of voting/polling technology for the large class, how is it used? Again this will not be discussed in detail since there are many variants adopted in practice. The most common is *Peer Instruction* (Mazur, 1997). The process is very simple but has been shown to be very effective. Firstly, a question (usually multiple choice) is posed to the class and shown on screen. The students are asked to think about it for a minute or so without any discussion, then send their response using their handsets – the results are displayed as a histogram. Very often the class are divided (and they react to this division amongst their peers). The second stage is to ask individual students to describe why they chose a particular answer (they are given time to think about this first) – this can be done by volunteers or by random selection. Other students are asked to comment on the responses: the lecturer can direct this, but does not reveal the correct answer at this stage. The third stage is to ask the students to discuss in groups (essentially those students who are sitting close to them, but this can be more structured): the class is then polled again, the results displayed and if necessary discussed again. The lecturer can give hints and continue with class discussion and further polling as necessary – it usually depends on the question. In fact, connected sets of questions are often used, but not necessarily all are presented to the class, again depending on what transpires. Classrooms are very lively (and noisy) but there is a freshness to large groups of students actively discussing mechanics concepts in class! Numerous studies on the effect of Peer Instruction are available: the writer's own (Banks, 2006; Nicol and Boyle, 2003; Boyle and Nicol, 2003) and those from Mazur and his colleagues (Crouch and Mazur, 2001; Crouch et al, 2007).

Finally, the discussion in the present section on teaching introductory mechanics does not give a complete picture of everything that happens in class. The class is not wholly based on Socratic Dialogue but uses a collection of pedagogical tools, for example *Just-In-Time Teaching* (Novak, 1999) and *Ranking Tasks* (O'Kuma et al, 2000). Extensive video resources are also used such as the *Mechanical Universe* series (Olenick et al, 2008), which gives an historical perspective and covers introductory mechanics, and *The Video Encyclopaedia of Physics* (The Education Group, 2019), which provides many physics demonstrations and can be coupled to multiple choice questions. These are supplemented by videos of practical problems (from sports, aerospace and structural engineering are good examples). The proper use of all of this technology must be underpinned by an individual instructor's reading of research in student learning. This is not the place to discuss this in any depth, but the reader can do worse than start with the US National Academies report *How People Learn*

(National Research Council, 2000). All of these tools combined help to structure the class away from a teaching format where a lecturer stands in front of a screen. It makes mechanics much more interesting and relevant.

2.5 More Advanced Topics: Continuum Mechanics

The reader should now be able to recognise that even basic concepts in mechanics can be problematic for the new learner while pedagogical techniques which can enhance deep understanding are available. It is now time to discuss (but in less detail) the consequences of this in a more advanced topic, namely continuum mechanics. The problems with teaching introductory mechanics arose through physics education. Continuum mechanics, and the concepts of stress and strain, elasticity, inelastic materials and strength of materials are not covered in physics education (but it has been argued they should be (Golub, 2003)). The immediate question is: are there conceptual issues here? Most instructors delivering introductory mechanics of materials courses would recognise that there are, but that these can be of a different nature, and sometimes subtler. There are two areas in introductory courses where conceptual issues must be addressed: *statics* and the concept of *stress*. There has been some work done relating to the former, less for the latter.

2.5.1 Teaching Introductory Statics

Most of the work done in this area has been carried out by Paul Steif (Steif and Dollár, 2005) and colleagues at Carnegie Mellon University. This work is an extension of the developments in introductory physics and mechanics. Certainly for an engineering student statics is one of the key supporting subjects for later studies of strength of materials, continuum mechanics and dynamics. It is normally introduced in the first year of a mechanics course usually after point mass dynamics. Yet it is essentially a new subject and requires different concepts such as distributed mass, centre of gravity, moments about a point due to a force, rotational equilibrium and couples (or torques). Students find the last two concepts quite confusing and mostly learn some trigonometric 'rule' to manage problem solving. Indeed, most introductory textbooks begin with revision of all the required mathematical tools from trigonometry and vector algebra and then treat statics as an exercise in applying these tools. Unfortunately engineering deals with real components, which are often in contact, have internal forces resulting from externally applied loads, connections and supports. Skill in drawing valid free body diagrams, now different from those required for connected point masses, becomes even more essential – this is an order of magnitude more difficult than introductory mechanics and students struggle. Supports, connections and joints are simplified and students similarly learn 'rules' to represent them in free body diagrams. The writer strongly believes that Freudenthal's notion of 'body experience' (Freudenthal, 1993) is even more crucial here if it is recognised that most students will not have that experience. As a result, classroom demonstrations with student participation are felt essential. For example, a student can be asked to balance a metre stick at a quarter point (with their finger, which is usually quite entertaining) with a known weight suspended from a light string at the closest end. Using a multiple-choice question, the class are asked to work out the weight of the measuring stick. Experience has shown that more than half the class will get this wrong. Further dialogue reveals that the difficulty lies in knowing where the centre of gravity of the stick is placed. Additionally, videos are also particularly useful: real engineering components, structural failures and so on. Numerous videos are available online and the UK Open University is a good source.

The style of multiple choice question required for most statics teaching using Socratic Dialogue differs slightly from introductory dynamics. The aim is to get the students to think about and discuss the nature of forces in some component, often components in contact and with various supports. There are typically a set of questions asking the class to think about what happens when the supports are removed and what's required in a free body diagram of the whole structure, then each component is taken on its own and the contact forces are introduced for the free body diagram and so on. Calculations may be done from step to step, but the emphasis must be on identifying the correct support forces, contact forces and internal forces. A formal problem solving strategy can be introduced using the multiple representation method discussed above and at that stage a complete mathematical analysis can be carried out. Many examples of conceptual questions for statics can be found throughout Steif's writings (Steif and Dollár, 2005); they are fairly straightforward to generate but it is important to always remember that the identification of forces and the sequence of free body diagrams must be emphasised. The importance of understanding concepts in statics should be clear to students by this time following the similar approach used in introductory dynamics.

Steif recommends that a particular *sequence* of introducing concepts in statics should be adopted: (1) 2-D translational and rotational equilibrium of bodies under the action of forces only, including the concept of centre of gravity, (2) couples and static equivalence, (3) equilibrium of bodies with forces and couples acting together, (4) separation of bodies, (5) contacting bodies and distributed forces, (6) frictional contact forces and their net effects and finally (7) equilibrium with normal and frictional forces. The concept of separation of bodies is felt to be critical since students are more likely to fail to appreciate that two concepts are involved: every force acts between two objects, typically in contact, and that the equations of equilibrium can apply to a single object with an associated free body diagram. Practical 'body experience' is essential. Friction is not introduced until later in the sequence so that students can see the importance of neglecting friction in examples used for previous concepts. The final concept in statics, (7) allows structured problem solving to be presented but with additional insight into the role of forces, moments and couples.

In the writer's experience, educators who adopt the more mathematically-based approaches found in traditional statics texts sometimes fail to appreciate the conceptual difficulties the students have with this topic. The concepts are usually lost in the practical obscurity of the mathematics. Steif and Dantzler (2005) also developed a *statics concept inventory* to assess students' prior misconceptions. Characteristic results from administering this test highlight conceptual problems with friction, static equivalence and in particular internal forces between contacting bodies and their associated free body diagrams.

2.5.2 Introducing Continuum Mechanics

The point at which the concepts of *stress* and *strain*, and their relationship for linear elastic bodies, is introduced can vary. In the writer's classes this is done as part of introductory mechanics rather than in a separate new course. It was felt that the notion of deformable bodies, and an idea of how to deal with them, was important at a very early stage. Students should understand that statics is not solely concerned with calculating support and connection forces: internal forces are also calculated for a reason. Since the science of deformable bodies is new to most students they arrive with few existing misconceptions. It is then very important not to introduce any confusing ideas about the nature of this new subject at this early stage.

One fascinating misconception which most students have is related not to mechanics, but rather lies in the mathematics of vectors. Most will claim that the principal characteristic of a vector, which differentiates it from a scalar, is that it has both *magnitude* and *direction*. Of course they have forgotten (or indeed never been told) that its definition requires a system of combining similar vectors (an algebra) and that addition must be commutative. This must be pointed out to avoid misconceptions when stress and strain tensors are introduced in later years. In relation to this many novice students will always write Newton's Second Law in scalar form even if they acknowledge that force and acceleration are vectors. They should also be forewarned that not all engineering quantities with 'directions' are vectors: finite rotations are a good example where addition is not commutative which can be easily demonstrated with a book in class.

Many introductory engineering mechanics textbooks which cover statics will often introduce stress and strain following the calculation of internal forces in members in trusses. Stress is immediately defined as the internal force divided by cross-sectional area and strain, for no apparent reason, as the change in length divided by the original length. After defining a linear elastic material, Young's modulus of elasticity is introduced and some graphs of tensile tests shown. Thus immediately students are led to believe this is straightforward, or 'obvious'; but the underlying conceptual difficulty is hidden. The writer has always found it quite advantageous to start the discussion of stress from a historical perspective:

To begin with the instructor can point out that until fairly recently in the development of mechanics, component *strength* could only be crudely characterized by *empirical* means – for example by measuring the deformation of a real structure under a given load. But as mathematics and engineering science developed numerous simple mathematical models relating internal force to deformation of specific structures (mostly beams and columns) were developed. For example, Galileo, in the second of his New Sciences (Lamar Crosby Jr., 2003), came up with a theory for the strength of cantilevered beams - but it was out by a factor of three, since he probably never tested the theory. Nonetheless, his theory was believed to be widely used. Although he didn't have the concepts of stress or strain Galileo, while discussing animal shape and form, did understand that there was a problem with *scaling*: '... (large) increase in height can be accomplished only by employing a material which is harder and stronger than usual, or by enlarging the size of the bones, thus changing their shape until the form and appearance of the animals suggests a monstrosity ...". The important concept here is that, when introducing strength and stiffness of components, students are made to realise that now the mechanics *doesn't scale*. In class this can be discussed using a simple question adapted from one of Galileo's dialogues: consider a rope suspending a heavy weight – how should the dimensions of the rope be changed if the dimensions of the weight are doubled? After considerable discussion students will finally see that the 'load bearing capacity' of the rope does not depend on its length. The notion of force divided by area as a significant quantity in mechanics can be appreciated.

When relating stress and strain Young's Modulus of Elasticity can be introduced. Thomas Young's (1773-1829) contribution was based on the observation that in a tensile test (in his case, as a physician, he was interested in the mechanics of the eye) it was more useful to plot load divided by area against elongation divided by original length than to plot simply load against elongation. Using this new method the plot was the same for all tested specimens. His work was published in 1807 but largely ignored; he later elaborated his ideas in his *Course of Lectures on Natural Philosophy and the Mechanical Arts* (Young, 1845). In fact, Young's experimental idea was described twenty-five years earlier by the Italian 'mechanician' Giordani Riccati (and can also be inferred in some studies by Euler in 1727 (Truesdell, 1960)). This justification for stress and strain is not really historically accurate (Truesdell, 1968): Young did not introduce his modulus as a characteristic of a material, but of the specimens he was testing and spoke in his initial 1807 lecture about fluxion rather than stress to add to the confusion. He used the works of the Bernoullis and Euler but did not reference them. Of course in 1822 Augustin-Louis Cauchy (1789-1857) introduced his *stress principle*, published in Cauchy (1827). At first sight this looks like a branch of advanced mathematics and would be familiar to anyone studying continuum mechanics today. One of Truesdell's essays (Truesdell, 1968) describes the historical development of the stress principle, yet once more this is dominated by mathematics. It is argued here that from a teaching perspective in an introductory course this should be avoided. The historical development described above at least provides some justification for the concepts of stress and strain and the associated idea of needing global parameters to describe material behaviour. This approach can be extended to introduce Poisson's ratio and the yield stress and the concept of an isotropic material before much of the mathematics of stress and strain is developed. From a conceptual perspective it has been found crucial to explain to students why some concept in mechanics is required and why it takes a specific form. The literature in science education has shown that for a deep understanding of a concept there must be a clear connection between the concept, its associated mathematical variables and what they represent (Laurillard, 2001). Unfortunately, most historical accounts of the development of the mechanics of deformable bodies emphasise the mathematical basis of continuum mechanics, even from the earliest by Todhunter and Pearson (1886). Later accounts (Todhunter and Pearson, 1953; Capecchi and Ruta, 2015), while updating the historical detail, use a similar approach - perhaps this is unavoidable although Heyman's history of structural analysis (Heyman, 1998) uses an approach which is more useful for teaching purposes.

Unfortunately, little has been done in terms of the development of a conceptual approach in strength/mechanics of materials or continuum mechanics beyond the introductory stages. The reasons for this are not entirely clear other than the effort required (and the perceived lack of reward in doing so). It is significant that attempts led by Paul Steif (Richardson et al, 2003), as part of the US Foundation Coalition funded by the Engineering Education Program of the National Science Foundation, to develop *a concept inventory for strength of materials* was not initially successful. The group reportedly found it difficult to come up with good 'distractor' options for multiple choice questions. Some follow-up studies are available in the literature but the approach does not seem to have been widely adopted, other than in Steif's 2011 *Mechanics of Materials* textbook (Steif, 2011), which didn't get beyond a first edition and appears to have only been released in the USA and Canada.

2.6 Conclusions

This Chapter has attempted to provide an overview of issues in the teaching of mechanics and describe some new, and old, pedagogical techniques which can be used in large classes in particular. Mechanics has been notorious as a subject afflicted by conceptual difficulties which lie in waiting to trap the unsuspecting novice student. It has been argued that these difficulties need to be highlighted as part of any teaching methodology which is adopted. Further, Socratic Dialogue, enabled today by technology in a large class, has proven effective in mechanics teaching from antiquity. It has also been argued that a historical approach in teaching is beneficial in highlighting conceptual problems and this should not be avoided in order to make the subject more 'modern'. While absolutely fundamental in mechanics, mathematical techniques should not direct the development of the subject to students as they strive to learn a difficult subject and leap Butterfield's hurdle. Students should see the association between concepts and their related mathematical variables. As underlined by Freudenthal (1993) students should also be taught to relate their body experience, and real engineering, to these fundamental concepts and what they represent in practical terms. Much of the discussion has related to introductory courses in mechanics, but this stage of students' learning is probably the most important. This does not infer that more advanced topics in later teaching years cannot benefit from the same approach, although this is rarely found in practice. In fact, a case can be made that discussion (Socratic Dialogue) is more relevant in advanced topics. As an example, consider how the student's first experience with inelastic material behaviour is usually presented. For metals the elastic limit or yield stress is their first encounter, taken from a graph of a tensile test. The students can then solve problems which require the maximum load to ensure elastic behaviour in some component to be calculated, which becomes a focus for examination preparation. Yet the elastic limit is perhaps not the most important concept in plastic behaviour of a metal: post-yield it is the unloading behaviour which is a more significant concept. Further the significance of plastic collapse for a hypothetical perfectly-plastic material for engineering design can be misleading. In time-dependent creep of metals students' first encounter centres around a standard creep curve from a tensile test when it is the behaviour of creeping components and the significance of different applied loads and/or fixed constraints which needs to be better understood in a conceptual framework. Looking at other topics in advanced material behaviour yields similar awkward concepts: for example, the Rule of Mixtures for composite materials. Of course nothing has been mentioned about other areas of mechanics beyond introductory statics or dynamics, but the interested reader can find similar studies in fluid dynamics and heat transfer – this Chapter can only cover so much.

Finally, nothing has been said about the role of advanced simulation in teaching mechanics. Despite a plea in 2006 from the US National Science Foundation Report on Simulation-Based Engineering Science (NSF/U.S., 2006) for engineering educators to reflect on possible consequences, there has been arguably little progress. It is the writer's view that this remains one of the most pressing unresolved questions in mechanics teaching. Simulation technology has considerable potential in teaching and learning, but it should not be forgotten that in the early stages of engineering design qualitative decisions have to be made and there is usually no time or resource for detailed finite element simulation. A deep conceptual understanding is then essential for practising engineers. This issue was perhaps best expressed by the historian Daniel Boorstin in an interview in the Washington Post published on 29th January 1984. Boorstin was quoted as saying '. . . the greatest obstacle to discovery is not ignorance – it is the illusion of knowledge . . . '.

References

Allchin D (1992) History as a tool in science education (resource centre for science teachers using sociology, history and philosophy of science). URL shipseducation.net/tool.htm, Accessed9May2019

Archimedes (1952) On the equilibrium of planes or the centres of gravity of planes I. In: Great Books of the Western World, Britannica, London, vol 11, pp 502–509

Aristotle (1936) Mechanical problems. In: Minor Works, Harvard University Press, Cambridge, Mass.

Arons A (1974) Toward wider public understanding of science: Addendum. Am J Phys 42:157–158

Arons A (1997) Teaching Introductory Physics. Wiley, New York

Banks D (2006) Audience Response Systems in Higher Education: Applications and Cases. IGI Global, Singapore

Barbour J (2001) The Discovery of Dynamics. Oxford University Press, Oxford

Bellman R (1984) Eye of the Hurricane: An Autobiography. World Scientific Publishing, Singapore

Boyle J, Nicol D (2003) Using classroom communication systems to support interaction and discussion in large class settings. Association for Learning Technology Journal (ALT-J) 11:43–57
Brougham H (1871) The Life and Times of Henry, Lord Brougham. W. Blackwood and Sons, Edinburgh, London
Brush S (1989) History of science and science education. In: Shortland M, Warwick A (eds) Teaching the History of Science, Blackwell, Oxford, pp 54–66
Butterfield H (1959) The Origins of Modern Science. The Macmillan Company, London
Capecchi D, Ruta G (2015) Strength of Materials and Theory of Elasticity in 19th Century Italy. Springer, Heidelberg
Cauchy A (1827) On the pressure or tension in a solid body. Exercices de Mathématiques 2:42
Clark W (2008) Academic Charisma and the Origins of the Research University. University of Chicago Press, Chicago
Cobban A (1971) Medieval student power. Past and Present 53:28–66
Crouch C, Mazur E (2001) Peer instruction: ten years of experience and results. Am J Phys 69:970–981
Crouch CH, Watkins J, Fagen AP, Mazur E (2007) Peer instruction: Engaging students one-on-one, all at once. In: Research-Based Reform of University Physics, vol 1, p 55
Dufresne RJ, Gerace WJ, Leonard WJ, Mestre JP, Wenk L (1996) Classtalk: A classroom communication system for active learning. Journal of Computing in Higher Education 7(2):3–47
Duhem P (1996) Essays in the History and Philosophy of Science. Hackett Pub. Co., Indianapolis
Duncan D (2004) Clickers in the Classroom. Benjamin Cummings, San Francisco
Emmerson G (1973) Engineering Education: A Social History. David & Charles, Newton Abbot
Epstein L (1981) Thinking Physics. Insight Press, San Francisco
Fawdry R (1920) The teaching of mechanics to beginners. The Mathematical Gazette 10:30–34
Ferguson E (1992) Engineering and the Mind's Eye. The MIT Press, Cambridge
Filon L (1926) Some points on the teaching of rational mechanics. The Mathematical Gazette 13:146–153
Fox R, Guagnini A (1993) Education, Technology and Industrial Performance in Europe, 1850-1939. Cambridge University Press, Cambridge
Freudenthal H (1993) Thoughts on teaching mechanics didactical phenomenology of the concept of force. In: Streefland L (ed) The Legacy of Hans Freudenthal, Springer, Dordrecht, Educational Studies in Mathematics, pp 71–87
Gerace WJ, Dufresne RJ, Leonard WJ, Mestre JP (2000) Minds-on-physics: materials for developing concept-based problem solving skills in physics. UMPERG report. URL www.srri.umass.edu/sites/srri/files/gerace-1999mdc/index.pdfAccessed9May2019
Goldstein DS, Wallis PD (eds) (2015) Clickers in the Classroom: Using Classroom Response Systems to Increase Student Learning. Stylus Publishing, Virginia
Golin G (2013) James Clerk Maxwell, a modern educator. Physics Today 66:8–10
Golub J (2003) Continuum mechanics in physics education. Physics Today 56:10
Hake R (1991) My conversion to the Arons-Advocated Method of science education. Teaching Education 3:109–111
Hake R (1998) Interactive-engagement vs traditional methods: A six-thousand-student survey of mechanics test data for introductory physics courses. American Journal of Physics 66:64–74
Hake R (2019) The Arons-Advocated Method. URL www.researchgate.net/publication/229002144_The_Arons-advocated_method
Halloun I, Hestenes D (1985) The initial knowledge state of college physics students. American Journal of Physics 53:1043–1055
Hannam J (ed) (2011) The Genesis of Science: How the Christian Middle Ages Launched the Scientific Revolution. Regenery Publishing, Washington
Hestenes D, Wells M, Swackhamer G (1992) Force concept inventory. The Physics Teacher 30:141–158
Hewitt P (1987) Conceptual Physics. Addison Wesley, San Francisco
Hewitt P (2002) Touch This! Conceptual Physics for Everyone: Mechanics. Addison-Wesley, San Francisco

Heyman J (1998) Structural Analysis: A Historical Approach. Cambridge University Press, Cambridge
Huntington E (1917) The logical skeleton of elementary dynamics. The American Mathematical Monthly 24:1–16
Lagrange J (1997) Analytical Mechanics (English translation by A. Boissonnade and V. Vagliente). Boston Studies in the Philosophy of Science, Springer, Netherlands, Dordrecht
Laird W (1986) The scope of renaissance mechanics. Osiris 2:43–68
Lamar Crosby Jr H (ed) (1955) Thomas of Bradwardine, His Tractatus de Proportionibus: Its Significance for the Development of Mathematical Physics. University of Wisconsin Press, Madison
Lamar Crosby Jr H (ed) (2003) Dialogues Concerning Two New Sciences. Reprint. Dover, Mineola, N.Y.
Laurillard D (2001) Rethinking University Teaching: A Conversational Framework for the Effective Use of Learning Technologies, 2nd edn. Taylor & Francis Ltd., London
Leonard WJ, Dufresne RJ, Gerace WJ, Mestre JP (1990) Minds On Physics: Motion, Teacher Guide, vol 1-4. Kendall Hunt Publishing
Mach E (1960) The Science of Mechanics: A Critical and Historical Account of its Development (English Translation by T. McCormack). Open Court, La Salle
Makdisi G (1974) The scholastic method in medieval education. Speculum 49:640–661
Mazur E (1997) Peer Instruction: A User's Manual. Prentice Hall, Upper Saddle River, New Jersey
Milne E (1954) The teaching of mechanics in school and university. The Mathematical Gazette 38:5–10
Moore T (2009) Six Ideas that Shaped Physics, 2nd edn. McGraw-Hill, New York
National Research Council (ed) (2000) How People Learn: Brain, Mind, Experience, and School (Expanded Edition). National Academies Press, Washington, DC
Navier C (1838) Leçons sur l'Application de la Méchanique. Carilian-Goeury, Paris
Newton I (1729) Philosophiæ Naturalis Principia Mathematica (English Trans by Andrew Motte: Mathematical Principles of Natural Philosophy)). Benjamin Motte, London
Nicol D, Boyle J (2003) Peer instruction versus class-wide discussion in large classes: a comparison of two interaction methods in the wired classroom. Studies in Higher Education 28:458–473
Novak G (ed) (1999) Just-In-Time Teaching. Prentice Hall, New Jersey
NSF/US (2006) Revolutionizing Engineering Science Through Simulation: A Report of the National Science Foundation Blue Ribbon Panel on Simulation-Based Engineering Science. National Science Foundation, URL www.nsf.gov/pubs/reports/sbes_final_report.pdf
O'Kuma TL, Maloney DP, Hieggelke CJ (eds) (2000) Ranking Task Exercises in Physics. Prentice Hall, New Jersey
Olenick RP, Apostol TM, Goodstein DL (eds) (2008) Beyond the Mechanical Universe: Introduction to Mechanics and Heat. Cambridge University Press, Cambridge
Rankine W (1858) Manual of Applied Mechanics. Griffin & Company, London
Rankine W, Bamber E (1873) A Mechanical Text-Book. Griffin & Company, London
Renn J, Damerow P, McLaughlin P (2003) Publication No. 239, Max Planck Institute for the History of Science, Berlin, chap Aristotle, Archimedes, Euclid, and the Origin of Mechanics: The Perspective of Historical Epistemology, pp 43–59
Richardson J, Steif P, Morgan J, Dantzler J (2003) Development of a concept inventory for strength of materials. In: 33rd Annual Frontiers in Education, 2003. FIE 2003., T3D, p Paper 29
Steif P (2011) Mechanics of Materials. Pearson, Boston
Steif PS, Dantzler JA (2005) A statics concept inventory: development and psychometric analysis. Journal of Engineering Education 94:363–371
Steif PS, Dollár A (2005) Reinventing the teaching of statics. Int J Engng Ed 21(4):723–729
The Education Group (2019) The video encyclopedia of physics. URL sales.physicsdemos.com/
Tobias S (1990) They're Not Dumb, They're Different: Stalking the Second Tier. Research Corporation, Tuscon

Todhunter I, Pearson K (1886) A History of the Theory of Elasticity and the Strength of Materials, vol I. Cambridge University Press, Cambridge
Todhunter I, Pearson K (1953) History of the Strength of Materials. McGraw-Hill, New York
Truesdell C (1960) The Rational Mechanics of Flexible or Elastic Bodies, 1638–1788. Orell Fussli, Turici
Truesdell C (1968) Essays in the History of Mechanics. Springer, Berlin
Truesdell C (1980) The Tragicomical History of Thermodynamics, 1822-1854. Springer, Berlin
Turning Technologies (2019) URL www.turningtechnologies.eu/
Walker M (2016) The Development of the Mechanics' Institute Movement in Britain and Beyond. Routledge, London
Wallace W (1971) Mechanics from Bradwardine to Galileo. Journal of the History of Ideas 32:15–18
Young T (1845) Course of Lectures on Natural Philosophy and the Mechanical Arts. Taylor and Walton, London

Chapter 3
Modeling of Damage of Ductile Materials

Michael Brünig, Moritz Zistl, and Steffen Gerke

Abstract The paper discusses a thermodynamically consistent anisotropic continuum damage model for ductile metals. It takes into account different elastic potential functions to simulate the effect of damage on elastic material behavior. In addition, a yield condition and a flow rule describe plastic behavior whereas a damage criterion and a damage rule characterize various damage processes in a phenomenological way. To validate the constitutive laws and to identify material parameters different experiments have been performed where specimens have been taken from thin metal sheets. As an example, the X0-specimen is tested under different biaxial loading conditions covering a wide range of stress states. Results for proportional and corresponding non-proportional loading histories are discussed. During the experiments strain fields in critical regions of the specimens are analyzed by digital image correlation (DIC) technique while the fracture surfaces are examined by scanning electron microscopy (SEM). Corresponding numerical simulations have been performed and numerical results are compared with available experimental ones. In addition, based on the numerical analyses stress states as well as plastic and damage fields can be predicted allowing explanation of different damage processes on the micro-level. The results also elucidate the effect of loading history on damage and fracture behavior in ductile metals.

Keywords: Anisotropic damage · Biaxial experiments · Non-proportional loading paths · Digital image correlation · Scanning electron microscopy · Numerical simulations

---

Michael Brünig · Moritz Zistl · Steffen Gerke
Institute for Mechanics and Structural Analysis, Bundeswehr University Munich, 85577 Neubiberg, Germany,
e-mail: michael.bruenig@unibw.de, moritz.zistl@unibw.de, steffen.gerke@unibw.de

H. Altenbach and A. Öchsner (eds.), *State of the Art and Future Trends in Material Modeling*, Advanced Structured Materials 100,
https://doi.org/10.1007/978-3-030-30355-6_3

3.1 Introduction

In various engineering disciplines the realistic and accurate modeling of inelastic deformation behavior as well as of damage and failure processes of ductile materials is essential for the solution of different boundary value problems. For example, large inelastic deformations caused by dislocations along preferred slip planes are often accompanied by nucleation, growth of defects on the micro-scale leading to reduction in strength of materials and to shortening the life time of engineering structures. Thus, these aspects demand to provide realistic information on stress distributions and localizations in material elements and to lay down safety factors against failure. On the other hand, due to testing techniques, statistical variations of material properties, machine precision, or manufacturing processes of specimens scatter of experimental data occurs and, therefore, phenomenological material models shall predict the essential features of experimentally observed behavior. Hence, proper understanding and mechanical modeling of stress-state-dependent damage mechanisms are of great importance in the analysis of the effects on deterioration of ductile materials and in elucidating the processes leading from microscopic defects to final fracture.

In order to provide information on stress states in loaded structures and on stress-state-dependent formation of micro-defects, modeling of damage and failure in ductile materials received considerable attention during the past decades. Different constitutive approaches have been published based on experiments or numerical simulations on the micro-level (Brünig, 2003a; Brünig et al, 2008; Chaboche, 1988a; Chow and Wang, 1987a; Chow and Yang, 2004; Lemaitre, 1996; Lu and Chow, 1990; Voyiadjis and Kattan, 1999). Within these continuum damage mechanics concepts the effect of micro-defects on material properties is analyzed in a phenomenological way. Critical values of proposed damage variables are used as major parameters describing the onset of fracture. In this context, an important aspect is the appropriate choice of the variables characterizing the damage state of the material. For example, scalar-valued damage parameters are often used due to their simplicity and numerical efficiency (Lemaitre, 1985a,b; Tai and Yang, 1986). However, in several engineering applications the scalar damage variables have been found to lead to inaccurate results (Chow and Lu, 1992; Chow and Wang, 1987a; Wang and Chow, 1989, 1990). Therefore, anisotropic continuum damage models using tensor-valued damage variables seem to be more suitable to simulate experimentally observed behavior (Brünig et al, 2015; Chow and Wang, 1987a; Chow and Yang, 2004; Lu and Chow, 1990; Murakami and Ohno, 1980).

Development of an accurate phenomenological constitutive model is based on specification of characteristic micro- and macroscopic behavior of the investigated ductile material. For example, damage and failure processes depend on the current stress state acting in a material point (Bao and Wierzbicki, 2004; Brünig et al, 2015). In particular, for high stress triaxialities damage is the result of nucleation, growth and coalescence of nearly spherical micro-voids whereas small positive or negative stress triaxialities lead to formation and growth of micro-shear-cracks. Combination of these basic damage modes on the micro-level occur for moderate positive stress triaxialities and for high negative stress triaxialities no damage has been observed in experiments with ductile metals. Therefore, detailed experimental and numerical analysis of these stress-state-dependent damage and failure processes are needed to develop and to validate accurate continuum damage models.

Numerical analyses on the micro-level examining deformation behavior of micro-defect-containing unit cells have been performed to get information on stress-state-dependent damage and fracture mechanisms (Brocks et al, 1995; Brünig et al, 2011, 2013, 2014, 2018b; Chew et al, 2006; Kim et al, 2003; Shen et al, 2014). Based on these numerical calculations individual behavior of micro-defects in unit cells under wide ranges of loading conditions can be investigated and the effects of their coalescence resulting in macro-cracks can be studied in detail. Systematic unit cell studies enable detection of stress-state-dependent damage and fracture processes which have not been revealed by experiments alone.

Furthermore, various experiments with carefully designed specimens have been discussed in the literature to develop and to validate damage and fracture models. In particular, uniaxial tension tests with unnotched and differently notched specimens have been performed to examine stress-state-dependent inelastic deformation behavior as well as damage and fracture modes (Bao and Wierzbicki, 2004; Bonora et al, 2005; Brünig et al, 2011, 2008; Driemeier et al, 2010). However, as has been shown by corresponding numerical simulations these tests with uniaxially loaded unnotched, differently notched and shear specimens only cover a small range of stress states (see Fig. 3.1) and, thus, further experiments with newly designed specimens have been proposed. For example, tests with butterfly specimens have been presented (Bai and Wierzbicki, 2008; Dunand and Mohr, 2011; Mohr and Henn, 2007) which are uniaxially loaded in different directions using special experimental equipment and, thus, these tests are able to cover a wide range of stress triaxialities. Alternatively, two-dimensional experiments with different biaxially loaded specimens have been developed to examine stress-state-dependent deformation, damage and failure behavior (Brünig et al, 2015; Brünig et al, 2016; Brünig et al, 2018a; Gerke et al, 2017).

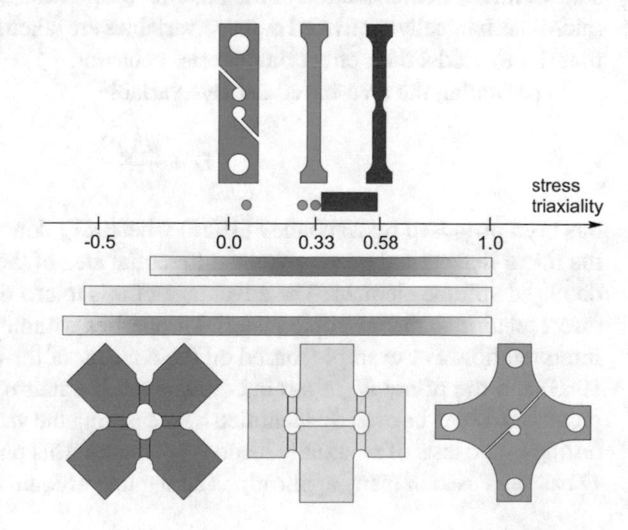

Fig. 3.1 Ranges of stress triaxialities for different specimens (unnotched specimen (green), notched specimens (red), shear specimen (blue), X0-specimen (orange), H-specimen (yellow), Z-specimen (grey))

Corresponding numerical simulations have elucidated that these specimens cover a wide range of stress triaxialities (Fig. 3.1) allowing investigation of various damage and failure processes on the micro-scale. In addition, proportional as well as different non-proportional loading paths have been used in these biaxial tests to examine the effect of the loading history on damage and failure (Brünig et al, 2019; Gerke et al, 2019).

In the present overview article a phenomenological continuum damage model is motivated and discussed. To validate the constitutive equations and to determine stress-state-dependent functions experiments with biaxially loaded specimens have been performed. Experimental results with proportional and different non-proportional loading conditions are discussed. In addition, corresponding numerical simulations are used to detect stress states as well as plastic and damage fields. Based on the numerical results damage and failure mechanisms on the micro-level are revealed which are also visualized by scanning electron microscopy of fracture surfaces.

3.2 Continuum Damage Model

3.2.1 Basic Ideas

Due to fast technological developments during the past decades modeling and numerical simulation of inelastic deformation behavior as well as of damage and fracture processes on the micro-level in materials and structural elements are highly important subjects in a large number of engineering disciplines. In this context, constitutive modeling of damage in ductile materials has received considerable attention with focus on the appropriate choice of efficient damage variables. In the open literature different ways have been proposed allowing characterization of the state of internal deterioration of the material properties. Often, phenomenological or micro-mechanically motivated damage variables are taken into account in mechanical theories to model their effect on material behavior.

In particular, the area-based damage variable

$$D = \frac{dA_d}{dA} \tag{3.1}$$

has been proposed by Kachanov (1958) where dA_d denotes the differential area of the micro-defects and dA means the differential area of the considered representative damaged volume element. The advantage of this micro-defect area fraction D is its direct relation to macroscopic material properties without further micro-mechanical interpretation. For example, based on the concept of the effective stress (Rabotnov, 1963) and the principle of strain equivalence (Lemaitre, 1996) the scalar damage parameter D can be directly identified by measuring the variation of Young's modulus during cyclic tests of uniaxially loaded specimens. This phenomenological parameter D has been used in many applications taking into account continuum damage models

(Bonora et al, 2005; Chaboche, 1988a,b; Lemaitre, 1985a,b; Tai and Yang, 1986). Of course, the scalar-valued damage parameter D averages many micro-defect-related parameters (i.e. the number of micro-defects, their shapes and sizes, the degree of adhesion between the defects, local variations of micro-defect densities, effects of local concentration of stresses on the micro-level) but it can be taken to be an adequate phenomenological variable characterizing the current state of damage. On the other hand, to analyze anisotropic damage behavior a corresponding damage tensor $\boldsymbol{D}$ has been proposed taking into account different micro-defect area fractions in the principal directions of the anisotropy (Chow and Wang, 1987a,b; Jie et al, 2011; Ju, 1990; Lu and Chow, 1990; Murakami, 1988; Murakami and Ohno, 1980; Voyiadjis and Kattan, 1992; Wang and Chow, 1990).

An alternative damage variable has been proposed by Gurson (1977) defining the micro-defect volume fraction

$$f = \frac{dV_d}{dV} \tag{3.2}$$

with the differential volume of the micro-defects dV_d and the differential volume of the representative damaged volume element dV. This continuous damage variable f is directly given by the microscopic damaged material geometry and, therefore, its current value can be directly determined by the use of microscopy (Landron et al, 2013, 2011). The scalar-valued damage parameter f also averages the micro-defect-related parameters discussed above and it is taken to be an adequate micro-mechanically motivated variable implemented in the Gurson yield condition (Gurson, 1977) (or respective extended yield criteria) characterizing the current state of damage (Benzerga and Leblond, 2010; Besson, 2010; Tvergaard, 1989).

Although modeling of damage processes provide a measure of material degradation at the micro-scale both scalar-valued damage parameters, D (3.1) and f (3.2), reflect average material degradation at the macro-level. They are based on different philosophies and their relation depends on the geometry of the examined representative volume element and the incorporated micro-defect. For example, the relation $D = \frac{2}{3} f$ is valid for a spherical volume element with a spherical micro-void. It should be noted that both damage parameters have advantages and disadvantages and both have been successfully taken into account in different damage approaches used in a wide range of engineering applications.

Since the scalar damage variable f can only simulate isotropic damage behavior, Brünig (2002, 2003a) proposed a kinematic concept to simulate anisotropic damage behavior and introduced the trace of the damage strain tensor

$$\operatorname{tr} \boldsymbol{A}^{da} = \ln \frac{dV}{dV - dV_d} = \ln(1 - f)^{-1} \tag{3.3}$$

as a function of the micro-defect volume fraction f. The damage strain tensor $\boldsymbol{A}^{da}$ represents irreversible macroscopic strains caused by formation, growth and coalescence of defects on the micro-scale. In addition, the rate of the isochoric damage strain rate tensor

$$\dot{\boldsymbol{H}}^{da}_{iso} = \frac{1}{3}(1-f)^{-1}\,\dot{f}\,\mathbf{1}\,. \tag{3.4}$$

is introduced. In the continuum model discussed in the present paper the damage strain tensor $\boldsymbol{A}^{da}$ as well as the damage strain rate tensor $\dot{\boldsymbol{H}}^{da}$ are generalized to model anisotropic damage behavior allowing a wide range of applications in various engineering fields.

3.2.2 Thermodynamically Consistent Model

The thermodynamically consistent continuum framework proposed by Brünig (2003a, 2016) is used to model anisotropic damage behavior in ductile metals. This phenomenological approach is based on the introduction of the micro-defect volume fraction f (3.2) as well as the corresponding tensors (3.3) and (3.4) and, thus, uses a kinematic description of damage caused by stress-state-dependent formation and growth of different micro-defects. This leads to the definition of damage strain tensors taking into account isotropic as well as anisotropic behavior providing realistic simulation of micro-failure induced degradation in ductile materials.

The continuum damage model is based on the introduction of damaged and corresponding fictitious undamaged configurations, both defined as respective initial, current and elastically unloaded ones. Based on this kinematic model, the strain rate tensor can be additively decomposed into the elastic, $\dot{\boldsymbol{H}}^{el}$, the effective plastic, $\dot{\bar{\boldsymbol{H}}}^{pl}$, and the damage, $\dot{\boldsymbol{H}}^{da}$, parts. In the respective undamaged and damaged configurations free energy functions are introduced to formulate elastic constitutive equations which are affected by increasing damage. In addition, a yield criterion and a flow rule characterize the plastic behavior with respect to the undamaged configurations. In a similar way, in the damaged configurations a damage condition and a damage rule govern different stress-state-dependent damage processes on the micro-level in a phenomenological macroscopic manner.

In particular, considering the undamaged configurations constitutive laws for the elastic-plastic behavior of the undamaged matrix material are formulated. In this context, plastic internal variables are introduced serving as a basic tool to carry forward information from the crystal lattice of ductile metals to the macro-level and to characterize the hardening behavior of the undamaged material. Based on observations in a large number of experiments with metals undergoing elastic-plastic deformations the elastic behavior of the undamaged material is not influenced by the plastic hardening behavior and the evolution of plastic deformations. Thus, in the undamaged configurations the free energy function is taken to be additively decomposed into the effective elastic and the effective plastic parts

$$\bar{\phi} = \bar{\phi}^{el}(\boldsymbol{A}^{el}) + \bar{\phi}^{pl}(\gamma) \tag{3.5}$$

where $\boldsymbol{A}^{el}$ is the logarithmic elastic strain tensor and γ represents the internal mechanical state variable representing the current amount of effective plastic strain. Thus, the variable γ is taken to be the equivalent plastic strain measure in the proposed continuum model. The additive decomposition of the effective energy function (3.5) leads to the hyper-elastic law and to the introduction of the effective Kirchhoff stress tensor

$$\bar{\boldsymbol{T}} = 2G\,\boldsymbol{A}^{el} + \left(K - \frac{2}{3}G\right) \mathrm{tr}\boldsymbol{A}^{el}\,\mathbf{1} \tag{3.6}$$

where G and K are the constant shear and bulk modulus of the undamaged matrix material, respectively.

Furthermore, isotropic plastic behavior is modeled by the yield condition

$$f^{pl}\left(\bar{I}_1, \bar{J}_2, c\right) = \sqrt{\bar{J}_2} - c\left(1 - \frac{a}{c}\bar{I}_1\right) = 0\,, \tag{3.7}$$

where $\bar{I}_1 = \mathrm{tr}\bar{\boldsymbol{T}}$ and $\bar{J}_2 = \frac{1}{2}\,\mathrm{dev}\bar{\boldsymbol{T}} \cdot \mathrm{dev}\bar{\boldsymbol{T}}$ represent the first and second deviatoric invariants of the effective Kirchhoff stress tensor (3.6) whereas c denotes the equivalent stress of the undamaged material. In Eq. (3.7) a is the hydrostatic stress coefficient based on experiments (Spitzig et al, 1975, 1976) indicating that the plastic behavior of ductile metals is affected by the hydrostatic stress state. It should be noted that in the present formulation isotropic plastic behavior is assumed which can be seen as a realistic assumption for many metals and alloys. However, especially in thin metal sheets, anisotropy must be taken into account. For example, this can be done by generalization of the invariants by projection tensors containing coefficients of the current state of plastic anisotropy (Chow and Wang, 1987a, 1988) or by extended yield conditions (Badreddine et al, 2010; Ha et al, 2018). Further generalization to take into account kinematic hardening in ductile metals is possible by implementing a back-stress tensor in the isotropic or anisotropic yield criterion (Badreddine et al, 2010; Chow and Yang, 2004).

Furthermore, the rate of the plastic strain tensor

$$\dot{\bar{\boldsymbol{H}}}^{pl} = \dot{\gamma}\,\bar{\boldsymbol{N}} \tag{3.8}$$

models the evolution of isochoric plastic deformations in the undamaged ductile material taking into account the normalized deviatoric stress tensor

$$\bar{\boldsymbol{N}} = \frac{1}{\sqrt{2\,\bar{J}_2}}\,\mathrm{dev}\bar{\boldsymbol{T}}$$

and the equivalent plastic strain rate measure $\dot{\gamma} = \bar{\boldsymbol{N}} \cdot \dot{\bar{\boldsymbol{H}}}^{pl}$. Since formation of volumetric plastic strains has not been measured in ductile metals (Spitzig et al, 1975, 1976) the isochoric flow rule (3.8) is taken to accurately simulate the plastic strain rate behavior. Plastic anisotropy in the flow rule can also be modeled by generalizing the normalized stress tensor by a projection tensor (Chow and Wang, 1987a, 1988).

Moreover, modeling of deformation and failure behavior of the anisotropically damaged material is based on the consideration of the damaged configurations. The framework of irreversible thermodynamics with internal state variables is used to develop the equations for the finite elastic-plastic-damage deformations of ductile materials. In the phenomenological continuum approach the Helmholtz free energy function ϕ is introduced as a basis for the formulation of the constitutive equations. The nonlinearities in ductile material behavior observed in experiments are caused by two distinct changes in the micro-structure, the plastic flow and the formation of micro-defects. For example, plastic flow leads to irreversible deformations caused by dislocation processes along preferred slip planes predominantly controlled by shear stresses on the micro-level. On the other hand, during loading of the material sample decohesion or fracture of large inclusions or precipitates lead to micro-voids or micro-shear-cracks destroying the band between material grains. These mechanisms also cause irreversible deformations on the macro-scale and, in contrast to the plastic flow, they affect the elastic material properties. Thus, the elastic part of the energy function depends on elastic and damage measures whereas the plastic and damage parts are formulated in terms of two sets of respective internal state variables corresponding to the formation of dislocations (plastic internal variables) and to nucleation, growth and coalescence of micro-defects (damage internal variables).

The Helmholtz free energy function of the anisotropically damaged material is additively decomposed into three parts:

$$\phi = \phi^{el}(\boldsymbol{A}^{el}, \boldsymbol{A}^{da}) + \phi^{pl}(\gamma) + \phi^{da}(\mu) \,. \tag{3.9}$$

The elastic part, ϕ^{el}, is written in terms of both the elastic and the damage strain tensor, $\boldsymbol{A}^{el}$ and $\boldsymbol{A}^{da}$, and the plastic and damage parts are expressed in terms of the respective plastic and damage internal variables, γ and μ.

It has been observed in experiments that damage affects Young's modulus (Lemaitre, 1985a) and Poisson's ratio (Chow and Wang, 1987b) as well as the shear and the bulk modulus (Spitzig et al, 1988). Hence, the Kirchhoff stress tensor of the damaged material

$$\begin{aligned} \boldsymbol{T} = {} & 2\left(G + \eta_2 \, \mathrm{tr}\boldsymbol{A}^{da}\right) \boldsymbol{A}^{el} \\ & + \left[\left(K - \frac{2}{3}G + 2\eta_1 \, \mathrm{tr}\boldsymbol{A}^{da}\right) \mathrm{tr}\boldsymbol{A}^{el} \; + \; \eta_3 \left(\boldsymbol{A}^{da} \cdot \boldsymbol{A}^{el}\right)\right] \boldsymbol{1} \\ & + \eta_3 \, \mathrm{tr}\boldsymbol{A}^{el} \boldsymbol{A}^{da} + \eta_4 \left(\boldsymbol{A}^{el}\boldsymbol{A}^{da} + \boldsymbol{A}^{da}\boldsymbol{A}^{el}\right) \end{aligned} \tag{3.10}$$

takes into account the additional constitutive parameters η_1, η_2, η_3 and η_4 modeling deterioration of elastic material behavior caused by damage. Based on the hyperelastic law (3.10) simulation of independent decrease of Young's modulus, Poisson's ratio, shear modulus and bulk modulus with increasing damage detected in experiments (Spitzig et al, 1988) can be realized (Brünig, 2003a).

The concept of damage surface is used to determine onset of damage in analogy to the yield surface concept employed in plasticity theory. Different ways to introduce a damage condition have been proposed in the literature. For example, Lemaitre (1985a) suggested a damage dissipation criterion expressed in terms of the damage strain energy release rate which is work-conjugate to the damage parameter D (3.1). In addition, a generalized version for anisotropic damage based on a damage strain energy release tensor has been developed by Chow and Wang (1987a) but some anomalies occurred during applications. Therefore, Chow and Wang (1987a) proposed an alternative damage criterion formulated in terms of the effective stress tensor. The stress space concept has also been used by Brünig (2003a); Brünig et al (2015, 2008) expressing the damage condition in terms of the Kirchhoff stress tensor (3.10) with respect to the damaged configurations. Following this concept, the damage condition can be written in the form

$$f^{da} = \alpha I_1 + \beta\sqrt{J_2} - \sigma = 0 \tag{3.11}$$

where the stress invariants $I_1 = \mathrm{tr}\tilde{\boldsymbol{T}}$ and $J_2 = \frac{1}{2}\,\mathrm{dev}\tilde{\boldsymbol{T}} \cdot \mathrm{dev}\tilde{\boldsymbol{T}}$ ($\tilde{\boldsymbol{T}}$ is work-conjugate to $\dot{\boldsymbol{H}}^{da}$ and simply related to $\boldsymbol{T}$, see Brünig (2003a) for further details) corresponding to the effects of hydrostatic and deviatoric stresses on damage due to shape and orientation of micro-defects. In Eq. (3.11) σ denotes the equivalent damage stress measure which can be seen as material toughness to micro-defect propagation. In addition, the stress-state-dependent variables α and β are associated to different damage and failure processes on the micro-level. In the current approach, stress state dependence is expressed in terms of the stress triaxiality

$$\eta = \frac{\sigma_m}{\sigma_{eq}} = \frac{I_1}{3\sqrt{3J_2}} \tag{3.12}$$

with the mean stress $\sigma_m = I_1/3$ and the equivalent von Mises stress $\sigma_{eq} = \sqrt{3J_2}$ as well as on the Lode parameter

$$\omega = \frac{2\tilde{T}_2 - \tilde{T}_1 - \tilde{T}_3}{\tilde{T}_1 - \tilde{T}_3} \quad \text{with } \tilde{T}_1 \geq \tilde{T}_2 \geq \tilde{T}_3 \tag{3.13}$$

written in terms of the principal Kirchhoff stress components $\tilde{T}_1$, $\tilde{T}_2$ and $\tilde{T}_3$.

Furthermore, formation of macroscopic strains caused by damage and failure processes on the micro-level are analyzed with the damage strain rate tensor

$$\dot{\boldsymbol{H}}^{da} = \dot{\mu}\left(\frac{\bar{\alpha}}{\sqrt{3}}\mathbf{1} + \frac{\bar{\beta}}{\sqrt{2}}\boldsymbol{N} + \bar{\delta}\boldsymbol{M}\right) \tag{3.14}$$

where $\dot{\mu}$ is a non-negative scalar factor and represents the equivalent damage strain rate characterizing the amount of increase in damage. In addition, the normalized stress related deviatoric tensors

$$\boldsymbol{N} = \frac{1}{\sqrt{2J_2}}\mathrm{dev}\tilde{\boldsymbol{T}}$$

and

$$\boldsymbol{M} = \frac{\text{dev}\tilde{\boldsymbol{S}}}{\|\text{dev}\tilde{\boldsymbol{S}}\|}$$

with

$$\text{dev}\tilde{\boldsymbol{S}} = \text{dev}\tilde{\boldsymbol{T}}\ \text{dev}\tilde{\boldsymbol{T}} - \frac{2}{3}J_2\boldsymbol{1} \tag{3.15}$$

have been introduced. The parameters $\bar{\alpha}, \bar{\beta}$ and $\bar{\delta}$ in Eq. (3.14) are kinematic parameters denoting the portion of volumetric and isochoric damage-based deformations also corresponding to different damage and failure mechanisms on the micro-level. These stress-state-dependent parameters will be identified by results of numerical simulations on the micro-scale.

3.2.3 Damage Mode Parameters Based on Numerical Simulations on the Micro-scale

It is difficult or nearly impossible to identify damage related material parameters by experiments alone. Thus, to get more inside in the complex damage and failure processes on the micro-level, Brünig et al (2013) performed three-dimensional numerical analyses of void-containing unit cells covering a wide range of stress triaxialities and Lode parameters. The results of the micro-mechanical numerical calculations are used to detect general trends as well as to propose stress-state-dependent functions for damage criteria and damage evolution laws.

In particular, for different stress triaxiality coefficients η and Lode parameters ω the damage mode parameter α is given by

$$\alpha(\eta) = \begin{cases} 0 \ \text{ for } \ \eta_{cut} \leq \eta \leq 0 \\ \\ \dfrac{1}{3} \ \text{ for } \ \eta > 0 \end{cases} \tag{3.16}$$

where η_{cut} denotes the cut-off value of stress triaxiality below which no damage occurs in ductile metals. Based on biaxial experiments with remarkable negative stress triaxialitites the value η_{cut} has been shown to be a stress-state-dependent function (Brünig et al, 2018a). In addition, the parameter β is taken to be the non-negative function

$$\beta(\eta, \omega) = \beta_0(\eta, \omega = 0) + \beta_\omega(\omega) \geq 0 \tag{3.17}$$

with

$$\beta_0(\eta) = \begin{cases} -0.45\,\eta + 0.85 \ \text{ for } \ \eta_{cut} \leq \eta \leq 0 \\ \\ -1.28\,\eta + 0.85 \ \text{ for } \ \eta > 0 \end{cases} \tag{3.18}$$

and

$$\beta_\omega(\omega) = -0.017\,\omega^3 - 0.065\,\omega^2 - 0.078\,\omega\,. \tag{3.19}$$

Furthermore, based on the results of the micro-mechanical numerical calculations the stress-state-dependent parameters $\bar{\alpha}$, $\bar{\beta}$ and $\bar{\delta}$ in the damage rule (3.14) have been developed. In particular, the non-negative parameter $\bar{\alpha} \geq 0$ corresponding to volumetric damage strain rates caused by isotropic growth of micro-defects is given by the relation

$$\bar{\alpha}(\eta) = \begin{cases} 0 & \text{for } \eta < 0.09864 \\ -0.07903 + 0.80117\,\eta & \text{for } 0.09864 \leq \eta \leq 1 \\ 0.49428 + 0.22786\,\eta & \text{for } 1 < \eta \leq 2 \\ 0.87500 + 0.03750\,\eta & \text{for } 2 < \eta \leq \dfrac{10}{3} \\ 1 & \text{for } \eta > \dfrac{10}{3} \end{cases} \tag{3.20}$$

and dependence on the Lode parameter ω has not been revealed by the numerical simulations on the micro-scale. In addition, the parameter $\bar{\beta}$ corresponding to anisotropic isochoric damage strain rates caused by evolution of micro-shear-cracks is given by the relation

$$\bar{\beta}(\eta,\omega) = \bar{\beta}_0(\eta) + f_\beta(\eta)\,\bar{\beta}_\omega(\omega) \tag{3.21}$$

with

$$\bar{\beta}(\eta,\omega) = \begin{cases} 0.94840 + 0.11965\,\eta + f_\beta(\eta)(1-\omega^2) & \text{for } \eta_{cut} \leq \eta \leq \dfrac{1}{3} \\ 1.14432 - 0.46810\,\eta + f_\beta(\eta)(1-\omega^2) & \text{for } \dfrac{1}{3} < \eta \leq \dfrac{2}{3} \\ 1.14432 - 0.46810\,\eta & \text{for } \dfrac{2}{3} < \eta \leq 2 \\ 0.52030 - 0.15609\,\eta & \text{for } 2 < \eta \leq \dfrac{10}{3} \\ 0 & \text{for } \eta > \dfrac{10}{3} \end{cases} \tag{3.22}$$

and

$$f_\beta(\eta) = -0.02520 + 0.03780\,\eta\,. \tag{3.23}$$

The additional parameter $\bar{\delta}$ also corresponding to the anisotropic damage strain rates caused by the formation of micro-shear-cracks is given by the relation

$$\bar{\delta}(\eta, \omega) = \begin{cases} (-0.12936 + 0.19404\,\eta)(1 - \omega^2) & \text{for } \eta_{cut} \leq \eta \leq \dfrac{2}{3} \\ 0 & \text{for } \eta > \dfrac{2}{3} \end{cases} . \tag{3.24}$$

It should be noted that these damage mode parameters (3.16)-(3.24) are based on numerical analysis using symmetry boundary conditions of the unit cells (Brünig et al, 2013). Additional numerical calculations have been performed using periodic boundary conditions (Brünig et al, 2018b) and they confirmed the functions (3.16)-(3.24).

3.3 Experiments and Corresponding Numerical Simulations

3.3.1 Experimental Equipment and Specimens

New biaxial experiments have been proposed (Gerke et al, 2017) to analyze inelastic deformation behavior as well as stress-state-dependent damage and fracture mechanisms in ductile metals. Different loading scenarios can be taken into account with proportional and associated non-proportional histories (Gerke et al, 2019) and results of additional experiments with alternative loading paths are analyzed in the present paper. The experiments are performed in a biaxial test machine type LFM-BIAX 20 kN (Walter + Bai) shown in Fig. 3.2(a) containing four individually driven cylinders with loads up to ±20 kN located in perpendicular axes. The specimens are fixed in the four heads of the cylinders using clamped or hinged boundary conditions, respectively. During the tests three-dimensional displacement fields are recorded by digital image correlation (DIC) technique in selected zones of the specimens. In the stereo setting four 6.0Mpx cameras with corresponding lighting system shown in Fig. 3.2(b) are used. In addition, after the tests fracture surfaces are investigated by scanning electron microscopy to reveal different fracture modes depending on the loading histories.

Tests are performed with the aluminum alloy AlSiMgMn (EN AW 6082-T6). Material parameters have been identified by numerical fitting of experimental stress–strain curves of uniaxial tension tests with smooth flat dog–bone–shape specimens.

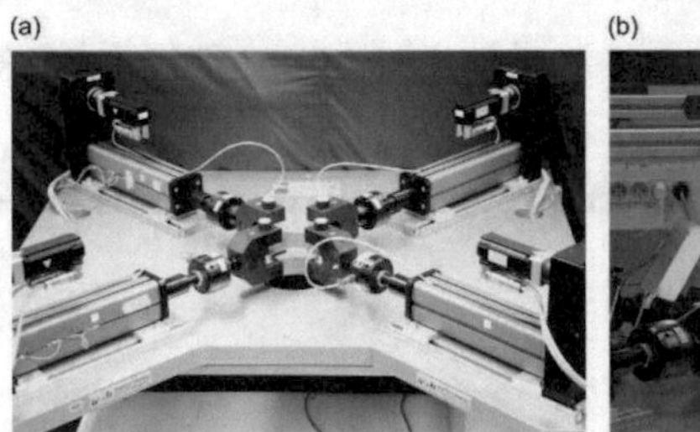

Fig. 3.2 Biaxial test machine

Initial elastic behavior is modeled by Young's modulus $E = 69{,}000$ MPa and Poisson's ratio $\nu = 0.29$. The effect of damage on elastic behavior is taken into account by the parameters $\eta_1 = -20{,}000$ MPa, $\eta_2 = -20{,}000$ MPa, $\eta_3 = -20{,}000$ MPa, and $\eta_4 = -20{,}000$ MPa for positive stress triaxialities ($\eta \geq 0$) whereas $\eta_1 = -4{,}000$ MPa, $\eta_2 = -300{,}000$ MPa, $\eta_3 = 600{,}000$ MPa, and $\eta_4 = -75{,}000$ MPa have been chosen for negative stress triaxialities. In addition, an isotropic hardening model is taken into account and plastic hardening behavior is modeled by the current yield stress

$$c = c_o \left(\frac{H_o \gamma}{n c_o} + 1 \right)^n \tag{3.25}$$

with the initial yield stress $c_o = 163.5$ MPa, the initial hardening modulus $H_o = 850$ MPa and the hardening exponent $n = 0.182$. Damage softening is simulated by the equivalent damage stress

$$\sigma = \sigma_o - H_1 \mu^2 \tag{3.26}$$

with the initial damage stress $\sigma_o = 242$ MPa and the modulus $H_1 = 400$ MPa.

Biaxial specimens are extracted from sheets with 4 mm thickness. In the present paper results with the X0-specimen shown in Fig. 3.3(a) undergoing proportional and different corresponding non-proportional loading paths are discussed. The X0-specimen is based on four crosswise arranged bars with a central opening and four notched parts arranged at 45° (Fig. 3.3(d)) where inelastic deformations as well as damage and fracture are expected to appear in localized bands. The dimensions of the X0-specimen are shown in Fig. 3.3(c): its length is 240 mm in each direction and the depth of the notches is 1 mm on each side leading to thickness reduction from 4

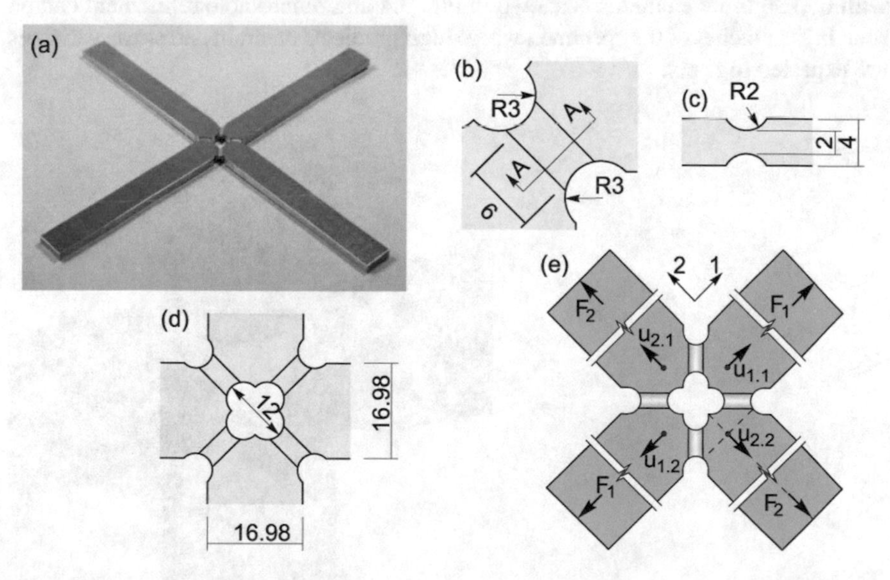

Fig. 3.3 X0-specimen (all dimensions in mm)

mm to 2 mm. These notched regions are 6 mm long (Fig. 3.3(b)) and their radii are 2 mm in thickness and 3 mm in plane directions, respectively.

The X0-specimen is individually loaded in two directions by the forces F_1 and F_2 in the corresponding axes (Fig. 3.3(e)) and different combinations of these loads lead to tension, compression and shear behavior in the notches. They cause different damage and fracture processes on the micro-level and their dependence on the loading history is analyzed in detail in the present paper. During the biaxial experiments the three-dimensional displacements of the red points shown in Fig. 3.3(e) are monitored by DIC. As a special case, the in-plane displacements $u_{1.1}$ and $u_{1.2}$ in axis 1 as well as $u_{2.1}$ and $u_{2.2}$ in axis 2 lead to the relative displacements $\Delta u_{ref.1} = u_{1.1} - u_{1.2}$ and $\Delta u_{ref.2} = u_{2.1} - u_{2.2}$ used in the analysis. The displacements normal to the plane are also controlled to detect possible buckling during compressive loading which must be avoided to receive comparable experimental results.

3.3.2 Numerical Aspects

The numerical simulations have been carried out using the finite element program ANSYS. It has been enhanced by a user-defined material subroutine taking into account the constitutive equations of the proposed continuum damage model. Integration of the constitutive rate equations (3.8) and (3.14) is performed by the inelastic predictor–elastic corrector method (Brünig, 2003b). Eight-node-elements of type Solid185 with linear displacement fields have been used to predict the three-dimensional displacement fields and to quantify strains and stresses in the specimen. The mesh with 65,736 finite elements is shown in Fig. 3.4 and remarkable refinement can be seen in the notches of the specimen where high gradients of strain and stress variables are expected to occur.

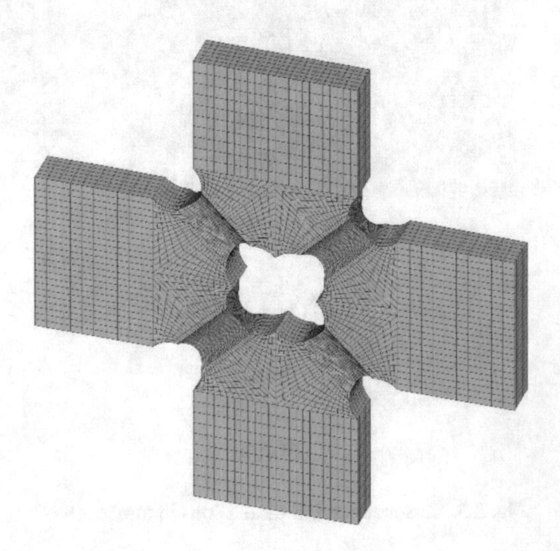

Fig. 3.4 Finite element mesh

3.3.3 Results of Biaxial Experiments and Corresponding Numerical Simulations

The investigated loading paths are shown in Fig. 3.5 taking into account the proportional as well as three corresponding non-proportional ones. In particular, in the proportional loading path (P 1/0) the X0-specimen is only loaded by the force F_1 and the final load at fracture is $F_1 = 6.5$ kN. In the second case (NP1 1/-1 to 1/0) the specimen is first loaded by $F_1 = -F_2 = 3.5$ kN and, then, the load F_1 is kept constant whereas F_2 is reduced to zero until the proportional path is reached. In the final step of this loading scenario the load F_1 is further increased until fracture of the specimen happens at $F_1 = 6.8$ kN. In the alternative non-proportional loading path (NP2 1/-1 to 1/0) the X0-specimen is again first loaded by $F_1 = -F_2$ but now up to $F_1 = -F_2 = 3.8$ kN before the axis switch takes place. In the next step again F_1 is kept constant and F_2 is reduced to zero and the final step is identical to the proportional case with only further loading by F_1 up to final fracture which in this alternative non-proportional loading history occurs at $F_1 = 7.0$ kN. During the last non-proportional loading path (NP 1/+1 to 1/0) the specimen is first loaded by $F_1 = F_2$ up to $F_1 = F_2 = 5.4$ kN, then again F_1 is kept constant and F_2 is driven to zero and after the proportional path is reached the load F_1 is further increased until fracture happens at $F_1 = 8.4$ kN. The loading-path-diagram (Fig. 3.5) clearly shows that the final load at fracture is remarkably affected by the loading history and compared to the proportional loading path the fracture loads differ up to about 30%.

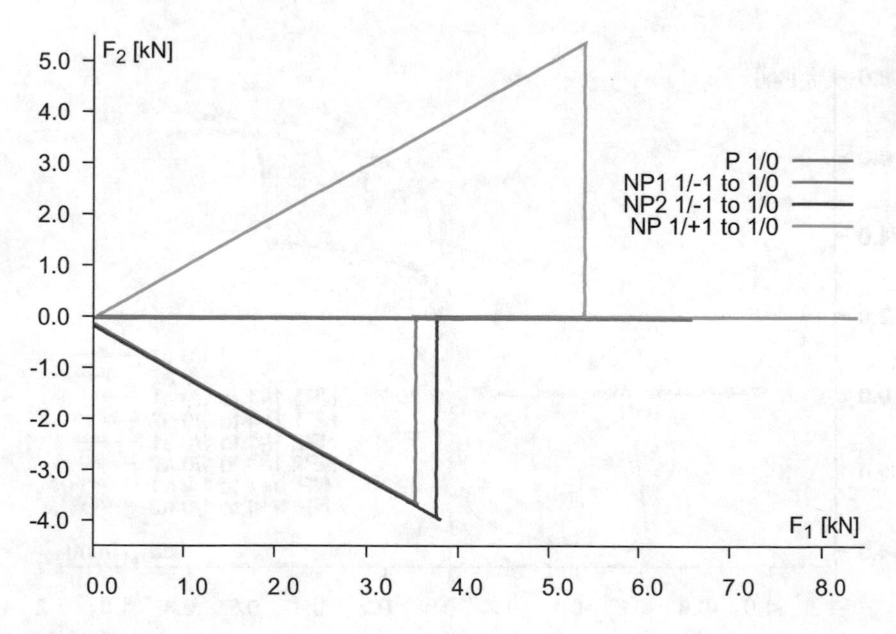

Fig. 3.5 Loading paths

Figure 3.6 shows corresponding load-displacement curves. In particular, in the proportional loading case (P 1/0) the load F_1 first nearly linearly increases and with subsequent inelastic deformation of the specimen up to fracture the relative displacement in axis 1 (a1) reaches $\Delta u_{ref.1}$ = 1.0 mm. During this proportional loading path the load F_2 is always zero whereas the relative displacement in axis 2 (a2) is $\Delta u_{ref.2}$ = -0.57 mm at the end of the test. In the first non-proportional loading path (NP1 1/-1 to 1/0) the load F_1 increases linearly in the elastic part and non-linearly in the subsequent short inelastic one and the displacement in axis 1 is $\Delta u ref.1$ = 0.26 mm before the axis switch takes place. Similar behavior can be seen in axis 2 with compressive loading and the displacement reaches $\Delta u_{ref.2}$ =-0.29 mm. In the next step, when load F_2 is reduced to zero and the displacement in axis 2 reduces to $\Delta u_{ref.2}$ = -0.24 mm whereas the load F_1 is kept constant and the corresponding displacement in axis 1 reduces to $\Delta u_{ref.1}$ = 0.24 mm. In the final proportional step F_1 further increases first linearly indicating elastic behavior with subsequent non-linear curve corresponding to inelastic behavior. At the final load at fracture the corresponding displacement in axis 1 is $\Delta u_{ref.1}$ = 0.87 mm and in axis 2 the final displacement is $\Delta u_{ref.2}$ = -0.60 mm. In the alternative non-proportional loading history (NP2 1/-1 to 1/0) with load switch at $F_1 = -F_2 = 3.8$ kN the corresponding displacements are $\Delta u_{ref.1}$ = 0.57 mm and $\Delta u_{ref.2}$ = -0.62 mm. Then after decrease of F_2 to zero the displacements are in axis 1 only marginally reduced to $\Delta u_{ref.1}$ = 0.55 mm whereas in axis 2 $\Delta u_{ref.2}$ = -0.56 mm is reached. After the final proportional step the displacements at fracture are $\Delta u_{ref.1}$ = 0.96 mm and $\Delta\ u_{ref.2}$ = -0.80 mm, respectively. In the last non-proportional loading path (NP

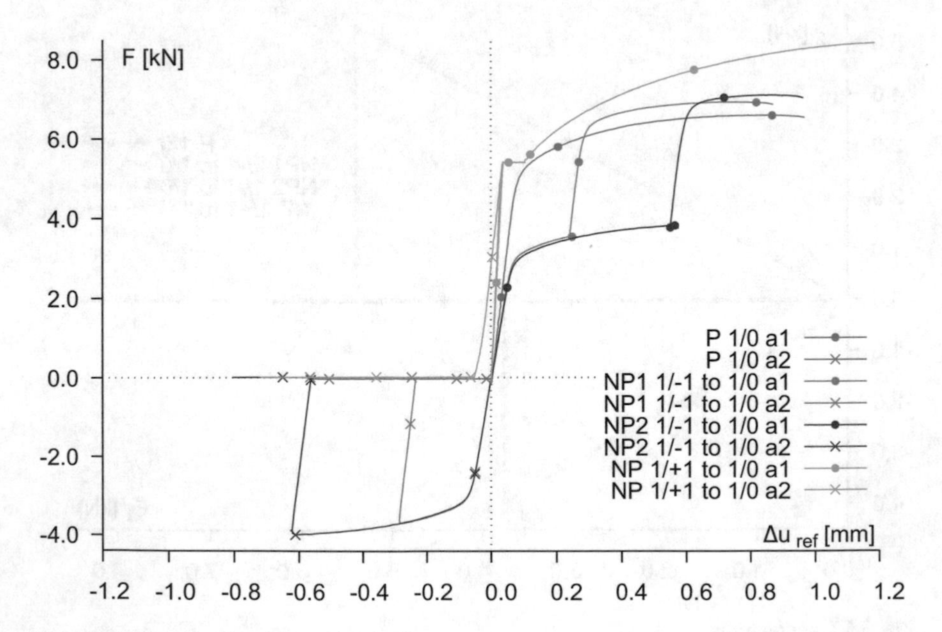

Fig. 3.6 Experimental load-displacement curves

1/+1 to 1/0) both loads F_1 and F_2 identically increase and before load switch the corresponding displacements are $\Delta u_{ref.1} = \Delta u_{ref.2} = 0.04$ mm. In the next unloading step, the load F_2 is reduced to zero and the displacement in axis 2 is $\Delta u_{ref.2}$ = -0.05 mm whereas the displacement in axis 1 reaches $\Delta u_{ref.1}$ = 0.10 mm. In the final proportional step F_1 further increases and the final displacement in axis 1 is $\Delta u_{ref.1}$ = 1.19 mm when fracture happens. In axis 2 the final displacement $\Delta u_{ref.2}$ = -0.72 mm is reached. These load-displacement curves clearly show that the loading path also remarkably affects the final displacements at fracture.

Comparison of experimental and numerical results for the load-displacement curves are shown in Fig. 3.7. Based on the material parameters given above the experimental and numerically predicted curves show good agreement. Only in the proportional case (Fig. 3.7(a)) slight over-prediction can be seen whereas for the non-proportional loading path NP 1/+1 to 1/0 (Fig. 3.7(d)) the numerically predicted load at fracture is about 10% smaller than the experimental one. For the other non-proportional loading histories (Figs. 3.7(b) and (c)) the differences are only marginal. Thus, the finite element program based on the proposed continuum damage model accurately predicts the load-displacement behavior of the X0-specimen for proportional and different non-proportional loading histories.

Figure 3.8 shows distribution of the maximum principal strains in the notched regions of the X0-specimen for various loading paths and at different load stages.

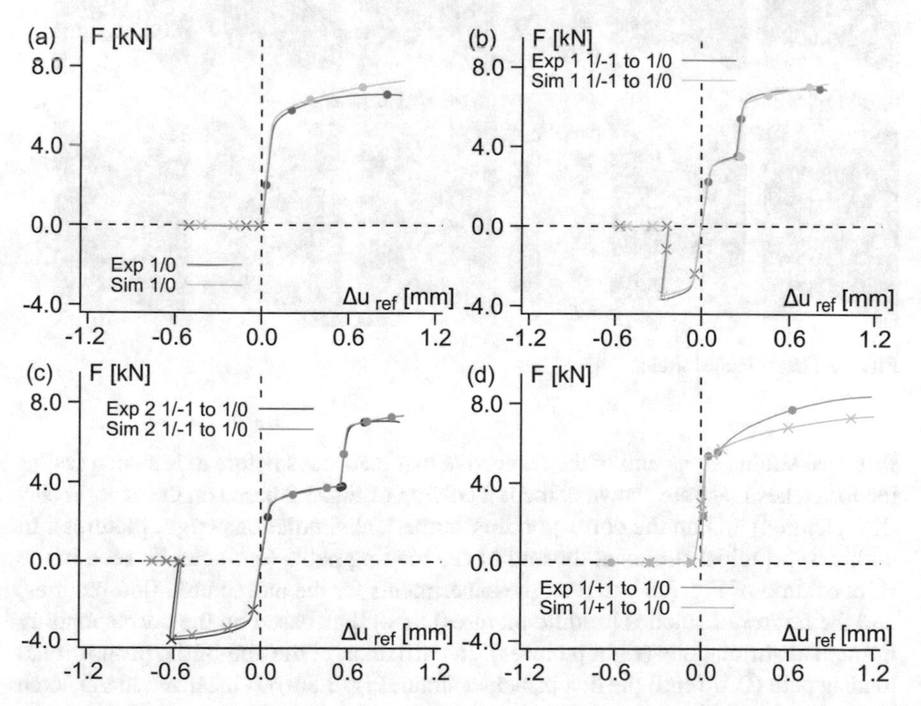

Fig. 3.7 Load-displacement curves: Experiments and numerical simulations

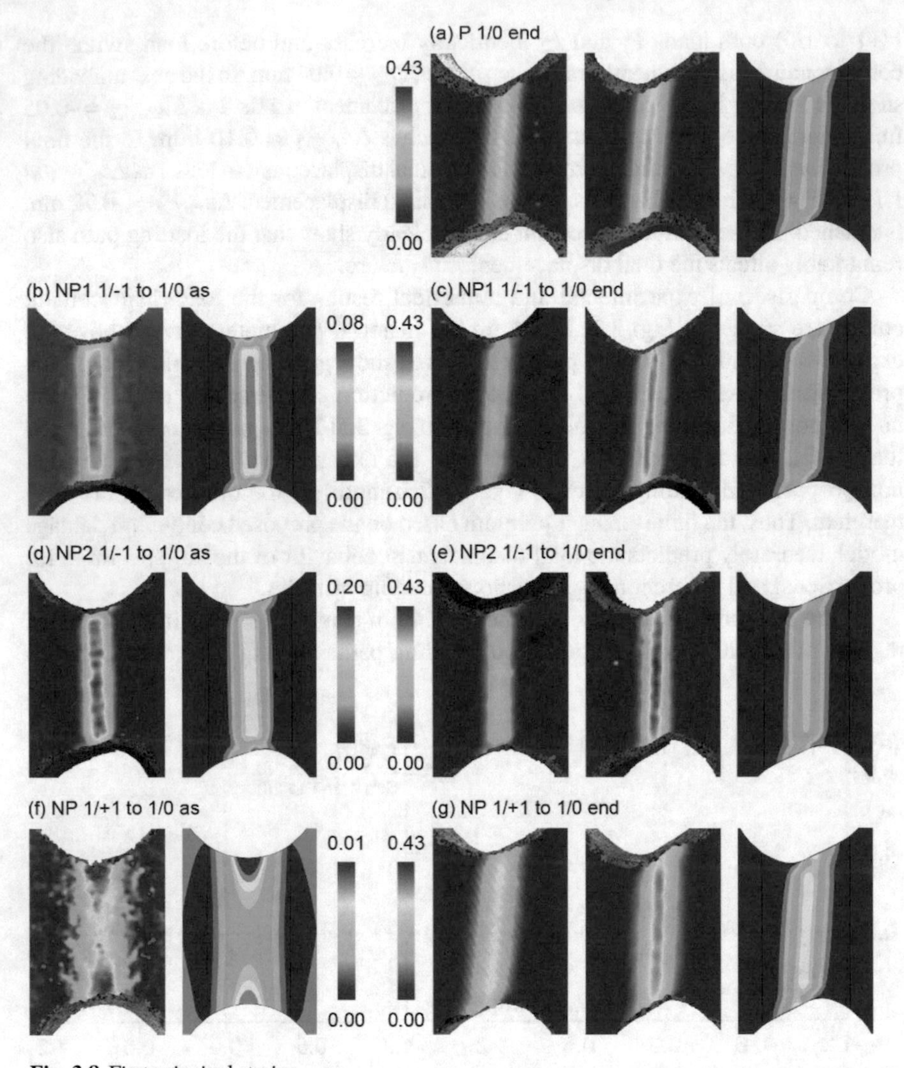

Fig. 3.8 First principal strains

Principal strains at the end of the respective first load steps before axis switch (as) of the load takes place are shown in the left column of Fig. 3.8 based on the experiments (left pictures) and on the corresponding numerical simulations (right pictures). In addition, principal strains at the end of the loading paths (end) can be seen in the right column of Fig. 3.8 based on the experiments for the unfractured (left pictures) and the fractured notches (middle pictures) as well as based on the corresponding numerical simulations (right pictures). In particular, at the end of the proportional loading path (P 1/0 end) the first principal strain (Fig. 3.8(a)) is localized in the notch in bands oriented from top-right to bottom-left with maximum values of 30% in the unfractured notch (left picture) and 34% in the fractured one (middle picture). In the corresponding numerical simulation the principal strain is slightly smaller with maximum value of 26% where the deformation of the notch as well as distribution of the localized principal strain field based on the experiment and the numerical simulation are very similar. In the first non-proportional loading path (NP1 1/-1 to 1/0) the principal strains remain small after the first load step before axis switch (as) of the loads. Nearly vertical bands can be seen in Fig. 3.8(b) with maxima of 7% in the experiment (left picture) and 8% in the numerical simulation. With further loading increase of the first principal strain occurs and at the end of the test (Fig. 3.8(c)) the strains are again localized in diagonal bands with maximum values of 21% in the unfractured notch (left picture) and 38% in the fractured one (middle picture) whereas in the numerical simulation 22% are reached. In the alternative non-proportional case (NP2 1/-1 to 1/0) a vertical band of the localized first principal strain can be seen in Fig. 3.8(d) after the first load step before axis switch (as) with maximum values of 20% in the experiment (left picture) and 15% in the numerical simulation (right picture). At the end of this loading process (Fig. 3.8(e)) the vertical band remains in the unfractured notch with values up to 26% whereas a slightly diagonal band with orientation from top-right to bottom-left occurs in the fractured notch with maximum values of 43%. In the numerical simulation also a vertical band of the localized first principal strain is predicted with values up to 26% corresponding well to the behavior in the unfractured notch in the experiment. In addition, evolution of the first principal strain for the final non-proportional loading path (NP 1/+1 to 1/0) are visualized. After the first load step before axis switch (as) only marginal strains of only 1% occur (Fig. 3.8(f)). At the end of the loading path, Fig. 3.8(g) shows localized strain bands with diagonal right-to-left orientation. In the experiments, maximum values of 33% in the unfractured notch (left picture) and 37% in the fractured notch (middle picture) are measured whereas in the numerical simulation 26% are predicted (right picture). It can be summarized that the loading history affects the final principal strain behavior but similar orientation of localized bands and maximum values are reached. The numerically predicted strains agree well with those in the unfractured notches where only damage processes on the micro-scale are active and no cracks occurred.

Distributions of the numerically predicted stress triaxiality η in the notched region of the X0-specimen for different loading histories are shown in Fig. 3.9. In particular, at the end of the proportional loading path P 1/0 end (Fig. 3.9(a)) on the surface of the notch (S) a wide band with η about 0.25 can be seen. In the longitudinal section (L) a wide area of these stress triaxiality values are predicted with some parts with slightly higher values. In the cross section (C) the stress triaxiality distribution is nearly homogeneous with values of about 0.25. These stress triaxialities are typical for a shear-tension stress state which occurs here during tensile loading in axis 1 only leading to shear-tension mechanisms in the notches of the X0-specimen. During the non-proportional path NP1 1/-1 to 1/0 the specimen is in the first step loaded by $F_1 = -F_2$ up to 3.5 kN and at the end of this load step before the axis switch (as) of the loads takes place a wide area of nearly zero stress triaxialities is numerically predicted on the surface (S), in the longitudinal section (L) and in the cross section (C) of the notches of specimen (Fig. 3.9(b)). The load ratio $F_1 : F_2 = -1$ leads to shear mechanisms which are nearly homogeneously distributed in the notched part

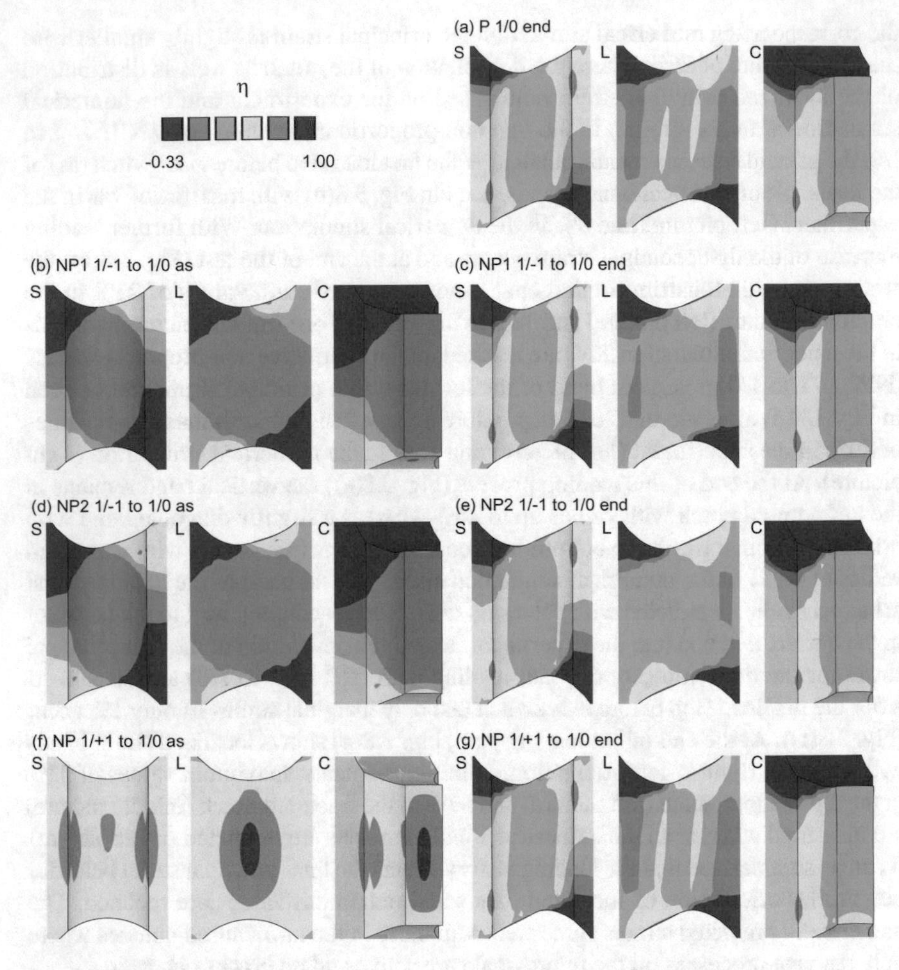

Fig. 3.9 Stress triaxialities η: S =surface, L = longitudinal section, C = cross section

of the specimen. After further loading up to the proportional part (NP1 1/-1 to 1/0 end) again stress triaxialities of about 0.25 occur in the notch (Fig. 3.9(c)) and the distribution is very similar to that one after the proportional loading path (Fig. 3.9(a)). In the second non-proportional path NP2 1/-1 to 1/0 with the first load step $F_1 = -F_2$ up to 3.8 kN before axis switch (as) again nearly zero stress triaxialities are predicted (Fig. 3.9(d)) and they agree well with the stress triaxialites shown in Fig. 3.9(b) for the same load ratio but earlier in the load step. This clearly indicates that the stress triaxialities remain nearly unchanged during loading with the same load ratio. At the end of this load path (NP2 1/-1 to 1/0 end) the stress triaxialities shown in Fig. 3.9(e) are again similar to those ones after the proportional loading path (Fig 3.9(a)). On the other hand, after the first step of the non-proportional path NP 1/+1 to 1/0 before axis switch (as) takes place the stress triaxiality shown in Fig. 3.9(f) is high with maximum of $\eta = 0.65$ on the surface (S) and higher maxima up to 1.0 in the center of the notch (see (L) and (C)) and remarkable gradients of the stress triaxiality can be seen in the notch. This stress triaxiality distribution is typical for high tensile loading with remarkable amount of positive hydrostatic stress caused by $F_1 = F_2$ and the geometry of the notches. After further loading up to the proportional path (NP 1/+1 to 1/0 end) the distribution of the stress triaxiality shown in Fig. 3.9(g) is again similar to all other states at the end of the respective experiments (Figs. 3.9(a), (c) and (e)). This clearly indicates that the final stress state is not affected by the loading histories and only depends on the final load ratio.

Corresponding distributions of the Lode parameter ω in the notched parts of the X0-specimen are shown in Fig. 3.10. At the end of the proportional and the different non-proportional loading paths very similar distributions and values are numerically predicted (see Figs. 3.10(a), (c), (e) and (f)) with values of about 0.5 and nearly

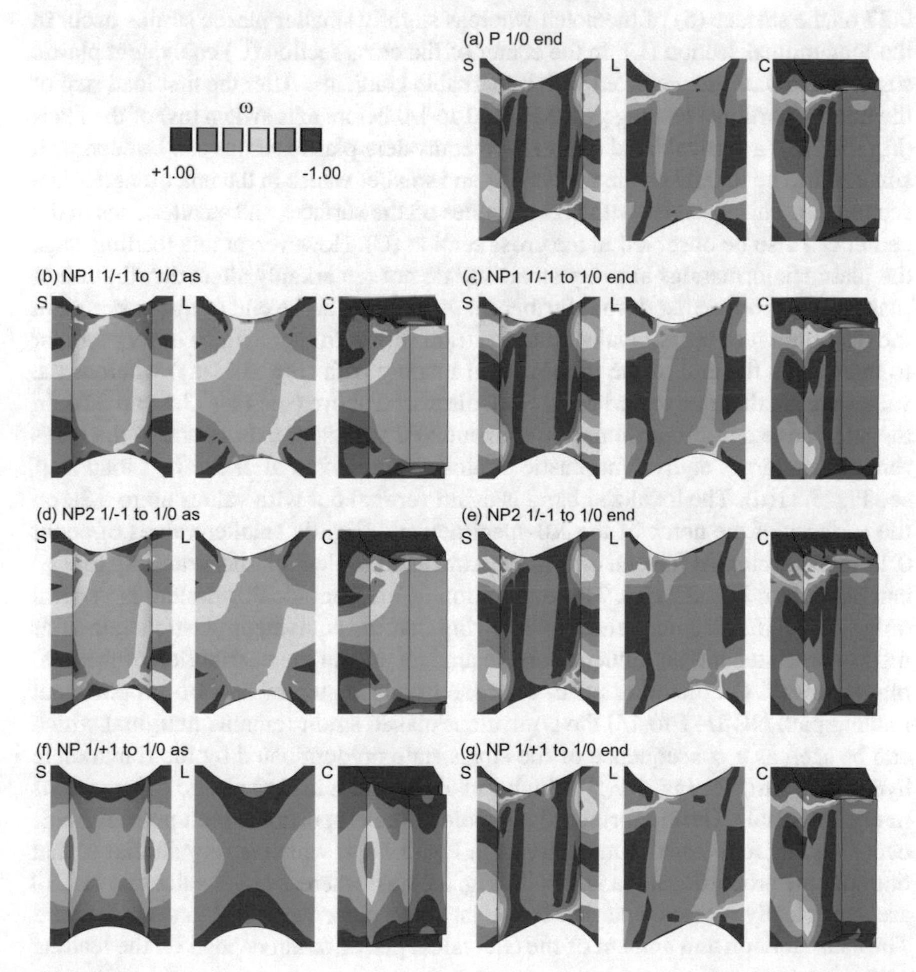

Fig. 3.10 Lode parameters ω: S =surface, L = longitudinal section, C = cross section

homogeneous fields in the notched region. In addition, after the first steps of the non-proportional paths NP1 and NP2 the Lode parameter is about zero with homogeneous distribution in the center of the notch (Figs. 3.10(b) and (d)) corresponding to shear stress behavior. However, after the first step of the non-proportional path NP (Fig. 3.10(f)) the distribution of the Lode parameter shows remarkable gradients in the notch with values up to +1.0 at the boundaries. Again it can be concluded that the final stress state is not affected by the loading histories and only depends on the final load ratio.

Based on the numerical simulations the equivalent plastic strains γ characterizing the amount of plastic deformations can be analyzed in detail where again load steps before axis switch of the loads and at the end of the experiments are considered. In particular, at the end of the proportional loading path (Fig. 3.11(a)) a slightly diagonal localized band of the equivalent plastic strain can be seen with values up to 0.27 on the surface (S) of the notch whereas slightly smaller plastic strains occur in the longitudinal section (L). In the center of the cross section (C) equivalent plastic strains up to 0.27 are predicted with remarkable gradients. After the first load step of the non-proportional loading path NP1 1/-1 to 1/0 before axis switch (as) of the loads (Fig. 3.11(b)) a vertical band of localized equivalent plastic strains can be seen with small values up to 0.09 on the surface (S) and smaller values in the inner longitudinal section (L). This behavior with higher values on the surfaces and smaller ones in the center can also be observed in the cross section (C). However, at this loading stage the plastic deformations are very small and do not remarkably affect distribution of plastic strains occurring during further loading. Thus, at the end of this experiment the distribution of the equivalent plastic strain shown in Fig. 3.11(c) is very similar to that one at the end of the proportional loading path (Fig. 3.11(a)) whereas the values are smaller due to earlier fracture discussed above (see Figs. 3.5 and 3.6). In the alternative non-proportional experiment NP2 with later axis switch of the loads remarkably larger equivalent plastic strains are predicted after the first load step, see Fig. 3.11(d). The localized band is again vertical but with values up to 0.20 on the surfaces of the notch of the X0-specimen and slightly smaller values of about 0.17 in the center. At the end of this experiment equivalent plastic strains up to 0.30 can be seen in Fig. 3.11(e). The orientation of the final localized band is vertical only with slight diagonal tendency showing that the equivalent plastic strain after the first load step has an influence on the amount and on the distribution of the final plastic strains. On the other hand, after the first load step of the non-proportional loading path NP 1/+1 to 1/0 the equivalent plastic strain remains marginal which can be seen as a consequence of the stress state predominated by the remarkable hydrostatic part, see Fig. 3.9(f), which usually does not lead to plastic deformations in ductile metals. During further loading along the proportional path plastic strains occur leading to the distribution shown in Fig. 3.11(g) which is very similar to that one after the proportional loading path (Fig. 3.11(a)) whereas larger values up to 0.30 are numerically predicted caused by the larger displacements and loads at fracture. Thus, distribution and amount of the equivalent plastic strain depend on the loading path.

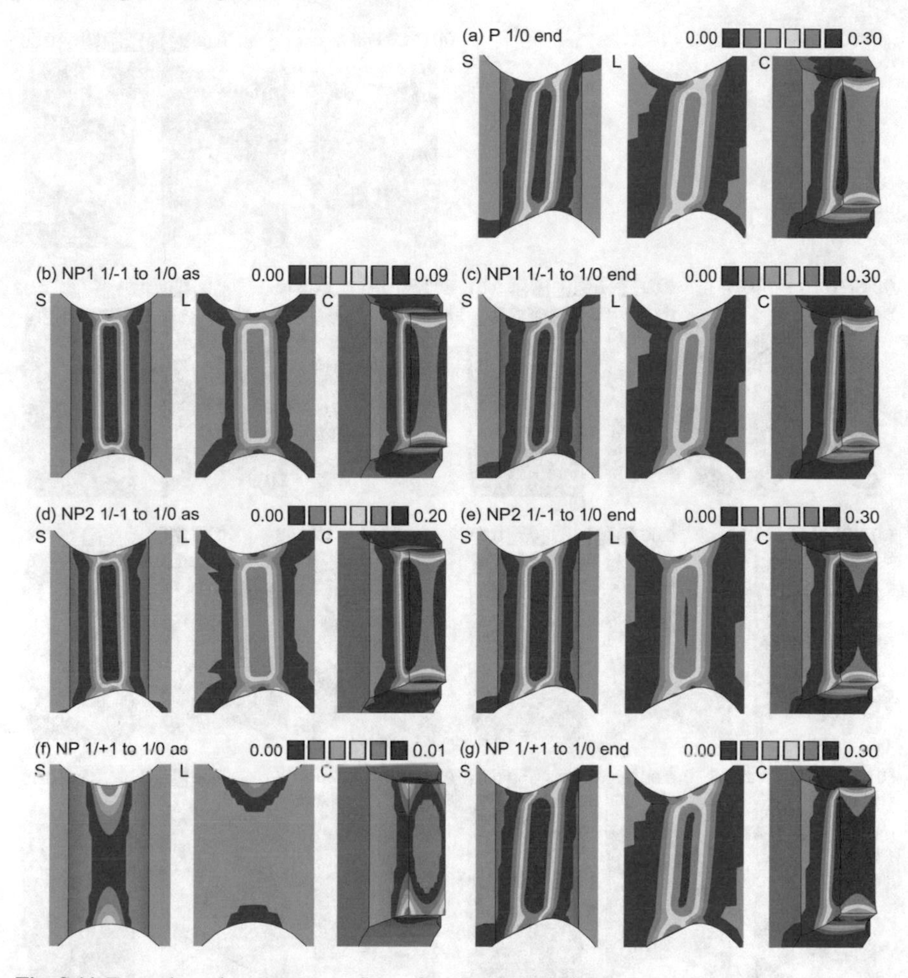

Fig. 3.11 Equivalent plastic strains γ: S =surface, L = longitudinal section, C = cross section

Furthermore, based on the numerical analysis the behavior of the equivalent damage strain μ occurring during proportional and various corresponding non-proportional loading histories can be studied in detail. In particular, after the proportional loading path a diagonally oriented localized band of equivalent damage strain can be seen in Fig. 3.12(a). On the surface of the notch (S) $\mu = 0.06$ is predicted whereas only 0.03 is reached in the longitudinal section (L). In the cross section (C) which is the cut in the mid-plane of the notch maximum values of $\mu = 0.06$ can only be seen on the left and right boundaries whereas smaller equivalent damage strains up to 0.03 appear in the center of the notch. During the non-proportional path NP1 after the first load step the equivalent damage strain remains marginal (Fig. 3.12(b)) and at the end of this step again a slightly diagonal localized band can be seen (Fig 3.12(c)). However, the values are only about 0.03 on the surface (S) and 0.02 in the center (L). In the cross section (C) a region with values of about 0.02 can be seen on the left

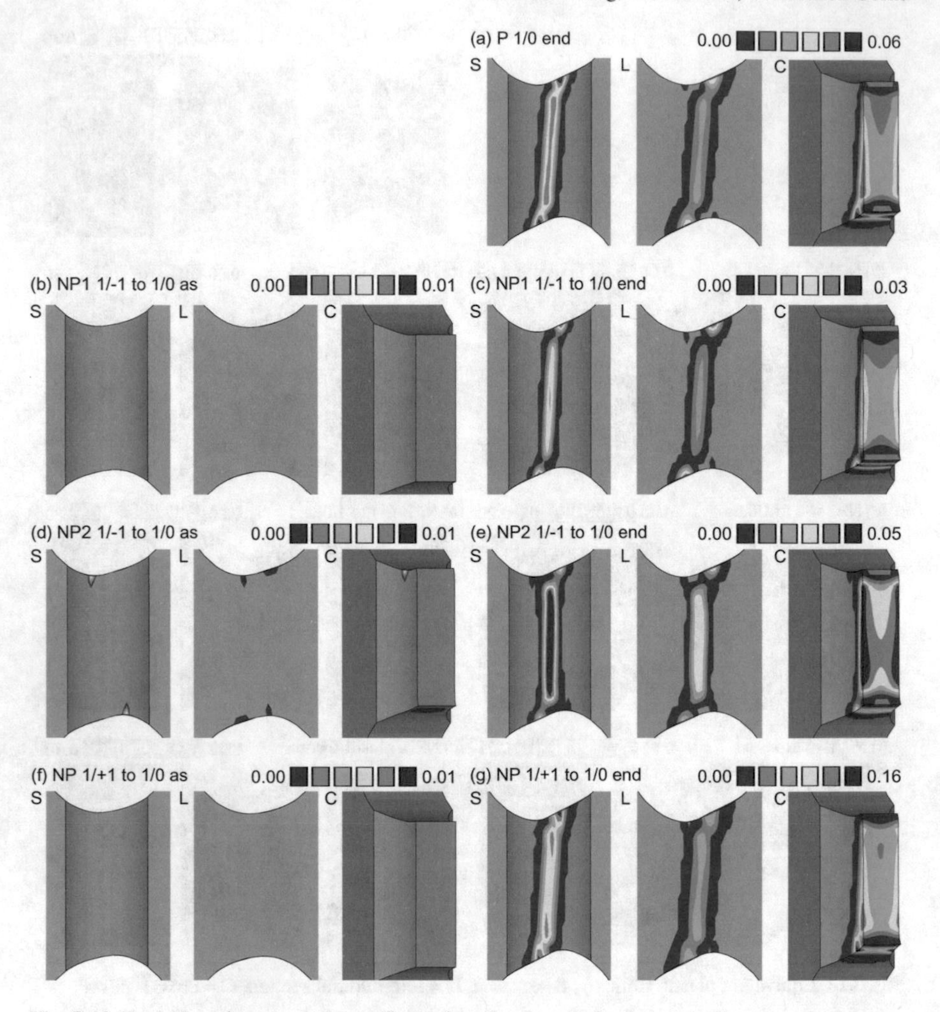

Fig. 3.12 Equivalent damage strains μ: S =surface, L = longitudinal section, C = cross section

and right boundaries whereas damage is predicted to be smaller in other parts. In the alternative non-proportional path NP2 few points with small equivalent damage strain occur after the first loading step before axis switch indicating onset of damage at this stage of loading (Fig. 3.12(d)). During further deformation of the X0-specimen a vertical localized band of equivalent damage strains occurs with values up to 0.05 on the surface (S) and 0.03 in the center (L) (Fig. 3.12(e)). In the cross section (C) damage reaches values of about 0.05 on the left and right boundaries and slightly smaller equivalent damage strains are numerically predicted in the center of the notch. During the non-proportional path NP 1/+1 to 1/0 no damage is predicted after the first load step (Fig. 3.12(f)) but at the end remarkable equivalent damage strains up to 0.16 can be seen in a localized band on the surface of the notch (S) whereas only 0.09 are reached in the the center (L) (Fig. 3.12(g)). In the cross section (C) the equivalent damage strains also reaches the maximum of 0.16 on the left and right boundaries whereas in the center $\mu = 0.09$ is predicted. It should be noted that the form of the localized bands of the equivalent plastic strains (Fig. 3.11) and the equivalent damage strains (Fig. 3.12) for the investigated loading histories are very similar indicating that damage mainly occurs in the regions where plastic deformations take place.

Figure 3.13 shows photos of the central parts of the fractured specimens as well as SEM pictures of the fracture surfaces. In the photos, the fracture lines correspond to the location and orientation of the bands of maximum equivalent damage strains (Fig. 3.13) and only show slight differences whereas effect of the loading history on the damage and fracture processes on the micro-scale can be clearly seen in the SEM pictures. In particular, at the end of the proportional loading path (P 1/0 end) growth of voids in combination with shear effects can be seen which is the typical damage and fracture mechanism for the tension-shear deformation behavior in the notched regions of the X0-specimen under the only load F_1. At the end of the first non-proportional loading path (NP1 1/-1 to 1/0 end) remarkable shear effects in combination with few voids are visualized by SEM. First loading up to $F_1 = -F_2 = 3.5$ kN causes inelastic shear deformation behavior in the notches leading to micro-shear-cracks. After unloading of F_2 additional final loading with F_1 leads to additional growth of some voids with simultaneous formation of further micro-shear-cracks leading to the shear-predominated fracture behavior shown in Fig. 3.13(b). Compared to the proportional loading history the shear effects are more predominant with less and smaller voids on the micro-level. On the other hand, at the end of the alternative non-proportional loading path (NP2 1/-1 to 1/0 end) more pronounced shear effects in combination with few voids are visualized by SEM which can be seen as a result of larger displacements at the load switch compared to the case (NP1 1/-1 to 1/0 end). During this non-proportional loading path first loading up to $F_1 = -F_2 = 3.8$ kN leads to these remarkable inelastic shear deformation behavior in the notches causing micro-shear-cracks. After unloading of F_2 additional final loading with F_1 causes growth of some voids with simultaneous formation of further micro-shear-cracks leading to the shear-predominated fracture behavior shown in Fig. 3.13(c). Compared to the proportional loading history the shear effects are more predominant with less and smaller voids on the micro-level. Furthermore, after the last non-proportional loading path (NP 1/+1 to 1/0 end) remarkable voids can be seen in Fig. 3.13(d) which are slightly sheared and superimposed by few micro-shear-cracks. In this case first loading with $F_1 = F_2$ causes tensile stresses with high portion of hydrostatic stress due to the notches which causes on the micro-level predominant growth of voids. After decrease of F_2 to zero further loading in axis 1 only leads to further growth of voids with small superimposed shear effects. Compared to the proportional loading path more and larger voids can be seen on the micro-scale which are less sheared. Based on these observations on the micro-level it can be concluded that the loading path remarkably affects the damage and fracture processes and the mechanisms occurring firstly are the predominant ones in the final fracture process.

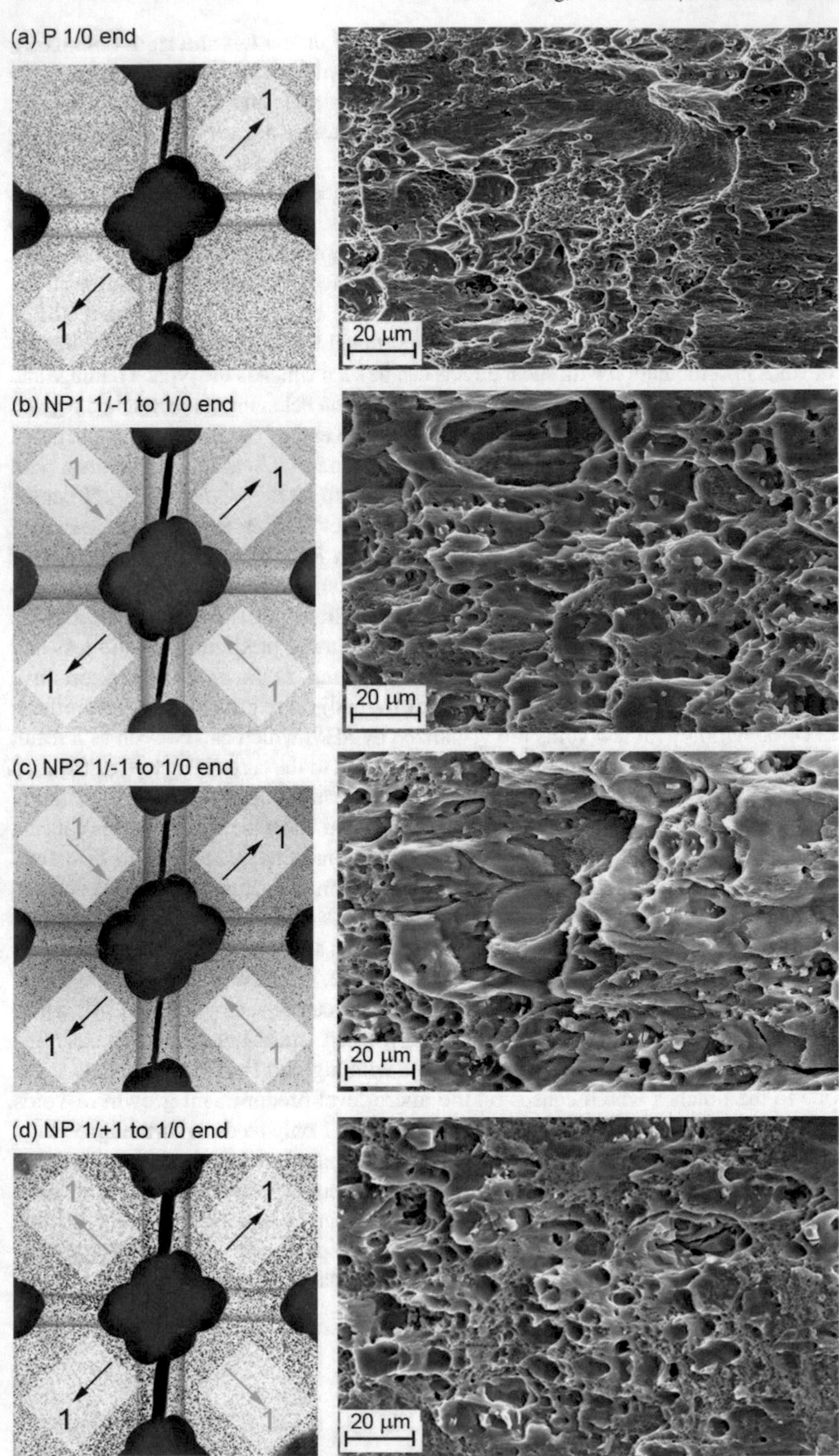

Fig. 3.13 Fractured specimens and SEM pictures of the fracture surfaces

3.4 Conclusions

The paper has discussed a continuum framework to model damage of ductile materials. The phenomenological continuum approach takes into account different branches in the damage criterion corresponding to different stress-state-dependent mechanisms on the micro-level. Evolution of plastic and damage strains is modeled by rate equations. Stress-state-dependent functions for micro-mechanically motivated parameters have been developed by numerical analysis on the micro-scale studying the behavior of three-dimensionally loaded void-containing representative volume elements. To validate the phenomenological continuum damage model and the proposed stress-satedependent functions new experiments with the biaxially loaded X0-specimen have been performed and results have been compared with those taken from corresponding numerical simulations. Focus was on different loading paths with the same final loading ratio. The experimental investigations revealed the effect of non-proportional loading paths on the damage and fracture behavior in ductile metals compared to proportional ones. In the critical notched regions of the specimen various shear-tension behaviors are caused by different proportional and non-proportional biaxial loading histories leading to different strain states as well as to different stress-state-dependent damage and fracture processes on the micro-scale. Thus, evolution of damage and fracture processes on the micro-level are remarkably affected by the loading history and have to be considered in validation of accurate material models predicting failure and life time of engineering structures.

Acknowledgements The project has been funded by the Deutsche Forschungsgemeinshaft (DFG, German Research Foundation) – project number 322157331, this financial support is gratefully acknowledged. The SEM images of the fracture surfaces presented in this paper were performed at the Institut für Werkstoffe im Bauwesen, Bundeswehr University Munich and the support of Wolfgang Saur is gratefully acknowledged.

References

Badreddine H, Saanouni K, Dogui A (2010) On non-associative anisotropic finite plasticity fully coupled with isotropic ductile damage for metal forming. International Journal of Plasticity 26(11):1541 – 1575, doi:10.1016/j.ijplas.2010.01.008

Bai Y, Wierzbicki T (2008) A new model of metal plasticity and fracture with pressure and Lode dependence. International Journal of Plasticity 24(6):1071 – 1096, doi:10.1016/j.ijplas.2007.09.004

Bao Y, Wierzbicki T (2004) On fracture locus in the equivalent strain and stress triaxiality space. International Journal of Mechanical Sciences 46(1):81 – 98, doi:10.1016/j.ijmecsci.2004.02.006

Benzerga AA, Leblond JB (2010) Ductile fracture by void growth to coalescence. In: Aref H, van der Giessen E (eds) Advances in Applied Mechanics, vol 44, Elsevier, pp 169 – 305, doi:10.1016/S0065-2156(10)44003-X

Besson J (2010) Continuum models of ductile fracture: A review. International Journal of Damage Mechanics 19(1):3–52, doi:10.1177/1056789509103482

Bonora N, Gentile D, Pirondi A, Newaz G (2005) Ductile damage evolution under triaxial state of stress: theory and experiments. International Journal of Plasticity 21(5):981 – 1007, doi:10.1016/j.ijplas.2004.06.003

Brocks W, Sun DZ, Hönig A (1995) Verification of the transferability of micromechanical parameters by cell model calculations with visco-plastic materials. International Journal of Plasticity 11(8):971 – 989, doi:10.1016/S0749-6419(95)00039-9

Brünig M, Gerke S, Schmidt M (2016) Biaxial experiments and phenomenological modeling of stress-state-dependent ductile damage and fracture. International Journal of Fracture 200:63–76, doi:10.1007/s10704-016-0080-3

Brünig M (2002) Numerical analysis and elastic–plastic deformation behavior of anisotropically damaged solids. International Journal of Plasticity 18(9):1237 – 1270, doi:10.1016/S0749-6419(01)00076-6

Brünig M (2003a) An anisotropic ductile damage model based on irreversible thermodynamics. International Journal of Plasticity 19(10):1679 – 1713, doi:10.1016/S0749-6419(02)00114-6

Brünig M (2003b) Numerical analysis of anisotropic ductile continuum damage. Computer Methods in Applied Mechanics and Engineering 192(26):2949 – 2976, doi:10.1016/S0045-7825(03)00311-6

Brünig M (2016) A thermodynamically consistent continuum damage model taking into account the ideas of CL Chow. International Journal of Damage Mechanics 25(8):1130–1141, doi:10.1177/1056789516639119

Brünig M, Chyra O, Albrecht D, Driemeier L, Alves M (2008) A ductile damage criterion at various stress triaxialities. International Journal of Plasticity 24(10):1731 – 1755, doi:10.1016/j.ijplas.2007.12.001, special Issue in Honor of Jean-Louis Chaboche

Brünig M, Albrecht D, Gerke S (2011) Numerical analyses of stress-triaxiality-dependent inelastic deformation behavior of aluminum alloys. International Journal of Damage Mechanics 20(2):299–317, doi:10.1177/1056789509351837

Brünig M, Gerke S, Hagenbrock V (2013) Micro-mechanical studies on the effect of the stress triaxiality and the Lode parameter on ductile damage. International Journal of Plasticity 50:49 – 65, doi:10.1016/j.ijplas.2013.03.012

Brünig M, Gerke S, Hagenbrock V (2014) Stress-state-dependence of damage strain rate tensors caused by growth and coalescence of micro-defects. International Journal of Plasticity 63:49 – 63, doi:10.1016/j.ijplas.2014.04.007, Deformation Tensors in Material Modeling in Honor of Prof. Otto T. Bruhns

Brünig M, Brenner D, Gerke S (2015) Stress state dependence of ductile damage and fracture behavior: Experiments and numerical simulations. Engineering Fracture Mechanics 141:152 – 169, doi:10.1016/j.engfracmech.2015.05.022

Brünig M, Gerke S, Schmidt M (2018a) Damage and failure at negative stress triaxialities: Experiments, modeling and numerical simulations. International Journal of Plasticity 102:70 – 82, doi:10.1016/j.ijplas.2017.12.003

Brünig M, Hagenbrock V, Gerke S (2018b) Macroscopic damage laws based on analysis of microscopic unit cells. ZAMM - Journal of Applied Mathematics and Mechanics / Zeitschrift für Angewandte Mathematik und Mechanik 98(2):181–194, doi:10.1002/zamm.201700188

Brünig M, Gerke S, Zistl M (2019) Experiments and numerical simulations with the H-specimen on damage and fracture of ductile metals under non-proportional loading paths. Engineering Fracture Mechanics p 106531, doi:10.1016/j.engfracmech.2019.106531

Chaboche JL (1988a) Continuum damage mechanics. Part I: General concepts. Trans ASME Journal of Applied Mechanics 55(1):59–64, doi:10.1115/1.3173661

Chaboche JL (1988b) Continuum damage mechanics. Part II: Damage growth, crack initiation, and crack growth. Trans ASME Journal of Applied Mechanics 55(2):65–72, doi:10.1115/1.3173662

Chew HB, Guo TF, Cheng L (2006) Effects of pressure-sensitivity and plastic dilatancy on void growth and interaction. International Journal of Solids and Structures 43(21):6380 – 6397, doi:10.1016/j.ijsolstr.2005.10.014

Chow CL, Lu TJ (1992) An analytical and experimental study of mixed-mode ductile fracture under nonproportional loading. International Journal of Damage Mechanics 1(2):191–236, doi:10.1177/105678959200100203

Chow CL, Wang J (1987a) An anisotropic theory of continuum damage mechanics for ductile fracture. Engineering Fracture Mechanics 27(5):547 – 558, doi:10.1016/0013-7944(87)90108-1

Chow CL, Wang J (1987b) An anisotropic theory of elasticity for continuum damage mechanics. International Journal of Fracture 33(1):3–16, doi:10.1007/BF00034895

Chow CL, Wang J (1988) A finite element analysis of continuum damage mechanics for ductile fracture. International Journal of Fracture 38(2):83–102, doi:10.1007/BF00033000

Chow CL, Yang XJ (2004) A generalized mixed isotropic-kinematic hardening plastic model coupled with anisotropic damage for sheet metal forming. International Journal of Damage Mechanics 13(1):81–101, doi:10.1177/1056789504039258

Driemeier L, Brünig M, Micheli G, Alves M (2010) Experiments on stress-triaxiality dependence of material behavior of aluminum alloys. Mechanics of Materials 42(2):207 – 217, doi:10.1016/j.mechmat.2009.11.012

Dunand M, Mohr D (2011) On the predictive capabilities of the shear modified Gurson and the modified Mohr-Coulomb fracture models over a wide range of stress triaxialities and Lode angles. Journal of The Mechanics and Physics of Solids 59:1374–1394, doi:10.1016/j.jmps.2011.04.006

Gerke S, Adulyasak P, Brünig M (2017) New biaxially loaded specimens for the analysis of damage and fracture in sheet metals. International Journal of Solids and Structures 110-111:209 – 218, doi:10.1016/j.ijsolstr.2017.01.027

Gerke S, Zistl M, Bhardwaj A, Brünig M (2019) Experiments with the X0-specimen on the effect of non-proportional loading paths on damage and fracture mechanisms in aluminum alloys. International Journal of Solids and Structures 163:157 – 169, doi:10.1016/j.ijsolstr.2019.01.007

Gurson AL (1977) Continuum theory of ductile rupture by void nucleation and growth: Part I – Yield criteria and flow rules for porous ductile media. Trans ASME Journal of Engineering Materials and Technology 99(1):2–15, doi:10.1115/1.3443401

Ha J, Baral M, Korkolis YP (2018) Plastic anisotropy and ductile fracture of bake-hardened AA6013 aluminum sheet. International Journal of Solids and Structures 155:123 – 139, doi:10.1016/j.ijsolstr.2018.07.015

Jie M, Chow CL, Wu X (2011) Damage-coupled FLD of sheet metals for warm forming and nonproportional loading. International Journal of Damage Mechanics 20(8):1243–1262, doi:10.1177/1056789510396331

Ju JW (1990) Isotropic and anisotropic damage variables in continuum damage mechanics. Journal of Engineering Mechanics 116(12):2764–2770, doi:10.1061/(ASCE)0733-9399(1990)116:12(2764)

Kachanov L (1958) On rupture time under condition of creep (in Russ.). Izvestija akademii nauk SSSROtd Techn Nauk (8):26–31

Kim J, Gao X, Srivatsan T (2003) Modeling of crack growth in ductile solids: a three-dimensional analysis. International Journal of Solids and Structures 40(26):7357 – 7374, doi:10.1016/j.ijsolstr.2003.08.022

Landron C, Maire E, Bouaziz O, Adrien J, Lecarme L, Bareggi A (2011) Validation of void growth models using X-ray microtomography characterization of damage in dual phase steels. Acta Materialia 59(20):7564 – 7573, doi:10.1016/j.actamat.2011.08.046

Landron C, Bouaziz O, Maire E, Adrien J (2013) Experimental investigation of void coalescence in a dual phase steel using X-ray tomography. Acta Materialia 61(18):6821 – 6829, doi:10.1016/j.actamat.2013.07.058

Lemaitre J (1985a) A continuous damage mechanics model for ductile fracture. Trans ASME Journal of Engineering Materials and Technology 107(1):83 – 89, doi:10.1115/1.3225775

Lemaitre J (1985b) Coupled elasto-plasticity and damage constitutive equations. Computer Methods in Applied Mechanics and Engineering 51(1):31 – 49, doi:10.1016/0045-7825(85)90026-X

Lemaitre J (1996) A Course on Damage Mechanics. Springer, Berlin, Heidelberg

Lu TJ, Chow CL (1990) On constitutive equations of inelastic solids with anisotropic damage. Theoretical and Applied Fracture Mechanics 14(3):187 – 218, doi:10.1016/0167-8442(90)90020-Z

Mohr D, Henn S (2007) Calibration of stress-triaxiality dependent crack formation criteria: A new hybrid experimental-numerical method. Experimental Mechanics 47(6):805 – 820, doi:10.1007/s11340-007-9039-7

Murakami S (1988) Mechanical modeling of material damage. Trans ASME Journal of Applied Mechanics 55(2):280 – 286, doi:10.1115/1.3173673

Murakami S, Ohno N (1980) A continuum theory of creep and creep damage. In: Ponter ARS, Hayhorst DR (eds) Creep in Structures, Springer, Berlin, pp 422–443

Rabotnov YN (1963) On the equations of state of creep. In: Progress in Applied Mechanics - The Prager Anniversary Volume, MacMillan, New York, pp 307–315

Shen J, Mao J, Boileau J, Chow CL (2014) Material damage evaluation with measured microdefects and multiresolution numerical analysis. International Journal of Damage Mechanics 23(4):537–566, doi:10.1177/1056789513501913

Spitzig WA, Sober RJ, Richmond O (1975) Pressure dependence of yielding and associated volume expansion in tempered martensite. Acta Metallurgica 23(7):885 – 893, doi:10.1016/0001-6160(75)90205-9

Spitzig WA, Sober RJ, Richmond O (1976) The effect of hydrostatic pressure on the deformation behavior of maraging and HY-80 steels and its implications for plasticity theory. Metallurgical Transactions A 7:1703–1710, doi:10.1007/BF02817888

Spitzig WA, Smelser RE, Richmond O (1988) The evolution of damage and fracture in iron compacts with various initial porosities. Acta Metallurgica 36(5):1201 – 1211, doi:10.1016/0001-6160(88)90273-8

Tai WH, Yang BX (1986) A new microvoid-damage model for ductile fracture. Engineering Fracture Mechanics 25(3):377 – 384, doi:10.1016/0013-7944(86)90133-5

Tvergaard V (1989) Material failure by void growth to coalescence. In: Hutchinson JW, Wu TY (eds) Advances in Applied Mechanics, vol 27, Elsevier, pp 83 – 151, doi:10.1016/S0065-2156(08)70195-9

Voyiadjis G, Kattan P (1999) Advances in Damage Mechanics: Metals and Metal Matrix Composites. Elsevier, Amsterdam

Voyiadjis GZ, Kattan PI (1992) A plasticity-damage theory for large deformation of solids — I. Theoretical formulation. International Journal of Engineering Science 30(9):1089 – 1108, doi:10.1016/0020-7225(92)90059-P

Wang J, Chow CL (1989) Mixed mode ductile fracture studies with nonproportional loading based on continuum damage mechanics. Trans ASME Journal of Engineering Materials Technology 111(2):204–209, doi:10.1115/1.3226455

Wang J, Chow CL (1990) A non-proportional loading finite element analysis of continuum damage mechanics for ductile fracture. International Journal for Numerical Methods in Engineering 29(1):197–209, doi:10.1002/nme.1620290113

Chapter 4
Creep in Heat-resistant Steels at Elevated Temperatures

Johanna Eisenträger and Holm Altenbach

Abstract Power generation is one of the most important applications for components made of heat-resistant steels. Here, creep deformations occur due to the prevailing high temperatures. In close connection with the aim to reduce the emissions and costs involved in power generation, precise constitutive modeling for creep has gained importance over the past years. Therefore, the current contribution provides a brief overview concerning the state-of-the-art in constitutive modeling for creep of heat-resistant steels. In its first part, basic notions about creep are introduced, and microstructural mechanisms are discussed. The second part presents commonly used models for creep in heat-resistant steels, such as the unified CHABOCHE-type models, nonunified approaches, as well as mixture models. Furthermore, the endochronic theories and multi-surface models are briefly introduced. The contribution concludes with a section on the constitutive modeling of creep damage, where the physically-based cavity growth mechanism models as well as approaches based on continuum damage mechanics are discussed.

Keywords: Creep · Heat-resistant steels · Constitutive modeling

4.1 Introduction

When modeling the mechanical behavior of engineering components in a high-temperature environment, it is crucial to consider creep, i.e. the inelastic deformation under sustained loads (below the yield stress σ_{y}) particularly at elevated temperatures (0.3–0.7 of the liquidus temperature T_{L}). This phenomenon can influence the mechanical behavior significantly and reduces the lifetime of components to a great extent.

Johanna Eisenträger · Holm Altenbach
Institut für Mechanik, Fakultät für Maschinenbau, Otto-von-Guericke-Universität Magdeburg, Universitätsplatz 2, 39106 Magdeburg, Germany,
e-mail: johanna.eisentraeger@ovgu.de, holm.altenbach@ovgu.de

H. Altenbach and A. Öchsner (eds.), *State of the Art and Future Trends in Material Modeling*, Advanced Structured Materials 100,
https://doi.org/10.1007/978-3-030-30355-6_4

One of the most important high-temperature applications for heat-resistant steels is power generation. In power plants, two major operating modes can be distinguished: stationary and intermittent service. During stationary service, prevailing high temperatures around 600°C (Masuyama, 2001; Breeze, 2005) induce creep deformations under constant loads. In order to account for the discontinuous power generation of renewable sources such as solar or wind energy, the output of conventional fossil-fuel and nuclear power plants is continuously adapted, resulting in frequent start-ups and shut-downs of these power plants. Thus, power plant components are also affected by fatigue. The combination of stationary and intermittent service results in creep-fatigue loads, i.e. cyclic loads with long holding times, on the components (Fournier et al, 2005; Röttger, 1997).

Within the last twenty years, the need to reduce emissions of power plants and to improve the thermodynamic efficiency in power generation has become obvious. These aims can be achieved by raising the operating temperatures – a measure, which would induce even higher loads on engineering components. In order to reduce costs and emissions, creep failure must be prevented, and the lifetime of components should be predicted precisely. This poses complex demands on employed constitutive models. As additional challenges, which corresponding constitutive models have to tackle, multiaxial, non-isothermal, and cyclic operating conditions of power systems are becoming the norm. Thus, not only long-term creep strains should be predicted accurately, but also short-term plastic strains as well as the interaction of both deformation modes.

The preceding considerations clearly show that although research in creep has been conducted for a long time, i.e. since the beginning of the 20th century, this topic has gained importance throughout the years. For this reason, the contribution at hand aims at providing a brief overview concerning the state-of-the-art in constitutive modeling for creep of heat-resistant steels. Note that it is impossible to provide a complete overview of this topic due to the great variety of developed constitutive models and the immense number of published works. Instead, this paper seeks to highlight a selection of commonly used constitutive models for creep in heat-resistant steels. Any inadvertent omission of relevant publications in the wide body of literature is not done on purpose and we wish to apologize in advance.

The contribution at hand is divided into five sections. After this first introductory part, basics about creep in heat-resistant steels, such as microstructural features of these alloys, a general definition of creep, and the classical creep curve, are discussed in Sect. 4.2. The third and major section of this contribution presents various constitutive models for creep, whereby we distinguish between early approaches, unified, and nonunified models. Section 4.4 provides a brief overview of two major classes of constitutive models for creep damage: the cavity growth mechanism models and approaches based on continuum damage mechanics. The final section gives a brief summary and identifies areas for further research.

4.2 Basics About Creep in Heat-resistant Steels

4.2.1 Microstructure of Heat-resistant Steels

Since creep is based on various microstructural processes, this section focuses on typical microstructural characteristics of heat-resistant steels. Note that the influence of the microstructure on the different creep stages is discussed in Sect. 4.2.3. Furthermore, the content of this section is primarily based on standard monographs from the field of material sciences, such as Czichos et al (2014); Weißbach et al (2015).

The microscale of metallic materials exhibits a regular structure, which is commonly referred to as lattice. Using a microscope, one can identify individual grains, i.e. areas with similar lattice orientation, and their boundaries. The grain boundaries are two-dimensional lattice defects, which separate adjacent grains. With respect to the magnitude of the disorientation of adjacent grains, one distinguishes high-angle and low-angle grain boundaries. In general, the boundary between adjacent grains with a disorientation angle lower than $\approx 15°$ is called "low-angle grain boundary", whereas boundaries between adjacent grains with a higher disorientation angle are referred to as "high-angle grain boundaries" (Priester, 2013). The lattice orientation might vary slightly inside a grain. Thus, one can distinguish subgrains separated by low-angle grain boundaries from other subgrains (Straub, 1995), cf. Fig. 4.1.

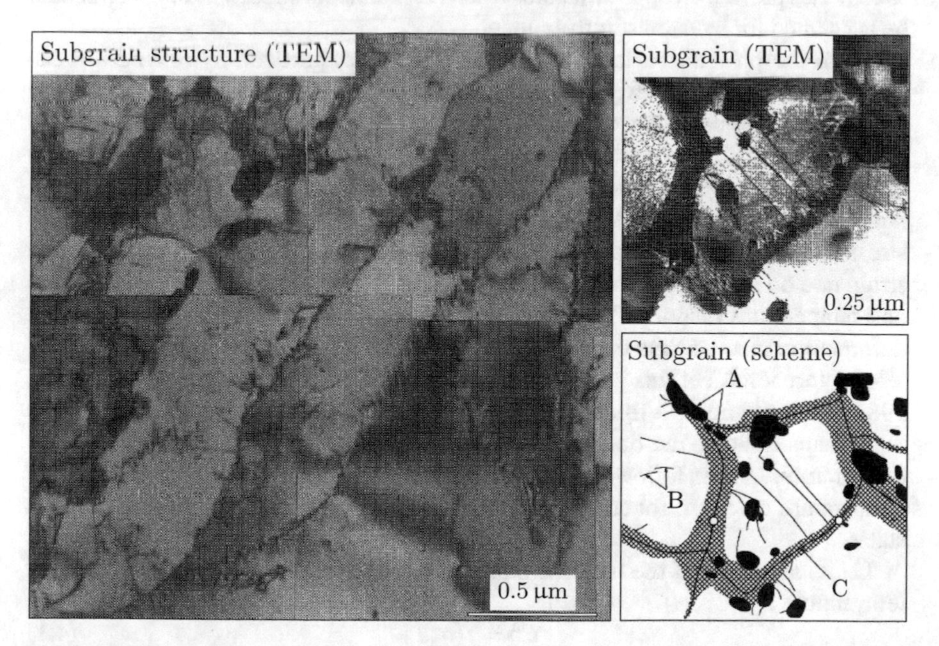

Fig. 4.1 Typical microstructure of heat-resistant steels (A carbides, B dislocations, C boundary), cf. Straub (1995); Polcik (1998); Eisenträger (2018)

Heat-resistant steels comprise several phases, such as ferrite, martensite, austenite, and precipitates. If the carbon content of an alloy does not exceed 0.2%, lath martensite represents the primary phase. Such alloys are not capable of forming an austenitic phase. Between and also inside the martensite laths, dislocations (one-dimensional lattice defects) concentrate and form subgrain boundaries. Furthermore, high-angle grain boundaries separate lath packs of different orientations. Moreover, carbides, i.e. the most common precipitates in heat-resistant steels, are an important part of the microstructure and tend to concentrate on the (sub)grain boundaries (Straub, 1995). This concentration of the carbides on the boundaries is also illustrated in Fig. 4.1.

4.2.2 Definition of Creep and Influence of Stress and Temperature

The phenomenon "creep" describes the continuous increase in deformation under constant loads, taking place particularly at elevated temperatures and under moderate load levels. Creep is also closely related to relaxation, which refers to the continuous decrease of the stress level in a material subjected to constant prescribed strains at high temperatures. Note that the time-dependent inelastic deformation of structures is not only affected by creep and relaxation, but stress redistribution occurs as well. Furthermore, in contrast to creep in homogeneous bulk materials, where uniaxial stress states prevail, creep in structures results in multiaxial stress states, which should be accounted for by a constitutive model.

Up to the present, various monographs on creep, particularly related to heat-resistant steels, have been published and could be consulted to find experimental data of creep tests and get a brief overview of established constitutive models for creep (Hult, 1966; Ilschner, 1973; Odqvist, 1974; Frost and Ashby, 1982; Nabarro and de Villiers, 1995; Penny and Marriott, 1995; Abe et al, 2008; Naumenko and Altenbach, 2007, 2016). An essential part of constitutive models for creep are stress and temperature response functions, which describe the dependence of the strain rate on stress and temperature, respectively. In order to find adequate stress and temperature response functions, one should take into account that different deformation mechanisms occur in a material, depending on the applied stress and temperature level. For this purpose, deformation mechanism maps (Frost and Ashby, 1982; Nabarro and de Villiers, 1995; François et al, 2012) can be consulted since these maps indicate the dominant deformation mechanism for given stresses and temperatures. In the following, let us briefly introduce commonly used stress and temperature response functions with respect to one-dimensional stress and strain states.

Let us assume that the inelastic strain rate is a function of the stress σ and the temperature T

$$\dot{\varepsilon}^{\text{in}} = f(\sigma, T). \tag{4.1}$$

This function can be extended to include further influences. However, the inelastic strain rate function of this type has a disadvantage: the identification is difficult if we have a common function for all influences. To solve this problem, a separation ansatz is frequently utilized to define the dependence of the inelastic strain rate $\dot{\varepsilon}^{\text{in}}$ on the stress σ and the temperature T. Thereby, the inelastic strain rate is approximated as a product of the stress response function f_σ and the temperature response function f_T (Naumenko and Altenbach, 2016):

$$\dot{\varepsilon}^{\text{in}} = f_\sigma(\sigma)\, f_T(T). \tag{4.2}$$

The NORTON power law represents one of the most commonly used choices for the stress response function (Frost and Ashby, 1982):

$$f_\sigma(\sigma) = A\sigma^n. \tag{4.3}$$

The material parameters A and n (creep exponent) should be estimated from experiments. Diffusional creep is usually described by employing a linear stress function (Herring, 1950; Harper and Dorn, 1957; Coble, 1963; Lifshitz, 1963), i.e. $n = 1$:

$$f_\sigma(\sigma) = B\sigma. \tag{4.4}$$

Another common choice to describe creep at low temperatures and wide stress ranges is a hyperbolic function proposed by Dyson and McLean (2001):

$$f_\sigma(\sigma) = C \sinh (D\sigma)\,. \tag{4.5}$$

Note that the material parameters B, C, and D must be determined based on experimental data. To express the dependence of the inelastic strain rate on temperature, the ARRHENIUS function is a popular choice (Dorn, 1955; Ilschner, 1973; Xiao and Guo, 2011):

$$f_T(T) = \alpha \exp\left(-\frac{Q}{\mathrm{R}T}\right) \tag{4.6}$$

with the material parameter α, the activation energy Q, and the universal gas constant $\mathrm{R} \approx 8.31696\,\mathrm{J\,(mol\,K)^{-1}}$.

As has already been pointed out, the temperature exerts a significant influence on creep and relaxation phenomena. This is due to the fact that creep is based on microstructural processes, which are highly dependent on temperature. However, because of the great variety of materials affected by creep, the classification of creep with respect to temperature is not uniform in literature, i.e. a clear and widespread definition of low and high-temperature creep does not exist. Note that temperature ranges in creep are often indicated with respect to the liquidus temperature T_L of the material under consideration. According to Naumenko and Altenbach (2016), high-temperature materials are used at a temperature range of 0.3–0.7 T_L. In Frost and Ashby (1982), it is stated that polycrystalline solids start to creep at 0.5 T_L, whereas a temperature level of 0.9 T_L is referred to as "very high" temperature level. In addition, according to McLean (1966), the temperature range 0.3–0.9 T_L is most important for engineering applications, while in this main creep range, two separate creep regimes are distinguished, low temperature (LT) creep at 0.3–0.5 T_L and high temperature (HT) creep at 0.5–0.9 T_L. Thus, in the paper at hand, we will define LT creep with respect to a temperature range of 0.3–0.5 T_L, while HT creep is assumed to occur in a temperature range of 0.5–0.7 T_L. Higher temperatures are related to other mechanisms and should be presented by the so-called power-break law.

4.2.3 Classical Creep Curve. Primary, Secondary, and Tertiary Creep

During a creep test, a tensile specimen is subjected to a sustained load under a constant temperature level. As main result of a creep test, one obtains a creep curve, as schematically depicted in Fig. 4.2. Further typical examples of creep curves can be found in several monographs (Odqvist, 1974; Odqvist and Hult, 1962; Penny and Marriott, 1995) and papers related to the experimental analysis of creep (El-Magd et al, 1996; Hyde et al, 1999; Kloc et al, 2001). ANDRADE, as one of the first researchers conducting creep tests, introduced the division of the classical creep curve into primary, secondary, and tertiary stage (da C. Andrade, 1910, 1914), as indicated by the symbols "I", "II", and "III" in Fig. 4.2. In the following, all three creep stages are explained, and the governing microstructural mechanisms are discussed. Further information on this topic can be found in Frost and Ashby (1982); Nabarro and de Villiers (1995); Rösler et al (2012).

After applying a load to a specimen at the beginning of a creep test, there is an instantaneous elastic deformation of the specimen. Afterwards, the *primary creep* stage prevails. This stage is marked by a decreasing strain rate, which is primarily due to hardening processes, i.e. the movement of dislocations is inhibited due to subgrain boundaries, precipitates, and the accumulation of dislocations. However, recovery and relaxation processes take place as well, e.g. the restructuring of lattice defects or thermally activated changes in the dislocation structure, such that dislocations annihilate, climb, and glide. Constitutive models for primary creep often involve a backstress (Orowan, 1934a,b,c; Malinin and Khadjinsky, 1972; Estrin and Mecking, 1984; Miller, 1987; Krempl, 1999; Dyson and McLean, 2001), cf. Sect. 4.3.

The *secondary creep* stage is marked by an approximately constant strain rate, cf. Fig. 4.2, since hardening and recovery processes are balanced. This is in contrast

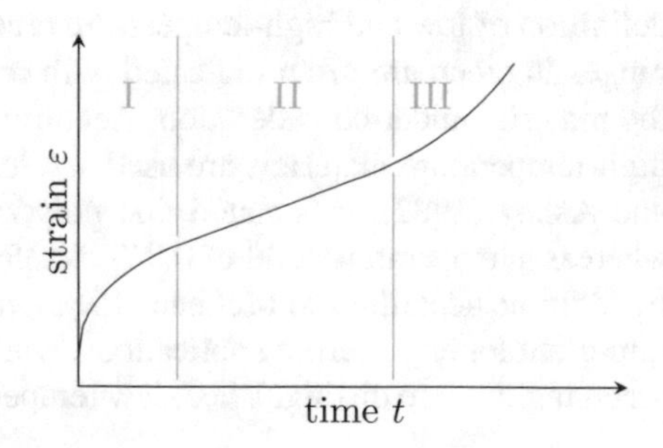

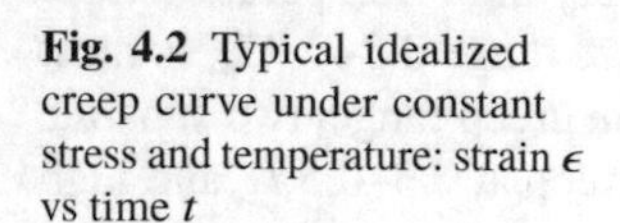
Fig. 4.2 Typical idealized creep curve under constant stress and temperature: strain ϵ vs time t

to the primary creep stage, which is dominated by hardening processes. Frequently, this stage is also referred to as "stationary" creep – a term, which should be used with caution, since it does not account for the various *instationary* processes and changes on the microstructural level. Note that the inelastic strain rate is not only constant, but also attains its minimum during this stage. Due to the constant strain rate, which simplifies the formulation of a constitutive model significantly, the first constitutive models for creep were related to secondary creep. Within this classical theory of creep, potentials are introduced (Odqvist, 1974; Odqvist and Hult, 1962). By deriving these creep potentials with respect to the stresses, one obtains a relation for the inelastic strain rate.

In addition, there is the *tertiary creep* stage, which marks the final phase of a creep test. During this stage, the deformation is accelerated by an increasing strain rate, cf. Fig. 4.2. This increase in strain rate is attributed to damage and softening processes, such as the formation, growth, and coalescence of voids, subgrain coarsening, the coarsening of precipitates such as carbides (particularly at the grain boundaries), and microstructural aging. The constitutive modeling of tertiary creep is closely related to the simulation of damage processes, cf. Sect. 4.4.

Above considerations are only valid for one-dimensional stress and deformation states. However, creep in structures is usually a multiaxial process. In ideal circumstances, constitutive models for multiaxial creep are calibrated with experimental data from multiaxial creep tests. Typical examples for experimental results of multiaxial creep tests can be found in Lin et al (2005b); Kowalewski (2001). For most materials, primary and secondary creep is independent of the type of stress state such that a good correlation of the von Mises equivalent stress $\sigma_{\text{vM}} = \sqrt{\frac{3}{2}\boldsymbol{\sigma}' : \boldsymbol{\sigma}'}$ with the strain rate can be established. Here, $\boldsymbol{\sigma}'$ denotes the deviator of the stress tensor. Thus, for primary and secondary creep, the results of uniaxial creep tests can also be used to calibrate the constitutive model. However, in the tertiary creep stage, the strain rate depends strongly on the stress state such that the von Mises stress is often not applicable (Naumenko and Altenbach, 2016) and a corresponding constitutive model should be formulated and calibrated based on multiaxial experimental creep data. Stress state dependent creep is discussed in Rabotnov (1969); Altenbach and Zolochevsky (1992, 1994); Altenbach et al (1995) and the references within.

Last but not least it is worth pointing out that the creep curve in Fig. 4.2 is idealized and serves only as an example to illustrate the three creep stages. One should keep in mind that the real shape of a creep curve varies strongly depending on the examined alloy. As one example, creep curves for 9–12% Cr steels, which are typical representatives of heat-resistant steels, do not usually exhibit a secondary stage (Simon et al, 2007; Ringel et al, 2004; Blum, 2008; Kimura, 2004). Instead of a constant strain rate, only the minimum strain rate can be extracted from the creep curves. Furthermore, for these alloys, the tertiary creep is particularly due to the coarsening of subgrains as main softening process (Straub, 1995). Nevertheless, other softening and damage processes, such as the coarsening of carbides and the formation of voids, also take place at the microstructure. In order to account for this pronounced softening effect, additional evolution equations can be incorporated into the constitutive models, cf. for example the constitutive model presented in Sect. 4.3.3.2.

4.3 Constitutive Modeling of Creep

This section presents commonly used constitutive models for creep in heat-resistant steels. It is composed of three parts, while the first part describes the so-called "early approaches", i.e. classical plasticity theory, the endochronic theory of VALANIS, and the multi-surface models initiated by MRÓZ. Section 4.3.2 discusses various nonunified models, which introduce separate strain measures to describe creep and plasticity. The unified models, using only one inelastic strain to simulate both creep and plasticity, are reviewed in Sect. 4.3.3.

Since the research on creep started already at the beginning of the 20th century with (da C. Andrade, 1910, 1914), an immense number of papers presenting experimental creep data and various constitutive models for rate-dependent inelasticity has been published, such that we will only name typical examples in the following. In Benallal et al (1989); Bruhns et al (1992); Lemaitre and Chaboche (1994); Haupt and Lion (1995); Khan and Jackson (1999); Krempl (1979); Krempl and Khan (2003), detailed experimental investigations of inelastic properties of metals can be found. Furthermore, reviews of various aspects of viscoplasticity are available, e.g. Perzyna (1966); Walker (1981). Typical examples for constitutive theories for the viscoplastic behavior of metals are the treatises (Haupt and Lion, 1995; Krempl, 1987; Perzyna, 1966; Lion, 2000).

The earliest constitutive models for creep in heat-resistant steels are theoretical approaches without internal variables (Stowell, 1957; Perzyna, 1963). Next, viscoplastic models including internal variables are presented, cf. Geary and Onat (1974); Bodner and Partom (1975); Hart (1976); Miller (1976a,b); Ponter and Leckie (1976); Krieg (1975). The concept of internal variables has been introduced by (Coleman and Gurtin, 1967). This concept states that the stress state depends on the strain as well as on internal variables, which represent a material state depending on the process history or, with other words, an evolution process. For the internal variables, ordinary differential equations (ODEs) need to be formulated with respect to time (creep behavior) or load (plasticity), the so-called "evolution equations". Up to the present, internal variables are frequently used to describe inelastic material behavior. The first models for creep in heat-resistant steels including internal variables have been refined in the 80s, cf. Stouffer and Bodner (1979); Walker (1981); Chaboche and Rousselier (1983a,b); Estrin and Mecking (1984); Krempl et al (1986); Lowe and Miller (1986).

Since a great variety of constitutive models on creep in heat-resistant steels is available, it is important to establish a systematic classification of these approaches. Although there are several criteria for classification, one often distinguishes two general classes of constitutive models: physically-based (microscopic) models and phenomenological (macroscopic) models (Charkaluk et al, 2002). Physically-based or microscopic models describe the macroscopic material behavior by considering the microstructural evolution of an alloy. Internal variables based on microstructural quantities, such as the dislocation density, are employed to reflect the micromechanics of deformation. That is why both microscopic phenomena, such as the evolution of microstructure, and macroscopic mechanical behavior, e.g. creep and relaxation, are represented. Typical examples are the models presented in Bodner and Partom (1975); Hart (1976); Miller (1976a,b); Stouffer and Bodner (1979); Sauzay et al (2005, 2008). On the other hand, phenomenological or macroscopic models are formulated based on the results of mechanical tests on a material. These models require significantly less effort compared to the physically-based approaches, which demand for extensive micrography to determine the involved parameters. For this reason, the majority of the presented models in Sect. 4.3.2 and 4.3.3 belongs to the group of phenomenological approaches.

4.3.1 Early Approaches

4.3.1.1 Classical Plasticity Theory

Since multiaxial creep behavior is similar to classical plasticity in certain aspects (in both theories, it is assumed that the continuum behaves as a fluid), classical plasticity theory has been directly employed for multiaxial creep analysis from 1900 to 1950. Historical reviews on this theory can be found in the monographs of Penny and Marriott (1995); Boyle and Spence (1983) or in the surveys Bruhns (2014, 2018). However, this theory is limited in practical application due to its derivation from the criteria of yielding failure. Furthermore, the physical damage process, cf. Sect. 4.2.3, is not accounted for (Yao et al, 2007).

Within this theory, the results of uniaxial creep tests are used to formulate effective stress criteria to compute the creep strains for multiaxial stress states. Thus, uniaxial constitutive relationships, as already introduced in Sect. 4.2.2, form the basis for modeling creep with classical plasticity theory. In Eq. (4.2), the separation ansatz to describe the dependence of the uniaxial strain rate on stress and temperature has already been introduced. In many cases, a specific time response function f_t is also taken into account, such that Eq. (4.2) can be modified as follows (Boyle and Spence, 1983; Penny and Marriott, 1995):

$$\dot{\varepsilon}^{\text{in}} = f_\sigma(\sigma)\, f_T(T)\, f_t(t). \tag{4.7}$$

This equation is a simplification of the generalized ansatz in Eq. (4.1). As already pointed out in Sect. 4.2.2, NORTON's power law is widely used for the stress response function, cf. Eq. (4.3), and the ARRHENIUS's law in Eq. (4.6) is often utilized for the temperature response function (Dorn, 1955). Furthermore, BAILEY's law represents a common choice for the time response function (Bailey, 1935):

$$f_t(t) = Et^m \tag{4.8}$$

with the material parameters E and m.

In order to formulate a constitutive model for multiaxial creep based on the previously introduced response functions with respect to the uniaxial stress state, one should keep in mind that creep is a shear-dominated process for isotropic and homogeneous materials (Boyle and Spence, 1983; Viswanathan, 1989). Based on this, the following three assumptions are usually formulated to derive multiaxial creep models (Boyle and Spence, 1983; Viswanathan, 1989):

1. Volume constancy is preserved during creep, such that the volumetric creep strain rate $\operatorname{tr}\left(\dot{\boldsymbol{\varepsilon}}^{\text{in}}\right)$ is zero.
2. The principal shear strain rates are proportional to the principal shear stresses.
3. The equivalent (VON MISES) strain rate is related to the equivalent (VON MISES) stress in the same way as for the uniaxial case.

There are models which do not assume that the volume constancy is preserved. Such models are presented in Altenbach et al (1995); Altenbach and Öchsner (2014); Altenbach et al (2014) and the references within. Models for steady-state multiaxial creep and the application to high-temperature components can be found in Boyle and Spence (1983); Penny and Marriott (1995).

4.3.1.2 Endochronic Theory

Another approach to model inelasticity in heat-resistant steels is the endochronic theory, which has been developed in the 70s by (Valanis, 1970). In contrast to phenomenological models with internal variables, where the present deformation and stress state depends only on the present value of observable variables, this theory requires the knowledge of the whole history of the deformation process. By introducing a so-called "intrinsic" time, the endochronic theory is able to explain phenomena, which classical plasticity theory could not cope with, such as cross and cyclic hardening as well as initial strain problems. In a later work by VALANIS, the theory is slightly modified by introducing an internal time which is related to the inelastic strain (Valanis, 1978). As further improvement, incremental or differential forms of the integral relation of stress and strain for inelasticity are formulated in Valanis and Fan (1983). Note that the obtained differential relation features a form, which significantly deviates from classical plasticity theory as presented in the previous section: no yield surface is introduced, and the typical distinction of elastic and inelastic processes is not required.

Based on the initial works by VALANIS, WATANABE and ATLURI derive a differential stress-strain relation based on the concept of intrinsic time (Watanabe and Atluri, 1986). Within this theory, the concept of a yield surface is retained, and elastic and plastic processes are defined in an analogous way to the concepts in classical plasticity. In order to allow for numerical solution of complex examples, the theory is implemented in the finite element method (FEM) by employing a generalized midpoint-radial-return algorithm, which enables computing the stress with respect to given strains.

4.3.1.3 Multi-surface Models

The multi-surface models represent another early approach to model inelasticity in steels and have been initiated by (Mróz, 1967). In the framework of this theory, several nested surfaces are introduced, whereby the smallest surface inside is the yield surface. The other surfaces are the bounding surfaces, since the yield surface always stays inside the bounding surfaces. If the yield surface meets the subsequent bounding surface, which now becomes the active surface, they will stay "attached" to each other at the current loading stress. The translation of the currently active surface is defined by MRÓZ's rule. Furthermore, the translation of the yield surface in the stress space is described by kinematic hardening and a corresponding backstress.

Based on the idea of the multi-surface model in Mróz (1967), (Dafalias and Popov, 1975, 1976) as well as (Krieg, 1975) developed two-surface models. Here, separate isotropic and kinematic hardening rules for the bounding surface and the yield surface are utilized. Note that CHABOCHE points out the great flexibility in modeling when using multi-surface models due to their additional degrees of freedom. However, these models exhibit significant drawbacks for complex nonproportional conditions (Chaboche, 2008).

4.3.2 Nonunified Models

Macromechanical models for creep and plasticity can be classified as unified and nonunified models. The notion of unified models has been established by (Chaboche and Rousselier, 1983b). This type of models accounts for only one time-dependent inelastic strain, whereas nonunified models introduce separate variables for instantaneous plastic strains and time-dependent inelastic deformation.

A very popular nonunified approach for creep in heat-resistant steels is the two-layer viscoplasticity model, which was originally developed by KICHENIN *et al.* to describe the visco-elastic behavior of polyethylene (Kichenin et al, 1996). As depicted in Fig. 4.3, two independent parallel networks are introduced: a viscous (time-dependent) and a plastic (time-independent) component. By introducing the stresses σ_V and σ_P in the viscous and plastic network, respectively, one can formulate the governing equations of the two-layer viscoplasticity model for one-dimensional stress and strain states Kichenin et al (1996):

$$\sigma = \sigma_V + \sigma_P \tag{4.9}$$

$$\sigma_V = \eta \dot{\varepsilon}_V = E_V (\epsilon - \epsilon_V) \tag{4.10}$$

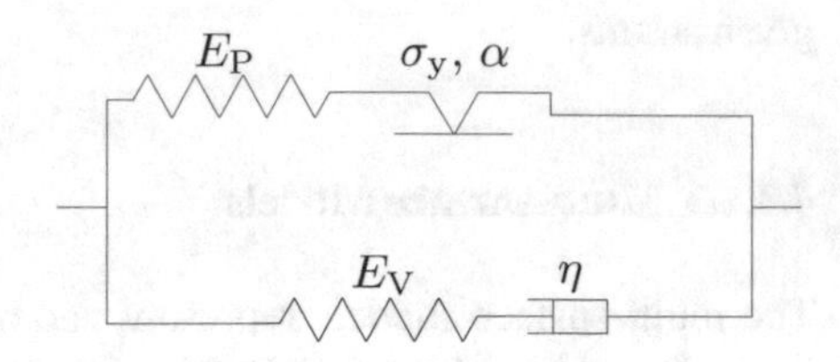

Fig. 4.3 Two-layer viscoplasticity model (Kichenin et al, 1996): YOUNG's moduli E_{P} and E_{V} in plastic and viscous network, viscosity η, threshold stress σ_{y} (yield stress), coefficient α for kinematic hardening

$$\sigma_{\mathrm{P}} = \begin{cases} E_{\mathrm{P}}\varepsilon & \text{if } \sigma_{\mathrm{P}} \leq \sigma_{\mathrm{y}} \\ \sigma_{\mathrm{y}} + \dfrac{\alpha E_{\mathrm{P}}}{\alpha + E_{\mathrm{P}}}\left(\varepsilon - \dfrac{\sigma_{\mathrm{y}}}{E_{\mathrm{P}}}\right) & \text{if } \sigma_{\mathrm{P}} > \sigma_{\mathrm{y}} \end{cases} \tag{4.11}$$

with the total strain ε, the strain ϵ_{V} in the viscous element, the YOUNG's moduli E_{P} and E_{V} in the plastic and viscous network, respectively, the viscosity η, the threshold stress σ_{y}, and the coefficient α for kinematic hardening.

Based on the initial work by KICHENIN *et al.*, FIGIEL and GÜNTHER implement a nonlinear isotropic and kinematic hardening model in the elastic-plastic network, whereas NORTON's power law, cf. Eq. (4.3), is used for the elastic-viscous network to describe creep deformations (Figiel and Günther, 2008). Note that the two-layer viscoplasticity model is also implemented in the commercial finite element code ABAQUS. Later, a similar two-layer viscoplasticity model is introduced in Leen et al (2010) with combined isotropic and kinematic hardening for plasticity and a power law creep model in order to describe the cyclic behavior of a high nickel-chromium alloy at 20–900°C. Furthermore, CHARKALUK and co-workers use the two-layer viscoplasticity model to represent the cyclic behavior of a cast iron under thermomechanical loading (Charkaluk et al, 2002). Temperatures up to 700°C are taken into account. In Solasi et al (2007), the analysis of polyelectrolyte membranes in fuel cells is performed based on the two-layer viscoplasticity model. With respect to heat-resistant steels, FARRAGHER *et al.* predict the thermo-mechanical cyclic behavior of a P91 steel at 400°C and 500°C with the two-layer viscoplasticity model (Farragher et al, 2013, 2014).

In the following, let us indicate additional, more recent examples of various non-unified models for heat-resistant steels. SHANG, LEEN, and HYDE formulate another nonunified constitutive model for creep incorporating isotropic and kinematic hardening models to describe the behavior of superplastic forming dies (Shang et al, 2006). The nonunified elasto-viscoplastic model of VELAY *et al.* incorporates several internal variables to simulate the cyclic behavior of the tempered martensitic steel 55NiCrMoV7 (Velay et al, 2006). Furthermore, another nonunified model is proposed in Wang et al (2015) in order to model the thermo-mechanical behavior of a high-chromium heat-resistant steel. Thereby, the authors decompose the total inelastic strain into a creep strain and a viscoplastic component.

However, several disadvantages of nonunified constitutive models have been pointed out, cf. e.g. Chaboche and Rousselier (1983b). Particularly for cyclic creep and the interaction of creep and plastic deformation, the separation of creep and plasticity gives unsatisfactory results when compared to experimental data (Krempl, 2000). Furthermore, the notion of "instantaneous" strains is not precisely defined. Additionally, numerical difficulties can occur when implementing different flow rules for instantaneous plastic strains and time-dependent inelastic strains. As a remedy, unified models are available and will be discussed in the next section.

4.3.3 Unified Models

4.3.3.1 Chaboche-type Models

The unified model by CHABOCHE represents a widespread approach to model creep and viscoplasticity in heat-resistant steels. General examples for unified CHABOCHE models can be found in Miller (1976a,b, 1987); Hart (1976); Cernocky and Krempl (1979, 1980); Walker (1981). For more specific applications of this type of constitutive model to creep at high temperatures, the interested reader is referred to the papers Moreno and Jordan (1986); Delobelle and Oytana (1984); Chaboche and Nouailhas (1989); Chan et al (1988, 1989). Within the unified viscoplasticity model of CHABOCHE (Chaboche and Rousselier, 1983a,b), plastic and creep strains are represented by only one parameter, the inelastic strain. Frequently, the description of the material behavior is refined by considering kinematic hardening.

Since the CHABOCHE model is usually formulated in combination with kinematic hardening, we will provide a short overview on this phenomenon in the following. Kinematic hardening is a commonly used method to implement the BAUSCHINGER effect in a constitutive model. This effect refers to the increase of yield strength in the direction of plastic flow during plastic deformation. In the reverse direction, the yield strength decreases such that the yield stress in compression is lower than the yield stress in tension (Zhang and Jiang, 2008). Kinematic hardening describes the translation of the elastic domain in the stress space (Lemaitre and Chaboche, 1994), such that the BAUSCHINGER effect can be taken into account. Therefore, all models for kinematic hardening employ a specific internal variable, the backstress tensor, which defines the position of the loading surface (Lemaitre and Chaboche, 1994; Chaboche and Rousselier, 1983a,b). One of the first models to describe kinematic hardening is the linear model by (Prager, 1949), where the following evolution equation is used for the backstress $\boldsymbol{\beta}$:

$$\dot{\boldsymbol{\beta}} = B_1 \dot{\boldsymbol{\varepsilon}}^{\text{in}} \tag{4.12}$$

with the material parameter B_1 and the inelastic strain rate tensor $\dot{\boldsymbol{\varepsilon}}^{\text{in}}$. However, a linear hardening behavior, i.e. the collinearity of the backstress with the inelastic strain, is rarely observed in experiments. Therefore, the nonlinear ARMSTRONG-FREDERICK model is now widely used (Armstrong and Frederick, 1966). This approach comprises a nonlinear function for kinematic hardening with a recall term to describe the dynamic recovery:

$$\dot{\boldsymbol{\beta}} = B_2 \dot{\boldsymbol{\varepsilon}}^{\text{in}} - B_3 \dot{\varepsilon}^{\text{in}}_{\text{vM}} \boldsymbol{\beta}, \tag{4.13}$$

whereby B_2 and B_3 are material parameters, and $\dot{\varepsilon}^{\text{in}}_{\text{vM}} = \sqrt{\frac{2}{3} \dot{\boldsymbol{\varepsilon}}^{\text{in}} : \dot{\boldsymbol{\varepsilon}}^{\text{in}}}$ denotes the von Mises equivalent inelastic strain rate. It has been shown in Chaboche (2008) and many other papers that the dynamic recovery term improves the simulation results significantly. As a further improvement of the simulation results, Chaboche suggests to superpose several backstresses (Chaboche and Rousselier, 1983a,b; Jiang and Kurath, 1996):

$$\boldsymbol{\beta} = \sum_{i=1}^{r} \boldsymbol{\beta}_i \tag{4.14}$$

with a separate evolution equation for each backstress:

$$\dot{\boldsymbol{\beta}}_i = B_{4_i} \dot{\boldsymbol{\varepsilon}}^{\text{in}} - B_{5_i} \dot{\varepsilon}^{\text{in}}_{\text{vM}} \boldsymbol{\beta}_i \tag{4.15}$$

with the material parameters B_{4_i} and B_{5_i}. Apart from the superposition of several backstresses in Eq. (4.14) and the corresponding evolution equations (4.15), the typical Chaboche model features the following governing equations, considering isothermal conditions and small deformations (Chaboche, 2008):

- the additive decomposition of the strain tensor into the elastic and plastic part:

$$\boldsymbol{\varepsilon} = \boldsymbol{\varepsilon}^{\text{el}} + \boldsymbol{\varepsilon}^{\text{in}}, \tag{4.16}$$

- Hooke's law for linear elasticity:

$$\boldsymbol{\sigma} = \boldsymbol{C} : \left(\boldsymbol{\varepsilon} - \boldsymbol{\varepsilon}^{\text{in}} \right) \tag{4.17}$$

  with the elastic stiffness tensor $\boldsymbol{C}$ of fourth rank,
- the definition of an elasticity domain based on the scalar function f:

$$f = ||\boldsymbol{\sigma} - \boldsymbol{\beta}||_{\text{H}} - k \leq 0, \tag{4.18}$$

  whereby k denotes the initial size of the yield surface, and Hill's criterion is employed introducing the fourth-rank tensor $\mathcal{H}$ to define the quadratic norm $||\boldsymbol{\sigma}||_{\text{H}} = \sqrt{\boldsymbol{\sigma} : \mathcal{H} : \boldsymbol{\sigma}}$,
- a flow rule to define the evolution of inelastic strains:

$$\dot{\boldsymbol{\varepsilon}}^{\text{in}} = \dot{\lambda} \frac{\partial f}{\partial \boldsymbol{\sigma}} \tag{4.19}$$

  with the plastic rate parameter $\dot{\lambda}$.

In addition, an evolution equation for the size of the yield surface k is required, which allows for the description of isotropic hardening. In the following, we will provide several examples of recent applications of the unified Chaboche model to creep in heat-resistant steels.

Yaguchi and Takahashi utilize a unified Chaboche-type viscoplastic model to simulate the cyclic behavior of a 9Cr-1Mo steel at 200–600°C. The cyclic softening effect is represented by a modified kinematic hardening equation, and the applied stress is divided into three components: a backstress, an effective stress, and an aging stress. This procedure enables a precise description of the inelastic behavior of the 9Cr-1Mo steel considering monotonic tension, stress relaxation, creep, and non-isothermal cyclic deformation. Later, this approach is refined by considering Ohno-Wang kinematic hardening to improve the prediction of ratcheting, i.e. the accumulation of inelastic strains under cyclic loads (Yaguchi and Takahashi, 2005a,b). In Zhao et al (2001), a unified Chaboche model is formulated to predict the stabilized cyclic loops of a nickel-based superalloy at various strain ranges and at high temperatures. Subsequently, the simulation results are improved by considering several types of test data such as monotonic, cyclic, relaxation, and creep tests (Tong and Vermeulen, 2003; Tong et al, 2004). Furthermore, static recovery terms as well as plastic strain memory terms have been included, cf. Zhan and Tong (2007); Zhan et al (2008).

In addition, a combined creep-fatigue damage approach based on the Chaboche model is presented in Chaboche and Gallerneau (2001). Here, the evolution of creep damage is coupled with time, while fatigue damage refers to the number of cycles. This model has been applied to nickel-based superalloys in Yeom et al (2007); Gharad et al (2006) as well as to a single crystal in Dunne and Hayhurst (1992a,b). Moreover, Hartrott and co-workers present a unified viscoplasticity model to simulate the thermo mechanical behavior of the heat-resistant P23 steel (von Hartrott et al, 2009). Another unified Chaboche model is developed in Hyde et al (2010) to predict the mechanical behavior of a 316 stainless steel in the temperature range 300–600°C. Last but not least, Saad *et al.* formulate a unified Chaboche model used for the thermo-mechanical fatigue of P91 and P92 steels (Saad, 2012; Saad et al, 2011a,b).

It is worth noting that the Chaboche model features certain similarities with the multi-surface models and the endochronic theory. In Chaboche and Rousselier (1983a,b), the equivalence between the nonlinear Armstrong-Frederick kinematic hardening rule and a simple two-surface model, cf. Sect. 4.3.1.3, based on bounding and yield surfaces is demonstrated. Furthermore, in Ohno and Wang (1991), the generalized nonlinear kinematic hardening rule by Chaboche (Chaboche and Rousselier, 1983a,b), cf. Eqs (4.14) and (4.15), is transformed into a multi-surface form. In addition, it is shown in Chaboche (1989) that the Chaboche model features similarities with a time-independent constitutive model Ohno and Kachi (1986), the unified viscoplastic models proposed in Walker (1981) and Krempl et al (1986), as well as the differential form derived in Watanabe and Atluri (1986) for the endochronic theory, cf. Sect. 4.3.1.2.

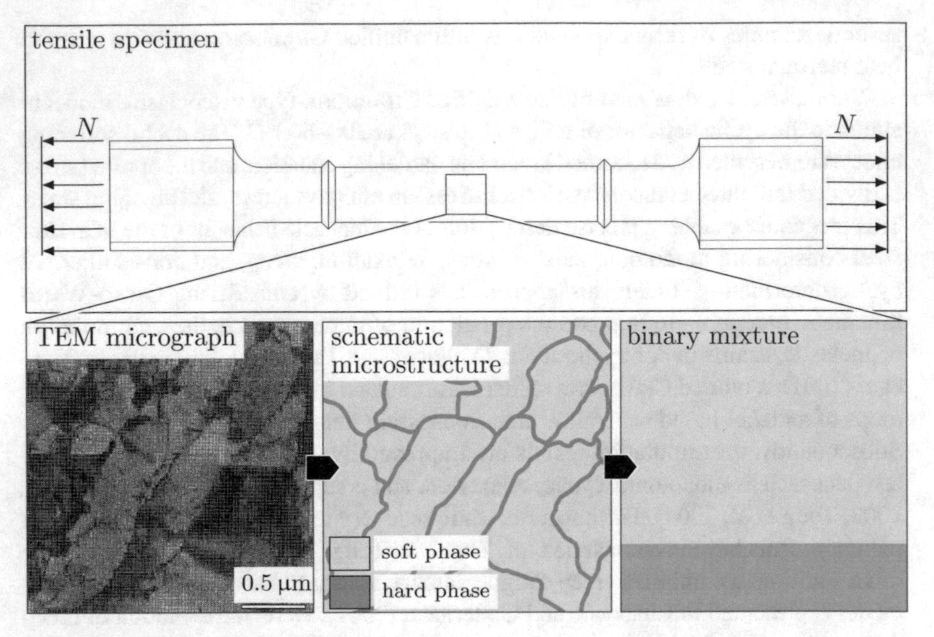

Fig. 4.4 Representation of the microstructure by means of the mixture model, cf. Straub (1995); Eisenträger (2018)

4.3.3.2 Mixture Models

Another important class of unified constitutive models for inelasticity are the so-called mixture models. Initially, these models have been developed by material scientists, e.g. Straub (1995); Polcik (1998); Barkar and Ågren (2005). Furthermore, the original derivation of the model is closely related to the microscale since the inelastic material behavior including hardening and softening is simulated by introducing an iso-strain composite with soft and hard constituents (Straub, 1995; Polcik, 1998). The hard constituent is related to the (sub)grain boundaries, i.e. regions with a high dislocation density and a large number of carbides, whereas the soft constituent represents the subgrain interior, i.e. regions with a low dislocation density and a small number of carbides (Blum, 2008). This division of the real microstructure into the two constituents is illustrated in Fig. 4.4. Note that the volume fraction of the hard constituent is closely related to microstructural features, such as the mean subgrain size, and assumed to decrease towards a saturation value to model softening due to the coarsening of subgrains (Naumenko et al, 2011). Usually, results from microscopic observations are used to calibrate these micromechanical models (Straub, 1995; Polcik, 1998; Barkar and Ågren, 2005).

Nevertheless, microscopy often requires a lot of effort and time. On the contrary, macroscopic material tests such as creep or HT tensile tests are straightforward to conduct and less time-consuming. Therefore, Naumenko *et al.* transform a micromechanical mixture model into a macroscopic one, cf. Naumenko et al (2011);

Naumenko and Gariboldi (2014). With the aim of simplifying the calibration procedure, a backstress of ARMSTRONG-FREDERICK-type and a softening variable are introduced as internal variables. In Eisenträger et al (2017, 2018a); Eisenträger (2018), a new calibration procedure based on creep and HT tensile tests is presented and the range of applicability of the model is significantly extended with respect to temperatures ($400°\mathrm{C} \leq T \leq 650°\mathrm{C}$) and stresses ($100\,\mathrm{MPa} \leq \sigma \leq 700\,\mathrm{MPa}$). Further details regarding the implementation in the FEM are discussed in Eisenträger et al (2018b). In the following, the governing equations of the mixture model are presented briefly. For further details on the derivation, the interested reader is referred to Eisenträger (2018).

As a starting point, we assume that the individual constituents exhibit an identical elastic behavior, while their inelastic behavior is different. To indicate the difference in the inelastic response of both constituents, one is referred to as "soft", while the other is labeled "hard". In the following equations, the index k is employed, taking the values s and h ($\Box_k \forall\, k \in \{\mathrm{s}, \mathrm{h}\}$). Since the following derivations are restricted to small strains, we postulate the equality of the linear strain tensor $\boldsymbol{\varepsilon}$ in the soft and the hard constituent (iso-strain assumption, Naumenko et al, 2011):

$$\boldsymbol{\varepsilon} = \boldsymbol{\varepsilon}_\mathrm{h} = \boldsymbol{\varepsilon}_\mathrm{s}. \tag{4.20}$$

In order to compute the overall stress $\boldsymbol{\sigma}$, a rule of mixture is applied:

$$\boldsymbol{\sigma} = \eta_\mathrm{s}\boldsymbol{\sigma}_\mathrm{s} + \eta_\mathrm{h}\boldsymbol{\sigma}_\mathrm{h} \tag{4.21}$$

with the volume fractions η_k. The following relation is valid due to the mass conservation constraint:

$$\eta_\mathrm{s} + \eta_\mathrm{h} = 1 \qquad \forall \quad 0 < \eta_k < 1.$$

In addition, we introduce the additive split of the strains into the elastic and the inelastic part, which are denoted by the superscripts $\Box^\mathrm{el}$ and $\Box^\mathrm{in}$, respectively:

$$\boldsymbol{\varepsilon} = \boldsymbol{\varepsilon}_k^\mathrm{el} + \boldsymbol{\varepsilon}_k^\mathrm{in}. \tag{4.22}$$

To describe the linear isotropic elastic behavior, the three-dimensional HOOKE's law is employed:

$$\boldsymbol{\sigma}_k = \boldsymbol{C} : \boldsymbol{\varepsilon}_k^\mathrm{el}. \tag{4.23}$$

Note that the same stiffness tensor $\boldsymbol{C}$ is used for both constituents because an identical elastic behavior is assumed. In addition to Eqs (4.20)–(4.23), evolution equations for the volume fractions η_k and the inelastic strains $\boldsymbol{\varepsilon}_k^\mathrm{in}$ need to be formulated. In a next step, two internal variables are introduced in order to enable the calibration of the model based on macroscopic material tests: an ARMSTRONG-FREDERICK-type backstress tensor $\boldsymbol{\beta}$ and a dimensionless softening variable Γ, which is closely related to the volume fraction of the hard constituent, cf. Naumenko et al (2011); Eisenträger (2018). By introducing the new internal variables, the following set of governing evolution equations for the inelastic strain in the mixture, the backstress, and the softening variable is derived (Eisenträger, 2018):

$$\dot{\boldsymbol{\varepsilon}}^{\mathrm{in}} = \frac{3}{2} f_\sigma(\tilde{\sigma}_{\mathrm{vM}})\, f_T(T)\, \frac{\tilde{\boldsymbol{\sigma}}'}{\tilde{\sigma}_{\mathrm{vM}}}, \tag{4.24}$$

$$\dot{\boldsymbol{\beta}} = \frac{1}{G(T)} \frac{\mathrm{d}G(T)}{\mathrm{d}T} \dot{T} \boldsymbol{\beta} + 2G(T) \frac{\eta_{\mathrm{h}_0}}{1-\eta_{\mathrm{h}_0}} \left[\dot{\boldsymbol{\varepsilon}}^{\mathrm{in}} - \frac{3}{2} \frac{\dot{\varepsilon}^{\mathrm{in}}_{\mathrm{vM}}}{\beta_{\mathrm{vM}_\star}(\sigma_{\mathrm{vM}})} \boldsymbol{\beta} \right], \tag{4.25}$$

$$\dot{\Gamma} = C_\Gamma \left[\Gamma_\star(\sigma_{\mathrm{vM}}) - \Gamma \right] \dot{\varepsilon}^{\mathrm{in}}_{\mathrm{vM}} \tag{4.26}$$

with the inelastic strain in the mixture $\boldsymbol{\varepsilon}^{\mathrm{in}} = \eta_{\mathrm{s}} \boldsymbol{\varepsilon}^{\mathrm{in}}_{\mathrm{s}} + \eta_{\mathrm{h}} \boldsymbol{\varepsilon}^{\mathrm{in}}_{\mathrm{h}}$, the stress and temperature response functions f_σ and f_T, the effective stress deviator $\tilde{\boldsymbol{\sigma}}' = \boldsymbol{\sigma}' - \Gamma\boldsymbol{\beta}$, and the corresponding von Mises equivalent quantities $\tilde{\sigma}_{\mathrm{vM}}, \sigma_{\mathrm{vM}}, \dot{\varepsilon}^{\mathrm{in}}_{\mathrm{vM}}, \beta_{\mathrm{vM}_\star}$. The variables G, η_{h_0}, $\Gamma_\star$ are temperature-dependent material parameters, which are determined during the calibration of the model.

In order to determine the model's response to a prescribed stress state $\boldsymbol{\sigma}$, the system of ODEs (4.24)–(4.26) must be solved, while providing initial conditions for the inelastic strain, the backstress, and the softening variable. In Eisenträger et al (2018a); Eisenträger (2018), the presented mixture model is calibrated based on creep and HT tensile tests with respect to the heat-resistant steel X20CrMoV12-1. The subsequent implementation in the FEM based on implicit Euler integration of the evolution equations (4.24)–(4.26) is discussed in Eisenträger et al (2018b). Overall, only 14 material parameters are required for robust simulations with respect to wide ranges of stress and temperature, i.e. $400°\mathrm{C} \le T \le 650°\mathrm{C}$ and $100\,\mathrm{MPa} \le \sigma \le 700\,\mathrm{MPa}$ (Eisenträger et al, 2018b). The model accounts for rate-dependent inelasticity in conjunction with nonlinear kinematic hardening. To describe the tertiary creep stage, a scalar damage variable can be incorporated as additional internal variable (Naumenko et al, 2011).

To conclude, the presented mixture model offers two main advantages in comparison to other approaches: a relatively small number of parameters for wide ranges of applicability and the possibility to calibrate the model based only on macroscopic material tests. In contrast to the previously discussed Chaboche models, only a single backstress and a softening variable are introduced. Since such a low number of internal variables is required, the resulting number of material parameters is relatively small, considering the wide range of applicability. The calibration of the model relies only on macroscopic tests such that time-consuming microscopic observations are not necessary.

4.4 Constitutive Modeling of Creep Damage

In order to describe the tertiary stage of the creep curve, cf. Sect. 4.2.3, constitutive models for creep damage are essential. According to Kassner and Hayes (2003); Goodall and Skelton (2004), the main cause of creep failure is the nucleation, growth, and coalescence of cavities (or voids) on the grain boundaries. It is important to note that macroscopic damage is closely related to microstructural processes. Thus, damage can be defined as the existence of micro defects, such as line defects, cf. Lemaitre (1996). In this sense, damage evolution summarizes the development of micro defects, to be precise the nucleation, growth, and coalescence of defects on the microstructure (Skrzypek and Ganczarski, 1999). The current section presents two major types of creep damage models: the microscopic cavity growth mechanism models and a more phenomenological approach, i.e. models based on continuum damage mechanics.

4.4.1 Cavity Growth Mechanism Models

Cavity growth mechanism (CGM) models predict the evolution of creep damage by accounting for microstructural damage processes. The first paper on CGM models was published in 1959 by HULL and RIMMER (Hull and Rimmer, 1959). In the 70s and 80s, this approach was further refined, cf. Rice and Tracey (1969); Hayhurst (1972); Manjoine (1975); Gurson (1977); Ashby et al (1978); Edward and Ashby (1979); Cane (1981); Manjoine (1982); Tvergaard and Needleman (1984); Huddleston (1985); Raj and Ashby (1975); Cocks and Ashby (1982b,a). More recent works on the CGM models have been published as well (Hales, 1994; Margolin et al, 1998; Spindler et al, 2001; Spindler, 2004; Ragab, 2002). According to Lin et al (2005b), typical applications for CGM models are HT creep, cold and hot metal forming, as well as superplastic forming. CGM-based models are widely used to predict the influence of multiaxial stress states on the creep failure strain or the time to rupture (Hull and Rimmer, 1959; McClintock, 1968; Rice and Tracey, 1969; Hayhurst, 1972; Cocks and Ashby, 1980; Manjoine, 1975, 1982; Margolin et al, 1998; Ragab, 2002; Spindler et al, 2001).

The CGM models are based on the physical process of creep cavity growth. Small voids or cavities are predominantly found at the grain boundaries. In addition, particles which lack cohesion with the matrix may act as voids. Furthermore, in Edward and Ashby (1979), grain boundary sliding is indicated as another important cause for void nucleation. Cavities nucleate particularly during the primary and secondary creep stage (Kassner and Hayes, 2003), whereas their growth and coalescence occurs primarily during the tertiary creep stage. This microstructural behavior is verified in Skleničk a et al (2003) for 9–12% Cr steels.

According to Hales (1994); Michel (2004); Yao et al (2007), three different types of cavity growth mechanisms can be distinguished: diffusion-controlled cavity growth, plasticity-controlled cavity growth, and constrained cavity growth. The diffusion-controlled cavity growth is proposed in Hull and Rimmer (1959); Raj and Ashby (1975) to predict the time to rupture under creep deformation. Thereby, the growth rate of cavities is primarily influenced by the shape of voids and the diffusion process. However, with increasing cavity size, the effect of diffusion-controlled growth decreases quickly, and the plasticity-controlled growth dominates the material behavior (Nicolaou et al, 2000). Within the plasticity-controlled growth of voids, the viscoplastic deformation of the surrounding material is the underlying cause for the increase in cavity size (Rice and Tracey, 1969; Hancock, 1976). Note that this mechanism influences the material behavior particularly under high strain rates. First models for the plasticity-controlled cavity growth were published in the 60s (McClintock, 1968; Rice and Tracey, 1969). In recent years, these approaches have been improved and applied to superplasticity in Khaleel et al (2001); Taylor et al (2002).

The first constrained cavity growth mechanism model has been proposed in Dyson (1976), and was refined gradually in further papers (Cocks and Ashby, 1982b,a; Edward and Ashby, 1979; Cocks and Ashby, 1980; Tvergaard, 1984; Yousefiani et al, 2000; Delph, 2002). According to this model, cavities grow when the local deformation rate exceeds the deformation rate of the surrounding material due to cavity growth, such that the rate of cavity growth is constrained to produce the local strain at the same rate as the deformation caused by the remote stress. Thus, the constrained cavity growth is due to the viscoplastic strains *and* diffusional processes, and this mechanism can be interpreted as coupled occurrence of the first two mechanisms, i.e. the diffusion-controlled cavity growth and the plasticity-controlled growth of voids.

4.4.2 Continuum Damage Mechanics Models

Within the continuum damage mechanics (CDM) models, a phenomenological description of damage is established. Here, damage variables are introduced based on the equivalence principle and the concept of representative volume elements (RVEs), such that the micro defects can be "smeared out", and the stress and strain state can be considered as homogeneous. Although the CDM models were initiated already during the 1950s, they have been highlighted again since the 1990s with the rapid development of computer technology and FEM since these models are straightforward to implement in numerical methods. In recent years, a multitude of papers on the CDM models has been published, cf. e.g. Othman et al (1993); Hayhurst et al (1994); Kowalewski et al (1994b,a); Perrin and Hayhurst (1996, 1999); Hyde et al (1996, 2006); Jing et al (2001, 2003); Xu (2001, 2004); Hayhurst et al (2005b,a); Lin et al (2005a).

The CDM model were initiated by Rabotnov, who developed a phenomenological model for creep damage with an evolution equation for a scalar damage variable ω, see Rabotnov (1959). One year before, the dual variable continuity $\psi = 1 - \omega$ was suggested in Kachanov (1958). A zero damage variable, i.e. $\omega = 0$, indicates the virgin material state (never existing in real materials), while $\omega = 1$ refers to the complete failure of the material (in real materials the failure occurs at $\omega = 0.3 \ldots 0.7$, Lemaitre and Chaboche (1994); Lemaitre (1996)). In Leckie and Hayhurst (1974), this concept was extended for the first time to multiaxial stress states. In addition to the damage variable concept, Rabotnov introduced the concept of effective stress (Rabotnov, 1959). Following this concept, the effective stress $\tilde{\sigma}$ in the uniaxial case is equal to the real stress σ, modified by the damage variable ω:

$$\tilde{\sigma} = \frac{\sigma}{1-\omega}. \tag{4.27}$$

Based on this, LEMAITRE established the equivalence principle (Lemaitre, 1971, 1985), which states that any constitutive equation for a damaged material may be derived in the same way as for a virgin material, except that the usual stress is replaced by the effective stress. In the following, CHABOCHE and LEMAITRE presented a method to derive constitutive laws based on the framework of irreversible thermodynamics and the principle of strain equivalence (Chaboche, 1981; Lemaitre, 1985). However, the concept of a scalar damage variable does not account for the anisotropic nature of damage. For this reason, several vector and tensor representations of damage variables have been developed later to account for anisotropy (Murakami and Ohno, 1981; Krajcinovic and Fonseka, 1981; Fonseka and Krajcinovic, 1981; Betten, 1982; Krajcinovic, 1983; Murakami, 1983; Chaboche, 1984). In addition, a large number of textbooks, monographs, and reviews on CDM models is available, e.g. Kachanov (1986); Chaboche (1988); Krajcinovic (1989); Lemaitre (1996); Voyiadjis and Kattan (1999); Skrzypek and Ganczarski (1999); Wohua and Valliappan (1998a,b); Lin et al (2005b); Yao et al (2007); Betten (2008); Murakami (2012). Further applications of CDM models to simulate creep damage are presented in Murakami et al (2000); Becker et al (2002); Hayhurst et al (2005b,a), and similar papers with respect to creep-fatigue damage can be found as well, cf. Jing et al (2003); Stolk et al (2004).

In the following, let us focus on several CDM models, while classifying these approaches into two groups: multiaxial models with a *single* damage variable and multiaxial frameworks incorporating *several* damage variables. Creep damage models with a single variable are straightforward to formulate. Nevertheless, the physical nature of the damage parameter is not examined, and different damage mechanisms cannot be taken into account. A commonly used approach is the LEMAITRE constitutive equation for damage (Lemaitre, 1985), which is based on the framework of irreversible thermodynamics and for example used in Jing et al (2001) for the multiaxial creep prediction of an aero-engine turbine disc. As an alternative, the KACHANOV-RABOTNOV constitutive equation (Leckie and Hayhurst, 1974) can be used to model creep damage. In Becker et al (2002), this approach is utilized for the numerical modeling of a titanium alloy at 650°C and a 0.5Cr0.5Mo0.25V steel at 640°C.

Considering several variables while simulating multiaxial creep damage results in a more elaborated formulation, but also requires significantly more effort with respect to the calibration and numerical implementation of the constitutive model. CDM models with several damage variables resolve the drawback of corresponding models with only one damage variable, which cannot account for the physical nature of the involved damage parameter. Furthermore, various studies show that the damage of high-temperature materials is due to different mechanisms, for example grain boundary sliding, void growth, diffusion of vacancies along the boundary, and the coarsening of precipitates, such as carbides, cf. also Sect. 4.2.3. The general form of equations for CDM models considering n different damage variables ω_n is given in Hayhurst (2005) as follows:

$$\dot{\boldsymbol{\varepsilon}} = f\left(\boldsymbol{\sigma}, \omega_1, \omega_2, \ldots, \omega_n, T\right), \tag{4.28}$$

$$\dot{\omega}_1 = g_1(\boldsymbol{\sigma}, \omega_1, \omega_2, \ldots, \omega_n, T), \tag{4.29}$$

$$\dot{\omega}_2 = g_2(\boldsymbol{\sigma}, \omega_1, \omega_2, \ldots, \omega_n, T), \tag{4.30}$$

$$\vdots$$

$$\dot{\omega}_n = g_n(\boldsymbol{\sigma}, \omega_1, \omega_2, \ldots, \omega_n, T). \tag{4.31}$$

In CDM models, the hyperbolic stress response function, cf. Eq. (4.5), is frequently used for the inelastic strain rates (Perrin and Hayhurst, 1996; Othman et al, 1993; Kowalewski et al, 1994a). The damage rate functions should account for the fact that the nucleation and growth of cavities reduce the load bearing section and accelerate creep damage, such that the effect of cavitation damage should be explicitly represented in the model. As one example, Dyson suggests a specific function for the cavitation damage, cf. Dyson (1988). In addition, precipitate coarsening should be taken into account since it represents another important cause for the occurrence of damage. Note that the evolution equation for a corresponding damage variable is given in Dyson (1988); Perrin and Hayhurst (1996). Moreover, damage in the tertiary creep stage is also due to the accumulation of dislocations, which could be implemented in a constitutive model by the rate function proposed in Dyson (1988); Othman et al (1993). Finally, many CDM models consider a primary creep state variable H, which is not explicitly related to damage, but allows for the simulation of strain hardening during the primary creep stage, cf. Kowalewski et al (1994a); Perrin and Hayhurst (1996) and Sect. 4.2.3.

In the sequel, let us provide specific examples for CDM models with several damage variables. In Othman et al (1993), a CDM model is presented to predict the multiaxial creep behavior of a nickel-based superalloy. Two damage variables are incorporated: a cavitation damage state variable and a dislocation multiplication state variable. A total of three damage variables are used in the constitutive model by Kowalewski and co-workers, which describe cavitation damage, precipitate coarsening, and hardening during primary creep. Here, the creep behavior of an aluminum alloy is simulated under multiaxial stress states. Dyson uses another CDM model to describe the evolution of creep strain based on microstructural processes, such that his framework involves dominant mechanisms, which determine the evolution of creep strain as normalized damage parameters. Note that in general, the Kachanov-Rabotnov constitutive equations with a single variable or several parameters are widespread in the prediction of multiaxial creep of ferritic steels, which are commonly used for components for high-temperature applications.

4.5 Conclusion and Outlook

The current contribution aims at providing a brief overview concerning the state-of-the-art in constitutive modeling for creep of heat-resistant steels. Since it is impossible to provide a complete overview of this topic due to the great variety of developed constitutive models and the immense number of published works in this field, this paper seeks to highlight a selection of commonly used constitutive models for creep in heat-resistant steels. At first, basic characteristics of creep in heat-resistant steels are described, typical microstructural properties of heat-resistant steels are introduced, the influence of stress and temperature on creep is discussed, and the classical creep curve including primary, secondary, and tertiary creep is presented. Special emphasis is placed on the important role of microstructural processes in creep.

The major section of this contribution presents various constitutive models for creep. As the so-called "early approaches", classical plasticity theory, the endochronic theory, and multi-surface models are briefly presented. Next, we discuss several nonunified models, which utilize separate variables for instantaneous plastic strains and time-dependent inelastic deformation. Among these models, the two-layer viscoplasticity model is commonly used. Due to several drawbacks of nonunified models, particularly due to difficulties while accounting for the interaction of creep and plasticity, unified models have become popular over the past years. Here, only one inelastic strain is used to describe creep and plasticity. The contribution at hand introduces two general approaches: the CHABOCHE-type models as well as mixture models. While CHABOCHE-type models are very popular due to their straightforward calibration based on macroscopic experimental data, they usually involve a large number of material parameters. In contrast, the proposed mixture model is formulated with respect to a microstructural background and allows for robust simulations with respect to large ranges of stress and temperature, whereas the number of parameters is relatively low. Nevertheless, one should note that the calibration procedure of the mixture model is more complex than the calibration of a CHABOCHE-type model.

Finally, two major classes of constitutive models for creep damage are discussed: the cavity growth mechanism models and approaches based on continuum damage mechanics. Note that creep damage models are essential to describe the tertiary creep stage. While cavity growth mechanism models are closely related to microstructural processes, continuum damage mechanics models are phenomenological approaches. However, since continuum damage mechanics models are straightforward to implement in numerical methods, such as the FEM, these models have become popular in recent years due to the widespread use of numerical methods.

Power generation represents one of the most important high-temperature applications for heat-resistant steels. In the future, it is expected that operating temperatures will raise to more than 600–650°C in order to increase thermodynamic efficiency. Thus, one should focus on the development of creep models accounting for this high temperature range. Furthermore, additional studies are necessary for modeling complex phenomena such as nonproportional hardening and flow, ratcheting, and thermo-mechanical behavior. For this purpose, it is crucial to estimate the reliability of material parameters, which constitutes a major challenge because of the limited availability of experimental data from creep tests. This is particularly true for recently developed alloys, such as the 9–12% Cr martensitic steels, so that the availability of long-term creep data is restricted due to the relatively low age of these steels. In order to solve this issue, one could extrapolate from short-term creep data to long-term predictions based on conventional empirical methods. However, further research should be conducted concerning the reliability of these extrapolation methods.

References

Abe F, Kern TU, Viswanathan R (eds) (2008) Creep-resistant steels. Woodhead Publishing Limited

Altenbach H, Öchsner A (eds) (2014) Plasticity of Pressure-Sensitive Materials. Engineering Materials, Springer, Berlin, Heidelberg, doi:10.1007/978-3-642-40945-5

Altenbach H, Zolochevsky A (1992) Energy version of creep and stress-rupture strength theory for anisotropic and isotropic materials which differ in resistance to tension and compression. Journal of Applied Mechanics and Technical Physics 33(1):101–106

Altenbach H, Zolochevsky AA (1994) Eine energetische Variante der Theorie des Kriechens und der Langzeitfestigkeit für isotrope Werkstoffe mit komplizierten Eigenschaften. ZAMM - Journal of Applied Mathematics and Mechanics / Zeitschrift für Angewandte Mathematik und Mechanik 74(3):189–199, doi:10.1002/zamm.19940740311

Altenbach H, Altenbach J, Zolochevsky A (1995) Erweiterte Deformationsmodelle und Versagenskriterien der Werkstoffmechanik. Deutscher Verlag für Grundstoffindustrie, Leipzig, Stuttgart

Altenbach H, Bolchoun A, Kolupaev VA (2014) Phenomenological Yield and Failure Criteria. In: Altenbach H, Öchsner A (eds) Plasticity of Pressure-Sensitive Materials, Engineering Materials, Springer, pp 49–152, doi:10.1007/978-3-642-40945-5_2

Armstrong PJ, Frederick CO (1966) A Mathematical Representation of the Multiaxial Bauschinger Effect

Ashby MF, Edward GH, Davenport J, Verrall RA (1978) Application of bound theorems for creeping solids and their application to large strain diffusional flow. Acta Metallurgica 26(9):1379–1388, doi:10.1016/0001-6160(78)90153-0

Bailey RW (1935) The utilization of creep test data in engineering design. Proceedings of The Institution of Mechanical Engineers 131(1):131–349

Barkar T, Ågren J (2005) Creep simulation of 9–12% Cr steels using the composite model with thermodynamically calculated input. Materials Science and Engineering: A 395(1–2):110–115, doi:10.1016/j.msea.2004.12.004

Becker AA, Hyde TH, Sun W, Andersson P (2002) Benchmarks for finite element analysis of creep continuum damage mechanics. Computational Materials Science 25(1–2):34–41, doi:10.1016/s0927-0256(02)00247-1

Benallal A, Gallo PL, Marquis D (1989) An experimental investigation of cyclic hardening of 316 stainless steel and of 2024 aluminium alloy under multiaxial loadings. Nuclear Engineering and Design 114(3):345–353, doi:10.1016/0029-5493(89)90112-x

Betten J (1982) Net-stress analysis in creep mechanics. Ingenieur-Archiv 52(6):405–419, doi:10.1007/bf00536211

Betten J (2008) Creep Mechanics, 3rd edn. Springer, Berlin, Heidelberg

Blum W (2008) Mechanisms of creep deformation in steel. In: Abe F, Kern TU, Viswanathan R (eds) Creep-resistant steels, Woodhead Publishing, pp 365–402

Bodner SR, Partom Y (1975) Constitutive Equations for Elastic-Viscoplastic Strain-Hardening Materials. Journal of Applied Mechanics 42(2):385–389, doi:10.1115/1.3423586

Boyle JT, Spence J (1983) Stress Analysis for Creep. Elsevier, doi:10.1016/c2013-0-00873-0

Breeze P (2005) Power Generation Technologies. Newnes, Oxford

Bruhns OT (2014) Some Remarks on the History of Plasticity – Heinrich Hencky, a Pioneer of the Early Years. In: Stein E (ed) The History of Theoretical, Material and Computational Mechanics - Mathematics Meets Mechanics and Engineering, Springer, Lecture Notes in Applied Mathematics and Mechanics, vol 1, pp 133–152, doi:10.1007/978-3-642-39905-3_9

Bruhns OT (2018) History of Plasticity. In: Altenbach H, Öchsner A (eds) Encyclopedia of Continuum Mechanics, Springer, pp 1–61, doi:10.1007/978-3-662-53605-6_281-1

Bruhns OT, Lehmann T, Pape A (1992) On the description of transient cyclic hardening behaviour of mild steel CK 15. International Journal of Plasticity 8(4):331–359, doi:10.1016/0749-6419(92)90054-g da C Andrade EN (1910) On the Viscous Flow in Metals, and Allied Phenomena. Proceedings of the Royal Society A: Mathematical, Physical and Engineering Sciences 84(567):1–12, doi:10.1098/rspa.1910.0050

da C Andrade EN (1914) The Flow in Metals under Large Constant Stresses. Proceedings of the Royal Society A: Mathematical, Physical and Engineering Sciences 90(619):329–342

Cane BJ (1981) Creep fracture of dispersion strengthened low alloy ferritic steels. Acta Metallurgica 29(9):1581–1591, doi:10.1016/0001-6160(81)90040-7

Cernocky EP, Krempl E (1979) A non-linear uniaxial integral constitutive equation incorporating rate effects, creep and relaxation. International Journal of Non-Linear Mechanics 14(3):183–203, doi:10.1016/0020-7462(79)90035-0

Cernocky EP, Krempl E (1980) A theory of viscoplasticity based on infinitesimal total strain. Acta Mechanica 36(3–4):263–289, doi:10.1007/bf01214636

Chaboche JL (1981) Continuous damage mechanics — A tool to describe phenomena before crack initiation. Nuclear Engineering and Design 64(2):233–247, doi:10.1016/0029-5493(81)90007-8

Chaboche JL (1984) Anisotropic creep damage in the framework of continuum damage mechanics. Nuclear Engineering and Design 79(3):309–319, doi:10.1016/0029-5493(84)90046-3

Chaboche JL (1988) Continuum Damage Mechanics: Part I—General Concepts. Journal of Applied Mechanics 55(1):59, doi:10.1115/1.3173661

Chaboche JL (1989) Constitutive equations for cyclic plasticity and cyclic viscoplasticity. International Journal of Plasticity 5(3):247–302, doi:10.1016/0749-6419(89)90015-6

Chaboche JL (2008) A review of some plasticity and viscoplasticity constitutive theories. International Journal of Plasticity 24(10):1642–1693, doi:10.1016/j.ijplas.2008.03.009

Chaboche JL, Gallerneau F (2001) An overview of the damage approach of durability modelling at elevated temperature. Fatigue and Fracture of Engineering Materials and Structures 24(6):405–418, doi:10.1046/j.1460-2695.2001.00415.x

Chaboche JL, Nouailhas D (1989) A Unified Constitutive Model for Cyclic Viscoplasticity and Its Applications to Various Stainless Steels. Journal of Engineering Materials and Technology 111(4):424–430, doi:10.1115/1.3226490

Chaboche JL, Rousselier G (1983a) On the Plastic and Viscoplastic Constitutive Equations: Part I: Rules Developed With Internal Variable Concept. Journal of Pressure Vessel Technology 105(2):153–158, doi:10.1115/1.3264257

Chaboche JL, Rousselier G (1983b) On the Plastic and Viscoplastic Constitutive Equations: Part II: Application of Internal Variable Concepts to the 316 Stainless Steel. Journal of Pressure Vessel Technology 105(2):159–164, doi:10.1115/1.3264258

Chan KS, Bodner SR, Lindholm US (1988) Phenomenological Modeling of Hardening and Thermal Recovery in Metals. Journal of Engineering Materials and Technology 110:1–8, doi:10.1115/1.3226003

Chan KS, Lindholm US, Bodner SR, Walker KP (1989) High Temperature Inelastic Deformation Under Uniaxial Loading: Theory and Experiment. Journal of Engineering Materials and Technology 111(4):345–353, doi:10.1115/1.3226478

Charkaluk E, Bignonnet A, Constantinescu A, Van KD (2002) Fatigue design of structures under thermomechanical loadings. Fatigue and Fracture of Engineering Materials and Structures 25(12):1199–1206, doi:10.1046/j.1460-2695.2002.00612.x

Coble RL (1963) A Model for Boundary Diffusion Controlled Creep in Polycrystalline Materials. Journal of Applied Physics 34(6):1679–1682, doi:10.1063/1.1702656

Cocks ACF, Ashby MF (1980) Intergranular fracture during power-law creep under multiaxial stresses. Metal Science 14(8-9):395–402, doi:10.1179/030634580790441187

Cocks ACF, Ashby MF (1982a) Creep fracture by coupled power-law creep and diffusion under multiaxial stress. Metal Science 16(10):465–474, doi:10.1179/msc.1982.16.10.465

Cocks ACF, Ashby MF (1982b) On creep fracture by void growth. Progress in Materials Science 27(3-4):189–244, doi:10.1016/0079-6425(82)90001-9

Coleman BD, Gurtin ME (1967) Thermodynamics with Internal State Variables. The Journal of Chemical Physics 47(2):597–613, doi:10.1063/1.1711937

Czichos H, Skrotzki B, Simon FG (2014) Das Ingenieurwissen: Werkstoffe. Springer, Berlin Heidelberg, doi:10.1007/978-3-642-41126-7

Dafalias YF, Popov EP (1975) A model of nonlinearly hardening materials for complex loading. Acta Mechanica 21(3):173–192, doi:10.1007/bf01181053

Dafalias YF, Popov EP (1976) Plastic Internal Variables Formalism of Cyclic Plasticity. Journal of Applied Mechanics 43(4):645–651, doi:10.1115/1.3423948

Delobelle P, Oytana C (1984) Experimental study of the flow rules of a 316 stainless steel at high and low stresses. Nuclear Engineering and Design 83(3):333–348, doi:10.1016/0029-5493(84)90126-2

Delph TJ (2002) Some selected topics in creep cavitation. Metallurgical and Materials Transactions A 33(2):383–390, doi:10.1007/s11661-002-0099-0

Dorn JE (1955) Some fundamental experiments on high temperature creep. Journal of the Mechanics and Physics of Solids 3(2):85–116, doi:10.1016/0022-5096(55)90054-5

Dunne FPE, Hayhurst DR (1992a) Continuum Damage Based Constitutive Equations for Copper under High Temperature Creep and Cyclic Plasticity. Proceedings of the Royal Society A: Mathematical, Physical and Engineering Sciences 437(1901):545–566, doi:10.1098/rspa.1992.0079

Dunne FPE, Hayhurst DR (1992b) Modelling of Combined High-Temperature Creep and Cyclic Plasticity in Components Using Continuum Damage Mechanics. Proceedings of the Royal Society A: Mathematical, Physical and Engineering Sciences 437(1901):567–589, doi:10.1098/rspa.1992.0080

Dyson BF (1976) Constraints on diffusional cavity growth rates. Metal Science 10(10):349–353, doi:10.1179/030634576790431417

Dyson BF (1988) Creep and fracture of metals: mechanisms and mechanics. Revue de Physique Appliquée 23(4):605–613, doi:10.1051/rphysap:01988002304060500

Dyson BF, McLean M (2001) Micromechanism-quantification for creep constitutive equations. In: Murakami S, Ohno N (eds) IUTAM Symposium on Creep in Structures, Kluwer, Dordrecht, pp 3–16

Edward GH, Ashby MF (1979) Intergranular fracture during power-law creep. Acta Metallurgica 27(9):1505–1518, doi:10.1016/0001-6160(79)90173-1

Eisenträger J (2018) A Framework for Modeling The Mechanical Behavior of Tempered Martensitic Steels at High Temperatures. PhD thesis, Otto von Guericke University Magdeburg

Eisenträger J, Naumenko K, Altenbach H, Gariboldi E (2017) Analysis of Temperature and Strain Rate Dependencies of Softening Regime for Tempered Martensitic Steel. The Journal of Strain Analysis for Engineering Design 52:226–238, doi:10.1177/0309324717699746

Eisenträger J, Naumenko K, Altenbach H (2018a) Calibration of a Phase Mixture Model for Hardening and Softening Regimes in Tempered Martensitic Steel Over Wide Stress and Temperature Ranges. The Journal of Strain Analysis for Engineering Design 53:156–177, doi:10.1177/0309324718755956

Eisenträger J, Naumenko K, Altenbach H (2018b) Numerical implementation of a phase mixture model for rate-dependent inelasticity of tempered martensitic steels. Acta Mechanica 229:3051–3068, doi:10.1007/s00707-018-2151-1

El-Magd E, Betten J, Palmen P (1996) Auswirkungen der Schädigungsanisotropie auf die Lebensdauer von Stählen bei Zeitstandbeanspruchung. Materialwissenschaft und Werkstofftechnik 27(5):239–245, doi:10.1002/mawe.19960270510

Estrin Y, Mecking H (1984) A unified phenomenological description of work hardening and creep based on one-parameter models. Acta Metallurgica 32(1):57–70, doi:10.1016/0001-6160(84)90202-5

Farragher TP, Scully S, O'Dowd NP, Leen SB (2013) Thermomechanical Analysis of a Pressurized Pipe Under Plant Conditions. Journal of Pressure Vessel Technology 135:011,204–1–011,204–9, doi:10.1115/1.4007287

Farragher TP, Scully S, O'Dowd NP, Hyde CJ, Leen SB (2014) High Temperature, Low Cycle Fatigue Characterization of P91 Weld and Heat Affected Zone Material. Journal of Pressure Vessel Technology 136(2):021,403–1–021,403–10, doi:10.1115/1.4025943

Figiel L, Günther B (2008) Modelling the high-temperature longitudinal fatigue behaviour of metal matrix composites (SiC/Ti-6242): Nonlinear time-dependent matrix behaviour. International Journal of Fatigue 30(2):268–276, doi:10.1016/j.ijfatigue.2007.01.056

Fonseka GU, Krajcinovic D (1981) The Continuous Damage Theory of Brittle Materials, Part 2: Uniaxial and Plane Response Modes. Journal of Applied Mechanics 48(4):816–824, doi:10.1115/1.3157740

Fournier B, Sauzay M, Mottot M, Brillet H, Monnet I, Pineau A (2005) Experimentally Based Modelling of Cyclically Induced Softening in a Martensitic Steel at High Temperature. In: Shibli IA, Holdsworth SR, Merckling G (eds) ECCC Creep Conference, DES tech publications, pp 649–661

François D, Pineau A, Zaoui A (2012) Mechanical Behaviour of Materials. Springer Netherlands, doi:10.1007/978-94-007-2546-1

Frost HJ, Ashby MF (1982) Deformation-Mechanism Maps: The Plasticity and Creep of Metals and Ceramics. Pergamon Press

Geary JA, Onat ET (1974) Representation of nonlinear hereditary mechanical behavior. Tech. rep., Oak Ridge National Lab.(ORNL), Oak Ridge, TN (United States), URL https://www.osti.gov/servlets/purl/4258567

Gharad AE, Zedira H, Azari Z, Pluvinage G (2006) A synergistic creep fatigue failure model damage (case of the alloy Z5NCTA at 550°C). Engineering Fracture Mechanics 73(6):750–770, doi:10.1016/j.engfracmech.2005.10.008

Goodall IW, Skelton RP (2004) The importance of multiaxial stress in creep deformation and rupture. Fatigue and Fracture of Engineering Materials and Structures 27(4):267–272, doi:10.1111/j.1460-2695.2004.00743.x

Gurson AL (1977) Continuum Theory of Ductile Rupture by Void Nucleation and Growth: Part I—Yield Criteria and Flow Rules for Porous Ductile Media. Journal of Engineering Materials and Technology 99(1):2–15, doi:10.1115/1.3443401

Hales R (1994) The Role of Cavity Growth Mechanisms in Determining Creep-Rupture under Multiaxial Stresses. Fatigue & Fracture of Engineering Materials and Structures 17(5):579–591, doi:10.1111/j.1460-2695.1994.tb00257.x

Hancock JW (1976) Creep cavitation without a vacancy flux. Metal Science 10(9):319–325, doi:10.1179/msc.1976.10.9.319

Harper J, Dorn J (1957) Viscous creep of aluminum near its melting temperature. Acta Metallurgica 5(11):654–665, doi:10.1016/0001-6160(57)90112-8

Hart EW (1976) Constitutive Relations for the Nonelastic Deformation of Metals. Journal of Engineering Materials and Technology 98(3):193–202, doi:10.1115/1.3443368

von Hartrott P, Holmström S, Caminada S, Pillot S (2009) Life-time prediction for advanced low alloy steel P23. Materials Science and Engineering: A 510-511:175–179, doi:10.1016/j.msea.2008.04.117

Haupt P, Lion A (1995) Experimental identification and mathematical modeling of viscoplastic material behavior. Continuum Mechanics and Thermodynamics 7(1):73–96, doi:10.1007/bf01175770

Hayhurst DR (1972) Creep rupture under multi-axial states of stress. Journal of the Mechanics and Physics of Solids 20(6):381–382, doi:10.1016/0022-5096(72)90015-4

Hayhurst DR (2005) CDM mechanisms-based modelling of tertiary creep: ability to predict the life of engineering components. Archives of Mechanics 57(2–3):103–132

Hayhurst DR, Dyson BF, Lin J (1994) Breakdown of the skeletal stress technique for lifetime prediction of notched tension bars due to creep crack growth. Engineering Fracture Mechanics 49(5):711–726, doi:10.1016/0013-7944(94)90035-3

Hayhurst DR, Goodall IW, Hayhurst RJ, Dean DW (2005a) Lifetime Predictions For High-Temperature Low-Alloy Ferritic Steel Weldments. The Journal of Strain Analysis for Engineering Design 40(7):675–701, doi:10.1243/030932405x30885

Hayhurst RJ, Mustata R, Hayhurst DR (2005b) Creep constitutive equations for parent, Type IV, R-HAZ, CG-HAZ and weld material in the range 565–640°C for Cr–Mo–V weldments. International Journal of Pressure Vessels and Piping 82(2):137–144, doi:10.1016/j.ijpvp.2004.07.014

Herring C (1950) Diffusional Viscosity of a Polycrystalline Solid. Journal of Applied Physics 21(5):437–445, doi:10.1063/1.1699681

Huddleston RL (1985) An Improved Multiaxial Creep-Rupture Strength Criterion. Journal of Pressure Vessel Technology 107(4):421–429, doi:10.1115/1.3264476

Hull D, Rimmer DE (1959) The growth of grain-boundary voids under stress. Philosophical Magazine 4(42):673–687, doi:10.1080/14786435908243264

Hult JAH (1966) Creep in Engineering Structures. John Wiley & Sons Canada

Hyde CJ, Sun W, Leen SB (2010) Cyclic thermo-mechanical material modelling and testing of 316 stainless steel. International Journal of Pressure Vessels and Piping 87(6):365–372, doi:10.1016/j.ijpvp.2010.03.007

Hyde TH, Xia L, Becker AA (1996) Prediction of creep failure in aeroengine materials under multi-axial stress states. International Journal of Mechanical Sciences 38(4):385–403, doi:10.1016/0020-7403(95)00063-1

Hyde TH, Sun W, Williams JA (1999) Creep behaviour of parent, weld and HAZ materials of new, service-aged and repaired 1/2Cr1/2Mo1/4V: 2 1/4Cr1Mo pipe welds at 640°C. Materials at High Temperatures 16(3):117–129, doi:10.1179/mht.1999.011

Hyde TH, Becker AA, Sun W, Williams JA (2006) Finite-element creep damage analyses of P91 pipes. International Journal of Pressure Vessels and Piping 83(11-12):853–863, doi:10.1016/j.ijpvp.2006.08.013

Ilschner B (1973) Hochtemperatur-Plastizität: Warmfestigkeit und Warmverformbarkeit metallischer und nichtmetallischer Werkstoffe. Reine und angewandte Metallkunde in Einzeldarstellungen, Springer

Jiang Y, Kurath P (1996) Characteristics of the Armstrong-Frederick type plasticity models. International Journal of Plasticity 12(3):387–415, doi:10.1016/s0749-6419(96)00013-7

Jing JP, Sun Y, Xia SB, Feng GT (2001) A continuum damage mechanics model on low cycle fatigue life assessment of steam turbine rotor. International Journal of Pressure Vessels and Piping 78(1):59–64, doi:10.1016/s0308-0161(01)00005-9

Jing JP, Meng G, Sun Y, Xia SB (2003) An effective continuum damage mechanics model for creep-fatigue life assessment of a steam turbine rotor. International Journal of Pressure Vessels and Piping 80(6):389–396, doi:10.1016/s0308-0161(03)00070-x

Kachanov LM (1958) O vremeni razrusheniya v usloviyakh polzuchesti (On the time to rupture under creep conditions, in Russ.). Izv AN SSSR Otd Tekh Nauk 8:26–31

Kachanov LM (1986) Introduction to continuum damage mechanics. Springer Science & Business Media, doi:10.1007/978-94-017-1957-5

Kassner ME, Hayes TA (2003) Creep cavitation in metals. International Journal of Plasticity 19(10):1715–1748, doi:10.1016/s0749-6419(02)00111-0

Khaleel MA, Zbib HM, Nyberg EA (2001) Constitutive modeling of deformation and damage in superplastic materials. International Journal of Plasticity 17(3):277–296, doi:10.1016/s0749-6419(00)00036-x

Khan AS, Jackson KM (1999) On the evolution of isotropic and kinematic hardening with finite plastic deformation Part I: compression/tension loading of OFHC copper cylinders. International Journal of Plasticity 15(12):1265–1275, doi:10.1016/s0749-6419(99)00037-6

Kichenin J, Dang KV, Boytard K (1996) Finite-element simulation of a new two-dissipative mechanisms model for bulk medium-density polyethylene. Journal of Materials Science 31(6):1653–1661, doi:10.1007/bf00357878

Kimura K (2004) 9Cr-1Mo-V-Nb steel. In: Yagi K, Merckling G, Kern TU, Irie H, Warlimont H (eds) Creep Properties of Heat Resistant Steels and Superalloys, Advanced Materials and Technologies, Springer Berlin Heidelberg, pp 126–133, doi:10.1007/10837344_27

Kloc L, Skienička V, Ventruba J (2001) Comparison of low stress creep properties of ferritic and austenitic creep resistant steels. Materials Science and Engineering: A 319–321:774–778, doi:10.1016/s0921-5093(01)00943-1

Kowalewski ZL (2001) Assessment of the Multiaxial Creep Data Based on the Isochronous Creep Surface Concept. In: IUTAM Symposium on Creep in Structures, Springer Netherlands, pp 401–410, doi:10.1007/978-94-015-9628-2_38

Kowalewski ZL, Hayhurst DR, Dyson BF (1994a) Mechanisms-based creep constitutive equations for an aluminium alloy. The Journal of Strain Analysis for Engineering Design 29(4):309–316, doi:10.1243/03093247v294309

Kowalewski ZL, Lin J, Hayhurst DR (1994b) Experimental and theoretical evaluation of a high-accuracy uni-axial creep testpiece with slit extensometer ridges. International Journal of Mechanical Sciences 36(8):751–769, doi:10.1016/0020-7403(94)90090-6

Krajcinovic D (1983) Constitutive Equations for Damaging Materials. Journal of Applied Mechanics 50(2):355–360, doi:10.1115/1.3167044

Krajcinovic D (1989) Damage mechanics. Mechanics of Materials 8(2–3):117–197, doi:10.1016/0167-6636(89)90011-2

Krajcinovic D, Fonseka GU (1981) The Continuous Damage Theory of Brittle Materials, Part 1: General Theory. Journal of Applied Mechanics 48(4):809–815, doi:10.1115/1.3157739

Krempl E (1979) An experimental study of room-temperature rate-sensitivity, creep and relaxation of AISI type 304 stainless steel. Journal of the Mechanics and Physics of Solids 27(5-6):363–375, doi:10.1016/0022-5096(79)90020-6

Krempl E (1987) Models of viscoplasticity some comments on equilibrium (back) stress and drag stress. Acta Mechanica 69(1-4):25–42, doi:10.1007/bf01175712

Krempl E (1999) Creep-Plasticity Interaction. In: Altenbach H, Skrzypek JJ (eds) Creep and Damage in Materials and Structures, Springer Vienna, pp 285–348, doi:10.1007/978-3-7091-2506-9_6

Krempl E (2000) Viscoplastic models for high temperature applications. International Journal of Solids and Structures 37(1-2):279–291, doi:10.1016/s0020-7683(99)00093-1

Krempl E, Khan F (2003) Rate (time)-dependent deformation behavior: an overview of some properties of metals and solid polymers. International Journal of Plasticity 19(7):1069–1095, doi:10.1016/s0749-6419(03)00002-0

Krempl E, McMahon JJ, Yao D (1986) Viscoplasticity based on overstress with a differential growth law for the equilibrium stress. Mechanics of Materials 5(1):35–48, doi:10.1016/0167-6636(86)90014-1

Krieg RD (1975) A Practical Two Surface Plasticity Theory. Journal of Applied Mechanics 42(3):641–646, doi:10.1115/1.3423656

Leckie FA, Hayhurst DR (1974) Creep Rupture of Structures. Proceedings of the Royal Society A: Mathematical, Physical and Engineering Sciences 340(1622):323–347, doi:10.1098/rspa.1974.0155

Leen SB, Deshpande A, Hyde TH (2010) Experimental and Numerical Characterization of the Cyclic Thermomechanical Behavior of a High Temperature Forming Tool Alloy. Journal of Manufacturing Science and Engineering 132(5):051,013–1–051,013–12, doi:10.1115/1.4002534

Lemaitre J (1971) Evaluation of Dissipation and Damage in Metals Submitted to Dynamic Loading. In: Proceedings I. C. M. 1

Lemaitre J (1985) A Continuous Damage Mechanics Model for Ductile Fracture. Journal of Engineering Materials and Technology 107:83–89

Lemaitre J (1996) A Course on Damage Mechanics. Springer Science & Business Media

Lemaitre J, Chaboche JL (1994) Mechanics of solid materials. Cambridge University Press

Lifshitz IM (1963) On the theory of diffusion-viscous flow of polycrystalline bodies. Soviet Physics JETP 17:909–920

Lin J, Kowalewski ZL, Cao J (2005a) Creep rupture of copper and aluminium alloy under combined loadings—experiments and their various descriptions. International Journal of Mechanical Sciences 47(7):1038–1058, doi:10.1016/j.ijmecsci.2005.02.010

Lin J, Liu Y, Dean TA (2005b) A Review on Damage Mechanisms, Models and Calibration Methods under Various Deformation Conditions. International Journal of Damage Mechanics 14(4):299–319, doi:10.1177/1056789505050357

Lion A (2000) Constitutive modelling in finite thermoviscoplasticity: a physical approach based on nonlinear rheological models. International Journal of Plasticity 16(5):469–494, doi:10.1016/s0749-6419(99)00038-8

Lowe TC, Miller AK (1986) Modeling Internal Stresses in the Nonelastic Deformation of Metals. Journal of Engineering Materials and Technology 108(4):365–373, doi:10.1115/1.3225896

Malinin NN, Khadjinsky GM (1972) Theory of creep with anisotropic hardening. International Journal of Mechanical Sciences 14(4):235–246, doi:10.1016/0020-7403(72)90065-3
Manjoine MJ (1975) Ductility Indices at Elevated Temperature. Journal of Engineering Materials and Technology 97(2):156–161, doi:10.1115/1.3443276
Manjoine MJ (1982) Creep-Rupture Behavior of Weldments. Welding J 61(2):50–57
Margolin BZ, Karzov GP, Shvetsova VA, Kostylev VI (1998) Modelling for transcrystalline and intercrystalline fracture by void nucleation and growth. Fatigue and Fracture of Engineering Materials and Structures 21(2):123–137, doi:10.1046/j.1460-2695.1998.00474.x
Masuyama F (2001) Advances in Physical Metallurgy and Processing of Steels. History of Power Plants and Progress in Heat Resistant Steels. The Iron and Steel Institute of Japan International 41(6):612–625, doi:10.2355/isijinternational.41.612
McClintock FA (1968) A Criterion for Ductile Fracture by the Growth of Holes. Journal of Applied Mechanics 35(2):363–371, doi:10.1115/1.3601204
McLean D (1966) The physics of high temperature creep in metals. Reports on Progress in Physics 29(1):1–33
Michel B (2004) Formulation of a new intergranular creep damage model for austenitic stainless steels. Nuclear Engineering and Design 227(2):161–174, doi:10.1016/j.nucengdes.2003.09.005
Miller A (1976a) An Inelastic Constitutive Model for Monotonic, Cyclic, and Creep Deformation: Part I - Equations Development and Analytical Procedures. Journal of Engineering Materials and Technology 98(2):97–105, doi:10.1115/1.3443367
Miller A (1976b) An Inelastic Constitutive Model for Monotonic, Cyclic, and Creep Deformation: Part II - Application to Type 304 Stainless Steel. Journal of Engineering Materials and Technology 98(2):106–113, doi:10.1115/1.3443346
Miller AK (ed) (1987) Unified Constitutive Equations for Creep and Plasticity. Springer Netherlands, doi:10.1007/978-94-009-3439-9
Moreno V, Jordan EH (1986) Prediction of material thermomechanical response with a unified viscoplastic constitutive model. International Journal of Plasticity 2(3):223–245, doi:10.1016/0749-6419(86)90002-1
Mróz Z (1967) On the description of anisotropic workhardening. Journal of the Mechanics and Physics of Solids 15(3):163–175, doi:10.1016/0022-5096(67)90030-0
Murakami S (1983) Notion of Continuum Damage Mechanics and its Application to Anisotropic Creep Damage Theory. Journal of Engineering Materials and Technology 105(2):99–105, doi:10.1115/1.3225633
Murakami S (2012) Continuum Damage Mechanics - A Continuum Mechanics Approach to the Analysis of Damage and Fracture, Solid Mechanics and its Applications, vol 185. Springer Netherlands
Murakami S, Ohno N (1981) A Continuum Theory of Creep and Creep Damage. In: Creep in Structures, Springer Berlin Heidelberg, pp 422–444, doi:10.1007/978-3-642-81598-0_28
Murakami S, Liu Y, Mizuno M (2000) Computational methods for creep fracture analysis by damage mechanics. Computer Methods in Applied Mechanics and Engineering 183(1-2):15–33, doi:10.1016/s0045-7825(99)00209-1
Nabarro FRN, de Villiers HL (1995) The Physics of Creep: Creep and Creep-resistant Alloys. Taylor & Francis
Naumenko K, Altenbach H (2007) Modeling of Creep for Structural Analysis. Springer Berlin Heidelberg, doi:10.1007/978-3-540-70839-1
Naumenko K, Altenbach H (2016) Modeling High Temperature Materials Behavior for Structural Analysis: Part I: Continuum Mechanics Foundations and Constitutive Models. Advanced Structured Materials, Springer International Publishing, doi:10.1007/978-3-319-31629-1
Naumenko K, Gariboldi E (2014) A phase mixture model for anisotropic creep of forged Al–Cu–Mg–Si alloy. Materials Science and Engineering: A 618:368–376, doi:10.1016/j.msea.2014.09.012
Naumenko K, Altenbach H, Kutschke A (2011) A Combined Model for Hardening, Softening, and Damage Processes in Advanced Heat Resistant Steels at Elevated Temperature. International Journal of Damage Mechanics 20(4):578–597, doi:10.1177/1056789510386851

Nicolaou PD, Semiatin SL, Ghosh AK (2000) An analysis of the effect of cavity nucleation rate and cavity coalescence on the tensile behavior of superplastic materials. Metallurgical and Materials Transactions A 31(5):1425–1434, doi:10.1007/s11661-000-0260-6

Odqvist FKG (1974) Mathematical Theory of Creep and Creep Rupture. Oxford University Press

Odqvist FKG, Hult J (1962) Kriechfestigkeit metallischer Werkstoffe. Springer, doi:10.1007/978-3-642-52432-5

Ohno N, Kachi Y (1986) A Constitutive Model of Cyclic Plasticity for Nonlinear Hardening Materials. Journal of Applied Mechanics 53(2):395–403, doi:10.1115/1.3171771

Ohno N, Wang JD (1991) Transformation of a nonlinear kinematic hardening rule to a multisurface form under isothermal and nonisothermal conditions. International Journal of Plasticity 7(8):879–891, doi:10.1016/0749-6419(91)90023-r

Orowan E (1934a) Zur Kristallplastizität. I. Zeitschrift für Physik 89(9-10):605–613, doi:10.1007/bf01341478

Orowan E (1934b) Zur Kristallplastizität. II. Zeitschrift für Physik 89(9-10):614–633, doi:10.1007/bf01341479

Orowan E (1934c) Zur Kristallplastizität. III. Zeitschrift für Physik 89(9-10):634–659, doi:10.1007/bf01341480

Othman AM, Hayhurst DR, Dyson BF (1993) Skeletal Point Stresses in Circumferentially Notched Tension Bars Undergoing Tertiary Creep Modelled with Physically Based Constitutive Equations. Proceedings of the Royal Society A: Mathematical, Physical and Engineering Sciences 441(1912):343–358, doi:10.1098/rspa.1993.0065

Penny RK, Marriott DL (1995) Design for Creep. Chapman & Hall

Perrin IJ, Hayhurst DR (1996) Creep constitutive equations for a 0.5Cr–0.5Mo–0.25V ferritic steel in the temperature range 600–675°C. The Journal of Strain Analysis for Engineering Design 31(4):299–314, doi:10.1243/03093247v314299

Perrin IJ, Hayhurst DR (1999) Continuum damage mechanics analyses of type IV creep failure in ferritic steel crossweld specimens. International Journal of Pressure Vessels and Piping 76(9):599–617, doi:10.1016/s0308-0161(99)00051-4

Perzyna P (1963) The constitutive equations for rate sensitive plastic materials. Quarterly of Applied Mathematics 20(4):321–332, doi:10.1090/qam/144536

Perzyna P (1966) Fundamental Problems in Viscoplasticity. In: Chernyi GG, Dryden HL, Germain P, Howarth L, Olszak W, Prager W, Probstein RF, Ziegler H (eds) Advances in Applied Mechanics, vol 9, Elsevier, pp 243–377, doi:10.1016/s0065-2156(08)70009-7

Polcik P (1998) Modellierung des Verformungsverhaltens der warmfesten 9-12% Chromstähle im Temperaturbereich von 550-650°C. PhD thesis, Friedrich-Alexander-Universität, Erlangen-Nürnberg

Ponter ARS, Leckie FA (1976) Constitutive Relationships for the Time-Dependent Deformation of Metals. Journal of Engineering Materials and Technology 98(1):47–51, doi:10.1115/1.3443336

Prager W (1949) Recent Developments in the Mathematical Theory of Plasticity. Journal of Applied Physics 20(3):235–241, doi:10.1063/1.1698348

Priester L (2013) Grain Boundaries. From Theory to Engineering. Springer, doi:10.1007/978-94-007-4969-6

Rabotnov YN (1959) O mechanizme dlitel'nogo razrusheniya (A mechanism of the long term fracture, in Russ.). Voprosy prochnosti materialov i konstruktsii, AN SSSR pp 5–7

Rabotnov YN (1969) Creep Problems in Structural Members. North-Holland, Amsterdam

Ragab AR (2002) Creep Rupture Due to Material Damage by Cavitation. Journal of Engineering Materials and Technology 124(2):199–205, doi:10.1115/1.1446076

Raj R, Ashby MF (1975) Intergranular fracture at elevated temperature. Acta Metallurgica 23(6):653–666, doi:10.1016/0001-6160(75)90047-4

Rice JR, Tracey DM (1969) On the ductile enlargement of voids in triaxial stress fields∗. Journal of the Mechanics and Physics of Solids 17(3):201–217, doi:10.1016/0022-5096(69)90033-7

Ringel M, Roos E, Maile K, Klenk A (2004) Advanced constitutive equations for 10 Cr forged and cast steel for steam turbines under creep fatigue and thermo-mechanical fatigue. In: 30. MPA-Seminar

'Safety and reliability in energy technology' in conjunction with the 9th German-Japanese seminar Vol 2 (Papers 27–53), pp 32.1–32.14

Rösler J, Harders H, Bäker M (2012) Mechanisches Verhalten der Werkstoffe. Springer Fachmedien Wiesbaden, doi:10.1007/978-3-8348-2241-3

Röttger DR (1997) Untersuchungen zum Wechselverformungs- und Zeitstandverhalten der Stähle X20CrMoV121 und X10CrMoVNb91. PhD thesis, Universität GH, Essen

Saad AA (2012) Cyclic plasticity and creep of power plant materials. PhD thesis, University of Nottingham, Nottingham, URL http://eprints.nottingham.ac.uk/id/eprint/12538

Saad AA, Hyde CJ, Sun W, Hyde TH (2011a) Thermal-mechanical fatigue simulation of a P91 steel in a temperature range of 400–600°C. Materials at High Temperatures 28(3):212–218, doi:10.3184/096034011X13072954674044

Saad AA, Sun W, Hyde TH, Tanner DWJ (2011b) Cyclic softening behaviour of a P91 steel under low cycle fatigue at high temperature. Procedia Engineering 10:1103–1108, doi:10.1016/j.proeng.2011.04.182

Sauzay M, Brillet H, Monneta I, Mottot M, Barcelo F, Fournier B, Pineau A (2005) Cyclically induced softening due to low-angle boundary annihilation in a martensitic steel. Materials Science and Engineering A 400–401:241–244, doi:10.1016/j.msea.2005.02.092

Sauzay M, Fournier B, Mottot M, Pineau A, Monnet I (2008) Cyclic softening of martensitic steels at high temperature: Experiments and physically based modelling. Materials Science and Engineering A 483–484:410–414, doi:10.1016/j.msea.2006.12.183

Shang J, Leen SB, Hyde TH (2006) Finite-element-based methodology for predicting the thermo-mechanical behaviour of superplastic forming tools. Proceedings of the Institution of Mechanical Engineers, Part L: Journal of Materials: Design and Applications 220(3):113–123, doi:10.1243/14644207jmda85

Simon A, Samir A, Scholz A, Berger C (2007) Konstitutive Beschreibung eines 10%Cr-Stahls zur Berechnung betriebsnaher Kriechermüdungsbeanspruchung. Materialwissenschaft und Werkstofftechnik 38(8):635–641, doi:10.1002/mawe.200600125

Sklenička V, Kuchařová K, Svoboda M, Kloc L, Buršík J, Kroupa A (2003) Long-term creep behavior of 9–12%Cr power plant steels. Materials Characterization 51(1):35–48, doi:10.1016/j.matchar.2003.09.012

Skrzypek JJ, Ganczarski A (1999) Modeling of material damage and failure of structures: theory and applications. Springer Science & Business Media

Solasi R, Zou Y, Huang X, Reifsnider K (2007) A time and hydration dependent viscoplastic model for polyelectrolyte membranes in fuel cells. Mechanics of Time-Dependent Materials 12(1):15–30, doi:10.1007/s11043-007-9040-7

Spindler MW (2004) The multiaxial creep ductility of austenitic stainless steels. Fatigue and Fracture of Engineering Materials and Structures 27(4):273–281, doi:10.1111/j.1460-2695.2004.00732.x

Spindler MW, Hales R, Skelton RP (2001) The multiaxial creep ductility of an ex-service type 316H stainless steel. In: 9th International Conference on Creep and Fracture of Engineering Materials and Structures, pp 679–688

Stolk J, Verdonschot N, Murphy BP, Prendergast PJ, Huiskes R (2004) Finite element simulation of anisotropic damage accumulation and creep in acrylic bone cement. Engineering Fracture Mechanics 71(4-6):513–528, doi:10.1016/s0013-7944(03)00048-1

Stouffer DC, Bodner SR (1979) A constitutive model for the deformation induced anisotropic plastic flow of metals. International Journal of Engineering Science 17(6):757–764, doi:10.1016/0020-7225(79)90050-8

Stowell EZ (1957) A Phenomenological Relation Between Stress, Strain Rate and Temperature for Metals at Elevated Temperatures. Tech. rep., REPORT 1343. National Advisory Committee for Aeronautics, URL https://ntrs.nasa.gov/archive/nasa/casi.ntrs.nasa.gov/19930091014.pdf

Straub S (1995) Verformungsverhalten und Mikrostruktur warmfester martensitischer 12%-Chromstähle. PhD thesis, Friedrich-Alexander-Universität, Erlangen-Nürnberg

Taylor MB, Zbib HM, Khaleel MA (2002) Damage and size effect during superplastic deformation. International Journal of Plasticity 18(3):415–442, doi:10.1016/s0749-6419(00)00106-6

Tong J, Vermeulen B (2003) The description of cyclic plasticity and viscoplasticity of waspaloy using unified constitutive equations. International Journal of Fatigue 25(5):413–420, doi:10.1016/s0142-1123(02)00162-7

Tong J, Zhan ZL, Vermeulen B (2004) Modelling of cyclic plasticity and viscoplasticity of a nickel-based alloy using Chaboche constitutive equations. International Journal of Fatigue 26(8):829–837, doi:10.1016/j.ijfatigue.2004.01.002

Tvergaard V (1984) On the creep constrained diffusive cavitation of grain boundary facets. Journal of the Mechanics and Physics of Solids 32(5):373–393, doi:10.1016/0022-5096(84)90021-8

Tvergaard V, Needleman A (1984) Analysis of the cup-cone fracture in a round tensile bar. Acta Metallurgica 32(1):157–169, doi:10.1016/0001-6160(84)90213-X

Valanis KC (1970) A Theory of Viscoplasticity without a Yield Surface. Tech. rep., Air Force Office of Scientific Research, Office of Aerospace Research, United States Air Force, URL www.dtic.mil/dtic/tr/fulltext/u2/725030.pdf

Valanis KC (1978) Fundamental Consequences of a New Intrinsic Time Measure. Plasticity as a Limit of the Endochronic Theory. Tech. rep., Division of Materials Engineering, The University of Iowa, Report G224-DME-78-001, URL https://apps.dtic.mil/docs/citations/ADA302661

Valanis KC, Fan J (1983) Endochronic Analysis of Cyclic Elastoplastic Strain Fields in a Notched Plate. Trans ASME Journal of Applied Mechanics 50(4a):789–794, doi:10.1115/1.3167147

Velay V, Bernhart G, Penazzi L (2006) Cyclic behavior modeling of a tempered martensitic hot work tool steel. International Journal of Plasticity 22(3):459–496, doi:10.1016/j.ijplas.2005.03.007

Viswanathan R (1989) Damage mechanisms and life assessment of high temperature components. ASM international

Voyiadjis GZ, Kattan PI (1999) Advances in Damage Mechanics: Metals and Metal Matrix Composites. Elsevier New York

Walker KP (1981) Research and Development Program for Nonlinear Structural Modeling with Advanced Time-Temperature Dependent Constitutive Relationships. Tech. rep., NASA CR-165533

Wang J, Steinmann P, Rudolph J, Willuweit A (2015) Simulation of creep and cyclic viscoplastic strains in high-Cr steel components based on a modified Becker–Hackenberg model. International Journal of Pressure Vessels and Piping 128:36–47, doi:10.1016/j.ijpvp.2015.02.003

Watanabe O, Atluri SN (1986) Internal time, general internal variable, and multi-yield-surface theories of plasticity and creep: A unification of concepts. International Journal of Plasticity 2(1):37–57, doi:10.1016/0749-6419(86)90015-x

Weißbach W, Dahms M, Jaroschek C (2015) Werkstoffkunde. Springer Fachmedien Wiesbaden, Wiesbaden, doi:10.1007/978-3-658-03919-6

Wohua Z, Valliappan S (1998a) Continuum Damage Mechanics Theory and Application-Part I: Theory. International Journal of Damage Mechanics 7(3):250–273, doi:10.1177/105678959800700303

Wohua Z, Valliappan S (1998b) Continuum Damage Mechanics Theory and Application-Part II: Application. International Journal of Damage Mechanics 7(3):274–297, doi:10.1177/105678959800700304

Xiao YH, Guo C (2011) Constitutive modelling for high temperature behavior of 1Cr12Ni3Mo2VNbN martensitic steel. Materials Science and Engineering: A 528(15):5081–5087, doi:10.1016/j.msea.2011.03.050

Xu Q (2001) Creep damage constitutive equations for multi-axial states of stress for 0.5Cr0.5Mo0.25V ferritic steel at 590°C. Theoretical and Applied Fracture Mechanics 36(2):99–107, doi:10.1016/s0167-8442(01)00060-x

Xu Q (2004) The development of validation methodology of multi-axial creep damage constitutive equations and its application to 0.5Cr0.5Mo0.25V ferritic steel at 590°C. Nuclear Engineering and Design 228(1-3):97–106, doi:10.1016/j.nucengdes.2003.06.021

Yaguchi M, Takahashi Y (2005a) Ratchetting of viscoplastic material with cyclic softening, part 1: experiments on modified 9Cr-1Mo steel. International Journal of Plasticity 21(1):43–65, doi:10.1016/j.ijplas.2004.02.001

Yaguchi M, Takahashi Y (2005b) Ratchetting of viscoplastic material with cyclic softening, part 2: application of constitutive models. International Journal of Plasticity 21(4):835–860, doi:10.1016/j.ijplas.2004.05.012

Yao HT, Xuan FZ, Wang Z, Tu ST (2007) A review of creep analysis and design under multi-axial stress states. Nuclear Engineering and Design 237(18):1969–1986, doi:10.1016/j.nucengdes.2007.02.003

Yeom JT, Lee CS, Kim JH, Lee DG, Park NK (2007) Continuum Damage Model of Creep-Fatigue Interaction in Ni-Base Superalloy. Key Engineering Materials 340–341:235–240, doi:10.4028/www.scientific.net/kem.340-341.235

Yousefiani A, Mohamed FA, Earthman JC (2000) Creep rupture mechanisms in annealed and overheated 7075 Al under multiaxial stress states. Metallurgical and Materials Transactions A 31(11):2807–2821, doi:10.1007/bf02830340

Zhan Z, Fernando US, Tong J (2008) Constitutive modelling of viscoplasticity in a nickel-based superalloy at high temperature. International Journal of Fatigue 30(7):1314–1323, doi:10.1016/j.ijfatigue.2007.06.010

Zhan ZL, Tong J (2007) A study of cyclic plasticity and viscoplasticity in a new nickel-based superalloy using unified constitutive equations. Part I: Evaluation and determination of material parameters. Mechanics of Materials 39(1):64–72, doi:10.1016/j.mechmat.2006.01.005

Zhang J, Jiang Y (2008) Constitutive modeling of cyclic plasticity deformation of a pure polycrystalline copper. International Journal of Plasticity 24(10):1890–1915, doi:10.1016/j.ijplas.2008.02.008

Zhao LG, Tong J, Vermeulen B, Byrne J (2001) On the uniaxial mechanical behaviour of an advanced nickel base superalloy at high temperature. Mechanics of Materials 33(10):593–600, doi:10.1016/s0167-6636(01)00071-0

Chapter 5
Surface Elasticity Models: Comparison Through the Condition of the Anti-plane Surface Wave Propagation

Victor A. Eremeyev

Abstract In order to discuss the peculiarities of few models of surface elasticity we consider here the dispersion relations for anti-plane surface waves. We show that the dispersion curves are quite sensitive to the choice of the model. We consider here the linear Gurtin-Murdoch model, strain- and stress-gradient surface elasticity models.

Keywords: Surface elasticity · Anti-plane surface wave propagation · Dispersion curves · Gurtin-Murdoch model · Strain-gradient surface elasticity model · Stress-gradient surface elasticity models

5.1 Introduction

The interest to generalized models of continua grows recently with respect to appearance of new microstructured materials as well as in order to describe new phenomena observed at the micro- and nano-scale, see, e.g., Forest et al (2011); Liebold and Müller (2015); Aifantis (2016). In particular, the surface elasticity models found various applications in micro- and nano-mechanics, see, e.g., Duan et al (2008); Wang et al (2011); Javili et al (2013b,a); Eremeyev (2016) and the reference therein. Having origin in the landscape works by Laplace (1805, 1806); Young (1805); Poisson (1831) and Gibbs, see Longley and Van Name (1928), the rational continual model of the surface elasticity was developed by Gurtin and Murdoch (1975, 1978). Later it was generalized by Steigmann and Ogden (1997, 1999) in order to take into account bending surface stiffness. As surface mechanics should describe quite

Victor A. Eremeyev
Faculty of Civil and Environmental Engineering, Gdańsk University of Technology, ul. Gabriela Narutowicza 11/12 80-233 Gdańsk, Poland,
e-mail: victor.eremeev@pg.edu.pl
Southern Federal University, Milchakova str. 8a, 344090 Rostov on Don & Southern Scientific Center of RASci, Chekhova str. 41, 344006 Rostov on Don, Russia,
e-mail: eremeyev.victor@gmail.com

H. Altenbach and A. Öchsner (eds.), *State of the Art and Future Trends in Material Modeling*, Advanced Structured Materials 100,
https://doi.org/10.1007/978-3-030-30355-6_5

different phenomena, in the literature are known various extensions of surface-related mechanics, see, e.g., dell'Isola and Seppecher (1997); dell'Isola et al (2012b); Placidi et al (2014); Lurie et al (2016, 2009); Belov et al (2019); Eremeyev (2019b) and the references therein.

The presence of surface stresses influences the effective (apparent) properties of nanostructured materials, such as nano-composites (Kushch et al, 2013; Nazarenko et al, 2016, 2018; Zemlyanova and Mogilevskaya, 2018; Han et al, 2018) or nano-plates and shells (Altenbach and Eremeyev, 2011; Altenbach et al, 2010, 2012; Ru, 2016). In addition, surface energy may result in new phenomena as the appearance surface/interfacial waves considered within the Gurtin-Murdoch approach (Xu et al, 2015; Eremeyev et al, 2016) and for certain generalizations of the Gurtin-Murdoch model (Eremeyev, 2017, 2019b,a). Let us note that this class of waves exist also for another type of media with surface energy such as strain-gradient media, see Vardoulakis and Georgiadis (1997); Georgiadis et al (2000); Yerofeyev and Sheshenina (2005); dell'Isola et al (2012a); Rosi et al (2015); Li et al (2015); Gourgiotis and Georgiadis (2015). The comparison of the Gurtin-Murdoch model with the Toupin-Mindlin strain gradient elasticity was given by Eremeyev et al (2018b), whereas the similarities with the dynamics of a square lattice was discussed by Eremeyev and Sharma (2019).

The aim of this paper is to compare the dispersion relations and condition of existence of anti-plane surface waves in various media with surface energy. The key-point of the surface elasticity is the presence of surface stresses $\boldsymbol{\tau}$. For the latter we assume additional constitutive equation. Here we consider the classic Gurtin-Murdoch model as well two extensions such as surface strain and surface stress gradient elasticity.

The paper is organized as follows. First, in Sect. 5.2 we present the basic equations for an elastic half-space with surface stresses. Then in Sect. 5.3 we consider various constitutive equations for $\boldsymbol{\tau}$. Here we introduce both the integral and differential constitutive equations. In other words, we consider both strongly and weak nonlocal models of surface elasticity. Finally, we discuss the dispersion relations in Sect. 5.4.

5.2 Anti-plane Motions of an Elastic Half-Space

In what follows we restrict ourselves by isotropic materials undergoing infinitesimal deformations. So in the bulk we have the Hooke law

$$\boldsymbol{\sigma} = 2\mu\boldsymbol{e} + \lambda\boldsymbol{I}\ \mathrm{tr}\,\boldsymbol{e}, \quad \boldsymbol{e} = \frac{1}{2}\left(\nabla\boldsymbol{u} + (\nabla\boldsymbol{u})^T\right), \tag{5.1}$$

where $\boldsymbol{\sigma}$ and $\boldsymbol{e}$ are the stress and strain tensors, respectively, λ and μ are Lamé elastic moduli, tr is the trace operator, the superscript T stands for the transpose operation, ∇ is the 3D nabla operator, and $\boldsymbol{I}$ is the 3D unit tensor. Hereinafter we use the direct (coordinate-free) tensor calculus as described in Simmonds (1994); Lebedev et al (2010); Eremeyev et al (2018a). As a result, the gradient of the displacement vector

$\boldsymbol{u} = \boldsymbol{u}(\boldsymbol{x}, t)$ is given by

$$\nabla \boldsymbol{u} = \frac{\partial u_j}{\partial x_i} \boldsymbol{i}_i \otimes \boldsymbol{i}_j,$$

where $\otimes$ denotes the dyadic product, x_1, x_2, x_3 are Cartesian coordinates with corresponding base vectors $\boldsymbol{i}_k$, $k = 1, 2, 3$, $\boldsymbol{x} = x_i \boldsymbol{i}_i$ is the position vector, t is time, and Einstein's summation rule is utilized. The equation of the motion is given by

$$\nabla \cdot \boldsymbol{\sigma} = \rho \frac{\partial^2 \boldsymbol{u}}{\partial t^2}, \tag{5.2}$$

where ρ is the mass density and the dot stands for scalar product. For a free surface with surface stresses we get the generalized Young-Laplace equation as a boundary condition

$$\boldsymbol{n} \cdot \boldsymbol{\sigma} = \nabla_s \cdot \boldsymbol{\tau} - m \frac{\partial^2 \boldsymbol{u}}{\partial t^2}, \tag{5.3}$$

where $\boldsymbol{n}$ is the unit outward vector of normal to the boundary, $\nabla_s \equiv \boldsymbol{P} \cdot \nabla$ is the surface nabla operator, $\boldsymbol{P} \equiv \boldsymbol{I} - \boldsymbol{n} \otimes \boldsymbol{n}$ is the surface unit second-order tensor, and m is the surface mass density, see Gurtin and Murdoch (1978).

Let us consider anti-plane motions of an elastic half-space given by the inequality $-\infty \le x_3 \le 0$. The displacement vector takes the form

$$\boldsymbol{u} = u(x_2, x_3, t) \boldsymbol{i}_1, \tag{5.4}$$

see Achenbach (1973). In this case the equation of motion (5.2) is reduced to the wave equation

$$\mu \Delta u = \rho \partial_t^2 u, \tag{5.5}$$

where $\Delta = \partial_2^2 + \partial_3^2$ is the 2D Laplace operator. For brevity, in what follows we will denote partial derivatives as $\partial_k = \partial / \partial x_k$ and $\partial_t = \partial / \partial t$. For the anti-plane motion $\boldsymbol{\tau}$ takes the form

$$\boldsymbol{\tau} = \tau (\boldsymbol{i}_1 \otimes \boldsymbol{i}_2 + \boldsymbol{i}_2 \otimes \boldsymbol{i}_1), \quad \tau = \tau(x_2, x_3, t)$$

with only one surface stress $\tau(x_2, x_3, t)$. As a result, the generalized Young-Laplace equation (5.3) can be transformed into

$$\sigma_{31} = \partial_2 \tau - m \partial_t^2 u$$

or, considering Hooke's law (5.1), into

$$\mu \partial_3 u = \partial_2 \tau - m \partial_t^2 u. \tag{5.6}$$

Thus, to complete the boundary-value problem statement one needs in the constitutive relations for τ.

5.3 Constitutive Relations Within the Surface Elasticity

After Gurtin and Murdoch (1975) in addition to constitutive equations in the bulk one should independently introduce constitutive relations for surface stresses $\boldsymbol{\tau}$. Here we consider the simplified linear Gurtin-Murdoch model and some of its extensions.

5.3.1 Simplified Linear Gurtin-Murdoch Model

Within this model we get the following constitutive relation

$$\boldsymbol{\tau} = 2\mu_s\boldsymbol{\epsilon} + \lambda_s\boldsymbol{P}\,\mathrm{tr}\,\boldsymbol{\epsilon}, \tag{5.7}$$

where the surface strain tensor is defined by the formula

$$\boldsymbol{\epsilon} = \frac{1}{2}\left[\boldsymbol{P}\cdot(\nabla_s\boldsymbol{u}) + (\nabla_s\boldsymbol{u})^T\cdot\boldsymbol{P}\right],$$

and λ_s and μ_s are the surface Lamé moduli.

For anti-plane deformations we get that

$$\boldsymbol{\epsilon} = \epsilon(\boldsymbol{i}_1\otimes\boldsymbol{i}_2 + \boldsymbol{i}_2\otimes\boldsymbol{i}_1), \quad \epsilon = \frac{1}{2}\partial_2 u$$

and

$$\tau = \mu_s\partial_2 u. \tag{5.8}$$

Let us note that as the anti-plane motions constitute a very specific class of deformations, in this case τ takes form (5.8) also for linearized (non-simplified) Gurtin-Murdoch model, see also discussion by Ru (2010), as well as for the linear Steigmann-Ogden model.

5.3.2 Linear Stress-gradient Surface Elasticity

Motivated by long range surface interactions as described by de Gennes (1981); de Gennes et al (2004); Israelachvili (2011), we recently proposed the integral-type constitutive relations of Eringen's type (Eremeyev, 2019a)

$$\boldsymbol{\tau}(\boldsymbol{x}) = \int_{-\infty}^{\infty}\int_{-\infty}^{\infty}\alpha(\|\boldsymbol{x}-\boldsymbol{x}'\|)\left[2\mu_s\boldsymbol{\epsilon}(\boldsymbol{x}') + \lambda_s\left(\mathrm{tr}\,\boldsymbol{\epsilon}(\boldsymbol{x}')\right)\boldsymbol{P}\right]\mathrm{d}x_1'\mathrm{d}x_2', \tag{5.9}$$

where $\alpha(s)$ is a kernel function, which can be taken as a fundamental solution of an elliptic differential equation. For example, introducing an elliptic differential operator

$\mathcal{L}$ we define α as the normalized solution of

$$\mathcal{L}(\partial_1, \partial_2)\alpha = \delta(\boldsymbol{x}), \quad \int_{-\infty}^{\infty}\int_{-\infty}^{\infty} \alpha(\|\boldsymbol{x} - \boldsymbol{x}'\|)\mathrm{d}x_1'\mathrm{d}x_2' = 1, \tag{5.10}$$

where $\delta(\boldsymbol{x})$ is the Dirac delta-function. In this case we can transform (5.9) into differential form

$$\mathcal{L}(\partial_1, \partial_2)\boldsymbol{\tau} = 2\mu_s\boldsymbol{\epsilon} + \lambda_s\boldsymbol{P}\,\mathrm{tr}\,\boldsymbol{\epsilon}. \tag{5.11}$$

After Eringen (2002) we can consider

$$\mathcal{L} = -q^{-2}\Delta + 1,$$

where the parameter q is a reciprocal length, as an example of proper strongly non-local model. Here we have

$$\alpha(s) = \frac{1}{2\pi}K_0(qs),$$

where K_0 is a modified Bessel function of the second kind. So we get the following stress-gradient constitutive equation

$$-q^{-2}\Delta\boldsymbol{\tau} + \boldsymbol{\tau} = 2\mu_s\boldsymbol{\epsilon} + \lambda_s\boldsymbol{P}\,\mathrm{tr}\,\boldsymbol{\epsilon}. \tag{5.12}$$

Other choices of the kernel functions are also possible, see Eringen (2002). For example, if we take $\alpha = \delta(\boldsymbol{x})$ we get (5.7).

In the case of anti-plane motions, Eq. (5.9) can be transformed into one scalar integral equation

$$\tau = \int_{-\infty}^{\infty}\int_{-\infty}^{\infty} \alpha(\|\boldsymbol{x} - \boldsymbol{x}'\|)\mu_s\partial_2 u(\boldsymbol{x}')\,\mathrm{d}x_1'\mathrm{d}x_2',$$

or into its differential counterpart

$$\mathcal{L}(0, \partial_2)\tau = \mu_s\partial_2 u.$$

For $\mathcal{L} = -q^{-2}\Delta + 1$, this becomes

$$-q^{-2}\partial_2^2\tau + \tau = \mu_s\partial_2 u. \tag{5.13}$$

Obviously, Eq. (5.13) transforms into (5.8) at $q \to \infty$. So for this limit we get the classic Gurtin-Murdoch model. The presented here model belongs to the class of strongly non-local materials according to Maugin's classification, see Maugin (2017) for the general framework and Eremeyev (2019a) for more detail.

5.3.3 Linear Strain-gradient Surface Elasticity

Another non-local generalization of the Gurtin-Murdoch model can be introduced considering higher order gradient terms in the surface energy density

$$W_s = W_s(\boldsymbol{\epsilon}, \nabla_s \nabla_s \boldsymbol{u}),$$

see Eremeyev (2017). Partially the model was motivated by consideration of hyperbolic metasurfaces, see Eremeyev (2019b) and the reference therein. As a result, we came to the constitutive relation

$$\boldsymbol{\tau} = \mu_s \boldsymbol{\epsilon} + \lambda_s \mathbf{P} \operatorname{tr} \boldsymbol{\epsilon} - \mu_2 \nabla_s \cdot (\nabla_s \nabla_s \boldsymbol{u}), \tag{5.14}$$

where μ_2 is an additional surface elastic modulus. Here in the model there also exist surface hyperstresses as in the 3D strain-gradient elasticity, given by the formula

$$\boldsymbol{\mu} = \mu_2 \nabla_s \nabla_s \boldsymbol{u}.$$

For anti-plane deformations, Eq. (5.14) takes the form

$$\tau = \mu_s \partial_2 u - \mu_2 \partial_2^3 u. \tag{5.15}$$

As a result, Eq. (5.6) becomes a forth-order differential equation with respect to the tangent derivative.

5.4 Dispersion Relations

Considering the models above, we came to the boundary-value problem in the half-space which consists of the wave equation (5.5) and the boundary condition (5.6) where τ was introduced within the Gurtin-Murdoch, stress- and strain-gradient models according to (5.8), (5.13), and (5.15), respectively. Assuming steady-state behaviour, we consider a solution of (5.5) in the form

$$u = U(x_2, x_3) \exp(-\mathrm{i}\omega t), \tag{5.16}$$

where U is an amplitude, ω is a circular frequency, and $\mathrm{i} = \sqrt{-1}$ is the imaginary unit. With (5.16), Eq. (5.5) transforms into

$$\mu \Delta U = -\rho \omega^2 U, \tag{5.17}$$

which has decaying at $x_3 \to -\infty$ solution

$$U = U_0 \exp(\varkappa x_3) \exp(\mathrm{i} k x_2), \tag{5.18}$$

where k is a wavenumber, U_0 is a constant, and $\varkappa$ is given by

$$\varkappa = \varkappa(k,\omega) \equiv \sqrt{k^2 - \frac{\omega^2}{c_T^2}}, \quad c_T = \sqrt{\frac{\mu}{\rho}},$$

where c_T is the phase velocity of transverse waves in the bulk (Achenbach, 1973).

A nontrivial solution of (5.18), that is with $U_0 \neq 0$, exists if and only if it satisfies the boundary conditions at $x_3 = 0$. The latter will lead to a dispersion relation, i.e., an equation relating k and ω.

The displacement field $u(x_2, x_3, t)$ according to (5.16) and (5.18) leads to a surface stress in the form

$$\tau = T \exp(\mathrm{i}kx_2) \exp(-\mathrm{i}\omega t),$$

where T is a constant. For (5.8), (5.13), and (5.15), T is given by

$$T = \mathrm{i}k\mu_s U_0, \tag{5.19}$$

$$T = \frac{\mathrm{i}k}{1 + q^{-2}k^2}\mu_s U_0, \tag{5.20}$$

$$T = \mathrm{i}k(\mu_s + \mu_2 k^2)U_0, \tag{5.21}$$

respectively. Substituting these dependencies into (5.6) we get the dispersion relations

$$\mu\varkappa(k,\omega) = m\omega^2 - \mu_s k^2, \tag{5.22}$$

$$\mu\varkappa(k,\omega) = m\omega^2 - q^2 \frac{\mu_s k^2}{k^2 + q^2}, \tag{5.23}$$

$$\mu\varkappa(k,\omega) = m\omega^2 - \mu_s k^2 - \mu_2 k^4. \tag{5.24}$$

Introducing the phase velocity $c = \omega/k$ and characteristic wavenumber $p = \rho/m$ we transform (5.22)-(5.24) into dimensionless forms

$$\frac{c^2}{c_T^2} = \frac{c_s^2}{c_T^2} + \frac{p}{|k|}\sqrt{1 - \frac{c^2}{c_T^2}}, \tag{5.25}$$

$$\frac{c^2}{c_T^2} = \frac{c_s^2}{c_T^2}\left(1 + \frac{k^2}{q^2}\right)^{-1} + \frac{p}{|k|}\sqrt{1 - \frac{c^2}{c_T^2}}, \tag{5.26}$$

$$\frac{c^2}{c_T^2} = \frac{c_s^2}{c_T^2} + \frac{\mathbb{K}}{p^4}k^4 + \frac{p}{|k|}\sqrt{1 - \frac{c^2}{c_T^2}}, \tag{5.27}$$

where $\mathbb{K} = \mu_2 p^4/(c_T^2 m)$ and $c_s = \sqrt{\mu_s/m}$ is the shear wave velocity in the thin film associated with the Gurtin-Murdoch model.

Typical dispersion curves for these models are shown in Fig. 5.1 for different values of parameters. Let us discuss some similarities in dispersion curves. All curves start from the point $(0, c_T)$ with a horizontal tangent. So for small k that is for long waves there is no significant difference in models as it should be. Indeed, surface nonlocality plays a role for short waves. Moreover, within a fixed range $0 \leq k \leq k_1$,

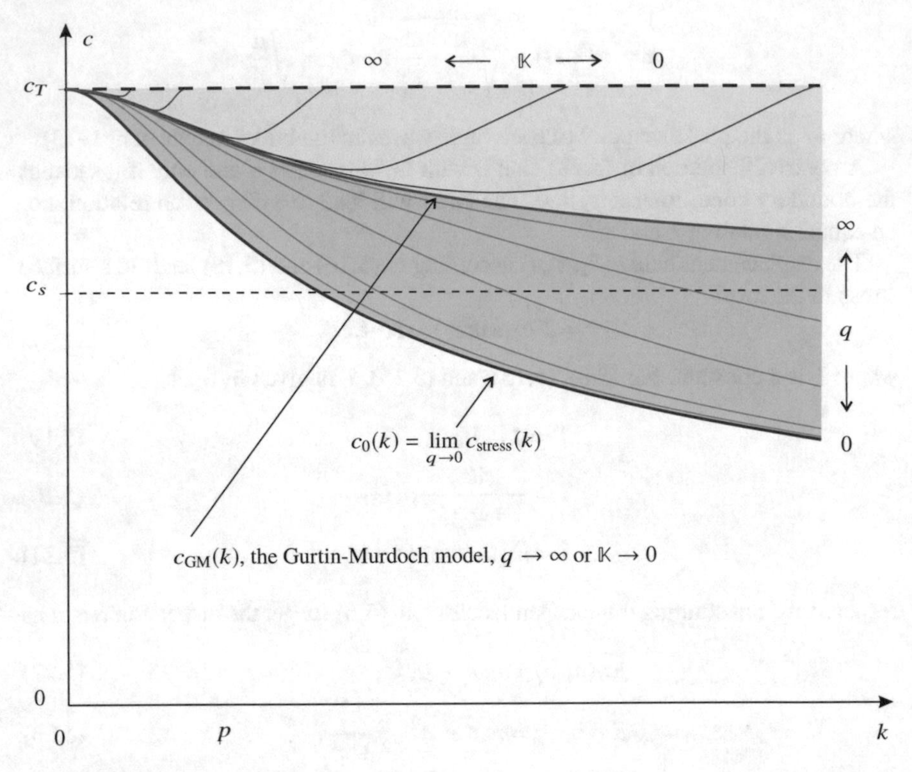

Fig. 5.1 Dispersion relations. c_{GM} curve corresponds to the Gurtin-Murdoch model. The dispersion curves c_{stress} for stress-gradient surface elasticity occupies the green area whereas dispersion curves c_{strain} for strain-gradient surface elasticity are in the yellow area. Here we assumed that $c_s = 3/4c_T$

the dispersion curve of the stress-gradient model for $q \to \infty$ will come arbitrarily close to the dispersion curve of the Gurtin-Murdoch model. The same behaviour demonstrate the dispersion curves for the strain-gradient model when $\mathbb{K} \to 0$. In what follows we assume the following notations: $c_{GM} = c_{GM}(k)$, $c_{\text{stress}} = c_{\text{stress}}(k)$, and $c_{\text{strain}} = c_{\text{strain}}(k)$ denote the phase velocity for the Gurtin-Murdoch, stress- and strain-gradient models.

For fixed q and $\mathbb{K}$ we have different behaviour of the dispersion curves for the stress- and strain gradient models at $k \to \infty$. For stress-gradient model we have that $c_{\text{stress}} \to 0$ when $k \to \infty$. Let us remind that the Gurtin-Murdoch dispersion curve tends to the finite velocity c_s at $k \to \infty$, see GM-curve in Fig. 5.1. For the strain-gradient model the dispersion curves approach the line $c_{\text{strain}} = c_T$ at $k = k_{\max}$, where $k_{\max}$ takes the value

$$k_{\max} = \sqrt{\frac{c_T^2 - c_s^2}{\mathbb{K}}}.$$

For the stress-gradient surface elasticity all dispersion curves are enclosed between the lower limiting curve for $q \to 0$, given by the formula

$$c_0^2 = \frac{c_T^2 p^2}{2k^2}\left(\sqrt{1+\frac{4k^2}{p^2}}-1\right), \tag{5.28}$$

and the dispersion curve of the Gurtin-Murdoch model, see Fig. 5.1. So we have the following bounds for c_{stress}

$$c_0(k) \le c_{\text{stress}}(k) \le c_{\text{GM}}(k). \tag{5.29}$$

Let us note that the dispersion curves for a square lattice lie also below the GM-curve, see Eremeyev and Sharma (2019). For the strain gradient model all dispersive curves are enclosed between the GM-curve and the line $c = c_T$, see Fig. 5.1,

$$c_{\text{GM}}(k) \le c_{\text{strain}}(k) \le c_T. \tag{5.30}$$

Thus, GM-curve separates dispersion curves for the stress- and stress-gradient model.

For all considered above models we consider the surface kinetic energy in the simplest form

$$K_s = \frac{1}{2} m \partial_t \boldsymbol{u} \cdot \partial_t \boldsymbol{u},$$

as was introduced by Gurtin and Murdoch (1978). Introduction of higher-order terms in the surface kinetic energy may significantly change the behaviour of the dispersion curves as in the case of the 3D models, see e.g. Askes and Aifantis (2011).

5.5 Conclusions

We have considered here the propagation of anti-plane surface waves in an elastic half-space with surfaces stresses within various models of surface elasticity. The linear Gurtin-Murdoch elasticity and strain- and stress-gradient surface elasticity models were compared. From the mathematical point of view the difference between the models consists of the boundary conditions at the half-space boundary. The analysis of dispersion relations was performed and the upper and lower bounds for the dispersion curves were found. In particular, it was shown that the dispersion curve for the Gurtin-Murdoch model separates the areas of dispersion curves for strain- and stress gradient surface elasticity.

Acknowledgements The author acknowledges financial support from the Russian Science Foundation under the grant "*Methods of microstructural nonlinear analysis, wave dynamics and mechanics of composites for research and design of modern metamaterials and elements of structures made on its base*" (No 15-19-10008-P).

References

Achenbach J (1973) Wave Propagation in Elastic Solids. North Holland, Amsterdam

Aifantis EC (2016) Internal Length Gradient (ILG) material mechanics across scales and disciplines. In: Bordas SPA, Balint DS (eds) Advances in Applied Mechanics, Elsevier, vol 49, pp 1–110

Altenbach H, Eremeyev VA (2011) On the shell theory on the nanoscale with surface stresses. International Journal of Engineering Science 49(12):1294–1301

Altenbach H, Eremeev VA, Morozov NF (2010) On equations of the linear theory of shells with surface stresses taken into account. Mechanics of Solids 45(3):331–342

Altenbach H, Eremeyev VA, Morozov NF (2012) Surface viscoelasticity and effective properties of thin-walled structures at the nanoscale. International Journal of Engineering Science 59:83–89

Askes H, Aifantis EC (2011) Gradient elasticity in statics and dynamics: An overview of formulations, length scale identification procedures, finite element implementations and new results. International Journal of Solids and Structures 48(13):1962–1990

Belov PA, Lurie SA, Golovina NY (2019) Classifying the existing continuum theories of ideal-surface adhesion. In: Adhesives and Adhesive Joints in Industry, IntechOpen

dell'Isola F, Seppecher P (1997) Edge contact forces and quasi-balanced power. Meccanica 32(1):33–52

dell'Isola F, Madeo A, Placidi L (2012a) Linear plane wave propagation and normal transmission and reflection at discontinuity surfaces in second gradient 3D continua. ZAMM - Journal of Applied Mathematics and Mechanics / Zeitschrift für Angewandte Mathematik und Mechanik 92(1):52–71

dell'Isola F, Seppecher P, Madeo A (2012b) How contact interactions may depend on the shape of Cauchy cuts in *n*th gradient continua: approach "á la d'alembert". ZAMP 63(6):1119–1141

Duan HL, Wang J, Karihaloo BL (2008) Theory of elasticity at the nanoscale. In: Aref H, van der Giessen E (eds) Advances in Applied Mechanics, Elsevier, vol 42, pp 1–68

Eremeyev VA (2016) On effective properties of materials at the nano-and microscales considering surface effects. Acta Mechanica 227(1):29–42

Eremeyev VA (2017) On nonlocal surface elasticity and propagation of surface anti-plane waves. In: Altenbach H, Goldstein RV, Murashkin E (eds) Mechanics for Materials and Technologies, Springer, Cham, Advanced Structured Materials, vol 46, pp 153–162

Eremeyev VA (2019a) On anti-plane surface wave propagation within the stress-gradient surface elasticity. In: Berezovski A, Soomere T (eds) Applied Wave Mathematics II, Mathematics of Planet Earth, vol 6, Springer, Cham

Eremeyev VA (2019b) Strongly anisotropic surface elasticity and antiplane surface waves. Philosophical Transactions of the Royal Society A pp 1–14, doi:10.1098/rsta.2019.0100

Eremeyev VA, Sharma BL (2019) Anti-plane surface waves in media with surface structure: Discrete vs. continuum model. International Journal of Engineering Science 143:33–38

Eremeyev VA, Rosi G, Naili S (2016) Surface/interfacial anti-plane waves in solids with surface energy. Mechanics Research Communications 74:8–13

Eremeyev VA, Cloud MJ, Lebedev LP (2018a) Applications of Tensor Analysis in Continuum Mechanics. World Scientific, New Jersey

Eremeyev VA, Rosi G, Naili S (2019) Comparison of anti-plane surface waves in strain-gradient materials and materials with surface stresses. Mathematics and Mechanics of Solids 24(8):2526–2535, doi:10.1177/1081286518769960

Eringen AC (2002) Nonlocal Continuum Field Theories. Springer, New York

Forest S, Cordero NM, Busso EP (2011) First vs. second gradient of strain theory for capillarity effects in an elastic fluid at small length scales. Computational Materials Science 50(4):1299–1304

de Gennes PG (1981) Some effects of long range forces on interfacial phenomena. J Physique Lettres 42(16):377–379

de Gennes PG, Brochard-Wyart F, Quéré D (2004) Capillarity and Wetting Phenomena: Drops, Bubbles, Pearls, Waves. Springer, New York

Georgiadis H, Vardoulakis I, Lykotrafitis G (2000) Torsional surface waves in a gradient-elastic half-space. Wave Motion 31(4):333–348

Gourgiotis P, Georgiadis H (2015) Torsional and SH surface waves in an isotropic and homogenous elastic half-space characterized by the Toupin–Mindlin gradient theory. International Journal of Solids and Structures 62(0):217 – 228

Gurtin ME, Murdoch AI (1975) A continuum theory of elastic material surfaces. Arch Ration Mech Analysis 57(4):291–323

Gurtin ME, Murdoch AI (1978) Surface stress in solids. International Journal of Solids and Structures 14(6):431–440

Han Z, Mogilevskaya SG, Schillinger D (2018) Local fields and overall transverse properties of unidirectional composite materials with multiple nanofibers and Steigmann–Ogden interfaces. International Journal of Solids and Structures 147:166 – 182

Israelachvili JN (2011) Intermolecular and Surface Forces, 3rd edn. Academic Press, Amsterdam

Javili A, dell'Isola F, Steinmann P (2013a) Geometrically nonlinear higher-gradient elasticity with energetic boundaries. Journal of the Mechanics and Physics of Solids 61(12):2381–2401

Javili A, McBride A, Steinmann P (2013b) Thermomechanics of solids with lower-dimensional energetics: on the importance of surface, interface, and curve structures at the nanoscale. a unifying review. Applied Mechanics Reviews 65(1):010,802

Kushch VI, Mogilevskaya SG, Stolarski HK, Crouch SL (2013) Elastic fields and effective moduli of particulate nanocomposites with the Gurtin-Murdoch model of interfaces. International Journal of Solids and Structures 50(7-8):1141–1153

Laplace PS (1805) Sur l'action capillaire. supplément à la théorie de l'action capillaire. In: Traité de mécanique céleste, vol 4. Supplement 1, Livre X, Gauthier–Villars et fils, Paris, pp 771–777

Laplace PS (1806) À la théorie de l'action capillaire. supplément à la théorie de l'action capillaire. In: Traité de mécanique céleste, vol 4. Supplement 2, Livre X, Gauthier–Villars et fils, Paris, pp 909–945

Lebedev LP, Cloud MJ, Eremeyev VA (2010) Tensor Analysis with Applications in Mechanics. World Scientific, New Jersey

Li Y, Wei PJ, Tang Q (2015) Reflection and transmission of elastic waves at the interface between two gradient-elastic solids with surface energy. European Journal of Mechanics A – Solids 52(C):54–71

Liebold C, Müller WH (2015) Are microcontinuum field theories of elasticity amenable to experiments? A review of some recent results. In: Chen GQ, Grinfeld M, Knops R (eds) Differential Geometry and Continuum Mechanics, Springer Proceedings in Mathematics & Statistics, vol 137, Springer, pp 255–278

Longley WR, Van Name RG (eds) (1928) The Collected Works of J. Willard Gibbs, PHD., LL.D., vol I Thermodynamics. Longmans, New York

Lurie S, Volkov-Bogorodsky D, Zubov V, Tuchkova N (2009) Advanced theoretical and numerical multiscale modeling of cohesion/adhesion interactions in continuum mechanics and its applications for filled nanocomposites. Computational Materials Science 45(3):709 – 714

Lurie S, Belov P, Altenbach H (2016) Classification of gradient adhesion theories across length scale. In: Altenbach H, Forest S (eds) Generalized Continua as Models for Classical and Advanced Materials, Advanced Structured Materials, vol 42, Springer, Cham, pp 261–277

Maugin GA (2017) Non-Classical Continuum Mechanics: A Dictionary. Springer, Singapore

Nazarenko L, Stolarski H, Altenbach H (2016) Effective properties of short-fiber composites with gurtin-murdoch model of interphase. International Journal of Solids and Structures 97:75–88

Nazarenko L, Stolarski H, Altenbach H (2018) Effective properties of particulate composites with surface-varying interphases. Composites Part B: Engineering 149:268–284

Placidi L, Rosi G, Giorgio I, Madeo A (2014) Reflection and transmission of plane waves at surfaces carrying material properties and embedded in second-gradient materials. Mathematics and Mechanics of Solids 19(5):555–578

Poisson SD (1831) Nouvelle théorie de l'action capillaire. Bachelier Père et Fils, Paris

Rosi G, Nguyen VH, Naili S (2015) Surface waves at the interface between an inviscid fluid and a dipolar gradient solid. Wave Motion 53(0):51–65

Ru CQ (2010) Simple geometrical explanation of Gurtin-Murdoch model of surface elasticity with clarification of its related versions. Science China Physics, Mechanics and Astronomy 53(3):536–544

Ru CQ (2016) A strain-consistent elastic plate model with surface elasticity. Continuum Mechanics and Thermodynamics 28(1-2):263–273

Simmonds JG (1994) A Brief on Tensor Analysis, 2nd edn. Springer, New Yourk

Steigmann DJ, Ogden RW (1997) Plane deformations of elastic solids with intrinsic boundary elasticity. Proceedings of the Royal Society A 453(1959):853–877

Steigmann DJ, Ogden RW (1999) Elastic surface-substrate interactions. Proceedings of the Royal Society A 455(1982):437–474

Vardoulakis I, Georgiadis HG (1997) SH surface waves in a homogeneous gradient-elastic half-space with surface energy. Journal of Elasticity 47(2):147–165

Wang J, Huang Z, Duan H, Yu S, Feng X, Wang G, Zhang W, Wang T (2011) Surface stress effect in mechanics of nanostructured materials. Acta Mech Solida Sinica 24:52–82

Xu L, Wang X, Fan H (2015) Anti-plane waves near an interface between two piezoelectric half-spaces. Mechanics Research Communications 67:8–12

Yerofeyev VI, Sheshenina OA (2005) Waves in a gradient-elastic medium with surface energy. Journal of Applied Mathematics and Mechanics 69(1):57 – 69

Young T (1805) An essay on the cohesion of fluids. Philosophical Transactions of the Royal Society of London 95:65–87

Zemlyanova AY, Mogilevskaya SG (2018) Circular inhomogeneity with Steigmann–Ogden interface: Local fields, neutrality, and Maxwell's type approximation formula. International Journal of Solids and Structures 135:85–98

Chapter 6
Anisotropic Material Behavior

Artur Ganczarski

Abstract This entry is focused on description of material anisotropy in elastic and plastic ranges. Concise classification of anisotropic materials with respect to symmetry of elastic matrices as referred to the crystal lattice symmetry is given, and extended analogy between symmetries of constitutive material matrices (elastic and yield/failure) is also discussed. In this entry basic features of anisotropic initial yield criteria are discussed. Two ways to account for anisotropy are presented: the explicit vs. implicit formulations. The explicit description of anisotropy is rigorously based on well established theory of common invariants (Sayir, Goldenblat–Kopnov, von Mises, Hill). The implicit approach involves linear transformation tensor of the Cauchy stress that accounts for anisotropy to enhance the known isotropic criteria to be able to capture anisotropy, hydrostatic pressure insensitivity and asymmetry of the yield surface (Barlat, Plunckett, Cazacu, Khan). The advantages and differences of both formulations are critically presented.

Keywords: Symmetry classes in elasticity · Anisotropic yield criteria · Anisotropic failure criteria

6.1 Elastic Anisotropy

„Material anisotropy means that the constitutive relation takes different forms depending on the Cartesian coordinate system we use."

Artur Ganczarski
Institute of Applied Mechanics, Cracow University of Technology, al. Jana Pawła II 37, 31-864 Kraków, Poland,
e-mail: artur.ganczarski@pk.edu.pl

H. Altenbach and A. Öchsner (eds.), *State of the Art and Future Trends in Material Modeling*, Advanced Structured Materials 100,
https://doi.org/10.1007/978-3-030-30355-6_6

This definition by Ottosen and Ristinmaa (2005) clearly shows that in the case of elasticity all information about anisotropy is included in the stiffness or compliance tensors. Hence, to study the elastic anisotropy means to study classes of symmetry of aforementioned tensors.

For the purpose of further considerations an analogy between the crystal lattice symmetry groups and classes and corresponding symmetry of the stiffness matrices defined for crystalline materials might occur useful (cf. e.g. Nye, 1957). Unit cells of the eight conventional crystal lattices are demonstrated based on Love (1944) and Jastrzebski (1987), whereas corresponding constitutive elasticity matrices are schematically sketched applying Nye's graphics (symbol ● refers to independent element, symbol ○ refers to dependent element, symbols ●—● or ○—○ represent pairs of identical matrix elements, symbols ●—⊙ stand for pairs of elements in which one is doubled (effect of engineering notation applied to shear strain $\gamma_{ij} = 2\varepsilon_{ij}$), whereas symbols ●—⊖ denote pairs of elements of the same absolute value but opposite signs, respectively.

6.1.1 Triclinic Symmetry

Deformation of representative cube taken of the generally anisotropic material of the triclinic symmetry subjected to exemplary axial tension along 3 axis is fully anisotropic. This means that it comprises both anisotropic axial strains (transformation of the cube to a rectangular prism) as well as anisotropic shear strains (transformation of the rectangular prism to a parallelepiped). In such a case of general deformation the elastic compliance matrix is fully populated. In other words, all components of the columnar stress vector depend on all 6 components of the columnar strain vector (36 combinations).

Final representation of compliance matrix for fully anisotropic (triclinic) material is as follows

$$
[\boldsymbol{E}^{-1}] = \left[\begin{array}{ccc|ccc}
\frac{1}{E_{11}} & -\frac{\nu_{21}}{E_{11}} & -\frac{\nu_{31}}{E_{11}} & \frac{\eta_{23(1)}}{E_{11}} & \frac{\eta_{31(1)}}{E_{11}} & \frac{\eta_{12(1)}}{E_{11}} \\
-\frac{\nu_{12}}{E_{22}} & \frac{1}{E_{22}} & -\frac{\nu_{32}}{E_{22}} & \frac{\eta_{23(2)}}{E_{22}} & \frac{\eta_{31(2)}}{E_{22}} & \frac{\eta_{12(2)}}{E_{22}} \\
-\frac{\nu_{13}}{E_{33}} & -\frac{\nu_{23}}{E_{33}} & \frac{1}{E_{33}} & \frac{\eta_{23(3)}}{E_{33}} & \frac{\eta_{31(3)}}{E_{33}} & \frac{\eta_{12(3)}}{E_{33}} \\
\hline
\frac{\eta_{(1)23}}{G_{23}} & \frac{\eta_{(2)23}}{G_{23}} & \frac{\eta_{(3)23}}{G_{23}} & \frac{1}{G_{23}} & \frac{\mu_{31(23)}}{G_{23}} & \frac{\mu_{12(23)}}{G_{23}} \\
\frac{\eta_{(1)31}}{G_{31}} & \frac{\eta_{(2)31}}{G_{31}} & \frac{\eta_{(3)31}}{G_{31}} & \frac{\mu_{(23)31}}{G_{31}} & \frac{1}{G_{31}} & \frac{\mu_{12(31)}}{G_{31}} \\
\frac{\eta_{(1)12}}{G_{12}} & \frac{\eta_{(2)12}}{G_{12}} & \frac{\eta_{(3)12}}{G_{12}} & \frac{\mu_{(23)12}}{G_{12}} & \frac{\mu_{(31)12}}{G_{12}} & \frac{1}{G_{12}}
\end{array}\right]
\left[\begin{array}{ccc|ccc}
\bullet & \bullet & \bullet & \bullet & \bullet & \bullet \\
 & \bullet & \bullet & \bullet & \bullet & \bullet \\
 & & \bullet & \bullet & \bullet & \bullet \\
\hline
 & & & \bullet & \bullet & \bullet \\
 & & & & \bullet & \bullet \\
 & & & & & \bullet
\end{array}\right] \quad (6.1)
$$

Symmetry of the elastic compliance matrix (6.1) results from symmetry of both stress and strain tensors, namely

Table 6.1 Classification of anisotropic elastic materials with respect to stiffness matrix symmetry referring to crystal lattice cf. Nye (1957)

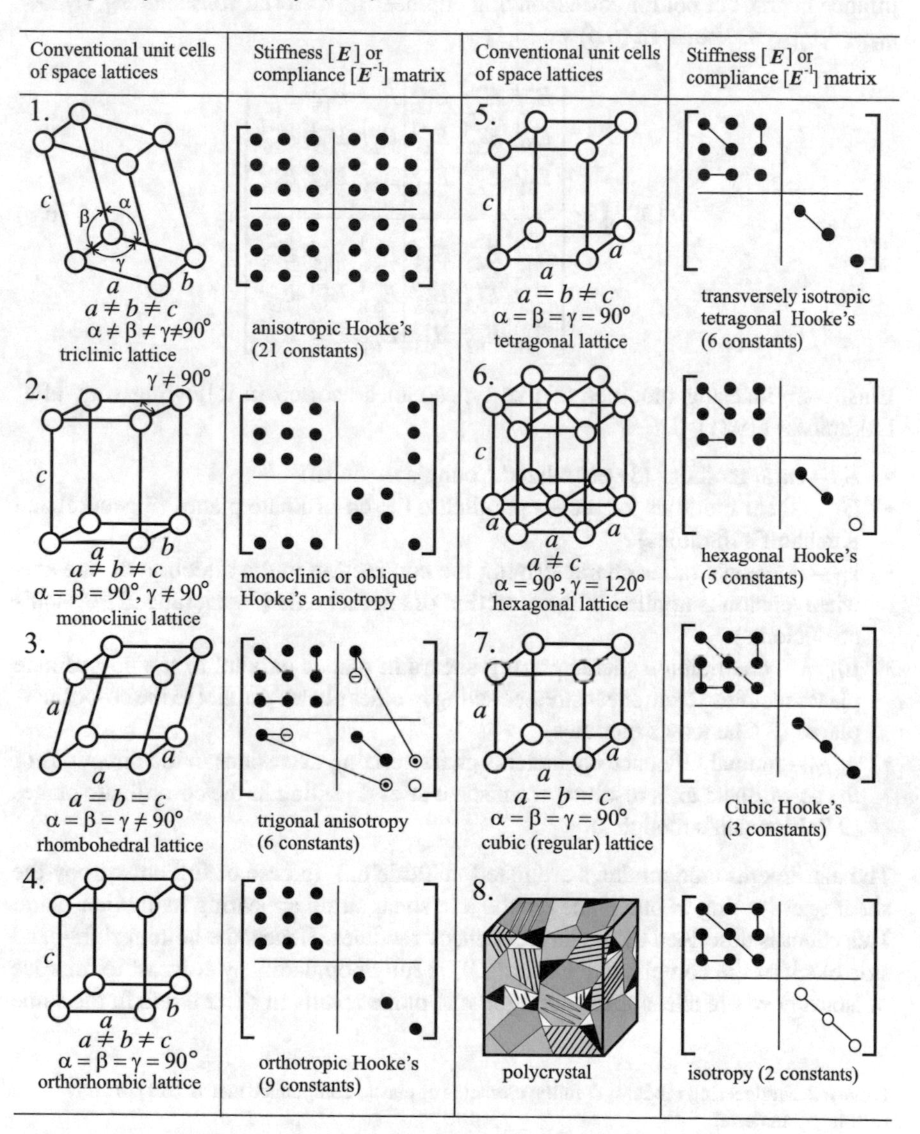

Conventional unit cells of space lattices	Stiffness $[\boldsymbol{E}]$ or compliance $[\boldsymbol{E}^{-1}]$ matrix	Conventional unit cells of space lattices	Stiffness $[\boldsymbol{E}]$ or compliance $[\boldsymbol{E}^{-1}]$ matrix
1. $a \neq b \neq c$, $\alpha \neq \beta \neq \gamma \neq 90°$, triclinic lattice	anisotropic Hooke's (21 constants)	5. $a = b \neq c$, $\alpha = \beta = \gamma = 90°$, tetragonal lattice	transversely isotropic tetragonal Hooke's (6 constants)
2. $\gamma \neq 90°$; $a \neq b \neq c$, $\alpha = \beta = 90°, \gamma \neq 90°$, monoclinic lattice	monoclinic or oblique Hooke's anisotropy	6. $a \neq c$, $\alpha = 90°, \gamma = 120°$, hexagonal lattice	hexagonal Hooke's (5 constants)
3. $a = b = c$, $\alpha = \beta = \gamma \neq 90°$, rhombohedral lattice	trigonal anisotropy (6 constants)	7. $a = b = c$, $\alpha = \beta = \gamma = 90°$, cubic (regular) lattice	Cubic Hooke's (3 constants)
4. $a \neq b \neq c$, $\alpha = \beta = \gamma = 90°$, orthorhombic lattice	orthotropic Hooke's (9 constants)	8. polycrystal	isotropy (2 constants)

$$
\begin{aligned}
\frac{\nu_{ij}}{E_{jj}} &= \frac{\nu_{ji}}{E_{ii}} \longrightarrow \nu_{ij}E_{ii} = \nu_{ji}E_{jj} \\
\frac{\eta_{ij(k)}}{E_{kk}} &= \frac{\eta_{(k)ij}}{G_{ij}} \longrightarrow \eta_{ij(k)}G_{ij} = \eta_{(k)ij}E_{kk} \\
\frac{\mu_{ij(ki)}}{G_{ki}} &= \frac{\mu_{(ki)ij}}{G_{ji}} \longrightarrow \mu_{ij(ki)}G_{ji} = \mu_{(ki)ij}G_{ki}
\end{aligned}
\tag{6.2}
$$

In should be pointed out that the symmetry $\boldsymbol{E}_{ij}^{-1} = \boldsymbol{E}_{ji}^{-1}$ holds for elements of compliance matrix but not for corresponding engineering material constants E_{ii}, ν_{ij}, G_{ij}, $\eta_{(i)jk}$, $\mu_{ij(ki)}$ as shown in (6.3) vs. (6.1)

$$[\boldsymbol{E}^{-1}] = \left[\begin{array}{ccc|ccc} E_{11}^{-1} & E_{12}^{-1} & E_{13}^{-1} & E_{14}^{-1} & E_{15}^{-1} & E_{16}^{-1} \\ E_{21}^{-1} & E_{22}^{-1} & E_{23}^{-1} & E_{24}^{-1} & E_{25}^{-1} & E_{26}^{-1} \\ E_{31}^{-1} & E_{32}^{-1} & E_{33}^{-1} & E_{34}^{-1} & E_{35}^{-1} & E_{36}^{-1} \\ \hline E_{41}^{-1} & E_{42}^{-1} & E_{43}^{-1} & E_{44}^{-1} & E_{45}^{-1} & E_{46}^{-1} \\ E_{51}^{-1} & E_{52}^{-1} & E_{53}^{-1} & E_{54}^{-1} & E_{55}^{-1} & E_{56}^{-1} \\ E_{61}^{-1} & E_{62}^{-1} & E_{63}^{-1} & E_{64}^{-1} & E_{65}^{-1} & E_{66}^{-1} \end{array}\right] \tag{6.3}$$

Elastic engineering modules of five types can be sorted in following way, after Lekhnitskii (1981):

- E_{ii} – axial modules (3 generalized Young's modules)
- G_{ij} – shear modules for planes parallel to the co-ordinate planes (3 generalized Kirchhoff's modules)
- ν_{ij} – Poisson's ratios characterizing the contraction in the direction of one axis when tension is applied in the direction of another axis (3 generalized Poisson's coefficients)
- $\mu_{ij(kl)}$ – coefficients characterizing shears in planes parallel to the co-ordinate planes resulting from shear stresses acting in other planes parallel to the co-ordinate planes (3 Chencov's modules)
- $\eta_{i(jk)}$ – mutual influence coefficients characterizing extensions in the directions of the co-ordinate axes resulting from shear stresses acting in the co-ordinate planes (9 Rabinovich's modules)

The aforementioned modules are listed in Table 6.2. In case of full anisotropy the shear stress acting in one plane results a in shear strain appearing in another plane. This effect is described by the three Chencov modules. Hence, the bottom right-hand side block of the compliance matrix (6.3) is fully populated, by contrast to the case of isotropy where shear stress acting in one plane results in shear strain in the same

Table 6.2 Engineering modules defining elements of elastic compliance matrix (6.1) of fully anisotropic material

Engineering elastic modules	Coupling between stress	strain	Corresponding axes or planes	Number of coefficients
E_{11}, E_{22}, E_{33}	axial	extension	the same axes: 1 → 1, etc.	3
G_{12}, G_{32}, G_{31}	shear	shear strain	the same planes: 12 → 12, etc.	3
$\nu_{21}, \nu_{31}, \nu_{32}$	axial	extension	different axes: 1 → 2, etc.	3
$\mu_{31(23)}, \mu_{12(23)}, \mu_{12(31)}$	shear	shear strain	different planes:13 → 23, etc.	3
$\eta_{23(1)}, \ldots, \eta_{12(3)}$	shear	extension	normal to shear plane: 23 → 1, etc.	9

plane exclusively. It means that in case of isotropy the considered blok of compliance matrix must have the diagonal form.

In order to describe effect of axial stresses on shear strains (upper right-hand side block), as well as effect of shear stresses on axial strains (lower left-hand side block), it is necessary to define 9 additional modules $\eta_{(i)jk}$, called Rabinovich's modules where the appropriate symmetry conditions hold (6.2). Total number of discussed modules is equal to 21. However, only 18 of them are truly independent because the compliance matrix $[E^{-1}]$ has to obey transformation with respect to 3 Euler's angles. It should be pointed out that in general case of anisotropy it is not possible to find any reference frame for which any elements of the compliance matrix can be equal to zero. The general case of anisotropy corresponds to the triclinic symmetry lattice cell in which all three edges differ each from the other and all three angles between them differ each from the other and none of them is equal 90°, as shown in item 1. of Table 6.1.

6.1.2 Monoclinic Symmetry

Among anisotropic materials the narrower group called monoclinic symmetry can be distinguished. Monoclinic or oblique symmetry corresponds to monoclinic space lattice cell symmetry in which all three edges differ each from the other whereas two angles are equal to 90° and one is different, as shown in item 2. of Table 6.1. The corresponding stiffness matrix symmetry characterizes through incomplete population in which only 13 elements are not equal to zero, as shown below

$$[E^{-1}] = \left[\begin{array}{ccc|ccc} \frac{1}{E_{11}} & -\frac{\nu_{21}}{E_{11}} & -\frac{\nu_{31}}{E_{11}} & & & \frac{\eta_{12(1)}}{E_{11}} \\ -\frac{\nu_{12}}{E_{22}} & \frac{1}{E_{22}} & -\frac{\nu_{32}}{E_{22}} & & & \frac{\eta_{12(2)}}{E_{22}} \\ -\frac{\nu_{13}}{E_{33}} & -\frac{\nu_{23}}{E_{33}} & \frac{1}{E_{33}} & & & \frac{\eta_{12(3)}}{E_{33}} \\ \hline & & & \frac{1}{G_{23}} & \frac{\mu_{31(23)}}{G_{23}} & \\ & & & \frac{\mu_{(23)31}}{G_{31}} & \frac{1}{G_{31}} & \\ \frac{\eta_{(1)12}}{G_{12}} & \frac{\eta_{(2)12}}{G_{12}} & \frac{\eta_{(3)12}}{G_{12}} & & & \frac{1}{G_{12}} \end{array}\right] \left[\begin{array}{ccc|ccc} \bullet & \bullet & \bullet & & & \bullet \\ & \bullet & \bullet & & & \bullet \\ & & \bullet & & & \bullet \\ \hline & & & \bullet & \bullet & \\ & & & & \bullet & \\ & & & & & \bullet \end{array}\right] \tag{6.4}$$

In other words in case of monoclinic symmetry only 3 of the Rabinovich modules and only 1 of the Chencov modules are different from zero.

6.1.3 Trigonal/Rhombohedral Symmetry

Another important narrower case of material anisotropy called trigonal anisotropy can be distinguished. The trigonal anisotropy corresponds the rhombohedral cell lattice in which all three edges are equal to each other and all three angles are equal but different from 90°, as shown in item 3. of Table 6.1. The corresponding compliance matrix takes the following representation

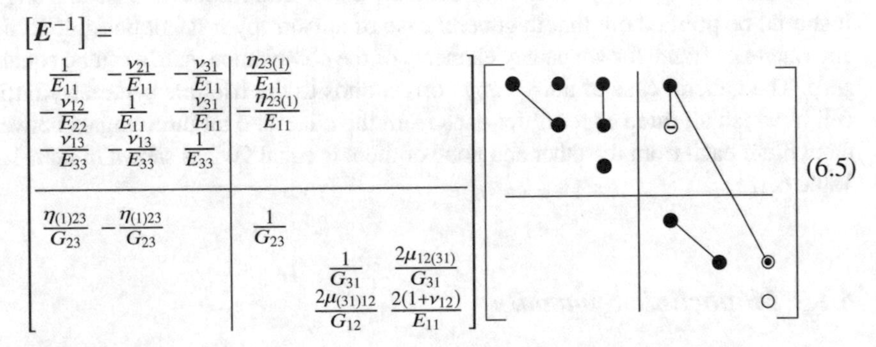

It is seen that in case of trigonal symmetry among Rabinovich's modules only 2 are non-zeroth but in fact only one of them is independent because they only differ in sign. Additionally, only one Chencov's modulus is different from zero but in fact it is the dependent modulus due to the specific coupling between components $2E^{-1}_{14} = E^{-1}_{56}$ and $E^{-1}_{24} = -E^{-1}_{14}$ as well as $E^{-1}_{11} = E^{-1}_{22}$, $E^{-1}_{44} = E^{-1}_{55}$, $E^{-1}_{13} = E^{-1}_{23}$ whereas $E^{-1}_{66} = (E^{-1}_{11} - E^{-1}_{12})/2$ must hold. Finally for trigonal symmetry only 6 elements of the compliance matrix are independent, see Berryman (2005).

6.1.4 Orthorhombic Symmetry

Majority of engineering materials exhibits a specific symmetry property, which may result in reduction of the number of non-zeroth elastic modules. It can be done when, for chosen symmetry group or class, some particular material directions are defined in such a way that transformation of the compliance matrix from an arbitrary co-ordinate frame to the given structural symmetry frame leads to the zeroth population of the top right-hand side and the bottom left-hand side blocks of the compliance matrix (6.1), and additionally the bottom right-hand side block possesses a diagonal form. In such practically important case both the nine Rabinovich $\eta_{(i)jk}$ and the three Chencov $\mu_{ij(kl)}$ modules are equal to zero, and consequently, coupling between the shear stresses and elongations does not exist such that shear strains are produced exclusively by the action on stresses at the same planes. In this particular symmetry, called orthotropy, there exist three mutually perpendicular axes (1, 2, 3) that determine the three material orthotropy planes. The orthotropy symmetry case corresponds to the orthorhombic lattice in which all three edges differ each from the other but all angles are equal to 90°, as presented in item 4. of Table 6.1.

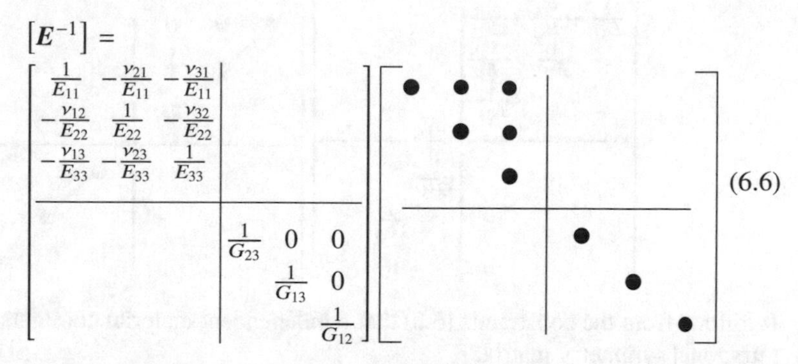

$$[E^{-1}] = \left[\begin{array}{ccc|ccc} \frac{1}{E_{11}} & -\frac{\nu_{21}}{E_{11}} & -\frac{\nu_{31}}{E_{11}} & & & \\ -\frac{\nu_{12}}{E_{22}} & \frac{1}{E_{22}} & -\frac{\nu_{32}}{E_{22}} & & & \\ -\frac{\nu_{13}}{E_{33}} & -\frac{\nu_{23}}{E_{33}} & \frac{1}{E_{33}} & & & \\ \hline & & & \frac{1}{G_{23}} & 0 & 0 \\ & & & & \frac{1}{G_{13}} & 0 \\ & & & & & \frac{1}{G_{12}} \end{array}\right] \left[\begin{array}{ccc|ccc} \bullet & \bullet & \bullet & & & \\ & \bullet & \bullet & & & \\ & & \bullet & & & \\ \hline & & & \bullet & & \\ & & & & \bullet & \\ & & & & & \bullet \end{array}\right] \tag{6.6}$$

The following conditions must hold to assure the matrix symmetry

$$\frac{\nu_{21}}{E_{11}} = \frac{\nu_{12}}{E_{22}} \qquad \frac{\nu_{13}}{E_{33}} = \frac{\nu_{31}}{E_{11}} \qquad \frac{\nu_{23}}{E_{33}} = \frac{\nu_{32}}{E_{22}} \tag{6.7}$$

Finally, in case of orthotropy number of independent material constants is 9, that is 3 generalized Hooke's modules E_{11}, E_{22}, E_{33}, 3 generalized Kirchhoff's modules G_{12}, G_{23}, G_{31} and three generalized Poisson's ratios $\nu_{21}, \nu_{23}, \nu_{31}$.

6.1.5 Tetragonal Transverse Isotropy

For several engineering applications the general orthotropic symmetry model occurs to be too complicated, since additional symmetry conditions frequently appear. In particular case when conditions of isotropy hold in selected orthotropy plane the so called transverse isotropy obeys.

In case of so called tetragonal symmetry material properties in the plane (1, 2) satisfy condition of cubic symmetry, see item 5. of Table 6.1

$$E_{11} = E_{22}, \quad G_{13} = G_{23}, \quad \nu_{31} = \nu_{32} \tag{6.8}$$

Hence, in case of transverse isotropy of tetragonal symmetry number of independent material constants is equal to 6: E_{11}, E_{33}, G_{23}, G_{12}, ν_{21}, ν_{31}. Corresponding crystal lattice is sketched in item 5. of Table 6.1, where tetragonal lattice being special case of the orthorhombic lattice with $a = b \neq c$ obeys.

When the constraints (6.8) are applied to compliance matrix (6.6) the transverse isotropy tetragonal symmetry case yields

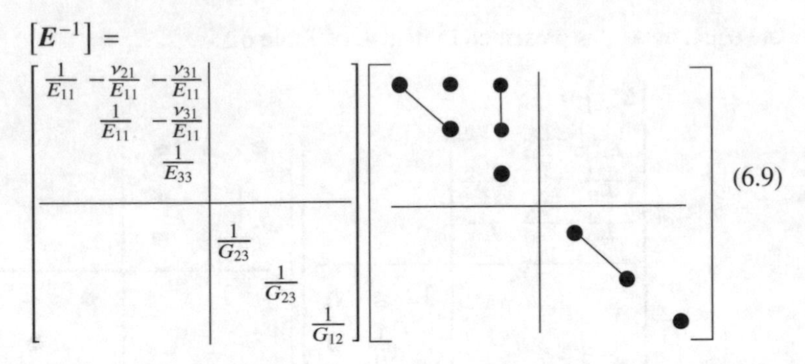

(6.9)

It follows from the constraints (6.8) that 6 independent material constants define the tetragonal symmetry matrix:

- E_{11}, E_{33} – two Young's modulus in the plane of isotropy and direction perpendicular to this plane,
- ν_{21}, ν_{31} – two Poisson's ratios referring to transverse contraction or swelling caused by tension or compression in direction perpendicular to isotropy plane,
- G_{12}, G_{23} – two different Kirchhoff's modules in the isotropy or orthotropy planes.

6.1.6 Hexagonal Transverse Isotropy

In special case of the transverse isotropy called hexagonal symmetry the additional constraint must obey for the shear modulus in the isotropy plane

$$G_{12} = \frac{E_{11}}{2(1+\nu_{21})} \qquad \text{or} \qquad E_{66}^{-1} = 2\left(E_{11}^{-1} - E_{12}^{-1}\right) \tag{6.10}$$

where modulus G_{12} is expressed in terms of the transverse Young's modulus E_{11} and transverse Poisson's ratio ν_{21}. Hence, in case of the transverse isotropy of hexagonal symmetry the number of independent constants is equal to 5: E_{11}, E_{33}, G_{23}, ν_{21}, ν_{31}. A choice of the five independent material constants from among six can be performed in optional way, for instance

$$[\boldsymbol{E}^{-1}] = \left[\begin{array}{ccc|ccc} \frac{1}{E_{11}} & -\frac{\nu_{21}}{E_{11}} & -\frac{\nu_{31}}{E_{11}} & & & \\ & \frac{1}{E_{11}} & -\frac{\nu_{31}}{E_{11}} & & & \\ & & \frac{1}{E_{33}} & & & \\ \hline & & & \frac{1}{G_{23}} & & \\ & & & & \frac{1}{G_{23}} & \\ & & & & & \frac{2(1+\nu_{21})}{E_{11}} \end{array}\right] \tag{6.11}$$

Rolled metals, some multi-phase composite materials, basalt or columnar ice are examples of transversely isotropic materials, however precise distinction between the tetragonal or hexagonal symmetry classes is often difficult (see for example Gan et al, 2000).

6.1.7 Cubic Symmetry

Further reduction of number of independent constants leads to the cubic symmetry for which the compliance matrix is characterized by 3 independent material constants $E_{11} = E_{22} = E_{33} = E$, $G_{23} = G_{31} = G_{12} = G$ and $\nu_{21} = \nu_{31} = \nu_{32} = \nu$. Hence, following form of the compliance matrix is furnished

$$[E^{-1}] = \left[\begin{array}{ccc|ccc} \frac{1}{E} & -\frac{\nu}{E} & -\frac{\nu}{E} & & & \\ & \frac{1}{E} & -\frac{\nu}{E} & & & \\ & & \frac{1}{E} & & & \\ \hline & & & \frac{1}{G} & & \\ & & & & \frac{1}{G} & \\ & & & & & \frac{1}{G} \end{array}\right] \tag{6.12}$$

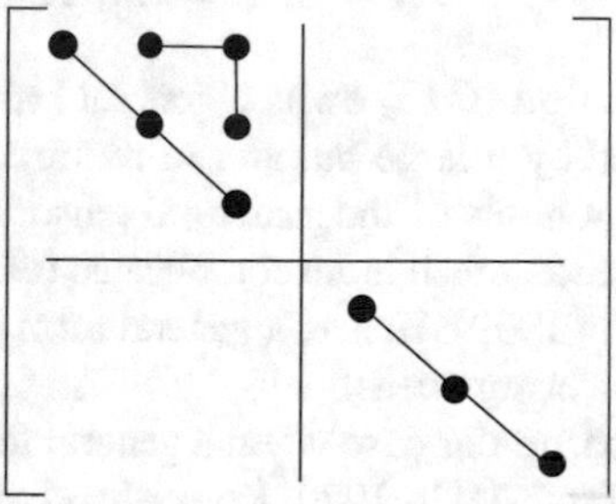

Note that in case of cubic symmetry the condition (6.10) does not hold. Corresponding cubic or regular lattice is shown in item 7. of Table 6.1. The particular example of the cubic symmetry material is the Nickel-based single crystal superalloys widely used in aircraft engines especially for turbine blades as discussed by Desmorat and Marull (2011). The cubic symmetry is the narrower symmetry case known from crystallography, see Jastrzebski (1987), since fully isotropic crystal lattices are unknown. Consequently, the material isotropy can be considered as a polycrystal (see item 8. of Table 6.1) being the assembly of sufficiently high number of monocrystals, randomly distributed and oriented such that constitutive relation does not depend of coordinate system (stiffness or compliance tensor can be defined by combination of only two independent material constants and second order unit tensors $\lambda\delta_{ij}\delta_{kl} + \mu(\delta_{ik}\delta_{jl} + \delta_{il}\delta_{jk})$).

6.2 Plastic Anisotropy

In the case of plasticity, analogously to the case of elasticity, all information about anisotropy is included in the constitutive law. However, this depends on the yield criterion and the hardening/softening rule. Hence, to study plastic anisotropy means at first to study classes of symmetry of yield criteria.

6.2.1 Goldenblat–Kopnov's Criterion

In a general case of material anisotropy, extension of the isotropic yield initiation criteria to the anisotropic yield/failure behaviour (Table 6.3), by the use of common invariants of the stress tensor and of the structural tensors of plastic anisotropy (cf. Hill, 1948; Sayir, 1970; Betten, 1988; Życzkowski, 2001), can be shown in a general fashion

$$f\left(\Pi, \Pi_{ij}\sigma_{ij}, \Pi_{ijkl}\sigma_{ij}\sigma_{kl}, \Pi_{ijklmn}\sigma_{ij}\sigma_{kl}\sigma_{mn}, \ldots\right) = 0 \tag{6.13}$$

where Einstein's summation convention holds.

In such a case, initiation of plastic flow or failure is governed by the structural tensors of material anisotropy of even-ranks:

$$\overset{<0>}{\mathbf{\Pi}} = \Pi,\ \overset{<2>}{\mathbf{\Pi}} = \Pi_{ij},\ \overset{<4>}{\mathbf{\Pi}} = \Pi_{ijkl},\ \overset{<6>}{\mathbf{\Pi}} = \Pi_{ijklmn}, \ldots,$$

etc. Equation (6.13) owns a general representation, but its practical identification is limited by a large number of required material tests and, additionally, because the components of the structural tensors are temperature dependent, which makes identification much more complicated (cf. e.g. Herakovich and Aboudi, 1999; Tamma and Avila, 1999). Hence, a general form (6.13) is usually more specified and limited for engineering needs.

In a particular case when a general tensorially-polynomial form of Eq. (6.13) is assumed (cf. Sayir, 1970; Kowalsky et al, 1999; Życzkowski, 2001; Ganczarski and Skrzypek, 2014) the polynomial anisotropic yield criterion is furnished

$$(\Pi_{ij}\sigma_{ij})^{\alpha} + (\Pi_{ijkl}\sigma_{ij}\sigma_{kl})^{\beta} + (\Pi_{ijklmn}\sigma_{ij}\sigma_{kl}\sigma_{mn})^{\gamma} + \ldots - 1 = 0 \tag{6.14}$$

where if the Voigt notation is used and the structural anisotropy tensors take corresponding matrix forms

$$[\overset{<2>}{\mathbf{\Pi}}] = \begin{bmatrix} \pi_{11} & \pi_{12} & \pi_{13} \\ & \pi_{22} & \pi_{23} \\ & & \pi_{33} \end{bmatrix}, \qquad [\overset{<4>}{\mathbf{\Pi}}] = \left[\begin{array}{ccc|ccc} \Pi_{11} & \Pi_{12} & \Pi_{13} & \Pi_{14} & \Pi_{15} & \Pi_{16} \\ & \Pi_{22} & \Pi_{23} & \Pi_{24} & \Pi_{25} & \Pi_{26} \\ & & \Pi_{33} & \Pi_{34} & \Pi_{35} & \Pi_{36} \\ \hline & & & \Pi_{44} & \Pi_{45} & \Pi_{46} \\ & & & & \Pi_{55} & \Pi_{56} \\ & & & & & \Pi_{66} \end{array}\right] \tag{6.15}$$

The even-rank structural anisotropy tensors $\Pi_{ij}, \Pi_{ijkl}, \Pi_{ijklmn}, \ldots$, in Eq. (6.14) are normalized by the common constant Π and $\alpha, \beta, \gamma \ldots$ etc., are arbitrary exponents of a polynomial representation. In a narrower case if $\alpha = 1, \beta = 1/2, \gamma = 1/3$, and limiting an infinite form (6.14) to the equation that contains only three common invariants, we arrive at the narrower form known as the Goldenblat and Kopnov criterion (Goldenblat and Kopnov, 1966)

$$\Pi_{ij}\sigma_{ij} + (\Pi_{ijkl}\sigma_{ij}\sigma_{kl})^{1/2} + (\Pi_{ijklmn}\sigma_{ij}\sigma_{kl}\sigma_{mn})^{1/3} - 1 = 0 \qquad (6.16)$$

which satisfies the dimensional homogeneity of three polynomial components.

Equation (6.16), when limited only to three common invariants of the stress tensor σ and structural anisotropy tensors of even orders: 2nd Π_{ij}, 4th Π_{ijkl} and 6th Π_{ijklmn} is not the most general one, in the meaning of the representation theorems, which determine the most general irreducible representation of the scalar and tensor functions that satisfy the invariance with respect to change of coordinates and material symmetry properties (cf. e.g. Spencer, 1971; Rymarz, 1993; Rogers, 1990). However, 2nd, 4th and 6th order structural anisotropy tensors, which are used in (6.16) or in case if $\alpha = 1, \beta = 1, \gamma = 1$ and the deviatoric stress representation is used by Kowalsky et al (1999)

$$h^{(1)}_{ij} s_{ij} + h^{(2)}_{ijkl} s_{ij} s_{kl} + h^{(3)}_{ijklmn} s_{ij} s_{kl} s_{mn} - h^{(0)} = 0 \qquad (6.17)$$

are found satisfactory for describing fundamental transformation modes of limit surfaces caused by plastic or failure processes, namely: isotropic change of size, kinematic translation and rotation, as well as surface distortion (Betten, 1988; Kowalsky et al, 1999, cf.).

Altenbach et al (1995); Altenbach and Kolupaev (2015) presented anisotropic yield criterion being simulteneous extension of the Altenbach–Zolochevsky isotropic criterion and the Goldenblat–Kopnov anisotropic criterion

$$\alpha a_{ij}\sigma_{ij} + \beta \left(b_{ijkl}\sigma_{ij}\sigma_{kl}\right)^{1/2} + \gamma \left(c_{ijkmnl}\sigma_{ij}\sigma_{kl}\sigma_{mn}\right)^{1/3} - 1 = 0 \qquad (6.18)$$

where α, β, γ are weight coefficients, whereas structural anisotropy tensors contain 6, 21 and 56 material constants.

6.2.2 Von Mises' Anisotropic Criterion

In what follows, we shall reduce class of the limit surface from the general tensorially-polynomial representation to the forms independent of both the first $\Pi_{ij}\sigma_{ij}$ and the third $\Pi_{ijklmn}\sigma_{ij}\sigma_{kl}\sigma_{mn}$ common invariants, but preserving the most general representation for the second common invariant, according to von Mises (1913, 1928). In such a case the 4th rank tensor of material anisotropy Π_{ijkl} is, in general, defined by 21 anisotropy modules (but 18 of them independent), since the anisotropy 6×6 matrix $[\boldsymbol{\Pi}]_{ij}$ (6.15) can completely be populated. Further reduction of the number of modules to 15 will be achieved, when the insensitivity of general von Mises quadratic form with respect to the change of hydrostatic stress will be assumed. In such a way the general tensorial von Mises criterion will be reduced to the deviatoric von Mises form defined by 15 anisotropy modules. A choice of 15 anisotropy modules considered as independent is, in general, not unique (cf. Szczepiński, 1993; Ganczarski and Skrzypek, 2013). However, the 15–parameter deviatoric von Mises criterion is sensitive to the change of sign of shear stresses, which may be considered as questionable (cf. e.g. Malinin and Rzysko, 1981). Simplest way to avoid a doubtful physical explanation for existence of terms linear for shear stresses τ_{ij}, a reduction of the 15–parameter von Mises equation to the 9–parameter orthotropic von Mises criterion can be done. This form does not satisfy the deviatoric property, but when the constraints of independence of the hydrostatic stress is consistently applied, it is easily reduced to the deviatoric form, known as orthotropic Hill's criterion, with only 6 independent moduli of orthotropy (cf. Hill, 1948).

Limiting ourselves to plastic yield initiation in ductile materials, a consecutive reduction of the general tensorially-polynomial anisotropic criterion (6.16) to the form dependent only on the 4th rank common invariant $\sigma_{ij}\Pi_{ijkl}\sigma_{kl}$ holds, as it was proposed in the von Mises criterion for anisotropic yield initiation (item #8 in Tab. 6.3) (cf. von Mises, 1913, 1928)

$$\sigma_{ij}\Pi_{ijkl}\sigma_{kl} - 1 = 0 \tag{6.19}$$

When the more convenient Voigt's vector-matrix notation is used, the form equivalent to (6.19) is obtained

$$\{\sigma\}^{\mathrm{T}}\,[\overset{<4>}{\Pi}\,]\{\sigma\} - 1 = 0 \tag{6.20}$$

where only one fourth-rank tensor of plastic anisotropy $\mathbf{\Pi}$ is saved.

Anisotropic von Mises criterion (6.19) or (6.20), being an initial yield criterion of anisotropic material is an extension of the isotropic Huber–von Mises criterion.

The structural 4th rank tensor of plastic anisotropy in equation (6.19) must be symmetric: $\Pi_{ijkl} = \Pi_{klij} = \Pi_{jikl} = \Pi_{ijlk}$, if stress tensor symmetry is assumed. Hence, in case if none other symmetry properties are implied, the von Mises plastic anisotropy tensor is defined by 21 modules. However, due to its invariance of the tensorial transformation rule, number of independent anisotropy modules is reduced to 18. Finally, the general anisotropic von Mises criterion can be furnished as

$$\begin{aligned}
&\Pi_{xxxx}\sigma_x^2 + \Pi_{yyyy}\sigma_y^2 + \Pi_{zzzz}\sigma_z^2 + 2\Pi_{xxyy}\sigma_x\sigma_y + \\
&2\Pi_{yyzz}\sigma_y\sigma_z + 2\Pi_{zzxx}\sigma_z\sigma_x + 4\Pi_{xxyz}\sigma_x\tau_{yz} + 4\Pi_{xxzx}\sigma_x\tau_{zx} + \\
&4\Pi_{xxxy}\sigma_x\tau_{xy} + 4\Pi_{yyyz}\sigma_y\tau_{yz} + 4\Pi_{yyzx}\sigma_y\tau_{zx} + 4\Pi_{yyxy}\sigma_y\tau_{xy} + \\
&4\Pi_{zzyz}\sigma_z\tau_{yz} + 4\Pi_{zzzx}\sigma_z\tau_{zx} + 4\Pi_{zzxy}\sigma_z\tau_{xy} + 8\Pi_{xyyz}\tau_{xy}\tau_{yz} + \\
&8\Pi_{yzzx}\tau_{yz}\tau_{zx} + 8\Pi_{zxxy}\tau_{zx}\tau_{xy} + 4\Pi_{yzyz}\tau_{yz}^2 + 4\Pi_{zxzx}\tau_{zx}^2 + \\
&4\Pi_{xyxy}\tau_{xy}^2 = 1
\end{aligned} \tag{6.21}$$

where Π_{ijkl} denote 21 components of the von Mises plastic anisotropy tensor.

The von Mises 6×6 matrix of plastic anisotropy, being symmetric and fully populated matrix representation of the 4th rank anisotropy tensor Π_{ijkl} shown in (6.19), is furnished as follows

$$
\overset{<4>}{[\,\mathbf{\Pi}\,]} = \left[\begin{array}{ccc|ccc} \Pi_{11} & \Pi_{12} & \Pi_{13} & \Pi_{14} & \Pi_{15} & \Pi_{16} \\ & \Pi_{22} & \Pi_{23} & \Pi_{24} & \Pi_{25} & \Pi_{26} \\ & & \Pi_{33} & \Pi_{34} & \Pi_{35} & \Pi_{36} \\ \hline & & & \Pi_{44} & \Pi_{45} & \Pi_{46} \\ & & & & \Pi_{55} & \Pi_{56} \\ & & & & & \Pi_{66} \end{array}\right] \left[\begin{array}{ccc|ccc} \bullet & \bullet & \bullet & \bullet & \bullet & \bullet \\ & \bullet & \bullet & \bullet & \bullet & \bullet \\ & & \bullet & \bullet & \bullet & \bullet \\ \hline & & & \bullet & \bullet & \bullet \\ & & & & \bullet & \bullet \\ & & & & & \bullet \end{array}\right] \tag{6.22}
$$

if engineering vectorial representation of the stress tensor $\{\sigma\}$ is chosen as

$$
\{\sigma\} = \left\{ \sigma_1\ \sigma_2\ \sigma_3\ \sigma_4\ \sigma_5\ \sigma_6 \right\}^{\mathrm{T}} = \left\{ \sigma_x\ \sigma_y\ \sigma_z\ \tau_{yz}\ \tau_{zx}\ \tau_{xy} \right\}^{\mathrm{T}} \tag{6.23}
$$

When the matrix coordinates Π_{ij} (6.22) are consistently defined by the tensorial coordinates Π_{ijkl}

$$
\begin{array}{lll}
\Pi_{11} = \Pi_{xxxx} & \Pi_{22} = \Pi_{yyyy} & \Pi_{33} = \Pi_{zzzz} \\
\Pi_{12} = \Pi_{xxyy} & \Pi_{13} = \Pi_{xxzz} & \Pi_{23} = \Pi_{yyzz} \\
\Pi_{14} = 2\Pi_{xxyz} & \Pi_{15} = 2\Pi_{xxzx} & \Pi_{16} = 2\Pi_{xxxy} \ \ldots \\
\Pi_{44} = 4\Pi_{yzyz} & \Pi_{55} = 4\Pi_{zxzx} & \Pi_{66} = 4\Pi_{xyxy} \\
\Pi_{45} = 4\Pi_{yzzx} & \Pi_{46} = 4\Pi_{xyyz} & \Pi_{56} = 4\Pi_{zxxy}
\end{array} \tag{6.24}
$$

we arrive at the general anisotropic von Mises equation equivalent to (6.21)

$$
\begin{array}{l}
\Pi_{11}\sigma_x^2 + \Pi_{22}\sigma_y^2 + \Pi_{33}\sigma_z^2 + 2(\Pi_{12}\sigma_x\sigma_y + \Pi_{23}\sigma_y\sigma_z + \Pi_{31}\sigma_z\sigma_x + \\
\Pi_{14}\sigma_x\tau_{yz} + \Pi_{15}\sigma_x\tau_{zx} + \Pi_{16}\sigma_x\tau_{xy} + \Pi_{24}\sigma_y\tau_{yz} + \Pi_{25}\sigma_y\tau_{zx} + \\
\Pi_{26}\sigma_y\tau_{xy} + \Pi_{34}\sigma_z\tau_{yz} + \Pi_{35}\sigma_z\tau_{zx} + \Pi_{36}\sigma_z\tau_{xy} + \Pi_{45}\tau_{yz}\tau_{zx} + \\
\Pi_{46}\tau_{xy}\tau_{yz} + \Pi_{56}\tau_{zx}\tau_{xy}) + \Pi_{44}\tau_{yz}^2 + \Pi_{55}\tau_{zx}^2 + \Pi_{66}\tau_{xy}^2 = 1
\end{array} \tag{6.25}
$$

Representation of the anisotropic von Mises condition (6.20) in deviatoric form is not trivial. The von Mises equation in the vector-matrix notation depends on both the deviatoric $\boldsymbol{s}$ and the hydrostatic part $\sigma_{\mathrm{h}}\mathbf{1}$, when stress decomposition $\boldsymbol{\sigma} = \boldsymbol{s} + \sigma_{\mathrm{h}}\mathbf{1}$ is applied, namely

$$
\{\boldsymbol{s}\}^{\mathrm{T}}\, \overset{<4>}{[\,\mathbf{\Pi}\,]}\, \{\boldsymbol{s}\} + \underline{\left(2\,\{\boldsymbol{s}\}^{\mathrm{T}} + \sigma_{\mathrm{h}}\,\{\mathbf{1}\}^{\mathrm{T}}\right)\left(\overset{<4>}{[\,\mathbf{\Pi}\,]}\,\{\mathbf{1}\}\,\sigma_{\mathrm{h}}\right)} - 1 = 0 \tag{6.26}
$$

The tensorial von Mises equation (6.26) can further be reduced to the deviatoric form independent of the hydrostatic pressure as follows

$$
\{\boldsymbol{s}\}^{\mathrm{T}}\, [{}_{\mathrm{dev}}\mathbf{\Pi}]\, \{\boldsymbol{s}\} - 1 = 0 \tag{6.27}
$$

only if the constraint

$$
\overset{<4>}{[\,\mathbf{\Pi}\,]}\, \{\mathbf{1}\} = 0 \tag{6.28}
$$

is consistently applied. The constraint (6.28) guarantees the deviatoric von Mises equation (6.27) be represented in the reduced 6–dimensional stress space by a cylindrical surface defined by 15 independent anisotropy modules, when 6 constraints are satisfied

$$\begin{array}{ll} \Pi_{11}+\Pi_{12}+\Pi_{13}=0 & \Pi_{12}+\Pi_{22}+\Pi_{23}=0 \\ \Pi_{13}+\Pi_{23}+\Pi_{33}=0 & \Pi_{14}+\Pi_{24}+\Pi_{34}=0 \\ \Pi_{15}+\Pi_{25}+\Pi_{35}=0 & \Pi_{16}+\Pi_{26}+\Pi_{36}=0 \end{array} \tag{6.29}$$

However, the final matrix representation (6.22) with (6.29) employed depends on a choice of independent elements. Two of such representations are of special importance.

In the first case, the elements of matrix (6.22) considered as independent are: $\Pi_{12},\Pi_{13},\Pi_{23};\Pi_{15},\Pi_{16},\Pi_{24},\Pi_{26},\Pi_{34},\Pi_{35}$ and $\Pi_{44},\Pi_{55},\Pi_{66};\Pi_{45},\Pi_{46},\Pi_{56}$, such that the following first representation for the deviatoric von Mises matrix is furnished

$$[{}_{\rm dev}\boldsymbol{\Pi}] = \left[\begin{array}{ccc|ccc} -\Pi_{12}-\Pi_{13} & \Pi_{12} & \Pi_{13} & -\Pi_{24}-\Pi_{34} & \Pi_{15} & \Pi_{16} \\ & -\Pi_{12}-\Pi_{23} & \Pi_{23} & \Pi_{24} & -\Pi_{15}-\Pi_{35} & \Pi_{26} \\ & & -\Pi_{13}-\Pi_{23} & \Pi_{34} & \Pi_{35} & -\Pi_{16}-\Pi_{26} \\ \hline & & & \Pi_{44} & \Pi_{45} & \Pi_{46} \\ & & & & \Pi_{55} & \Pi_{56} \\ & & & & & \Pi_{66} \end{array}\right] \left[\begin{array}{ccc|ccc} \circ & \bullet & \bullet & \circ & \bullet & \bullet \\ & \circ & \bullet & \bullet & \circ & \bullet \\ & & \circ & \bullet & \bullet & \circ \\ \hline & & & \bullet & \bullet & \bullet \\ & & & & \bullet & \bullet \\ & & & & & \bullet \end{array}\right] \tag{6.30}$$

if constraints (6.29) are applied as follows

$$\begin{array}{ll} \Pi_{11}=-\Pi_{12}-\Pi_{13}, & \Pi_{14}=-\Pi_{24}-\Pi_{34} \\ \Pi_{22}=-\Pi_{12}-\Pi_{23}, & \Pi_{25}=-\Pi_{15}-\Pi_{35} \\ \Pi_{33}=-\Pi_{13}-\Pi_{23}, & \Pi_{36}=-\Pi_{16}-\Pi_{26} \end{array} \tag{6.31}$$

In the second case, the elements of matrix (6.22) chosen as independent are: $\Pi_{11},\Pi_{22},\Pi_{33};\Pi_{15},\Pi_{16},\Pi_{24},\Pi_{26},\Pi_{34},\Pi_{35}$ and $\Pi_{44},\Pi_{55},\Pi_{66};\Pi_{45},\Pi_{46},\Pi_{56}$, hence we arrive at the second representation of the deviatoric von Mises matrix as follows

$$
[{}_{\mathrm{dev}}\boldsymbol{\Pi}] = \left[\begin{array}{ccc|ccc}
\Pi_{11} & \frac{1}{2}(\Pi_{33} - \Pi_{11} - \Pi_{22}) & \frac{1}{2}(\Pi_{22} - \Pi_{11} - \Pi_{33}) & -\Pi_{24} - \Pi_{34} & \Pi_{15} & \Pi_{16} \\
 & \Pi_{22} & \frac{1}{2}(\Pi_{11} - \Pi_{22} - \Pi_{33}) & \Pi_{24} & -\Pi_{15} - \Pi_{35} & \Pi_{26} \\
 & & \Pi_{33} & \Pi_{34} & \Pi_{35} & -\Pi_{16} - \Pi_{26} \\
\hline
 & & & \Pi_{44} & \Pi_{45} & \Pi_{46} \\
 & & & & \Pi_{55} & \Pi_{56} \\
 & & & & & \Pi_{66}
\end{array}\right]
\left[\begin{array}{ccc|ccc}
\bullet & \circ & \circ & \circ & \bullet & \bullet \\
 & \bullet & \circ & \bullet & \circ & \bullet \\
 & & \bullet & \bullet & \bullet & \circ \\
\hline
 & & & \bullet & \bullet & \bullet \\
 & & & & \bullet & \bullet \\
 & & & & & \bullet
\end{array}\right] \tag{6.32}
$$

if, instead of (6.31), other substitution is used

$$
\begin{aligned}
\Pi_{12} &= \frac{1}{2}(\Pi_{33} - \Pi_{11} - \Pi_{22}) & \Pi_{14} &= -\Pi_{24} - \Pi_{34} \\
\Pi_{13} &= \frac{1}{2}(\Pi_{22} - \Pi_{11} - \Pi_{33}) & \Pi_{25} &= -\Pi_{15} - \Pi_{35} \\
\Pi_{23} &= \frac{1}{2}(\Pi_{11} - \Pi_{22} - \Pi_{33}) & \Pi_{36} &= -\Pi_{16} - \Pi_{26}
\end{aligned} \tag{6.33}
$$

A choice of 15 elements in the von Mises matrix (6.22) considered as independent is not a unique procedure and can result in the different deviatoric von Mises equation forms. In particular, when a more convenient representation (6.30) is substituted for $[{}_{\mathrm{dev}}\boldsymbol{\Pi}]$ in (6.27) we arrive at the following von Mises equation expressed in the deviatoric stress space

$$
\begin{aligned}
&-\Pi_{12}\left(s_x - s_y\right)^2 - \Pi_{13}\left(s_x - s_z\right)^2 - \Pi_{23}\left(s_y - s_z\right)^2 + \\
&\quad \underline{2\left\{\tau_{yz}\left[\Pi_{24}\left(s_y - s_x\right) + \Pi_{34}\left(s_z - s_x\right)\right] + \right.} \\
&\quad \underline{\tau_{zx}\left[\Pi_{15}\left(s_x - s_y\right) + \Pi_{35}\left(s_z - s_y\right)\right] +} \\
&\quad \underline{\tau_{xy}\left[\Pi_{16}\left(s_x - s_z\right) + \Pi_{26}\left(s_y - s_z\right)\right] +} \\
&\quad \underline{\left. \Pi_{45}\tau_{yz}\tau_{zx} + \Pi_{46}\tau_{xy}\tau_{yz} + \Pi_{56}\tau_{zx}\tau_{xy}\right\} +} \\
&\quad \Pi_{44}\tau_{yz}^2 + \Pi_{55}\tau_{zx}^2 + \Pi_{66}\tau_{xy}^2 = 1
\end{aligned} \tag{6.34}
$$

It is visible that above equation owns the clear deviatoric structure hence, when the tensorial stress space is used instead of the deviatoric one, the analogous equivalent to (6.34) representation of the deviatoric von Mises equation is also true in terms of stress components (cf. Szczepiński, 1993)

$$
\begin{aligned}
&-\Pi_{12}\left(\sigma_x-\sigma_y\right)^2-\Pi_{13}\left(\sigma_x-\sigma_z\right)^2-\Pi_{23}\left(\sigma_y-\sigma_z\right)^2+\\
&\underline{2\left\{\tau_{yz}\left[\Pi_{24}\left(\sigma_y-\sigma_x\right)+\Pi_{34}\left(\sigma_z-\sigma_x\right)\right]+\right.}\\
&\underline{\tau_{zx}\left[\Pi_{15}\left(\sigma_x-\sigma_y\right)+\Pi_{35}\left(\sigma_z-\sigma_y\right)\right]+}\\
&\underline{\tau_{xy}\left[\Pi_{16}\left(\sigma_x-\sigma_z\right)+\Pi_{26}\left(\sigma_y-\sigma_z\right)\right]+}\\
&\underline{\left.\Pi_{45}\tau_{yz}\tau_{zx}+\Pi_{46}\tau_{xy}\tau_{yz}+\Pi_{56}\tau_{zx}\tau_{xy}\right\}+}\\
&\Pi_{44}\tau_{yz}^2+\Pi_{55}\tau_{zx}^2+\Pi_{66}\tau_{xy}^2=1
\end{aligned}
\tag{6.35}
$$

Note, that equations (6.34) or (6.35) are defined by 15 elements Π_{ij}. However, the underlined terms are sensitive to change of sign of shear stresses, e.g. $\tau_{yz}(\sigma_y-\sigma_x)$ etc., which is physically questionable and, finally, such terms are consequently omitted is some cases (cf. e.g. Malinin and Rzysko, 1981). Nevertheless, the full representation (6.35), might occur useful when the von Mises–Tsai–Wu extension to the brittle-like material is sought for (cf. Tsai and Wu, 1971).

6.2.3 Von Mises' Orthotropic Criterion and Hill's Deviatoric Criterion

General form of the 21–parameter anisotropic von Mises criterion (6.25) involves none material symmetry property. In a particular case if plastic orthotropy is assumed for the initial yield criterion (6.20), when represented in principal orthotropy axes, the 9–parameter orthotropic von Mises matrix (6.26) takes the form

$$
[{}_{\rm ort}\mathbf{\Pi}]=\left[\begin{array}{ccc|ccc}
\Pi_{11} & \Pi_{12} & \Pi_{13} & 0 & 0 & 0\\
 & \Pi_{22} & \Pi_{23} & 0 & 0 & 0\\
 & & \Pi_{33} & 0 & 0 & 0\\
\hline
 & & & \Pi_{44} & 0 & 0\\
 & & & & \Pi_{55} & 0\\
 & & & & & \Pi_{66}
\end{array}\right]
\left[\begin{array}{ccc|ccc}
\bullet & \bullet & \bullet & & & \\
 & \bullet & \bullet & & & \\
 & & \bullet & & & \\
\hline
 & & & \bullet & & \\
 & & & & \bullet & \\
 & & & & & \bullet
\end{array}\right]
\tag{6.36}
$$

In such a case the general anisotropic von Mises equation (6.25) is reduced to the narrower 9–parameter orthotropic von Mises criterion

$$
\begin{aligned}
\Pi_{11}\sigma_x^2+\Pi_{22}\sigma_y^2+\Pi_{33}\sigma_z^2+2(\Pi_{12}\sigma_x\sigma_y+\Pi_{23}\sigma_y\sigma_z+\Pi_{31}\sigma_z\sigma_x)+&\\
\Pi_{44}\tau_{yz}^2+\Pi_{55}\tau_{zx}^2+\Pi_{66}\tau_{xy}^2=1&
\end{aligned}
\tag{6.37}
$$

When the Voigt notation is used, the 9–parameter orthotropic von Mises criterion takes the form

$$
\{\sigma\}^{\rm T}\,[{}_{\rm ort}\mathbf{\Pi}]\,\{\sigma\}-1=0 \tag{6.38}
$$

that involves definition (6.36). Note that equation (6.38) belongs to the class of hydrostatic pressure sensitive criteria (cf. item #4 in Tab. 6.3 Khan and Liu, 2012; Khan et al, 2012).

In order to achieve pressure insensitive orthotropic criterion we apply a procedure described in Sect. 6.2.2. If we decompose again the stress tensor into deviatoric and volumetric parts $\boldsymbol{\sigma} = \boldsymbol{s} + \sigma_{\rm h}\mathbf{1}$ in the orthotropic von Mises equation (6.38) we arrive at the equation analogous to (6.26)

$$\{\boldsymbol{s}\}^{\rm T}\,[_{\rm ort}\boldsymbol{\Pi}]\,\{\boldsymbol{s}\} + \underline{\left(2\,\{\boldsymbol{s}\}^{\rm T} + \sigma_{\rm h}\,\{\mathbf{1}\}^{\rm T}\right)\left([_{\rm ort}\boldsymbol{\Pi}]\,\{\mathbf{1}\}\,\sigma_{\rm h}\right)} - 1 = 0 \tag{6.39}$$

Assuming further hydrostatic pressure insensitive form the following holds

$$[_{\rm ort}\boldsymbol{\Pi}]\,\{\mathbf{1}\} = 0 \tag{6.40}$$

which leads to three constraints instead of six in general case of von Mises anisotropic equation (6.29)

$$\begin{array}{c} \Pi_{11} + \Pi_{12} + \Pi_{13} = 0 \\ \Pi_{12} + \Pi_{22} + \Pi_{23} = 0 \\ \Pi_{13} + \Pi_{23} + \Pi_{33} = 0 \end{array} \tag{6.41}$$

In this way the orthotropic von Mises criterion (6.38) reduces to the pressure insensitive criterion called Hill's criterion Hill (1948, 1950) that contains 6 independent modules

$$\{\boldsymbol{s}\}^{\rm T}\,[\boldsymbol{\Pi}^{\rm H}]\,\{\boldsymbol{s}\} - 1 - 0 \tag{6.42}$$

Hill's matrix $[\boldsymbol{\Pi}^{\rm H}]$ appearing in equation (6.42) contains 6 independent modules. A choice of the three independent modules form six involved in equations (6.41) is not unique. In what follows two of them are discussed (see two aforementioned forms (6.30) and (6.32)).

In this way we arrive at the following Hill's matrices

$$[\boldsymbol{\Pi}^{\rm H}] = \left[\begin{array}{ccc|ccc} -\Pi_{12}-\Pi_{13} & \Pi_{12} & \Pi_{13} & & & \\ & -\Pi_{12}-\Pi_{23} & \Pi_{23} & & & \\ & & -\Pi_{13}-\Pi_{23} & & & \\ \hline & & & \Pi_{44} & & \\ & & & & \Pi_{55} & \\ & & & & & \Pi_{66} \end{array}\right] \left[\begin{array}{ccc|ccc} \circ & \bullet & \bullet & & & \\ & \circ & \bullet & & & \\ & & \circ & & & \\ \hline & & & \bullet & & \\ & & & & \bullet & \\ & & & & & \bullet \end{array}\right] \tag{6.43}$$

or

$$[\Pi^{\rm H}] = \left[\begin{array}{ccc|ccc} \Pi_{11} & \frac{\Pi_{33}-\Pi_{11}-\Pi_{22}}{2} & \frac{\Pi_{22}-\Pi_{11}-\Pi_{33}}{2} & & & \\ & \Pi_{22} & \frac{\Pi_{11}-\Pi_{22}-\Pi_{33}}{2} & & & \\ & & \Pi_{33} & & & \\ \hline & & & \Pi_{44} & & \\ & & & & \Pi_{55} & \\ & & & & & \Pi_{66} \end{array}\right] \left[\begin{array}{ccc|ccc} \bullet & \circ & \circ & & & \\ & \bullet & \circ & & & \\ & & \bullet & & & \\ \hline & & & \bullet & & \\ & & & & \bullet & \\ & & & & & \bullet \end{array}\right] \tag{6.44}$$

When the engineering notation is used, corresponding representations of the Hill's criterion are

$$\begin{gathered} -\left[\Pi_{23}\left(\sigma_y-\sigma_z\right)^2+\Pi_{13}\left(\sigma_z-\sigma_x\right)^2+\Pi_{12}\left(\sigma_x-\sigma_y\right)^2\right]+ \\ \Pi_{44}\tau_{yz}^2+\Pi_{55}\tau_{zx}^2+\Pi_{66}\tau_{xy}^2=1 \end{gathered} \tag{6.45}$$

or

$$\begin{gathered} \Pi_{11}\sigma_x^2+\Pi_{22}\sigma_y^2+\Pi_{33}\sigma_z^2+\left(\Pi_{33}-\Pi_{11}-\Pi_{22}\right)\sigma_x\sigma_y+ \\ \left(\Pi_{22}-\Pi_{11}-\Pi_{33}\right)\sigma_x\sigma_z+\left(\Pi_{11}-\Pi_{22}-\Pi_{33}\right)\sigma_y\sigma_z+ \\ \Pi_{44}\tau_{yz}^2+\Pi_{55}\tau_{zx}^2+\Pi_{66}\tau_{xy}^2=1 \end{gathered} \tag{6.46}$$

Both representations (6.45) or (6.46) describe the same Hill's limit surface, but applying two different choices of six independent elements of the Hill matrices (6.43) or (6.44). In order to calibrate Hill's criterion in the form (6.45) or (6.46) three tests of uniaxial tension $\sigma_x = k_x, \sigma_y = k_y, \sigma_z = k_z$ and three tests of pure shear $\tau_{xy} = k_{xy}, \tau_{yz} = k_{yz}, \tau_{zx} = k_{zx}$, in directions and planes of material orthotropy, must be performed. These tests allow to express 6 modules of material orthotropy in equations (6.45) and (6.46) in terms of 3 independent plastic tension limits k_x, k_y, k_z (in directions of orthotropy), and 3 independent plastic shear limits k_{yz}, k_{zx}, k_{xy} (in planes of material orthotropy). Hence,

$$\begin{aligned} -\Pi_{23} &= \frac{1}{2}\left(\frac{1}{k_y^2}+\frac{1}{k_z^2}-\frac{1}{k_x^2}\right) & \Pi_{44} &= \frac{1}{k_{yz}^2} \\ -\Pi_{13} &= \frac{1}{2}\left(\frac{1}{k_z^2}+\frac{1}{k_x^2}-\frac{1}{k_y^2}\right) & \Pi_{55} &= \frac{1}{k_{zx}^2} \\ -\Pi_{12} &= \frac{1}{2}\left(\frac{1}{k_x^2}+\frac{1}{k_y^2}-\frac{1}{k_z^2}\right) & \Pi_{66} &= \frac{1}{k_{xy}^2} \end{aligned} \tag{6.47}$$

such that orthotropic Hill's criteria equivalent to (6.45) or (6.46) can be furnished in terms of plastic anisotropy limits as follows

$$\frac{1}{2}\left(\frac{1}{k_y^2}+\frac{1}{k_z^2}-\frac{1}{k_x^2}\right)(\sigma_y-\sigma_z)^2+\frac{1}{2}\left(\frac{1}{k_z^2}+\frac{1}{k_x^2}-\frac{1}{k_y^2}\right)(\sigma_z-\sigma_x)^2+$$
$$\frac{1}{2}\left(\frac{1}{k_x^2}+\frac{1}{k_y^2}-\frac{1}{k_z^2}\right)(\sigma_x-\sigma_y)^2+\left(\frac{\tau_{yz}}{k_{yz}}\right)^2+\left(\frac{\tau_{zx}}{k_{zx}}\right)^2+\left(\frac{\tau_{xy}}{k_{xy}}\right)^2=1 \tag{6.48}$$

or

$$\left(\frac{\sigma_x}{k_x}\right)^2+\left(\frac{\sigma_y}{k_y}\right)^2+\left(\frac{\sigma_z}{k_z}\right)^2-\left(\frac{1}{k_x^2}+\frac{1}{k_y^2}-\frac{1}{k_z^2}\right)\sigma_x\sigma_y-$$
$$\left(\frac{1}{k_y^2}+\frac{1}{k_z^2}-\frac{1}{k_x^2}\right)\sigma_y\sigma_z-\left(\frac{1}{k_z^2}+\frac{1}{k_x^2}-\frac{1}{k_y^2}\right)\sigma_z\sigma_x+ \tag{6.49}$$
$$\left(\frac{\tau_{yz}}{k_{yz}}\right)^2+\left(\frac{\tau_{zx}}{k_{zx}}\right)^2+\left(\frac{\tau_{xy}}{k_{xy}}\right)^2=1$$

Note that under a particular plane stress condition, e.g. in the x, y plane, when $\sigma_z = \tau_{zx} = \tau_{yz} = 0$, both formulas (6.48) and (6.49) reduce to the 4–parameter orthotropic Hill's condition

$$\frac{\sigma_x^2}{k_x^2}+\frac{\sigma_y^2}{k_y^2}-\left(\frac{1}{k_x^2}+\frac{1}{k_y^2}-\frac{1}{k_z^2}\right)\sigma_x\sigma_y+\frac{\tau_{xy}^2}{k_{xy}^2}=1 \tag{6.50}$$

where initiation of plastic flow in the x, y plane is controlled not only by the in-plane limits k_x, k_y and k_{xy}, but also by the out-of-plane limit k_z, which may finally lead to inadmissible loss of convexity by the yield surface.

The Hill criterion (6.45) is formulated in the space of principal material directions of orthotropy which in general do not coincide with directions of principal stresses. In the particular case when the coaxiality holds $\sigma_x = \sigma_1$, $\sigma_y = \sigma_2$, $\sigma_z = \sigma_3$, $\tau_{xy} = \tau_{yz} = \tau_{zx} = 0$ we arrive at simplified

$$-\Pi_{23}(\sigma_2-\sigma_3)^2-\Pi_{13}(\sigma_3-\sigma_1)^2-\Pi_{12}(\sigma_1-\sigma_2)^2=1 \tag{6.51}$$

or when calibration (6.47) is used the explicit form of (6.51) is finally furnished

$$\frac{1}{2}\left(\frac{1}{k_2^2}+\frac{1}{k_3^2}-\frac{1}{k_1^2}\right)(\sigma_2-\sigma_3)^2+\frac{1}{2}\left(\frac{1}{k_3^2}+\frac{1}{k_1^2}-\frac{1}{k_2^2}\right)(\sigma_3-\sigma_1)^2+$$
$$\frac{1}{2}\left(\frac{1}{k_1^2}+\frac{1}{k_2^2}-\frac{1}{k_3^2}\right)(\sigma_1-\sigma_2)^2=1 \tag{6.52}$$

Hill's condition (6.52) represents cylindrical elliptic surface the axis of which coincides with the hydrostatic axis. Nevertheless in some cases the limit surface looses closed form for high othotropy degree which may occur when one of following expressions

$$\frac{1}{k_2^2}+\frac{1}{k_3^2}-\frac{1}{k_1^2}$$

elsewhere

$$\frac{1}{k_3^2}+\frac{1}{k_1^2}-\frac{1}{k_2^2}$$

or

$$\frac{1}{k_1^2}+\frac{1}{k_2^2}-\frac{1}{k_3^2}$$

changes the sign.

6.2.4 Barlat–Khan's Implicit Formulations

In this subsection another approach (implicit formulation) is discussed based on a series of papers developed by Barlat, Planckett, Cazacu and Khan to mention some names only. The implicit formulation involves the linear transformation of the Cauchy stress tensor $\boldsymbol{\sigma}$ to the transformed stress $\boldsymbol{\Sigma} = \boldsymbol{L} : \boldsymbol{\sigma}$ by use of transformation tensor $\boldsymbol{L}$ responsible for orthotropy. Such linear transformation concept of the stress tensor was first introduced by Sobotka (1969) and Boehler and Sawczuk (1970)

$$\widehat{\sigma}_{ij} = A_{ijkl}\sigma_{kl} \tag{6.53}$$

where A_{ijkl} stands for a certain dimensionless tensor of anisotropy that satisfies general symmetry conditions $A_{ijkl} = A_{jikl} = A_{ijlk} = A_{klij}$ and the well known isotropic yield conditions to hold for anisotropic materials as well if σ_{ij} are replaced by $\widehat{\sigma}_{ij}$. This approach is not directly based on the theory of common invariants in sense of Sayir, Goldenblat, Kopnov, Spencer, Boehler, Betten, etc. formalism (explicit formulation). According to this implicit approach an extension of isotropic initial yield/failure criteria is performed to account for the tension/compression asymmetry property and to material anisotropy frame (usually orthotropy) by applying the linear transformation to the stress tensor and inserting this transformed stress tensor into the originally isotropic yield/failure criteria.

In Cazacu et al (2006) the authors consider both the isotropic yield criterion for description of asymmetric yielding

$$\begin{gathered} f(J_{2s}, J_{3s}) = (|s_1| - \widehat{k}s_1)^a + (|s_2| - \widehat{k}s_2)^a + (|s_3| - \widehat{k}s_3)^a = 2k^a \\ \widehat{k} = \frac{1-h(\frac{k_t}{k_c})}{1+h(\frac{k_t}{k_c})} \qquad h\left(\frac{k_t}{k_c}\right) = \left[\frac{2^a - 2(\frac{k_t}{k_c})^a}{(2\frac{k_t}{k_c})^a - 2}\right]^{1/a} \end{gathered} \tag{6.54}$$

where s_i, $i = 1,\ldots,3$ are the principal values of the stress deviator and f gives the size of the yield locus (isotropic hardening), as well as its extension to include orthotropy by the use of linear transformation of the Cauchy stress deviator $\boldsymbol{\Sigma} = \boldsymbol{C} : \boldsymbol{s}$ through

$$\boldsymbol{C}=\left[\begin{array}{ccc|ccc} C_{11} & C_{12} & C_{13} & & & \\ C_{12} & C_{22} & C_{23} & & & \\ C_{13} & C_{23} & C_{33} & & & \\ \hline & & & C_{44} & & \\ & & & & C_{55} & \\ & & & & & C_{66} \end{array}\right] \tag{6.55}$$

which lead to following anisotropic equation

$$(|\Sigma_1| - \widehat{k}\Sigma_1)^a + (|\Sigma_2| - \widehat{k}\Sigma_2)^a + (|\Sigma_3| - \widehat{k}\Sigma_3)^a = 2k^a \tag{6.56}$$

Authors proved convexity of the isotropic yield form (6.54) as well as pressure insensitivity of its orthotropic form (6.56) obtained through the linear transformation to the transformed stress frame. However the question of convexity of the orthotropic form (6.56) remains open.

The proposed yield function appears to be suitable for description of the strong asymmetry and anisotropy observed in textured Mg-Th and Mg-Li binary alloy sheets and for titanium 4Al-1/4O_2, see Cazacu et al (2006). The orthotropic yield criterion proposed by Cazacu et al (2006) was also investigated in a series of multiaxial loading experiments on Ti-6Al-4V titanium alloy by Khan et al (2007).

Extension of Drucker's isotropic yield criterion to anisotropy by use of common invariants J_2^0 and J_3^0 is due to Cazacu and Barlat (2004), and investigated by Yoshida et al (2013), also discussed in details by Kolupaev (2018)

$$(J_2^0)^{3/2} - cJ_3^0 - k^3 = 0 \tag{6.57}$$

The constant c in the equation (6.57) accounts for the tension/compression asymmetry defined as

$$c = \frac{3\sqrt{3}(k_\mathrm{t}^3 - k_\mathrm{c}^3)}{2(k_\mathrm{t}^3 + k_\mathrm{c}^3)} \tag{6.58}$$

and belongs to two ranges

$$c \in \begin{cases} \left(0, \dfrac{3\sqrt{3}}{2}\right) & \text{for } k_\mathrm{t} > k_\mathrm{c} > 0 \\ \left(-\dfrac{3\sqrt{3}}{2}, 0\right) & \text{for } 0 < k_\mathrm{t} < k_\mathrm{c} \end{cases} \tag{6.59}$$

The second and third common invariants of orthotropy are defined as

$$
\begin{aligned}
J_2^0 &= \frac{1}{6}\left[a_1(\sigma_x-\sigma_y)^2 + a_2(\sigma_y-\sigma_z)^2 + a_3(\sigma_z-\sigma_x)^2\right] \\
&+ a_4\tau_{xy}^2 + a_5\tau_{xz}^2 + a_6\tau_{zy}^2 \\
J_3^0 &= \frac{1}{27}\left\{(b_1+b_2)\sigma_x^3 + (b_3+b_4)\sigma_y^3 + [2(b_1+b_4)-b_2-b_3]\sigma_z^3\right\} \\
&+ 2b_{11}\tau_{xy}\tau_{yz}\tau_{zx} + \frac{1}{9}\left\{2(b_1+b_2)\sigma_x\sigma_y\sigma_z - (b_1\sigma_y+b_2\sigma_z)\sigma_x^2\right. \\
&\left. -(b_3\sigma_z+b_2\sigma_x)\sigma_y^2 - [(b_1-b_2+b_4)\sigma_x + (b_1+b_3+b_4)\sigma_y]\sigma_z^2\right\} \\
&- \frac{1}{3}\left\{\tau_{yz}^2[(b_6+b_7)\sigma_x - b_6\sigma_y - b_7\sigma_z]\right. \\
&- \tau_{zx}^2[2b_9\sigma_y - b_8\sigma_z - (2b_9-b_8)\sigma_x] \\
&\left. - \tau_{xy}^2[2b_{10}\sigma_z - b_5\sigma_y - (2b_{10}-b_5)\sigma_x]\right\}
\end{aligned}
\tag{6.60}
$$

The discussed anisotropic criterion was successfully verified for textured magnesium Mg-Th and Mg-Li alloy sheets. Authors proved convexity of the enhanced isotropic yield criterion only for $c(k_t/k_c)$ belonging to the range

$$
[-\frac{3\sqrt{3}}{2}, \frac{3\sqrt{3}}{2}]
$$

In case of the anisotropic form of Cazacu and Barlat's criterion (6.57) the general proof of convexity for the wide class of highly tension/compression asymmetric and anisotropic materials may not be possible.

More complete representation of J_2^0 and J_3^0 common invariants as well as the extended model (6.57) verification for high-purity α-titanium is done by Nixon et al (2010).

Korkolis and Kyriakides (2008) applied anisotropic extension of Hosford's isotropic criterion in terms of principal stress deviator s_1, s_2 in case of plane stress state

$$
|s_1 - s_2|^n + |2s_1 + s_2|^n + |s_1 + 2s_2|^n = 2k^n \tag{6.61}
$$

Following Barlat et al (2003) they introduced anisotropy by use of a concept of two linear transformations $\boldsymbol{S}' = \boldsymbol{L}' : \boldsymbol{s}$ and $\boldsymbol{S}'' = \boldsymbol{L}'' : \boldsymbol{s}$ where $\boldsymbol{L}'$ and $\boldsymbol{L}''$ are transformation tensors introducing anisotropy

$$
|S_1' - S_2'|^n + |2S_1'' + S_2''|^n + |S_1'' + 2S_2''|^n = 2k^n \tag{6.62}
$$

Experimental validation of (6.62) is due to Korkolis and Kyriakides (2008) applied to Al-6260-T4 as well as due to Dunand et al (2012); Luo et al (2012) applied to AA6260-T6 alloys under classical tensile and butterfly shear tests.

Comparison of two different approaches: explicit formulation based of common invariants and implicit formulation composed as extension of isotropic criteria to anisotropy and tension/compression asymmetry leads to the following characteristic features.

The implicit formulation is very advantageous and fruitful in order to build numerical models able to capture experimental evidence for broad class of innovative metallic materials (mainly metal based alloys) that simultaneously exhibit tension/compression asymmetry, anisotropy and hydrostatic pressure insensitivity. Apart from these advantages some open questions may be highlighted. Among them there might be mentioned not obvious physical interpretation for the extended criteria based on known isotropic forms enhanced through strength differential sensitivity and orthotropic linear transformation of stress. The general proof of convexity is rather cumbersome and not attached in a complete and convinced form. Although the isotropic equations have understanding physical interpretations and satisfy convexity requirements the transposition of these equations to the transformed stress frame may lead to the loss of convexity.

By contrast use of the explicit approach based on well established theory of common invariants is more rigorous and so leads to more clear physical interpretation (energy) and convexity of quadratic or poly-quadratic forms. However this consistent approach leads to major difficulties when numerical implementation and experimental validation are considered. Additional difficulties arise when implementing the explicit approach to more general cases if the material orthotropy frame does not coincide with the principal stress frame. Such more general problem was discussed by Ganczarski and Lenczowski (1997) in case of Hill's and orthotropic Hosford's criteria. In such a case it is necessary to transform tensor of structural orthotropy to the frame of principal stress resulting in a possible loss of convexity and even degeneration of an initially closed surface to two-fold surface (non closed).

6.2.5 Brief Survey of Anisotropic Yield Criteria

In this section a brief survey of the selected commonly used pressure sensitive and insensitive initial yield criteria is presented. Chosen anisotropic yield criteria are collected in Table 6.3. In the item #5 two examples of implementation of implicit anisotropic extension of the isotropic Drucker yield criterion (dependent on the second and the third deviatoric stress invariants) referring to works by Cazacu and Barlat (2004) and Nixon et al (2010) are presented. By contrast to original notation used by authors the criterion is rewritten in a frame of transformed stress $\boldsymbol{\Sigma} = \boldsymbol{L} : \boldsymbol{\sigma}$ instead of the Cauchy stress frame $\boldsymbol{\sigma}$. Due to this concept the second J_2^0 and the third J_3^0 transformed invariants are expressed in term of only one fourth-rank transformation tensor $\boldsymbol{L}$ instead of the second-rank

$$\boldsymbol{s} : {}_{\text{dev}}\overset{<4>}{\boldsymbol{\Pi}} : \boldsymbol{s}$$

and the third-rank common invariants

Table 6.3 Survey of pressure sensitive and insensitive anisotropic yield criteria

#	Author(s)	Limit criterion
1	Goldenblat and Kopnov (1966) and Sayir (1970)	$(\Pi_{ij}\sigma_{ij})^{\alpha} + (\Pi_{ijkl}\sigma_{ij}\sigma_{kl})^{\beta} + (\Pi_{ijklmn}\sigma_{ij}\sigma_{kl}\sigma_{mn})^{\gamma} + \ldots = 1$
2	Kowalsky et al (1999)	$h^{(0)} + h^{(1)}_{ij} s_{ij} + s_{ij} h^{(2)}_{ijkl} s_{kl} + s_{ij} s_{kl} h^{(3)}_{ijklmn} s_{mn} = 0$
	Altenbach et al (1995, 2014)	$\alpha a_{ij}\sigma_{ij} + \beta \left(b_{ijkl}\sigma_{ij}\sigma_{kl}\right)^{1/2} + \gamma \left(c_{ijkmnl}\sigma_{ij}\sigma_{kl}\sigma_{mn}\right)^{1/3} - 1 = 0$
3	von Mises (1913, 1928)	$\Pi_{ijkl}\sigma_{ij}\sigma_{kl} = 1$
4	Khan et al (2012)	${}_{\mathrm{ort}}\Pi_{ijkl}\sigma_{ij}\sigma_{kl} = 1$
5	Cazacu and Barlat (2004) and Nixon et al (2010)	$\left\{\frac{1}{2}\mathrm{tr}\left[(\boldsymbol{L}:\boldsymbol{\sigma})\cdot(\boldsymbol{L}:\boldsymbol{\sigma})\right]\right\}^{3/2} - c\frac{1}{3}\mathrm{tr}\left[(\boldsymbol{L}:\boldsymbol{\sigma})\cdot(\boldsymbol{L}:\boldsymbol{\sigma})\cdot(\boldsymbol{L}:\boldsymbol{\sigma})\right] = k^3$
6	Szczepiński (1993)	$\boldsymbol{s} : {}_{\mathrm{dev}}\overset{<4>}{\boldsymbol{\Pi}} : \boldsymbol{s} = 1$
7	Hill (1948, 1950)	$\boldsymbol{s} : \overset{<4>}{\boldsymbol{\Pi}^{\mathrm{H}}} : \boldsymbol{s} = 1$
8	Cazacu et al (2006) and Khan et al (2007)	$(\lvert\Sigma_1\rvert - \widehat{k}\Sigma_1)^a + (\lvert\Sigma_2\rvert - \widehat{k}\Sigma_2)^a + (\lvert\Sigma_3\rvert - \widehat{k}\Sigma_3)^a = 2k^a$
9	Ganczarski and Lenczowski (1997)	$a_1\lvert\sigma_y - \sigma_z\rvert^m + a_2\lvert\sigma_z - \sigma_x\rvert^m + a_3\lvert\sigma_x - \sigma_y\rvert^m + a_4\lvert\tau_{yz}\rvert^m + a_5\lvert\tau_{zx}\rvert^m + a_6\lvert\tau_{xy}\rvert^m = 1$
10	Korkolis and Kyriakides (2008)	$\lvert S'_1 - S'_2\rvert^n + \lvert 2S''_1 + S''_2\rvert^n + \lvert S''_1 + 2S''_2\rvert^n = 2k^n$

$$\boldsymbol{s} : {}_{\mathrm{dev}}\overset{<6>}{\boldsymbol{\Pi}} : \boldsymbol{s} : \boldsymbol{s}$$

necessary to be implemented when the Goldenblat–Kopnov explicit formulation would be used. The discussed implicit formulation shows essential reduction of the number of material constants that have to be identified in order to capture experimental data. Note that the transformation tensor $\boldsymbol{L}$ exhibits format of the Hill orthotropy matrix however it is dimensionless. When compare items #6 and #7 corresponding to the deviatoric von Mises criterion (6.35) written in the form suggested by Szczepiński (1993) and to the Hill criterion (6.45) (Hill, 1948, 1950) different population of corresponding plastic matrices is applied. In case of Hill's format the terms which are sensitive to change of sign of shear stresses, for instance $\tau_{yz}(s_y - s_z), \ldots, \tau_{yz}\tau_{zx}, \ldots$ are omitted. It is equivalent to the reduction of a number of independent plastic modules from 15 to 6.

To describe both the asymmetry between tension and compression and the anisotropy observed in hexagonal closed packed metal sheets Cazacu et al (2006) and Khan et al (2007) proposed extension of isotropic criterion (6.54) to the case of orthotropy represented by item #8. It consists in application of fourth order linear transformation operator on the Cauchy stress tensor expressed by its principal values. The proposed anisotropic criterion was successively applied to the description of the anisotropy and asymmetry of the yield loci of textured polycrystalline magnesium and binary Mg–Th, Mg–Li alloys and α titanium.

Orthotropic generalization of the Hosford criterion in which principal axes of material orthotropy do not coincide with principal stress axes was proposed by Ganczarski and Lenczowski (1997) in the form of item #9. Next the convexity check of the yield condition was performed in case of the brass sheet of Russian commercial symbol Ł22, that is material of strong orthotropy slightly different than transverse isotropy. The last criterion item #10 is another anisotropic generalization of Hosford's isotropic criterion done by Korkolis and Kyriakides (2008) and addressed to Al-6260-T4 tubes inflated under combined internal pressure and axial load.

References

Altenbach H, Kolupaev V (2015) Classical and non-classical failure criteria. In: Altenbach H, Sadowski T (eds) Failure and Damage Analysis of Advanced Materials, Springer, Wien, Heidelberg, New York, Dordrecht, London, CISM International Centre for Mechanical Sciences Courses and Lectures, vol 560, pp 1–66

Altenbach H, Altenbach J, Zolochevsky A (1995) A generalized constitutive equation for creep of polymers at multiaxial loading. Mechanics of Composite Materials 31(6):511–518

Altenbach H, Bolchoun A, Kolupaev VA (2014) Phenomenological yield and failure criteria. In: Altenbach H, Öchsner A (eds) Plasticity of Pressure-Sensitive Materials, Springer, Heidelberg, New York, Dordrecht, London, Engineering Materials, pp 49–152

Barlat F, Brem JC, Yoon JW, Chung K, Dick RE, Lege DJ, Pourboghrat F, Choi SH, Chu E (2003) Plane stress function for aluminium alloy sheets – part I: theory. Int J Plast 19:1297–1319

Berryman G (2005) Bounds and self-consistent estimates for elastic constants of random polycrystals with hexagonal, trigonal, and tetragonal symmetries. J Mech Phy Solids 53:2141–2173

Betten J (1988) Applications of tensor functions to the formulation of yield criteria for anisotropic materials. Int J Plast 4:29–46

Boehler JP, Sawczuk A (1970) Équilibre limite des sols anisotropes. J Mécanique 9:5–33

Cazacu O, Barlat F (2004) A criterion for description of anisotropy and yield differential effects in pressure-insensitive materials. Int J Plast 20:2027–2045

Cazacu O, Planckett B, Barlat F (2006) Orthotropic yield criterion for hexagonal close packed metals. Int J Plast 22:1171–1194

Desmorat R, Marull R (2011) Non-quadratic Kelvin modes based plasticity criteria for anisotropic materials. Int J Plast 27:327–351

Dunand M, Maertens AP, Luo M, Mohr D (2012) Experiments and modeling of anisotropic aluminum extrusions under multi-axial loading - part I: Plasticity. Int J Plast 36:34–49

Gan H, Orozco CE, Herkovich CT (2000) A strain-compatible method for micromechanical analysis of multi-phase composites. Int J Solids Struct 37:5097–5122

Ganczarski A, Lenczowski J (1997) On the convexity of the Goldenblatt-Kopnov yield condition. Arch Mech 49:461–475

Ganczarski A, Skrzypek J (2013) Mechanics of Novel Materials. Iss. Cracow Univ. Technol., Cracow

Ganczarski AW, Skrzypek JJ (2014) Constraints on the applicability range of Hill's criterion: strong orthotropy or transverse isotropy. Acta Mechanica 225(9):2563–2582

Goldenblat II, Kopnov VA (1966) Obobshchennaya teoriya plasticheskogo techeniya anizotropnyh sred. Sbornik Stroitelńaya Mehanika pp 307–319

Herakovich CT, Aboudi J (1999) Thermal effects in composites. In: Hetnarski RB (ed) Thermal Stresses V, Lastran Corp. Publ. Division, Rochester

Hill R (1948) A theory of the yielding and plastic flow of anisotropic metals. Proc Roy Soc London A193:281–297

Hill R (1950) The Mathematical Theory of Plasticity. Oxford University Press, Oxford

Jastrzebski ZD (1987) The Nature and Properties of Engineering Materials. John Wiley & Sons Inc., New York

Khan AS, Liu H (2012) Strain rate and temperature dependent fracture criteria for isotropic and anisotropic metals. Int J Plast 37:1–15

Khan AS, Kazmi R, Farrokh B (2007) Multiaxial and non-proportional loading responses, anisotropy and modeling of Ti-6Al-4V titanium alloy over wide ranges of strain rates and temperatures. Int J Plast 23:931–950

Khan AS, Yu S, Liu H (2012) Deformation enhanced anisotropic responses of Ti-6Al-4V alloy, Part II: A stress rate and temperature dependent anisotropic yield criterion. Int J Plast 38:14–26

Kolupaev VA (2018) Equivalent Stress Concept for Limit State Analysis, Advanced Structured Materials, vol 86. Springer, Cham

Korkolis YP, Kyriakides S (2008) Inflation and burst of aluminum tubes. Part II: An advanced yield function including deformation-induced anisotropy. Int J Plast 24:1625–1637

Kowalsky UK, Ahrens H, Dinkler D (1999) Distorted yield surfaces - modeling by higher order anisotropic hardening tensors. Comput Mat Sci 16:81–88

Lekhnitskii SG (1981) Theory of Elasticity of an Anisotropic Body. Mir Publishers, Moscow

Love AEH (1944) A Treatise on the Mathematical Theory of Elasticity. Dover Publ., New York

Luo M, Dunand M, Moth D (2012) Experiments and modeling of anisotropic aluminum extrusions under multi-axial loading - Part II: Ductile fracture. Int J Plast 32–33:36–58

Malinin NN, Rzysko J (1981) Mechanics of Materials. PWN, Warsaw

von Mises R (1913) Mechanik der festen Körper im plastisch deformablen Zustand. Nachrichten von der Gesellschaft der Wissenschaften zu Göttingen, Mathematisch-Physikalische Klasse 4:582–592

von Mises R (1928) Mechanik der plastischen Formänderung von Kristallen. ZAMM 8(13):161–185

Nixon ME, Cazacu O, Lebensohn RA (2010) Anisotropic response of high-purity α-titanium: experimental characterization and constitutive modeling. Int J Plast 26:516–532

Nye JF (1957) Physical Properties of Crystals: Their Representations by Tensors and Matrices. Clarendon Press, Oxford

Ottosen NS, Ristinmaa M (2005) The Mechanics of Constitutive Modeling. Elsevier, Amsterdam

Rogers TG (1990) Yield criteria, flow rules, and hardening in anisotropic plasticity. In Boehler J.P. (ed) Yielding, damage and failure of anisotropic solids. Mech. Eng. Publ., London

Rymarz C (1993) Continuum Mechanics. PWN, Warsaw

Sayir M (1970) Zur Fließbedingung der Plastizitätstheorie. Ingenierurarchiv 39:414–432

Sobotka Z (1969) Theorie des plastischen Fliessens von anisotropen Körpern. ZAMM 49:25–32

Spencer AJM (1971) Theory of invariants. In: Eringen C (ed) Continuum Physics, Academic Press, New York

Szczepiński W (1993) On deformation-induced plastic anisotropy of sheet metals. Arch Mech 45(1):3–38

Tamma KK, Avila A (1999) An integrated micro/macro modelling and computational methodology for high temperature composites. In Hetnarski RB (ed) Thermal Stresses V. Lastran Corp. Publ. Division, Rochester

Tsai ST, Wu EM (1971) general theory of strength for anisotropic materials. Int J Numer Methods Engng 38:2083–2088

Yoshida F, Hamasaki HM, Uemori T (2013) A user-friendly 3D yield function to describe anisotropy of steel sheets. Int J Plast 45:119–139

Życzkowski M (2001) Anisotropic yield conditions. In: Lemaitre J (ed) Handbook of Materials Behavior Models, Academic Press, San Diego

Chapter 7
Coupled Problems in Thermodynamics

Elena A. Ivanova and Dmitry V. Matias

Abstract We consider three basic methods adopted in modern thermodynamics. We discuss the state of the art, current problems and development prospects. We also discuss the possibility and necessity of constructing mechanical models of thermal processes and models of other processes of "non-mechanical nature". Next, we consider one of the possible mechanical models of thermal and electromagnetic processes. In order to illustrate the consequences of this model, we analyze the mutual influence of thermal and electromagnetic waves at the interface between two materials.

Keywords: Micropolar continuum · Cosserat continuum · Rotational degrees of freedom · Thermodynamics · Electrodynamics · Wave propagation

7.1 Introduction

In book "The aim and structure of physical theory", Pierre Duhem writes (Duhem, 1954, p. 19): *"A physical theory is not an explanation. It is a system of mathematical propositions, deduced from a small number of principles, which aim to represent as simply, as completely, and as exactly as possible a set of experimental laws."* Classical thermodynamics fully satisfies this definition. However, modern thermodynamics is

Elena A. Ivanova
Peter the Great St. Petersburg Polytechnic University, Department of Theoretical Mechanics, Institute of Applied Mathematics and Mechanics, Politechnicheskaya 29, 195251 St. Petersburg & Laboratory of Mechatronics, Institute for Problems in Mechanical Engineering of Russian Academy of Sciences, Bolshoy pr. V.O. 61, 199178 St. Petersburg, Russia,
e-mail: elenaivanova239@gmail.com

Dmitry V. Matias
Peter the Great St. Petersburg Polytechnic University, Department of Theoretical Mechanics, Institute of Applied Mathematics and Mechanics, Politechnicheskaya, 29, 195251, St. Petersburg, Russia,
e-mail: dvmatyas@gmail.com

H. Altenbach and A. Öchsner (eds.), *State of the Art and Future Trends in Material Modeling*, Advanced Structured Materials 100,
https://doi.org/10.1007/978-3-030-30355-6_7

not only a physical theory in the sense of Duhem's definition. Modern thermodynamics is something unique in science.

First of all, we note that modern thermodynamics consists of three branches, which are not actually associated with each other. They differ from each other both by interpretations of the basic concepts and by mathematical methods. Discussing the three branches, we mean, first, thermodynamics that is developed in the frame of continuum mechanics and is a logical extension of classical thermodynamics, second, non-equilibrium thermodynamics, which is also an extension of classical thermodynamics, which is originated by Prigogine and De Groot, and third, statistical thermodynamics, which arose from the kinetic theory. These three sciences study the same issues and use the same scientific terms. But there is a non trivial question of whether the concepts of temperature, entropy and other thermodynamic quantities used in continuum mechanics, non-equilibrium thermodynamics and statistical thermodynamics are identical. This question is important since, in modern thermodynamics, there is a tendency of ideas and methods that developed independently for a long time, to mutually influence each other and to unite with each other.

Modern thermodynamics long ago went beyond the purely thermal problems, which resulted in the creation of classical thermodynamics some time ago. With the development progress, the application field of thermodynamics broadens and new fields of science such as gas dynamics, thermoelasticity, thermoviscoelasticity, thermoelectricity, thermomagnetism, the theory of structural and phase transformations are created. Modern thermodynamics is used in almost all fields of natural sciences. It is used in describing chemical reactions, studying the structure of matter and modeling radiation. It plays an important role both in creating the theories that describe the processes in the microcosm and in creating cosmogonic theories. Thermodynamics is a link in modeling the mutual influence of processes of different nature. Such a unique role of thermodynamics is due to the fact that it became not just a theory describing a specific physical process, but turned into a research method that can be used to describe a wide variety of the physical processes. We note that the aforesaid mainly concerns non-equilibrium thermodynamics and statistical thermodynamics. Unfortunately, in the most cases, continuum mechanics only considers either purely mechanical problems or problems at the intersection of mechanics and other natural sciences. At the present stage of its development, continuum mechanics rarely claims on the creation of models of processes of "a non-mechanical nature". Such a position of continuum mechanics seems to be not promising either for the development of continuum mechanics itself or for the development of science in general. Continuum mechanics, as well as non-equilibrium thermodynamics and statistical thermodynamics is a set of the physical theories describing some specific processes, on the one hand, and it is a research method that can and should be used to model processes of different nature, on the other hand. We are deeply convinced that continuum mechanics should adhere to the similar position with respect to its models and methods as non-equilibrium thermodynamics and statistical thermodynamics. This makes it possible for continuum mechanics to play an important role in the study of all physical processes without exceptions.

A surprising feature of modern thermodynamics is as follows. On the one hand, thermodynamics as a phenomenological science is involved in the modeling many physical processes. On the other hand, in thermodynamics, there is no a generally accepted model of thermal processes that can be used always and everywhere. A simple and clear mechanical model, which is the basis of the kinetic theory, has a limited range of applicability even within the framework of statistical thermodynamics. In non-equilibrium thermodynamics and continuum mechanics, this model is not used at all. In continuum mechanics, there are several different mechanical models of thermal processes. However, none of them can be compared in its popularity to the model adopted in the kinetic theory. Non-equilibrium thermodynamics is a purely phenomenological science, in which there are only mathematical models. Do modern science in general and modern thermodynamics in particular need the mechanical models? Certainly, this question is controversial. And it has something in common with another debatable question, which is worded in the above quotation from Duhem's book, namely, whether a physical theory should be an explanation?

7.2 Historical Remarks and the State of the Art

7.2.1 Preliminary Remarks

In the short article, we cannot give an extensive historical overview. But there is no need whatsoever since the history of thermodynamics is well expounded in Müller (2007). In addition, detailed overviews of different models of thermal conductivity and thermoelasticity can be found in Ignaczak and Ostoja-Starzewski (2009); Straughan (2011). In this section, we carry out a comparative analysis of the three basic approaches, which are being developed in thermodynamics over a long period of time, and outline the main directions of the development of modern thermodynamics. The review does not claim to be exhaustive. We purpose only to pay attention to some facts that, in our opinion, are important for understanding the state of things in modern thermodynamics.

7.2.2 Statistical Thermodynamics and Continuum Mechanics

For the first time, the idea to reduce thermal phenomena to mechanical phenomena was implemented in a mathematical form by Daniel Bernoulli. Bernoulli argued that heat is a manifestation of molecules vibrations, assumed all the molecular velocities to be equal each other and interpreted the gas pressure to be the result of the action of the molecules colliding with the vessel wall. The kinetic theory was further developed in the works of Clausius and Maxwell. In 1857, Clausius derived the basic formula of the kinetic theory of gases, according to which the gas pressure is equal to two thirds of the average kinetic energy of all molecules per unit volume. The derivation of this formula can be found in many books, see, e.g., Feynman et al (1963). A comparison of the expression for pressure obtained from the kinetic theory with the expression given by the ideal gas equation led to the idea of identifying the temperature with the average kinetic energy of translational motion $\bar{K}$:

$$\bar{K} = \frac{3}{2}\, k_B T, \tag{7.1}$$

where k_B is the Boltzmann constant, T is the absolute temperature. In 1859, Maxwell replaced Bernoulli's hypothesis of velocity equality with the formula for velocity distribution, which was later named after him. Thus, beginning with the works of Clausius and Maxwell, the kinetic theory has become a statistical one. In 1872, Boltzmann derived the kinetic equation for the distribution function, which describes non-equilibrium processes (viscous flow, thermal conduction, diffusion) in gases of low density, and, in 1877, he suggested the famous relation between entropy S and the thermodynamic probability W, see Boltzmann (1974):

$$S = k_B \ln W. \tag{7.2}$$

A modern account of the Maxwell–Boltzmann statistics can be found, e.g., in Berkley Physics Course (Reif, 1967). It is important to pay attention to how the concept of temperature is introduced in Reif (1967). Let us denote by $W(E)$ the number of possible states of a macroscopic system when its energy is in the range from E to $E + \delta E$. Following Reif (1967), we introduce quantity $\beta(E)$ as

$$\beta(E) = \frac{\partial \ln W}{\partial E}. \tag{7.3}$$

The dimension of parameter β is equal to unit divided by the dimension of energy. Let us represent β^{-1} as

$$\beta^{-1} = k_B T, \tag{7.4}$$

where the new parameter T is called the absolute temperature. In view of Eqs. (7.3), (7.4), it is easy to get the formula relating entropy, energy and temperature:

$$T^{-1} = \frac{\partial S}{\partial E}. \tag{7.5}$$

It can be proved that, in the case of an ideal monatomic gas, the definition of temperature (7.3)–(7.5) is equivalent to the definition (7.1). To be exact, the definitions (7.1) and (7.3)–(7.5) are equivalent under the condition that the Maxwell distribution is used in Eqs. (7.3)–(7.5).

In the case of any distribution, different from the Maxwell distribution, the definition of temperature (7.3)–(7.5) cannot be reduced to Eq. (7.1), and hence the temperature loses its simple and intuitive mechanical interpretation. This fact is very important since the further development of statistical thermodynamics consists in creating and using a number of new statistics, based on the distributions different from the Maxwell distribution. In 1924, Bose originated a statistics describing systems of particles with zero spin or integer spin. These particles are called bosons. Photons and certain nuclei are bosons. Subsequently, this statistics was sophisticated by Einstein. Now it is called the Bose–Einstein statistics. In 1926, Fermi originated a statistics describing systems of particles with half-integer spin, which are called fermions. Electrons, protons and neutrons are fermions. In the same year, Dirac found out a quantum-mechanical meaning of this statistics. That is why, now it is called the Fermi–Dirac statistics. In the case of a strongly rarefied gas at temperatures higher than several tens of kelvins, both the statistics turn into the Maxwell–Boltzmann statistics. For classical and quantum statistics, the distribution functions can be expressed in one general form:

$$\langle n_i \rangle = \frac{n_i}{N_i} = \frac{1}{e^{(E_i - \mu)/k_B T} + \delta}, \tag{7.6}$$

where E_i is the energy of state, μ is the chemical potential, n_i is the number of particles in the given state, N_i is the multiplicity of state of the particles with energy E_i. For the Bose–Einstein quantum distribution, $\delta = -1$, for the Maxwell–Boltzmann classical distribution, $\delta = 0$ and $\mu = 0$, for the the Fermi–Dirac quantum distribution, $\delta = +1$. A detailed account of statistical thermodynamics, both classical and quantum, can be found, e.g., in Kubo (1965); Huang (1963); Kittel (1970). Statistical thermodynamics of non-equilibrium processes is discussed in Eisenschitz (1958); Röpke (2013). A substantiation of the formalism of statistical thermodynamics is given in Ruelle (1978); Krylov (2003).

We believe that statistical physics and continuum mechanics have more traits in common than it seems ex facte. First of all, we note that both statistical physics and continuum mechanics are the research methods. Nevertheless they differ from each other from a mathematical point of view, but, in essence, they have the same structure. Both the sciences are based on certain fundamental equations. The fundamental equations are used in solving all the problems. But there is a certain freedom of choice, which allows us to consider various models in the framework of the given concept. We mean the choice of expression for the internal energy in continuum mechanics and the choice of distribution function in statistical physics. Certainly, this is a very schematic representation. There are many nuances, but a detailed discussion of these issues is beyond the scope of this paper. We aim only to pay attention to a couple of important facts. First, the choice of distribution function in statistical physics is the same thing as the choice of expression for the internal energy in continuum mechanics. Second, in modern statistical physics, as in modern continuum mechanics, there is no generally accepted mechanical interpretation of temperature and there is no generally accepted mechanical model of thermal processes.

7.2.3 Non-equilibrium Thermodynamics and Continuum Mechanics

A purely phenomenological approach is used for the description of thermal processes in non-equilibrium thermodynamics. The detailed consideration of the method of non-equilibrium thermodynamics can be found in works of its originators, namely Prigogine and De Groot — see Kondepudi and Prigogine (1998); Prigogine (1955); De Groot (1951). Here we briefly outline the main ideas of the method. In order to do this, we consider matter with density ρ occupying volume V. Let some property of the matter be characterized by an extensive thermodynamic parameter B. Quantity B, as any extensive quantity, satisfies the following relations:

$$\lim_{\Delta V\to 0}\frac{\Delta B}{\Delta V}=\rho b, \qquad B=\int\limits_V \rho b\, dV, \tag{7.7}$$

where b is the mass-specific density of B. A change of B in volume V is caused by two factors. The first factor is a flux of this quantity into volume V from outside. The second factor is a production of this quantity inside volume V. Let $\mathbf{J}_B$ denote the flux density of B through surface Σ, bounding volume V, and σ_B denote the production rate of B inside volume V. Then the balance equation for quantity b takes the form:

$$\int\limits_V \frac{d(\rho b)}{dt}\, dV = -\int\limits_\Sigma \mathbf{n}\cdot\mathbf{J}_B\, d\Sigma + \int\limits_V \sigma_B\, dV, \tag{7.8}$$

where $\mathbf{n}$ denotes the unit outer normal vector to the surface Σ. Since the integrands are considered to be continuous, and Eq. (7.8) is formulated for an arbitrary volume V, from this equation it follows the local form of the balance equation:

$$\rho\frac{db}{dt} = -\nabla\cdot\mathbf{J}_B + \sigma_B. \tag{7.9}$$

Equation (7.9) is written for the scalar quantity b. However, in accordance with the ideology of non-equilibrium thermodynamics, analogous equations can be formulated for vector or tensor quantities having any physical meaning. This is one of the key differences between non-equilibrium thermodynamics and continuum mechanics. In classical continuum mechanics, there are only four balance equations: the mass balance, the momentum balance, the angular momentum balance and the energy balance. It should be noted that, in modern continuum mechanics, there is a tendency to include new balance equations, and this is undoubtedly a result of the influence of non-equilibrium thermodynamics. The breadth of views on the balance equations allows non-equilibrium thermodynamics to describe almost all physical processes, which are associated with thermal effects in one way or another. In addition, in the so-called extended non-equilibrium thermodynamics (Jou et al, 2001), it is considered acceptable to write the balance equations not only for the basic quantities, but also for their fluxes, for fluxes of fluxes, etc. Due to this, extended non-equilibrium thermodynamics is able to describe any relaxation processes.

The system of balance equations in non-equilibrium thermodynamics, as in continuum mechanics, is not closed. In order to close this system thermodynamic forces and fluxes are introduced as conjugate quantities. This is done as follows. Quantity σ_B in Eq. (7.9) is represented as

$$\sigma_B = \sum_\alpha J_\alpha X_\alpha, \tag{7.10}$$

where fluxes J_α and thermodynamic forces X_α are not necessarily scalar quantities. For example, thermodynamic force $\nabla\left(T^{-1}\right)$ corresponds to heat flux $\mathbf{h}$. The relations between thermodynamic forces and fluxes in non-equilibrium thermodynamics are analogies of the constitutive equations in continuum mechanics. These relations have the following form:

$$J_\alpha = \sum_\beta L_{\alpha\beta} X_\beta, \qquad L_{\alpha\beta} = \left(\frac{\partial J_\alpha}{\partial X_\beta}\right)_{eq}. \tag{7.11}$$

Quantities $L_{\alpha\beta}$ are called phenomenological coefficients. They satisfy the Onsager reciprocal relations

$$L_{\alpha\beta} = L_{\beta\alpha}. \tag{7.12}$$

The Onsager reciprocal relations play an important role in non equilibrium thermodynamics. They are considered to be the macroscopic equalities that result from the microscopic reversibility. The proof of the Onsager reciprocal relations is a nontrivial problem and this issue is often discussed in literature. We note that in continuum mechanics the relations similar to Eq. (7.12) are obtained naturally. This is a feature of the continuum mechanics method, which allows us to obtain the constitutive equations by using the energy balance equation and the second law of thermodynamics. Equation (7.11) are postulated, not derived in non-equilibrium thermodynamics. This is the reason why the Onsager reciprocal relations have to be proved.

In non-equilibrium thermodynamics, as in continuum mechanics, temperature and entropy are introduced by the equation relating these quantities to the internal energy, and in addition, temperature is considered to be a quantity measured by a thermometer, and entropy is defined as a quantity conjugate to the temperature. However, in non-equilibrium thermodynamics the concept of entropy flux and the concept of entropy production are introduced, and the entropy balance equation is considered to be one of the basic equations. This approach is not the same as compared to the classical Coleman–Noll procedure based on using the Clausius–Duhem inequality.

Thus, the method of non-equilibrium thermodynamics and the method of continuum mechanics are very similar. However, the postulates, on which these sciences are based, differ from each other. In addition, because of some historical reasons, these two sciences have evolved independently of each other. As a result, both the sciences have acquired such a structure that their contradictionless integration is extremely difficult. In fact, any attempt to combine the ideas of non-equilibrium thermodynamics and continuum mechanics leads to the constracting of an original theory. We are convinced that the influence of non-equilibrium thermodynamics on continuum mechanics is useful, primarily because it contributes to an expansion of the application field of continuum mechanics.

At present, the following situation arises in that part of thermodynamics, which is a purely phenomenological science. Classical equilibrium thermodynamics, which is substantially based on experimental studies, is described by the nonlinear equations. Non-equilibrium thermodynamics, which is increasingly based on theoretical premises and to a lesser extent on experiments, is a well-developed science only in its linear version. With regard to the creation of a nonlinear theory, one can only speak of more or less successful attempts made by different authors. The reason is as follows. Based on formal mathematical considerations, a linear model can be constructed quite easily and completely unambiguously. At the same time, from the formal mathematical point of view, this linear model can be generalized to a nonlinear case by many ways. Concerning the development prospects of nonlinear thermodynamics, we note that continuum mechanics with its well-developed methods of nonlinear theory of elasticity has certain advantages over non-equilibrium thermodynamics.

7.2.4 A Brief Overview of Current Research

In the modern literature on thermodynamics, nonlinear and coupled problems are actively studied and discussed. Without claiming to be an exhaustive literature review, we indicate the main research areas in this field. Many papers covering only mathematical questions are regularly published for several decades. Various aspects of constructing analytical, semi-analytic and numerical solutions of the nonlinear heat conduction equations are discussed in such papers — see, e.g., Campo (1982); Jordan et al (1987); Polyanin et al (2000); Ebadian and Darania (2008); Habibi et al (2015). Such works usually deal with the simplest nonlinear heat conduction equations. The nonlinearity of these equations consists in the fact that the material constants of the linear equations are replaced by some functions of temperature (more often by polynomials). Among the mathematical works, it is worth mentioning the papers where the authors consider laser heat sources (see, e.g., Fong et al, 2010), as this type of thermal influences is most often found in the modern literature. We also refer to works where methods based on Lagrangian mechanics and variational principles are developed — see, e.g., Biot (1970, 1984); Lebon and Casas-Vazquez (1974); Lebon and Dauby (1990); Gay-Balmaz and Yoshimura (2019). Another large group of publications consists of applied works, which are devoted to modeling thermal processes in technical devices — see, e.g., Grudinin et al (2011); Chaibi et al (2012); Huang et al (2012); Markides et al (2013). In such works, the mutual influence of thermal processes and processes of other physical nature (optical, electrical, magnetic) is usually taken into account. Other distinctive features of applied works are the use of numerical methods, the use of parameters of specific technical devices in calculations and the comparison of modeling results with the experimental data. We note that the models of nonlinear thermal processes in the widest scale range, from geophysical processes (see, e.g., Mottaghy and Rath, 2006), up to biological processes at the molecular level (see, e.g., LeMesurier, 2008), are presented in the modern literature. There are a large number of papers devoted to studying nonlinear effects associated with thermal radiation — see, e.g., Khandekar et al (2015); Ananth et al (2015). There exist a variety of mathematical models used to describe various thermal processes. Some of them are based on classical concepts, and others are based on quantum-mechanical concepts. However, purely empirical relations, which are not based on any models, play an important role in all nonlinear theories.

The most of modern theoretical studies are devoted to the discussion of the entropy concept. The interest in this subject arose as soon as it was established that some models of thermal conductivity, different from the classical Fourier model, contradict the second law of thermodynamics. Different opinions were expressed and different approaches were proposed to solve the problem, see, e.g., Rubin (1992); Baik and Lavine (1995); Dugdale (1996); Barletta and Zanchini (1997); Zanchini (1999); Čápek and Sheehan (2005); Lieb and Yngvason (1997); Ostoja-Starzewski (2016, 2017). However, at present this question remains controversial. We believe that the discussion of this interesting and important topic will continue for a long time. It seems to us that the problem lies in the ambiguity of the concept of entropy, the absence of its generally accepted definition and the incomprehensibility of its physical meaning. In the kinetic theory, where there are clear mechanical models of thermal processes and temperature has a clear mechanical meaning, the basic equations are derived without the use of entropy. Boltzmann's probabilistic definition of entropy has, in our opinion, a purely mathematical meaning. In modern statistical thermodynamics, where entropy plays a very important role, not only the physical meaning of entropy does not become clearer, but also the physical meaning of temperature loses its clarity. We believe that entropy has a clear physical meaning only in classical equilibrium thermodynamics, where an adiabatic process is actually an isentropic process. In more complex models, which are studied in non-equilibrium thermodynamics and continuum mechanics, the adiabatic process need not to be isentropic and vice versa. That is why the intuitively clear definition of entropy given in classical thermodynamics can be extended to the case of non-equilibrium processes in various ways and all definitions of entropy appropriate for non-equilibrium processes are more formal and less understandable than the classical definition. It is obvious that the entropy introduced in classical equilibrium thermodynamics must satisfy the second law of thermodynamics in its classical form. But it is not obvious at all, whether the entropy introduced for non-equilibrium processes should satisfy one or another formulation of the second law. It is quite possible that for this reason many scientists continue to consider various models of thermal conductivity that conflict with the second law of thermodynamics, and all attempts to solve the problem of the contradiction are, in fact, reduced to new interpretations of the second law of thermodynamics.

7.3 Mechanical Models for Studying Coupled Problems in Thermodynamics

7.3.1 Preliminary Remarks

The idea of describing magneto-, electro- and thermomechanical processes by means of continuum mechanics models is well known. Continuum mechanics models based on rotational degrees of freedom have been suggested in Dixon and Eringen (1964, 1965); Treugolov (1989); Grekova and Zhilin (2001); Grekova (2001); Zhilin (2006b,a, 2012); Ivanova and Kolpakov (2013, 2016); Ivanova (2019a), micromorphic continuum models have been considered in Bardeen et al (1957); Eringen (2003); Galeş et al (2011), two-component continuum models have been discussed in Tiersten (1964); Maugin (1988); Eringen and Maugin (1990); Fomethe and Maugin (1996), continuum models with microstructure based on rotational degrees of freedom have been constructed in Zhilin (2012); Shliomis and Stepanov (1993); Ivanova (2015).

The presented study continues and develops the research carried out in Ivanova (2010, 2011, 2012, 2014, 2015, 2017, 2018, 2019a,b); Vitokhin and Ivanova (2017), where an original approach to constructing theories of thermo- and thermoviscoelasticity, as well as theories of electromagnetism, have been worked out. This approach is based on the idea to introduce mechanical analogies of physical quantities in the framework of the suggested model, to show that under a number of simplifying assumptions the equations of the suggested model coincide with well-known equations of thermodynamics and electrodynamics and then to explore the properties of the model in its general form.

7.3.2 The Cosserat Continuum of Special Type

Now we give a brief account of the basic equations of the Cosserat continuum of special type. Let vector $\mathbf{r}$ identify the position of some point of space. We use the following notations: $\mathbf{v}(\mathbf{r},t)$ is the velocity vector field; $\mathbf{u}(\mathbf{r},t)$ is the displacement vector field; $\mathbf{P}(\mathbf{r},t)$ is the rotation tensor field, $\boldsymbol{\theta}(\mathbf{r},t)$ is the rotation vector field, and $\boldsymbol{\omega}(\mathbf{r},t)$ is the angular velocity vector field. For simplicity's sake, we consider a linear theory. We assume that in the reference configuration tensor $\mathbf{P}(\mathbf{r},t)$ is equal to the unit tensor $\mathbf{E}$ and rotation vector $\boldsymbol{\theta}(\mathbf{r},t)$ is equal to zero. Then, the kinematic relations have the form

$$\mathbf{v}(\mathbf{r},t) = \frac{d\mathbf{u}(\mathbf{r},t)}{dt}, \quad \mathbf{P}(\mathbf{r},t) = \mathbf{E} + \boldsymbol{\theta}(\mathbf{r},t) \times \mathbf{E}, \quad \boldsymbol{\omega}(\mathbf{r},t) = \frac{d\boldsymbol{\theta}(\mathbf{r},t)}{dt}. \tag{7.13}$$

We consider an isotropic continuum and therefore we assume the mass-specific densities of kinetic energy, the linear momentum vector and the angular momentum vector to be

$$\mathscr{K} = \frac{1}{2}\mathbf{v}\cdot\mathbf{v} + \frac{1}{2}J\,\boldsymbol{\omega}\cdot\boldsymbol{\omega}, \quad \mathscr{K}_1 = \mathbf{v}, \quad \mathscr{K}_2 = \mathbf{r}\times\mathbf{v} + J\,\boldsymbol{\omega}, \tag{7.14}$$

where constant J is the mass-specific density of moments of inertia. The angular momentum density $\mathscr{K}_2$ is calculated with respect to the origin of the reference frame. We note that the mass balance equation in the linear approximation takes the form $\rho = \varrho\left(1 - \nabla\cdot\mathbf{u}\right)$, where ϱ is the density of mass in the reference configuration and ρ is the current value of the mass density.

Now we consider the force vector $\boldsymbol{\tau}_n$ and the moment vector $\mathbf{T}_n$ modeling the surrounding medium influence on the surface $\mathscr{S}$ of the elementary volume $\mathscr{V}$. Next, by standard reasoning, we introduce the concept of stress tensor $\boldsymbol{\tau}$ associated with the stress vector $\boldsymbol{\tau}_n$ and the concept of moment stress tensor $\mathbf{T}$ associated with the moment stress vector $\mathbf{T}_n$. These tensors are defined by the Cauchy relations $\boldsymbol{\tau}_n = \mathbf{n}\cdot\boldsymbol{\tau}$ and $\mathbf{T}_n = \mathbf{n}\cdot\mathbf{T}$ where $\mathbf{n}$ denotes the unit outer normal vector to the surface $\mathscr{S}$. Further, we assume the external forces per unit mass to be equal to zero. In this case, the dynamics equations are written as

$$\nabla\cdot\boldsymbol{\tau} = \varrho\frac{d\mathbf{v}}{dt}, \qquad \nabla\cdot\mathbf{T} + \boldsymbol{\tau}_\times + \varrho\mathbf{L}_f = \varrho J\,\frac{d\boldsymbol{\omega}}{dt}. \tag{7.15}$$

Here $\mathbf{L}_f$ is the mass-specific density of external moments, $(\,)_\times$ denotes the vector invariant of a tensor that is defined for an arbitrary dyad as $(\mathbf{ab})_\times = \mathbf{a}\times\mathbf{b}$.

We consider the continuum to be isolated. In this case, by standard reasoning, the energy balance equation can be transformed to the form

$$\varrho\frac{dU}{dt} = \boldsymbol{\tau}^T\cdot\cdot\left(\nabla\mathbf{v} + \mathbf{E}\times\boldsymbol{\omega}\right) + \mathbf{T}^T\cdot\cdot\nabla\boldsymbol{\omega}, \tag{7.16}$$

where U is the internal energy per unit mass. The double scalar product is defined as $\mathbf{ab}\cdot\cdot\mathbf{cd} = (\mathbf{b}\cdot\mathbf{c})(\mathbf{a}\cdot\mathbf{d})$.

Further, we accept several simplifying assumptions that are the basis of the theory under consideration.

Hypothesis 1. *The moment stress tensor* $\mathbf{T}$ *has the following form:*

$$\mathbf{T} = T\mathbf{E} - \mathbf{M}\times\mathbf{E}. \tag{7.17}$$

In view of assumption (7.17) and the fact that the linear theory is considered, the energy balance equation (7.16) can be reduced to the form (Ivanova, 2019a):

$$\varrho\frac{dU}{dt} = \boldsymbol{\tau}^T\cdot\cdot\frac{d\mathbf{e}}{dt} + \varrho T\,\frac{d\Theta_\rho}{dt} + \varrho\mathbf{M}\cdot\frac{d\boldsymbol{\Psi}_\rho}{dt}, \tag{7.18}$$

where the following notations are used:

$$\mathbf{e} = \nabla\mathbf{u} + \mathbf{E}\times\boldsymbol{\theta}, \quad \Theta_\rho = \varrho^{-1}\mathrm{tr}\,\boldsymbol{\Theta}, \quad \boldsymbol{\Psi}_\rho = \varrho^{-1}\boldsymbol{\Theta}_\times, \quad \boldsymbol{\Theta} = \nabla\boldsymbol{\theta}. \tag{7.19}$$

Here $\mathbf{e}$ and $\boldsymbol{\Theta}$ are the strain tensors, Θ_ρ and $\boldsymbol{\Psi}_\rho$ are the strain measures corresponding to the spherical and antisymmetric parts of tensor $\mathbf{T}$, respectively. Factor ϱ^{-1} is introduced for convenience. Since we consider the elastic continuum, from the energy balance equation (7.18) it follows that the mass-specific density of internal energy is a function of three arguments, namely $U = U(\mathbf{e}, \Theta_\rho, \boldsymbol{\Psi}_\rho)$.

Hypothesis 2. *The mass-specific density of internal energy is a function of only two arguments, namely*

$$U = U(\Theta_\rho, \boldsymbol{\Psi}_\rho). \tag{7.20}$$

Substituting Eq. (7.20) into Eq. (7.18), we obtain the Cauchy–Green relations

$$\boldsymbol{\tau} = \mathbf{0}, \qquad T = \frac{\partial U(\Theta_\rho, \boldsymbol{\Psi}_\rho)}{\partial \Theta_\rho}, \qquad \mathbf{M} = \frac{\partial U(\Theta_\rho, \boldsymbol{\Psi}_\rho)}{\partial \boldsymbol{\Psi}_\rho}. \tag{7.21}$$

Specifying the mass-specific density of internal energy as

$$U = \frac{1}{2} C_\Theta \Theta_\rho^2 + \frac{1}{2} C_\Psi \Psi_\rho^2, \qquad \Psi_\rho = \sqrt{\boldsymbol{\Psi}_\rho \cdot \boldsymbol{\Psi}_\rho}, \tag{7.22}$$

and substituting Eq. (7.22) into Eq. (7.21), we obtain the constitutive equations for T and $\mathbf{M}$. Stiffness parameters C_Θ and C_Ψ in Eq. (7.22) are considered to be independent of time.

The angular momentum balance equation contains a vector of external moment $\mathbf{L}_f$. In the considered model, vector $\mathbf{L}_f$ characterizes a viscous damping. For more detail on the physical meaning of the moment vector of viscous damping, see Ivanova (2010, 2011, 2012, 2014, 2017). The solutions of several model problems, which allow us to determine the structure the moment vector of viscous damping, can be found in Ivanova (2011, 2012, 2014). Here we specify vector $\mathbf{L}_f$ taking into account the results obtained in the mentioned papers.

Hypothesis 3. *The moment vector of viscous damping has the form*

$$\mathbf{L}_f = -\beta J \boldsymbol{\omega}, \tag{7.23}$$

where constant $\boldsymbol{\beta}$ *has the meaning of a damping coefficient.*

It is easy to show that the system of the basic equations of the continuum under study can be separated into two independent systems. The system describing the translational motion is trivial. The system describing the rotational motion is

$$\begin{gathered} \nabla T - \nabla \times \mathbf{M} - \beta \varrho J \boldsymbol{\omega} = \varrho J \frac{d\boldsymbol{\omega}}{dt}, \quad T = C_\Theta \Theta_\rho, \quad \mathbf{M} = C_\Psi \boldsymbol{\Psi}_\rho, \\ \Theta_\rho = \frac{\Theta}{\varrho}, \quad \Theta = \nabla \cdot \boldsymbol{\theta}, \quad \boldsymbol{\Psi}_\rho = \frac{\boldsymbol{\Psi}}{\varrho}, \quad \boldsymbol{\Psi} = \nabla \times \boldsymbol{\theta}, \quad \boldsymbol{\omega} = \frac{d\boldsymbol{\theta}}{dt}. \end{gathered} \tag{7.24}$$

Thus, the system of partial differential equations (7.24) describes the behavior of the continuum based only on rotational degrees of freedom. Further, we use the continuum for simulating thermodynamic and electromagnetic processes in matter. We note that, for simplicity's sake, we neglect all mechanical processes in matter and we also neglect heat supply from any external sources.

7.3.3 Mechanical Analogies of Physical Quantities

We suppose that quantities T and Θ_ρ are related to the absolute temperature T_a that can be measured by a thermometer and the mass-specific density of entropy Θ_a by the formulas

$$T = aT_a, \qquad \Theta_\rho = \frac{1}{a}\,\Theta_a, \tag{7.25}$$

where a is the normalization factor. We also suppose that the moment stress vector $\mathbf{M}$ is the analogy of the electric field vector $\mathscr{E}$ and the volume density of proper angular momentum $\varrho J\,\omega$ is the analogy of the magnetic induction vector $\mathscr{B}$, namely

$$\mathbf{M} = \chi\mathscr{E}, \qquad \varrho J\,\boldsymbol{\omega} = \chi\mathscr{B}, \tag{7.26}$$

where χ is the normalization factor. A detailed consideration of prerequisites for the choice of the analogies of thermodynamic and electromagnetic quantities in the framework of the suggested model can be found in Ivanova (2019a).

7.3.4 Simulating Thermodynamic and Electromagnetic Processes in Matter

Now we consider a homogeneous isotropic matter possessing thermodynamic and electromagnetic properties. We assume that electric charges and electric currents are absent. In this case, in order to simulate thermodynamic and electromagnetic processes in the matter we can use the mechanical model considered in Sect. 7.3.2 and the mechanical analogies of the physical quantities considered in Sect. 7.3.3. It is important to note that since the linear theory is considered the quantity T_a has the meaning of a small deviation of the temperature from its reference value T_a^* and quantity Θ_a has the meaning of a small deviation of the mass-specific density of entropy from its reference value Θ_a^*.

Next, we assume that the material constants characterizing properties of the continuum are related to the known physical constants by the formulas

$$J = \frac{\chi^2\mu\mu_0}{\varrho}, \qquad C_\Psi = \frac{\chi^2\varrho}{\varepsilon_0\varepsilon}, \qquad C_\Theta = \frac{a^2 T_a^*}{c_v}, \qquad \beta = \frac{a^2}{\chi^2}\frac{T_a^*}{\lambda\,\mu_0\mu}, \tag{7.27}$$

where ε is the relative permittivity, μ is the relative permeability, c_v is the specific heat at constant volume, λ is the heat conduction coefficient. We note that the first and second formulas in Eq. (7.27) are the exactly same as in Ivanova (2015, 2019a,b), and the third formula in Eq. (7.27) is in agrement with the corresponding formula previously obtained for the case of a heat-conductive elastic material, see Ivanova (2010, 2011, 2012, 2014, 2017).

In view of the adopted physical analogies given by Eqs. (7.25), (7.26) and (7.27), it is not difficult to show that from Eq. (7.24) it follows that

$$\nabla \times \mathscr{B} = \varepsilon_0 \varepsilon \mu_0 \mu \, \frac{d\mathscr{E}}{dt}, \qquad \frac{a}{\chi} \, \nabla T_a - \nabla \times \mathscr{E} - \frac{a^2}{\chi^2} \frac{T_a^*}{\lambda \, \mu_0 \mu} \, \mathscr{B} = \frac{d\mathscr{B}}{dt},$$
$$\frac{d\Theta_a}{dt} = \frac{a}{\chi} \, \frac{1}{\rho_* \mu_0 \mu} \, \nabla \cdot \mathscr{B}, \qquad T_a = \frac{T_a^*}{c_v} \, \Theta_a. \tag{7.28}$$

We note that the first equation in Eq. (7.28) coincides with one of Maxwell's equations and the last equation in Eq. (7.28) is well known in the linear theory of thermoelacticity.

Now we transform Eq. (7.28) to the equations for the electric field vector $\mathscr{E}$ and the vortex part of the magnetic induction vector $\mathscr{B}$:

$$c^2 \, \Delta(\nabla \times \mathscr{E}) = \beta \, \frac{d(\nabla \times \mathscr{E})}{dt} + \frac{d^2(\nabla \times \mathscr{E})}{dt^2}, \qquad \nabla \cdot \mathscr{E} = 0,$$
$$c^2 \, \Delta(\nabla \times \mathscr{B}) = \beta \, \frac{d(\nabla \times \mathscr{B})}{dt} + \frac{d^2(\nabla \times \mathscr{B})}{dt^2}, \qquad c^2 = \frac{1}{\varepsilon_0 \varepsilon \mu_0 \mu}, \tag{7.29}$$

and the equations for temperature T_a, the mass-specific density of entropy Θ_a and the potential part of the magnetic induction vector $\mathscr{B}$:

$$c_r^2 \, \Delta T_a = \beta \, \frac{dT_a}{dt} + \frac{d^2 T_a}{dt^2}, \qquad c_r^2 \Delta \Theta_a = \beta \, \frac{d\Theta_a}{dt} + \frac{d^2 \Theta_a}{dt^2},$$
$$c_r^2 \, \Delta(\nabla \cdot \mathscr{B}) = \beta \, \frac{d(\nabla \cdot \mathscr{B})}{dt} + \frac{d^2(\nabla \cdot \mathscr{B})}{dt^2}, \qquad c_r^2 = \frac{a^2}{\chi^2} \frac{T_a^*}{\mu_0 \mu \, \varrho c_v}. \tag{7.30}$$

It is easy to see that Eq. (7.29) are similar to Maxwell's equations in a conductive medium. The only difference is the physical meaning of the terms that ensure damping of electromagnetic waves. Three equations in Eq. (7.30) are equivalent and coincide with a hyperbolic heat conduction equation, see Cataneo (1958); Chandrasekharaiah (1998); Jou et al (2001); Babenkov and Ivanova (2014); Babenkov and Vitokhin (2018). This equation can be rewritten in the more customary form:

$$\Delta T_a = \frac{\rho_* c_v}{\lambda} \frac{dT_a}{dt} + \frac{1}{c_r^2} \frac{d^2 T_a}{dt^2}. \tag{7.31}$$

According to Debye's law, the specific heat is proportional to temperature cubed at temperatures close to absolute zero and is close to constant value at temperatures above the Debye temperature. Therefore, as it follows from the last relation in Eq. (7.30), the propagation velocity of thermal waves c_r tends to infinity like $1/T_a^*$ as $T_a^* \to 0$ and it tends to infinity like $\sqrt{T_a^*}$ as $T_a^* \to \infty$.

7.3.5 Analysis of the Wave Behavior at the Interface

Now we consider an incident plane wave propagating in a material. This may be a torsional wave, which is an analogy of a longitudinal wave in the case of translational degrees of freedom, or this may be a bending wave, which is an analogy of a transverse wave in the case of translational degrees of freedom. Further, we study wave processes that take place when the incident wave reaches an interface between two materials possessing different physical properties. We note that, in the case of rotational degrees of freedom, the wave processes at the interface are similar to the wave processes that take place at the interface in the case of translational degrees of freedom. Thus, when a torsional wave or a bending wave reaches the interface, a different number of waves (from one to four) can occur. However, in the case of continua based on translational degrees of freedom, the transverse wave propagation velocity is less than the longitudinal wave propagation velocity. This restriction does not inherent to the continua based on rotational degrees of freedom. Below, considering the behavior of torsional and bending waves, we focus our attention on the conditions that are impossible in the case of longitudinal and transverse waves. We do not aim to study in detail the dependence of the wave behavior on the parameters. The results below are illustrative.

In order to study the wave processes occurring near the interface between two materials, we can use Eq. (7.28) written in terms of the electromagnetic and thermodynamic quantities. However, it is more convenient to use the corresponding equations written in terms of the mechanical quantities, namely Eq. (7.24). It is easy to show that Eq. (7.24) can be reduced to the following one:

$$\frac{C_\Theta}{\varrho}\nabla\nabla\cdot\boldsymbol{\theta} - \frac{C_\Psi}{\varrho}\nabla\nabla\times\boldsymbol{\theta} - \beta\varrho J\frac{d\boldsymbol{\theta}}{dt} - \varrho J\frac{d^2\boldsymbol{\theta}}{dt^2} = 0. \tag{7.32}$$

Further, we denote the parameters of the first and second materials by indexes 1 and 2, respectively. We introduce the dimensionless time $\tilde{t}$ and the dimensionless gradient operator $\tilde{\nabla}$ as

$$\tilde{t} = \beta_1 t, \qquad \tilde{\nabla} = \frac{1}{\varrho_1\beta_1}\sqrt{\frac{C_\Psi^{(1)}}{J_1}}\,\nabla. \tag{7.33}$$

In view of Eq. (7.33) we can write down Eq. (7.32) for the first and second materials as

$$\begin{aligned} &\alpha_1\tilde{\nabla}\tilde{\nabla}\cdot\boldsymbol{\theta} - \tilde{\nabla}\tilde{\nabla}\times\boldsymbol{\theta} - \frac{d\boldsymbol{\theta}}{d\tilde{t}} - \frac{d^2\boldsymbol{\theta}}{d\tilde{t}^2} = 0, \\ &\alpha_2\tilde{\nabla}\tilde{\nabla}\cdot\boldsymbol{\theta} - \alpha\tilde{\nabla}\tilde{\nabla}\times\boldsymbol{\theta} - \alpha_3\frac{d\boldsymbol{\theta}}{d\tilde{t}} - \frac{d^2\boldsymbol{\theta}}{d\tilde{t}^2} = 0, \end{aligned} \tag{7.34}$$

where the dimensionless parameters are calculated by the formulas

$$\alpha = \frac{C_\Psi^{(2)}\varrho_1^2 J_1}{\varrho_2^2 J_2 C_\Psi^{(1)}}, \quad \alpha_1 = \frac{C_\Theta^{(1)}}{C_\Psi^{(1)}}, \quad \alpha_2 = \frac{C_\Theta^{(2)}\varrho_1^2 J_1}{\varrho_2^2 J_2 C_\Psi^{(1)}}, \quad \alpha_3 = \frac{\beta_2}{\beta_1}. \tag{7.35}$$

We note that constants α, α_1 and α_2 can be expressed in terms of the wave propagation velocities, namely

$$\alpha = \left(\frac{c^{(2)}}{c^{(1)}}\right)^2, \quad \alpha_1 = \left(\frac{c_r^{(1)}}{c^{(1)}}\right)^2, \quad \alpha_2 = \left(\frac{c_r^{(2)}}{c^{(1)}}\right)^2. \tag{7.36}$$

A numerical analysis of the wave propagation problem has been carried out by using the finite-volume method. Results of the analysis are presented in Figs. 7.1–7.5 where distributions of the rotation vector magnitude are shown. Dashed lines in the figures separate the media with different physical properties. Numbers in the figures indicate waves of different types.

Certainly, the analysis of Figs. 7.1–7.5 does not allow us to distinguish torsional waves and bending waves. In order to determine the types of waves, the analysis of the projections of the rotation vector has been performed. Figures 7.1–7.5 correspond to moments of time close to the moments when the incident waves reach the second

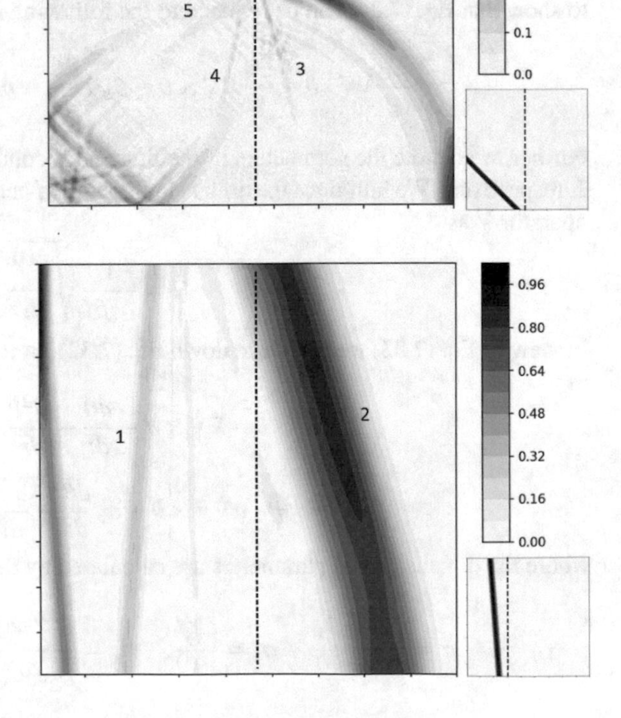

Fig. 7.1 The case of the incident torsional wave and $\alpha_1 = 15$, $\alpha_2 = 10$. Notation: 1 is the incident torsional wave, 2 is the refracted torsional wave, 3 is the refracted bending wave, 4 is the reflected bending wave, 5 is the reflected torsional wave

Fig. 7.2 The case of the incident torsional wave with the incident angle close to zero and $\alpha_1 = 10$, $\alpha_2 = 100$. Notation: 1 is the reflected torsional wave, 2 is the refracted torsional wave

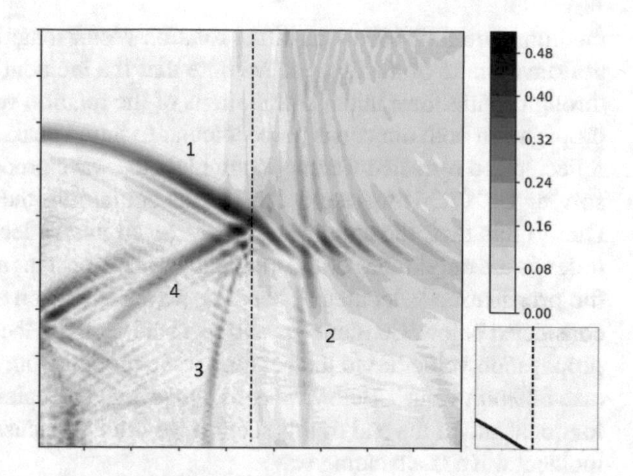

Fig. 7.3 The case of the incident torsional wave with the large incident angle and $\alpha_1 = 10$, $\alpha_2 = 100$. Notation: 1 is the incident torsional wave, 2 is the refracted bending wave, 3 is the reflected bending wave, 4 is the reflected torsional wave

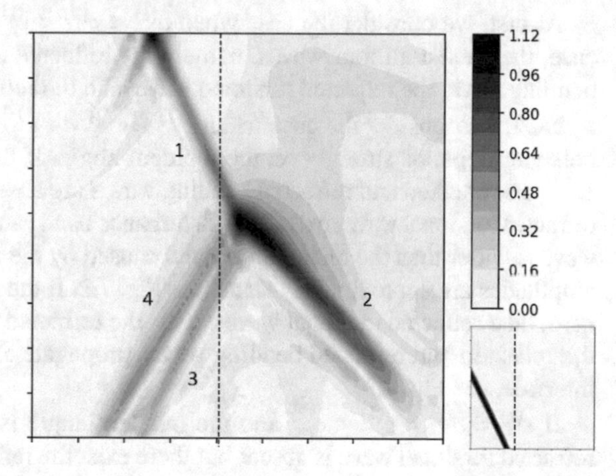

Fig. 7.4 The case of the incident torsional wave with the large incident angle and $\alpha_1 = 2/3$, $\alpha_2 = 10$. Notation: 1 is the incident torsional wave, 2 is the refracted bending wave, 3 is the reflected torsional wave, 4 is the reflected bending wave

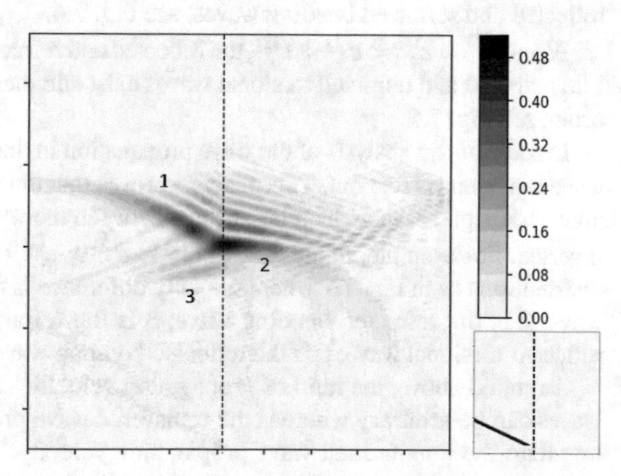

Fig. 7.5 The case of the incident torsional wave and $\alpha_1 = 1/15$, $\alpha_2 = 1/10$. Notation: 1 is the incident torsional wave, 2 is the refracted torsional wave, 3 is the reflected torsional wave

medium. Initial distributions of the rotation vector magnitude are shown in the right bottom corners of the figures. We note that the incident plane waves are generated through nonuniform initial distributions of the rotation vector. As a result, the waves propagate in both directions perpendicular to initial peak. In order to easily recognize reflected and refracted waves, we simulate the wave processes without damping, i.e., solving Eq. (7.34) we neglect the terms containing the first time derivatives of $\boldsymbol{\theta}$. Due to this fact, the amplitudes of refracted and reflected waves are of the same order as the amplitudes of the incident waves. Constant α characterizing the ratio of the propagation velocities of bending waves is chosen as $\alpha = 1.5$ for all examples considered below. Constants α_1 and α_2 characterizing the ratios of the torsional wave propagation velocities to the bending wave propagation velocity in the first material take different values. Below we consider several examples where the incident wave is torsional and at the end of this section we briefly discuss what should change if the incident wave is a bending wave.

At first, we consider the case when $c^{(1)} \approx c^{(2)} < c_r^{(1)} \approx c_r^{(2)}$, see Fig. 7.1. In this case, there exist all four waves, namely the reflected torsional wave, the reflected bending wave, the refracted torsional wave, and the refracted bending wave.

Next, we consider the case when $c^{(1)} \approx c^{(2)} < c_r^{(1)} < c_r^{(2)}$. In this case, the wave behavior depends strongly on the incident angle. If the incident angle is close to zero, the reflected and refracted bending waves are absent whereas the reflected and refracted torsional waves exist. The disturbance band caused by the refracted torsional wave is wider than the disturbance band caused by the incident wave, however their amplitudes are approximately same, see Fig. 7.2. If the incident angle is not close to zero, then reflected torsional wave exists, the refracted torsional wave is absent and the reflected and refracted bending waves propagate almost perpendicularly to the interface, see Fig. 7.3.

If $c^{(1)} \approx c^{(2)} \approx c_r^{(1)} < c_r^{(2)}$ and the incident angle is not close to zero, then the refracted torsional wave is absent but there exist the reflected torsional wave and the reflected and refracted bending waves, see Fig. 7.4.

When $c_r^{(1)} \approx c_r^{(2)} < c^{(1)} \approx c^{(2)}$, the reflected and refracted bending waves are absent. The reflected and refracted torsional waves exist and their amplitudes are of the same order, see Fig. 7.5.

In addition, an analysis of the wave propagation in the case of the bending incident wave has been carried out. This analysis proves that the wave behavior similar to that above takes place when the relationships between the wave propagation velocities are inverted. For example, in the case of $c^{(1)} \approx c^{(2)} > c_r^{(1)} \approx c_r^{(2)}$, we have the amplitude distributions as in Fig. 7.1 where the only difference is that 1 is the incident bending wave, 2 is the refracted bending wave, 3 is the refracted torsional wave, 4 is the reflected torsional wave, 5 is the reflected bending wave.

As stated above, the ratio of propagation velocities of the bending and torsional waves can be arbitrary whereas the transverse wave propagation velocity is always less than the longitudinal wave propagation velocity. This difference is important since, considering the classical continuum with translational degrees of freedom, in the case of the incident longitudinal wave we could not obtain the results similar to that presented in Fig. 7.5 and in the case of the incident transverse wave we could not obtain the results similar to that presented in Figs. 7.1–7.4. Thus, on the one hand, mechanical models and mechanical analogues allows us to predict what processes are possible and what processes are not possible and give a hint about a direction of experimental research. On the other hand, if predictions obtained on the basis of a mechanical model contradict experimental data this is a reason to change the mechanical model but this is not a reason to assert that it is impossible to describe the given physical process using the continuum mechanics methods.

7.4 Whether Modern Thermodynamics Needs Mechanical Models

Summarizing the above analysis of trends in the development of modern thermodynamics, we note, that along with the significant progress in developing mathematical methods, there is a completely different situation with mechanical models of thermal processes. In modern statistical thermodynamics, the mechanical model no longer plays an important role that it played at the initial stage of the development of this science. Non-equilibrium thermodynamics is a purely phenomenological science. In this science, the question of mechanical interpretations of temperature and other thermodynamic quantities has never been raised. In fact, now mechanical models of thermal processes are developed only in the framework of continuum mechanics. In this regard, we return to the issues raised in the introduction, namely, whether modern science needs mechanical models and whether a physical theory should be an explanation. We are convinced that the answer to both questions should be affirmative. In our opinion, the main and almost the only incentive to create a new theory is the scientist's desire to understand the essence of the phenomenon, i.e., to explain this phenomenon, at least to himself. People tend to think using analogies. If a person finds some kind of analogy between a new phenomenon for him and the one he knows well or can imagine, he has a feeling that he has found an explanation for this phenomenon and understands the essence of it. Mechanical models are useful due to several reasons. Firstly, they create vivid visual images that stimulate intuitive thinking. Secondly, they allow us to draw analogies with the well-studied phenomena. Thirdly, the use of mechanical models allows for a more fundamental study, since it involves the derivation of differential equations from the first principles through logical reasoning.

References

Ananth P, Dinesh A, Sugunamma V, Sandeep N (2015) Effect of nonlinear thermal radiation on stagnation flow of a casson fluid towards a stretching sheet. Ind Eng Lett 5(8):70–79

Babenkov MB, Ivanova EA (2014) Analysis of the wave propagation processes in heat transfer problems of the hyperbolic type. Continuum Mech Thermodyn 26(4):483–502

Babenkov MB, Vitokhin EY (2018) Thermoelastic waves in a medium with heat-flux relaxation. In: Altenbach, H and Öchsner, A (ed) Encyclopedia of Continuum Mechanics, Springer, Berlin, Heidelberg

Baik C, Lavine AS (1995) On hyperbolic heat conduction equation and the second law of thermodynamics. Trans ASME J Heat Transf 117:256 – 263

Bardeen J, Cooper LN, Schrieffer JR (1957) Micromorphic theory of superconductivity. The Phys Rev 106(1):162–164

Barletta A, Zanchini E (1997) Hyperbolic heat conduction and local equilibrium: A second law analysis. Int J Heat Mass Transf 40(5):1007 – 1016

Biot MA (1970) Variational Principles in Heat Transfer: A Unified Lagrangian Analysis of Dissipative Phenomena. Oxford Mathematical Monographs, Clarendon Press, Oxford

Biot MA (1984) New variational-Lagrangian irreversible thermodynamics with application to viscous flow, reaction–diffusion, and solid mechanics. Advances in Applied Mechanics 24:1 – 91

Boltzmann L (1974) Theoretical Physics and Philosophical Problems: Selected Writings. D. Reidel Publishing Company, Boston

Campo A (1982) Estimate of the transient conduction of heat in materials with linear thermal properties based on the solution for constant properties. Heat and Mass Transf 17(1):1–9

Čápek V, Sheehan DP (2005) Challenges to the Second Law of Thermodynamics. Theory and Experiment. Springer

Cataneo C (1958) A form of heat conduction equation which eliminates the paradox of instantaneous propagation. Compte Rendus 247:431–433

Chaibi M, Fernández T, Mimouni A, Rodriguez-Tellez J, Tazón A, Mediavilla A (2012) Nonlinear modeling of trapping and thermal effects on GaAs and GaN MESFET/HEMT devices. Progr Electromagn Res 124:163–186

Chandrasekharaiah DS (1998) Hyperbolic thermoelasticity: A review of recent literature. Appl Mech Rev 51:705–729

De Groot SR (1951) Thermodynamics of Irreversible Processes. North Holland, Amsterdam

Dixon RC, Eringen AC (1964) A dynamical theory of polar elastic dielectrics — I. Int J Engng Sci 2:359–377

Dixon RC, Eringen AC (1965) A dynamical theory of polar elastic dielectrics — II. Int J Engng Sci 3:379–398

Dugdale JS (1996) Entropy and its Physical Meaning. Taylor & Francis, London

Duhem P (1954) The Aim and Structure of Physical Theory. Princeton Univercity Press, Princeton, New Jersey

Ebadian A, Darania P (2008) Study of exact solutions of nonlinear heat equations. Comput Appl Math 27(2):107–121

Eisenschitz R (1958) Statistical Theory of Irreversible Processes. Oxford University Press, London

Eringen AC (2003) Continuum theory of micromorphic electromagnetic thermoelastic solids. Int J Engng Sci 41:653–665

Eringen AC, Maugin GA (1990) Electrodynamics of Continua. Springer–Verlag, New York

Feynman RP, Leighton RB, Sands M (1963) The Feynman Lectures on Physics. Mainly Mechanics, Radiation, and Heat, vol 1. Addison Wesley Publishing Company, London

Fomethe A, Maugin GA (1996) Material forces in thermoelastic ferromagnets. Continuum Mech Thermodyn 8:275–292

Fong E, Lam TT, Davis SE (2010) Nonlinear heat conduction in isotropic and orthotropic materials with laser heat source. J Thermophys Heat Transf 24(1):104–111

Galeş C, Ghiba ID, Ignătescu I (2011) Asymptotic partition of energy in micromorphic thermopiezoelectricity. Journal of Thermal Stresses 34:1241–1249

Gay-Balmaz F, Yoshimura H (2019) From Lagrangian mechanics to nonequilibrium thermodynamics: A variational perspective. Entropy 21(1), doi:10.3390/e21010008, 8

Grekova E, Zhilin P (2001) Basic equations of Kelvin's medium and analogy with ferromagnets. Journal of Elasticity 64:29–70

Grekova EF (2001) Ferromagnets and Kelvin's medium: Basic equations and wave processes. Journal of Computational Acoustics 9(2):427–446

Grudinin I, Lee H, Chen T, Vahala K (2011) Compensation of thermal nonlinearity effect in optical resonators. Opt Express 19(8):7365–7372

Habibi M, Oloumi M, Hosseinkhani H, Magidi S (2015) Numerical investigation into the highly nonlinear heat transfer equation with Bremsstrahlung emission in the inertial confinement fusion plasmas. Contrib Plasma Phys 55(9):677–684

Huang C, Fan J, Zhu L (2012) Dynamic nonlinear thermal optical effects in coupled ring resonators. AIP Adv 2(032131):1–8

Huang K (1963) Statistical Mechanics. John Wiley and Sons, New York

Ignaczak J, Ostoja-Starzewski M (2009) Thermoelasticity with Finite Wave Speeds. Oxford Science Publications, Oxford

Ivanova EA (2010) Derivation of theory of thermoviscoelasticity by means of two-component medium. Acta Mech 215:261–286

Ivanova EA (2011) On one model of generalized continuum and its thermodynamical interpretation. In: Altenbach H, Maugin GA, Erofeev V (eds) Mechanics of Generalized Continua, Springer, Berlin, Heidelberg, Advanced Structured Materials, vol 7, pp 151–174

Ivanova EA (2012) Derivation of theory of thermoviscoelasticity by means of two-component Cosserat continuum. Tech Mech 32:273–286

Ivanova EA (2014) Description of mechanism of thermal conduction and internal damping by means of two-component Cosserat continuum. Acta Mech 225:757–795

Ivanova EA (2015) A new model of a micropolar continuum and some electromagnetic analogies. Acta Mech 226:697–721

Ivanova EA (2017) Description of nonlinear thermal effects by means of a two-component Cosserat continuum. Acta Mech 228:2299–2346

Ivanova EA (2018) Thermal effects by means of two-component cosserat continuum. In: Altenbach H, Öchsner A (eds) Encyclopedia of Continuum Mechanics, Springer, Berlin, pp 1–12

Ivanova EA (2019a) On micropolar continuum approach to some problems of thermo- and electrodynamics. Acta Mech 230:1685–1715

Ivanova EA (2019b) Towards micropolar continuum theory describing some problems of thermo- and electrodynamics. In: Altenbach H, Irschik H, Metveenko V (eds) Contributions to Advanced Dynamics and Continuum Mechanics, Springer, Cham, Advanced Structured Materials, vol 114, pp 1–19

Ivanova EA, Kolpakov YE (2013) Piezoeffect in polar materials using moment theory. J Appl Mech Tech Phys 54(6):989–1002

Ivanova EA, Kolpakov YE (2016) A description of piezoelectric effect in non-polar materials taking into account the quadrupole moments. Z Angew Math Mech 96(9):1033–1048

Jordan A, Khaldi S, Benmouna M, Borucki A (1987) Study of non-linear heat transfer problems. Revue de Physique Appliqueé 22(1):101–105

Jou D, Casas-Vazquez J, Lebon G (2001) Extended Irreversible Thermodynamics. Springer, Berlin

Khandekar C, Pick A, Johnson SG, Rodriguez AW (2015) Radiative heat transfer in nonlinear Kerr media. Phys Rev B 91(115406):1–9

Kittel C (1970) Thermal Physics. John Wiley and Sons, New York

Kondepudi D, Prigogine I (1998) Modern Thermodynamics: From Heat Engines to Dissipative Structures. Wiley, Chichester

Krylov NS (2003) Works on the Substantiation of Statistical Physics (in Russ.). Editorial URSS, Moscow

Kubo R (1965) Statistical Mechanics. Elsevier Science Publishers B.V., Amsterdam

Lebon G, Casas-Vazquez J (1974) Lagrangian formulation of unsteady non-linear heat transfer problems. J Enging Math 8(1):31 – 44

Lebon G, Dauby PC (1990) Heat transport in dielectric crystals at low temperature: a variational formulation based on extended irreversible thermodynamics. Phys Rev A 42:4710 – 4715

LeMesurier B (2008) Modeling thermal effects on nonlinear wave motion in biopolymers by a stochastic discrete nonlinear Schrödinger equation with phase damping. Discrete and Contin Dyn Syst S 1(2):317–327

Lieb EH, Yngvason J (1997) A guide to entropy and the second law of thermodynamics. Notices of the AMS (May):571 – 581

Markides CN, Osuolale A, Solanki R, Stan GBV (2013) Nonlinear heat transfer processes in a two-phase thermofluidic oscillator. Appl Energy 104:958–977

Maugin GA (1988) Continuum Mechanics of Electromagnetic Solids. Elsevier Science Publishers, Oxford

Mottaghy D, Rath V (2006) Latent heat effects in subsurface heat transport modeling and their impact on palaeotemperature reconstructions. Geophys J Int 164:236–245

Müller I (2007) A History of Thermodynamics. The Doctrine of Energy and Entropy. Springer, Berlin, Heidelberg, doi:10.1007/978-3-540-46227-9

Ostoja-Starzewski M (2016) Second law violations, continuum mechanics, and permeability. Continuum Mechanics and Thermodynamics 28(1-2):489–501, doi:10.1007/s00161-015-0451-4

Ostoja-Starzewski M (2017) Admitting spontaneous violations of the second law in continuum thermomechanics. Entropy 19(78):1–10, doi:10.3390/e19020078

Polyanin AD, Zhurov AI, Vyaz'min AV (2000) Exact solutions of nonlinear heat- and mass-transfer equations. Theor Found of Chem Eng 34(5):403–415

Prigogine I (1955) Introduction to Thermodynamics of Irreversible Processes. Charles C. Thomas Publishers, Springfield

Reif F (1967) Berkeley Physics Course, vol 5. Statistical Physics. McGraw-Hill Book Company, New York

Röpke G (2013) Nonequilibrium Statistical Physics. Wiley-VCH, Weinheim

Rubin MB (1992) Hyperbolic heat conduction and the second law. Int J Eng Sci 30(11):1665 – 1676

Ruelle D (1978) Thermodynamic Formalism. The Mathematical Structures of Classical Equilibrium Statistical Mechanics. Addison-Wesley Publishing Company, London

Shliomis MI, Stepanov VI (1993) Rotational viscosity of magnetic fluids: contribution of the Brownian and Néel relaxational processes. J Magnetism Magnetic Mat 122:196–199

Straughan B (2011) Heat Waves, Applied Mathematical Sciences, vol 177. Springer, New York

Tiersten HF (1964) Coupled magnetomechanical equations for magnetically saturated insulators. J Math Phys 5(9):1298–1318

Treugolov IG (1989) Moment theory of electromagnetic effects in anisotropic solids. Journal of Applied Mathematics and Mechanics 53(6):786–790

Vitokhin EY, Ivanova EA (2017) Dispersion relations for the hyperbolic thermal conductivity, thermoelasticity and thermoviscoelasticity. Continuum Mech Thermodyn 29(6):1219–1240

Zanchini E (1999) Hyperbolic heat conduction theories and nondecreasing entropy. Phys Rev B Condens Matter Mater Phys 60(2):991 —- 997

Zhilin PA (2006a) Advanced Problems in Mechanics, vol 2. Institute for Problems in Mechanical Engineering, St. Petersburg

Zhilin PA (2006b) Advanced Problems in Mechanics (In Russ.), vol 1. Institute for Problems in Mechanical Engineering, St. Petersburg

Zhilin PA (2012) Rational Continuum Mechanics (in Russ.). Polytechnic University Publishing House, St. Petersburg

Chapter 8
Estimation of Energy of Fracture Initiation in Brittle Materials with Cracks

Ruslan L. Lapin, Nikita D. Muschak, Vadim A. Tsaplin, Vitaly A. Kuzkin, and Anton M. Krivtsov

Abstract We study deformation and fracture of a brittle material under mixed quasi-static loading. Numerical simulations of deformation of a cubic sample containing a single crack are carried out using the particle dynamics method. Effect of ratio of compressive and shear loads on energy of fracture initiation is investigated for two crack shapes and various crack orientations. The energy of fracture initiation in a material containing multiple cracks is estimated using the non-interaction approximation. It is shown that in the case of mixed loading (compression and shear) the energy is significantly lower than in the case of pure compression. Presented results may serve for minimization of energy consumption during disintegration of solid minerals.

Keywords: Brittle fracture · Fracture initiation · Cracks · Particle dynamics method · Desintegration of rocks · Energy consumption

8.1 Introduction

One of the key technological challenges for mining industry is minimization of energy consumption during disintegration (fracture) of solid minerals (Vaisberg and

---

Ruslan L. Lapin · Nikita D. Muschak
Peter the Great St. Petersburg Polytechnic University, Department of Theoretical Mechanics, Institute of Applied Mathematics and Mechanics, Politechnicheskaya 29, 195251 St. Petersburg, Russia,
e-mail: lapruslan@gmail.com, niky-m@yandex.ru

Vadim A. Tsaplin · Vitaly A. Kuzkin · Anton M. Krivtsov
Peter the Great St. Petersburg Polytechnic University, Department of Theoretical Mechanics, Institute of Applied Mathematics and Mechanics, Politechnicheskaya 29, 195251 St. Petersburg & Laboratoty "Discrete models in mechanics", Institute for Problems in Mechanical Engineering of Russian Academy of Sciences, Bolshoy pr. V.O. 61, 199178 St. Petersburg, Russia,
e-mail: vtsaplin@yandex.ru, kuzkinva@gmail.com, akrivtsov@bk.ru

H. Altenbach and A. Öchsner (eds.), *State of the Art and Future Trends in Material Modeling*, Advanced Structured Materials 100,
https://doi.org/10.1007/978-3-030-30355-6_8

Kameneva, 2014; Vaisberg et al, 2018a,b). The energy required for fracture of rocks strongly depends on their heterogeneous internal structure. Development of scanning technologies, for example, computer microtomography, makes it possible to determine shapes and sizes of heterogeneities (Vaisberg and Kameneva, 2014; Vaisberg et al, 2018a; Vesga et al, 2008). However, finding relation between microstructure of a rock and its mechanical properties is still a challenging problem for mechanics (Kachanov and Sevostianov, 2018; Altenbach and Sadowski, 2015; Altenbach and Öchsner, 2011).

Influence of heterogeneities on effective elastic properties of materials is studied in many works on micromechanics (Torquato, 1991). Materials with pores (Shafiro and Kachanov, 1997; Kumar and Han, 2005; Bîrsan and Altenbach, 2011), cracks (Sayers and Kachanov, 1991; Saenger, 2008; Min and Jing, 2003; Grechka and Kachanov, 2006), and inclusions (Shafiro and Kachanov, 2000) of different shapes are considered. Proper microstructural parameters determining contribution of heterogeneities to effective properties are introduced (Kachanov and Sevostianov, 2005; Kachanov, 1999). State of the art in calculation of effective elastic properties is summarized in the recent book (Kachanov and Sevostianov, 2018). In particular, the non-interaction approximation allowing to calculate effective properties analytically is discussed in detail.

Success of micromechanics in prediction of effective elastic properties is caused by the fact that these properties are insensitive to many features of real microstructure. Estimation of influence of heterogeneities, e.g. cracks, on strength properties is more challenging. Complexity of estimation of strength properties is related to the fact that strength is determined by local stress fields. Therefore many works are devoted to development of numerical schemes for accurate calculation of the local fields (Linkov, 2002; Krivtsov, 2007; Kuna, 2013). In particular, efficient methods for calculation of stress intensity factors in materials containing multiple cracks are proposed in Rejwer et al (2014); Kushch et al (2009); Jaworski et al (2016). Relation between distribution of stress intensity factor and effective strength of a material is discussed in Rejwer et al (2014).

From practical point of view, it is important to find relation between loading type and energy required for fracture of a material. This problem is not fully covered in literature. In Bratov and Krivtsov (2016), a simple two-dimensional model for estimation of energy of fracture initiation is proposed. Influence of loading type on the energy is investigated. It is shown that mixed loading (compression and shear) is energetically more efficient than pure compression. In the present paper, we generalize the results of Bratov and Krivtsov (2016) for the three-dimensional case.

The paper is organized as follows. In Sect. 8.2, discrete model of a brittle material is presented. In Sect. 8.3, simulation of deformation and fracture of a sample containing an infinite rectangular crack is carried out. Energy of fracture initiation under various loads for different crack orientations is calculated. Numerical results are compared with analytical estimates (Bratov and Krivtsov, 2016). In Sect. 8.4, the energy of fracture initiation is calculated for a penny-shaped crack. Generalization for the case of multiple non-interacting penny-shaped cracks is carried out in Sect. 8.5.

8.2 Discrete Model of a Brittle Material

In this section, a discrete model of deformation and fracture of a brittle material (e.g. rock) is presented. A material is simulated using the particle dynamics method (Krivtsov, 2007, 2004, 2003). In this method, a material is represented as a set of interacting particles (~material points) connected by bonds. Cubic sample of a material is considered. Number of particles in the sample is of order of $5 \cdot 10^5$. Particles form a quasi-random lattice (Tsaplin and Kuzkin, 2017). Positions of the particles are calculated using the following algorithm, proposed in Tsaplin and Kuzkin (2017). A perfect face-centered cubic lattice (FCC) is created. Particles are located at nodes of the lattice (Fig. 8.1A). Then particles get random displacements (Fig. 8.1B). Magnitudes of the displacements are of order of $0.4d$. Here d is the step of the FCC lattice. Each pair of particles at the distance less than $1.9d$ is connected by a linear elastic spring (bond). On average, each particle has 20 bonds. The equilibrium bond length is equal to the initial distance between connected particles. In Tsaplin and Kuzkin (2017) it is shown that resulting material has isotropic elastic and strength properties.

During the simulation, the following equations of motion for particles are solved numerically:

$$m\dot{\mathbf{v}}_i = \sum_j \mathbf{F}_{ij} - \beta \mathbf{v}_i. \tag{8.1}$$

Here, summation is carried out over all particles j connected with particle i; m is particle mass; $\mathbf{v}_i$ is particle velocity; $\mathbf{F}_{ij}$ is force in the bond connecting particles i and j; β is coefficient of artificial dissipation, which is introduced in order to suppress vibrations caused by deformation of the sample. Forces, $\mathbf{F}_{ij}$, arising in bonds are calculated as

$$\mathbf{F}_{ij} = c_{ij}(r_{ij} - r_{ij}^0)\mathbf{e}_{ij}, \qquad c_{ij} = c_0 \frac{d}{r_{ij}^0}, \tag{8.2}$$

where c_{ij} is bond stiffness; r_{ij} is distance between particles i, j; r_{ij}^0 is initial bond length; $\mathbf{e}_{ij}$ is unit vector directed along the line connecting the particles; c_0 is

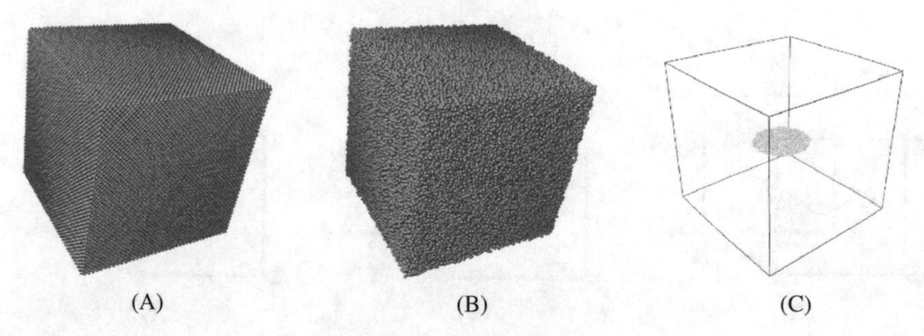

Fig. 8.1 (A) FCC lattice; (B) quasi-random lattice; (C) an example of a crack in the sample (particles near the crack are shown only)

stiffness of a bond with initial length equal to d. Equations of motion (8.1) are solved numerically using the symplectic leap-frog integration scheme with time step $2 \cdot 10^{-2} T_*$, where

$$T_* = 2\pi\sqrt{\frac{m}{c_0}}.$$

Periodic boundary conditions in all space directions are used.

Macroscopic elastic moduli of the considered material are calculated in Tsaplin and Kuzkin (2017). It is shown that the Young modulus and the Poisson ratio are related with microparameters as

$$\nu = 0.255, \qquad E = 1.48\frac{c_0}{d}. \tag{8.3}$$

An initial crack is created by removing bonds between the particles that cross a crack surface (Fig. 8.1C). Contact between crack faces is neglected. Crack propagation is simulated by removing bonds, satisfying the following inequality:

$$\frac{r_{ij} - r_{ij}^0}{r_{ij}^0} > \varepsilon_{cr}, \tag{8.4}$$

where ε_{cr} is the critical bond deformation. In further calculations ε_{cr} is equal to $2 \cdot 10^{-4}$.

8.3 An Infinite Rectangular Crack

In this section, we study initiation of fracture (crack propagation) in a sample containing an infinite rectangular crack under various loads. An initial crack is shown in Fig. 8.2A. Angle α is a parameter defining orientation of the crack with respect to direction of loading. Crack length is equal to 0.35 of periodic cell size. During the simulation, every $20T_*$ a periodic cell is subjected to a uniform strain ($\Delta\varepsilon_{zz}$ or $\Delta\varepsilon_{yz}$ or both). Three cases are considered: (a) compression along z axis with increment

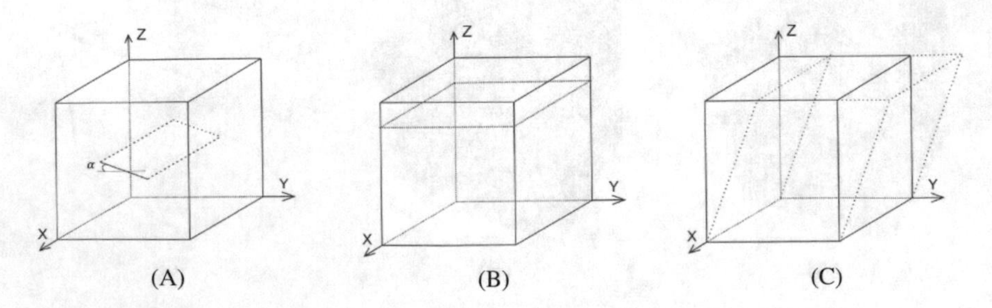

Fig. 8.2 Periodic cell containing an infinite crack (A). Change of the periodic cell under compression (B) and shear (C)

$\Delta\varepsilon_{zz} = 5 \cdot 10^{-6}$ (see Fig. 8.2B); (b) pure shear with increment $\Delta\varepsilon_{yz} = 5 \cdot 10^{-6}$ (Fig. 8.2C) and (c) mixed loading with increments $\Delta\varepsilon_{zz} = \Delta\varepsilon_{yz} = 5 \cdot 10^{-6}$.

Deformation of the sample leads to breakage of bonds. A moment of fracture initiation is tracked by the number of broken bonds. It is assumed that the fracture begins when the number of broken bonds increases by 5% compared to the number of bonds removed for creation of the initial crack. When the criterion is satisfied, the strain energy density is calculated as a sum of potential energies of all bonds in the periodic cell:

$$U = \frac{1}{2V} \sum c_{ij}(r_{ij} - r_{ij}^{0})^2, \tag{8.5}$$

where V is volume of the periodic cell. Further U is referred to as the *energy of fracture initiation*.

We compare simulation results with analytical estimates obtained in Bratov and Krivtsov (2016). In Bratov and Krivtsov (2016), a single crack under compression and shear loads applied at infinity is considered in two-dimensional formulation. Solution of corresponding elasticity problem yields stresses near the crack tip. Neuber-Novozhilov fracture criterion (Novozhilov, 1969) is used. The criterion is used for estimation of energy of fracture initiation.

Comparison of analytical estimates Bratov and Krivtsov (2016) with the results of particle dynamics simulations is presented in Fig. 8.3. Every point on the plot corresponds to average over 5 simulations with different realizations of a quasi-random lattice. Figure 8.3 shows that numerical and analytical results are in a qualitative agreement. Deviations are caused by different fracture criteria and material models. In Bratov and Krivtsov (2016), linear fracture mechanics is used. In the framework of this approach, at certain crack orientations the fracture criterion is never satisfied. For example, uniaxial compression of a sample along crack direction does not lead to fracture. Therefore energy of fracture initiation, U, formally tends to infinity (see Fig. 8.3). In contrast, presented discrete model yields finite energy of fracture initiation for any crack orientation. Moreover, fracture is possible even in the absence of a

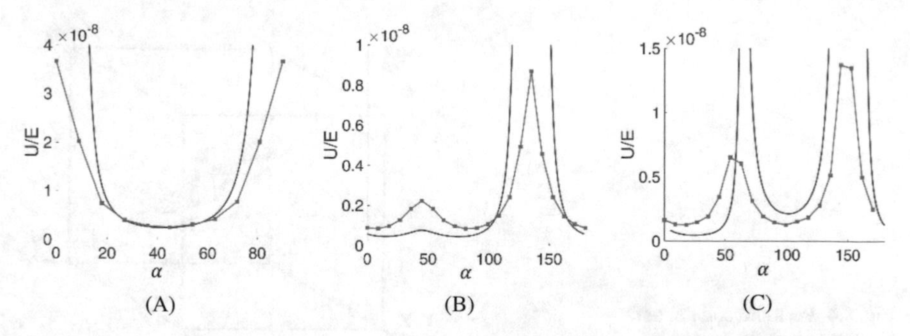

Fig. 8.3 Energy of fracture initiation, U, as a function of rectangular crack orientation under uniaxial compression (A), pure shear (B), and compression with a shear (C). In the latter case $\Delta\varepsilon_{zz} = \Delta\varepsilon_{yz}$. Results of particle dynamics modeling (squares) and analytical estimates (Bratov and Krivtsov, 2016) (solid line) are shown

crack. Additionally, we note that relation between bond breakage criterion (8.4) and the Neuber-Novozhilov criterion is not straightforward.

Figure 8.3 shows that the energy of fracture initiation strongly depends on crack orientation and loading type. In particular, the energy has clear minima corresponding to the most energetically beneficial crack orientations. In the following section, this fact is considered in detail for a penny-shaped crack.

8.4 A Penny-shaped Crack

In this section, initiation of fracture in a sample containing a penny-shaped (circular) crack (Fig. 8.4) is considered. As in the previous section, the sample is subjected to compressive and shear strains with increments $\Delta\varepsilon_{zz}$ and $\Delta\varepsilon_{yz}$ every $20T_*$. Crack orientation is specified by angle α (see Fig. 8.4). The ratio of crack diameter to size of the periodic cell is equal to 0.35.

Influence of crack orientation and ratio of strain increments,

$$k = \frac{\Delta\varepsilon_{yz}}{\Delta\varepsilon_{zz}},$$

on energy of fracture initiation is investigated. Results of numerical simulations are shown in Fig. 8.5. Each point on the plot corresponds to average over 5 realizations of a quasi-random lattice. Figure 8.5 shows that under uniaxial compression ($k = 0$) energy of fracture initiation, U, has minima at $\alpha = 45°; 135°$ and maxima at $\alpha = 0°; 90°; 180°$. Adding shear leads to decrease of minimum and maximum values of fracture initiation energy. Moreover, for $k > 0.5$ and for all crack orientations, the energy of fracture initiation is less than in the case of uniaxial compression. Therefore mixed loading is energetically more efficient than uniaxial compression.

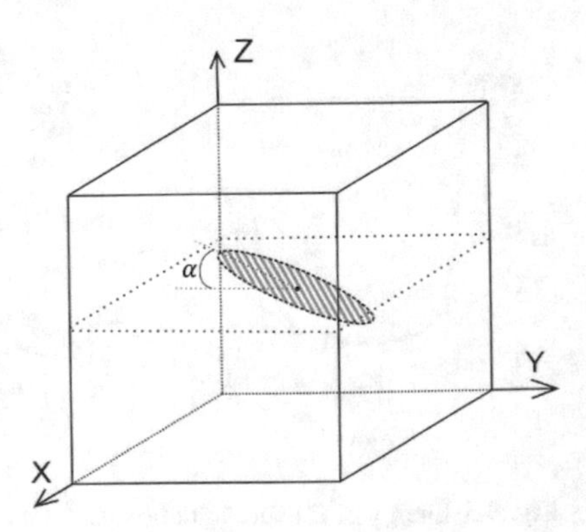

Fig. 8.4 Periodic cell containing a penny-shaped crack

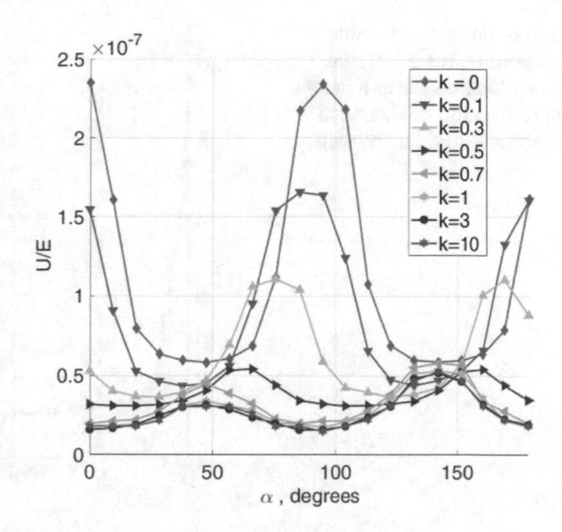

Fig. 8.5 Dependence of the energy of fracture initiation, U, on crack orientation for various ratios of shear and compressive strain increments, k (uniaxial compression corresponds to $k = 0$)

8.5 Multiple Randomly Oriented Penny-shaped Cracks (Non-interaction Approximation)

In this section, energy of fracture initiation in a material containing multiple randomly oriented penny-shaped cracks is estimated using the results obtained above and the non-interaction approximation (Kachanov and Sevostianov, 2018). Consider a material containing randomly located and oriented cracks. Suppose that the cracks are located far from each other. In the framework of the non-interaction approximation, we assume that mutual influence of cracks can be neglected. Then for each crack, the energy of fracture initiation can be estimated using Fig. 8.5. We assume that fracture starts at cracks with orientations, corresponding to minimum of function $U(\alpha)$. For each ratio of shear and compressive strains increments, k, the minima are calculated using results shown in Fig. 8.5. Resulting energy of fracture initiation for a material containing multiple cracks is presented in Fig. 8.6.

Figure 8.6 shows that the energy of fracture initiation has maximum at $k = 0$ (uniaxial compression) and monotonically decreases with increasing shear load. Note that adding a small shear load, significantly decreases energy of fracture initiation. These results are in a good qualitative agreement with experimental observations (Vaisberg et al, 2016) and analytical estimates (Bratov and Krivtsov, 2016).

8.6 Conclusions

Energy of fracture initiation for a material with a single crack under mixed loading (compression and shear) was calculated numerically. Rectangular and penny-shaped cracks were considered. Dependencies of the energy of fracture initiation on crack orientation for various ratios of compression and shear loads were obtained.

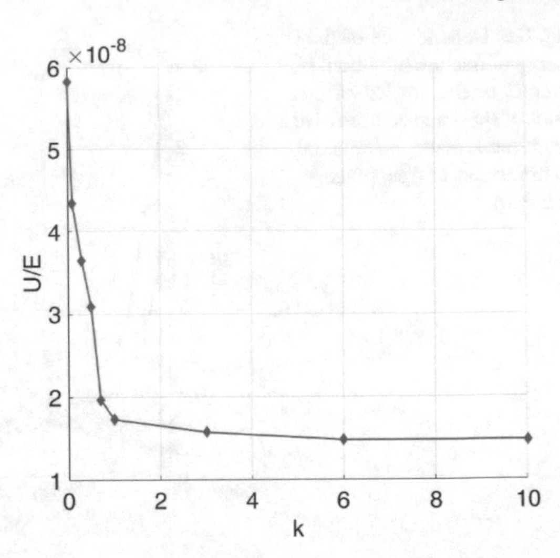

Fig. 8.6 Energy of fracture initiation, U, for a material with multiple cracks as a function of the ratio of shear and compressive strain increments, k

The dependencies were employed for estimation of energy of fracture initiation in a material containing multiple cracks under the non-interaction approximation. It was shown that the energy strongly depends on loading type. It has maximum in the case of uniaxial compression and it decreases monotonically with increasing shear load. It was shown that adding a small shear load yields significant decrease of the energy of fracture initiation.

Presented results may serve for minimization of energy consumption during disintegration of rocks, for example, in vibrational crushers (Vaisberg et al, 2018b). In particular, the results suggest that energy consumption under mixed loading is several times less than under uniaxial compression. This fact is in a good qualitative agreement with experimental observations (Vaisberg et al, 2018b, 2016).

Acknowledgements This work was supported by the Russian Science Foundation (Grant No. 17-79-30056). The authors are deeply grateful L.A Vaisberg for formulation of the problem and useful discussions. The work was initiated in the course of joint investigation of technological processes of vibration disintegration of materials, the main developer of which is REC "Mekhanobr-Tekhnika". Numerical modeling was performed using the Polytechnic supercomputer center at Peter the Great St. Petersburg Polytechnic University.

References

Altenbach H, Öchsner A (eds) (2011) Cellular and Porous Materials in Structures and Processes, CISM International Centre for Mechanical Sciences, vol 521. Springer, Vienna

Altenbach H, Sadowski T (eds) (2015) Failure and Damage Analysis of Advanced Materials, CISM International Centre for Mechanical Sciences, vol 560. Springer, Vienna

Bîrsan M, Altenbach H (2011) On the theory of porous elastic rods. International Journal of Solids and Structures 48(6):910 – 924

Bratov VA, Krivtsov AM (2016) Analysis of energy required for initiation of inclined crack under uniaxial compression and mixed loading. Engineering Fracture Mechanics 219:106,518

Grechka V, Kachanov M (2006) Effective elasticity of rocks with closely spaced and intersecting cracks. GEOPHYSICS 71(3):D85–D91

Jaworski D, Linkov A, Rybarska-Rusinek L (2016) On solving 3d elasticity problems for inhomogeneous region with cracks, pores and inclusions. Computers and Geotechnics 71(6):295–309

Kachanov M (1999) Solids with cracks and non-spherical pores: proper parameters of defect density and effective elastic properties. International Journal of Fracture 97(1-4):1 – 32

Kachanov M, Sevostianov I (2005) On quantitative characterization of microstructures and effective properties. International Journal of Solids and Structures 42(2):309 – 336

Kachanov M, Sevostianov I (2018) Micromechanics of Materials, with Applications, Solid Mechanics and Its Applications, vol 249. Springer

Krivtsov AM (2003) Molecular dynamics simulation of impact fracture in polycrystalline materials. Meccanica 38(01):61–70

Krivtsov AM (2004) Molecular dynamics simulation of plastic effects upon spalling. Physics of the Solid State 46(6):1055–1060

Krivtsov AM (2007) Deformation and Fracture of Solids with a Microstructure (in Russ.). Fizmatlit, Moscow

Kumar M, Han D (2005) Pore shape effect on elastic properties of carbonate rocks, Society of Exploration Geophysicists, pp 1477–1480

Kuna M (2013) Finite Elements in Fracture Mechanics, Solid Mechanics and Its Applications, vol 213. Springer Netherlands

Kushch VI, Sevostianov I, Mishnaevsky L (2009) Effect of crack orientation statistics on effective stiffness of mircocracked solid. International Journal of Solids and Structures 46(6):1574 – 1588

Linkov AM (2002) Boundary Integral Equations in Elasticity Theory. Kluwer Academic Publishers, Dordrecht-Boston-London

Min KB, Jing L (2003) Numerical determination of the equivalent elastic compliance tensor for fractured rock masses using the distinct element method. International Journal of Rock Mechanics and Mining Sciences 40(6):795 – 816

Novozhilov VV (1969) On a necessary and sufficient criterion for brittle strength. Journal of Applied Mathematics and Mechanics 33(2):201 – 210

Rejwer E, Rybarska-Rusinek L, Linkov A (2014) The complex variable fast multipole boundary element method for the analysis of strongly inhomogeneous media. Engineering Analysis with Boundary Elements 43:105–116

Saenger EH (2008) Numerical methods to determine effective elastic properties. International Journal of Engineering Science 46(6):598 – 605

Sayers CM, Kachanov M (1991) A simple technique for finding effective elastic constants of cracked solids for arbitrary crack orientation statistics. International Journal of Solids and Structures 27(6):671 – 680

Shafiro B, Kachanov M (1997) Materials with fluid-filled pores of various shapes: Effective elastic properties and fluid pressure polarization. International Journal of Solids and Structures 34(27):3517 – 3540

Shafiro B, Kachanov M (2000) Anisotropic effective conductivity of materials with nonrandomly oriented inclusions of diverse ellipsoidal shapes. Journal of Applied Physics 87(12):8561–8569

Torquato S (1991) Random heterogeneous media: microstructure and improved bounds on effective properties. Applied Mechanics Reviews 44(2):37–76

Tsaplin VA, Kuzkin VA (2017) On using quasi-random lattices for simulation of isotropic materials. Materials Physics and Mechanics 32(12):321–327

Vaisberg LA, Kameneva EE (2014) X-ray computed tomography in the study of physico-mechanical properties of rocks. Gornyi Zhurnal 9:85–90

Vaisberg LA, Baldaeva TM, Ivanov TM, Otroshchenko AA (2016) Screening efficiency with circular and rectilinear vibrations. Obogashchenie Rud 1:1–12

Vaisberg LA, Kameneva EE, Nikiforova VS (2018a) Microtomographic studies of rock pore space as the basis for rock disintegration technology improvements. Obogashchenie Rud 3:51–55

Vaisberg LA, Kruppa PI, Baranov VF (2018b) Microtomographic studies of rock pore space as the basis for rock disintegration technology improvements. Obogashchenie Rud 3:51–55

Vesga LF, Vallejo LE, Lobo-Guerrero S (2008) DEM analysis of the crack propagation in brittle clays under uniaxial compression tests. International Journal for Numerical and Analytical Methods in Geomechanics 32(11):1405–1415

Chapter 9
Effective Elastic Properties Using Maxwell's Approach for Transversely Isotropic Composites

Leandro Daniel Lau Alfonso, Reinaldo Rodríguez-Ramos, Jose A. Otero, Frédéric Lebon, Federico J. Sabina, Raul Guinovart-Díaz, and Julián Bravo-Castillero

Abstract In this contribution an analysis of static properties of transversely isotropic, porous and nano-composites is considered. Present work features explicit formulas for effective coefficient in these types of composites. The reinforcements of the

---

Leandro Daniel Lau Alfonso
Instituto de Cibernética, Matemática y Física, ICIMAF, Calle 15 No. 551, entre C y D, Vedado, Habana 4, CP–10400, Cuba,
e-mail: leandro@icimaf.cu

Reinaldo Rodríguez-Ramos
Facultad de Matemática y Computación, Universidad de la Habana, San Lázaro y L, Vedado, Habana 4, CP–10400, Cuba & Visiting Professor from February 15 to August 15, 2019 at Instituto de Investigaciones en Matemáticas Aplicadas y en Sistemas, Universidad Nacional Autónoma de México, Alcaldía Álvaro Obregón, 01000 CDMF, México,
e-mail: reinaldo@matcom.uh.cu

Jose A. Otero
Instituto Tecnológico de Estudios Superiores de Monterrey, CEM, E.M. CP 52926, México,
e-mail: j.a.otero@tec.mx

Frédéric Lebon
Aix-Marseille University, CNRS, Centrale Marseille, LMA, 4 Impasse Nikola Tesla, CS 40006, 13453 Marseille Cedex 13, France,
e-mail: lebon@lma.cnrs-mrs.fr

Federico J. Sabina
Instituto de Investigaciones en Matemáticas Aplicadas y en Sistemas, Universidad Nacional Autónoma de México, Alcaldía Álvaro Obregón, 01000 CDMF, México,
e-mail: fjs@mym.iimas.unam.mx

Raul Guinovart-Díaz
Facultad de Matemática y Computación, Universidad de la Habana, San Lázaro y L, Vedado, Habana 4, CP–10400, Cuba,
e-mail: guino@matcom.uh.cu

Julián Bravo-Castillero
Instituto de Investigaciones en Matemáticas Aplicadas y en Sistemas, Universidad Nacional Autónoma de México, Alcaldía Álvaro Obregón, 01000 CDMF, México,
e-mail: julian@mym.iimas.unam.mx

H. Altenbach and A. Öchsner (eds.), *State of the Art and Future Trends in Material Modeling*, Advanced Structured Materials 100,
https://doi.org/10.1007/978-3-030-30355-6_9

composites are a set of spheroidal inclusions with identical size and shape. The center is randomly distributed and the inclusions are embedded in an homogeneous infinite medium (matrix). An study of theoretical predictions obtained by Maxwell approach using two different density distribution functions, which describe the alignment inclusions is done. The method allows to report the static effective elastic coefficients in composites ensemble with inclusions of different geometrical shapes and configurations embedded into a matrix. The effective properties of composites are computed using the Maxwell homogenization method in Matlab software. Another novelty of this contribution is the calculation of new explicit analytical formulas for the control of the alignment tensors $\mathbf{N}^{*}$ and $\mathbf{N}^{s*}$ which is in charged of the alignment distribution of inclusions within matrix through disorder parameters λ and s, respectively. The alignment tensors $\mathbf{N}^{*}$ and $\mathbf{N}^{s*}$ are obtained as average of all possible alignments of the inclusions inside the composite. Numerical results are obtained and compared with some other theoretical approaches reported in the literature.

Keywords: Maxwell's scheme · Inhomogeneities · Homogenization · Transversely isotropic composites

9.1 Introduction

It was the scientist James Clerck Maxwell, who proposed a method to calculate the effective conductivity of a homogeneous spherical material that contained a finite amount of inclusions of spherical type in Maxwell (1954). This aroused the interest of the scientific community since in its method Maxwell did not consider interactions between inclusions and arrived, in the case of spherical inclusions, to the same predictions as other methods that take them into account.

In Kanaun (2016); Levin and Kanaun (2012) recent results are presented in this area. In Kanaun (2016) four methods are compared, the original and generalized Maxwell schemes and the one-particle and multi-particle effective field methods (EFM). Those approaches give closed predictions for small volume fractions of inclusions ($p < 0.3$), furthermore, another method such as (MT) method is mentioned, concluding that in the case of isotropic materials with spherical inclusions of equal behaviour, the original Maxwell and the one-particle effective field methods, coincide in their results, and in the case of spheroidal inclusions the original Maxwell and the one-particle EFM deviate substantially from the multi-particle EFM and generalized Maxwell scheme, being remarkable that both last approaches coincide practically for all the values of the volume fraction. The Maxwell method is extended in Levin and Kanaun (2012) to an homogeneous anisotropic medium containing an arbitrary set of homogeneous anisotropic ellipsoidal inclusions, where it is shown that the explicit equations obtained by the Maxwell method for isotropic materials coincide in the case of spherical inclusions with the MT, and for spheroidal inclusions oriented in a parallel direction, the equations for the effective elastic modulus tensor are the same, which are obtained by Maxwell and MT methods.

In addition, in Sevostianov (2014), Maxwell's scheme is reformulated from the rigidity and flexibility contribution tensors, being remarkable that the form of the fictitious building, affects the predictions of the effective coefficients. Moreover, explicit formula for choosing the aspect ratio of fictitious building is given when it presents spheroidal shape, and the advantages of the Maxwell method are proposed. It shows that the reformulation of its scheme as a function of the tensor of contribution of flexibility, is equivalent to the reformulation as a function of the tensor of contribution of rigidity. In Sevostianov and Giraud (2013), the Maxwell method reformulated as a function of the rigidity contribution tensor is illustrated, through four examples: the first is a material that contains multiple pores of identical form, the second, a material that contains three families of inclusions that have different properties and forms, the third, a material that contains circular cracks with certain orientations and, finally, a material containing randomly oriented pores with no ellipsoidal shapes.

In Martinez-Ayuso et al (2017) a study is presented on the homogenization of piezoelectric materials with pores, through numerical and analytical methods. The results obtained in Martinez-Ayuso et al (2017) are compared by two different methods: (MT) method and self-consistent scheme of Hershey (1954) and Kröner (1958). Besides, these results are contrasted with two classical bounds known in the homogenization theory, the Hashin-Strikman and Halpin-Tsai bounds. In addition, a numerical model (FEM) of representative volume element is developed, based on the analysis of finite element for different percentage of inclusions in the material. The locally exact homogenization theory for unidirectional composites with square periodicity and isotropic phases proposed by Drago and Pindera (2008) is extended in Wang and Pindera (2016) to architectures with hexagonal symmetry and transversely isotopic phases, through a numerical method that uses Fourier transformation.

In McCartney and Kelly (2008) the far-field methodology developed by Maxwell (1954), is used to estimate effective thermoelastic properties in multi-phase isotropic composites. Furthermore, effective bulk and shear moduli are estimated, as well as thermal expansion coefficients in these types of composites. Besides, results are compared with formulas and dimensions, known in the literature. The generalized Maxwell method is developed by Levin et al (2012) for the calculation of effective parameters in poroelastic composites. This method is compared with other self-consistent methods existing in the literature. Moreover, examples of applications of the generalized method of Maxwell for the calculation of effective parametric parameters for heterogeneous materials constituted by rocks are reported. The Maxwell homogenization scheme (Kushch and Sevostianov, 2016) is formulated in terms of moments of dipole and of the tensor of contribution of element of representative volume, also deals with the problem of effective conductivity in a composite with spheroidal inclusions aligned, analyzing the convergence of that solution.

The novelty of the present contribution is the derivation of explicit analytical formulae for the control of the alignment tensors $\mathbf{N}^{*}$ and $\mathbf{N}^{s*}$. These functions distribute the alignment of inclusions inside the matrix material through disorder parameters λ and s, respectively, obtained as an average of all possible alignments of the inclusions within the nano transversely isotropic composite. Another novelty of this contribution consists in the study of Maxwell's approach predictions using two different density distribution functions (Sevostianov, 2014; Giraud et al, 2007) for alignment inclusions inside the composite. The different types of inclusions taken into account in the model are spheroidals. Spheroidal fibre, spherical inclusion and spheroidal disk are considered for different ranges of the aspect ratio parameter. Moreover, explicit formulae for effective elastic tensor are given for porous, nano and composites with global transversely isotropic behaviour formed by constituents with transversely isotropic symmetry as well. Comparisons with other theoretical approaches, such as, closed relations reported by Christensen model and FEM (Dong et al, 2005), Locally-exact homogenization theory (LEHT) (Wang and Pindera, 2016), among others are given.

9.2 Statement of Fundamental Equations

A solid material of volume V that posses linear elastic behaviour is considered. In this case, the constitutive equations for a linear elastic solid can be written in terms of stress tensor σ_{ij} and strain tensor ε_{ij} through Hooke law

$$\sigma_{ij} = C_{ijkl}\varepsilon_{kl}, \tag{9.1}$$

where C_{ijkl} is the stiffness fourth order tensor. In previous equation, the indexes i, j, k, l go from 1 to 3. The elastic constants satisfy the following symmetry relationships

$$C_{ijkl} = C_{jikl} = C_{ijlk}. \tag{9.2}$$

The symmetry of elastic constants (9.2) reduces the number of elastic independent constants from 81 to 36 (see Qu and Cherkaoui, 2006; Roger and Dieulesaint, 2000). For crystals, stiffness tensor C_{ijkl}, formed by 36 components, is also symmetric respect to permutation of pairs of indexes

$$C_{ijkl} = C_{klij}. \tag{9.3}$$

The existence of equality (9.3) in the general case, lead to a reduction of the number of independent components for the stiffness tensor from 36 to 21, that is, the number of constants of a solid without symmetry.

The strain $\boldsymbol{\varepsilon}$ and displacement $\boldsymbol{u}$ are related by the Cauchy linear relationship

$$\varepsilon_{ij} = \frac{1}{2}(u_{i,j} + u_{j,i}). \tag{9.4}$$

The components of elastic tensor moduli, strain and stress in matrix notation, is useful to write using the abbreviate notation. The binary combinations $ij = m$ $(i, j = 1,2,3)$ and $kl = n$ $(k,l = 1,2,3)$ are substituted by an index from 1 to 6 following the next scheme $(m,n = 1,\ldots,6)$

$$(11) \to 1;\ (22) \to 2;\ (33) \to 3;\ (23,32) \to 4;\ (31,13) \to 5;\ (12,21) \to 6. \quad (9.5)$$

Equation (9.1) can be written in matrix notation

$$\begin{pmatrix} \sigma_1 \\ \sigma_2 \\ \sigma_3 \\ \sigma_4 \\ \sigma_5 \\ \sigma_6 \end{pmatrix} = \begin{pmatrix} C_{11} & C_{12} & C_{13} & C_{14} & C_{15} & C_{16} \\ C_{21} & C_{22} & C_{23} & C_{24} & C_{25} & C_{26} \\ C_{31} & C_{32} & C_{33} & C_{34} & C_{35} & C_{36} \\ C_{41} & C_{42} & C_{43} & C_{44} & C_{45} & C_{46} \\ C_{51} & C_{52} & C_{53} & C_{54} & c_{55} & C_{56} \\ C_{61} & C_{62} & C_{63} & C_{64} & C_{65} & C_{66} \end{pmatrix} \begin{pmatrix} \varepsilon_1 \\ \varepsilon_2 \\ \varepsilon_3 \\ \varepsilon_4 \\ \varepsilon_5 \\ \varepsilon_6 \end{pmatrix}. \quad (9.6)$$

Material's behavior is described through 21 constants of the tensor C_{ijkl}.

The elastic fourth rank stiffness tensor for composites with transversely isotropic symmetry, oriented along the x_3 symmetry axis, is given by

$$\mathbf{C} = \begin{pmatrix} C_{11} & C_{12} & C_{13} & 0 & 0 & 0 \\ C_{12} & C_{11} & C_{13} & 0 & 0 & 0 \\ C_{13} & C_{13} & C_{33} & 0 & 0 & 0 \\ 0 & 0 & 0 & C_{44} & 0 & 0 \\ 0 & 0 & 0 & 0 & C_{44} & 0 \\ 0 & 0 & 0 & 0 & 0 & \dfrac{C_{11}-C_{12}}{2} \end{pmatrix}. \quad (9.7)$$

As a particular case, if the material has isotropic symmetry, the independent constants are reduced to 2. Thus, constitutive equation (9.6) can be written

$$\begin{pmatrix} \sigma_1 \\ \sigma_2 \\ \sigma_3 \\ \sigma_4 \\ \sigma_5 \\ \sigma_6 \end{pmatrix} = \begin{pmatrix} C_{11} & C_{12} & C_{12} & 0 & 0 & 0 \\ C_{12} & C_{11} & C_{12} & 0 & 0 & 0 \\ C_{12} & C_{12} & C_{11} & 0 & 0 & 0 \\ 0 & 0 & 0 & \frac{1}{2}(C_{11}-C_{12}) & 0 & 0 \\ 0 & 0 & 0 & 0 & \frac{1}{2}(C_{11}-C_{12}) & 0 \\ 0 & 0 & 0 & 0 & 0 & \frac{1}{2}(C_{11}-C_{12}) \end{pmatrix} \begin{pmatrix} \varepsilon_1 \\ \varepsilon_2 \\ \varepsilon_3 \\ \varepsilon_4 \\ \varepsilon_5 \\ \varepsilon_6 \end{pmatrix}. \quad (9.8)$$

In some cases, it is convenient to represent stiffness tensor C_{ijkl} for isotropic materials in the form

$$C_{ijkl} = \lambda\delta_{ij}\delta_{kl} + \mu(\delta_{ik}\delta_{jl} + \delta_{il}\delta_{jk}) = K\delta_{ij}\delta_{kl} + G\left(\delta_{ik}\delta_{jl} + \delta_{il}\delta_{jk} - \frac{2}{3}\delta_{ij}\delta_{kl}\right), \quad (9.9)$$

where K and G are the bulk and shear modulus, respectively, and λ and μ are Lame's constants, which are related with the constants of tensor C_{ijkl} of Eq. (9.8) by

$$\begin{aligned} C_{11} &= C_{22} = C_{33} = \lambda + 2\mu, \\ C_{12} &= C_{13} = C_{23} = \lambda, \\ C_{44} &= C_{55} = C_{66} = \mu = (C_{11} - C_{12})/2. \end{aligned} \quad (9.10)$$

The stress field given by stress tensor represents an equilibrium state with volume forces f_i in all the points of the volume V, it means that stress field satisfies the equilibrium equation

$$\frac{\partial \sigma_{ij}}{\partial x_j} + f_i = 0. \tag{9.11}$$

Equation (9.11) is valid for any material point inside a continuum medium.

Introducing **T** basis (Sevostianov, 2014) for transversely isotropic tensor

$$T^1_{ijkl} = \Theta_{ij}\Theta_{kl}, \quad T^2_{ijkl} = \frac{\Theta_{ik}\Theta_{lj} + \Theta_{il}\Theta_{kj} - \Theta_{ij}\Theta_{kl}}{2}, \quad T^3_{ijkl} = \Theta_{ij}\xi_k\xi_l, \tag{9.12}$$

$$T^4_{ijkl} = \xi_i\xi_j\Theta_{kl}, \quad T^5_{ijkl} = \frac{\Theta_{ik}\xi_l\xi_j + \Theta_{il}\xi_k\xi_j + \Theta_{jk}\xi_l\xi_i + \Theta_{jl}\xi_k\xi_i}{4}, \tag{9.13}$$

$$T^6_{ijkl} = \xi_i\xi_j\xi_k\xi_l, \quad \Theta_{ij} = \delta_{ij} - \xi_i\xi_j, \tag{9.14}$$

$$\boldsymbol{\xi} = (\xi_1, \xi_2, \xi_3) = (\sin\psi\cos\theta, \sin\psi\sin\theta, \cos\psi), \quad \psi \in [0,\pi], \theta \in [0,2\pi], \tag{9.15}$$

where δ_{ij} is the Kronecker delta, and using decomposition of stiffness tensor (9.7) in **T** basis oriented along x_3 symmetry axis (Sevostianov, 2014), it holds

$$\begin{aligned} \mathbf{C} &= k\mathbf{T}^1 + 2m\mathbf{T}^2 + l(\mathbf{T}^3 + \mathbf{T}^4) + 4\mu\mathbf{T}^5 + n\mathbf{T}^6, \\ \mathbf{C} &= (k, 2m, l, l, 4\mu, n), \end{aligned} \tag{9.16}$$

where

$$\begin{aligned} k &= \frac{C_{1111} + C_{1122}}{2}, \quad m = \frac{C_{1111} - C_{1122}}{2}, \\ l &= C_{1133}, \quad \mu = C_{2323}, \quad n = C_{3333}, \end{aligned} \tag{9.17}$$

where k is the plane-strain bulk modulus for lateral dilatation without longitudinal extension, m is the rigidity modulus for shearing in any transverse direction, l is the associated cross-modulus, μ is the longitudinal or axial shear modulus, and n is the modulus for longitudinal uniaxial strain.

9.3 Geometry of Inclusions

Initially a composite material is considered on the framework of Maxwell approach, with a fictitious building (the fictitious building is an arbitrary region that is framed within the composite) of spheroidal shape inside the homogeneous material (matrix). The composite can be ensemble, in principle, for a single type of inclusion or different types of spheroidal inclusions. In Fig. 9.1 it is featured a sample of the composite with the fictitious building (region in dark gray) inside matrix material (region in normal gray). Three different types of spheroidal inclusions accounted in the present model are described in a mathematical form through the set

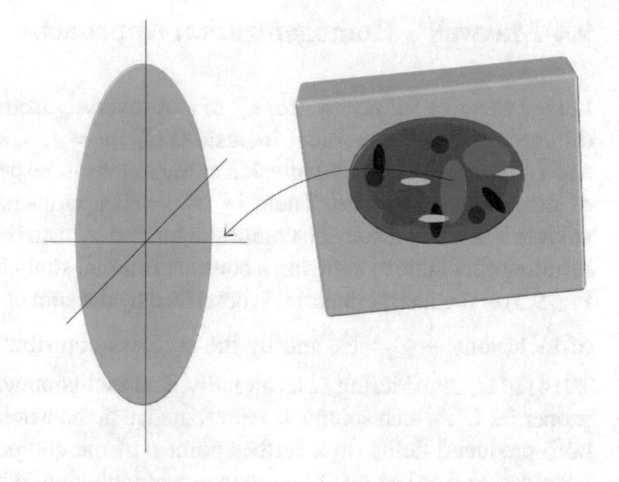

Fig. 9.1 Representative nano transversal isotropic material with spheroidal inclusions and fictitious building

$$\mathfrak{I} = \left\{ (x, y, z) \in \mathbb{R}^3 : \frac{x^2}{(\gamma a)^2} + \frac{y^2}{(\gamma a)^2} + \frac{z^2}{a^2} \leq 1, a \in \mathbb{R} \right\},$$

where the parameter γ, is denominated aspect ratio of the inclusions. In the present model, the parameter γ is taken as $\gamma = \frac{x_1}{x_3}$, taking the value $\gamma = 1$ for spherical inclusions, $\gamma < 1$ for fibre cylindrical inclusions and $\gamma > 1$ for spheroidal disk inclusions, Fig. 9.2. Analogously, the aspect ratio of the fictitious building within matrix material is assigned to the parameter Γ and it describes the geometrical form of this construction inside the composite through the set

$$F_B = \left\{ (x, y, z) \in \mathbb{R}^3 : \frac{x^2}{(\Gamma b)^2} + \frac{y^2}{(\Gamma b)^2} + \frac{z^2}{b^2} \leq 1, b \in \mathbb{R} \right\}.$$

The different types of inclusions taken into account in the model are spheroidals. In Fig. 9.1 all possible spheroidal inclusions embedded into the fictitious body is featured and in Fig. 9.2 spheroidal fibre, spherical and spheroidal disk inclusions are shown for different ranges of the aspect ratio γ.

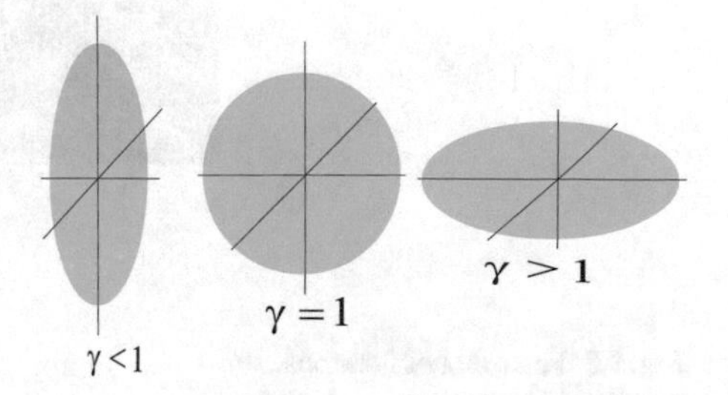

Fig. 9.2 Description of three different types of spheroidal inclusions taking into account in the present model

9.4 Maxwell's Homogenization Approach

Let the effective stiffness tensor $\mathbf{C}^*$ of a transversely isotropic elastic composite, with different types of spheroidal inclusions $\Omega_m, m = 1, ..., n$ of volume V_m, and be $\mathbf{C}^0$ and $\mathbf{C}^i$ the matrix and ith-inclusion stiffness tensor, respectively, being V the volume of the whole composite. Then, by Maxwell approach, a fictitious building Ω of volume $\bar{V}$ is taken inside the composite formed by matrix material and inclusions, the resulting effect due to applying a constant external strain field ε^0 (this fact is shown in Fig. 9.3) to the matrix material is described by the sum of stiffness contribution tensor of inclusions $\frac{1}{V}\sum_i V_i\mathbf{N}_i$ and by the stiffness contribution tensor $\bar{\mathbf{N}}$ (Sevostianov, 2014) of Ω, considering Ω as an individual inclusion with homogeneous unknown properties $\mathbf{C}^*$, which volume is representative in the whole composite. Later, equating both produced fields (in a further point ρ of the composite), by inclusions and Ω, considering this last one like independent inclusion, it holds the effective equation obtained by Maxwell approach

$$\frac{\bar{V}}{V}\bar{\mathbf{N}} = \frac{1}{V}\sum_i V_i\mathbf{N}_i. \tag{9.18}$$

Given in the formula (9.36)-(9.38) of the Appendix, tensor $\mathbf{N}$ in the Maxwell approach contains the information about the geometrical shape and elastic properties of the inclusions, it also depends on the elastic properties of the matrix material through the components of tensor $\mathbf{P}$ which contains the aspect ratio γ in its integral expressions (9.27). The tensor $\mathbf{P}$ (Hill, 1965) describes the geometrical shape and properties of fictitious building and it contains the elastic properties of matrix material and the aspect ratio of the fictitious building Γ, and for ellipsoidal inclusions the following relation holds

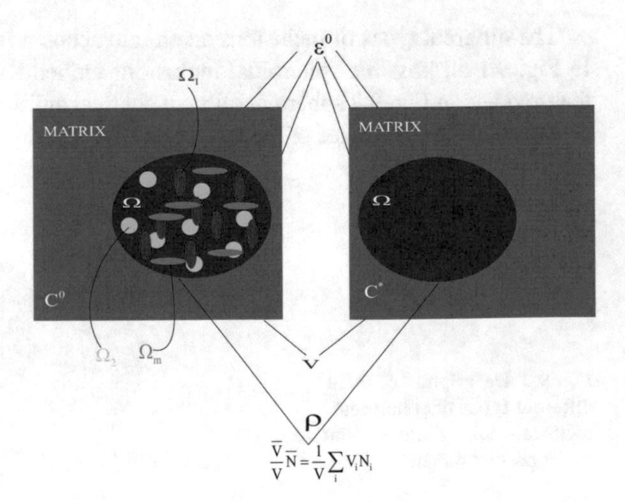

Fig. 9.3 Featuring of fictitious building in matrix material

$$\mathbf{N} = \left[\left(\mathbf{C}^i - \mathbf{C}^0 \right)^{-1} + \mathbf{P} \right]^{-1}, \tag{9.19}$$

and tensor $\mathbf{P}$ is calculated

$$\mathbf{P}_{ijkl} = \int_\Psi K^*_{ijkl}(a^{-1}k)d\Psi, \quad K_{ijkl}(x) = -\left[\nabla_j \nabla_l G_{ik}(x) \right]_{(ij)(kl)}, \tag{9.20}$$

$$\mathbf{a} = \begin{pmatrix} 1/a_1 & 0 & 0 \\ 0 & 1/a_2 & 0 \\ 0 & 0 & 1/a_3 \end{pmatrix}, \tag{9.21}$$

where Ψ is the unit sphere, a_1, a_2, a_3 are the axis of the ellipsoidal volume V, G_{ik} is the static Green function of operator $\nabla_j C^0_{ijkl} \nabla_l$ and K^*_{ijkl} the Fourier transform of function K_{ijkl}, given by

$$K^*_{ijkl} = \frac{1}{4} \left[\xi_j (\mathbf{G}_{ik})^{-1} \xi_l + \xi_i (\mathbf{G}_{jk})^{-1} \xi_l + \xi_j (\mathbf{G}_{il})^{-1} \xi_k + \xi_i (\mathbf{G}_{jl})^{-1} \xi_k \right], \tag{9.22}$$

where ξ_j are given by (9.15). From (9.19) it holds that when Ω has ellipsoidal form

$$\tilde{\mathbf{N}} = \left[\left(\mathbf{C}^* - \mathbf{C}^0 \right)^{-1} + \mathbf{P} \right]^{-1}, \tag{9.23}$$

and then by substitution (9.23) into (9.18), it yields the final expression for effective stiffness tensor of composite

$$\mathbf{C}^* = \mathbf{C}^0 + \left[\left(\frac{1}{V} \sum_i V_i \mathbf{N}_i \right)^{-1} - \mathbf{P} \right]^{-1}. \tag{9.24}$$

Equation (9.24) is the most important formula of Maxwell method because it allows to write the explicit expression for computing effective stiffness tensor of composites. Then, following the idea of (Sevostianov, 2014), one can replace in eq. (9.24) the sum $\sum_i \mathbf{N}_i$ by the quantity $\mathbf{N}^*$ given by formulae (9.49)-(9.54) of the Appendix, and the quantity $\sum_i \frac{V_i}{V}$ by parameter v_f which denotes the volume fraction of inclusion in the model, consequently, for two-phase composites Eq. (9.24) becomes

$$\mathbf{C}^* = \mathbf{C}^0 + v_f \cdot \left[\left(\mathbf{N}^* \right)^{-1} - v_f \cdot \mathbf{P} \right]^{-1}, \tag{9.25}$$

Moreover, the Maxwell approach allows us to replace tensor $\mathbf{N}^*$ of Eq. (9.25) by tensor $\mathbf{N}^{S*}$ given by formulae (9.65)-(9.76) of the Appendix for obtaining two Maxwell homogenization approaches that differ one from the other in the density distribution function chosen for modeling the alignment of inclusions within matrix materials.

9.5 Analysis of Numerical Results

In this section, present model (PM) using the Maxwell method applied to nanocomposites and transversely isotropic composite materials is validated with other theoretical approaches reported in different works. In Table 9.1 are shown the material properties used in the calculations.

The present model is compared with Christensen, FEM approach (Table 1 of Dong et al (2005)) and Locally Exact Homogenization Theory (Wang and Pindera, 2016) for validating the case of composite with isotropic matrix and transversely isotropic inclusion. Besides, a comparison with Mogilevskaya et al (2014) for effective tetragonal elastic moduli of two phase fiber reinforced composite is done. Furthermore, a validation for the porous inclusion case with respect to Sevostianov (2014); Vilchevskaya and Sevostianov (2015) approaches is performed. The present model (PM) is compared with the approach reported in Selmi et al (2007) for the case of two phase nanocomposite. Moreover, a study about the influence on the effective coefficient through PM using two different density distribution functions (DDF) for describing parallel alignment inclusions within matrix is reported. From now on, DDF given by (9.47) and (9.63) are identified by DDF1 and DDF2, respectively. The procedure given by the Eq. (9.25) using tensor $\mathbf{N}^*$ reported in (9.48) and $\mathbf{N}^{s*}$ given in (9.64) is denoted by PM-DDF1 and PM-DDF2, respectively. The numerical results shown in the figures are taken from the mentioned literature in each study. Present model's predictions are based in the effective equation (9.25) obtained in the last section.

Table 9.1 Mechanical properties of the constituents used for the computation

material properties (GPa)	C_{1111}	C_{1122}	C_{1133}	C_{3333}	C_{2323}
Epoxy	6.64444	3.57778	3.57778	6.64444	1.53333
Boron inclusions	459.667	114.917	114.917	459.667	172.375
3501-6 Epoxy	6.464	3.33	3.33	6.464	1.576
AS4 Graphite	15.6879	3.18792	3.77517	226.51	15
matrix	1	0.5	0.5	1	0.25
moderate inclusion	18	15	15	18	1.5
high inclusion	1200	1000	1000	1200	100
matrix1	130.27	56.0962	56.0962	130.27	37.0869
matrix2	17.9482	10.1082	10.6614	32.2627	8.75362
porous	10.1^{-7}	10.12^{-7}	10.13^{-7}	10.14^{-7}	10.15^{-7}
LaRC-SI	8.14286	5.42857	5.42857	8.14286	1.35714
continuum Graphene	3024	1008	1008	3024	1008
Aluminum	101.9	50.022	50.022	101.9	25.94
SiC	474.2	98	98	474.2	188.1

9.5.1 Density Distribution Functions

The influence on effective properties obtained through Maxwell approach using density distribution function given by (9.47) in Appendix denoted by DDF1 for modeling different inclusion alignment cases inside matrix material, is studied by Sevostianov (2014) for porous composite. Now, the purpose is to show the results of Maxwell approach using DDF2 given by (9.63) in Appendix (PM-DDF2) and (PM-DDF1) for modeling inclusion's alignment inside composite, and compare these two approaches between them and other predictions reported in other research works not only for porous composites but for a more general variety of different composites like Epoxy/Boron, 3501-6 Epoxy/AS4 Graphite, matrix/moderate inclusion, matrix/high inclusion, among others.

In Fig. 9.4 is shown the dependence on the alignment parameter s of six alignment functions τ_i given by formulae (9.71)-(9.76) of the Appendix when disorder parameter $s \to \infty$, s controls the alignment of inclusions inside the matrix material through tensor $\mathbf{N}^{s*}$. Present model takes into account the random or non random alignment of every type of inclusion inside the matrix through parameter s. Figure 9.4 reflects that functions $\tau_i(s)$ have horizontal asymptotes at different values as $s \to \infty$. It is remarkable that these six functions remain constants as the disorder parameter s approaches to infinity.

In Table 9.2 are shown the predictions of the tensor $\mathbf{N}^*$ numerical values (in GPa) given by formulae (9.49)-(9.54) for different values of disorder parameter λ, for modeling cylindrical inclusions within the matrix in a Epoxy/Boron composite (see Table 9.1), taking into account aspect ratios of inclusions and fictitious building $\gamma = \Gamma = 0.01$. From Table 9.2 one can observe that for values of disorder parameter λ higher than 100, the components of tensor $\mathbf{N}^*$ remain close one each other, it implies that the components of effective elastic tensor are also closed through PM-DDF1 approach.

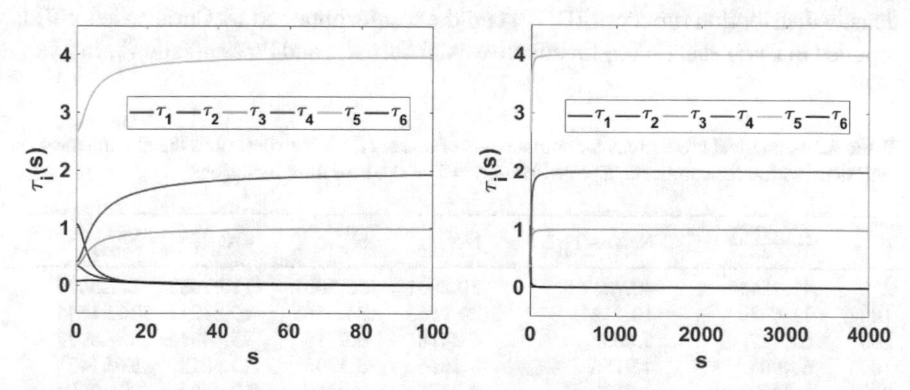

Fig. 9.4 Dependence of alignment functions τ_i given by Eq. (9.71)-(9.76) of the Appendix related to disorder parameter s

Table 9.2 Numerical predictions for components of tensor $\mathbf{N}^*$ which describes the alignment of inclusions inside the composite Epoxy/Boron, $\gamma = \Gamma = 0.01$ for fibre inclusions

λ	$\frac{\mathbf{N}^*_{1111}+\mathbf{N}^*_{1122}}{2}$	$\mathbf{N}^*_{1111} - \mathbf{N}^*_{1122}$	$\mathbf{N}^*_{1133}$	$\mathbf{N}^*_{3311}$	$4\mathbf{N}^*_{2323}$	$\mathbf{N}^*_{3333}$
2	33.3135	32.5450	29.7264	29.7264	118.2359	156.5058
10	7.2516	5.7731	11.2025	11.2025	45.5601	334.8491
100	6.4908	4.9134	2.8862	2.8862	12.4927	371.1578
1000	6.4911	4.9126	2.7783	2.7783	12.0634	371.5879
3700	6.4911	4.9126	2.7773	2.7773	12.0594	371.5920
5000	6.4911	4.9126	2.7773	2.7773	12.0593	371.5921

Results are given in GPa.

In Table 9.3 are shown numerical values (in GPa) of tensor $\mathbf{N}^{s*}$ given by formulae (9.65)-(9.70) of Appendix, for different values of disorder parameter s, taking into account $\gamma = \Gamma = 0.01$ for modeling cylindrical Boron inclusions within epoxy matrix (see the material properties in Table 9.1). From Table 9.3 one can observe that for values of disorder parameter s higher than 3000, the components of tensor $\mathbf{N}^{s*}$ remain close one each other, it implies that the components of effective elastic tensor are also closed through PM-DDF2 approach. From Fig.9.4 and Tables 9.2 and 9.3 one can conclude that best value of parameter s in PM-DDF2 for estimating PM-DDF1 approach with $\lambda = 100$ (aligned inclusion case reported by Sevostianov (2014)) is the value $s = 3700$.

9.5.2 Study of Composites Constituted by Isotropic Matrix and Isotropic Inhomogeneities

In Table 9.4 it is shown comparisons between present model through two different density distribution functions (DDF) and the results obtained by Christensen, FEM reported in Dong et al (2005) for effective axial bulk K^*_{12} and Poisson ratio ν^*_{31} in a two

Table 9.3 Numerical predictions for components of tensor $\mathbf{N}^{s*}$ which describes the alignment of inclusions within the composite Epoxy/Boron, $\gamma = \Gamma = 0.01$ for fibre inclusions

s	$\frac{\mathbf{N}^{s*}_{1111}+\mathbf{N}^{s*}_{1122}}{2}$	$\mathbf{N}^{s*}_{1111} - \mathbf{N}^{s*}_{1122}$	$\mathbf{N}^{s*}_{1133}$	$\mathbf{N}^{s*}_{3311}$	$4\mathbf{N}^{s*}_{2323}$	$\mathbf{N}^{s*}_{3333}$
2	41.3244	40.7217	30.2001	30.2001	119.7988	122.5679
10	11.6640	10.4124	24.9563	24.9563	100.1217	262.1844
100	6.5477	5.0052	6.2418	6.2418	25.8454	357.5077
1000	6.4904	4.9156	3.1396	3.1396	13.5013	370.1455
3700	6.4907	4.9133	2.8755	2.8755	12.4503	371.2004
5000	6.4908	4.9131	2.8500	2.8500	12.3487	371.3022

Results are given in GPa.

Table 9.4 Comparisons with Table 1 of Dong et al (2005), $\gamma = \Gamma = 0.001$ for fibre inclusions

Effective properties | v_f	0.1	0.3	0.5	0.7	0.9
K_{12}^* (Christensen)	5.83056	7.86606	11.4567	19.4865	53.5138
K_{12}^* (FEM)	5.83056	7.8661	11.457	19.487	53.514
K_{12}^* (PM-DDF1)	5.8305	7.8658	11.4559	19.4835	53.4872
K_{12}^* (PM-DDF2)	5.8305	7.8658	11.4559	19.4834	53.4868
ν_{31}^* (Christensen)	0.331156	0.296514	0.265414	0.237339	0.211869
ν_{31}^* (FEM)	0.33116	0.29651	0.26541	0.23734	0.21187
ν_{31}^* (PM-DDF1)	0.3323	0.2998	0.2707	0.2443	0.2203
ν_{31}^* (PM-DDF2)	0.3322	0.2995	0.2701	0.2436	0.2195

Results are given in GPa.

phase composite formed by Boron inclusions and Epoxy matrix whose properties are shown in Table 9.1. Both phases, matrix and inclusions exhibit isotropic symmetry. The effective coefficients K_{12}^*, ν_{31}^* are given by formulae (10) of Dong et al (2005). In the calculations are using for both type of DDF, Boron inclusions aspect ratio $\gamma = 0.001$, the aspect ratio of fictitious building $\Gamma = 0.001$, which is the same to the inclusions. The disorder parameter $\lambda = 100$ for modeling parallel aligned inclusion case using the PM-DDF1 approach, and disorder parameter $s = 3700$ for PM-DDF2. From Table 9.4 is remarkable that predictions through present model give good agreement with those obtained by Christensen, FEM approaches published in Dong et al (2005).

A comparison with Mogilevskaya et al (2014) and semi-analytical finite element method (SAFEM) reported in Otero et al (2013, 2017) implemented for this type of composite under perfect contact is given in Tables 9.5 and 9.6 for tetragonal elastic moduli of two phase fibre reinforced composites. Numerical predictions for the effective coefficient C_{1313}^* normalized by the shear modulus μ_0 of matrix material for two different composites made of matrix/moderate inclusion and matrix/high inclusion, are shown in Tables 9.5 and 9.6, respectively. All the constituents in both composites are in Table 9.1 and have isotropic symmetry. This comparison corresponds to numerical results obtained by the present approach and different predictions obtained by the expressions (6), (20) and RUC (Repetitive square unit cell) model reported in Mogilevskaya et al (2014). The computations are done for $\gamma = \Gamma = 0.001$, $\lambda = 100$ and parameter $s = 3700$ for PM-DDF1 and PM-DDF2 approaches, respectively. Table 9.5 shows good agreement between PM-DDF1, PM-DDF2, Eqs. (6), (20) and RUC (Mogilevskaya et al, 2014) as well as SAFEM approaches for matrix/moderate inclusion composite. On the other hand, Table 9.6 reveals good comparison between PM-DDF1, PM-DDF2, RUC and SAFEM approaches, however, it is not the same with the results reported by Eqs. (6) and (20) of Mogilevskaya et al (2014) for the case of matrix/high inclusion composite.

Additionaly, the composites matrix/moderate inclusion and matrix/high inclusions are considered and numerical calculations for the effective coefficient $(C_{1111}^* - C_{1122}^*)/2$ normalized by the shear modulus of matrix μ_0 and the effective coefficient C_{1133}^*

Table 9.5 Comparison of the effective coefficient C^*_{1313} normalized by the shear modulus of matrix μ_0 with Table 5 of Mogilevskaya et al (2014), $\gamma = \Gamma = 0.001$ for fibre inclusions in the composite matrix/moderate inclusion

v_f	Eq. (6)	Eq. (20)	RUC	PM-DDF1	PM-DDF2	SAFEM
0.05	1.074	1.075	1.074	1.0742	1.0742	1.0741
0.1	1.154	1.155	1.154	1.1542	1.1542	1.1538
0.15	1.240	1.242	1.240	1.2406	1.2405	1.2400
0.2	1.333	1.335	1.333	1.3341	1.3341	1.3334
0.25	1.435	1.437	1.435	1.4359	1.4358	1.4351
0.3	1.545	1.548	1.546	1.5469	1.5468	1.5463
0.35	1.667	1.670	1.669	1.6685	1.6683	1.6688
0.4	1.800	1.804	1.805	1.8023	1.8021	1.8045
0.45	1.947	1.952	1.956	1.9503	1.9500	1.9564
0.5	2.111	2.117	2.128	2.1147	2.1144	2.1283
0.55	2.294	2.302	2.326	2.2986	2.2981	2.3256
0.6	2.500	2.511	2.556	2.5055	2.5049	2.5561
0.65	2.733	2.750	2.832	2.7401	2.7394	2.8319
0.7	3.000	3.024	3.173	3.0083	3.0075	3.1731
0.75	3.308	3.344	3.620	3.3180	3.3170	3.6197

Results are given in GPa.

Table 9.6 Comparison of the effective coefficient C^*_{1313} normalized by the shear modulus of matrix μ_0 with Table 8 of Mogilevskaya et al (2014), $\gamma = \Gamma = 0.001$ for fibre inclusions in the composite matrix/high inclusion

v_f	Eq. (6)	Eq. (20)	RUC	PM-DDF1	PM-DDF2	SAFEM
0.05	1.105	1.106	1.105	1.1239	1.1220	1.1047
0.1	1.221	1.224	1.221	1.2641	1.2598	1.2209
0.15	1.351	1.355	1.351	1.4241	1.4168	1.3509
0.2	1.497	1.502	1.497	1.6085	1.5972	1.4972
0.25	1.662	1.668	1.663	1.8232	1.8067	1.6633
0.3	1.851	1.858	1.854	2.0765	2.0529	1.8540
0.35	2.069	2.077	2.076	2.3797	2.3466	2.0762
0.4	2.322	2.333	2.340	2.7492	2.7027	2.3396
0.45	2.622	2.635	2.659	3.2094	3.1437	2.6590
0.5	2.980	2.999	3.058	3.7983	3.7039	3.0583
0.55	3.418	3.446	3.578	4.5789	4.4392	3.5779
0.6	3.963	4.008	4.294	5.6628	5.4471	4.2939
0.65	4.662	4.737	5.372	7.2694	6.9133	5.3719
0.7	5.590	5.720	7.274	9.8970	9.2429	7.2737
0.75	6.882	7.123	12.227	14.9721	13.5163	12.2267

The results are given in GPa.

normalized by the matrix coefficient C^m_{1133} are shown in Tables 9.7 and 9.8, respectively. Comparisons are given by present model, Mogilevskaya et al (2014), SAFEM reported in Otero et al (2013, 2017) and asymtotic homogenization method (AHM) proposed in Bravo-Castillero et al (2012); Guinovart-Diaz et al (2001). Aspect ratio parameters

Table 9.7 Comparison of the effective coefficient $(C^*_{1111} - C^*_{1122})/2$ normalized by the shear modulus of matrix μ_0 with Table 4 of Mogilevskaya et al (2014), $\gamma = \Gamma = 0.001$ for fibre inclusions in the composite matrix/moderate inclusion

v_f	Eq. (4)	Eq. (13)	Eq. (18)	RUC	PM-DDF1	PM-DDF2	SAFEM	AHM
0.05	1.066	1.069	1.069	1.069	1.063	1.063	1.0653	1.0653
0.1	1.137	1.148	1.150	1.148	1.1312	1.1312	1.1409	1.1409
0.15	1.214	1.242	1.244	1.241	1.2052	1.2052	1.2281	1.2281
0.2	1.298	1.349	1.352	1.349	1.2858	1.2857	1.3285	1.3285
0.25	1.389	1.474	1.477	1.474	1.3739	1.3739	1.4438	1.4438
0.3	1.489	1.616	1.621	1.617	1.4707	1.4706	1.5755	1.5755
0.35	1.600	1.777	1.783	1.780	1.5774	1.5774	1.7257	1.7257
0.4	1.722	1.958	1.965	1.965	1.6958	1.6958	1.8963	1.8963
0.45	1.858	2.157	2.166	2.172	1.8277	1.8277	2.0894	2.0893
0.5	2.010	2.374	2.385	2.403	1.9758	1.9758	2.3070	2.3069
0.55	2.181	2.604	2.616	2.656	2.1431	2.143	2.5516	2.5512
0.6	2.376	2.841	2.856	2.929	2.3336	2.3336	2.8253	2.8243
0.65	2.599	3.078	3.096	3.218	2.5526	2.5525	3.1309	3.1282
0.7	2.857	3.306	3.327	3.517	2.8069	2.8068	3.4718	3.4650
0.75	3.158	3.514	3.538	3.823	3.1058	3.1057	3.8532	3.8369

Results are given in GPa.

Table 9.8 Comparison of the effective coefficient C^*_{1133} normalized by the coefficient C^m_{1133} of the matrix with Table 9 of Mogilevskaya et al (2014), $\gamma = \Gamma = 0.001$ for fibre inclusions in the composite matrix/high inclusion

v_f	Eq. (7)	Eq. (21)	RUC	PM-DDF1	PM-DDF2	SAFEM	AHM
0.05	1.032	1.033	1.032	1.1045	1.1036	1.0956	1.0956
0.1	1.068	1.069	1.068	1.2206	1.2188	1.2019	1.2019
0.15	1.109	1.11	1.109	1.3505	1.3475	1.3206	1.3206
0.2	1.154	1.156	1.154	1.4965	1.4923	1.4542	1.4542
0.25	1.205	1.208	1.205	1.662	1.6564	1.6058	1.6058
0.3	1.263	1.267	1.264	1.8512	1.844	1.7796	1.7796
0.35	1.33	1.335	1.331	2.0695	2.0605	1.9810	1.9810
0.4	1.409	1.415	1.411	2.3242	2.313	2.2184	2.2184
0.45	1.501	1.509	1.506	2.6252	2.6114	2.5038	2.5037
0.5	1.611	1.621	1.623	2.9863	2.9694	2.8569	2.8564
0.55	1.746	1.757	1.772	3.4276	3.4069	3.3119	3.3101
0.6	1.913	1.927	1.974	3.9792	3.9537	3.9348	3.9278
0.65	2.127	2.142	2.275	4.6881	4.6564	4.8737	4.8447
0.7	2.411	2.427	2.804	5.6329	5.593	6.5509	6.4123
0.75	2.804	2.819	4.177	6.9548	6.9033	11.0435	9.9304

Results are given in GPa.

γ and Γ in the calculations are taken the same as in Tables 9.5 and 9.6, $\lambda = 100$ and $s = 3700$.

A validation of the present model (PM) with experimental data reported by Premkumar et al (1992); Liaw et al (1995); El-Eskandarany (1998); Kumar et al

(2009); Upadhyay and Singh (2012); Qu et al (2016) is shown in Fig. 9.5. The experimental and theoretical studies are for the effective Young modulus for different volume fraction of SiC isotropic spherical inclusions into Aluminum isotropic matrix, where the properties are given in Table 9.1. The experimental data is reported for different percentages of volume fraction. PM predictions is consistent with Liaw et al (1995); Qu et al (2016); Upadhyay and Singh (2012)

$$E_t^* = \frac{1}{S_{1111}^*}, \quad E_a^* = \frac{1}{S_{3333}^*}, \quad \mathbf{S}^* = (\mathbf{C}^*)^{-1} \tag{9.26}$$

For isotropic composites the effective axial E_a^* and tangential E_t^* Young modulus are the same and are given by Eq. (9.26) taking into account that $S_{1111}^* = S_{3333}^*$.

9.5.3 Study of Composites Constituted by Isotropic Matrix and Transversely Isotropic Inhomogeneities

Figure 9.6 displays another validation of the present model approach for the effective axial shear modulus of a material composed by isotropic matrix 3501-6 Epoxy and AS4 Graphite with transversely isotropic symmetry inclusions. The inclusions are embedded into the matrix. The elastic properties of the constituents are given in Table 9.1. This validation assures good concordance with numerical predictions

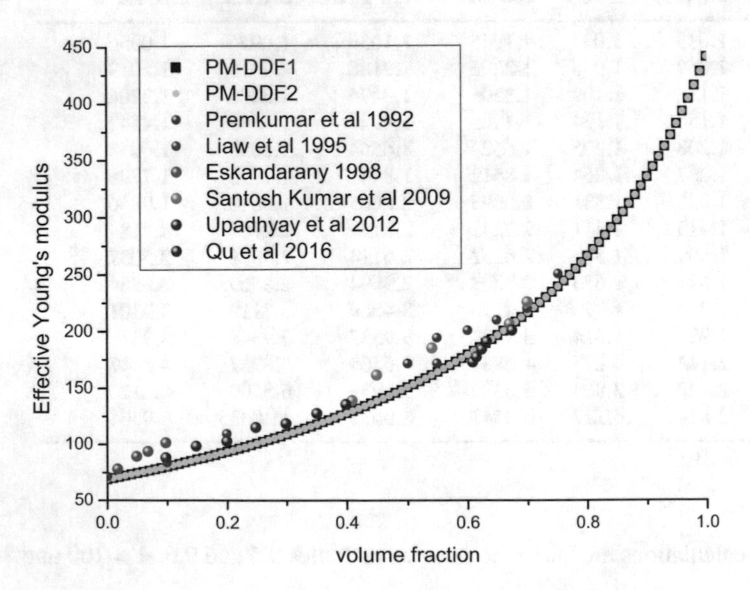

Fig. 9.5 Effective Young's modulus prediction between PM and experimental data reported in Premkumar et al (1992); Liaw et al (1995); El-Eskandarany (1998); Kumar et al (2009); Upadhyay and Singh (2012); Qu et al (2016), $\gamma = \Gamma = 1$ for spherical inclusions

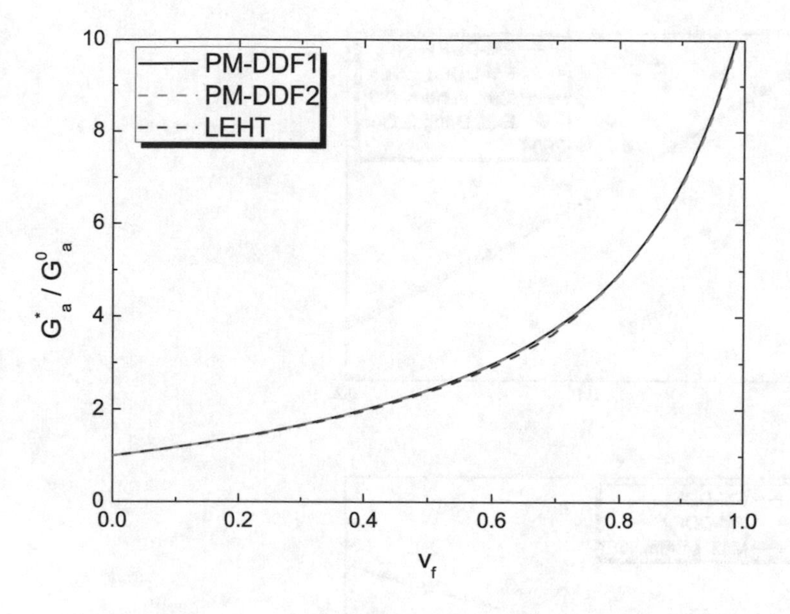

Fig. 9.6 Comparison between PM and LEHT (Wang and Pindera, 2016) approaches for the effective axial shear modulus in a composite made of isotropic matrix 3501-6 Epoxy and AS4 Graphite with transversely isotropic symmetry, $\gamma = \Gamma = 0.01$ for fibre inclusions

reported by LEHT homogenization approach (Wang and Pindera, 2016). In Fig. 9.6 are shown numerical predictions for effective axial shear modulus of composite 3501-6 Epoxy/AS4 Graphite (Table 9.1) related to the dependence of the volume fraction of fibre inclusions. A comparison between PM using both density distribution functions (DDF) and Locally Exact Homogenization Theory (LEHT) is done. In both models PM-DDF1 and PM-DDF2, the aspect ratio of fictitious building and fibre inclusions are $\gamma = \Gamma = 0.01$, disorder parameters in DDF1 and DDF2 are $\lambda = 100$ and $s = 3700$, respectively for modeling the case of aligned fibres inside the matrix material. From Fig. 9.6, PM and LEHT give very close results for this effective property of the composite in the whole range of volume fraction.

9.5.4 Porous Composite with Isotropic and Transversely Isotropic Matrix

The next study focus on numerical prediction of porous composites. Validation of PM with Maxwell approach reported by Sevostianov (2014) (as SMM), Vilchevskaya and Sevostianov (2015) (as Max. schem.) and experimental results reported in Dong and Guo (2004) is shown in Fig. 9.7. The effective transversal shear modulus of two phase composite matrix2/porous (see Table 9.1) is compared between PM, Max. schem.

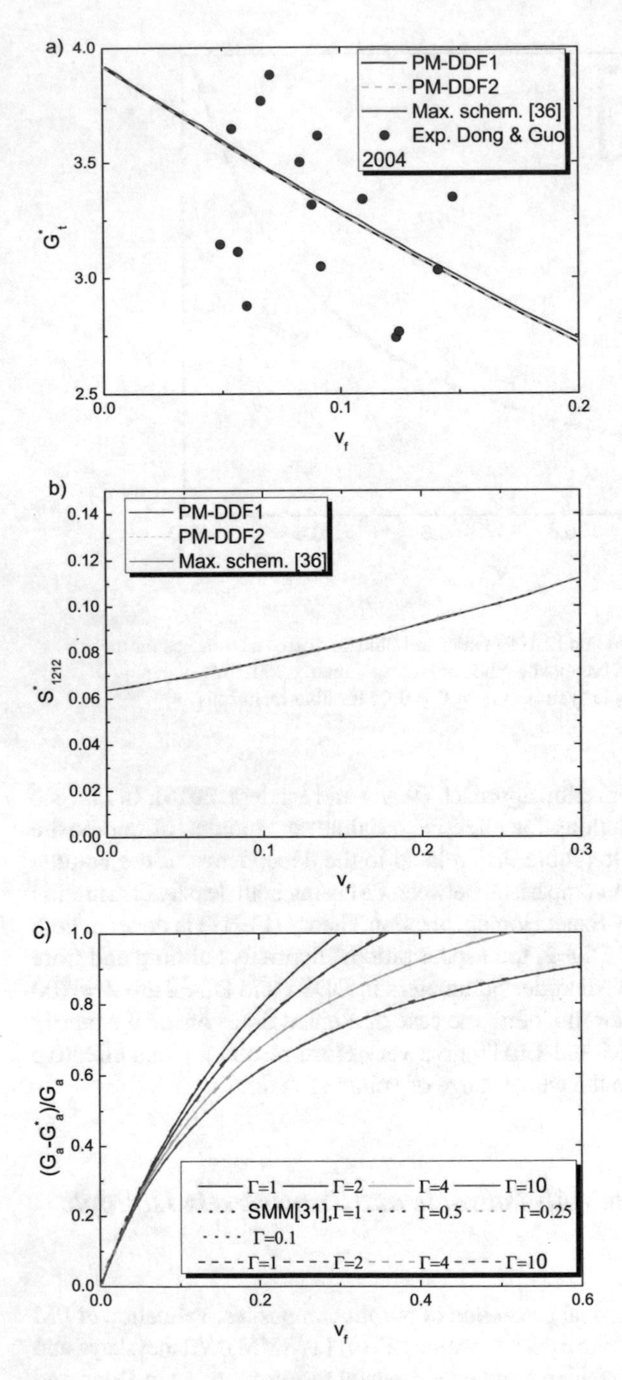

Fig. 9.7 Effective coefficient a) G_t^*, b) S_{1212}^* for matrix2/porous composite, $\gamma = \Gamma = 1/1.39$ for fibre spheroidal porous inclusions, and c) $G_a - G_a^*$ for matrix1/porous composite normalized by axial shear modulus G_a of matrix material, $\gamma = 10$ for disk spheroidal porous inclusions at four fixed values of Γ

Vilchevskaya and Sevostianov (2015) and the experimental data reported by Dong and Guo (2004) in Fig. 9.7a). The matrix2 exhibits transversely isotropic symmetry. In both approaches of PM, i.e. PM-DDF1 and PM-DDF2, are taken $\gamma = \Gamma = 1/1.39$, $\lambda = 100$ for PM-DDF1 and $s = 3700$ for PM-DDF2 as disorder parameters for modeling the porous inclusion parallel aligned case. From Fig. 9.7a) one can observe that PM-DDF1 and PM-DDF2 approaches give close results, while both approach of the present model (PM-DDF1 and PM-DDF2) have good correspondence with Max. schem. Vilchevskaya and Sevostianov (2015) approach and experimental data of Dong and Guo (2004). In Fig. 9.7b) it is shown a numerical comparison for the effective compliance coefficient S^*_{1212} between PM and Max. schem. reported in Vilchevskaya and Sevostianov (2015), for two phase composite matrix2/porous of Table 9.1. In both predictions of PM approach (PM-DDF1 and PM-DDF2), are taken the same values of γ, Γ, λ and s like in Fig.9.7a). From Fig. 9.7b) one can observe that PM-DDF1, PM-DDF2 and Max. schem. Vilchevskaya and Sevostianov (2015) approaches practically coincide in the whole range of volume fraction. In Fig. 9.7c) are shown the numerical comparison of the effective axial coefficient $G_a - G^*_a$ normalized by the matrix material axial shear modulus G_a for two phase composite constituted by matrix1/porous of Table 9.1 between PM and SMM reported in Sevostianov (2014). The solid and dash curves indicate PM-DDF1 and PM-DDF2 approaches, respectively. The constituent matrix1 exhibit isotropic symmetry. The porous aspect ratio are taken as $\gamma = 10$ and the aspect ratio of fictitious building Γ is fixed as $\Gamma = 1, 2, 4, 10$ to see the influence of different geometrical shape of fictitious building on the behavior of this effective coefficient. From Fig. 9.7c), it is observed that when the effective coefficient $(G_a - G^*_a)/G_a > 1$ then $G^*_a < 0$, therefore, this fact conduces to lost of physical meaning, thus the only curve with real physical meaning is that none having value $\Gamma = 10$, i.e., the same aspect ratio for inclusions and fictitious building. For PM-DDF1 is taken $\lambda = 100$ and for PM-DDF2 $s = 3700$. Comparisons between the two approaches of PM (PM-DDF1 and PM-DDF2) and numerical results (SMM) shown in Fig. 9.7b) of Sevostianov (2014) give close results. Furthermore, in Table 9.9, numerical comparisons are reported using the present model and SAFEM for the effective transversal shear modulus G^*_t in the composite matrix2/porous. There is a good match between results obtained through both approaches.

Table 9.9 Comparison between the present model and SAFEM approach for the effective transversal G^*_t shear modulus in the composite made by fibre spheroidal porous inhomogeneities embedded into the matrix2, $\gamma = \Gamma = 1/1.39$, $\lambda = 100$, $s = 3700$

v_f	SAFEM G^*_t (GPa)	PM-DDF1 G^*_t (GPa)	PM-DDF2 G^*_t (GPa)
0.1	3.3354	3.2749	3.275
0.2	2.7686	2.7161	2.7162
0.3	2.2516	2.2273	2.2274

9.5.5 Two-phase Nano-composites

The last validation of the present model (PM) is done for the case of a two phase nano-composite constituted by isotropic matrix LaRC-SI and continuum graphene as inclusions with isotropic symmetry. The properties of these materials are shown in Table 9.1. In Fig. 9.8 are shown numerical predictions of PM-DDF1 and PM-DDF2 approaches for the effective axial G_a^* and transversal G_t^* shear modulus normalized by the corresponding axial G_a and transversal G_t shear modulus of matrix material in a two phase nanocomposite LaRC-SI/continuum graphene. The effective axial G_a^* and transversal G_t^* shear modulus calculated by PM are validated with three different methods (the Sequential, the Two-level based on Mori-Tanaka method and the FE approaches) reported in Selmi et al (2007) for nanocomposites. The parameters for PM predictions are taken $\gamma = \Gamma = 0.1$, $\lambda = 100$ and $s = 3700$ for PM-DDF1 and PM-DDF2, respectively. Figure 9.8 evidences that PM-DDF1 and PM-DDF2 match very good with Sequential, Two-level and FE approaches reported in Selmi et al (2007).

9.6 Conclusions

From the research that has been carried out, it is possible to conclude that based on the results obtained, Maxwell homogenization approach has been very successful as a procedure to estimate effective elastic properties of composites constituted by phases of transversely isotropic symmetry. A methodology for computing the overall properties is reported from the outcome of our investigation. Closed formulas for effective stiffness tensor of composites with different arrangement of the inclusions have been deduced. Current homogenization technique is compared with other theoretical approaches, getting low computational cost of the implemented algorithm

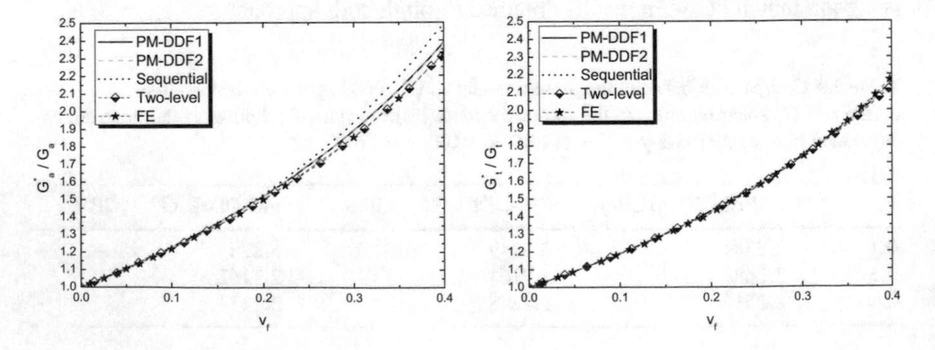

Fig. 9.8 Effective axial G_a^* and transversal G_t^* shear modulus normalized by the matrix material axial G_a and transversal G_t shear modulus, respectively, for a composite LaRC-SI/continuum Graphene, $\gamma = \Gamma = 0.1$ for fibre inclusions

derived from the present model and correlating satisfactorily with the results reported in the studied references . The used method is especially reachable and forthright procedure for the calculation of effective properties of heterogeneous media with more of two different types of constituents. As stated in the introduction our main target was to hand over new analytical formulae for the control alignment tensors $\mathbf{N}^*$ and $\mathbf{N}^{s*}$ which are emphasized in the appendix of the current contribution. Besides, an study about Maxwell approach using two different density distribution functions is done for predicting effective mechanical properties of transversely isotropic, nano and porous composites.

Appendix

Tensor $\mathbf{P}$ of Maxwell explicit effective formula (9.24) for transversely isotropic spheroidal inclusions of aspect ratio $\gamma = \frac{x_1}{x_3}$, where x_1, x_2, x_3 are the inclusion's axis, is given by

$$
\begin{aligned}
\mathbf{P}_{1111} &= \frac{1}{16}\left(\int_{-1}^{1}\frac{(1-u^2)\cdot\Upsilon(u)}{b}du\right) + \frac{3}{16}\left(\int_{-1}^{1}\frac{d(1-u^2)\cdot\Upsilon(u)}{h_0+h_1u^2+h_2u^4}du\right),\\
\mathbf{P}_{1122} &= -\frac{1}{16}\left(\int_{-1}^{1}\frac{(1-u^2)\cdot\Upsilon(u)}{b}du\right) + \frac{1}{16}\left(\int_{-1}^{1}\frac{d(1-u^2)\cdot\Upsilon(u)}{h_0+h_1u^2+h_2u^4}du\right),\\
\mathbf{P}_{1133} &= -\frac{(C_{1133}+C_{2323})}{4}\left(\int_{-1}^{1}\frac{u^2\cdot\Upsilon(u)}{h_0+h_1u^2+h_2u^4}du\right)+\\
&\quad+\frac{(C_{1133}+C_{2323})}{4}\left(\int_{-1}^{1}\frac{u^4\cdot\Upsilon(u)}{h_0+h_1u^2+h_2u^4}du\right),\\
\mathbf{P}_{3333} &= \frac{C_{1111}}{2}\left(\int_{-1}^{1}\frac{u^2\cdot\Upsilon(u)}{h_0+h_1u^2+h_2u^4}du\right)+\\
&\quad+\frac{(C_{2323}-C_{1111})}{2}\left(\int_{-1}^{1}\frac{u^4\cdot\Upsilon(u)}{h_0+h_1u^2+h_2u^4}du\right),\\
\mathbf{P}_{2323} &= -\frac{(C_{1133}+C_{1111})}{8}\cdot\left(\int_{-1}^{1}\frac{u^2\cdot\Upsilon(u)du}{h_0+h_1u^2+h_2u^4}\right)+\frac{1}{16}\cdot\left(\int_{-1}^{1}\frac{u^2\cdot\Upsilon(u)du}{b}\right)+\\
&\quad+\frac{C_{1111}}{16}\cdot\left(\int_{-1}^{1}\frac{\Upsilon(u)du}{h_0+h_1u^2+h_2u^4}\right)+\\
&\quad+\frac{[2C_{1133}+C_{3333}+C_{1111}]}{16}\cdot\left(\int_{-1}^{1}\frac{u^4\cdot\Upsilon(u)du}{h_0+h_1u^2+h_2u^4}\right),
\end{aligned}
\tag{9.27}
$$

$$\mathbf{P}_{1212} = \frac{\mathbf{P}_{1111} - \mathbf{P}_{1122}}{2} = \frac{1}{16} \cdot \left(\int_{-1}^{1} \frac{(1-u^2) \cdot \Upsilon(u)}{b} du \right) + $$
$$+ \frac{1}{16} \cdot \left(\int_{-1}^{1} \frac{d(1-u^2) \cdot \Upsilon(u)}{(h_0 + h_1 u^2 + h_2 u^4)} du \right),$$

where

$$\Upsilon(u) = \frac{\gamma^2}{\left[\gamma^2 + u^2(1-\gamma^2)\right]^{3/2}}, \tag{9.28}$$

and

$$b = \frac{(C_{1111} - C_{1122})}{2}(1-u^2) + C_{2323}u^2, \quad d = C_{2323}(1-u^2) + C_{3333}u^2, \tag{9.29}$$

$$h_0 = C_{1111}C_{2323}, \quad h_1 = C_{1111}C_{3333} - C_{1133}^2 - 2C_{1133}C_{2323} - 2C_{1111}C_{2323}, \tag{9.30}$$
$$h_2 = C_{3333}C_{2323} + C_{1111}C_{2323} + 2C_{1133}C_{2323} + C_{1133}^2 - C_{1111}C_{3333}. \tag{9.31}$$

Using decomposition (9.16), (9.17) and eq. (9.27), it holds

$$\mathbf{P} = (p_1, p_2, p_3, p_4, p_5, p_6), \tag{9.32}$$

where

$$p_1 = \frac{\mathbf{P}_{1111} + \mathbf{P}_{1122}}{2}, \; p_2 = \mathbf{P}_{1111} - \mathbf{P}_{1122}, \; p_3 = p_4 = \mathbf{P}_{1133}, \tag{9.33}$$
$$p_5 = 4\mathbf{P}_{2323}, \; p_6 = \mathbf{P}_{3333}.$$

Tensor $\mathbf{N}$ of Maxwell explicit effective formula (9.24) for transversely isotropic spheroidal inclusions, is given by the relation

$$\mathbf{N} = \left[\left(\mathbf{C}^1 - \mathbf{C}^0 \right)^{-1} + \mathbf{P} \right]^{-1}, \tag{9.34}$$

where $\mathbf{C}^0$ and $\mathbf{C}^1$ are the stiffness tensor of matrix and inclusions material, respectively. Explicitly

$$\mathbf{N} = (n_1, n_2, n_3, n_4, n_5, n_6), \tag{9.35}$$

where

$$n_1 = \frac{1}{2\blacklozenge} \left[\frac{d_1 + d_2}{d_5(d_1 + d_2) - 2d_3^2} + p_6 \right], \quad n_2 = \frac{d_1 - d_2}{1 + p_2(d_1 - d_2)}, \tag{9.36}$$

$$n_3 = n_4 = -\frac{1}{\blacklozenge} \left[p_3 - \frac{d_3}{d_5(d_1 + d_2) - 2d_3^2} \right], \quad n_5 = \frac{4d_4}{1 + p_5 d_4}, \tag{9.37}$$

$$n_6 = \frac{1}{\blacklozenge}\left[\frac{d_5}{d_5(d_1+d_2)-2d_3^2}+2p_1\right], \tag{9.38}$$

$$\blacklozenge = \frac{1+p_6 d_5+2p_1(d_1+d_2)+4p_3 d_3}{d_5(d_1+d_2)-2d_3^2}+2p_1 p_6-2p_3^2,$$

$$d_1 = C^1_{1111}-C^0_{1111},\ d_2 = C^1_{1122}-C^0_{1122},\ d_3 = C^1_{1133}-C^0_{1133},$$
$$d_4 = C^1_{2323}-C^0_{2323},\ d_5 = C^1_{3333}-C^0_{3333}, \tag{9.39}$$

where $p_i, i = 1,2,\cdots,6$ are given by eq. (9.33). Remark that this explicit expressions for components $n_i, i = 1,2,...,6$ used in the decomposition of tensor **N** into basis **T** given by (9.12)-(9.14) are different and more general than explicit expressions given by formulae (A.9) and (A.10) of Sevostianov (2014), because they are for a material with transversely isotropic symmetry.

Expression of tensor **Q** for transversely isotropic spheroidal inclusions, is given by the relation

$$\mathbf{Q} = \mathbf{C}^0\cdot\left(\mathbf{I}-\mathbf{P}\cdot\mathbf{C}^0\right), \tag{9.40}$$

where $\mathbf{C}^0$ is the stiffness tensor of matrix material. Explicitly

$$\mathbf{Q} = (q_1,q_2,q_3,q_4,q_5,q_6), \tag{9.41}$$

and

$$q_1 = \frac{1}{2}D_a(1-2p_1D_a-4p_3C^0_{13})-p_6D_c,\quad q_2 = D_b(1-p_2D_b), \tag{9.42}$$

$$q_3 = q_4 = C^0_{13}W_a-D_aW_b-2p_3D_c,\quad q_5 = 4C^0_{44}W_c, \tag{9.43}$$

$$q_6 = C^0_{33}W_a-4C^0_{13}(W_b-p_1C^0_{13}), \tag{9.44}$$

$$D_a = C^0_{11}+C^0_{12},\quad D_b = C^0_{11}-C^0_{12},\quad D_c = (C^0_{13})^2, \tag{9.45}$$

$$W_a = 1-p_6C^0_{33},\quad W_b = 2p_1C^0_{13}+p_3C^0_{33},\quad W_c = 1-p_5C^0_{44}, \tag{9.46}$$

where $p_i, i = 1,...6$, are the components of tensor **P** of eq. (9.33). Remark that this explicit expressions for components $q_i, i = 1,2,...,6$ used in the decomposition of tensor **Q** into basis **T** given by (9.12)-(9.14) are different and more general than explicit expressions given by formulae (A.6) of Sevostianov (2014), because they are for a material with transversely isotropic symmetry.

Then, introducing parameter λ Sevostianov (2014) that controls the random or not random aligned inclusions cases, by means of density distribution function

$$P_\lambda(\psi) = \frac{1}{2\pi}\left[(1+\lambda^2)e^{-\lambda\psi}+\lambda e^{-\lambda\frac{\pi}{2}}\right], \tag{9.47}$$

using eq. (9.36)-(9.38), thus, tensor $\mathbf{N}^*$ can be calculated by

$$\mathbf{N}^*_{ijkl} = \int_0^{\frac{\pi}{2}} P_\lambda(\psi) \sin\psi \, d\psi \int_0^{2\pi} \left(\sum_{p=1}^{6} n_p \mathbf{T}^p_{ijkl} \right) d\theta. \tag{9.48}$$

Explicitly,

$$N^*_{1111} = \frac{1}{2}\left[2(w_1 + w_2) + 2(w_3 + w_4) g_1(\lambda) + \frac{3}{4} w_5 g_2(\lambda) \right], \tag{9.49}$$

$$N^*_{1122} = \frac{1}{2}\left[2w_1 + 2w_3 g_1(\lambda) + \frac{1}{4} w_5 g_2(\lambda) \right], \tag{9.50}$$

$$N^*_{3333} = \frac{1}{2}\left[2(w_1 + w_2) + 2(w_3 + w_4) g_3(\lambda) + 2w_5 g_4(\lambda) \right], \tag{9.51}$$

$$N^*_{1133} = \frac{1}{2}\left[2w_1 + 2w_3 g_5(\lambda) + 2w_5 g_6(\lambda) \right], \tag{9.52}$$

$$N^*_{1313} = \frac{1}{2}\left[w_2 + w_4 g_5(\lambda) + 2w_5 g_6(\lambda) \right], \tag{9.53}$$

$$N^*_{1212} = \frac{N^*_{1111} - N^*_{1122}}{2}, \tag{9.54}$$

with

$$w_1 = n_1 - \frac{n_2}{2}, \quad w_2 = n_2, \quad w_3 = 2n_3 + n_2 - 2n_1, \tag{9.55}$$

$$w_4 = n_5 - 2n_2, \quad w_5 = n_6 + n_1 + \frac{n_2}{2} - 2n_3 - n_5, \tag{9.56}$$

and functions $g_j(\lambda), j = 1, 2, \cdots, 6$ are given by formulas

$$g_1(\lambda) = \frac{18 - \lambda(\lambda^2 + 3)e^{-\frac{\lambda\pi}{2}}}{6(\lambda^2 + 9)}, \tag{9.57}$$

$$g_2(\lambda) = \frac{120}{(\lambda^2 + 9)(\lambda^2 + 25)} - \lambda \frac{7\lambda^4 + 178\lambda^2 + 435}{15(\lambda^2 + 9)(\lambda^2 + 25)} e^{-\frac{\lambda\pi}{2}}, \tag{9.58}$$

$$g_3(\lambda) = \frac{(\lambda^2 + 3)(3 + \lambda e^{-\lambda\pi/2})}{3(\lambda^2 + 9)}, \tag{9.59}$$

$$g_4(\lambda) = \frac{24 + (\lambda^2 + 1)(\lambda^2 + 21)}{(\lambda^2 + 9)(\lambda^2 + 25)} + \lambda e^{-\lambda\pi/2} \frac{[(\lambda^2 + 9)(\lambda^2 + 25) - 120]}{5(\lambda^2 + 9)(\lambda^2 + 25)}, \tag{9.60}$$

$$g_5(\lambda) = \frac{3}{2(\lambda^2 + 9)} + \frac{(\lambda^2 + 3)(6 + \lambda e^{-\lambda\pi/2})}{12(\lambda^2 + 9)}, \tag{9.61}$$

$$g_6(\lambda) = \frac{3(\lambda^2 + 5)}{(\lambda^2 + 9)(\lambda^2 + 25)} + \frac{\lambda[(\lambda^2 + 1)(\lambda^2 + 18) + 12]}{15(\lambda^2 + 9)(\lambda^2 + 25)} e^{-\lambda\pi/2}. \tag{9.62}$$

Remark that explicitly expressions obtained here for functions $g_j(\lambda), j = 1, 2, ..., 6$ are different of explicit expressions obtained for functions $g_i(\lambda)$ reported in formula (3.6) of Sevostianov (2014), except for functions $g_1(\lambda), g_4(\lambda)$ that coincides in both contributions.

On the other hand, one can replace density distribution function given by (9.47) by density distribution function

$$W_s(\psi) = \frac{1}{4\pi} \frac{s \cdot \cosh(s \cdot \cos\psi)}{\sinh s}, \tag{9.63}$$

in which the parameter s is the disorder parameter, and similar to Eq. (9.48), one can calculate tensor $\mathbf{N}^{s*}$ given by

$$\mathbf{N}^{s*}_{ijkl} = \int_0^{\pi} W_s(\psi) \sin\psi \, d\psi \int_0^{2\pi} \left(\sum_{p=1}^{6} n_p T^p_{ijkl} \right) d\theta, \tag{9.64}$$

which explicitly is written

$$\mathbf{N}^{s*}_{1111} = \frac{1}{4}\left[4(w_1 + w_2) + 4(w_3 + w_4)\tau_1(s) + \frac{3}{4} w_5 \tau_2(s) \right], \tag{9.65}$$

$$\mathbf{N}^{s*}_{1122} = \frac{1}{4}\left[4w_1 + 4w_3\tau_1(s) + \frac{1}{4} w_5 \tau_2(s) \right], \tag{9.66}$$

$$\mathbf{N}^{s*}_{3333} = \frac{1}{4}\left[4(w_1 + w_2) + 4(w_3 + w_4)\tau_3(s) + 2w_5\tau_4(s) \right], \tag{9.67}$$

$$\mathbf{N}^{s*}_{1133} = \frac{1}{4}\left[4w_1 + \frac{w_3}{2}\tau_5(s) + 4w_5\tau_6(s) \right], \tag{9.68}$$

$$\mathbf{N}^{s*}_{1313} = \frac{1}{4}\left[2w_2 + \frac{w_4}{4}\tau_5(s) + 4w_5\tau_6(s) \right], \tag{9.69}$$

$$\mathbf{N}^{s*}_{1212} = \frac{\mathbf{N}^{s*}_{1111} - \mathbf{N}^{s*}_{1122}}{2}, \tag{9.70}$$

where $w_i, i = 1, \cdots, 5$ are given by Eq. (9.55)-(9.56), and functions $\tau_i, i = 1, \cdots, 6$ are,

$$\tau_1(s) = \frac{1}{s}\left[\frac{s - \tanh s}{s \tanh s} \right], \tag{9.71}$$

$$\tau_2(s) = \frac{16}{s^4}\left[\frac{s^2 \tanh s - 3(s - \tanh s)}{\tanh s} \right], \tag{9.72}$$

$$\tau_3(s) = \frac{1}{s^2}\left[\frac{s^2 \tanh s - 2(s - \tanh s)}{\tanh s} \right], \tag{9.73}$$

$$\tau_4(s) = \frac{2}{s^4}\left[\frac{s^2(s^2+8)\tanh s - 4(s-\tanh s)(6+s^2)}{\tanh s}\right], \tag{9.74}$$

$$\tau_5(s) = \frac{4}{s^2}\left[\frac{s^2\tanh s - (s-\tanh s)}{\tanh s}\right], \tag{9.75}$$

$$\tau_6(s) = \frac{1}{s^4}\left[\frac{(s-\tanh s)(s^2+12) - 4s^2\tanh s}{\tanh s}\right]. \tag{9.76}$$

Acknowledgements Authors thank to PHC Carlos J. Finlay 2018 project No. 39142TA (France-Cuba), to Proyecto Nacional de Ciencias Básicas titulado Métodos físico-matemáticos para el estudio de nuevos materiales y la propagación de ondas. Aplicaciones, Project No. 7515. The author RR thanks to Departamento de Matemáticas y Mecánica, IIMAS and PREI-DGAPA at UNAM, for the support to his research project and Ramiro Chávez Tovar and Ana Pérez Arteaga for computational assistance.

References

Bravo-Castillero J, Guinovart-Diaz R, Rodriguez-Ramos R, Sabina FJ, Brenner R (2012) Unified analytical formulae for the effective properties of periodic fibrous composites. Materials Letters 73:68 – 71, doi:10.1016/j.matlet.2011.12.106

Dong XN, Guo XE (2004) The dependence of transversely isotropic elasticity of human femoral cortical bone on porosity. Journal of Biomechanics 37(8):1281–1287, doi:10.1016/j.jbiomech.2003.12.011

Dong XN, Zhang X, Huang YY, Guo XE (2005) A generalized self-consistent estimate for the effective elastic moduli of fiber-reinforced composite materials with multiple transversely isotropic inclusions. International Journal of Mechanical Sciences 47(6):922 – 940, doi:10.1016/j.ijmecsci.2005.01.008

Drago AS, Pindera MJ (2008) A locally-exact homogenization theory for periodic microstructures with isotropic phases. Trans ASME Journal of Applied Mechanics 75(5):051,010–051,010–14, doi:10.1115/1.2913043

El-Eskandarany MS (1998) Mechanical solid state mixing for synthesizing of SiCp/Al nanocomposites. Journal of Alloys and Compounds 279(2):263 – 271, doi:10.1016/S0925-8388(98)00658-6

Giraud A, Huynh Q, Hoxha D, Kondo D (2007) Effective poroelastic properties of transversely isotropic rock-like composites with arbitrarily oriented ellipsoidal inclusions. Mechanics of Materials 39(11):1006 – 1024, doi:10.1016/j.mechmat.2007.05.005

Guinovart-Diaz R, Bravo-Castillero J, Rodriguez-Ramos R, Sabina FJ, Martinez-Rosado R (2001) Overall properties of piezocomposite materials 1–3. Materials Letters 48(2):93 – 98, doi:10.1016/S0167-577X(00)00285-8

Hershey AV (1954) The elasticity of an isotropic aggregate of anisotropic cubic crystals. Trans ASME Journal of Applied Mechanics 21(3):236–240, doi:10.1115/1.2913043

Hill R (1965) A self-consistent mechanics of composite materials. Journal of the Mechanics and Physics of Solids 13(4):213 – 222, doi:10.1016/0022-5096(65)90010-4

Kanaun S (2016) Efficient homogenization techniques for elastic composites: Maxwell scheme vs. effective field method. International Journal of Engineering Science 103:19 – 34, doi:10.1016/j.ijengsci.2016.03.004

Kröner E (1958) Berechnung der elastischen Konstanten des Vielkristalls aus den Konstanten des Einkristalls. Zeischrift für Physik 151(4):504–518, doi:10.1007/BF01337948

Kumar SS, Bai VS, Rajkumar KV, Sharma GK, Jayakumar T, Rajasekharan T (2009) Elastic modulus of Al-Si/SiC metal matrix composites as a function of volume fraction. Journal of Physics D: Applied Physics 42(17):175,504, doi:10.1088/0022-3727/42/17/175504

Kushch VI, Sevostianov I (2016) Maxwell homogenization scheme as a rigorous method of micromechanics: Application to effective conductivity of a composite with spheroidal particles. International Journal of Engineering Science 98:36 – 50, doi:10.1016/j.ijengsci.2015.07.003, special Issue Dedicated to Sergey Kanaun´s 70th Birthday

Levin V, Kanaun S, Markov M (2012) Generalized Maxwell's scheme for homogenization of poroelastic composites. International Journal of Engineering Science 61:75 – 86, doi:10.1016/j.ijengsci.2012.06.011, the special issue in honor of Nikita F. Morozov

Levin VM, Kanaun SK (2012) Application of Maxwell method in solution of homogenization problem for anizotropic elastic media with ellipsoidal inclusions. Proceedings of Petrozavodsk State University 1(8):86–89

Liaw PK, Shannon RE, Clark WG, Harrigan WC, Jeong H, Hsu DK (1995) Nondestructive characterization of material properties of metal-matrix composites. Materials Chemistry and Physics 39(3):220 – 228, doi:10.1016/0254-0584(94)01431-F

Martinez-Ayuso G, Friswell MI, Adhikari S, Khodaparast HH, Berger H (2017) Homogenization of porous piezoelectric materials. International Journal of Solids and Structures 113-114:218 – 229, doi:10.1016/j.ijsolstr.2017.03.003

Maxwell J (1954) A Treatise on Electricity and Magnetism, 3rd edn. Dover, New York

McCartney LN, Kelly A (2008) Maxwell's far-field methodology applied to the prediction of properties of multi-phase isotropic particulate composites. Proceedings of the Royal Society A: Mathematical, Physical and Engineering Sciences 464(2090):423–446, doi:10.1098/rspa.2007.0071

Mogilevskaya SG, Kushch VI, Nikolskiy D (2014) Evaluation of some approximate estimates for the effective tetragonal elastic moduli of two-phase fiber-reinforced composites. Journal of Composite Materials 48(19):2349–2362, doi:10.1177/0021998313498103

Otero JA, Rodriguez-Ramos R, Bravo-Castillero J, Guinovart-Diaz R, Sabina FJ, Monsivais G (2013) Semi-analytical method for computing effective properties in elastic composite under imperfect contact. International Journal of Solids and Structures 50(3):609 – 622, doi:10.1016/j.ijsolstr.2012.11.001

Otero JA, Rodriguez-Ramos R, Monsivais G (2017) Computation of effective properties in elastic composites under imperfect contact with different inclusion shapes. Mathematical Methods in the Applied Sciences 40(9):3290–3310, doi:10.1002/mma.3956

Premkumar MK, Hunt Jr WH, Sawtell RR (1992) Aluminum composite materials for multichip modules. JOM 44(7):24–28, doi:10.1007/BF03222271

Qu J, Cherkaoui M (2006) Fundamentals of Micromechanics of Solids. John Wiley & Sons, New Jersey

Qu SG, Lou HS, Li XQ (2016) Influence of particle size distribution on properties of sic particles reinforced aluminum matrix composites with high sic particle content. Journal of Composite Materials 50(8):1049–1058, doi:10.1177/0021998315586864

Roger D, Dieulesaint E (2000) Elastic Waves in Solids, vol I. Springer, New York

Selmi A, Friebel C, Doghri I, Hassis H (2007) Prediction of the elastic properties of single walled carbon nanotube reinforced polymers: A comparative study of several micromechanical models. Composites Science and Technology 67(10):2071 – 2084, doi:10.1016/j.compscitech.2006.11.016

Sevostianov I (2014) On the shape of effective inclusion in the Maxwell homogenization scheme for anisotropic elastic composites. Mechanics of Materials 75:45 – 59, doi:10.1016/j.mechmat.2014.03.003

Sevostianov I, Giraud A (2013) Generalization of Maxwell homogenization scheme for elastic material containing inhomogeneities of diverse shape. International Journal of Engineering Science 64:23 – 36, doi:10.1016/j.ijengsci.2012.12.004

Upadhyay A, Singh R (2012) Prediction of effective elastic modulus of biphasic composite materials. Modern Mechanical Engineering 2(1):6–13, doi:10.4236/mme.2012.21002

Vilchevskaya E, Sevostianov I (2015) Effective elastic properties of a particulate composite with transversely-isotropic matrix. International Journal of Engineering Science 94:139 – 149, doi:10.1016/j.ijengsci.2015.05.006

Wang G, Pindera MJ (2016) Locally-exact homogenization theory for transversely isotropic unidirectional composites. Mechanics Research Communications 78:2 – 14, doi:10.1016/j.mechrescom.2015.09.011, Recent Advances in Multiscale, Multifunctional and Functionally Graded Materials

Chapter 10
Advanced Numerical Models for Predicting the Load and Environmentally Dependent Behaviour of Adhesives and Adhesively Bonded Joints

Eduardo André de Sousa Marques, Alireza Akhavan-Safar, Raul Duarte Salgueiral Gomes Campilho, Ricardo João Camilo Carbas, and Lucas Filipe Martins da Silva

Abstract Adhesive bonding is a very flexible and efficient joining method which has been extensively adopted in applications where low structural weight and high mechanical performance is a required. However, the design process and strength prediction methods for bonded joints is still a topic where intensive research is being carried out, especially when taking into consideration the recent advances in adhesive formulations. Several methods, both analytical and numerical, are currently available for accurately modelling the behaviour of adhesive joints under quasi-static loads and in unaged conditions. However, experimental testing has also demonstrated that adhesives and adhesive joints exhibit a large degree of sensitivity to the cyclic loading and the environmental conditions, which can result in drastic changes to the mechanical behaviour of a joint. Such changes pose significant challenges when designing bonded structures and modelling the long-term behaviour of an adhesive joint, necessitating the development of advanced models, able to introduce a variable degree of degradation as a function of several environmental variables. This work presents a description of several cohesive zone models, able to simulate damage, by first discussing the state of the art techniques available for modelling of adhesives and adhesive joints, followed by the description of the specific approaches that can be employed for studying the behaviour under impact rates, fatigue and hygrothermal ageing.

Eduardo André de Sousa Marques · Alireza Akhavan-Safar · Ricardo João Camilo Carbas
Instituto de Ciência e Inovação em Engenharia Mecânica e Engenharia Industrial (INEGI), Rua Dr. Roberto Frias, 4200-465 Porto, Portugal,
e-mail: emarques@fe.up.pt, akhavan101@gmail.com, carbas@fe.up.pt

Raul Duarte Salgueiral Gomes Campilho
Departamento de Engenharia Mecânica, Instituto Superior de Engenharia do Porto, Instituto Politécnico do Porto, Porto, Portugal,
e-mail: rds@isep.ipp.pt

Lucas Filipe Martins da Silva
Departamento de Engenharia Mecânica, Faculdade de Engenharia (FEUP), Universidade do Porto, Rua Dr. Roberto Frias, 4200-465 Porto, Portugal,
e-mail: lucas@fe.up.pt

H. Altenbach and A. Öchsner (eds.), *State of the Art and Future Trends in Material Modeling*, Advanced Structured Materials 100,
https://doi.org/10.1007/978-3-030-30355-6_10

Keywords: Adhesive joint · Modelling · Durability · Loading rate · Ageing · Fatigue · Finite element analysis

10.1 Introduction

Many key players in the manufacturing sector, such as the major automakers, have significantly increased the use of adhesives for joining load-bearing components in an effort to reduce vehicle weight, improve fuel economy and reduce emissions. However, the design process of a large scale bonded structure can be extremely complex, requiring detailed knowledge of both the substrate properties and those of the adhesives, which can vary widely from brittle epoxies to highly deformable rubbers (da Silva et al, 2011). Adhesives, which are mostly polymer based, are viscoelastic, and their properties greatly depend on several factors such humidity and loading rate. While some design criteria consider these effects (Kinloch, 1987), there is still difficulty (Adams and Harris, 1996), in creating models that can accurately represent the mechanical behaviour of adhesives and adhesive joints under a wide variety of conditions. This process can be performed by various methods, which are typically divided into analytical and numerical methods. Analytical methods are simple to implement but lack the inherent flexibility of the analytical methods. The shear lag analysis or analytical method of Volkersen (1938) (capable of modelling the elasticity of the adhesive material) was one of the first analytical methods proposed for the analysis of single lap joints. It was followed by other methods which of increasingly higher complexity such as the Goland and Reissner first approximation (Goland and Reissner, 1944) (which does not neglect the joint rotation), the Generalized Failure Criterion (Hart-Smith, 1973) (which takes in account of the ability of the adhesive material to withstand plastic deformation) and the adherend failure criterion (Adams et al, 1997) which is based on the Goland and Reissner theory and suitable for adherends that deform plastically. The use of analytical methods, in contrast, has grown exponentially with the increased availability of commercial finite element (FE) codes and software. The first authors that used FE models for joints were Wooley and Carver (1971), followed by Adams and Peppiatt (1974). Despite the work of Wooley and Carver (1971) being considered as a significant evolution in the prediction of the failure load in Single lap joints (SLJs), these models still do not consider fracture mechanics. This method has been recently combined with fracture mechanics to create cohesive damage models, which offer the possibility of predicting the crack propagation as a result of simulated degradation of the material. Recently, numerical models have been developed where the strength of materials and fracture mechanics approaches have been combined, creating cohesive zone models (CZM). CZMs attempt to minimize the limitations of the strength of materials and fracture mechanics approaches (Cavalli and Thouless, 2001; Liljedahl et al, 2006a). These models can be successfully employed for thin planes of materials, making them especially well-suited for adhesive joints (Campilho et al, 2007). The evaluation of new bonded designs for vehicle and aircraft structures may be stream-lined using

FE analysis, reducing the large costs inherent in building or testing prototypes. This work presents a detailed explanation of the current trends that govern the research on modelling the mechanical behaviour of adhesively bonded joints.

This chapter is divided into four major sections, each one describing an area of study in the numerical modelling of adhesive joints. The first section will describe in detail the advanced techniques that can be used to predict failure in adhesive joints, such as the use of CZMs and the use extended finite elements (XFEM). The following section is focused on modelling the loading rate dependent behaviour of adhesives and adhesive joints, providing examples on how to model the behaviour of joints under impact. Another chapter is devoted to modelling the moisture uptake and the changes in the adhesive behaviour induced by this moisture. Lastly, the final section provides information on how to simulate fatigue induced damage in adhesives and adhesive joints.

10.2 Modelling Adhesives and Adhesive Joints Using Cohesive Zone Models and Extended Finite Element Method

Conventional techniques for strength prediction of bonded joints such as continuum mechanics or linear elastic fracture mechanics (LEFM) based techniques are usually limited to small-scale yielding ahead of the crack tip. However, for modern toughened adhesives the plastic zones that develop in the adhesive layer can lead to much larger plastic yielding than elastic yielding (Ji et al, 2010). CZMs were developed in the late 1950's/early 1960's by Barenblatt (1959) and Dugdale (1960) for static applications. CZMs are based on spring or more typically cohesive elements (Feraren and Jensen, 2004), connecting two-dimensional (2D) or three-dimensional (3D) solid elements of structures. The CZM laws can be easily incorporated in conventional FE software to simulate crack propagation in various materials, including adhesively bonded joints (Ji et al, 2010). This method defines cohesive laws to model interfaces or entire finite regions.

The CZM laws are established between paired nodes of cohesive elements. They can be used to connect superimposed nodes of elements representing different materials or different plies in composites, to simulate a zero thickness interface (local approach; Fig. 10.1a; Pardoen et al, 2005), or they can be applied directly between two non-contacting materials to simulate a thin strip of finite thickness between them, e.g. to simulate an adhesive bond (continuum approach; Fig. 10.1b; Xie and Waas, 2006). Some numerical works where strength prediction of bonded joints is achieved by CZMs take advantage of the local approach (Liljedahl et al, 2006a; Campilho et al, 2005). With this methodology, the plastic dissipations of the adhesive bond are simulated by the solid finite elements, whilst the cohesive elements simulate damage growth (Fig. 10.1a). The "intrinsic fracture energy" should be considered for the CZM laws instead of the fracture toughness (G_c), while the plastic dissipations of ductile materials take place at the solid elements representative of the adhesive bond (Liljedahl et al, 2006a). Thus, G_c is the sum of these two components. The

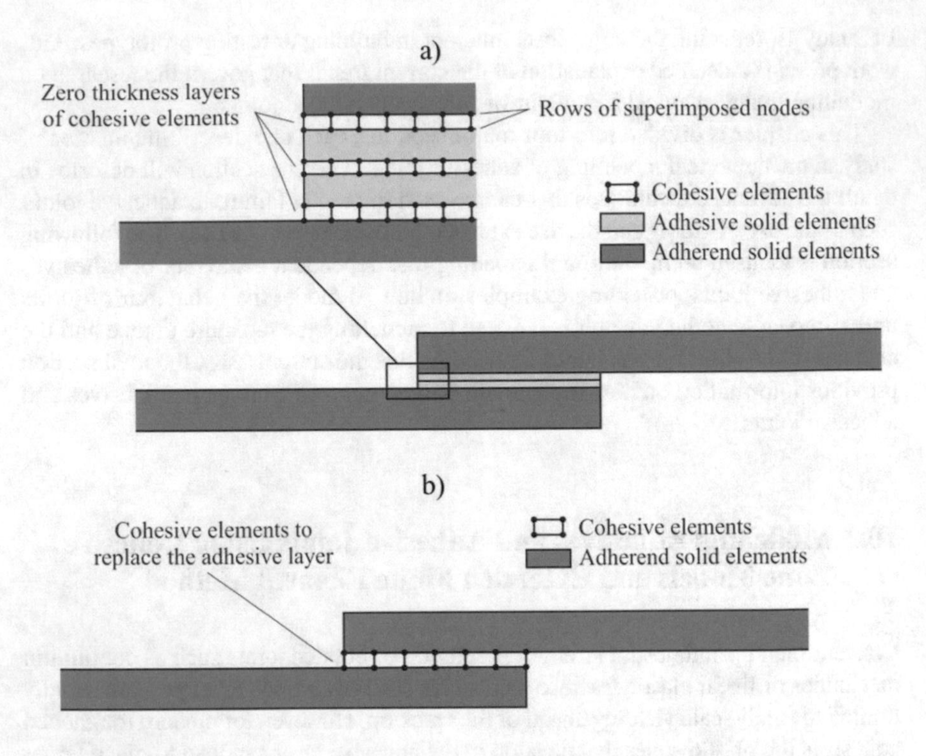

Fig. 10.1 Cohesive elements to simulate zero thickness failure paths – local approach (a) and to model a thin adhesive bond between the adherends – continuum approach (b) in an adhesive joint

effects of external and internal constraints on the plastic dissipations of an adhesive bond are thus accountable for in the local approach. Application of the continuum approach (Fig. 10.1b) involves the replacement of the entire adhesive bond by a single row of cohesive elements with the representative behaviour of the adhesive bond (Campilho et al, 2008a; Kafkalidis and Thouless, 2002). Oppositely to the local approach, the CZM elements' stiffness represents the adhesive layer stiffness in each mode of loading. This approach has been widely used in the simulation of bonded joints, with accurate results after proper calibrations are undertaken for the CZM laws (Campilho et al, 2009a). The main disadvantage of this approach is that CZM become dependent on the joint geometry, which affects the size of the fracture process zone (FPZ) and plasticity ahead of the crack tip (Ji et al, 2010).

CZMs simulate stress evolution and subsequent softening up to complete failure, which corresponds to material degradation. The CZM laws are usually represented by linear relations at each one of the loading stages (Yang and Thouless, 2001). Figure 10.2 represents the 2D triangular CZM model actually implemented in Abaqus (Providence, RI, USA) for static damage growth. The 3D additionally includes the tearing component. More details on the 3D CZM are available in Campilho et al (2008b) or in Abaqus (2013). The subscripts n and s relate to pure normal (tension)

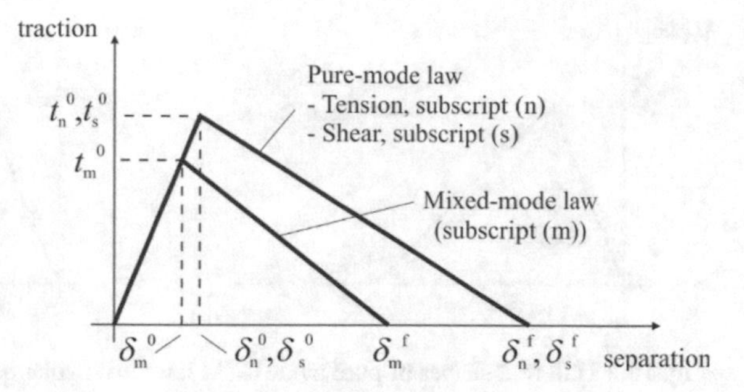

Fig. 10.2 Triangular CZM law (adapted from Abaqus, 2013)

and shear behaviours, respectively. t_n and t_s are defined as current stresses in tension and shear, respectively, δ_n^0 and δ_s^0 are the peak strength displacements, and δ_n^f and δ_s^f the failure displacements (defined by G_{Ic} or G_{IIc}, respectively, as these represent the area under the CZM laws). In the mixed-mode CZM law, t_m^0 is the mixed-mode cohesive strength, δ_m^0 the corresponding displacement, and δ_m^f the mixed-mode failure displacement. Under mixed loading, stress and/or energetic criteria are used to combine the pure mode laws. With this procedure, the complete failure response of structures can be simulated (Zhu et al, 2009). Typically, continuum mechanics criteria are used for the onset of damage and energy criteria for propagation (Chen et al, 2011). This allows the simulation of onset and non-self-similar growth of damage without user intervention and not requiring an initial flaw, unlike what occurs with conventional fracture mechanics criteria. On the other hand, compared to continuum mechanics techniques, CZMs are mesh independent if enough integration points are simultaneously under softening during the failure process (Campilho et al, 2011, 2009b). This technique also allows combining multiple failure possibilities, and the knowledge of the damage onset site is not required as input, since damages initiates at any CZM element when the damage onset criterion is attained. The main limitation of CZMs is that cohesive elements must exist at the planes where damage is prone to occur, which, in several applications, can be difficult to know in advance. However, in bonded joints that damage propagation is restricted to the adhesive layer or the adhesive/adherend interfaces, which turns the analysis procedure easier. Developed CZMs include triangular (Alfano and Crisfield, 2001), linear-parabolic (Allix and Corigliano, 1996), polynomial (Chen, 2002), exponential (Chandra et al, 2002) and trapezoidal laws (Campilho et al, 2008a). The triangular, linear-exponential and trapezoidal shapes are the most commonly used CZM shapes (Fig. 10.3). In the trapezoidal law, δ_n^s and δ_s^s are the stress softening onset displacements.

CZMs can also be adapted to simulate ductile adhesive layers by using trapezoidal laws (Campilho et al, 2009a). Although it is always advised the use of the most suitable CZM shape and to perform accurate parameter estimations, some works have demonstrated acceptable predictions for small variations to the optimal CZM parameters and shapes (Liljedahl et al, 2006a; Biel and Stigh, 2008). Nonetheless, the effect of the CZM law shape on the strength predictions may significantly vary depending on the structure geometry and post-elastic behaviour of the materials.

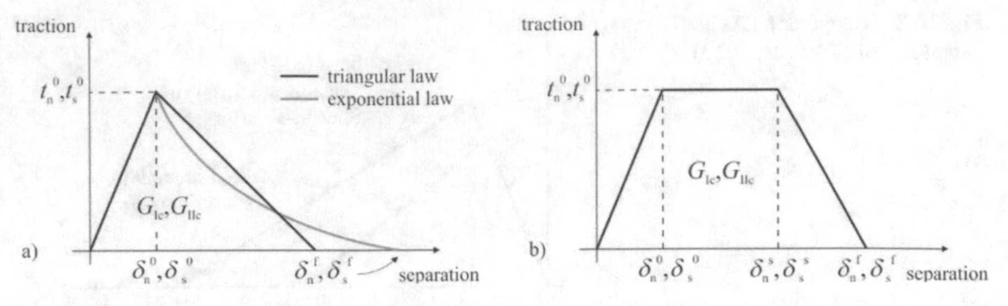

Fig. 10.3 Different shapes of pure mode CZM laws: triangular or linear-exponential (a) and trapezoidal (b)

The CZM law effects became evident in the experimental and FE study of Pinto et al (2009), whose objective was the strength comparison of single-lap joints with similar and dissimilar adherends and values of t_P bonded with the adhesive 3M DP-8005. The accurate shape of the CZM law was considered fundamental for the strength prediction and $P - \delta$ response of the structure when using stiff adherends. Under these conditions, peel stresses are minimal and, due to the large longitudinal stiffness, shear stresses distribute more evenly along the bond length. Thus, the $P - \delta$ curve is very similar in shape to the chosen shear CZM law. On the other hand, compliant adherends led to large shear and peel stress gradients. Since this implies different damage states along the adhesive layer, using an inaccurate CZM law gives adhesive stresses that are over predicted at some elements and under predicted at others. Thus, by using compliant adherends the overall behaviour gave smaller errors. Ridha et al (2011) considered scarf repairs on composite panels bonded with the high elongation epoxy adhesive FM 300M (Cytec). CZM laws with linear, exponential and trapezoidal softening were compared, and linear degradation resulted in under predictions of the repairs strength of nearly 20%, on account of excessive plastic degradation at the bond edges that was not observed in the real joints. Regarding the application of CZM for strength prediction of adhesive bonds, trapezoidal laws are recommended for ductile adhesives (Feraren and Jensen, 2004; Campilho et al, 2010), and this is particularly critical when considering stiff adherends, due to the practically absence of differential deformation effects in these components along the overlap (Pinto et al, 2009; Alfano, 2006). In contrast, triangular CZMs are efficient for brittle materials that do not plasticize by a significant amount after yielding (Campilho et al, 2011), and also for the intralaminar fracture of composite adherends in bonded structures, due to their intrinsic brittleness (Xie et al, 2006). For adhesives that exhibit a relatively brittle behaviour in tension while showing large plastic flow in shear, the proper selection of the CZM parameters and also the minimization of the constant stress (plastic flow) region in the tensile law result on a good representation of the adhesive behaviour. The material/interfacial behaviour that the CZM law is simulating should always be the leading decision factor to select the most appropriate shape. Despite this fact, other issues should be taken into account (da Silva and Campilho, 2012). In fact, the CZM law shape also influences the iterative solving procedure and the time required to attain the solution of a given engineering problem:

larger convergence difficulties in the iterative solving procedure usually take place for trapezoidal rather than triangular CZM laws, due to the more abrupt change of stiffness in the cohesive elements during stress softening. Actually, for a fixed value of the material properties G_{Ic} and G_{IIc}, the larger the constant stress length of the trapezoidal law, the bigger is the descending slope. Additionally, exponential and trapezoidal CZM are more difficult to formulate and implement in FE software.

The CZM parameter effects are also detailed in several works. The main joint geometry parameters that affect the CZM parameters of adhesive layers are t_A (thickness of the adhesive) and t_P (thickness of the substrate), which emphasizes the importance of the t_A and t_P consistency between the fracture tests and the structures to be simulated (Leffler et al, 2007; Chai, 1986; Bascom and Cottington, 1976). In Carlberger and Stigh (2010), the CZM parameters in tension and shear were determined for a thin layer of adhesive using the DCB and ENF test configurations, respectively, considering $0.1 \leq t_A \leq 1.6$ mm. It was concluded that the CZM parameters significantly vary with t_A, namely an increase of G_{Ic} and G_{IIc} with this parameter. Corroboration of the adhesive restraining effects was equally accomplished by Ji et al (2010), which studied the influence of t_A on t_n^0 and G_{Ic} for a brittle epoxy adhesive, by using the DCB specimen and the direct method for parameter estimation. Results clearly showed a reduction of tn0 and increase of G_{Ic} with bigger values of t_A. On the other hand, a few studied showed variations of G_{Ic} and G_{IIc} by modification of t_P. In Mangalgiri et al (1987), symmetric and asymmetric DCB specimens were experimentally tested with different values of t_P (by considering 8, 16 or 24 plies of carbon-fibre adherends). The static tests showed a large improvement of G_{Ic} between composites with 8 and 16 plies. Devitt et al (1980) equally used the DCB test to investigate this effect and found a 9% increase in the value of GIc of bonded joints made of glass-epoxy composites by duplicating the number of plies of the adherends. From these studies, it is clear that the differences take place at relatively low t_P values. Since most bonded joints are made between thin adherends/sheets, the understanding of how t_P affects the fracture toughness is highly relevant.

The eXtended Finite Element Method (XFEM) is a recent improvement of the FE method for modelling damage growth in structures. It uses damage laws for the prediction of fracture that are based on the bulk strength of the materials for the initiation of damage and strain for the assessment of failure (defined by G_{Ic}), rather than the cohesive tractions and tensile/shear relative displacements used in CZM. XFEM gains an advantage over CZM modelling as it does not require the crack to follow a predefined path. Actually, cracks are allowed to grow freely within a bulk region of a material without the requirement of the mesh to match the geometry of the discontinuities neither remeshing near the crack (Mohammadi, 2008). This method is an extension of the FE method, whose fundamental features were firstly presented in the late 90's by Belytschko and Black (1999). The XFEM relies on the concept of partition of unity, that can be implemented in the traditional FE by the introduction of local enrichment functions for the nodal displacements near the crack to allow its growth and separation between the crack faces (Moës et al, 1999). Due to crack growth, the crack tip continuously changes its position and orientation depending on the loading conditions and structure geometry, simultaneously to the creation of the necessary enrichment functions for the nodal points of the finite elements around the crack path/tip. It uses damage laws for the prediction of fracture that are based on the bulk strength of the materials for the initiation of damage and strain for the assessment of failure (defined by G_{Ic}), rather than the values of t_n^0/t_s^0 or w/v used in CZM.

Varying applications to this innovative technique were proposed to simulate different engineering problems. Sukumar et al (2000) updated the method to three-dimensional damage simulation. Modelling of intersecting cracks with multiple branches, multiple holes and cracks emanating from holes was addressed by Daux et al (2000). The problem of cohesive propagation of cracks in concrete structures was studied by Moës and Belytschko (2002), considering three-point bending and four-point shear scaled specimens. More advanced features such as plasticity, contacting between bodies and geometrical non-linearities, which show a particular relevance for the simulation of fracture in structures, are already available within the scope of XFEM. The employment of plastic enrichments in XFEM modelling is accredited to Elguedj et al (2006), which used a new enriched basis function to capture the singular fields in elasto-plastic fracture mechanics. Modelling of contact by the XFEM was firstly introduced by Dolbow et al (2001) and afterwards adapted to frictional contact by Khoei and Nikbakht (2006). Fagerström and Larsson (2006) implemented geometrical nonlinearities within the XFEM. Fatigue applications for XFEM were proposed recently (Xu and Yuan, 2009; Sabsabi et al, 2011), but these have not yet been applied to the mixed mode fracture of bonded joints.

10.3 Modelling of Adhesives and Adhesive Joints Under Varying Loading Rates and Impact Conditions

The process of modelling the impact behaviour of adhesive joints is highly relevant for the automotive industry, as it allows the simulation of the crash behaviour of bonded vehicle structures, enabling optimization processes to be carried out and reducing the need for costly full scale or component scale crash tests. This approach usually demands the use of dynamical models, as for large strain rates the influence of inertial effects becomes significant and introduces added stresses in the joint. Some authors, such as Harris and Adams (1985) have employed models without any type of inertial effects but with strain rate dependent properties, which are valid for smaller rates of loading and small masses. Alternatively, if the materials are not shown to be strain rate sensitive and the impact speeds are high, models with quasi-static property descriptions are used in conjunction with inertial modelling to improve accuracy in stress predictions. This is also valid in joints where adherend yield occurs under impact, as in these joints the adhesive does not have a significant contribution on the behaviour of the joint and most metallic adherends do not exhibit strain rate dependency. Higuchi et al (2002a,b); Sawa et al (2003) used the DYNA3D software package to model the stress distribution propagation in SLJ specimens under diverse impact loadings. This type of dynamic models is formulated according to Eq. (10.1)

$$[M][A] + [K][U] = [F], \tag{10.1}$$

where $[M]$ is the mass matrix, $[A]$ is the acceleration vector, $[K]$ is the stiffness matrix, $[U]$ is the displacement vector and $[F]$ is the external load vector.

Challita and Othman (2010) also employed a three-dimensional FE dynamical model to assess the stresses present in split Hopkinson pressure bar (SHPB) testing of double lap joint (DLJ) specimens with metal substrates. For both substrates and adhesive, an elastic behaviour was assumed. Similar modelling work was performed by Hazimeh et al (2015) but using composite substrates in DLJ specimens.

More complex models combine the inertial effects with strain rate dependent properties, either for the adhesive, the adherends or both. In order to achieve reliable and accurate predictions of impact behaviour of adhesive joints using numerical models, the material properties used in the model should preferably be determined at the appropriate strain rates values (Duncan and Pearce, 1999; Charalambides and Dean, 1997; Dean and Charalambides, 1997). In Xia et al (2009a,b) the authors aimed to understand in depth the dynamic failure of weld-bonded joints. The models created by the authors used a fully dynamical analysis that treated the adhesive layer as a rigid link between the two strain rate dependent substrates. Yang et al (2012) evaluated the application of a simplified finite element for modelling of a toughened adhesively bonded joint. The numerical model employed strain rate dependent data for a toughened epoxy adhesive and for the steel substrates by defining curves of the failure parameters versus the effective strain rate. This was complemented by the addition to the model of strain rate dependent data for the pre-failure properties.

As an alternative to the use of experimentally derived strain rate dependent data, the model can directly employ a constitutive model. Zaera et al (2000) used a Cowper-Symmonds based model in a finite difference simulation of the impact behaviour of a bonded ceramic/metal armour. The authors were able to use this relatively simple constitutive model to model the experimental data sufficiently well for preliminary design calculations. The work of Sawa et al (2008) and Liao et al (2011, 2013) further demonstrated that the use of a Cowper-Symonds constitutive model for the adhesive could model the interface stress distributions in various types of joint geometries.

Dean et al (1997) conducted a study with the intention to compare the measured and predicted performance of SLJ specimens under impact conditions (strain rates ranging from $2 \cdot 10^{-5}$ to $115\ s^{-1}$), through the use of a drop weight apparatus. Von Mises and Drucker-Prager models implemented in a FEA were used to predict joint failure. The results indicated that the von Mises yield criterion was not suitable for toughened adhesives, but they indicated that the linear Drucker-Prager model seemed applicable. The authors concluded that for toughened adhesives, an elastic-plastic material model is needed and that this model should employ a yield criterion with sensitivity to the hydrostatic component stress.

Zgoul and Crocombe (2004) employed a rate dependent von Mises mode and a rate dependent Drucker-Prager to model the numerical behaviour of SLJ and thick adherent shear testing (TAST) specimens. All models were implemented in a FEA and the results compared with experimental data. The results were not very satisfactory as the von Mises model was found to be inaccurate as it did not account for hydrostatic sensitivity. The Drucker-Prager model was able to accommodate hydrostatic sensitivity but was reported to have convergence problems.

Currently, most of the academic and industrial research on the impact behaviour is focused on using CZMs, applied to complex models (Machado et al, 2018). Carlberger and Stigh (2007) demonstrated the validity of using this type of approach to predict impact strength. Authors such as Haufe et al (2010); May et al (2014); Clarke et al (2011); Avendaño et al (2016); Neumayer et al (2016) have shown that modern commercial software packages support the prediction accurate failure load predictions using complex dynamical cohesive models with strain rate dependent data.

In these cases, the simplest approach to model the behaviour of an adhesive joint under impact using CZM is to determine the average strain rate that an adhesive joint will be subjected and create a cohesive law that will reflect the properties of the adhesive at that given strain rate. This approach is somewhat rough as it does not take into account the fact that an adhesive layer will be almost certainly be subjected to different local strain rates due to the differential straining an adhesive joint typically experiences. However, the results can be sufficiently accurate for obtaining satisfactory load predictions in a design process.

To improve on this case, CZMs can instead be specifically developed to reflect the properties of adhesive layers as a function of the strain rate. This type of functionality is available in some commercial finite element software packages and it can also be implemented in custom designed cohesive elements, for example, using an user defined material model (UMAT or VUMAT) subroutine of Abaqus. In these elements, a material law is first necessary, which will provide a method to calculate the value for each value of strain rate. This can be achieved through a function or table, construed using experimental data and extrapolation.

For the case of a solution implemented in Abaqus, an UMAT subroutine (user material) is used to define the Jacobian matrix, $\partial\Delta\sigma/\partial\Delta\varepsilon$, for the mechanical constitutive model, updating stress. A scheme of the integration process of UMAT for each increment can be seen in Fig. 10.4.

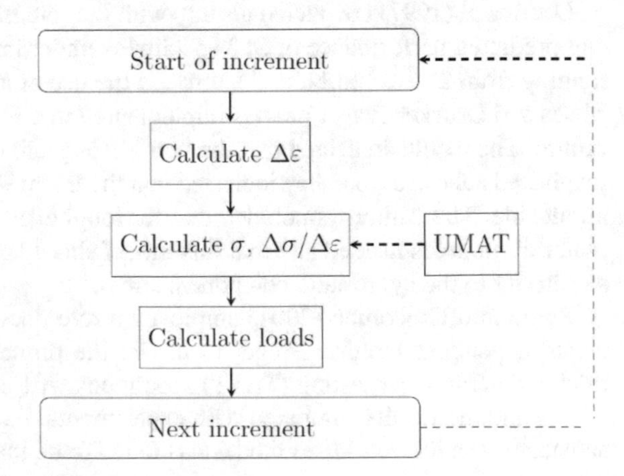

Fig. 10.4 Schematic representation of the UMAT integration process in Abaqus

For modelling a basic triangular cohesive law for a CZM, the necessary properties of the adhesive that should be introduced on the model are the Young's modulus, E, shear modulus, G_s, as well as expression that characterize the strain rate dependent tensile strength σ_{max}, shear strength, τ_{max}, critical energy release rate in mode I, G_{IC}, and critical energy release rate in mode II, G_{IIC}. The ultimate stress in mode I, σ_{Imax}, is the maximum tensile strength σ_{max} and ultimate stress in mode II σ_{IImax} is the maximum tensile strength τ_{max}.

It is important to note that, even if a code is developed to analyse cases in which the loading is mainly in mode I, it should be able to accommodate minor loadings in mode II that may appear during the test. Therefore, an UMAT element must be developed considering the possibility of mixed mode loads. In this way, if the loading is in pure mode I, only mode I properties will be considered but, if small mode II loads emerge, their contribution can still be accounted for.

Abaqus provides, for each time increment, the strain increment and its duration, which allows the direct determination of strain rate. As mentioned above, if the strain rate is known, the critical energy release rate in mode I can be determined, based on the imputed expression. Afterwards, and considering a triangular traction-separation law, the strain when maximum stress is reached, ε_{i0} (i = I, II), and the maximum strain, ε_{if} (i = I, II), are determined through the area of the triangle represented in Fig. 10.5. The stiffness k represents the Young's modulus and the shear modulus for mode I and II, respectively.

The current strain in the direction of mode I and II, ε_i (i = I, II), is also provided by Abaqus, enabling the calculation of the mixed mode of the loading. Afterwards, a new triangular law can be established for that mixed mode, combining the properties for pure mode I and pure mode II, as clarified in Fig. 10.6. The strain when maximum stress is reached, ε_{m0}, maximum strain, ε_{mf}, and current strain, ε_m, are determined for the mode under analysis.

It is considered that, when the strain in one element reaches ε_{m0}, damage initiates, and as strain progresses from ε_{m0} to ε_{mf} a damage coefficient, d, progresses linearly from 0 to 1 reaching 1 (full damage) when the strain reaches ε_{mf}. When ε_m is less than ε_{m0}, the Jacobian matrix is calculated, and stress is determined. When ε_m is higher than ε_{m0}, damage is calculated and then stress is determined. The degradation of the properties when damage occurs can be seen in Fig. 10.7.

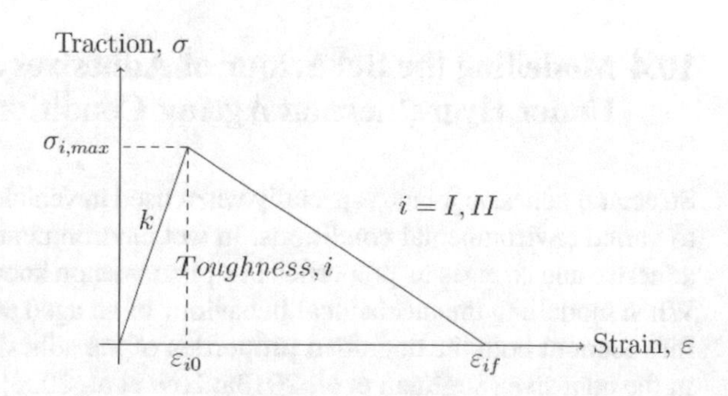

Fig. 10.5 Triangular law used in the UMAT for each pure mode

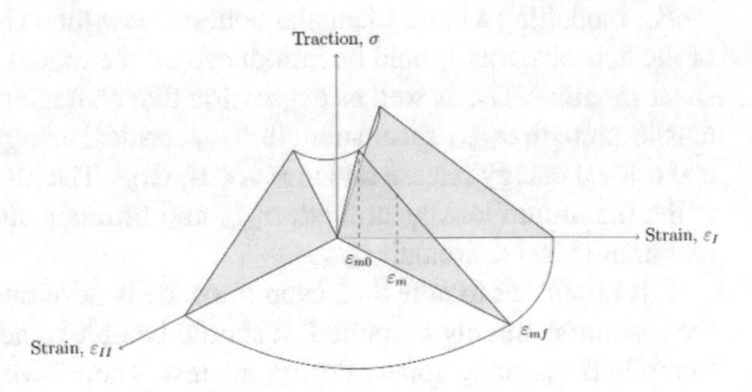

Fig. 10.6 Geometrical representation of the pure mode laws under mixed mode conditions

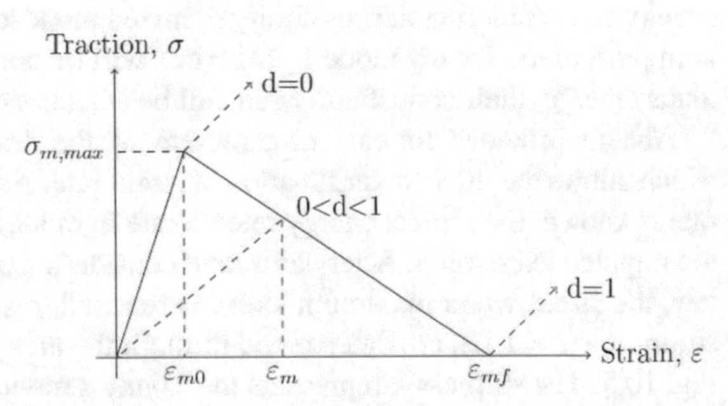

Fig. 10.7 Triangular law used in UMAT, showing three different levels of damage degradation, under mixed mode

Overall, the use of cohesive zone modelling for this purpose enables the development of robust models, which can be applied to structures with complex geometry. Two main challenges arise when adopting this procedure, however. The first is related to the difficulty in determining accurate strain rate dependent properties of the materials, which require the use of extensive testing procedures using equipment such as drop weight testers and SHPB. The second challenge is the complexity of these models, which is not entirely suitable for directly modelling very large structures due to the large computational costs involved. Considering these challenges, further research is still advised in both simplifying the testing procedures and the models, to enable a faster and yet still accurate path for the determination of reliable failure prediction for adhesive bonded structures under impact conditions.

10.4 Modelling the Behaviour of Adhesives and Adhesive Joints Under Hygrothermal Ageing Conditions

Structural adhesive joints, especially when used in vehicle construction, are subjected to varied environmental conditions. In wet environments, moisture ingresses in the adhesive and changes its properties in a phenomenon known as ageing of the adhesive. When modelling the mechanical behaviour of an aged adhesive joint, one must take into account both the degraded properties of the adhesive if the failure is cohesive in the adhesive (Sugiman et al, 2013a; Hua et al, 2006; Han et al, 2014a; Sugiman et al, 2013b) and the degraded properties of the interface (Sugiman et al, 2013b; Liljedahl et al, 2006b; Crocombe et al, 2006; Liljedahl et al, 2005) if interfacial failure occurs. Adhesives generally become more ductile and weaker when exposed to moist environments and the interface is prone to lose its toughness. In an adhesive joint, this frequently means that there is a gradient in the mechanical properties of the adhesive layer and on the interface as the exposed faces of the adhesive joint always absorb water faster than the joint's centre, which will cause the edges of the joint to degrade its properties much faster than in the centre if the joint has been exposed to hygrothermal ageing for a limited time and has not reached saturation yet (Ilioni et al, 2019). Therefore, in practice the numerical simulation must simulate a joint with graded material properties. In order to assess the gradient of mechanical properties in the adhesive layer or in the interface, it is necessary to have complete knowledge the amount of water at each point, which means that the water uptake into the adhesive joint must be computed, either using an analytical (Crocombe, 1997) or numerical method (Hua et al, 2006, 2008; Sugiman et al, 2013b; Han et al, 2014b). Crocombe (1997) was the first to include such simulation. The strength of single-lap adhesive joints with and without adhesive fillets was predicted using the adhesive failure strain as a criterion. It was found that, after 30 days of immersion in tap water, the joints were more prone to fail at their centre, where the adhesive ductility was lower due to the reduced water uptake. Later, Hua et al (2006) made a similar analysis. In this work, the von Mises yielding criterion was used as the failure criterion. This critical strain was calculated using dry and partially moist mixed-mode flexure tests. It was found that the critical strain determined using bulk specimens was higher than the actual critical strain in the adhesive joint, indicating that the wet-adhesive–adherend interface was more susceptible to moisture degradation. Carrere et al (2018) developed a method for predicting the mechanical behaviour of carbon-epoxy laminates using a finite fracture mechanics approach. In this model, the damage threshold is influenced by the aging, but the kinetics of the crack propagation remains almost constant. Usually, because water penetrates into the adhesive bondline through two directions, a 3D analysis must be undertaken in order to consider the gradient in the mechanical properties in both dimensions of the adhesive layer (Hua et al, 2006).

However, there are two cases in which a simpler 2D analysis may be enough to accurately predict the mechanical behaviour of the adhesive joint:

1. A rectangular adhesive layer, in which the length is considerably smaller than the width. In this case, the gradient in the width direction will be negligible. Only the gradient in the length direction will be important;
2. When permeable adherends, such as fibre reinforced plastic, are used. These adherends allow water to be absorbed through its thickness, allowing for a more uniform water absorption by the adhesive layer. Most adhesive joints degrade under service conditions. They absorb water while supporting a mechanical load. It is known that mechanical loading enhances degradation and water absorption of adhesives (Liljedahl et al, 2005; Han et al, 2014b).

Some authors have modelled the mechanical behaviour of adhesive joints using sequentially coupled analyses. This kind of analysis is normally made in two steps:

1. Calculating the moisture profile using a diffusion analysis;
2. Calculating the parameters used in the model, which are a function of the moisture amount in the adhesive, predicted in the previous step.

Graded properties are attributed to the adhesive, as the moisture concentration is usually not constant along the entire overlap. Usually the moisture uptake of the adhesive is determined using unstressed bulk specimens. However, some studies state (Liljedahl et al, 2005) that the diffusion of water into adhesives is affected by the stress state of the adhesive. The sequentially coupled analysis does not consider the stress enhanced diffusion. To overcome this setback, fully coupled models (Han et al, 2014a) have been developed. Using these models, the stress state of the adhesive is influenced by the water uptake, which will in turn be influenced by the stress state of the adhesive. In practice, this means that real adhesive joints, which are usually stressed during its work life, will absorb more water and their properties will be more degraded. In order to model the degradation of stressed adhesive joints subjected to moisture environments and to obtain their residual strength, Han et al (2014a) used two steps:

Step 1: Modelling the long-term aging process in the adhesive joint under combined thermal-hygro-mechanical service loading conditions with a fully coupled methodology, an analogy between moisture diffusion and conduction of heat was made and thermal-displacement-coupled elements were used in the adhesive layer. The von Mises stress was used to characterize the stress dependence of the moisture uptake. In this step, a constant creep load was applied to the adhesive joint. The moisture uptake and equivalent creep strain were defined as field variable and used in step 2.

Step 2: Simulation of the quasi-static tensile loading process in adhesive joints using cohesive zone models that had been previously aged (in step 1). The properties of the adhesive were set to be a function of the field variables defined in step 1. When an adhesive is aged under a moist environment, swelling occurs due to the absorbed bond water. However, it has been shown by several authors (Sugiman et al, 2013b; Reedy and Guess, 1996) that in an adhesive joint, no significant residual stresses arise due to relaxation of the adhesive and the strength of the joint remains almost unchanged.

Viana et al (2017c) present a practical example of the modelling of the water uptake process and its use for predicting the time necessary for achieving saturation. In this work, the water uptake in the adhesive joint was modelled using FE analysis. In the studied case, the bondline is very long, and considering that diffusion only occurs in the width direction is enough to predict the water absorption of the adhesive joint (Hua et al, 2006). The adhesive joints reached their maximum moisture uptake sooner than expected if only sorption in the bulk adhesive was considered, taking into account the adhesive diffusion properties determined in a previous study (Viana et al, 2017a).This is thought to be due to interfacial diffusion of water. In order to model this phenomenon, two types of models are usually considered:

1. A one dimensional model, in which the overall water uptake of the joint was modelled. The diffusion along the width of the joint was modelled using unidi-

mensional beam elements. The diffusion coefficient attributed to these elements were fitted so that the numerical prediction would match the moisture uptake that was calculated, considering the experimentally measured toughness. This way it was possible to compute the overall diffusion coefficient ($D_{average}$) of the joint through an inverse method;
2. A two-dimensional model, in which the water uptake of the bulk adhesive and the water uptake in the adhesive-adherend interface were modelled separately. In this model, the increase in the diffusion speed is attributed to capillary diffusion happening at the interface between the adhesive and the adherends. In order to model this phenomenon, two layers were considered (as shown in Fig. 10.8)

One layer of adhesive, whose diffusion properties are shown, as an example, in Table 10.1. Depending on the ageing environment, these adhesives may present Fickian or dual Fickian behaviour. For this reason, two diffusion coefficients and two equilibrium moisture uptakes are presented.

A very thin layer that represents the interface. The diffusion coefficient of this layer was fitted so that the water uptake of the adhesive would match the water uptake that was calculated from the experimentally measured toughness.

This way it is possible to compute the diffusion coefficient of the interface ($D_{interface}$) through an inverse method. As the coefficient of diffusion was set to be higher at the interface, diffusion of water occurs preferentially in this region, which is then responsible for bringing moisture deeper into the adhesive layer. This moisture

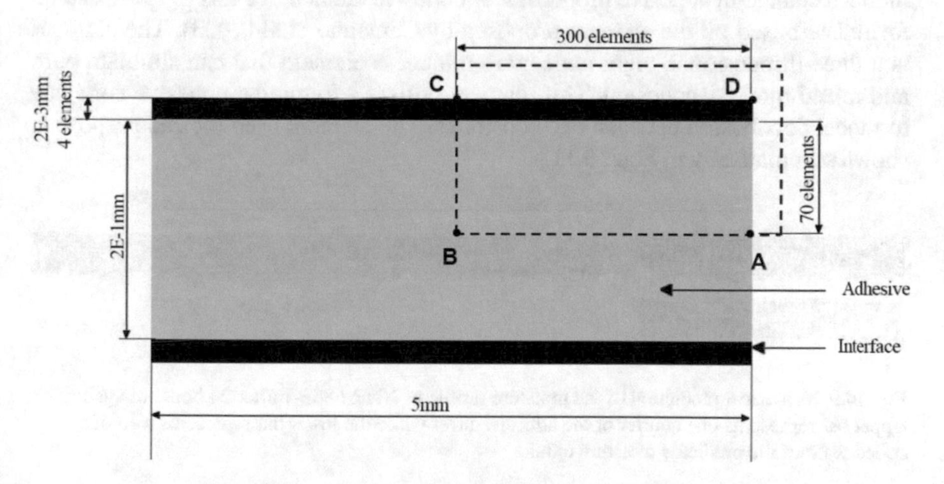

Fig. 10.8 Geometry of the model used to predict interfacial water uptake

Table 10.1 Moisture diffusion parameters of an adhesive

	Ageing Environment	D_1 (m^2/s)	mwt_1	D_1 (m^2/s)	mwt_2
XNR	Distilled Water	6.0E-13	0.0095	8.0E-14	0.0023
6852-1	Salt Water	6.0E-13	0.0080	8.0E-14	0.0006

is quickly absorbed by the adhesive. Fig. 10.9 shows an example of the computed moisture uptake of XNR 6852-1 adhesive after 24 hours of immersion.

The phenomenon of diffusion shares mathematics with the phenomenon of heat conduction and it is possible to model the moisture uptake of the adhesive simply as a heat transfer problem. The equivalent parameters to permeability coefficient, diffusion coefficient and solubility coefficient are thermal conductivity, thermal diffusivity and heat capacity respectively. In both the one-dimensional model and the two-dimensional model, the heat transfer elements available at the Abaqus library DC1D2 and DC2D4 for the 1D and 2D analyses, respectively, were used. Due to the symmetry of the problem, in order to reduce the computation effort and to increase the speed of the analysis only one quarter of the section of the specimens must be modelled. Considering the geometry shown Fig. 10.8, across line segments [AB], [BC] and [CD], no mass transfer occurs. In the line segment [AD] equilibrium moisture uptake is attained instantly because it is in contact with the ageing environment. Using this methodology, both the average diffusion coefficient and the diffusion coefficient of the interface can be determined. They are shown in Table 10.2.

The comparison between the experimental diffusion and the numerical predictions for one of the cases under study is shown in Fig. 10.10. Although there is some dispersion in the results, which is expected given the method used, the numerical prediction fits the experimental results well.

Similarly to what was presented for the strain rate dependence of adhesives, it also possible to formulate a CZM element, which takes into account the moisture induced change in adhesive properties. A cohesive element for this purpose can be formulate based on the element proposed by Camanho et al (2003). The element is a three-dimensional, eight node, zero thickness element that can simulate pure and mixed mode decohesion. This element utilises a triangular cohesive zone law to model decohesion between two substrates. The element used for this purpose is shown schematically in Fig. 10.11.

Fig. 10.9 Numerical prediction of the moisture profile of XNR 6852-1 after 24 hours of ageing. Upper bar represents one quarter of the adhesive layer while the lower bar represents a colour coded scale of the predicted moisture uptake

Table 10.2 Moisture diffusion parameters of the joints bonded the adhesive studied

	Ageing Environment	$D_{average}$ (m^2/s Obtained with 1D model	$D_{interface}$ (m^2/s Obtained with 2D model
XNR 6852-1	Destilled Water	6.0E-13	8.0E-14
	Salt Water	6.0E-13	8.0E-14

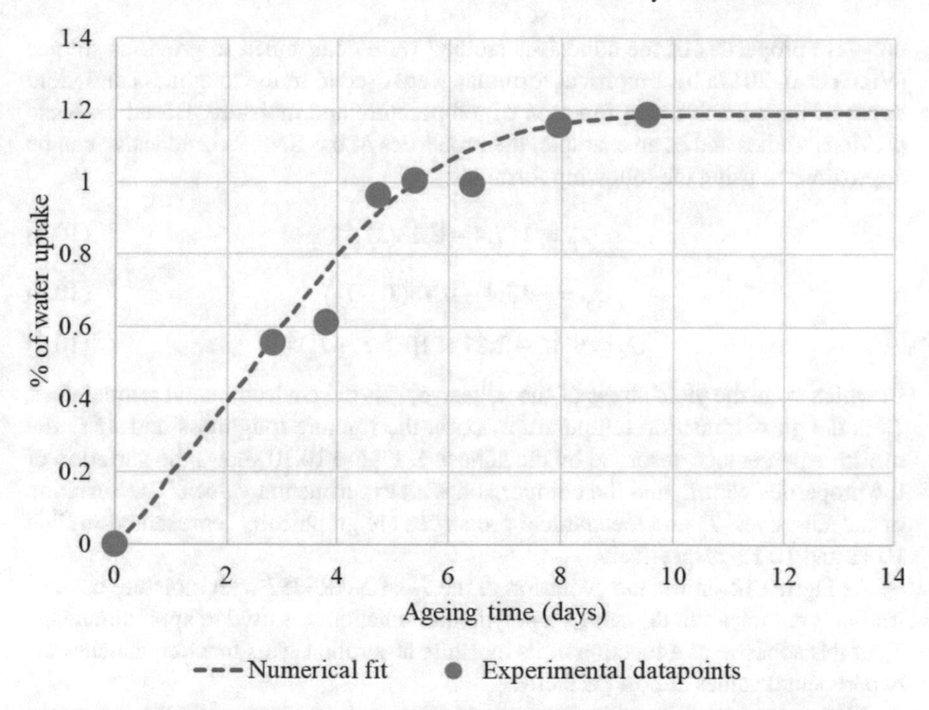

Fig. 10.10 Example of experimental data and numerical prediction for one of the cases under study (XNR 6852-1 in distilled water)

A high initial stiffness (K) is used to hold the top and bottom surfaces of decohesion element together in the linear elastic range until the yield stress is reached. After this, softening starts and the load decreases linearly until zero. The toughness of the adhesive is given by the area of the triangle. Modifications were made to this element to make it take into consideration the environmental temperature and absorbed moisture by the adhesive. The element reads the moisture field of the adhesive layer and attributes the yield stress and toughness of the adhesive according to the read moisture and environmental temperature. The moisture and temperature dependent

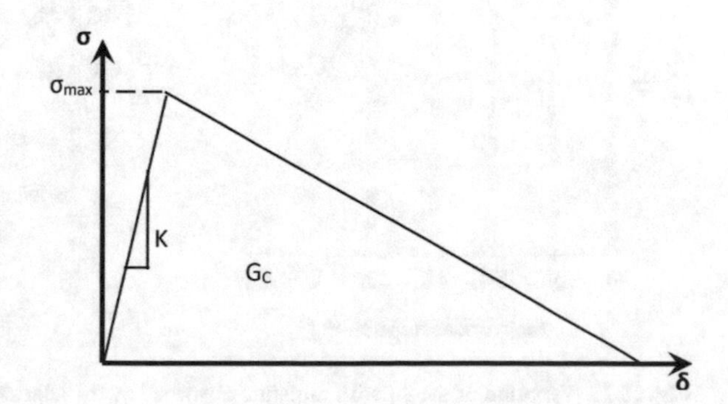

Fig. 10.11 Pure mode cohesive law used in the element proposed in this study

cohesive properties of the adhesives studied were determined in previous studies (Viana et al, 2017a,b). Empirical formulas were used to fit the toughness and yield stress of the adhesive as a function of temperature and moisture. Based on these previous studies and as an example, the properties of the XNR6852 adhesive can be approximated using the following formulas:

$$T_g = 117.4 - 8.23(M)^4 \tag{10.2}$$

$$\sigma_y = -15.4 - 0.75(T - T_g) \tag{10.3}$$

$$G_c = 8.46 - 2.27 \times 10^{-4}(T - T_g)^2 \tag{10.4}$$

In which σ_y is the yield stress of the adhesive, T is the environmental temperature, T_g is the glass transition temperature, G_c is the fracture toughness and M is the moisture percentage absorbed by the adhesive. Figure 10.10 shows the variation of the properties with T_g and the comparison with experimental values. The variation of the adhesives' T_g and mechanical properties are graphically represented in Figs. 10.12 and 10.13, respectively.

As Fig. 10.12 shows, the evolution of the T_g of XNR6852 with moisture concentration is not linear. In this study, a polynomial function was used to approximate the T_g of this adhesive as a function of its moisture absorption. This function matches the experimental values almost perfectly.

The properties of the adhesive were assumed to be dependent on the difference between the test temperature and T_g. Moisture absorbed by the adhesive will have the effect of decreasing the adhesive's T_g and indirectly affecting the adhesive's properties. This is in line with other studies (Jurf and Vinson, 1985). In this study, the evolution of the yield stress of both adhesives was considered to be linear, while the evolution of the fracture toughness was modelled with a fourth-degree polynomial equation. It is important to notice that the values obtained for the fracture toughness correspond only to cohesive fracture of the adhesive.

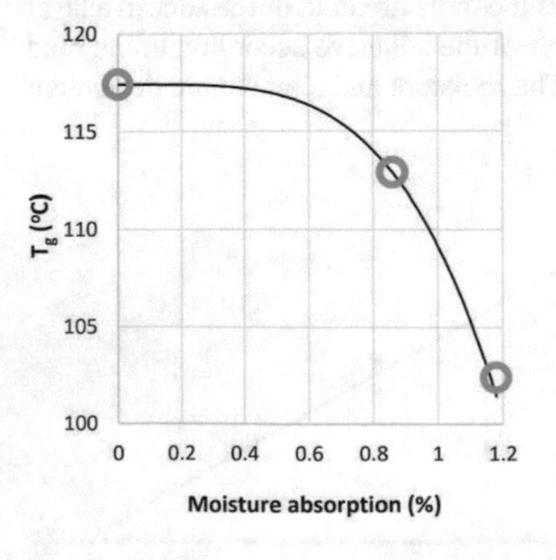

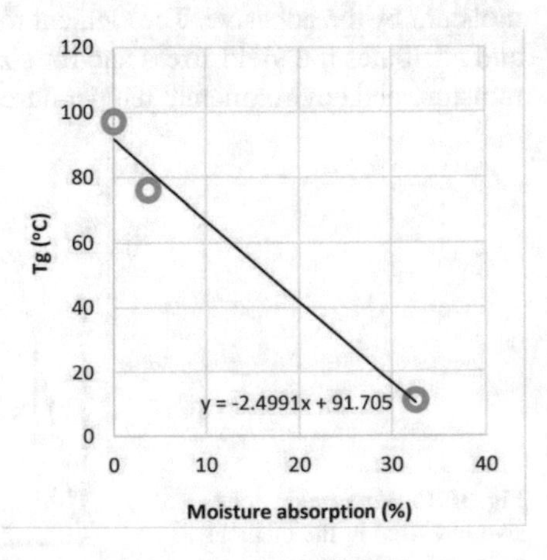

Fig. 10.12 Variation of the T_g with moisture absorbed by the adhesive

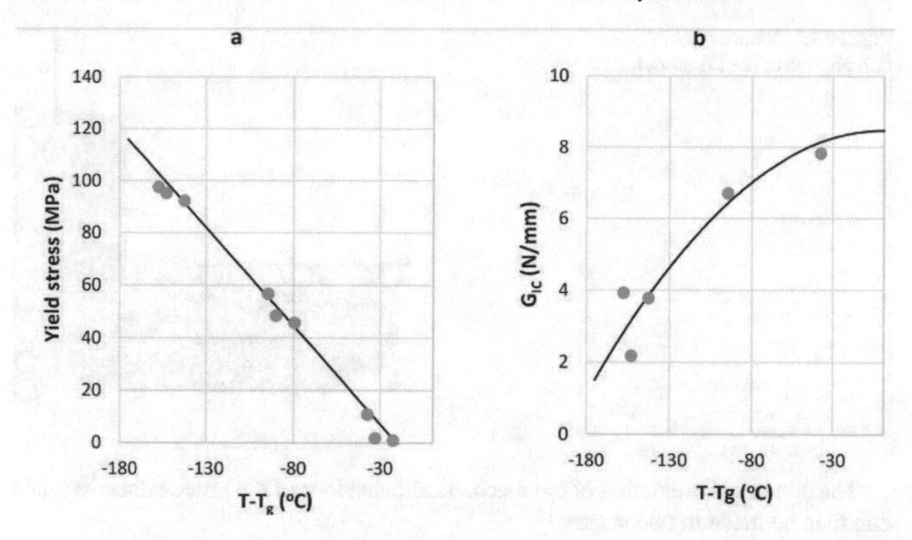

Fig. 10.13 Variation of the yield stress and mode I fracture toughness of the XNR6852 adhesive with $T - T_g$: column a: Variation of the strain rate, column b: Variation of mode I fracture toughness

The geometry of the specimen was discretised using the commercial FE software Abaqus. The substrates were modelled using C3D8 full integration elements, available in Abaqus library, to avoid hourglass effects. The bondline has two layers of elastic elements, also discretised using C3D8 elements. Between these two layers of elastic adhesive, the developed cohesive element was placed. The 4 mm wide bondline was modelled with a refined mesh of 20 elements. The rest of the specimen received a coarser mesh. Figure 10.14 shows the mesh used in this study.

Due to the symmetry of the specimen, in order to decrease the computational effort, only half of the specimen was modelled. The corresponding boundary conditions are shown in Fig. 10.15:

1. Every displacement in the lower substrate was set to zero;
2. A displacement of 0.5 mm was applied to the loading area of the upper substrate;
3. Displacements in the y and x directions of the middle plane were set to zero.

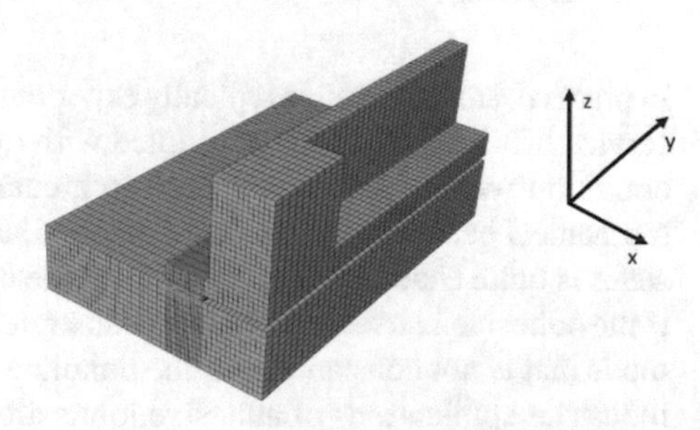

Fig. 10.14 Mesh of the complex joint geometry used

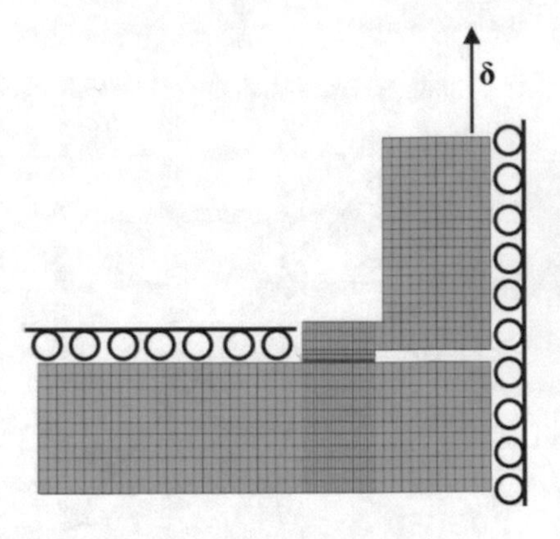

Fig. 10.15 Boundary condition applied to the FE model

The numerical prediction of the mechanical behaviour of the tested adhesive joints can then be made in two stages:

1. Prediction of the moisture gradient across the width of the adhesive taking into account the properties of each adhesive, described above.
2. Taking into account the moisture of each element, the moisture and temperature dependent properties were attributed to each element. This resulted into each element being assigned a distinct set of properties corresponding to a bondline with graded properties.

Overall, the published results and the novel techniques available in the literature demonstrate that it is indeed possible to model the loss in adhesive joint performance induced by the ingress of moisture in the adhesive. The necessary scientific background is established, enabling the creation of more powerful model formulations that are able to replicate experimentally determined behaviour. The main obstacle to the use of these techniques is, however, the time-consuming process of determining the parameters related to the water ingress in adhesives.

10.5 Modelling of Adhesives and Adhesive Joints Under Cyclic Loads

In practice, adhesive joints typically experience a multiaxial state of stress during their service life. This is typically coupled with cycling loading, a very common loading condition which sometimes result in a catastrophic failure of the structures. The mechanical behaviour of bonded structures subjected to these multiaxial, cyclic stress states is quite challenging to model and the analysis of the problem is more complex if the adhesive is used for bonding substrates with a configuration that leads mixed mode that is not constant along the bondline. However, despite these difficulties, the industrial applications of adhesive joints are increasing because of their significant advantages and precise numerical simulation methods are being requested by the industrial users of adhesive bonding. The goal of this section is to generally explain the procedure for numerically simulate the behaviour of adhesive joints subjected to cyclic loadings. A short survey on some already published methods on fatigue simulation of adhesive joints is also presented.

The total fatigue life of the joints can be divided into two different stages including the fatigue crack initiation and the fatigue crack propagation lives. For adhesive joints, especially when the bondline is very thin, is not easy to separate the initiation and propagation steps. The former corresponds to the life where the fracture energy is less than the threshold value. As soon as the accumulated energy reaches the threshold energy, the crack start to initiate and propagate where the fatigue crack propagation life starts. Depending on the material properties and the loading conditions, the total life may be governed by the crack initiation life or the crack propagation part. When the life is mainly spent before the crack initiation, the total life approaches such as strain based, stress based or a combination of stress and strain should be employed for fatigue life analysis of adhesive bonding. Using this approach, a good estimation of fatigue life can usually be obtained. In this condition, usually performing a simple linear elastic finite element analysis is sufficient to estimate the fatigue life. However, depending on the fatigue model, extra material (or geometry) constants maybe needed which should be obtained experimentally. Therefore, the main concern with these methods is the way that these parameters should be numerically measured. Critical plan and critical distance approaches are the two recent methods which are considered for the total fatigue life estimation of the materials.

However, when the crack propagation is an important part of the life or even when the damage tolerance is the concept considered for designing the joint, the fatigue crack propagation part should be analysed. Although stress or strain based approaches can be used for fatigue crack growth simulation (Ishii et al, 1999; Poursartip and Chinatambi, 1989), the most commonly used method for fatigue crack propagation analysis is the Paris law relation. Recently, several studies (Masaki et al, 1994; Alderliesten, 2016; Allegri et al, 2011; Andersons et al, 2004) have dealt with the numerical analysis of adhesive joints using different Paris law relations. By plotting the variation of crack length for each cycle as a function of the fracture energy in a log-log scale diagram a line with a constant slope will be appeared which corresponds to the stable fatigue crack propagation part of the test (see Fig. 10.16).

The Paris law is a fracture mechanics based approach, which links the strain energy release rate (or the stress intensity factor) to the fatigue crack growth (FCG) rate. Based on the Paris law curve shown in Fig. 10.16, the total life of the joints is divided to three regions including the initiation life, the stable crack propagation stage and the unstable fatigue crack propagation part. The slope of the line (m) for the second region, gives the exponent of the Paris law and determines the fatigue crack growth rate of the tested material. Several studies (Allegri et al, 2011; Andersons et al, 2004; Atodaria et al, 1997; Ramkumar and Whitcomb, 1985) have dealt with the effect of different Paris law functions on the slope of the Paris law curve. However, it should be noted that by changing the Paris law parameters, the slope of the curves may change as well. Using the Paris law values, it would be possible to simulate the fatigue crack propagation of adhesive joints. However, as the m is a function of

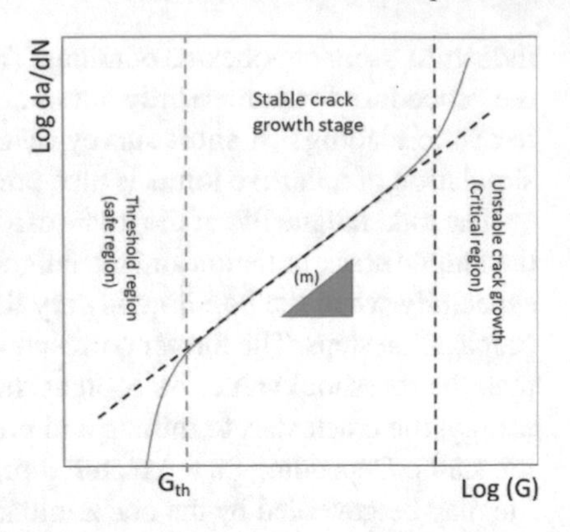

Fig. 10.16 Schematic view of the Paris law curve

loading conditions and of the geometries of the joint, to simulate the FCG behaviour, the m should be first obtained experimentally for a variety of loading conditions and geometries. To solve this issue, several authors (Masaki et al, 1994; Allegri et al, 2011; Andersons et al, 2004; Atodaria et al, 1997; Ramkumar and Whitcomb, 1985) have modified the Paris law function. The best relation is the one which could be able to collapse different Paris law curves into one. In this condition, by considering the experimental results for a specific configuration, the fatigue crack growth can be simulated for joints with different loading conditions.

Several researchers (Muñoz et al, 2006; Monteiro et al, 2019; Costa et al, 2018) have employed the Paris law relations to numerically analyse the fatigue behaviour of adhesive joints. They have considered CZM for FE analysis of adhesive joints subjected to cyclic loadings. The application of cohesive elements for fatigue life simulation of the adhesive materials is explained in the following section. The CZM technique can be used for fatigue numerical simulation of adhesive joints as well as quasi static analysis. In parallel with the static conditions described previously in this chapter, for a numerical analysis of fatigue using CZM, a damage initiation method and a damage evolution model should be defined first. In addition, a mixed mode criterion should be also employed to combine the pure mode I and pure mode II cohesive properties of the element based on the mode ratio.

Fatigue failure is a cumulative damage mechanism. Consequently, the mechanical properties of the adhesive degrade cycle by cycle until a specific number of cycles is reached and failure takes place. Accordingly, to simulate the fatigue response of adhesive joints, a degradation method should be also considered. A recently developed degradation approach is explained in the following section.

Usually the failure models already included in the commercial FE software cannot perform such a complex analysis for adhesive joints. To solve this issue writing a subroutine is unavoidable. Using Abaqus, there are different approaches to perform numerical fatigue life analysis using different types of subroutines. In this section the more advanced routines called user element (UEL) and user material (UMAT) will be introduced.

Different degradation approaches have been proposed by authors. Khoramishad et al (2010) proposed a fatigue damage criterion based on the strain values. They used CZM to simulate the fatigue response of adhesive joints.Muñoz et al (2006) also considered a damage mechanics based approach for degradation of the properties of the cohesive elements. A Paris like model is also proposed by some authors (Moroni and Pirondi, 2011) for degradation of cohesive properties of the cohesive elements due to the cyclic loadings. In this section one of the most recent models developed by Costa et al (2018) is introduced

$$y(N) = y_0 \left(1 - \frac{N}{N_f}\right)^k \tag{10.5}$$

In Eq. (10.5) y gives the value of the cohesive parameter (e.g. fracture energy) when the element has experienced N number of cycles. y_0 is the initial value of the corresponded cohesive parameter and the exponent k is a fitting parameter which should be obtained experimentally. N_f is the total fatigue life of the joint which can be estimated using Eq. (10.6)

$$N_f = \frac{\Delta a}{(\mathrm{d}a/\mathrm{d}N)_a} \tag{10.6}$$

Costa et al (2018) found that a higher degradation rate leads to an increase in the m and result in a lower threshold energy and of the intercept (C). Naturally, the CZM parameters were found to decrease faster for higher values of k. The total number of cycles at failure (N_f) is a function of the total length of the bonded area and of the average FCG rate. Based on the results presented by Costa et al (2018) using Eq. (10.6) a good estimation of the total life of the joints can be obtained. However, more experimental data, especially at lower load levels, is required to validate this approach for total fatigue life estimation of adhesive joints.

Figure 10.17 shows a schematic triangular shaped CZM where degradation has been applied on the cohesive properties of the element. Damage initiates when the separation is more than δ^0 and is completed when the separation reaches δ^f. The area below the triangular shape is defined by G_c. By replacing y in Eq. (10.5) by fracture energy or traction, the degradation of fracture energy and traction as a function of the number of cycles can be obtained. Regardless of the degradation approach, the remaining procedure is similar for numerical simulation of adhesive joints. The next section presents the constitutive equations of cohesive elements which can be employed for finite element analysis of joints subjected to cyclic loadings.

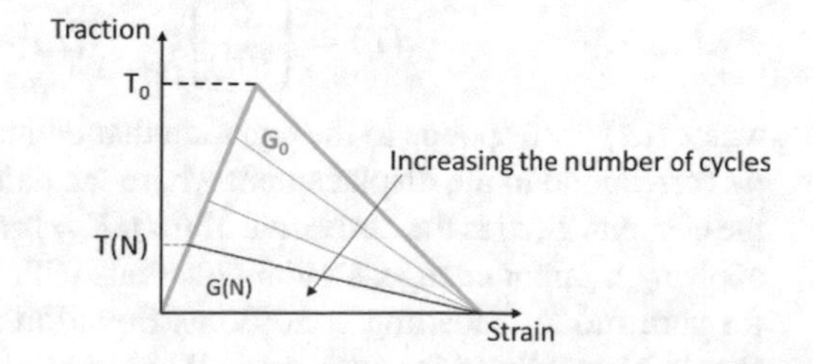

Fig. 10.17 The effect of data degradation on the triangular shape of the CZM

Different CZM shapes can be considered for numerical analysis of the fatigue behaviour of adhesive materials. The bilinear or triangular shape is the most commonly used model. The constitutive equations of the cohesive elements are as follows:

$$\sigma = \begin{bmatrix} \sigma_{\mathrm{I}} \\ \sigma_{\mathrm{II}} \\ \sigma_{\mathrm{III}} \end{bmatrix} = (1-d)K \begin{bmatrix} \delta_{\mathrm{I}} \\ \delta_{\mathrm{II}} \\ \delta_{\mathrm{III}} \end{bmatrix} - dK \begin{bmatrix} < -\delta_{\mathrm{I}} > \\ 0 \\ 0 \end{bmatrix} \tag{10.7}$$

It should be noted that the compressive normal load does not introduce any damage for the cohesive elements. K in Eq. (10.7) is the initial stiffness of the cohesive elements and d is the damage value, which is a function of the shape of the CZM. Based on Eq. (10.7) traction (stress) at each load increment and for each element can be calculated by knowing the values of the initial stiffness, the amount of damage and the values of the current separation (strain). Subroutines use the constitutive laws of cohesive elements for numerical simulation of adhesive joints. Different subroutines can be employed for fatigue life simulation of adhesive joints. User element (UEL) and user material (UMAT) are the two advanced subroutines which have been already employed for fatigue analysis of adhesive materials.

UEL uses the concepts of FE mechanics to calculate the displacement of the cohesive elements. To achieve this, the stiffness matrix of the element should be defined first. Equation (10.8) gives the general relation used to estimate the behaviour of the cohesive elements.

$$[K] \times \{d\} = \{f\}, \tag{10.8}$$

where $[K]$, $\{d\}$ and $\{f\}$ are the stiffness matrix, the displacement vector which is obtained by FE mechanics and the vector of the external forces, respectively. The stiffness matrix and the vector of the external forces for the cohesive elements can be calculated using the following matrix formulations:

$$[K] = w[B]^T[T_d][B], \qquad \{f\} = w[B]^T\{T\}, \tag{10.9}$$

where w and $[B]$ are the element width and the matrix of the global displacement-separation relation. $\{T\}$ and $[T_d]$ are also vector and matrix, respectively and include the traction-separation laws. Based on the behaviour of the material, the shape of the traction-separation law will change and consequently will lead to different formulations for the $\{T\}$ and $[T_d]$.

For mode I loadings in the local coordinate system $\{T\}$ and $[T_d]$ should be defined as:

$$\{T\} = \begin{Bmatrix} 0 \\ t(d) \end{Bmatrix}, \qquad [T_d] = \begin{bmatrix} 0 & 0 \\ 0 & t'(d) \end{bmatrix} \tag{10.10}$$

where $t(d)$ corresponds to the equation that define the triangular shape CZM. d_0 and d_f correspond to the displacement where the damage initiate and the final failure of the element. $t'(d)$ is the derivative of $t(d)$. K_{eff} is also the effective initial stiffness. By applying a minor change, similar equations with a minor change can be also applied for pure mode II loading conditions. Equation (10.11) shows the relations which should be employed for pure mode II analysis

$$\{T\} = \begin{Bmatrix} t(d) \\ 0 \end{Bmatrix}, \qquad [T_d] = \begin{bmatrix} t'(d) & 0 \\ 0 & 0 \end{bmatrix} \tag{10.11}$$

The basics of the FE mechanics, shape functions, the shape of the CZM and the cycle by cycle degradation of the cohesive properties of the element should be defined in UEL implemented in Abaqus. Figure 10.18 shows a schematic of the mode II loading conditions and the cohesive zone element in a global (x, y) and a local (ξ, η) coordinate.

4-node, 6-node or even 8-node element can be employed in this analysis. For a 4-node element, 4 shape functions should be defined. Assuming that the height of the cohesive element is null ($el_y = 0$ mm, see Fig. 10.18), subsequently the shape functions of node 1,4 and 2,3 should be only defined. At a specific ξ coordinate, the same functions apply to both nodes:

$$N_{1,4} = \frac{1}{2}(1 - \xi), \quad N_{2,3} = \frac{1}{2}(1 + \xi) \tag{10.12}$$

From Eq. (10.12) the matrix of the shape $[N]$ functions can be determined as follows:

$$[N] = \begin{bmatrix} N_{14} & N_{23} & N_{23} & N_{14} \\ N_{14} & N_{23} & N_{23} & N_{14} \end{bmatrix} \tag{10.13}$$

The strain displacement is represented by matrix $[B]$:

$$[B] = \lfloor R \rfloor \lfloor N \rfloor, \tag{10.14}$$

where matrix $[R]$ is the transformations matrix from global to local coordinates, defined as:

$$[R] = \begin{bmatrix} \cos\alpha & \sin\alpha \\ -\sin\alpha & \cos\alpha \end{bmatrix} \tag{10.15}$$

α is the angle between the coordinate systems (see Fig. 10.18).

Two different zones (zones 1 and 2 shown in Fig. 10.19) can be considered for determining the values of $t_i(d)$, $t_i'(d)$ and G_i. $t_1(d)$, $t_1'(d)$ are defined as follows:

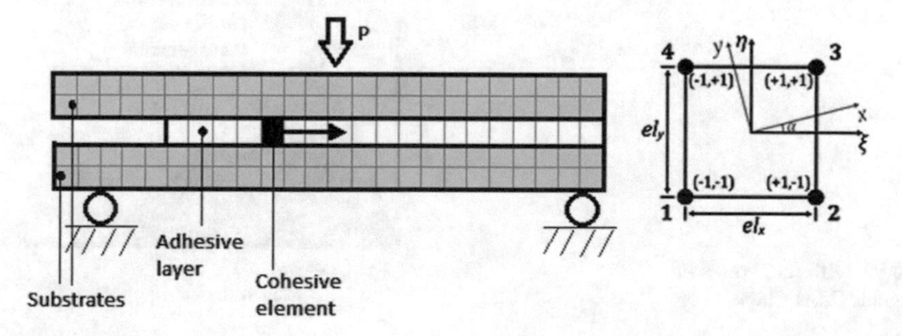

Fig. 10.18 Schematic of the ENF specimen and the cohesive zone element in global (x, y) and local (ξ, η)

$$t_1(d) = \frac{t_m d}{d_0},\ t_1'(d) = \frac{t_m}{d_0},\quad t_2(d) = t_m\left(1 - \frac{d - d_0}{d_f - d_0}\right),\ t_2'(d) = \frac{-t_m}{d_f - d_0} \tag{10.16}$$

where the exponents 1 and 2 denote the zone 1 and zone 2, respectively. By knowing the value of fracture energy, the value of d_{f} can be easily obtained. Using the above procedure and by employing a data degradation approach, the values of fracture energy and separation and subsequently the damage values can be calculated as a function of number of cycles. More details about the procedure can be found in the literature (Monteiro et al, 2019; Costa et al, 2018).

UMAT is another advanced subroutine which is considered for fatigue failure analysis of adhesive joints. To employ UMAT, in addition to the data degradation approach, a mixed mode criterion should be also employed. Power law and Benzeggagh-Kenane (BK) are the two widely used criteria for mixed mode loading conditions. Equation (10.17) shows the BK relation, respectively,

$$G_{\mathrm{c}} = G_{\mathrm{Ic}} + (G_{\mathrm{shear\,c}} - G_{\mathrm{Ic}})\left(\frac{G_{\mathrm{shear}}}{G_{\mathrm{T}}}\right)^{\lambda}, \tag{10.17}$$

where in the BK approach, $G_{\mathrm{shear}} = G_{\mathrm{II}} + G_{\mathrm{III}}$ and $G_{\mathrm{T}} = G_{\mathrm{I}} + G_{\mathrm{shear}}$. The energy release rate is calculated at each integration point for a given increment using the reduction factor given in Eq. (10.5). The conditions for damage onset and final failure can be expressed in terms of the displacement as shown in Eq. (10.18) repectively,

$$d = \frac{\delta^{\mathrm{f}}(\delta_{\mathrm{eq}} - \delta^0)}{\delta_{\mathrm{eq}}(\delta^{\mathrm{f}} - \delta^0)}, \tag{10.18}$$

where:

$$\delta_{\mathrm{eq}} = \sqrt{\delta_{\mathrm{I}}^2 + \delta_{\mathrm{shear}}^2},\qquad \delta_{\mathrm{shear}} = \sqrt{\delta_{\mathrm{II}}^2 + \delta_{\mathrm{III}}^2},$$

$$\delta^{0^2} = \delta_{\mathrm{I}}^2 + \delta_{\mathrm{II}}^2 + \delta_{\mathrm{III}}^2 = \delta_I^{0^2} + \left[\delta_{\mathrm{shear}}^{0\ \ 2} - \delta_I^{0^2}\right]\omega^{\lambda},$$

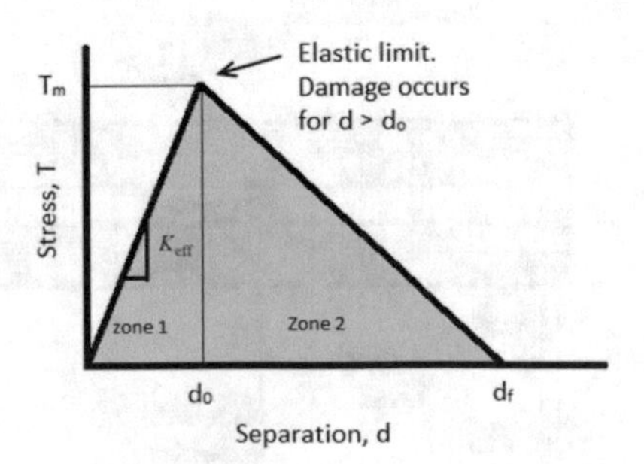

Fig. 10.19 Different zones in a triangular CZM shape

$$\delta^{\mathrm{f}} = \frac{\delta_I^0 \delta_I^{\mathrm{f}} + (\delta_{\mathrm{shear}}^0 \delta_{\mathrm{shear}}^{\mathrm{f}} - \delta_{\mathrm{I}}^0 \delta_{\mathrm{I}}^{\mathrm{f}})\omega^{\lambda}}{\delta^0},$$

$$\omega = \frac{G_{\mathrm{shear}}}{G_{\mathrm{T}}} = \frac{\beta^2}{1 + 2\beta^2 - 2\beta}, \qquad \beta = \frac{\delta_{\mathrm{shear}}}{\delta_{\mathrm{shear}} + < \delta_{\mathrm{I}} >}$$

where δ_{I}^0 and $\delta_{\mathrm{shear}}^0$ are the damage initiation conditions for normal and shear directions, respectively, and $\delta_{\mathrm{I}}^{\mathrm{f}}$ and $\delta_{\mathrm{shear}}^{\mathrm{f}}$ are displacements at final failure. ω represents the ratio of G_{shear} to G_{T} which is a function of the mode ratio. This process should be followed for each integration point of each element and for all the elements at each time increment. Whenever the damage (d) reaches 1 or when the fracture energy or traction is zero the point is fully failed. By this approach the size of the damage zone increase cycle by cycle. Using the second approach (UMAT) a DCB joint subjected to cyclic loading was analysed. Table 10.3 gives the material properties of the adhesive and the substrates. The dimensions of the analysed joint are shown in

Table 10.3 Mechanical properties of the adhesive and substrates (Monteiro et al, 2019)

Properties	Substrate	Adhesive
Maximum tensile strength (MPa)	-	31
Maximum tensile strain (%)	-	10.4
Young's modulus (MPa)	210000	1159
Poisson's ratio (ν)	0.3	-

Fig. 10.20, while Fig. 10.21 shows a typical damage evolution in adhesive layer for a joint subjected to mixed mode cyclic loading. The results are obtained based on the second approach (UMAT).

Acknowledgements The authors wish to acknowledge the funding provided by Ministério da Ciência, Tecnologia e Ensino Superior – Fundação para a Ciência e a Tecnologia (Portugal), under the projects POCI-01-0145-FEDER-028351, POCI-01-0145-FEDER-028473 and POCI-01-0145-FEDER-029839. The authors also wish to thank Project NORTE-01-0145-FEDER-000022 – "SciTech - Science and Technology for Competitive and Sustainable Industries", co-financed by Programa Operacional Regional do Norte (NORTE2020), through Fundo Europeu de Desenvolvimento Regional (FEDER) and PhD grant SFRH/BD/139341/2018 "Impact strength optimization with cohesive zone elements of multi-material bonded structures used in the automotive industry".

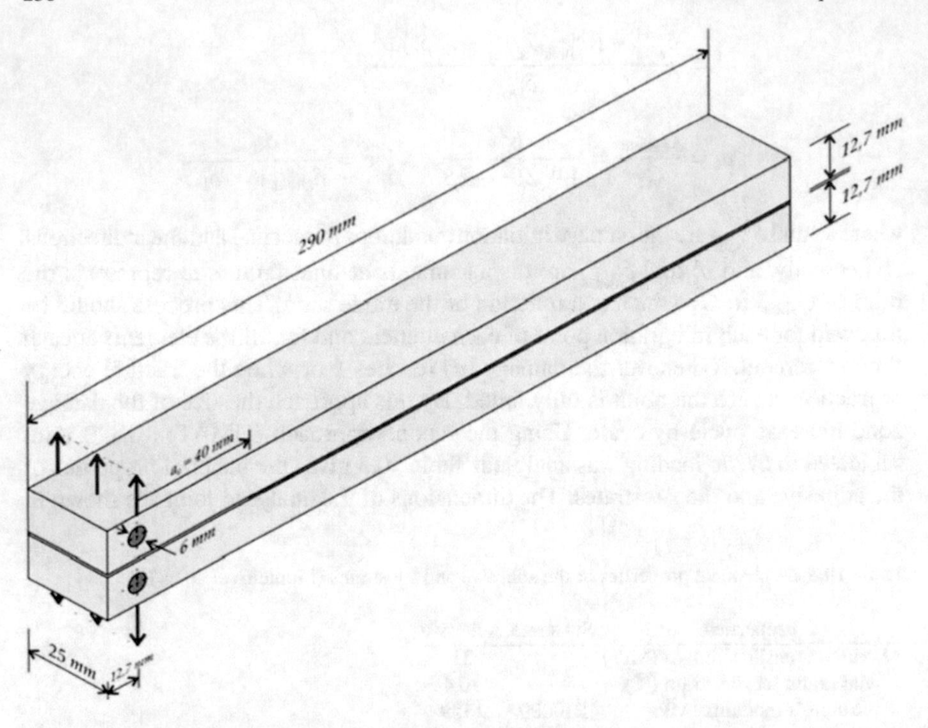

Fig. 10.20 Dimensions of the analysed DCB (mm)

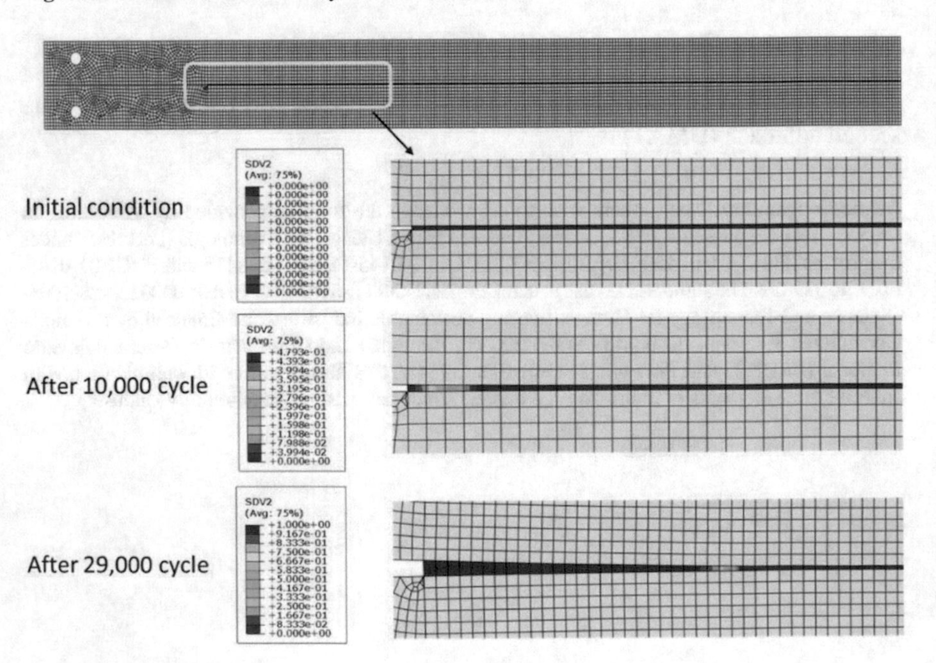

Fig. 10.21 Fatigue life analysis of a DCB joint using UMAT

References

Abaqus (2013) Documentation of the software Abaqus. Manual, Dassault Systèmes, Vélizy-Villacoublay

Adams R, Comyn J, Wake W (eds) (1997) Structural Adhesive Joints in Engineering. Springer Science & Business Media

Adams RD, Harris JA (1996) A critical assessment of the block impact test for measuring the impact strength of adhesive bonds. International Journal of Adhesion and Adhesives 16(2):61–71, special Issue: In honour of Dr K. W. Allen on the occasion of his 70th birthday

Adams RD, Peppiatt NA (1974) Stress analysis of adhesive-bonded lap joints. Journal of Strain Analysis 9(3):185–196

Alderliesten RC (2016) How proper similitude can improve our understanding of crack closure and plasticity in fatigue. International Journal of Fatigue 82:263–273, 10th Fatigue Damage of Structural Materials Conference

Alfano G (2006) On the influence of the shape of the interface law on the application of cohesive-zone models. Composites Science and Technology 66(6):723–730, Advances in statics and dynamics of delamination

Alfano G, Crisfield MA (2001) Finite element interface models for the delamination analysis of laminated composites: mechanical and computational issues. International Journal for Numerical Methods in Engineering 50(7):1701–1736

Allegri G, Jones MI, Wisnom MR, Hallett SR (2011) A new semi-empirical model for stress ratio effect on mode ii fatigue delamination growth. Composites Part A: Applied Science and Manufacturing 42(7):733–740

Allix O, Corigliano A (1996) Modeling and simulation of crack propagation in mixed-modes interlaminar fracture specimens. International Journal of Fracture 77(2):111–140

Andersons J, Hojo M, Ochiai S (2004) Empirical model for stress ratio effect on fatigue delamination growth rate in composite laminates. International Journal of Fatigue 26(6):597–604

Atodaria DR, Putatunda SK, Mallick PK (1997) A fatigue crack growth model for random fiber composites. Journal of Composite Materials 31(18):1838–1855

Avendaño R, Carbas RJC, Chaves FJP, Costa M, da Silva LFM, Fernandes A (2016) Impact loading of single lap joints of dissimilar lightweight adherends bonded with a crash-resistant epoxy adhesive. Trans ASME Journal of Engineering Materials and Technology 116(4):041,019

Barenblatt GI (1959) The formation of equilibrium cracks during brittle fracture. general ideas and hypotheses. axially-symmetric cracks. Journal of Applied Mathematics and Mechanics 23(3):622–636

Bascom WD, Cottington RL (1976) Effect of temperature on the adhesive fracture behavior of an elastomer-epoxy resin. The Journal of Adhesion 7(4):333–346

Belytschko T, Black T (1999) Elastic crack growth in finite elements with minimal remeshing. International Journal for Numerical Methods in Engineering 45(5):601–620

Biel A, Stigh U (2008) Effects of constitutive parameters on the accuracy of measured fracture energy using the dcb-specimen. Engineering Fracture Mechanics 75(10):2968–2983

Camanho PP, Davila CG, de Moura MF (2003) Numerical simulation of mixed-mode progressive delamination in composite materials. Journal of Composite Materials 37(16):1415–1438

Campilho RDSG, de Moura MFSF, Domingues JJMS (2005) Modelling single and double-lap repairs on composite materials. Composites Science and Technology 65(13):1948–1958

Campilho RDSG, de Moura MFSF, Domingues JJMS (2007) Stress and failure analyses of scarf repaired CFRP laminates using a cohesive damage model. Journal of Adhesion Science and Technology 21(9):855–870

Campilho RDSG, de Moura MFSF, Domingues JJMS (2008a) Using a cohesive damage model to predict the tensile behaviour of CFRP single-strap repairs. International Journal of Solids and Structures 45(5):1497–1512

Campilho RDSG, de Moura MFSF, Domingues JJMS, Morais JJL (2008b) Computational modelling of the residual strength of repaired composite laminates using a cohesive damage model. Journal of Adhesion Science and Technology 22(13):1565–1591

Campilho RDSG, de Moura MFSF, Pinto AMG, Morais JJL, Domingues JJMS (2009a) Modelling the tensile fracture behaviour of CFRP scarf repairs. Composites Part B: Engineering 40(2):149–157

Campilho RDSG, de Moura MFSF, Ramantani DA, Morais JJL, Domingues JJMS (2009b) Tensile behaviour of three-dimensional carbon-epoxy adhesively bonded single- and double-strap repairs. International Journal of Adhesion and Adhesives 29(6):678–686, special Issue on Durability of Adhesive Joints

Campilho RDSG, de Moura MFSF, Ramantani DA, Morais JJL, Barreto AMJP, Domingues JJMS (2010) Adhesively bonded repair proposal for wood members damaged by horizontal shear using carbon-epoxy patches. The Journal of Adhesion 86(5-6):649–670

Campilho RDSG, Banea MD, Pinto AMG, da Silva LFM, de Jesus AMP (2011) Strength prediction of single- and double-lap joints by standard and extended finite element modelling. International Journal of Adhesion and Adhesives 31(5):363–372, special Issue on Joint Design 2

Carlberger T, Stigh U (2007) An explicit FE-model of impact fracture in an adhesive joint. Engineering Fracture Mechanics 74(14):2247–2262

Carlberger T, Stigh U (2010) Influence of layer thickness on cohesive properties of an epoxy-based adhesive-an experimental study. The Journal of Adhesion 86(8):816–835

Carrere N, Tual N, Bonnemains T, Lolive E, Davies P (2018) Modelling of the damage development in carbon/epoxy laminates subjected to combined seawater ageing and mechanical loading. Proceedings of the Institution of Mechanical Engineers, Part L: Journal of Materials: Design and Applications 232(9):761–768

Cavalli MN, Thouless MD (2001) The effect of damage nucleation on the toughness of an adhesive joint. The Journal of Adhesion 76(1):75–92

Chai H (1986) On the correlation between the mode I failure of adhesive joints and laminated composites. Engineering Fracture Mechanics 24(3):413–431

Challita G, Othman R (2010) Finite-element analysis of SHPB tests on double-lap adhesive joints. International Journal of Adhesion and Adhesives 30(4):236–244

Chandra N, Li H, Shet C, Ghonem H (2002) Some issues in the application of cohesive zone models for metal-ceramic interfaces. International Journal of Solids and Structures 39(10):2827–2855

Charalambides M, Dean G (1997) Constitutive models and their data requirements for use in finite element analysis of adhesives under impact loading. Report (a)59, National Physical Laboratory. Great Britain, Centre for Materials Measurement and Technology

Chen J (2002) Predicting progressive delamination of stiffened fibre-composite panel and repaired sandwich panel by decohesion models. Journal of Thermoplastic Composite Materials 15(5):429–442

Chen Z, Adams RD, da Silva LFM (2011) Prediction of crack initiation and propagation of adhesive lap joints using an energy failure criterion. Engineering Fracture Mechanics 78(6):990–1007

Clarke M, Broughton JG, Hutchinson AR, Buckley M (2011) An investigation into the use of an embedded process zone model for predicting the structural behaviour of adhesive bonded joints. International Journal of Vehicle Structures & Systems 3(3)

Costa M, Viana G, Créac'hcadec R, da Silva LFM, Campilho RDSG (2018) A cohesive zone element for mode I modelling of adhesives degraded by humidity and fatigue. International Journal of Fatigue 112:173–182

Crocombe AD (1997) Durability modelling concepts and tools for the cohesive environmental degradation of bonded structures. International Journal of Adhesion and Adhesives 17(3):229–238

Crocombe AD, Hua YX, Loh WK, Wahab MA, Ashcroft IA (2006) Predicting the residual strength for environmentally degraded adhesive lap joints. International Journal of Adhesion and Adhesives 26(5):325–336

Daux C, Moës N, Dolbow J, Sukumar N, Belytschko T (2000) Arbitrary branched and intersecting cracks with the extended finite element method. International Journal for Numerical Methods in Engineering 48(12):1741–1760

Dean G, Lord G, Duncan B (1997) Comparison of the measured and predicted performance of adhesive joints under impact. Report (a)206, National Physical Laboratory. Great Britain, Centre for Materials Measurement and Technology

Dean GD, Charalambides MN (1997) Data requirements for the use of finite element methods to predict the impact performance of plastics. Report (a)58, National Physical Laboratory. Great Britain, Centre for Materials Measurement and Technology

Devitt DF, Schapery RA, Bradley WL (1980) A method for determining the mode i delamination fracture toughness of elastic and viscoelastic composite materials. Journal of Composite Materials 14(4):270–285

Dolbow J, Moës N, Belytschko T (2001) An extended finite element method for modeling crack growth with frictional contact. Computer Methods in Applied Mechanics and Engineering 190(51):6825–6846

Dugdale DS (1960) Yielding of steel sheets containing slits. Journal of the Mechanics and Physics of Solids 8(2):100–104

Duncan BC, Pearce A (1999) Comparison of impact and high rate tests for determining properties of adhesives and polymers needed for design under impact loading. Performance of adhesives joints programme project paj2 - report (a)134, National Physical Laboratory. Great Britain, Centre for Materials Measurement and Technology

Elguedj T, Gravouil A, Combescure A (2006) Appropriate extended functions for X-FEM simulation of plastic fracture mechanics. Computer Methods in Applied Mechanics and Engineering 195(7):501–515

Fagerström M, Larsson R (2006) Theory and numerics for finite deformation fracture modelling using strong discontinuities. International Journal for Numerical Methods in Engineering 66(6):911–948

Feraren P, Jensen HM (2004) Cohesive zone modelling of interface fracture near flaws in adhesive joints. Engineering Fracture Mechanics 71(15):2125–2142

Goland M, Reissner E (1944) The stresses in cemented joints. Journal of Applied Mechanics 11(1):A17–A27

Han X, Crocombe AD, Anwar SNR, Hu P (2014a) The strength prediction of adhesive single lap joints exposed to long term loading in a hostile environment. International Journal of Adhesion and Adhesives 55:1–11

Han X, Crocombe AD, Anwar SNR, Hu P, Li WD (2014b) The effect of a hot-wet environment on adhesively bonded joints under a sustained load. The Journal of Adhesion 90(5-6):420–436

Harris JA, Adams RD (1985) An assessment of the impact performance of bonded joints for use in high energy absorbing structures. Proceedings of the Institution of Mechanical Engineers, Part C: Journal of Mechanical Engineering Science 199(2):121–131

Hart-Smith L (1973) Adhesive-bonded double-lap joints. Technical Report NASACR-112235, National Aeronautics and Space Administration - Langley Research Center, Hampton, Virginia

Haufe A, Pietsch G, Graf T, Feucht M (2010) Modelling of weld and adhesive connections in crashworthiness applications with LS-DYNA. In: Proceedings of the NAFEMS seminar "simulation of connections and joints in structures", Wiesbaden, pp 1–6

Hazimeh R, Challita G, Khalil K, Othman R (2015) Finite element analysis of adhesively bonded composite joints subjected to impact loadings. International Journal of Adhesion and Adhesives 56:24–31, special Issue on Impact Phenomena of Adhesively Bonded Joints

Higuchi I, Sawa T, Suga H (2002a) Three-dimensional finite element analysis of single-lap adhesive joints subjected to impact bending moments. Journal of Adhesion Science and Technology 16(10):1327–1342

Higuchi I, Sawa T, Suga H (2002b) Three-dimensional finite element analysis of single-lap adhesive joints under impact loads. Journal of Adhesion Science and Technology 16(12):1585–1601

Hua Y, Crocombe AD, Wahab MA, Ashcroft IA (2006) Modelling environmental degradation in EA9321-bonded joints using a progressive damage failure model. The Journal of Adhesion 82(2):135–160

Hua Y, Crocombe AD, Wahab MA, Ashcroft IA (2008) Continuum damage modelling of environmental degradation in joints bonded with EA9321 epoxy adhesive. International Journal of Adhesion and Adhesives 28(6):302–313

Ilioni A, Gac PYL, Badulescu C, Thévenet D, Davies P (2019) Prediction of mechanical behaviour of a bulk epoxy adhesive in a marine environment. The Journal of Adhesion 95(1):64–84

Ishii K, Imanaka M, Nakayama H, Kodama H (1999) Evaluation of the fatigue strength of adhesively bonded CFRP/metal single and single-step double-lap joints. Composites Science and Technology 59(11):1675–1683

Ji G, Ouyang Z, Li G, Ibekwe S, Pang SS (2010) Effects of adhesive thickness on global and local Mode-I interfacial fracture of bonded joints. International Journal of Solids and Structures 47(18):2445–2458

Jurf R, Vinson JR (1985) Effect of moisture on the static and viscoelastic shear properties of epoxy adhesives. Journal of Materials Science 20(8):2979–2989

Kafkalidis MS, Thouless MD (2002) The effects of geometry and material properties on the fracture of single lap-shear joints. International Journal of Solids and Structures 39(17):4367–4383

Khoei AR, Nikbakht M (2006) Contact friction modeling with the extended finite element method (X-FEM). Journal of Materials Processing Technology 177(1):58–62, Proc. 11th Intern. Conf. on Metal Forming 2006

Khoramishad H, Crocombe AD, Katnam KB, Ashcroft IA (2010) Predicting fatigue damage in adhesively bonded joints using a cohesive zone model. International Journal of Fatigue 32(7):1146–1158

Kinloch AJ (ed) (1987) Adhesion and Adhesives. Springer Netherlands

Leffler K, Alfredsson KS, Stigh U (2007) Shear behaviour of adhesive layers. International Journal of Solids and Structures 44(2):530–545

Liao L, Kobayashi T, Sawa T, Goda Y (2011) 3-D FEM stress analysis and strength evaluation of single-lap adhesive joints subjected to impact tensile loads. International Journal of Adhesion and Adhesives 31(7):612–619

Liao L, Sawa T, Huang C (2013) Experimental and fem studies on mechanical properties of single-lap adhesive joint with dissimilar adherends subjected to impact tensile loadings. International Journal of Adhesion and Adhesives 44:91–98

Liljedahl CDM, Crocombe AD, Wahab MA, Ashcroft IA (2005) The effect of residual strains on the progressive damage modelling of environmentally degraded adhesive joints. Journal of Adhesion Science and Technology 19(7):525–547

Liljedahl CDM, Crocombe AD, Wahab MA, Ashcroft IA (2006a) Damage modelling of adhesively bonded joints. International Journal of Fracture 141(1-2):147–161

Liljedahl CDM, Crocombe AD, Wahab MA, Ashcroft IA (2006b) Modelling the environmental degradation of the interface in adhesively bonded joints using a cohesive zone approach. The Journal of Adhesion 82(11):1061–1089

Machado JJM, Marques EAS, da Silva LFM (2018) Adhesives and adhesive joints under impact loadings: An overview. The Journal of Adhesion 94(6):421–452

Mangalgiri PD, Johnson WS, Everett Jr RA (1987) Effect of adherend thickness and mixed mode loading on debond growth in adhesively bonded composite joints. The Journal of Adhesion 23(4):263–288

Masaki H, Shojiro O, Gustafson CG, Keisuke T (1994) Effect of matrix resin on delamination fatigue crack growth in CFRP laminates. Engineering Fracture Mechanics 49(1):35–47

May M, Voß H, Hiermaier S (2014) Predictive modeling of damage and failure in adhesively bonded metallic joints using cohesive interface elements. International Journal of Adhesion and Adhesives 49:7–17

Moës N, Belytschko T (2002) Extended finite element method for cohesive crack growth. Engineering Fracture Mechanics 69(7):813–833

Moës N, Dolbow J, Belytschko T (1999) A finite element method for crack growth without remeshing. International Journal for Numerical Methods in Engineering 46(1):131–150

Mohammadi S (2008) Extended Finite Element Method for Fracture Analysis of Structures. Blackwell Publishing, New Jersey

Monteiro J, Akhavan-Safar A, Carbas R, Marques E, Goyal R, El-zein M, da Silva LFM (2019) Mode II modeling of adhesive materials degraded by fatigue loading using cohesive zone elements. Theoretical and Applied Fracture Mechanics 103:102,253

Moroni F, Pirondi A (2011) A procedure for the simulation of fatigue crack growth in adhesively bonded joints based on the cohesive zone model and different mixed-mode propagation criteria. Engineering Fracture Mechanics 78(8):1808–1816, multiaxial Fracture

Muñoz JJ, Galvanetto U, Robinson P (2006) On the numerical simulation of fatigue driven delamination with interface elements. International Journal of Fatigue 28(10):1136–1146, the Third International Conference on Fatigue of Composites

Neumayer J, Koerber H, Hinterhölzl R (2016) An explicit cohesive element combining cohesive failure of the adhesive and delamination failure in composite bonded joints. Composite Structures 146:75–83

Pardoen T, Ferracin T, Landis C, Delannay F (2005) Constraint effects in adhesive joint fracture. Journal of the Mechanics and Physics of Solids 53(9):1951–1983

Pinto AMG, oes AGM, Campilho RDSG, de Moura MFSF, Baptista APM (2009) Single-lap joints of similar and dissimilar adherends bonded with an acrylic adhesive. The Journal of Adhesion 85(6):351–376

Poursartip A, Chinatambi N (1989) Fatigue damage development in notched $(0^2/\pm45)^s$ laminates. In: Lagace P (ed) Composite Materials: Fatigue and Fracture, ASTM International, West Conshohocken, PA, vol 2, pp 45–65

Ramkumar R, Whitcomb J (1985) Characterization of mode I and mixed-mode delamination growth in T300/5208 graphite/epoxy. In: Johnson W (ed) STP876-EB Delamination and Debonding of Materials, ASTM International, West Conshohocken, PA, pp 315–335

Reedy ED, Guess TR (1996) Butt joint strength: effect of residual stress and stress relaxation. Journal of Adhesion Science and Technology 10(1):33–45

Ridha M, Tan VBC, Tay TE (2011) Traction-separation laws for progressive failure of bonded scarf repair of composite panel. Composite Structures 93(4):1239–1245

Sabsabi M, Giner E, Fuenmayor FJ (2011) Experimental fatigue testing of a fretting complete contact and numerical life correlation using X-FEM. International Journal of Fatigue 33(6):811–822

Sawa T, Higuchi I, Suga H (2003) Three-dimensional finite element stress analysis of single-lap adhesive joints of dissimilar adherends subjected to impact tensile loads. Journal of Adhesion Science and Technology 17(16):2157–2174

Sawa T, Nagai T, Iwamoto T, Kuramoto H (2008) A study on evaluation of impact strength of adhesive joints subjected to impact shear loadings. In: ASME 2008 International Mechanical Engineering Congress and Exposition, American Society of Mechanical Engineers, pp 55–61

da Silva LFM, Campilho RDSG (2012) Advances in Numerical Modelling of Adhesive Joints. SpringerBriefs in Applied Sciences and Technology, Springer, Berlin, Heidelberg

da Silva LFM, Öchsner A, Adams RD (eds) (2011) Handbook of Adhesion Technology. Springer, Berlin, Heidelberg

Sugiman S, Crocombe AD, Aschroft IA (2013a) Experimental and numerical investigation of the static response of environmentally aged adhesively bonded joints. International Journal of Adhesion and Adhesives 40:224–237

Sugiman S, Crocombe AD, Aschroft IA (2013b) Modelling the static response of unaged adhesively bonded structures. Engineering Fracture Mechanics 98:296–314

Sukumar N, Moës N, Moran B, Belytschko T (2000) Extended finite element method for three-dimensional crack modelling. International Journal for Numerical Methods in Engineering 48(11):1549–1570

Viana G, Costa M, Banea MD, da Silva LFM (2017a) Behaviour of environmentally degraded epoxy adhesives as a function of temperature. The Journal of Adhesion 93(1-2):95–112

Viana G, Costa M, Banea MD, da Silva LFM (2017b) Moisture and temperature degradation of double cantilever beam adhesive joints. Journal of Adhesion Science and Technology 31(16):1824–1838

Viana G, Costa M, Banea MD, da Silva LFM (2017c) Water Diffusion in Double Cantilever Beam Adhesive Joints. Latin American Journal of Solids and Structures 14(2):188–201

Volkersen O (1938) Die Nietkraftverteilung in zugbeanspruchten Nietverbindungen mit konstanten Laschenquerschnitten. Luftfahrtforschung 15(1-2):41–47

Wooley GR, Carver DR (1971) Stress concentration factors for bonded lap joints. J Aircraft 8(10):817–820

Xia Y, Zhou Q, Wang P, Johnson N, Gayden X, Fickes J (2009a) Development of a high-efficiency modeling technique for weld-bonded steel joints in vehicle structures, part ii: Dynamic experiments and simulations. International Journal of Adhesion and Adhesives 29(4):427–433

Xia Y, Zhou Q, Wang P, Johnson N, Gayden X, Fickes J (2009b) Development of high-efficiency modeling technique for weld-bonded steel joints in vehicle structures - part i: Static experiments and simulations. International Journal of Adhesion and Adhesives 29(4):414–426

Xie D, Waas AM (2006) Discrete cohesive zone model for mixed-mode fracture using finite element analysis. Engineering Fracture Mechanics 73(13):1783–1796

Xie D, Salvi AG, Sun C, Waas AM, Caliskan A (2006) Discrete cohesive zone model to simulate static fracture in 2D triaxially braided carbon fiber composites. Journal of Composite Materials 40(22):2025–2046

Xu Y, Yuan H (2009) Computational analysis of mixed-mode fatigue crack growth in quasi-brittle materials using extended finite element methods. Engineering Fracture Mechanics 76(2):165–181

Yang QD, Thouless MD (2001) Mixed-mode fracture analyses of plastically-deforming adhesive joints. International Journal of Fracture 110(2):175–187

Yang X, Xia Y, Zhou Q, Wang PC, Wang K (2012) Modeling of high strength steel joints bonded with toughened adhesive for vehicle crash simulations. International Journal of Adhesion and Adhesives 39:21–32

Zaera R, Sánchez-Sáez S, Pérez-Castellanos JL, Navarro C (2000) Modelling of the adhesive layer in mixed ceramic/metal armours subjected to impact. Composites Part A: Applied Science and Manufacturing 31(8):823–833

Zgoul M, Crocombe AD (2004) Numerical modelling of lap joints bonded with a rate-dependent adhesive. International Journal of Adhesion and Adhesives 24(4):355–366

Zhu Y, Liechti KM, Ravi-Chandar K (2009) Direct extraction of rate-dependent traction-separation laws for polyurea/steel interfaces. International Journal of Solids and Structures 46(1):31–51

Chapter 11
A Short Review of Electromagnetic Force Models for Matter - Theory and Experimental Evidence

Wilhelm Rickert and Wolfgang H. Müller

Abstract From Maxwell's equations balance laws for the electromagnetic linear momentum, angular momentum, and energy can be found after recasting and using several identities of vector calculus. Therefore, the obtained equations are not "new results" but rather identities having the form of a balance law. However, there is some degree of freedom, (a) during construction of a particular identity and (b) for the choice of the to-be-balanced quantity, the non-convective flux, and the production term. In short, one is insecure which of the various forms is correct under which circumstances. This conundrum is referred to as the Abraham-Minkowski controversy, who first proposed different expressions for the electromagnetic linear momentum. The proper choice of electromagnetic force and torque expressions is of particular importance in matter where the mechanical and electromagnetic fields couple. The question arises as to whether a comparison between the predicted deformation behavior and the observed one can help to decide which electromagnetic force model is suitable for a material of interest. In this paper we shall briefly review the controversy and suggest new approaches for its solution on the continuum level.

Keywords: Electromagnetic force models · Magnetostriction · Electrostriction · Total forces and torques

11.1 Compilation of Relevant Force Models

Our starting point are the local balances of linear momentum for ponderable matter in regular points,

Wilhelm Rickert · Wolfgang H. Müller
Institut für Mechanik, Kontinuumsmechanik und Materialtheorie, Technische Universität Berlin, Sek. MS. 2, Einsteinufer 5, 10587 Berlin, Germany,
e-mail: rickert@tu-berlin.de, wolfgang.h.mueller@tu-berlin.de

H. Altenbach and A. Öchsner (eds.), *State of the Art and Future Trends in Material Modeling*, Advanced Structured Materials 100,
https://doi.org/10.1007/978-3-030-30355-6_11

$$\frac{\partial}{\partial t}(\rho \boldsymbol{v}) + \nabla \cdot (\rho \boldsymbol{v} \otimes \boldsymbol{v} - \boldsymbol{\sigma}) = \rho \boldsymbol{f} + \boldsymbol{f}^{\mathrm{EM}} , \tag{11.1}$$

and at a singular interface I with the normal $\boldsymbol{n}$ showing no intrinsic properties,

$$\boldsymbol{n} \cdot [\![\boldsymbol{\sigma} + \rho(\boldsymbol{w} - \boldsymbol{v}) \otimes \boldsymbol{v}]\!] = -\boldsymbol{f}_I^{\mathrm{EM}} , \tag{11.2}$$

where double brackets denote the jump across the interface. We denote by ρ the mass density, by $\boldsymbol{v}$ the particle velocity, by $\boldsymbol{w}$ the mapping velocity, by $\boldsymbol{f}$ the gravitational body force, and by $\boldsymbol{\sigma}$ the stress tensor. Unusual symbols worthy of a more detailed discussion are the volumetric electromagnetic force, $\boldsymbol{f}^{\mathrm{EM}}$, and its counterpart on a singular interface, $\boldsymbol{f}_I^{\mathrm{EM}}$. For these various expressions can be found in the literature, at least after some algebraic effort, and we will cite the pertinent references in what follows.

Probably the best known force model is the one attributed to Lorentz. We write:

$$\boldsymbol{f}^{\mathrm{L}} = q\boldsymbol{E} + \boldsymbol{J} \times \boldsymbol{B} , \qquad \boldsymbol{f}_I^{\mathrm{L}} = q_I \langle \boldsymbol{E} \rangle + \boldsymbol{J}_I \times \langle \boldsymbol{B} \rangle . \tag{11.3}$$

Pointed brackets refer to arithmetic averages of the right and left limit field values. q is the total charge, $\boldsymbol{J}$ is the total current, $\boldsymbol{E}$ and $\boldsymbol{B}$ are the electric and magnetic field vectors, respectively. The index I refers to the corresponding interface characteristics. All quantities are explained in detail for example in Müller (2014, Chapter 13). The correctness of these expressions is demonstrated indirectly in Müller (1985, Section 9.5) by construction of the standard Maxwell stress tensor and Poynting vector via Maxwell's equations. A more explicit proof is presented in Reich et al (2018, Appendix C). Suffice it to say that in the derivation and in those of the following force models ample use is made of the Maxwell-Lorentz-Aether relations, which hold true in an inertial system.

A second set of force models goes back to the work of Abraham (1909). Starting from the electromagnetic momentum density presented in that work and manipulating it similarly as the expression leading to the Lorentz force in Reich et al (2018) two different sets of expressions will result due to the intrinsic arbitrariness in the balance equations (see the example outlined in Reich et al, 2018, Section 2, for this issue):

$$\begin{aligned} \boldsymbol{f}^{\mathrm{A}_1} &= q\boldsymbol{E} + \boldsymbol{J} \times \mu_0 \boldsymbol{\mathfrak{H}} + (\nabla \times \boldsymbol{B}) \times \boldsymbol{M} + \mu_0 \boldsymbol{D} \times \frac{\partial \boldsymbol{M}}{\partial t} , \\ \boldsymbol{f}_I^{\mathrm{A}_1} &= q_I \langle \boldsymbol{E} \rangle + \boldsymbol{J}_I \times \mu_0 \langle \boldsymbol{\mathfrak{H}} \rangle - \mu_0 w_\perp \langle \boldsymbol{D} \rangle \times [\![\boldsymbol{M}]\!] + (\boldsymbol{n} \times [\![\boldsymbol{B}]\!]) \times \langle \boldsymbol{M} \rangle , \end{aligned} \tag{11.4}$$

and

$$\begin{aligned} \boldsymbol{f}^{\mathrm{A}_2} &= q\boldsymbol{E} + \boldsymbol{J} \times \mu_0 \boldsymbol{\mathfrak{H}} - \nabla \cdot (\boldsymbol{M} \otimes \boldsymbol{B}) + \mu_0 \boldsymbol{D} \times \frac{\partial \boldsymbol{M}}{\partial t} , \\ \boldsymbol{f}_I^{\mathrm{A}_2} &= q_I \langle \boldsymbol{E} \rangle + \boldsymbol{J}_I \times \mu_0 \langle \boldsymbol{\mathfrak{H}} \rangle - \mu_0 w_\perp \langle \boldsymbol{D} \rangle \times [\![\boldsymbol{M}]\!] + \\ &\quad + (\boldsymbol{n} \times [\![\boldsymbol{B}]\!]) \times \langle \boldsymbol{M} \rangle - \boldsymbol{n} \cdot [\langle \boldsymbol{M} \rangle \otimes [\![\boldsymbol{B}]\!] + [\![\boldsymbol{M}]\!] \otimes \langle \boldsymbol{B} \rangle] . \end{aligned} \tag{11.5}$$

We denote by $\boldsymbol{\mathfrak{H}}$ the electric current potential in matter, by $\boldsymbol{M}$ the magnetization, $\boldsymbol{D}$ is the total charge potential, μ_0 is the vacuum permeability. Note that these two choices are somewhat arbitrary and we could obtain even more force models from one electromagnetic momentum density.

In the same spirit the electromagnetic momentum density shown in the work by Minkowski (1910) leads to the following expressions:

$$\begin{aligned} \boldsymbol{f}^{\mathrm{M}_1} &= q\boldsymbol{E} + \boldsymbol{J}^{\mathrm{f}} \times \boldsymbol{B} + (\nabla \times \boldsymbol{M}) \times \boldsymbol{B} - (\nabla \times \boldsymbol{E}) \times \boldsymbol{P} \,, \\ \boldsymbol{f}_I^{\mathrm{M}_1} &= q_I \langle \boldsymbol{E} \rangle + \boldsymbol{J}_I^{\mathrm{f}} \times \langle \boldsymbol{B} \rangle + \langle \boldsymbol{P} \rangle \times (\boldsymbol{n} \times [\![\boldsymbol{E}]\!]) + (\boldsymbol{n} \times [\![\boldsymbol{M}]\!]) \times \langle \boldsymbol{B} \rangle \,, \end{aligned} \tag{11.6}$$

and

$$\begin{aligned} \boldsymbol{f}^{\mathrm{M}_2} &= q^{\mathrm{f}}\boldsymbol{E} + \boldsymbol{J}^{\mathrm{f}} \times \boldsymbol{B} - (\nabla \otimes \boldsymbol{M}) \cdot \boldsymbol{B} + (\nabla \otimes \boldsymbol{E}) \cdot \boldsymbol{P} \,, \\ \boldsymbol{f}_I^{\mathrm{M}_2} &= q_I^{\mathrm{f}} \langle \boldsymbol{E} \rangle + \boldsymbol{J}_I^{\mathrm{f}} \times \langle \boldsymbol{B} \rangle + \boldsymbol{n}(\langle \boldsymbol{P} \rangle \cdot [\![\boldsymbol{E}]\!] - \langle \boldsymbol{B} \rangle \cdot [\![\boldsymbol{M}]\!]) \,. \end{aligned} \tag{11.7}$$

$\boldsymbol{J}^{\mathrm{f}}$ and $\boldsymbol{J}_I^{\mathrm{f}}$ are the free total currents in regular and singular points, respectively, and q^{f} is the free charge density.

Finally, from the work of Einstein and Laub (1908) we find:

$$\begin{aligned} \boldsymbol{f}^{\mathrm{EL}} &= q^{\mathrm{f}}\boldsymbol{E} + \boldsymbol{J}^{\mathrm{f}} \times \mu_0 \mathfrak{H} + \boldsymbol{P} \cdot (\nabla \otimes \boldsymbol{E}) + \frac{\partial \boldsymbol{P}}{\partial t} \times \mu_0 \mathfrak{H} + \\ &\quad + \mu_0 \boldsymbol{M} \cdot (\nabla \otimes \mathfrak{H}) - \mu_0 \frac{\partial \boldsymbol{M}}{\partial t} \times \boldsymbol{D} \,, \\ \boldsymbol{f}_I^{\mathrm{EL}} &= \boldsymbol{f}_I^{\mathrm{M}_2} + \boldsymbol{n} [\![\boldsymbol{B} \cdot \boldsymbol{M} - \frac{\mu_0}{2} \boldsymbol{M} \cdot \boldsymbol{M}]\!] - w_\perp [\![\mathfrak{D} \times \mu_0 \boldsymbol{M} + \boldsymbol{P} \times \mu_0 \mathfrak{H}]\!] \,, \end{aligned} \tag{11.8}$$

where $\mathfrak{D}$ denotes the free charge potential.

11.2 Intermezzo

After the various force models have been presented the natural question arises as to which of them is the correct one? The answer is that all models are correct on the continuum scale for matter. There is not "the one" that will describe all situations correctly. Depending on the material one will be more realistic than the other. Experiments must decide which one this is. Various experiments come to mind. A first idea could be to measure the total force exerted on a body made of a material susceptible to external electromagnetic fields and to compare it with the total forces predicted by the various models. Second, it might be useful to study the deformation of that body. This requires us to solve a complex boundary value problem, because then we need to couple mechanics and electrodynamics, in particular we need to look at the stress-strain correlation. Third, besides forces it might be useful to study torques resulting during electromagnetic force interaction.

We shall outline the corresponding procedures in what follows in several case studies, which have been published by us before. Hence we just repeat the results, compare and comment on them. Our first example concerns a permanent magnetic, linear elastic sphere, resulting in magnetostriction. The second example is a silicone oil drop in castor oil (so that the oils do not mix) subjected to an external electric field.

Due to the different electric polarization in the two media deformation will result. The third example is the deformation of a spherical linear-elastic electret, similarly to magnetostriction, but with different types in poralization. And finally we shall report on the force and torque interaction between two rigid permanent spherical magnets.

11.3 Case I: Magnetostriction of a Spherical Permanent Magnet

This problem was analyzed before in detail in Reich et al (2018, Sect. 6). We considered the static case of a permanent magnetic sphere of radius R with uniform magnetization, $\boldsymbol{M} = M_0 \boldsymbol{e}_z$, which is treated as isotropically linear-elastic, with Lamé parameters λ and μ, mechanics-wise. In order to evaluate the expressions for the various force densities the following information for the electro-magnetic quantities are required:

$$\begin{aligned} &\boldsymbol{B}^{\mathrm{I}} = \tfrac{2}{3}\mu_0 \boldsymbol{M}\,, \quad \boldsymbol{\mathfrak{H}}^{\mathrm{I}} = -\frac{1}{3}\boldsymbol{M}\,, \quad q = q^{\mathrm{f}} - \nabla\cdot\boldsymbol{P} = 0\,, \\ &\boldsymbol{J} = \boldsymbol{J}^{\mathrm{f}} + \frac{\partial \boldsymbol{P}}{\partial t} + \nabla\times\boldsymbol{M} = \boldsymbol{0}\,, \quad q_I = q_I^{\mathrm{f}} - \boldsymbol{n}\cdot[\![\boldsymbol{P}]\!] = 0\,, \\ &\boldsymbol{J}_I = \boldsymbol{J}_I^{\mathrm{f}} - [\![\boldsymbol{P}]\!]w_\perp + \boldsymbol{n}\times[\![\boldsymbol{M}]\!] = \boldsymbol{n}\times[\![\boldsymbol{M}]\!]\,. \end{aligned} \tag{11.9}$$

Hence all of the volumetric force densities of all presented models vanish and the corresponding force densities on the interface are given by:

$$\begin{aligned} \boldsymbol{f}_I^{\mathrm{L}} &= \frac{1}{6}\mu_0 M_0^2(\sin^2\vartheta\,\boldsymbol{e}_r + 4\sin\vartheta\cos\vartheta\,\boldsymbol{e}_\vartheta) = \boldsymbol{f}_I^{(1)}\,, \\ \boldsymbol{f}_I^{\mathrm{A}_1} &= \frac{1}{6}\mu_0 M_0^2(\sin^2\vartheta\,\boldsymbol{e}_r + 4\sin\vartheta\cos\vartheta\,\boldsymbol{e}_\vartheta) = \boldsymbol{f}_I^{(1)}\,, \\ \boldsymbol{f}_I^{\mathrm{A}_2} &= \frac{1}{6}\mu_0 M_0^2(1 + 3\cos^2\vartheta)\boldsymbol{e}_r = \boldsymbol{f}_I^{(2)}\,, \\ \boldsymbol{f}_I^{\mathrm{M}_1} &= \frac{1}{6}\mu_0 M_0^2(\sin^2\vartheta\,\boldsymbol{e}_r + 4\sin\vartheta\cos\vartheta\,\boldsymbol{e}_\vartheta) = \boldsymbol{f}_I^{(1)}\,, \\ \boldsymbol{f}_I^{\mathrm{M}_2} &= \frac{1}{6}\mu_0 M_0^2(1 + 3\cos^2\vartheta)\boldsymbol{e}_r = \boldsymbol{f}_I^{(2)}\,, \\ \boldsymbol{f}_I^{\mathrm{EL}} &= \frac{1}{2}\mu_0 M_0^2\cos^2\vartheta\,\boldsymbol{e}_r = \boldsymbol{f}_I^{(3)}\,. \end{aligned} \tag{11.10}$$

It is noteworthy that in this case all models having (the same) symmetric electromagnetic stress measure (not shown here explicitly) yield the same surface force density. Interestingly, the non-symmetric Abraham and Minkowski models (also not detailed here) coincide for this magnetic problem. The Einstein-Laub model is distinct from the others. However, it can be seen that

$$\boldsymbol{f}_I^{\mathrm{EL}} = \boldsymbol{f}_I^{\mathrm{A}_2} - \frac{1}{6}\mu_0 M_0^2\boldsymbol{e}_r,$$

hence these models differ only by a constant radial (pressure) offset. Qualitative representations of the surface force densities indicating the direction of deformation are shown in Fig. 11.1. The deformation field is now determined based on Hooke's law, mechanical equilibrium, and traction boundary conditions containing the various surface force densities. The method of Hiramatsu and Oka was applied for solving the resulting Lamé-Navier equations. It allows to obtain closed-form solutions in terms of Legendre polynomials. Sketches of the sphere deforming into different types of spheroids is shown in Fig. 11.2. Naively speaking one would now think that a simple measurement of the surface contour of an originally spherical object deforming after its magnetization would suffice to identify the "most realistic" surface force density. Unfortunately for a typical solid and reasonably high values of magnetization

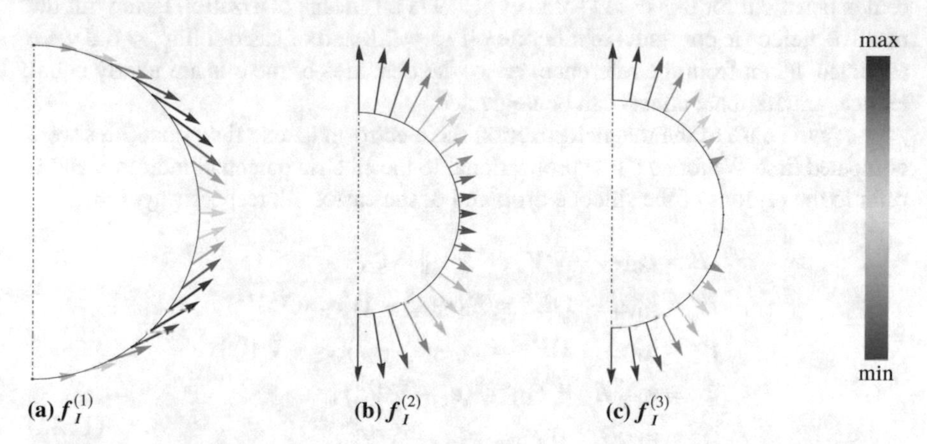

Fig. 11.1 Qualitative representations of the surface force densities. In (c), arrows are suppressed for small force magnitudes

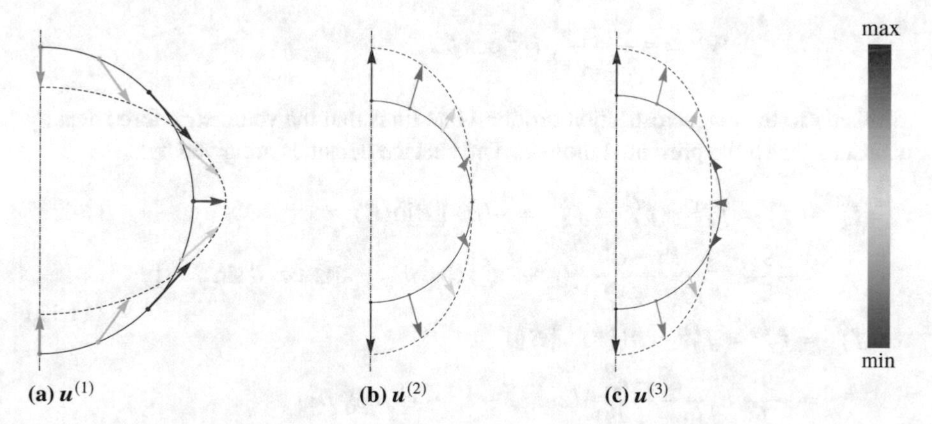

Fig. 11.2 Qualitative visualization of the surface displacements for the three electromagnetic force results. The ratio $\lambda/\mu = 1.27$ was used in order to model steel

the displacements are still extremely small (nanometer range), which renders a trustworthy measurement impossible. A more deformable object is required. Such a case will be presented in the next section.

11.4 Case II: Deformation of a Spherical Droplet due to Electric Polarization

This problem was analyzed before in detail in Reich et al (2018, Section 7). We considered the static case of a spherical silicone oil droplet of radius R in oxidized castor oil, placed in an homogeneous electric field $\boldsymbol{E}_0 = E_0 \boldsymbol{e}_z$. In fact there exists a real experiment for this case (Torza et al, 1971). Linear polarization laws with the relative dielectric constants of silicone oil $\epsilon_\mathrm{r}^\mathrm{S} \approx 2.8$ and of castor oil $\epsilon_\mathrm{r}^\mathrm{C} \approx 6.3$ were assumed, taken from the reference. Also, the densities of the oils are nearly equal. Hence, gravitational effects can be neglected.

As in the case of the magnetostriction the electro-magnetic field quantities were computed first. We found ($\mathcal{V}$ is proportional to the electric potential, indices S and C refer to the regions of the silicone drop and of the castor oil, respectively, $\tilde{r} = \frac{r}{R}$):

$$
\begin{aligned}
\boldsymbol{E} &= E_0(\boldsymbol{e}_z - \tilde{\nabla}\mathcal{V}) \,, \quad [\![\mathcal{V}]\!] = 0 \,, \\
\boldsymbol{P}^\mathrm{S} &= \epsilon_0(\epsilon_\mathrm{r}^\mathrm{S} - 1)\boldsymbol{E}^\mathrm{S} = E_0\epsilon_0(\epsilon_\mathrm{r}^\mathrm{S} - 1)(\boldsymbol{e}_z - \tilde{\nabla}\mathcal{V}^\mathrm{S}) \,, \\
\boldsymbol{P}^\mathrm{C} &= \epsilon_0(\epsilon_\mathrm{r}^\mathrm{C} - 1)\boldsymbol{E}^\mathrm{C} = E_0\epsilon_0(\epsilon_\mathrm{r}^\mathrm{C} - 1)(\boldsymbol{e}_z - \tilde{\nabla}\mathcal{V}^\mathrm{C}) \,, \\
\mathfrak{D}^\mathrm{S} &= \epsilon_0\epsilon_\mathrm{r}^\mathrm{S}\boldsymbol{E}^\mathrm{S} = E_0\epsilon_0\epsilon_\mathrm{r}^\mathrm{S}(\boldsymbol{e}_z - \tilde{\nabla}\mathcal{V}^\mathrm{S}) \,, \\
\mathfrak{D}^\mathrm{C} &= \epsilon_0\epsilon_\mathrm{r}^\mathrm{C}\boldsymbol{E}^\mathrm{C} = E_0\epsilon_0\epsilon_\mathrm{r}^\mathrm{C}(\boldsymbol{e}_z - \tilde{\nabla}\mathcal{V}^\mathrm{C}) \,, \\
\mathcal{V}^\mathrm{S} &= -\frac{\epsilon_\mathrm{r}^\mathrm{C} - \epsilon_\mathrm{r}^\mathrm{S}}{2\epsilon_\mathrm{r}^\mathrm{C} + \epsilon_\mathrm{r}^\mathrm{S}}\tilde{r}\cos\vartheta \,, \\
\mathcal{V}^\mathrm{C} &= -\frac{\epsilon_\mathrm{r}^\mathrm{C} - \epsilon_\mathrm{r}^\mathrm{S}}{2\epsilon_\mathrm{r}^\mathrm{C} + \epsilon_\mathrm{r}^\mathrm{S}}\tilde{r}^{-2}\cos\vartheta \,.
\end{aligned}
\tag{11.11}
$$

Similarly to the magnetostriction problem one finds that the volumetric force density vanishes for all the presented models. The surface densities are given by:

$$
\begin{aligned}
\boldsymbol{f}_I^{(4)} &= \boldsymbol{f}_I^\mathrm{L} = \boldsymbol{f}_I^{\mathrm{A}_1} = \boldsymbol{f}_I^{\mathrm{A}_2} = \boldsymbol{f}_I^{\mathrm{M}_1} = -(\boldsymbol{n}\cdot[\![\boldsymbol{P}]\!])\langle\boldsymbol{E}\rangle \\
&= -\frac{9}{2}\epsilon_0 E_0^2 \frac{\epsilon_\mathrm{r}^\mathrm{C} - \epsilon_\mathrm{r}^\mathrm{S}}{(2\epsilon_\mathrm{r}^\mathrm{C} + \epsilon_\mathrm{r}^\mathrm{S})^2}[(\epsilon_\mathrm{r}^\mathrm{C} + \epsilon_\mathrm{r}^\mathrm{S})\cos^2\vartheta\,\boldsymbol{e}_r - 2\epsilon_\mathrm{r}^\mathrm{C}\cos\vartheta\sin\vartheta\,\boldsymbol{e}_\vartheta] \,, \\
\boldsymbol{f}_I^{(5)} &= \boldsymbol{f}_I^{\mathrm{M}_2} = \boldsymbol{f}_I^\mathrm{EL} = \boldsymbol{n}(\langle\boldsymbol{P}\rangle\cdot[\![\boldsymbol{E}]\!]) \\
&= -\frac{9}{2}\epsilon_0 E_0^2 \frac{\epsilon_\mathrm{r}^\mathrm{C} - \epsilon_\mathrm{r}^\mathrm{S}}{(2\epsilon_\mathrm{r}^\mathrm{C} + \epsilon_\mathrm{r}^\mathrm{S})^2}(2\epsilon_\mathrm{r}^\mathrm{C}\epsilon_\mathrm{r}^\mathrm{S} - \epsilon_\mathrm{r}^\mathrm{C} - \epsilon_\mathrm{r}^\mathrm{S})\cos^2\vartheta\,\boldsymbol{e}_r \,.
\end{aligned}
\tag{11.12}
$$

Sketches of the forces are depicted in Fig. 11.3. In order to compute the deformation

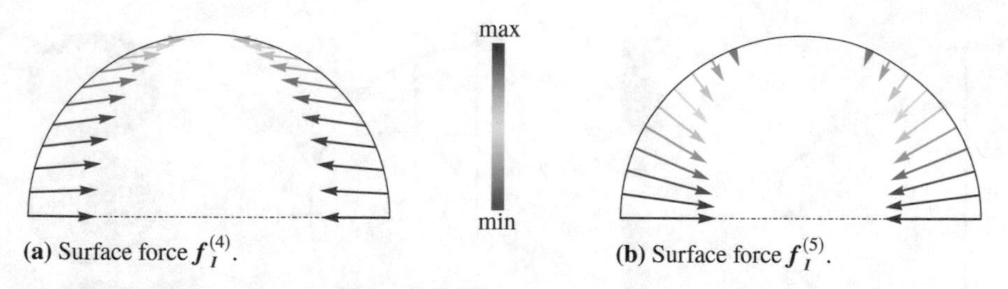

(a) Surface force $\boldsymbol{f}_I^{(4)}$. **(b)** Surface force $\boldsymbol{f}_I^{(5)}$.

Fig. 11.3 Qualitative representation of the computed force results for the oil drop experiment. For visualization, the values $\epsilon_r^S = 2.8$ and $\epsilon_r^C = 6.3$ were chosen. Note that the external electric field points in horizontal direction

the droplet was treated as a fluid at rest in terms of a hydrostatic pressure acting on it. A linear relationship between the pressure and the volume change was assumed and linked to the following (normalized) displacement *ansatz* on the interface

$$\boldsymbol{u}_I = \hat{u}(\tilde{u}_r(\vartheta)\boldsymbol{e}_r + \tilde{u}_\vartheta(\vartheta)\boldsymbol{e}_\vartheta) \,. \tag{11.13}$$

The deformation results of the experiments from Torza et al (1971) are depicted in Fig. 11.4c. For increasing electric field strength the drop deforms as an oblate spheroid. In Fig. 11.3 the surface force predictions of the various force models are qualitatively shown. $\boldsymbol{f}_I^{(4)}$ and $\boldsymbol{f}_I^{(5)}$ both suggest that the droplet should assume an oblate shape. Intuitively speaking, the surface force $\boldsymbol{f}_I^{(4)}$ might result in the correct deformation figure since the force $\boldsymbol{f}_I^{5}$ may cause dimples at poles, deviating from a spheroid form. Moreover, the magnitudes of the displacements differ. The models with the force $\boldsymbol{f}_I^{(4)}$ yields the smooth deformation figure depicted in Fig. 11.4a which is in good agreement with the experimental results in Fig. 11.4c. The deformation figure due to the models with the force $\boldsymbol{f}_I^{(5)}$ possesses a different curvature near the poles. Hence, the deformed body is not an oblate spheroid. However, this form is not observable in the experimental photographs. Therefore, it is reasonable to conclude that the models with the force $\boldsymbol{f}_I^{(5)}$ yield unphysical results, *i.e.*, the asymmetric Minkowski and the Einstein-Laub models are unlikely.

11.5 Case III: Elastic Deformation of Spherical Electrets due to Electric Polarization and Surface Charges

The analysis of this problem will be published in detail in Rickert et al (2019). Analogously to the droplet problem we shall consider the static case of spherical electrets. In (I), a linear dielectric in an externally applied electric field $\boldsymbol{E}_0$ is considered. Then, a real charge electret with surface charge $q_I^{\mathrm{f}} = \mathcal{Q}/A_{\mathrm{sph}}$ is analyzed, (II). From these solutions the cases of an oriented dipole electret (III) and a real

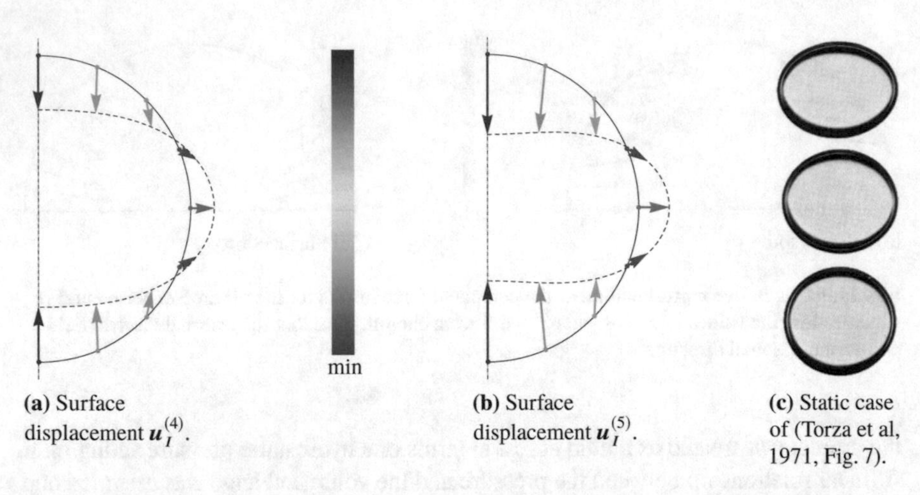

(a) Surface displacement $\boldsymbol{u}_I^{(4)}$.

(b) Surface displacement $\boldsymbol{u}_I^{(5)}$.

(c) Static case of (Torza et al, 1971, Fig. 7).

Fig. 11.4 Deformation figures of the oil droplet. (a) and (b): Predicted surface displacement using parameters $\epsilon_r^S = 2.8$, $\epsilon_r^C = 6.3$, $\gamma = 1$ and $\lambda_I/\mu_I = 1$. Scaling of the displacements was applied. (c): Experimental photos from Torza et al (1971, Fig. 7). The electric field points vertically. In (c), the electric field strength increases from the bottom to the top

charge electret with linear dielectric material behavior (IV) are readily obtained. We wish to calculate the various electromagnetic force densities and use them for predicting the deformation in situations (I)–(IV). As before the electromagnetic field quantities need to be determined first. By using the following scaling factors,

$$\frac{q_I^{\mathrm{f}}}{\epsilon_0} = \alpha\mathcal{E}\,,\ \frac{P_0}{\epsilon_0} = \beta\mathcal{E}\,,\ E_0 = \gamma\mathcal{E} \quad \Rightarrow\ a_0 = \frac{\alpha}{\epsilon_{\mathrm{r}}}\mathcal{E}\,,\ a_1 = \frac{\beta + (\epsilon_{\mathrm{r}} - 1)\gamma}{2\epsilon_{\mathrm{r}} + 1}\mathcal{E} \tag{11.14}$$

it can be shown that ($a_0 = \tilde{a}_0\mathcal{E}, a_1 = \tilde{a}_1\mathcal{E}$):

$$\begin{aligned}
\boldsymbol{E}^{\mathrm{I}} &= (\gamma - \tilde{a}_1)\mathcal{E}[\cos\vartheta\,\boldsymbol{e}_r - \sin\vartheta\,\boldsymbol{e}_\vartheta] = (\gamma - \tilde{a}_1)\mathcal{E}\boldsymbol{e}_z = \text{const.}\\
\boldsymbol{E}^{\mathrm{O}} &= \left\{\left[\tilde{a}_0\tilde{r}^{-2} + (2\tilde{a}_1\tilde{r}^{-3} + \gamma)\cos\vartheta\right]\boldsymbol{e}_r + \left(\tilde{a}_1\tilde{r}^{-3} - \gamma\right)\sin\vartheta\,\boldsymbol{e}_\vartheta\right\}\mathcal{E}\,,\\
\boldsymbol{P}^{\mathrm{I}} &= \kappa\epsilon_0\mathcal{E}[\cos\vartheta\,\boldsymbol{e}_r - \sin\vartheta\,\boldsymbol{e}_\vartheta] = \kappa\epsilon_0\mathcal{E}\boldsymbol{e}_z = \text{const.},\ \boldsymbol{P}^{\mathrm{O}} = \boldsymbol{0},\\
q_I &= \epsilon_0\left(\frac{q_I^{\mathrm{f}}}{\epsilon_0} - \boldsymbol{n}\cdot\frac{[\![\boldsymbol{P}]\!]}{\epsilon_0}\right) = \epsilon_0\mathcal{E}[\alpha + \kappa\cos\vartheta]\,,\\
\boldsymbol{J}_I &= -[\![\boldsymbol{P}]\!]w_\perp + \boldsymbol{n}\times[\![\boldsymbol{M}]\!] = \boldsymbol{0}\,,
\end{aligned} \tag{11.15}$$

where different scaling factors apply for the different situations:

$$
\begin{aligned}
&\text{(I)} \quad \mathcal{E} = E_0\,, \quad \alpha = 0\,, \quad \beta = 0\,, \quad \gamma = 1\,, \quad (\epsilon_\mathrm{r} \neq 1)\\
&\text{(II)} \quad \mathcal{E} = \frac{q_I^\mathrm{f}}{\epsilon_0}\,, \quad \alpha = 1\,, \quad \beta = 0\,, \quad \gamma = 0\,, \quad (\epsilon_\mathrm{r} = 1)\\
&\text{(III)} \quad \mathcal{E} = \frac{P_0}{\epsilon_0}\,, \quad \alpha = 0\,, \quad \beta = 1\,, \quad \gamma = 0\,, \quad (\epsilon_\mathrm{r} = 1)\\
&\text{(IV)} \quad \mathcal{E} = \frac{q_I^\mathrm{f}}{\epsilon_0}, \quad \alpha = 1\,, \quad \beta = 0\,, \quad \gamma = 0\,, \quad (\epsilon_\mathrm{r} \neq 1)
\end{aligned}
\tag{11.16}
$$

and $\kappa = \beta + (\gamma - \tilde{a}_1)(\epsilon_\mathrm{r} - 1)$. Now the non-vanishing surface force densities result:

$$
\begin{aligned}
\boldsymbol{f}_I^{(1)} &:= \boldsymbol{f}_I^{\mathrm{L}} = \boldsymbol{f}_I^{\mathrm{A}_1} = \boldsymbol{f}_I^{\mathrm{A}_2} = \boldsymbol{f}_I^{\mathrm{M}_1}\\
&= \epsilon_0 \mathcal{E}^2 \Bigg[\left(c_0^{(1)} \mathcal{P}_0(x) + c_1^{(1)} \mathcal{P}_1(x) + c_2^{(1)} \mathcal{P}_2(x) \right) \boldsymbol{e}_r +\\
&\quad + \left(d_1^{(1)} \frac{\mathrm{d}\mathcal{P}_1(x)}{\mathrm{d}\vartheta} + d_2^{(1)} \frac{\mathrm{d}\mathcal{P}_2(x)}{\mathrm{d}\vartheta} \right) \boldsymbol{e}_\vartheta \Bigg]\,,\\
\boldsymbol{f}_I^{(2)} &:= \boldsymbol{f}_I^{\mathrm{M}_2} = \boldsymbol{f}_I^{\mathrm{EL}}\\
&= \epsilon_0 \mathcal{E}^2 \left[\left(c_0^{(2)} \mathcal{P}_0(x) + c_1^{(2)} \mathcal{P}_1(x) + c_2^{(2)} \mathcal{P}_2(x) \right) \boldsymbol{e}_r + d_1^{(2)} \frac{\mathrm{d}\mathcal{P}_1(x)}{\mathrm{d}\vartheta} \boldsymbol{e}_\vartheta \right].
\end{aligned}
\tag{11.17}
$$

$\mathcal{P}_i(x)$ denote Legendre polynomials and the coefficients are given by:

$$
\begin{aligned}
c_0^{(1)} &= \frac{1}{2}\alpha\tilde{a}_0 + \tfrac{1}{6}\kappa(\tilde{a}_1 + 2\gamma)\,, & c_1^{(1)} &= \frac{1}{2}\alpha(\tilde{a}_1 + 2\gamma) + \tfrac{1}{2}\kappa\tilde{a}_0\,,\\
d_1^{(1)} &= \alpha(\gamma - \tilde{a}_1)\,, & d_2^{(1)} &= \frac{1}{3}\kappa(\gamma - \tilde{a}_1)\,,\\
c_0^{(2)} &= \frac{1}{2}\alpha\tilde{a}_0 + \tfrac{1}{2}\kappa\tilde{a}_1\,, & c_1^{(2)} &= \frac{1}{2}\alpha(\tilde{a}_1 + 2\gamma) + \frac{1}{2}\kappa\tilde{a}_0\,,\\
c_2^{(1)} &= \frac{1}{3}\kappa(\tilde{a}_1 + 2\gamma)\,, & d_1^{(2)} &= \alpha(\gamma - \tilde{a}_1)\,, \quad c_2^{(2)} = \kappa\tilde{a}_1\,.
\end{aligned}
\tag{11.18}
$$

The various force densities are illustrated in Fig. 11.5. They may serve as first indication for the deformation pattern. Now in complete analogy to Case I the elastic deformation response can be calculated in closed form for (I)–(IV) as will be shown in Rickert et al (2019). The resulting forms are shown in Fig. 11.6.

Unfortunately, as in the case of magnetostriction, the displacements are very small making experimental investigations difficult.

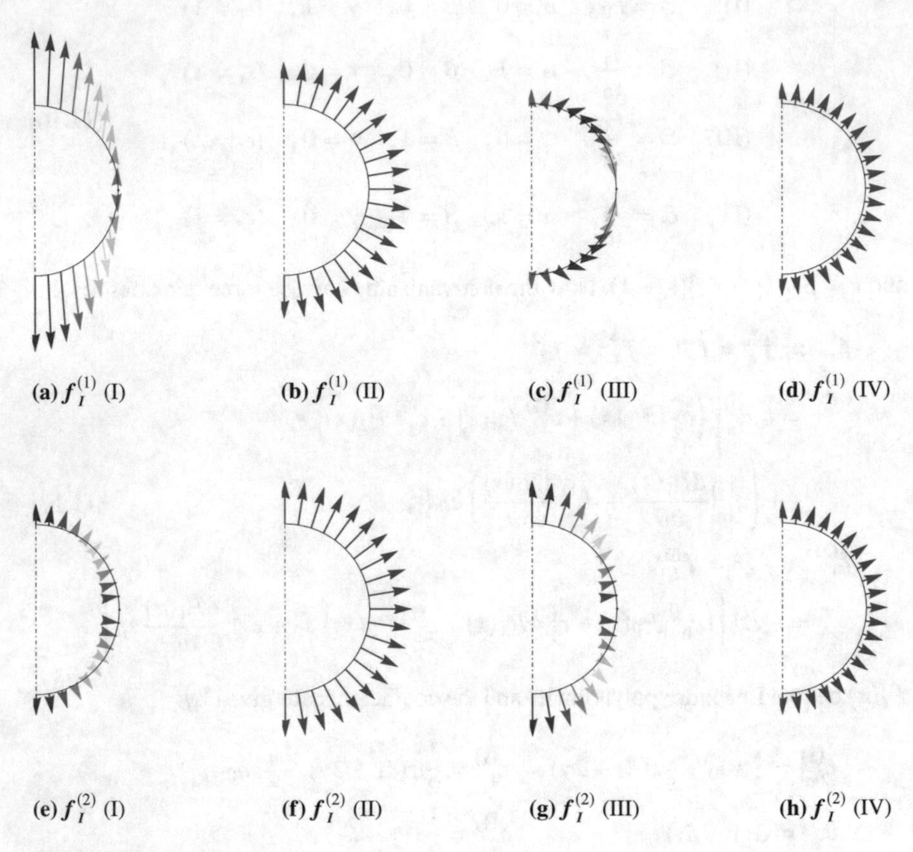

Fig. 11.5 Electret force densities predicted

11.6 Case IV: Force and Torque Interaction Between Spherical Magnets

The details of this problem will be published in Rickert and Müller (2019). We consider the interaction between two spherical rigid permanent magnets, homogeneously magnetized by $M_0^{(\mathrm{I})}$ and $M_0^{(\mathrm{II})}$ of radii $R_{(\mathrm{I})}$ and $R_{(\mathrm{II})}$, respectively: Fig. 11.7.

As one can show the dimensionless surface force densities $\tilde{\boldsymbol{f}}_I^{(\mathrm{EM})} = \boldsymbol{f}_I^{(\mathrm{EM})}/\hat{f}$ are given by:

$$\tilde{\boldsymbol{f}}_I^{\mathrm{L}} = \sin\vartheta' \boldsymbol{e}'_\varphi \times \tilde{\boldsymbol{B}}_{(\mathrm{I})} + \left\{ \frac{1}{6} \frac{M_0^{(\mathrm{II})}}{M_0^{(\mathrm{I})}} (\sin^2\vartheta' \boldsymbol{e}'_r + 4\sin\vartheta' \cos\vartheta' \boldsymbol{e}'_\vartheta) \right\} ,$$

$$\tilde{\boldsymbol{f}}_I^{\mathrm{A}_1} = \sin\vartheta' \boldsymbol{e}'_\varphi \times \tilde{\mathfrak{H}}_{(\mathrm{I})} + \left\{ \frac{1}{6} \sin\vartheta' \frac{M_0^{(\mathrm{II})}}{M_0^{(\mathrm{I})}} (\sin\vartheta' \boldsymbol{e}'_r + 4\cos\vartheta' \boldsymbol{e}'_\vartheta) \right\} ,$$

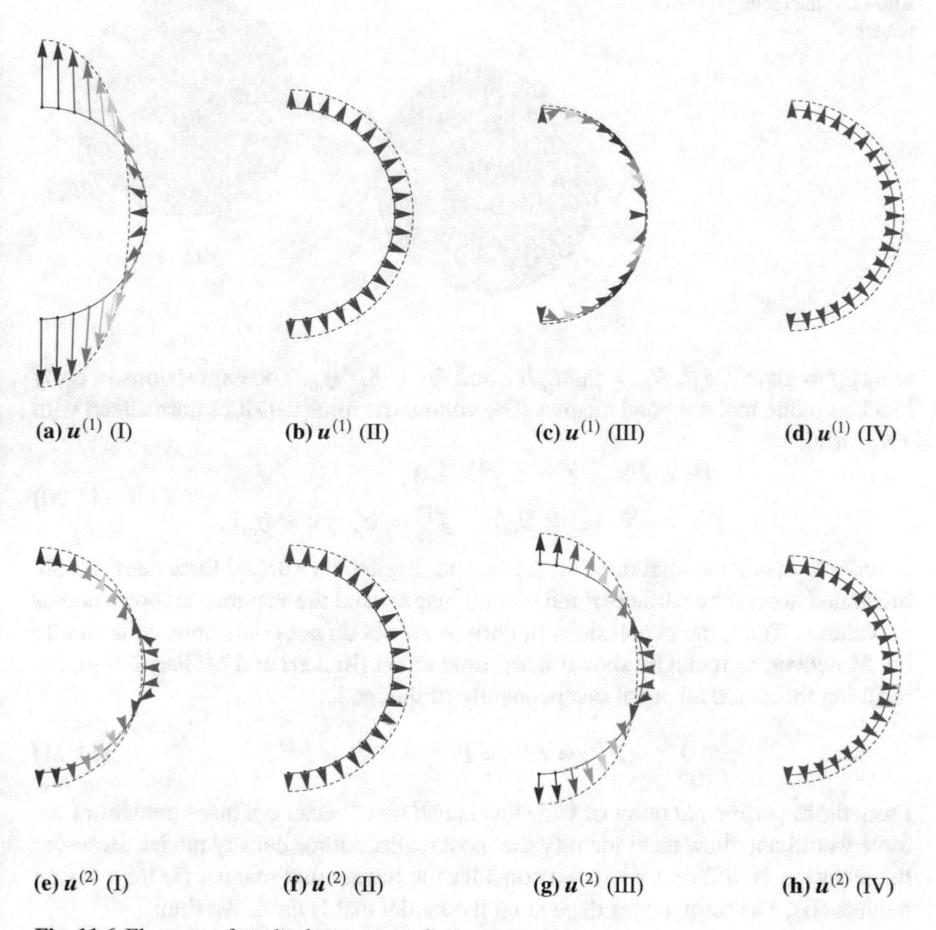

Fig. 11.6 Electret surface displacement predictions

$$
\begin{aligned}
\tilde{\boldsymbol{f}}_I^{\mathrm{A}_2} &= \sin\vartheta' \boldsymbol{e}'_\varphi \times \tilde{\mathfrak{H}}_{(\mathrm{I})} + \left\{ \cos\vartheta' \tilde{\boldsymbol{B}}_{(\mathrm{I})} + \frac{1}{6}\frac{M_0^{(\mathrm{II})}}{M_0^{(\mathrm{I})}}(1 + 3\cos^2\vartheta')\boldsymbol{e}'_r \right\}, \qquad (11.19) \\
\tilde{\boldsymbol{f}}_I^{\mathrm{M}_1} &= \sin\vartheta' \boldsymbol{e}'_\varphi \times \tilde{\boldsymbol{B}}_{(\mathrm{I})} + \left\{ \frac{1}{6}\frac{M_0^{(\mathrm{II})}}{M_0^{(\mathrm{I})}}(\sin^2\vartheta' \boldsymbol{e}'_r + 4\sin\vartheta' \cos\vartheta' \boldsymbol{e}'_\vartheta) \right\}, \\
\tilde{\boldsymbol{f}}_I^{\mathrm{M}_2} &= \boldsymbol{n}(\tilde{\boldsymbol{B}}_{(\mathrm{I})} \cdot \boldsymbol{e}'_z) + \left\{ \frac{1}{6}\frac{M_0^{(\mathrm{II})}}{M_0^{(\mathrm{I})}}(1 + 3\cos^2\vartheta')\boldsymbol{e}'_r \right\}, \\
\tilde{\boldsymbol{f}}_I^{\mathrm{EL}} &= \left\{ \frac{1}{6}\frac{M_0^{(\mathrm{II})}}{M_0^{(\mathrm{I})}}(1 + 3\cos^2\vartheta')\boldsymbol{e}'_r - \boldsymbol{n}\left(\frac{1}{2}\frac{M_0^{(\mathrm{II})}}{M_0^{(\mathrm{I})}}\sin^2\vartheta' + \frac{1}{6}\frac{M_0^{(\mathrm{II})}}{M_0^{(\mathrm{I})}}[4 + 3\cos^2\vartheta'] \right) \right\},
\end{aligned}
$$

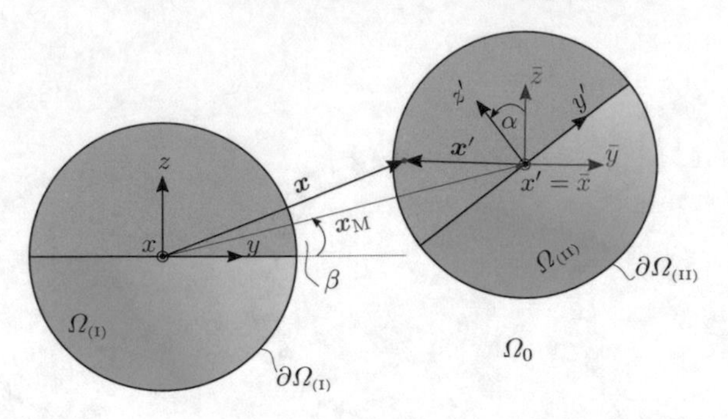

Fig. 11.7 Interacting spherical magnets

where $\hat{f} = \mu_0 M_0^{(I)} M_0^{(II)}$, $\boldsymbol{B}_{(I)} = \mu_0 M_0^{(I)} \tilde{\boldsymbol{B}}_{(I)}$ and $\boldsymbol{\mathfrak{H}}_{(I)} = M_0^{(I)} \tilde{\boldsymbol{\mathfrak{H}}}_{(I)}$. The expressions in curly brackets relate to the second magnet. The volumetric force densities normalized with $\hat{f} R_{(II)}^{-1}$ read:

$$\begin{aligned}
\tilde{\boldsymbol{f}}^{\mathrm{L}} &= \tilde{\boldsymbol{f}}^{\mathrm{A}_1} = \tilde{\boldsymbol{f}}^{\mathrm{M}_1} = \tilde{\boldsymbol{f}}^{\mathrm{M}_2} = \boldsymbol{0}\,, \\
\tilde{\boldsymbol{f}}^{\mathrm{A}_2} &= -\tilde{\nabla} \cdot (\boldsymbol{e}_z' \otimes \tilde{\boldsymbol{B}}_{(I)})\,, \quad \tilde{\boldsymbol{f}}^{\mathrm{EL}} = \boldsymbol{e}_z' \cdot (\tilde{\nabla} \otimes \tilde{\boldsymbol{\mathfrak{H}}}_{(I)})\,.
\end{aligned} \tag{11.20}$$

In order to obtain the total force on the second magnet, the surface force densities are integrated across the surface of the second magnet and the volumetric forces across its volume. Then, the expressions in curly brackets do not contribute, as it should be. Moreover, as it can be shown with some effort (Rickert and Müller, 2019), the resulting forces are all equal independently of the model:

$$\boldsymbol{F}^{\mathrm{L}} = \boldsymbol{F}^{\mathrm{A}_1} = \boldsymbol{F}^{\mathrm{A}_2} = \boldsymbol{F}^{\mathrm{M}_1} = \boldsymbol{F}^{\mathrm{M}_2} = \boldsymbol{F}^{\mathrm{EL}}\,. \tag{11.21}$$

From the experimental point of view this is bad news because a measurement of the force would not allow us to identify the most realistic force density model. However, the situation is different when we consider the torque that magnet (I) imposes on magnet (II). The torque does depend on the model that is used. We find:

$$\begin{aligned}
\boldsymbol{M}^{\mathrm{L}} &= \boldsymbol{M}^{\mathrm{A}_1} = \boldsymbol{M}^{\mathrm{M}_1} = \hat{f} R_{(II)}^3 \int\limits_{\partial\Omega_{(II)}} \sin\vartheta' \boldsymbol{e}_\varphi' (\boldsymbol{e}_r' \cdot \tilde{\boldsymbol{B}}_{(I)})\, \mathrm{d}\tilde{A}\,, \\
\boldsymbol{M}^{\mathrm{A}_2} &= -\hat{f} R_{(II)}^3 \int\limits_{\Omega_{(II)}} \boldsymbol{e}_r' \times \left[\boldsymbol{e}_z' \cdot (\tilde{\nabla} \otimes \tilde{\boldsymbol{B}}_{(I)})\right] \mathrm{d}\tilde{V} + \\
&\quad + \hat{f} R_{(II)}^3 \int\limits_{\partial\Omega_{(II)}} \boldsymbol{e}_r' \times \left[(\sin\vartheta' \boldsymbol{e}_\varphi' \times \tilde{\boldsymbol{\mathfrak{H}}}_{(I)} + \cos\vartheta' \tilde{\boldsymbol{B}}_{(I)})\right] \mathrm{d}\tilde{A}\,, \\
\boldsymbol{M}^{\mathrm{M}_2} &= \hat{f} R_{(II)}^3 \int\limits_{\partial\Omega_{(II)}} \boldsymbol{e}_r' \times \left[\boldsymbol{e}_r' (\tilde{\boldsymbol{B}}_{(I)} \cdot \boldsymbol{e}_z')\right] \mathrm{d}\tilde{A} = \boldsymbol{0}\,, \\
\boldsymbol{M}^{\mathrm{EL}} &= \hat{f} R_{(II)}^3 \int\limits_{\Omega_{(II)}} \boldsymbol{e}_r' \times \left[\boldsymbol{e}_z' \cdot (\tilde{\nabla} \otimes \tilde{\boldsymbol{B}}_{(I)})\right] \mathrm{d}\tilde{V}.
\end{aligned} \tag{11.22}$$

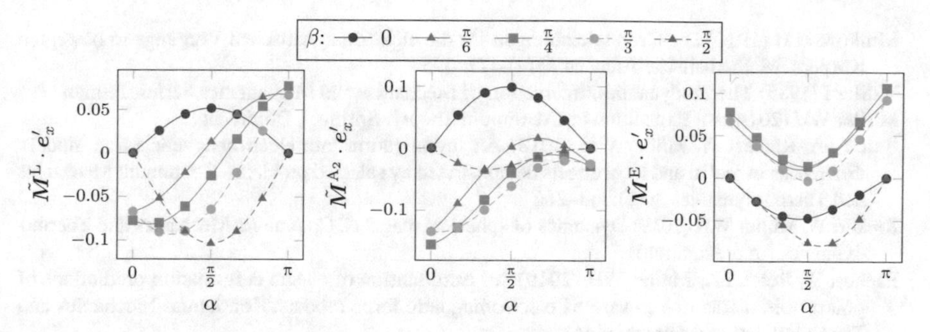

Fig. 11.8 Torques on the magnet (II) due to the magnetic field of the first one for different models and different configurations of the two magnets

Most interestingly, the second version of the Minkowski model yields no torque in any configuration of the two magnets. Therefore, we may conclude that this model is unrealistic. The non-vanishing torques are depicted in Fig. 11.8. From the figure it is clear, that the total torque on the second magnet is different for the distinct models and hence, only by measurement the correct force model for this situation can be found.

11.7 Conclusions and Outlook

The main objective of this paper was to draw attention to the fact that there exist different electromagnetic force models for ponderable matter. Which one is applicable depends on the concrete material that is subjected to electromagnetic fields. For a decision experiments must be performed and compared with theoretical predictions for the total force, the moment, and the deformation of the body in question. Four examples were presented to illustrate this complex situation. It is fair to say that in this context very few experiments have been performed and that the resulting deformations are usually very small, which makes a decision difficult. In conclusion one may say that until today the description of processes for bodies with a coupling between their thermo-mechanical and electromagnetic fields is still far from a complete rational understanding.

References

Abraham M (1909) Zur Elektrodynamik bewegter Körper. Rendiconti del Circolo Matematico di Palermo (1884-1940) 28(1):1–28

Einstein A, Laub J (1908) Über die im elektromagnetischen Felde auf ruhende Körper ausgeübten ponderomotorischen Kräfte. Annalen der Physik 331(8):541–550

Minkowski H (1910) Die Grundgleichungen für die elektromagnetischen Vorgänge in bewegten Körpern. Mathematische Annalen 68(4):472–525

Müller I (1985) Thermodynamics. Interaction of Mechanics and Mathematics Series, Pitman

Müller WH (2014) An Expedition to Continuum Theory. Springer, Dordrecht

Reich FA, Rickert W, Müller WH (2018) An investigation into electromagnetic force models: differences in global and local effects demonstrated by selected problems. Continuum Mechanics and Thermodynamics 30(2):233–266

Rickert W, Müller WH (2019) Dynamics of spherical magnets. Continuum Mechanics and Thermodynamics (in preparation)

Rickert W, Reich FA, Müller WH (2019) An examination of elastic deformation predictions of polarizable media due to various electromagnetic force models. Continuum Mechanics and Thermodynamics (in preparation)

Torza S, Cox RG, Mason SG (1971) Electrohydrodynamic deformation and burst of liquid drops. Philosophical Transactions of the Royal Society A: Mathematical, Physical and Engineering Sciences 269(1198):295–319

Chapter 12
Extreme Yield Figures for Universal Strength Criteria

Philipp L. Rosendahl, Vladimir A. Kolupaev, and Holm Altenbach

Abstract We propose a universal, generally applicable yield criterion that describes a single convex surface in principal stress space encompassing extreme yield figures as convexity limits. The novel criterion is derived phenomenologically exploiting geometrical properties of yield surfaces in principal stress space. It is systematically compared with known yield criteria using different forms of visualization.

Using a I_1 - substitution the criterion is applicable to materials with pressure-sensitive behavior and contains well-known strength criteria. Introducing appropriate parameter restrictions, it can be applied for the modeling of ductile and brittle material behavior. The implementation of the present criterion eliminates the necessity of choosing a specific yield criterion for a particular material.

The proposed criterion allows for excellent approximation of experimental data. It is applied to measured data of concrete and provides better accuracy than existing criteria from literature.

Keywords: Equivalent stress · Multi-axial loading · Isogonal hexagon · Isotoxal hexagon · Deviatoric plane · Concrete

Philipp L. Rosendahl
Technische Universität Darmstadt, Department of Mechanical Engineering, Darmstadt, Germany,
e-mail: rosendahl@fsm.tu-darmstadt.de

Vladimir A. Kolupaev
Fraunhofer Institute for Structural Durability and System Reliability (LBF), Schloßgartenstr. 6, D-64289 Darmstadt, Germany,
e-mail: Vladimir.Kolupaev@lbf.fraunhofer.de

Holm Altenbach
Otto-von-Guericke-Universität Magdeburg, Institut für Mechanik (IFME), Universitätsplatz 2, D-39106 Magdeburg, Germany,
e-mail: holm.altenbach@ovgu.de

H. Altenbach and A. Öchsner (eds.), *State of the Art and Future Trends in Material Modeling*, Advanced Structured Materials 100,
https://doi.org/10.1007/978-3-030-30355-6_12

12.1 Introduction

Phenomenological criteria provide a simple way to describe the onset of yielding, damage or brittle failure of a certain material without consideration of the microstructure. The critical state of the material is represented only by the stresses at which the limit of a material is reached. This set of stress states is denoted as limit surface. Consequently, a corresponding criterion is the mathematical description of all points on this limit surface.

In order to formulate such phenomenological criteria, the equivalent stress concept is typically used (Timoshenko, 1953). The concept suggests, that for each multi-axial stress state represented by a stress tensor $\boldsymbol{\sigma}$ a scalar value σ_{eq}, called equivalent stress, can be computed. This value can be compared to a uniaxial limit stress value σ_{T} where the index T denotes tension. Uniaxial tensile properties of engineering materials can be readily measured in experiments. Because of its simplicity and clarity, this method has become widely accepted and found practical use. However, selecting a specific criterion for a particular material requires certain knowledge of the material behavior.

Classical yield criteria for isotropic materials such as the VON MISES, TRESCA, and SCHMIDT-ISHLINSKI hypotheses are applicable to ductile materials and the RANKINE criterion (normal stress hypothesis) to brittle materials. They are mainly used for didactic purposes. The denomination "hypothesis" with regard to these criteria expresses the historical character of the concept.

Classical criteria are often too primitive to accurately represent experimental data. Hence, generalizations of these criteria were proposed. The MARIOTTE-ST. VENANT (strain) criterion (Filonenko-Borodich, 1960; Kolupaev, 2018), the BURZYŃSKI-YAGN criterion (Burzyński, 1928; Yagn, 1931), the MOHR-COULOMB criterion (Mohr, 1900a,b), and the PISARENKO-LEBEDEV criterion (Lebedev, 1965; Pisarenko and Lebedev, 1976) comprise one or two of the classical criteria. Owing to the lack of experimental data they are still seen as the standard in recent applications. However, applying new materials in critical components and design optimization procedures requires more comprehensive criteria. Even though several sophisticated generalized yield and strength criteria were formulated (Altenbach et al, 1995; Kolupaev, 2018; Pisarenko and Lebedev, 1976; Yu, 2018; Życzkowski, 1981), choosing an appropriate criterion for a particular material remains challenging because of generally incomplete data sets and scattering of the available measured data. Trying to fit different existing criteria is laborious and the optimal evaluation cannot be guaranteed. In order to eliminate the need to choose a specific criterion, a universally applicable criterion is necessary.

In the present work, we propose a universally applicable yield criterion that describes a single convex surface. The criterion exploits geometrical properties of yield surfaces in principal stress space. It contains extreme yield figures as the convexity restriction. Convex criteria beyond these limitations are unavailable. The criterion includes all well-known yield criteria like VON MISES, TRESCA, SCHMIDT-ISHLINSKY, etc. Using a I_1-substitution as a function of the trace of the stress tensor the criterion is applicable to pressure-sensitive material behavior. It incorporates various conditions to obtain special theories. The criterion is applied to experimental data of concrete and provides better accuracy than existing criteria from literature.

The present work is organized as follows. Section 12.2 presents requirements for the formulation of yield and strength criteria. Section 12.3 recalls methods to formulate the limit surfaces. Section 12.4 derives geometrical properties which can be used to compare different criteria, approximations, and measured data. In Sect. 12.5 the limits of convexity are discussed. Section 12.6 reviews existing criteria and proposes a universal yield criterion. In Sect. 12.7 experimental data of concrete are used to assess the performance of existing criteria and the novel proposition.

12.2 Requirements for Yield Criteria

Yield surfaces for pressure-insensitive isotropic materials are described by a prismatic or cylindrical body centered around the hydrostatic axis

$$\sigma_{\mathrm{I}} = \sigma_{\mathrm{II}} = \sigma_{\mathrm{III}}, \tag{12.1}$$

in principal stress space. Cross sections orthogonal to the hydrostatic axis are called deviatoric planes or π-planes sometimes restricted to the π_0-plane through the coordinate origin. Owing to isotropy, cross sections in the π-plane must be of trigonal symmetry (this includes rotational and hexagonal symmetry), see Fig. 12.1. Further, we require yield surfaces to be convex. Thus, basic cross sections may be described by a circle or regular polygons of trigonal or hexagonal symmetries: e.g., equilateral triangles, hexagons, enneagons (nine-sided polygons), dodecagons (twelve-sided polygons), among others. Each criterion described by a regular polygon in the π-plane has a counterpart which is obtained by its rotation by π/n in the π-plane where n is the number of vertices.

Criteria discussed in the present work are phenomenological. No sufficient conditions for their formulation can be given. However, the quality of a certain yield criterion may be assessed considering the following plausibility assumptions (Kolupaev, 2018):

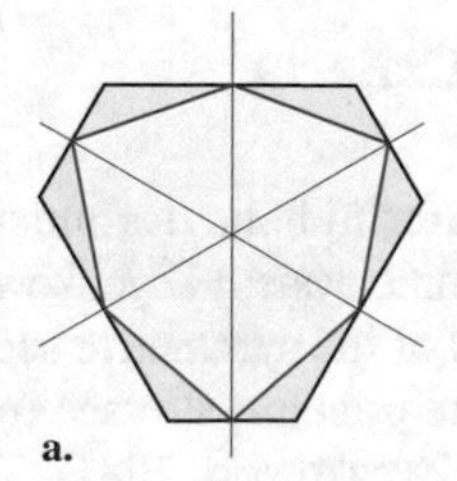

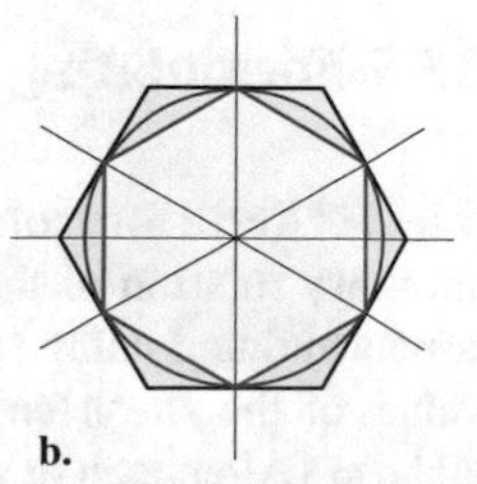

Fig. 12.1 Yield criteria in the π-plane normalized with respect to the equivalent stress $\sigma_{\mathrm{eq}} = \sigma_{\mathrm{T}}$: **a.** Isogonal (black) and isotoxal (blue) hexagons of trigonal symmetry, **b.** Regular hexagons of the SCHMIDT-ISHLINSKY (black) and TRESCA (blue) hypotheses of hexagonal symmetry and the circle of the VON MISES hypothesis (red) of rotational symmetry

- explicit solvability of the criterion with respect to the equivalent stress σ_{eq},
- only independent and as few parameters as possible,
- a wide range of possible convex shapes in the π-plane,
- a wide range of possible convex shapes in the meridian cross section (cross section containing the hydrostatic axis),
- once continuous differentiability at the hydrostatic nodes (apices of the limit surface),
- once continuous differentiability of the criterion in the π-plane except at the border of the convexity limits,
- a single surface in principal stress space without any additional outer contours and plane intersections, and
- unique assignment of the limit surface to parameters of the criterion.

These assumptions help to select user-friendly criteria with a wide range of applications. Yet, to our knowledge, criteria satisfying all of the above assumptions are not known. Up to now, the PODGÓRSKI yield criterion (Podgórski, 1984; Podgórski, 1985), the modified ALTENBACH-ZOLOCHEVSKY yield criterion (Kolupaev, 2017, 2018), and the modified YU strength criterion (Kolupaev, 2017, 2018) meet the plausibility assumptions in the best way possible and are recommended for application. The first two criteria with appropriate I_1-substitution and the YU strength criterion with straight meridian line include almost all known criteria and approximate the remaining criteria in the best way known. Still, different criteria must be chosen to describe specific shapes of the limit surface.

Possible shapes of yield criteria in the π-plane are limited by the requirement of convexity. The upper and lower convexity limits may be referred to as extreme yield figures (Bigoni and Piccolroaz, 2004; Marti, 1980; Sayir and Ziegler, 1969). The aim of this work is to generalize the description of extreme yield figures using a universal yield criterion which should satisfy as many plausibility assumptions as possible.

Extreme yield figures may take the shape of isogonal and isotoxal polygons of trigonal and hexagonal symmetry. Isogonal polygons are equiangular. An isotoxal polygon is equilateral, that is, all sides are of the same length (Koca and Koca, 2011; Tóth, 1964). In general, isogonal and isotoxal hexagons are of trigonal symmetry (Fig. 12.1a). The regular hexagons of the TRESCA and SCHMIDT-ISHLINSKY criteria have an additional symmetry axis and are of hexagonal symmetry (Fig. 12.1b). Isogonal and isotoxal dodecagons (twelve-sided polygons) are of hexagonal symmetry, too.

12.3 Formulating Yield Criteria

Yield criteria for isotropic material behavior must be invariant with respect to arbitrary rotation of the coordinate system (Życzkowski, 1981). Therefore, criteria are formulated using invariants of the symmetric second-rank stress tensor. Eigenvalues of the stress tensor are the principal stresses (or principal invariants) σ_{I}, σ_{II}, and σ_{III} (Altenbach et al, 1995; Życzkowski, 1981). The following order is assumed (Burzyński, 1928)

$$\sigma_{\rm I} \geq \sigma_{\rm II} \geq \sigma_{\rm III}. \tag{12.2}$$

Functions of invariants are also invariants (Appendix A.1). Thus we may also use the trace I_1 of the stress tensor and the invariants I_2', I_3' of the stress deviator (12.62) – (12.64) or cylindrical invariants (NOVOZHILOV's invariants) ξ, ρ, and θ (12.65) – (12.67) in order to formulate criteria. Yield and strength criteria describe limit surfaces in stress space which can be formulated for instance according to:

$$\Phi\left(\sigma_{\rm I}, \sigma_{\rm II}, \sigma_{\rm III}, \sigma_{\rm eq}\right) = 0 \;\text{ or }\; \Phi\left(I_1, I_2', I_3', \sigma_{\rm eq}\right) = 0 \;\text{ or }\; \Phi\left(\xi, \rho, \theta, \sigma_{\rm eq}\right) = 0, \tag{12.3}$$

where ξ (12.65) is the scaled invariant I_1 and describes the coordinate of the loading on the hydrostatic axis, ρ (12.66) is the scaled root of the second invariant I_2' and describes the radius in the π-plane and θ is the corresponding stress angle in the π-plane. ρ may be replaced by the stress triaxiality (inclination) factor ψ (12.69) which yields a description in spherical invariants. The stress angle θ is sometimes replaced by φ (12.68). All these formulations are equivalent. One or the other may be preferred depending on the modeling concept or desired application.

In the case of pressure-insensitive material behavior, the first invariant I_1 does not influence failure. For this behaviour, the formulations (12.3) can be reduced to

$$\Phi\left(I_2', I_3', \sigma_{\rm eq}\right) = 0 \qquad \text{or} \qquad \Phi\left(\rho, \theta, \sigma_{\rm eq}\right) = 0. \tag{12.4}$$

Formulations in terms of I_2' and I_3' in polynomial form cannot be recommended for application because of additional outer contours around the physically meaningful surface or plane intersections (Kolupaev, 2018).

In order to satisfy the first plausibility assumption, the equivalent stress $\sigma_{\rm eq}$ must be split from the yield function:

$$\sigma_{\rm eq} = \Phi(\rho, \theta) \qquad \text{or} \qquad \sigma_{\rm eq} = \Phi(\rho, \varphi). \tag{12.5}$$

Such formulations are advantageous for iterative computations, e.g., in FEM codes. We may further postulate a multiplicative split of yield criteria into a function of radius and a function of stress angle according to

$$\sigma_{\rm eq} = \Psi(\rho)\,\Omega(\theta) \qquad \text{or} \qquad \sigma_{\rm eq} = \Psi(\rho)\,\Omega(\varphi). \tag{12.6}$$

This split is motivated by its practical use. In order to highlight the deviations of the geometry of the criterion in the π-plane from the circle of the VON MISES hypothesis (Fig. 12.1b)

$$\sigma_{\rm eq} = \sqrt{3 I_2'} \qquad \text{with} \qquad \Omega(\theta) = 1 \;\text{ or }\; \Omega(\varphi) = 1, \tag{12.7}$$

the function of $\Psi(\rho)$ is often replaced by $\sqrt{3 I_2'}$ (Kolupaev, 2017; Kolupaev et al, 2018) which yields

$$\sigma_{\rm eq} = \sqrt{3 I_2'}\,\Omega(\theta) \qquad \text{or} \qquad \sigma_{\rm eq} = \sqrt{3 I_2'}\,\Omega(\varphi). \tag{12.8}$$

Normalizing criteria with respect to the uniaxial tensile limit loading, e.g., the tensile strength

$$\sigma_{\mathrm{eq}} = R^{\mathrm{T}}, \tag{12.9}$$

where the index T denotes tension, yields

$$\sigma_{\mathrm{eq}} = \sqrt{3I_2'}\,\frac{\Omega(\theta)}{\Omega(0)} \quad \text{or} \quad \sigma_{\mathrm{eq}} = \sqrt{3I_2'}\,\frac{\Omega(\varphi)}{\Omega(-\pi/6)}. \tag{12.10}$$

Reintroducing the first invariant of the stress tensor I_1 in (12.4) using the linear substitution (Kolupaev, 2018), see also Botkin (1940a,b); Drucker and Prager (1952); Mirolyubov (1953); Sandel (1919); Sayir (1970)

$$\sigma_{\mathrm{eq}} \rightarrow \frac{\sigma_{\mathrm{eq}} - \gamma_1 I_1}{1-\gamma_1} \quad \text{with} \quad \gamma_1 \in [0, 1[, \tag{12.11}$$

yields a pressure-sensitive generalization of, e.g., the criteria (12.10):

$$\frac{\sigma_{\mathrm{eq}} - \gamma_1 I_1}{1-\gamma_1} = \sqrt{3I_2'}\,\frac{\Omega(\theta)}{\Omega(0)} \quad \text{or} \quad \frac{\sigma_{\mathrm{eq}} - \gamma_1 I_1}{1-\gamma_1} = \sqrt{3I_2'}\,\frac{\Omega(\varphi)}{\Omega(-\pi/6)}, \tag{12.12}$$

that does not violate the first plausibility assumption.

The visualization of the criteria (12.12) in the Burzyński-plane $(I_1, \sqrt{3I_2'})$ is then obvious and a direct comparison with the von Mises hypothesis (12.7) is possible. The reciprocal value of the parameter γ_1 describes the intersection of the limit surface with the I_1-axis. This parameter does not interact with other parameters of the criterion and thus does not influence the geometry of cross sections in the π-plane.

It is to note that although the linear substitution, Eq. (12.11) provides good results in several cases, it produces an additional conical surface in the region

$$\frac{I_1}{R^{\mathrm{T}}} \geq \frac{1}{\gamma_1},$$

without physical meaning and the apex at hydrostatic tensile loading is C^0-continuously differentiable, cf. the fifth plausibility assumption. Further I_1-substitutions for a range of possible convex shapes in the meridian cross section (parabola, hyperbola, ellipsis) are given in Kolupaev (2018).

12.4 Comparing Different Yield Criteria

For analyses of measured data and comparison of approximations for various materials, measured data will be normalized by the appropriate limit tensile loading, e.g., by R^{T} (12.9)

$$\left(\frac{\sigma_{\mathrm{I}}}{R^{\mathrm{T}}}, \frac{\sigma_{\mathrm{II}}}{R^{\mathrm{T}}}, \frac{\sigma_{\mathrm{III}}}{R^{\mathrm{T}}}\right), \tag{12.13}$$

so that material properties get geometric meaning and different surfaces Φ can be compared in the same diagrams.

Let us distinguish pressure-insensitive yield criteria, which are comprehensively described in the π-plane and pressure-sensitive strength criteria. In the π-plane of the yield criterion (Fig. 12.2) certain types of loading with the stress angle θ share the same radius ρ (12.66) and collapse onto one point. Introducing the corresponding nomenclature these are:

- $\theta = 0$: T (uniaxial tension), CC (equibiaxial compression),
- $\theta = \pi/6$: K (torsion), Tt (biaxial tension with $I_3' = 0$), and Cc (biaxial compression with $I_3' = 0$),
- $\theta = \pi/3$: TT (equibiaxial biaxial tension), C (uniaxial compression).

Radii ρ at these stress angles θ are characteristic properties of the limit surface. Because of their I_1-dependence pressure-sensitive strength criteria have additional characteristic values which will later be related to the above. In order to visualize pressure-sensitive criteria certain cross sections I_1 = const. in the π-plane and the BURZYŃSKI-plane are needed.

12.4.1 Geometry of Limit Surfaces in the π-plane

Cross sections of pressure-insensitive criteria Eq. (12.4) may be described in polar coordinates as functions $\rho(\theta)$ where ρ and θ correspond to radius (12.66) and stress angle (12.67) in the π-plane, respectively. Let us introduce geometrical properties as relations of radii at the angles $\theta = \{\pi/12, \pi/6, \pi/4, \pi/3\}$ to the radius $\rho(0)$ as

$$r_{15} = \frac{\rho(\pi/12)}{\rho(0)}, \quad r_{30} = \frac{\rho(\pi/6)}{\rho(0)}, \quad r_{45} = \frac{\rho(\pi/4)}{\rho(0)}, \quad \text{and} \quad r_{60} = \frac{\rho(\pi/3)}{\rho(0)}. \tag{12.14}$$

The indices 0, 15, 30, 45, and 60 correspond to the stress angle of the loading point in degrees. With these values (12.14) as coordinate axes, different criteria can be easily compared in appropriate diagrams. The chosen angles θ are fractions of the angle $\pi/3$ between the symmetry axes in the π-plane (Figs. 12.3 and 12.4). According to

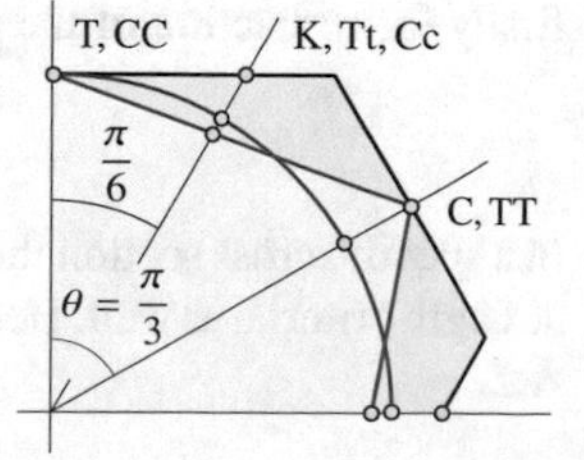

Fig. 12.2 Isogonal (black) and isotoxal (blue) hexagons in the π-plane normalized with respect to the equivalent stress $\sigma_{eq} = R^T$ (Fig. 12.1a): Enlarged detail with the VON MISES hypothesis (red) and the stress states (T, CC on the 0-meridian, K, Tt, Cc on the $\pi/6$-meridian, and C, TT on the $\pi/3$-meridian) for comparison

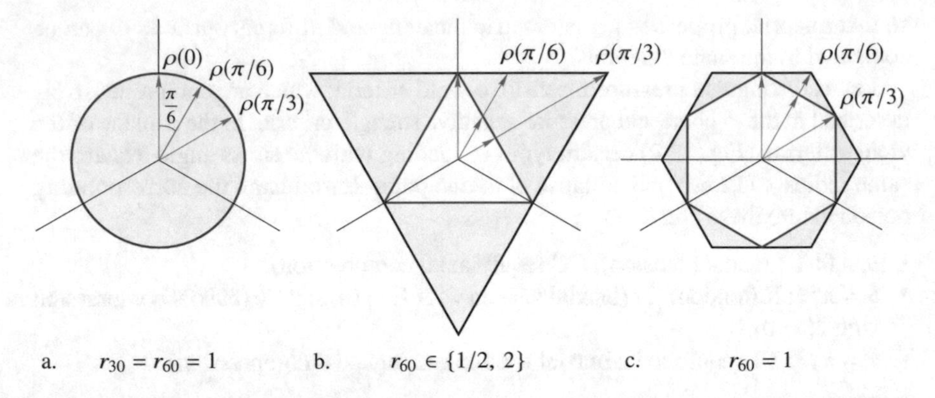

Fig. 12.3 Basic surfaces with the same radius $\rho(0)$ in the π-plane: **a.** Rotationally symmetric von Mises criterion (12.7), **b.** Equilateral triangles R_1 and R_2, and **c.** Equilateral hexagons H_1 and H_2. The values r_{30} and r_{60} are given for comparison

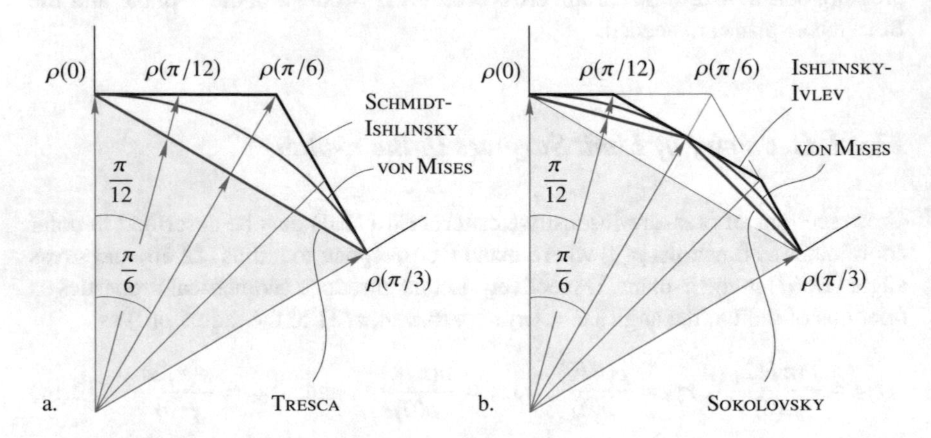

Fig. 12.4 Basic surfaces of hexagonal symmetry in the π-plane: **a.** Equilateral hexagons H_1 and H_2 (Tresca and Schmidt-Ishlinsky criteria) and **b.** Equilateral dodecagons D_1 and D_2 (Sokolovsky and Ishlinsky-Ivlev criteria) with the von Mises hypothesis (12.7). Because of hexagonal symmetry a cut-out of the angle $\theta \in [0, \pi/3]$ is representative (Kolupaev, 2018)

Betten-Troost (Betten, 1979, 2001; Bolchoun et al, 2011; Troost and Betten, 1974), convexity of the criterion is most critical at these angels and needs to be checked firstly for restriction of the parameters. With

$$I_1 = \text{const.} \tag{12.15}$$

in a specific cross section the relations (12.14) can be introduced for pressure-sensitive strength criteria, as well. Details on the calculation of the ratios r are given in Appendix A.2.

For the rotationally symmetric VON MISES hypothesis (12.7) all radii are equal (Fig. 12.3a) and we obtain

$$r_{15} = r_{30} = r_{45} = r_{60} = 1. \tag{12.16}$$

For criteria of hexagonal symmetry (e.g. TRESCA, SCHMIDT-ISHLINSKY, SOKOLOVSKY, and ISHLINSKY-IVLEV (Figs. 12.3c and 12.4)) the radii at the angels $\theta = 0$ and $\pi/3$ are equal $\rho(0) = \rho(\pi/3)$ and because of hexagonal symmetry we obtain $\rho(\pi/12) = \rho(\pi/4)$ which yields

$$r_{60} = 1 \qquad \text{and} \qquad r_{15} = r_{45}. \tag{12.17}$$

Ratios r at additional fractions of the $\pi/3$ can be given but are not required for the characterization of surfaces.

Using, the relations of (12.14) as coordinate axes, Figs. 12.5 and 12.6 show convexity restrictions for criteria of trigonal symmetry in the $r_{60} - r_{30}$ diagram and for criteria of hexagonal symmetry ($r_{60} = 1$) in the $r_{15} - r_{30}$ diagram, respectively. These diagrams allow a comparison of all criteria for isotropic material behavior. In Figs. 12.5 and 12.6 basic yield figures are labeled according to their shape: equilateral triangles are denoted R, equilateral hexagons H, equilateral dodecagons D, equilateral icositetragons C and the circular VON MISES criterion M. Indices 1 and 2 refer to an upward pointing tip or upward facing flat base, respectively. The yield figures M, D_1, D_2, C_1, and C_2 coincide in the $r_{60} - r_{30}$ diagram (Fig. 12.5) while the yield figures M, C_1, and C_2 coincide in the $r_{15} - r_{30}$ diagram (Fig. 12.6).

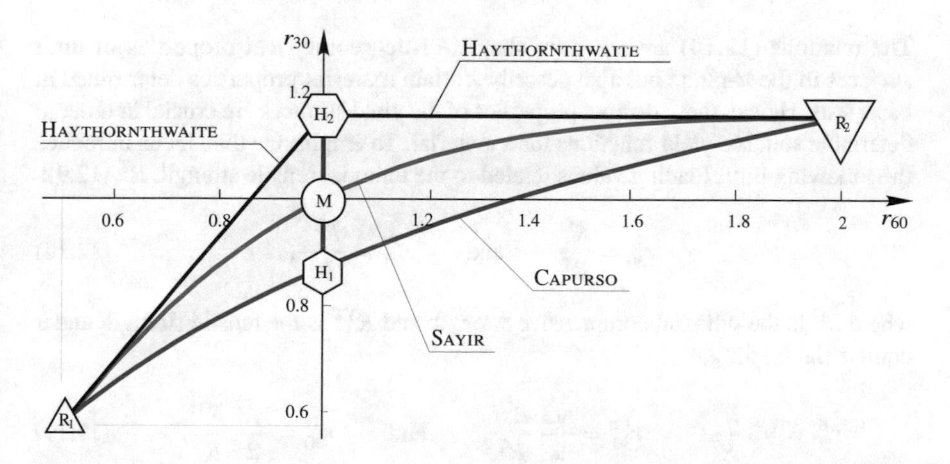

Fig. 12.5 Diagram $r_{60} - r_{30}$ for convex criteria of trigonal symmetry compared to the VON MISES hypothesis with $r_{30} = r_{60} = 1$ (Kolupaev, 2018). Basic cross sections are visualized (Table 12.1). Denotation according Kolupaev (2018): CAPURSO criterion (Capurso, 1967; Sayir, 1970), SAYIR criterion (Sayir, 1970; Sayir and Ziegler, 1969), and HAYTHORNTHWAITE criterion (Haythornthwaite, 1961, 1962)

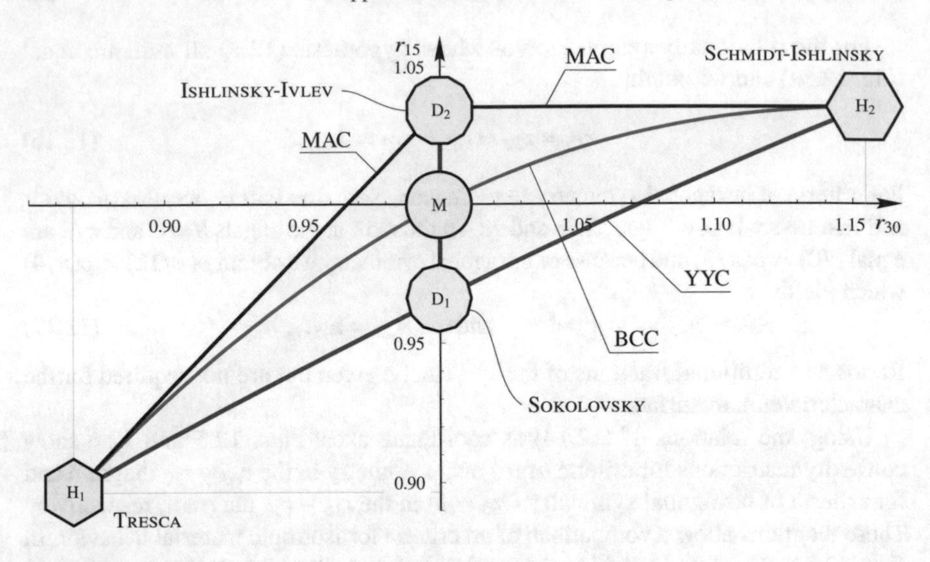

Fig. 12.6 Diagram $r_{30} - r_{15}$ for convex criteria of hexagonal symmetry ($r_{60} = 1$) compared to the VON MISES hypothesis with $r_{15} = r_{30} = 1$. Basic cross sections are visualized (Table 12.1). Abbreviations: YYC – YU yield criterion, BCC – bicubic criterion, and MAC – multiplicative ansatz criterion (Kolupaev, 2018)

12.4.2 Material Properties and Basic Experiments

The relations (12.14) are not only characteristic geometrical properties of limit surfaces in the π-plane but also describe certain material properties determined in basic tests. Hence, these distinct properties of the yield surfaces are crucial in order to determine suitable yield functions for a material. To emphasize this, let us introduce the following limit loading values related to the uniaxial tensile strength R^{T} (12.9):

$$r_{60}^{\mathrm{C}} = \frac{R^{\mathrm{C}}}{R^{\mathrm{T}}} \qquad \text{and} \qquad r_{60}^{\mathrm{TT}} = \frac{R^{\mathrm{TT}}}{R^{\mathrm{T}}}, \tag{12.18}$$

where R^{C} is the uniaxial compressive strength and R^{TT} is the tensile strength under equibiaxial loading,

$$r_{30}^{\mathrm{K}} = \sqrt{3}\,\frac{R^{\mathrm{K}}}{R^{\mathrm{T}}}, \qquad r_{30}^{\mathrm{Cc}} = \frac{\sqrt{3}}{2}\,\frac{R^{\mathrm{Cc}}}{R^{\mathrm{T}}}, \qquad \text{and} \qquad r_{30}^{\mathrm{Tt}} = \frac{\sqrt{3}}{2}\,\frac{R^{\mathrm{Tt}}}{R^{\mathrm{T}}}, \tag{12.19}$$

where R^{K} is the torsional strength and R^{Tt} and R^{Cc} are ultimate loadings of thin-walled tube specimens with closed ends under inner (Tt) and outer pressure (Cc), respectively, and

$$r_{0}^{\mathrm{CC}} = \frac{R^{\mathrm{CC}}}{R^{\mathrm{T}}}, \tag{12.20}$$

where R^{CC} is the compressive strength under equibiaxial compression. While a hydrostatic tensile test with $\sigma_{\mathrm{I}} = \sigma_{\mathrm{II}} = \sigma_{\mathrm{III}} > 0$ can hardly be realized (Kolupaev et al, 2014), the corresponding property for the hydrostatic tensile strength is important for the comparison of extrapolations. We may introduce

$$r^{\mathrm{TTT}} = \frac{R^{\mathrm{TTT}}}{R^{\mathrm{T}}} = \frac{1}{3\,\gamma_1} \qquad \text{and} \qquad r^{\mathrm{CCC}} = \frac{R^{\mathrm{CCC}}}{R^{\mathrm{T}}}. \tag{12.21}$$

where R^{TTT} and R^{CCC} are hydrostatic strengths under tension and compression, respectively. Except for porous and granular media, hydrostatic compressive failure does typically not occur for relevant loadings and $r^{\mathrm{CCC}} \to \infty$ can be assumed.

Now, r_{15}, r_{30}, and r_{60} describe the π-plane geometry at a chosen cross section (12.15) and r^{C}_{60}, r^{TT}_{60}, r^{K}_{30}, r^{Cc}_{30}, r^{Tt}_{30}, r^{CC}_{0}, r^{CCC}, and r^{TTT} describe corresponding material properties. When $\gamma_1 = 0$, pressure-sensitive strength criteria generally converge to pressure-insensitive yield criteria and the values on the same meridians (characterized by the angle θ) coincide (Fig. 12.2):

$$r_{60} = r^{\mathrm{C}}_{60} = r^{\mathrm{TT}}_{60}, \qquad r_{30} = r^{\mathrm{K}}_{30} = r^{\mathrm{Cc}}_{30} = r^{\mathrm{Tt}}_{30} \qquad \text{and} \qquad r^{\mathrm{CC}}_{0} = 1. \tag{12.22}$$

Hence, certain tests immediately determine specifics points of the yield surfaces. Convex pressure-insensitive criteria with the properties (12.17) do not distinguish between tensile and compressive behavior. Because of rotational or hexagonal symmetry, we additionally obtain

$$r_{60} = r^{\mathrm{C}}_{60} = r^{\mathrm{TT}}_{60} = r^{\mathrm{CC}}_{0} = 1. \tag{12.23}$$

Classical strength criteria such as the normal stress hypothesis, von Mises, Tresca, and Schmidt-Ishlinsky hypotheses, and also the criteria of Mohr-Coulomb and Pisarenko-Lebedev describe material behavior with the properties $R^{\mathrm{T}} = R^{\mathrm{TT}}$ and $R^{\mathrm{C}} = R^{\mathrm{CC}}$ which yields

$$r^{\mathrm{TT}}_{60} = 1 \qquad \text{and} \qquad r^{\mathrm{C}}_{60} = r^{\mathrm{CC}}_{0}, \tag{12.24}$$

which can be used for the comparison of approximations or for the formulation of fitting restrictions. Details on parameter identification and fitting procedures for pressure-sensitive material behavior are given in Appendix A.3.

12.5 Extreme Yield Figures

Lower and upper bounds of convexity for isotropic criteria in the $r_{60} - r_{30}$ diagram (Fig. 12.5) are obtained with extreme yield figures of isotoxal and isogonal hexagons (Figs. 12.7 and 12.8). The polynomial formulations (12.4) of these hexagons are known as Capurso and Haythornthwaite criteria, respectively (Kolupaev, 2018). However, their polynomial forms feature additional contours or intersections

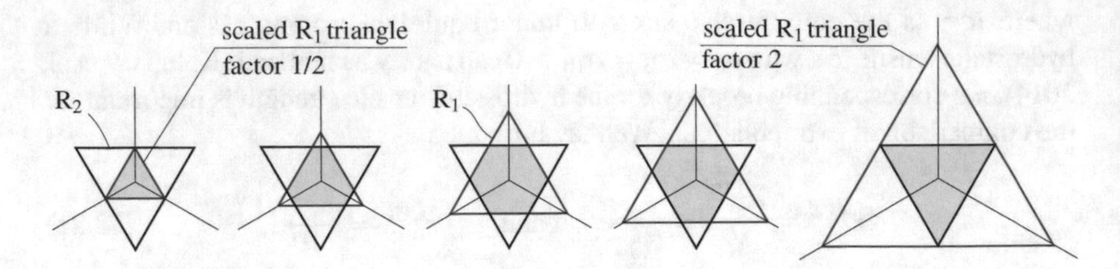

Fig. 12.7 Isogonal (equiangular) hexagons (upper convexity limit) are formed by the intersection of two triangles in the π-plane: the scaled R_1 triangle (blue) and the R_2 triangle (black)

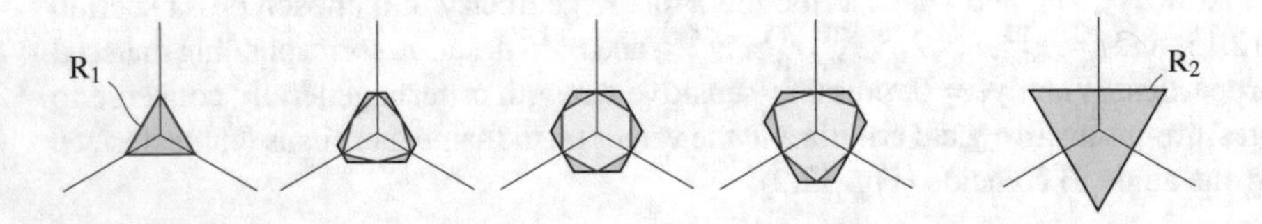

Fig. 12.8 Isotoxal (equilateral, lower convexity limit, blue) and isogonal hexagons (equiangular, upper convexity limit, black) in the π-plane with the R_1 (blue) and R_2 (black) triangle as limit cases

surrounding the physically reasonable shape of the surface Φ which makes the application involved.

Isotoxal hexagons (lower bound, transition R_1–H_1–R_2 in Fig. 12.5) as function of stress angle (12.10) can be formulated using the PODGÓRSKI criterion (Table 12.1) which describes the geometry of the CAPURSO criterion as single surface among others. A criterion for isogonal hexagons (upper bound, transition R_1–H_2–R_2) as function of stress angle without case discrimination is missing. Both hexagons degenerate to the same equilateral triangles R_1 and R_2 in limit cases (Figs. 12.3b and 12.8) with

$$r_{60} \in \left[\frac{1}{2}, 2\right].$$

These hexagons in the π-plane extended with the linear I_1-substitution (12.11) represent pyramides in principal stress space, which are important strength criteria for practical applications (Paul, 1968; Wronski and Pick, 1977; Yu, 2018).

The lower and upper bounds of the convexity restriction for isotropic criteria in the $r_{15} - r_{30}$ diagram (Fig. 12.6) are obtained with extremal yield figures of isotoxal and isogonal dodecagons. Isotoxal dodecagons (lower bound, transition H_1–D_1–H_2) as function of the stress angle can be described with the modified YU (mYu) criterion (Kolupaev, 2018). Only the polynomial formulation for isogonal dodecagons (upper bound, transition H_1–D_2–H_2) is known (Fig. 12.6, MAC). Both dodecagons degenerate to the same equilateral hexagons H_1 and H_2 in limit cases (Figs. 12.3c and 12.4a) with

$$r_{30} \in \left[\frac{\sqrt{3}}{2}, \frac{2}{\sqrt{3}}\right].$$

Although the I_1-substitution (12.11) is possible here, dodecagons are typically used as pressure-insensitive criteria. The differences between the regular dodecagons D_1 and D_2 (Figs. 12.4b) with $r_{30} = r_{60} = 1$ (Table 12.1) are

$$r_{15} \in \left[\frac{1}{2}\sqrt{2+\sqrt{3}}, \sqrt{2}\left(\sqrt{3}-1\right)\right].$$

This deviation below ±3.5% between the MAC and YYC includes all criteria of hexagonal symmetry (Fig. 12.6) compared in Kolupaev (2018). Differences between the regular icositetragons C_1 and C_2 (Table 12.1) are minor. They have the properties (12.16) and coincide with the von Mises hypothesis (12.7) in the $r_{15} - r_{30}$ diagram (Fig. 12.6). The icositetragons C_1 and C_2 will be obtained as a result of the generalization of the criteria of hexagonal symmetry.

Regular enneagons (9-gons), pentadecagons (15-gons), octadecagon (18-gons), and icositetragons (24-gons) are conceivable as yield criteria. However, because of increasing complexity in their formulation and low practical importance, they have found no applications and are only mentioned for the sake of completeness.

12.6 Generalized Strength Criteria

The phenomenological nature of yield and strength criteria has caused an unmanageable number of possible formulations. Having to choose an appropriate criterion can leave users confused. Connections between material behavior and the criterion are often not available. Selecting a criterion for a particular application is usually not based on objective arguments. The best possible approximation of measured data with physical background (convexity of the meridian and convexity in the π-plane, range of the inelastic Poisson's ratios at tension and compression, restriction of the hydrostatic strength R^{TTT} (12.21) among others) cannot be guaranteed.

Plausibility assumptions reduce the number of applicable criteria significantly. Up to now, the Podgórski and the modified Altenbach-Zolochevsky criteria with I_1-substitution (12.11) and the modified Yu strength criterion meet the plausibility assumptions in the best way and are recommended for application. Various materials can be described using these criteria.

Unfortunately, the listed criteria can not describe all extreme yield figures of trigonal and hexagonal symmetry (Figs. 12.5 and 12.6). An attempt to formulate a universal yield criterion will be shown in this section. If a universal criterion is present, further criteria have rather historical significance. With such criterion it can be checked whether an optimal approximation of the measured data with the convex shape in the π-plane is possible. Different approximations are easy to compare.

Table 12.1 Basic criteria in the π-plane: equilateral triangles (R), hexagons (H), dodecagons (D), icositetragon (C), the von Mises hypothesis (M). Further references of the criteria are given in Kolupaev (2018)

π-plane		R₁ Ivlev	R₂ Mariotte	H₁ Tresca	H₂ Schmidt-Ishlinsky	D₁ Sokolovsky	D₂ Ishlinsky-Ivlev	C₁ Rosendahl	C₂ Rosendahl	M von Mises
Geometry	r_{15}	$\sqrt{2}/2$	$\sqrt{2}\left(\sqrt{3}-1\right)$	$\sqrt{3/2}\left(\sqrt{3}-1\right)$	$\sqrt{2}\left(\sqrt{3}-1\right)$	$\frac{1}{2}\sqrt{2+\sqrt{3}}$	$\sqrt{2}\left(\sqrt{3}-1\right)$	1	1	1
	r_{30}	$1/\sqrt{3}$	$2/\sqrt{3}$	$\sqrt{3}/2$	$2/\sqrt{3}$	1	1	1	1	1
	r_{60}	1/2	2	1	1	1	1	1	1	1
mYu	r_{60}^{C}	–	–	1	1	1	–	–	–	–
	χ	–	–	0	1	1/2	–	–	–	–
Podg.	β	{0, 1}	{1, 0}	1/2	–	–	–	–	–	{0, 1}
	γ	{−1, 1}	{−1, 1}	{−1, 1}	–	–	–	–	–	0
mAZ	r_{60}	1/2	2	1	1	1	–	–	–	–
	ξ_{m}	0	0	0	1	1/2	–	–	–	–
CTS	α	0	0	[0, 1]	1	1	–	–	–	[0, 1]
	β	1	0	1/2	{0, 1}	{1/4, 3/4}	–	–	–	[0, 1]
	γ	1	1	1	1	1	–	–	–	0
UDF	α	0	0	0	0	0	0	0	0	[0, 1]
	β	1	0	1	0	1	0	1	0	[0, 1]
	γ	1	1	1	1	1	1	1	1	0
	n	1	1	2	2	4	4	6	6	$\mathbb{N}^+$
Ref.		Cicala (1961) Ivlev (1959)	Benvenuto (1991) Mariotte (1718)	Coulomb (1776) Tresca (1868)	Ishlinsky (1940) Schmidt (1932)	Yu (1961) Billington (1986)	Ivlev (1960) Shesterikov (1960)	– –	– –	Huber (1904) von Mises (1913)

Let us examine the modified YU strength criterion, the modified ALTENBACH-ZOLOCHEVSKY criterion, the PODGÓRSKI criterion, and the universal yield criterion in detail. Characteristic geometrical properties and values for basic tests will be computed. In the next step these criteria will be applied to measured data of concrete.

12.6.1 Modified Yu Strength Criterion

The modified YU strength criterion (mYU) is derived from a linear combination of the YU yield criterion (YYC) (12.98)–(12.99) and the normal stress hypothesis (12.102) as functions of the stress angle θ. It can be solved for the equivalent stress σ_{eq} which reads

$$\sigma_{\mathrm{eq}} = \frac{1}{3\, r_{60}^{\mathrm{C}}} \left[\sqrt{3 I_2'} \left[3\, \frac{\Omega(\theta, \chi)}{\Omega(0, \chi)} + 2\, (r_{60}^{\mathrm{C}} - 1) \cos\theta \right] + I_1\, (r_{60}^{\mathrm{C}} - 1) \right], \qquad (12.25)$$

with the shape function of the YYC

$$\Omega(\theta, \chi) = \sin\left(\chi\, \frac{\pi}{6} + \arcsin\left[\cos\left(\frac{1}{3} \arcsin\left[\cos 3\,\theta \right] \right) \right] \right), \qquad \chi \in [0, 1]. \quad (12.26)$$

The modified YU strength criterion contains:

- the MOHR-COULOMB criterion (Mohr, 1914) with the parameter $\chi = 0$,
- a continuous analogy of the PISARENKO-LEBEDEV criterion (Lebedev, 1965; Pisarenko and Lebedev, 1976) with $\chi = 1/2$,
- an analogon of the twin-shear criterion (TST) of YU (Yu, 2018) with $\chi = 1$,
- the normal stress hypothesis with $r_{60}^{\mathrm{C}} \to \infty$ for any $\chi \in [0, 1]$, and
- the YYC (12.98)–(12.99) with $r_{60}^{\mathrm{C}} = 1$ (Yu, 2018).

It is C^0-continuously differentiable which leads to ambiguities in the calculation of the gradient at edges. Like the original YU strength criterion (YSC) the mYU criterion contains three classical criteria and can be considered a generalized classical criterion with the properties (12.24). The mYU with two parameters r_{60}^{C} and χ is convenient for different applications as a yield and strength criterion. A detailed derivation of the mYU criterion and its geometric properties is given in Appendix A.4.

12.6.2 Podgórski Criterion

Normalized with respect to the uniaxial tensile stress (12.9), the PODGÓRSKI criterion (Podgórski, 1984; Podgórski, 1985; Kolupaev, 2017, 2018) reads

$$\sigma_{\mathrm{eq}} = \sqrt{3\, I_2'}\, \frac{\Omega(\theta, \beta, \eta)}{\Omega(0, \beta, \eta)}, \qquad (12.27)$$

with the shape function

$$\Omega(\theta,\beta,\eta) = \cos\left[\frac{1}{3}\,(\pi\beta - \arccos[\,\eta\cos 3\,\theta\,])\right], \qquad \beta \in [0,\,1], \quad \eta \in [-1,\,1]. \tag{12.28}$$

Replacing the parameter η by

$$\eta = \sin\left(\frac{\gamma\,\pi}{2}\right), \qquad \gamma \in [-1,1], \tag{12.29}$$

yields improved parameter sensitivity and numerical stability for fitting (Fig. 12.9), cf. Kolupaev (2018). The criterion involves several known criteria (Table 12.1: R_1, M, H_1, R_2 and Fig. 12.5, Capurso and Sayir ctiteria). For a comparison with other generalized criteria curves r_{15} = const. are shown in Fig. 12.10.

The Podgórski criterion (12.27) is C^1-continuously differentiable except at the border R_1–H_1–R_2 of the Capurso criterion. Convex π-plane geometries in the region between R_1–M–R_2 and R_1–H_2–R_2 (Fig. 12.5) are not covered with real-valued parameters β and η (Kolupaev, 2018). However, the criterion is convenient for many applications as a yield criterion. With the linear I_1-substitution (12.11) we obtain the conical criterion

$$\frac{\sigma_{\text{eq}} - \gamma_1\,I_1}{1-\gamma_1} = \sqrt{3\,I_2}\,\frac{\Omega(\theta,\beta,\gamma)}{\Omega(0,\beta,\gamma)}, \qquad \gamma_1 \in [0,\,1[. \tag{12.30}$$

which becomes a pyramid at the border R_1–H_1–R_2.

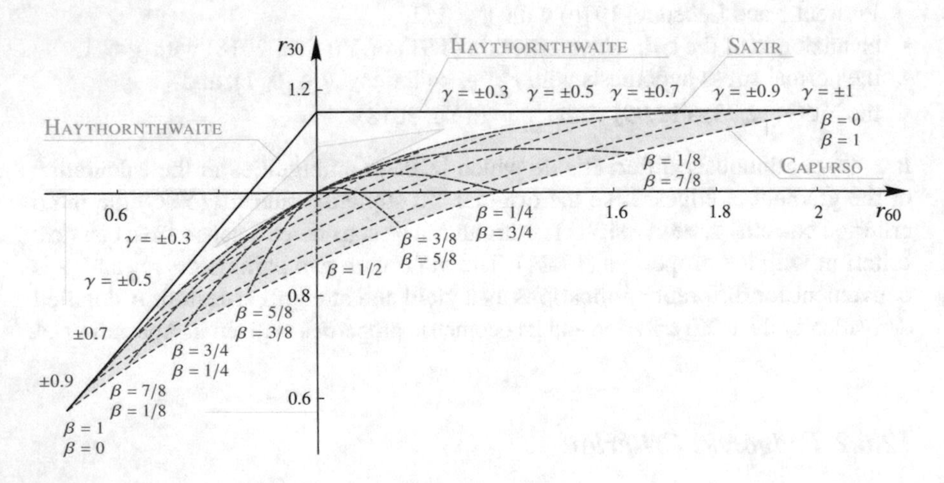

Fig. 12.9 Podgórski criterion (12.27) in the $r_{30} - r_{60}$ diagram (Fig. 12.5). The lines β = const., $\gamma \in [-1,\,1]$ (solid red) and γ = const., $\beta \in [0,\,1]$ (dashed blue) are shown, cf. Podgórski (1984); Podgórski (1985), adapted from Kolupaev (2017, 2018)

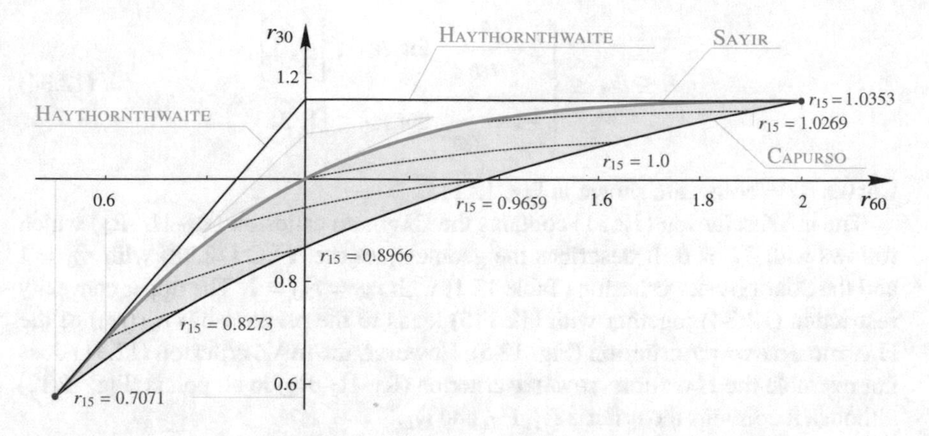

Fig. 12.10 PODGÓRSKI criterion (12.27) with the curves r_{15} = const. in the $r_{30} - r_{60}$ diagram (Fig. 12.9). The fixed values r_{15} are taken from Table 12.1. Additionally, the value $r_{15} = 0.8273$ computed with $r_{60} = 0.6, \beta = 1, \gamma = 0.7804$ and the value $r_{15} = 1.0269$ with $r_{60} = 1.9$, $\beta = 0.0290, \gamma = 1$ are shown

12.6.3 Modified Altenbach-Zolochevsky Criterion

The present modification of the ALTENBACH-ZOLOCHEVSKY criterion (Altenbach et al, 1995) was introduced in Kolupaev (2017, 2018) in order to describe all points in the $r_{60} - r_{30}$ diagram (Fig. 12.5). It reads

$$\sigma_{\text{eq}} = \sqrt{3I'_2}\,\frac{\Omega(\varphi, r_{60}, \xi_{\text{m}})}{\Omega(-\pi/6, r_{60}, \xi_{\text{m}})}, \tag{12.31}$$

with the shape function

$$\begin{aligned}\Omega(\varphi, r_{60}, \xi_{\text{m}}) = &\left(\frac{1}{r_{60}} - 1\right) \sin\varphi \\ &+ \frac{1 + r_{60} - 2\,r_{60}\,\xi_{\text{m}}}{\sqrt{3}\,r_{60}} \cos\varphi \\ &+ \xi_{\text{m}} \sin\left[\frac{\pi}{6} + \arcsin[\cos\varphi]\right],\end{aligned} \tag{12.32}$$

and

$$\Omega(-\pi/6, r_{60}, \xi_{\text{m}}) = 1. \tag{12.33}$$

The parameter ξ_{m} is a function of the geometrical values r_{30} and r_{60} (12.115) and is limited by the convexity condition

$$0 \leq \xi_{\mathrm{m}} \leq \begin{cases} 2 - \dfrac{1}{r_{60}} & \text{for } r_{60} \in \left[\dfrac{1}{2}, 1\right], \\ -1 + \dfrac{2}{r_{60}} & \text{for } r_{60} \in [1, 2]. \end{cases} \tag{12.34}$$

Curves r_{15} = const. are shown in Fig. 12.11.

The mAZ criterion (12.31) contains the CAPURSO criterion (R_1–H_1–R_2) which follows with $\xi_{\mathrm{m}} = 0$. It describes the geometry of the YYC (12.98) with $r_{60} = 1$ and the SOKOLOVSKY criterion (Table 12.1) with $r_{60} = r_{30} = 1$. The upper convexity restriction (12.34) together with (12.115) leads to the relationship $r_{60}\,(r_{30})$ of the HAYTHORNTHWAITE criterion (Fig. 12.5). However, the mAZ criterion (12.31) does not resemble the HAYTHORNTHWAITE criterion (R_1–H_2–R_2) in all points (Fig. 12.12) although it contains the criteria R_1, H_2, and R_2.

The mAZ criterion is C^0-continuously differentiable and can be recommended as a strength criterion. Extended with the linear I_1-substitution (12.11)

$$\frac{\sigma_{\mathrm{eq}} - \gamma_1 I_1}{1 - \gamma_1} = \sqrt{3 I_2'}\, \frac{\Omega(\varphi, r_{60}, \xi_{\mathrm{m}})}{\Omega(-\pi/6, r_{60}, \xi_{\mathrm{m}})}, \tag{12.35}$$

the mAZ has greater fitting capabilities than the mYU criterion (12.25).

12.6.4 Universal Yield Criterion of Trigonal Symmetry

A universal yield criterion must be capable of representing arbitrary C^1-continuous shapes in the $r_{30} - r_{60}$ diagram (Fig. 12.5) between the convexity limits described by

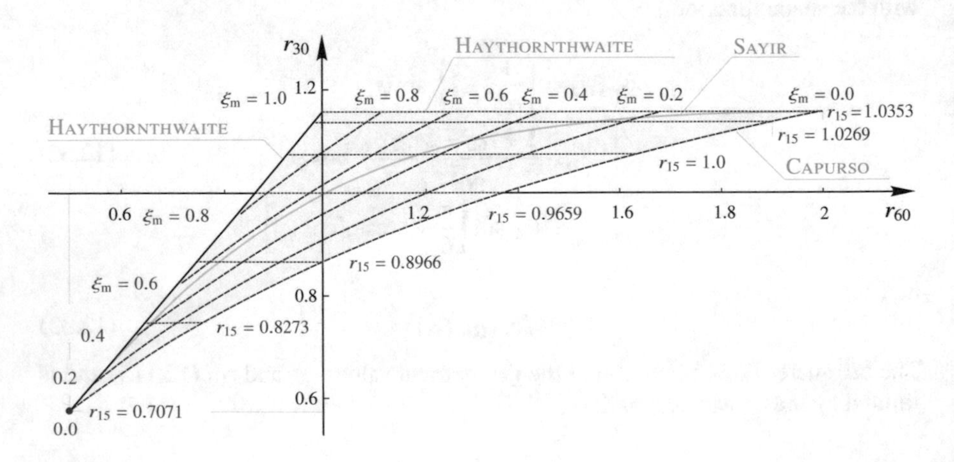

Fig. 12.11 Modified ALTENBACH-ZOLOCHEVSKY criterion (12.31) in the $r_{30} - r_{60}$ diagram (cf. Fig. 12.10). Lines ξ_{m} = const. (dashed, blue) and r_{15} = const. (dashed, red) are shown

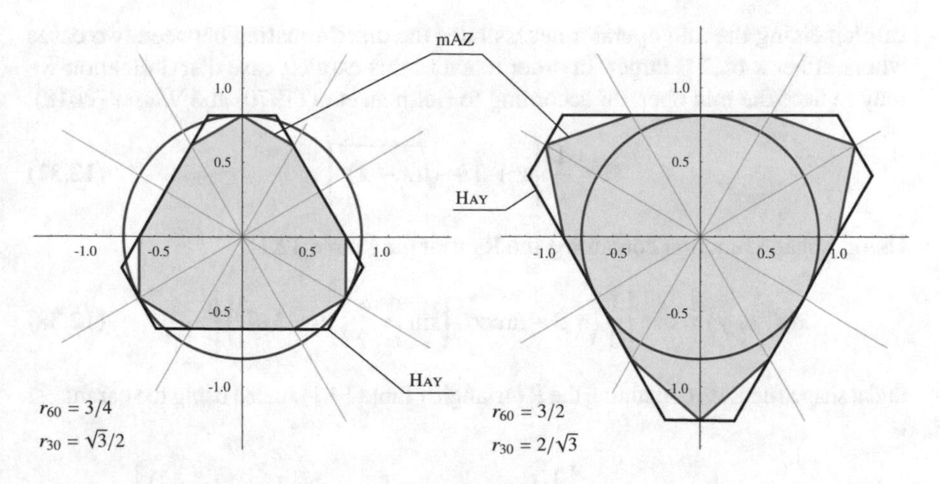

Fig. 12.12 HAYTHORNTHWAITE criterion (HAY: black line) and modified ALTENBACH-ZOLOCHEVSKY criterion (12.31) (mAZ: blue line) with $\xi_m = 2/3$ (left) and $\xi_m = 1/3$ (right) in the π-plane for the same values r_{60} and r_{30} in each case. The circle of the VON MISES hypothesis (red line) is shown for comparison (Kolupaev, 2017, 2018). Reproduced with permission from American Society of Civil Engineers ASCE

the HAYTHORNTHWAITE and CAPURSO criteria. In accordance with the plausibility conditions, it should describe a single continuous surface in stress space exhibiting no additional outer contours and plane intersections. Present generalizations are still limited in this regard. For instance, the modified ALTENBACH-ZOLOCHEVSKY criterion (12.31) is only C^0-continuously differentiable and the PODGÓRSKI criterion (12.27) is limited to shapes bounded by the SAYIR and the CAPURSO criteria (Fig. 12.9).

In order to allow a description of any convex shape in the $r_{30} - r_{60}$ diagram (Fig. 12.5), we propose a universal criterion exploiting properties of PODGÓRSKI-like yield figures. To this end, consider the HAYTHORNTHWAITE criterion which constitutes the upper limit case in the $r_{30} - r_{60}$ diagram. This criterion can be described by the minimum of an intersection of a triangle R_2 and a scaled triangle R_1 (Fig. 12.7). Because of the normalization (12.10) the choice of which triangle to scale is arbitrary and may be inverted. Both triangles can be described using the PODGÓRSKI shape function (12.28). However, the PODGÓRSKI shape function does not only allow for the description of both triangles but also for the description of any intermediate shape between the SAYIR and the CAPURSO criteria (Fig. 12.9). The intersection of other shapes than triangles can now represent any intermediate shape between the HAYTHORNTHWAITE and SAYIR criteria. The conventional PODGÓRSKI criterion (12.27) can be obtained in limit cases. Hence, any intermediate shapes between the HAYTHORNTHWAITE and the SAYIR criteria as well as any yield figure between the SAYIR and the CAPURSO criteria can be described in a single equation.

The intersection of two shape functions κ and λ is obtained using the minimum function

$$\Omega = \min\{\kappa, \lambda\}. \tag{12.36}$$

Implementing the min operator necessitates the discrimination between two cases where either κ or λ is larger. In order to avoid this explicit case discrimination we may replace the min operator according to Bellman et al (1970) and Walser (2018)

$$\Omega = \frac{1}{2}\left[\kappa + \lambda + \sqrt{(\kappa - \lambda)^2}\right]. \tag{12.37}$$

Using a shape function containing the R_2 triangle (Table 12.1)

$$\kappa(\theta, \beta, \gamma) = \cos\left[\frac{1}{3}\left(\pi\,\beta - \arccos\left[\sin\left[\gamma\,\frac{\pi}{2}\right]\,\cos[3\,\theta]\right]\right)\right], \tag{12.38}$$

and a shape function containing the R_1 triangle (Table 12.1) scaled using the parameter α

$$\lambda(\theta, \alpha, \beta, \gamma) = \frac{1}{2}\,(\alpha + 1)\,\cos\left[\frac{1}{3}\left(\pi\,\beta - \arccos\left[-\sin\left[\gamma\,\frac{\pi}{2}\right]\,\cos[3\,\theta]\right]\right)\right], \tag{12.39}$$

we obtain the universal yield criterion of trigonal symmetry (CTS)

$$\sigma_{\text{eq}} = \sqrt{3I_2}\,\frac{\Omega(\theta, \alpha, \beta, \gamma)}{\Omega(0, \alpha, \beta, \gamma)}, \tag{12.40}$$

normalized with respect to the unidirectional tensile stress $\sigma_{\text{eq}} = R^{\text{T}}$ at the angle $\theta = 0$ (12.10). The parameters α, β, and γ, which adjust the shape of the CTS criterion, are restricted as follows:

$$\alpha \in [0, 1], \qquad \beta \in [0, 1], \qquad \gamma \in [0, 1]. \tag{12.41}$$

For $\alpha = 0$ we obtain the Podgórski criterion (12.27)–(12.28). In this case, the function λ is scaled such that it does not intersect with κ anymore. When κ remains as the only shape function in effect, the same representation of curves $\beta = \text{const.}$, $\gamma \in [0, 1]$ as shown in Fig. 12.9 is obtained. Further basic yield functions described by (12.40) are summarized in Table 12.1. Because the CTS is the intersection of two convex figures in the π-plane with the same symmetry axes, its resulting yield figures are convex. Such figures are C^0-continuously differentiable.

The parameter α describes a transition from the triangles R_1 and R_2 (at $\alpha = 0$) to the hexagon H_2 (at $\alpha = 1$), as shown in Fig. 12.13a. That is, it corresponds to moving along the r_{60}-axis in the $r_{30} - r_{60}$ diagram (Fig. 12.14). The parameter β models the transition from shapes whose tip points up (e.g. R_1) to those who feature a flat topside (e.g. R_2), see Fig. 12.13b. This corresponds to moving along the r_{30}-axis (Fig. 12.15). The parameter $\gamma < 1$ smooths the yield surface and provides a C^1-continuous function as shown in Fig. 12.13c. The parameter controls the transition to the von Mises criterion at $\gamma = 0$ (Fig. 12.16).

To visualize the impact of different shape parameters, consider Figs. 12.14–12.17. Figure 12.14 shows CTS (12.40) curves for different $\alpha = \text{const.}$ at $\gamma = 1$ in the $r_{30} - r_{60}$ diagram, where $\beta \in [0, 1]$ is the curve parameter. When $\alpha < 1$, the curves

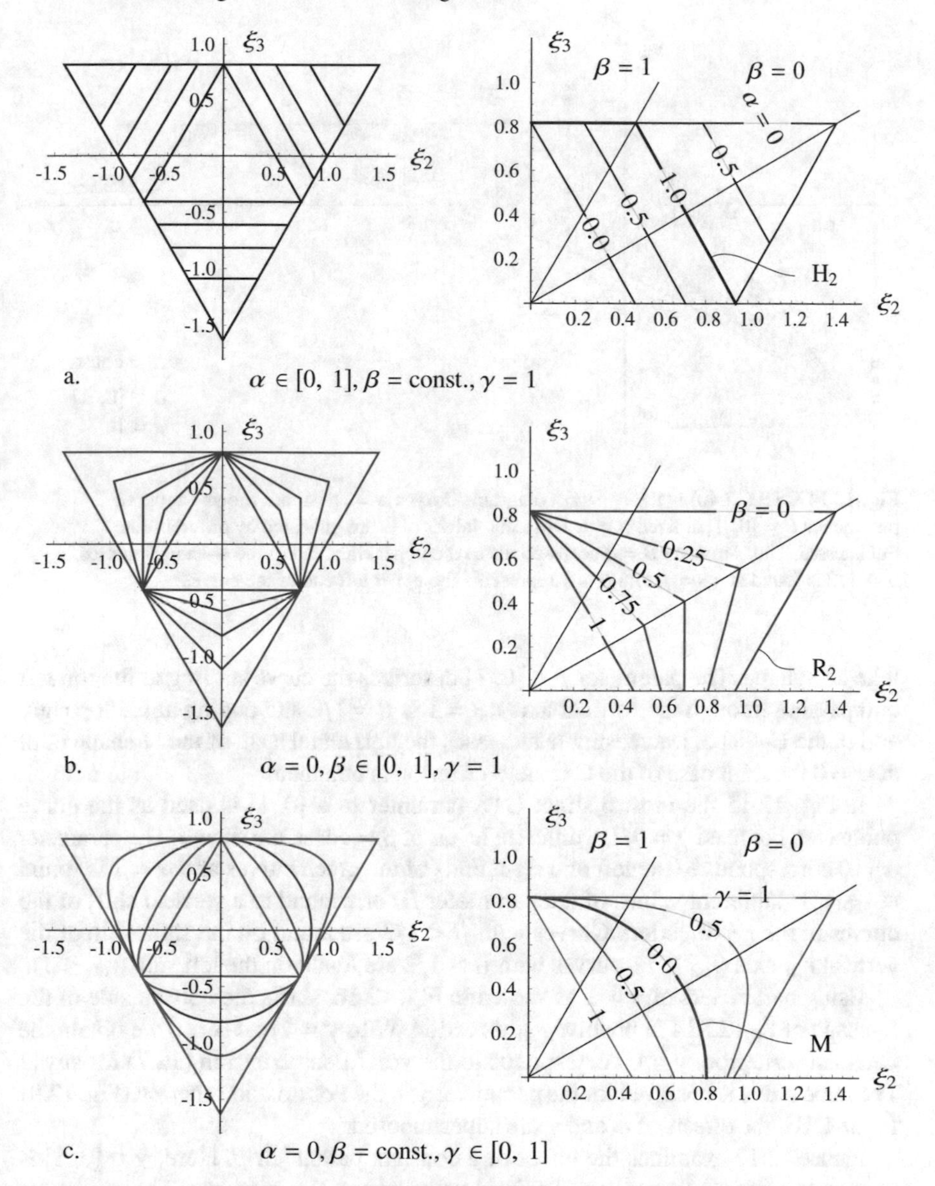

Fig. 12.13 Influence of parameters of the universal criterion of trigonal symmetry (CTS) (12.40) on the shape of yield surfaces the π-plane. **a.** α describes a transition from the triangles R_1 and R_2 ($\alpha = 0$) to the hexagon H_2 ($\alpha = 1$), **b.** β describes the transition of shapes whose tips point down (e.g. R_2 at $\beta = 0$) to shapes with a flat base (e.g. R_1 at $\beta = 1$), **c.** γ smooths the yield surface and provides a C^1-continuously differentiable function for $\gamma < 1$

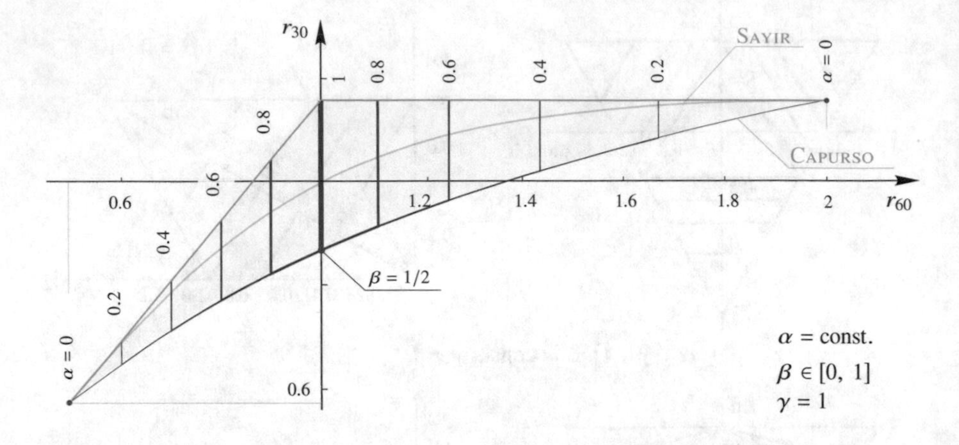

Fig. 12.14 CTS (12.40) in the $r_{30} - r_{60}$ diagram. Curves α = const. are shown for curve parameters $\beta = [0,\ 1]$ at fixed $\gamma = 1$. Different values of α are indicated by different line thicknesses. The parameter $\beta = 0$ corresponds to the upper right end of the U-shaped curves, $\beta = 1/2$ is found at $r_{60} = 1$, and $\beta = 1$ represents the upper left end of the curves

take a U-shape. The parameter $\beta \in [0,\ 1]$ describes the curve starting at the top left end passing through the vertical axis $r_{60} = 1$ at $\beta = 1/2$ and ending at the top right end of the U-shape. Decreasing α increases the horizontal span of the U-shape until at $\alpha = 0$ the limit case of the CAPURSO criterion is obtained.

In Fig. 12.15 the most distinct CTS paramter $\alpha \in [0,\ 1]$ is used as the curve parameter. For fixed $\gamma = 6/10$, different levels of β = const. are shown. The parameter $\alpha = 0$ corresponds to the left and right ends of the green curves and $\alpha = 1$ is found at $r_{60} = 1$. Different values of the parameter β correspond to a vertical shift of the curves in the r_{30} direction. Curves with $\beta < 1/2$ are found on the right side of the vertical r_{30}-axis ($r_{60} \geq 1$), curves with $\beta > 1/2$ are found on the left side ($r_{60} \leq 1$).

Using parameters of $\gamma \neq 1$ as shown in Fig. 12.16, shifts the bottom side of the U-shape of Fig. 12.14 in positive r_{30}-direction. With $\gamma = 1$ and $\alpha = 0$ we obtain the CAPURSO criterion. $\gamma = 0$ corresponds to the VON MISES criterion (12.7) for any α. The same effect is observed for the parameter γ in the PODGÓRSKI criterion (Fig. 12.9). In the CTS, the effects of α and γ are superimposed.

Figure 12.17 examines the effect of a constant parameter β. Here, $\gamma \in [0,\ 1]$ is used as the curve parameter and a fixed value of $\alpha = 0.6$ is shown. The parameter $\beta < 1/2$ describes shapes on the left side of the vertical r_{30}-axis while $\beta > 1/2$ describes shapes on its right side. The parameter $\beta = 1/2$ represents shapes on the r_{30}-axis at $r_{30} < 1$. Increasing β from 0 to 1/2 describes a vertical top to bottom transition on a parallel to the r_{30}-axis. The equivalent behavior is observed for β between 1/2 and 1 on the right side of the r_{30}-axis.

Fitting capabilities of the CTS Eq. (12.40) are illustrated using curves r_{15} = const. (Fig. 12.18). For $r_{15} = \sqrt{2}/2 \approx 0.7071$ only the triangle R_1 can be represented. Geometries with $r_{15} = 1.0353$ contain the shapes of the HAYTHORNTHWAITE criterion

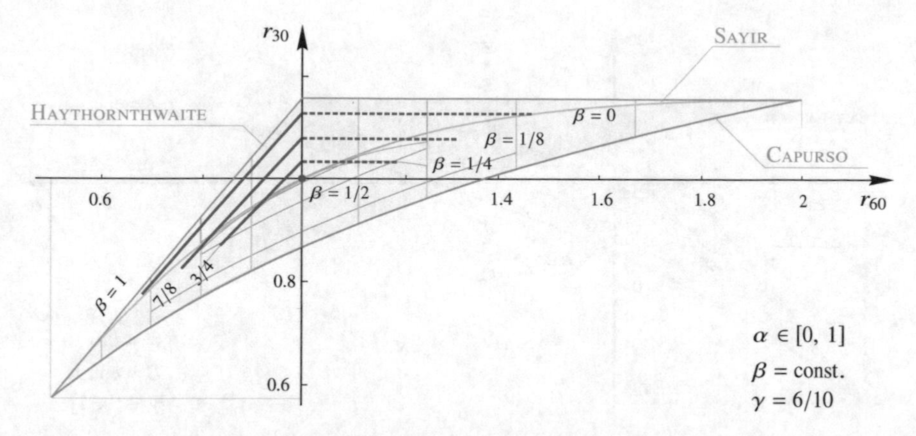

Fig. 12.15 CTS (12.40) in the $r_{30} - r_{60}$ diagram. Curves β = const. are shown for curve parameters $\alpha = [0, 1]$ at fixed $\gamma = 6/10$ (dark green). $\alpha = 0$ corresponds to the left ($\beta > 1/2$) or right ($\beta < 1/2$) ends of the curves. $\alpha = 1$ is found at $r_{60} = 1$. Lines of Figs. 12.14, 12.16, and 12.17 are shown for comparison (pale red, blue and black)

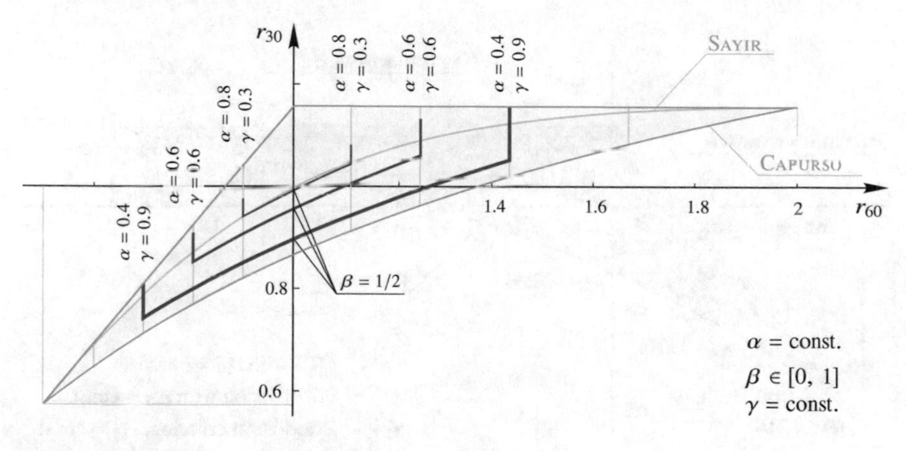

Fig. 12.16 CTS (12.40) in the $r_{30} - r_{60}$ diagram. Curves α = const. are shown for curve parameters $\beta \in [0, 1]$ each at a different constant value for γ = const. (blue). Different (α, γ) pairs are indicated by different line thicknesses. $\beta = 0$ corresponds to the upper right end of the U-shaped curves. $\beta = 1/2$ is found at $r_{60} = 1$ and $\beta = 1$ represents the upper left end of the curves. Lines of Fig. 12.14 are shown for comparison (pale red)

at $r_{60} \geq 1$ and some forms at $r_{60} \leq 1$. For in-between values of r_{15} = const. several convex geometries are possible. The CTS provides better fitting capabilities than PODGÓRSKI (12.27) or ALTENBACH-ZOLOCHEVSKY (12.31) criteria since for one r_{15} = const. entire $r_{30} - r_{60}$-patches bounded the other two criteria are covered. The comparison of the properties of the different criteria, e.g. shown in Figs. 12.10 and 12.11 based on these curves is straightforward.

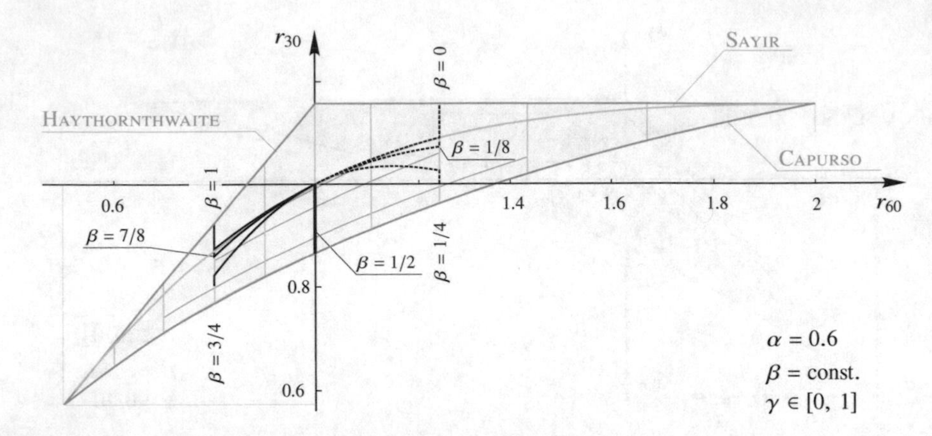

Fig. 12.17 CTS (12.40) in the $r_{30} - r_{60}$ diagram. Curves $\beta =$ const. are shown for curve parameters $\gamma = [0, 1]$ at fixed $\alpha = 0.6$ (black). $\gamma = 1$ corresponds to the left ($\beta > 1/2$) or right ($\beta < 1/2$) end of the curves. $\gamma = 0$ is found at $r_{30} = r_{60} = 1$. Lines of Fig. 12.14 and 12.16 are shown for comparison (pale red and blue)

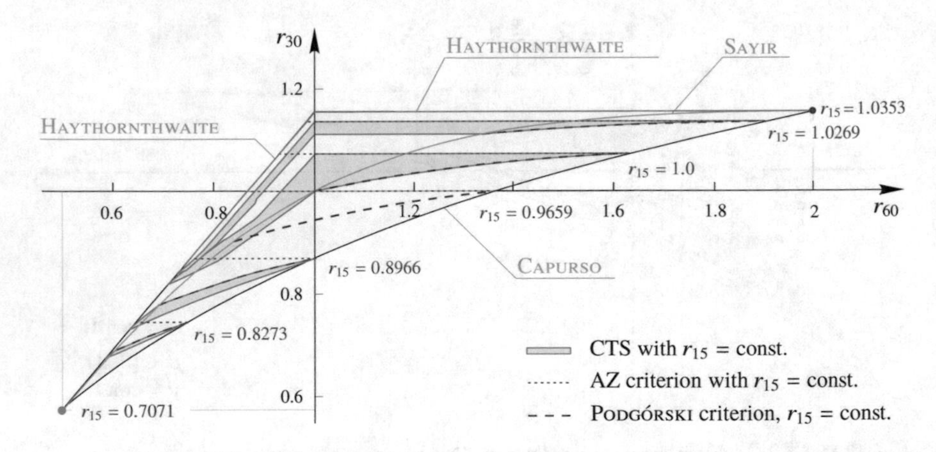

Fig. 12.18 CTS (12.40) with curves $r_{15} =$ const. in the $r_{30} - r_{60}$ diagram. Curves of the PODGÓRSKI criterion (12.27) and the ALTENBACH-ZOLOCHEVSKY criterion (12.31) are shown for comparison (Figs. 12.10 and 12.11). The curve $r_{15} = 0.9659$ of the CTS is excluded for better illustration

Applying the linear I_1-substitution (12.11) to the CTS (12.40) we obtain a universal conical (pyramidal) criterion

$$\frac{\sigma_{\text{eq}} - \gamma_1 I_1}{1 - \gamma_1} = \sqrt{3 I_2}\, \frac{\Omega(\theta, \alpha, \beta, \gamma)}{\Omega(0, \alpha, \beta, \gamma)}, \qquad \gamma_1 \in [0, 1[. \tag{12.42}$$

The geometrical values r_{15}, r_{30}, and r_{60} are obtained from (12.93). The basic test value r_0^{CC} can be computed according to (12.94). Further basic test values are obtained with (12.95)–(12.97). Please note that owing to the use of the operator (12.36) the assignment

$$(r_{60}, r_{30}, r_{15}) \to (\alpha, \beta, \gamma)$$

is not unique violating the last plausibility assumption.

12.6.5 Universal Deviatoric Function

The properties of the universal yield criterion of trigonal symmetry (CTS) allow for further generalization of the criterion to incorporate arbitrary yield surfaces of hexagonal symmetry, as well. To this end, let us introduce an integer parameter $n \in \mathbb{N}^+$. We will use this parameter to determine the number of

$$m = 3\,n$$

edges of equilateral m-gons obtained from κ and λ for $\alpha = 0$, $\beta \in \{0, 1\}$ and $\gamma = 1$. For this set of parameters, $n = 1$ yields triangles R_1 and R_2 and thus, the yield criterion of trigonal symmetry (12.40) and $n = 2$ yields hexagons H_1 and H_2 for the chosen set of parameters and constitutes a criterion of hexagonal symmetry. Higher values of n provide yield surfaces with additional edges. The criteria are of hexagonal symmetry for even n and of trigonal symmetry for odd n.

Again, we exploit reformulation of the min operator (12.37) and shape function scaling to obtain the general yield criterion with

$$\kappa(\theta, \beta, \gamma, n) = \cos\left[\frac{1}{3n}\left(\pi\,\beta - \arccos\left[\sin\left[\gamma\,\frac{\pi}{2}\right]\,\cos[3\,n\,\theta]\right]\right)\right], \tag{12.43}$$

and

$$\begin{aligned}\lambda(\theta, \alpha, \beta, \gamma, n) = &\left(\alpha + (1 - \alpha)\cos\left[\frac{\pi}{3n}\right]\right) \\ &\cos\left[\frac{1}{3n}\left(\pi\,\beta - \arccos\left[-\sin\left[\gamma\,\frac{\pi}{2}\right]\,\cos[3\,n\,\theta]\right]\right)\right].\end{aligned} \tag{12.44}$$

The equivalent stress reads

$$\sigma_{\text{eq}} = \sqrt{3\,I_2}\,\frac{\Omega(\theta, \alpha, \beta, \gamma, n)}{\Omega(0, \alpha, \beta, \gamma, n)}, \tag{12.45}$$

with the following parameter restrictions ensuring convexity:

$$\alpha \in [0, 1], \qquad \beta \in [0, 1], \qquad \gamma \in [0, 1], \qquad n \in \mathbb{N}^+. \tag{12.46}$$

The geometrical values r_{15}, r_{30}, and r_{60} follow with (12.93). The linear I_1-substitution (12.11) leads to the conical (pyramidal) surface (12.12)

$$\frac{\sigma_{eq} - \gamma_1 I_1}{1 - \gamma_1} = \sqrt{3 I_2} \frac{\Omega(\theta, \alpha, \beta, \gamma, n)}{\Omega(0, \alpha, \beta, \gamma, n)}, \qquad \gamma_1 \in [0, 1[. \tag{12.47}$$

with the basic test values (12.94)–(12.97). This convex criterion comprises all other criteria discussed in the present work. In Table 12.1 similarities of the parameters α, β, and γ for $n = 1$, 2, 4, and 6 are shown. Regular geometries with $n = 2$ are often used in theory of plasticity (Table 12.2). The geometries with $n > 2$ have not found application, yet.

Figure 12.19 shows the influence of the parameters α, β, and γ on yield figures of hexagonal symmetry obtained for $n = 2$. The parameter α represents the transition from the hexagons H_1 and H_2 ($\alpha = 0$) to the dodecagon D_2 ($\alpha = 1$) shown in Fig. 12.19a. This corresponds to moving along the r_{30}-axis in the $r_{30} - r_{15}$ diagram (Fig. 12.20). As shown in Fig. 12.19b the parameter β describes the transition from shapes with a flat base (e.g. H_2 at $\beta = 0$) to shapes standing on their tips (e.g. H_1 at $\beta = 1$). This corresponds to moving along the r_{15} axis in the $r_{30} - r_{15}$ diagram (Fig. 12.23). Again, γ smooths the yield surface providing a C^1-continuous function and describing the transition from any shape to the von Mises criterion at $\gamma = 0$ (Figs. 12.19c and 12.22).

Figure 12.20 shows curves $\alpha =$ const. at $\gamma = 1$ of the universal criterion of hexagonal symmetry (CHS) obtained for $n = 2$ in the $r_{30} - r_{15}$ diagram. The parameter $\beta \in [0, 1]$ describes a U-shaped curve starting at its upper right, passing through $r_{30} = 1$ at $\beta = 1/2$ and ending at its upper left corner. A smaller α increases the horizontal span of the U-curves. $\alpha = 0$ represents the YYC as convexity limit.

Using $\gamma < 1$, as shown in Fig. 12.21, translates the U-shapes of Fig. 12.20 in positive r_{15}-direction along the vertical axis. Again, $\gamma = 1$ and $\alpha = 0$ yields the YYC and $\gamma = 0$ corresponds to von Mises criterion for any α.

Figure 12.22 shows the influence of β in the criterion of hexagonal symmetry ($n = 2$). Curves with $\alpha = 0.6$ and $\gamma \in [0, 1]$ as the curve parameter are shown. $\beta < 1/2$ corresponds to curves on the left side of the vertical r_{15}-axis while $\beta > 1/2$ describes shapes on its right side. Shapes with $\beta = 1/2$ are found on the r_{15}-axis at $r_{30} = 1$. Increasing β from 0 to 1/2 or decreasing it from 1 to 1/2 shifts curves in negative r_{15}-direction.

Figure 12.23 shows different curves with $\beta =$ const. and $\gamma = 6/10$. Again, β shifts the curves along the r_{15}-axis. $\alpha = 0$ represents the left ($\beta > 1/2$) and right ($\beta < 1/2$) ends of the curves and $\alpha = 1$ is found on the r_{15}-axis at $r_{30} = 1$.

Table 12.2 Criterion of hexagonal symmetry (CHS): Universal yield function (UDF), Eq. (12.45), with $n = 2$ in the π-plane, cf. Table 12.1

π-plane	H_1	H_2	D_1	D_2	C_1	C_2	M
α	0	0	$[0, 1]$	1	1	–	$[0, 1]$
β	1	0	1/2	$\{0, 1\}$	$\{1/4, 3/4\}$	–	$[0, 1]$
γ	1	1	1	1	1	–	0

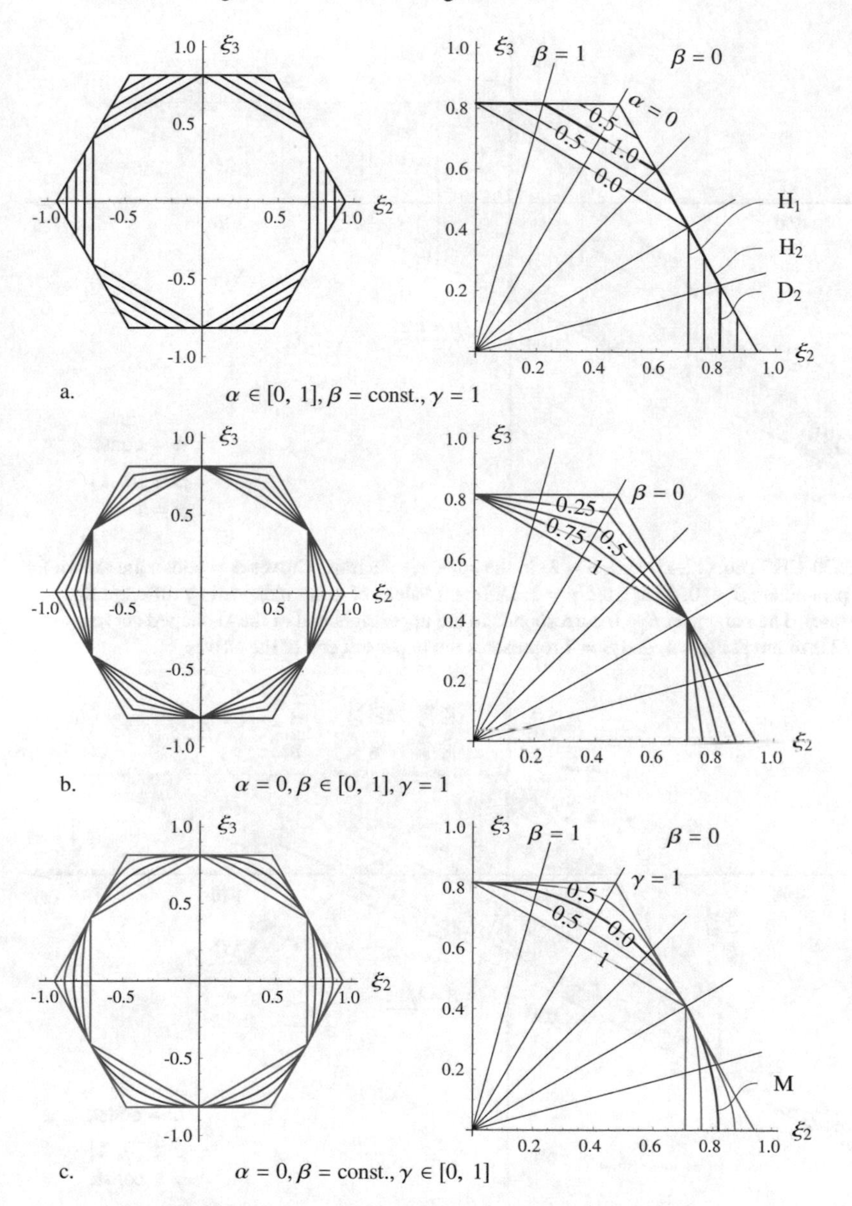

Fig. 12.19 Influence of parameters of the criterion of hexagonal symmetry (CHS) (Eq. (12.45) with $n = 2$) on the shape of yield surfaces in the π-plane. **a.** α describes a transition from the hexagon H_1 and H_2 ($\alpha = 0$) to the dodecagon D_2 ($\alpha = 1$), **b.** β describes the transition of shapes with a flat base (e.g. H_2 at $\beta = 0$) to shapes standing on their tips (e.g. H_1 at $\beta = 1$), **c.** γ smooths the yield surface and provides a C^1-continuously differentiable function for $\gamma < 1$

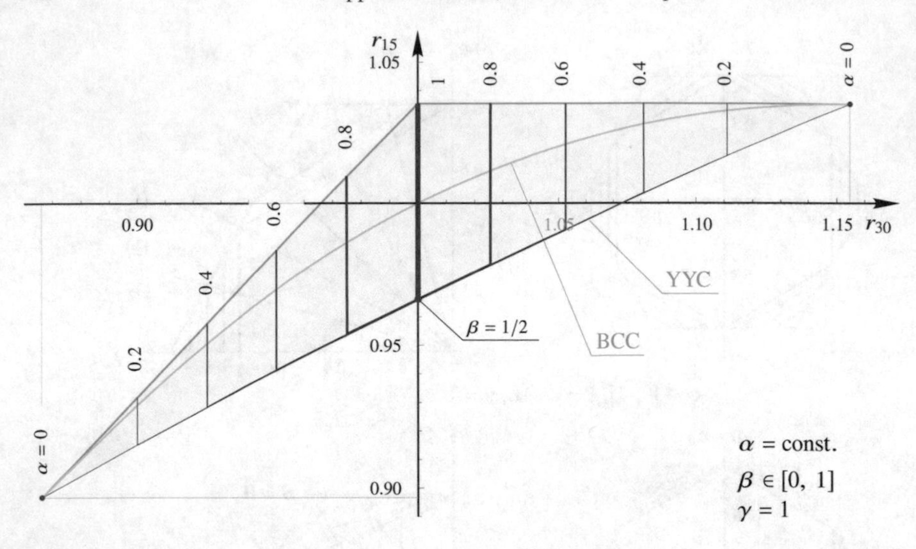

Fig. 12.20 CHS (Eq. (12.45) with $n = 2$) in the $r_{30} - r_{15}$ diagram. Curves α = const. are shown for curve parameters $\beta \in [0, 1]$ at fixed $\gamma = 1$. Different values of α are indicated by different line thicknesses. The parameter $\beta = 0$ corresponds to the upper right end of the U-shaped curves, $\beta = 1/2$ is found at $r_{30} = 1$, and $\beta = 1$ represents the upper left end of the curves

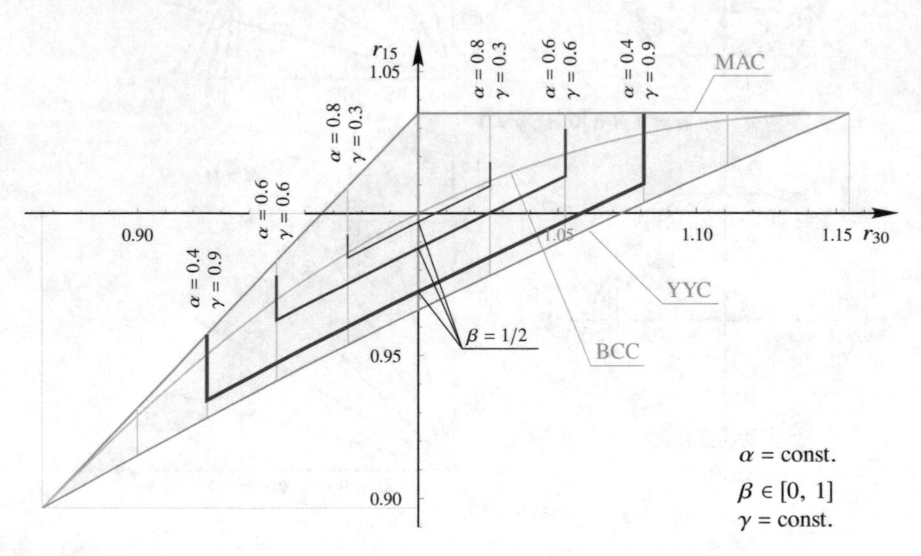

Fig. 12.21 CHS (Eq. (12.45) with $n = 2$) in the $r_{30} - r_{15}$ diagram. Three curves α = const. are shown for curve parameters $\beta \in [0, 1]$ each at a different constant value γ = const. (blue). Different (α, γ) pairs are indicated by different line thicknesses. $\beta = 0$ corresponds to the upper right end of the U-shaped curves. $\beta = 1/2$ is found at $r_{30} = 1$ and $\beta = 1$ represents the upper left end of the curves. Lines of Fig. 12.20 are shown for comparison (pale red)

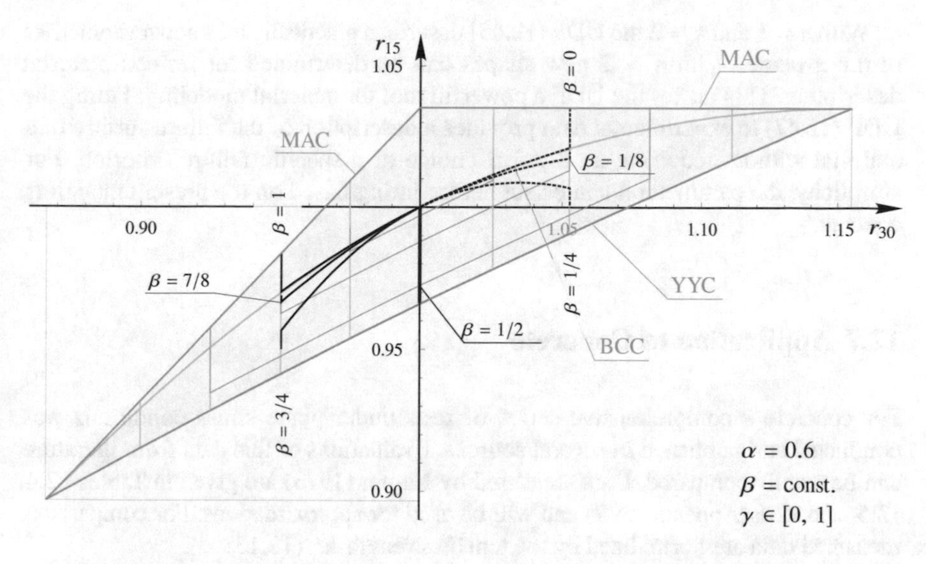

Fig. 12.22 CHS (Eq. (12.45) with $n = 2$) in the $r_{30} - r_{15}$ diagram. Curves β = const. are shown for curve parameters $\gamma = [0, 1]$ at fixed $\alpha = 0.6$ (black). $\gamma = 1$ corresponds to the left ($\beta > 1/2$) or right ($\beta < 1/2$) end of the curves. $\gamma = 0$ is found at $r_{30} = r_{15} = 1$. Lines of Fig. 12.20 and 12.21 are shown for comparison (pale red and blue)

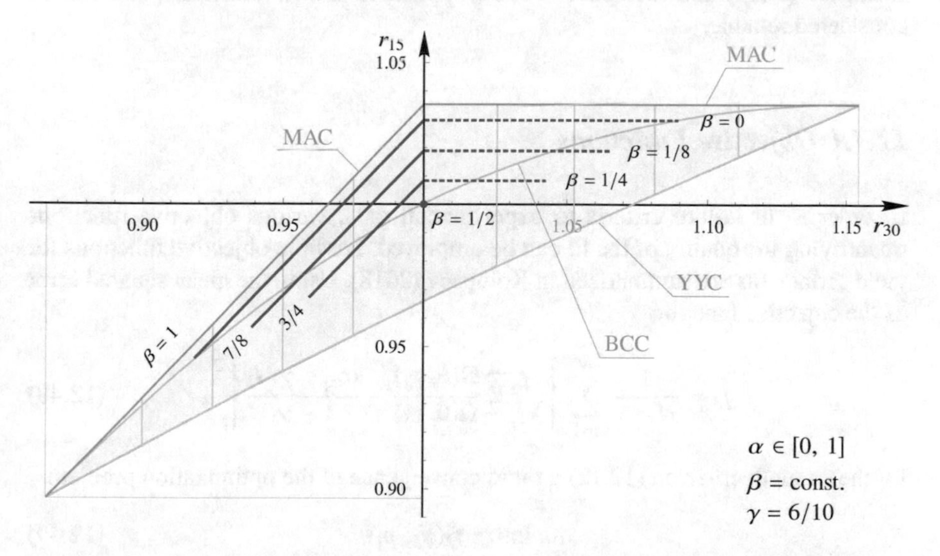

Fig. 12.23 CHS (Eq. (12.45) with $n = 2$) in the $r_{30} - r_{15}$ diagram. Curves β = const. are shown for curve parameters $\alpha = [0, 1]$ at fixed $\gamma = 6/10$ (dark green). $\alpha = 0$ corresponds to the left ($\beta > 1/2$) or right ($\beta < 1/2$) ends of the curves. $\alpha = 1$ is found at $r_{30} = 1$. Lines of Figs. 12.20, 12.21, and 12.22 are shown for comparison (pale red and black)

With $n = 1$ and $n = 2$ the UDF (12.45) describes practically all known geometries in the π-plane. With $n > 2$ new shapes can be determined for refined material description. This makes the UDF a powerful tool for material modeling. Fitting the UDF (12.47) to experimental data provides a description of the failure surface of a material without requiring an a priori choice of a specific failure criterion. For simplicity, the parameter n can be set before fitting based on the present modeling concept.

12.7 Application to Concrete

For concrete a comprehensive series of tests under plane stress conditions was conducted and published in several sources. Evaluations of this data from literature can be readily compared. Data measured by Kupfer (1973) are given in Tables 12.4, 12.5, and 12.6 (Appendix A.7) and will be used for approximations. For comparison, measured data are normalized by the tensile strength R^{T} (12.13).

The method shown in this section is not limited to concrete and can be transferred to any isotropic material with $R^{\mathrm{T}} > 0$. The fitting methods and visualization techniques used in the following are summarized in Altenbach et al (2014); Altenbach and Kolupaev (2014) and Kolupaev (2018), applied to several materials, and can be considered reliable.

12.7.1 Objective Functions

In order to fit failure criteria to experimental data, various objective functions quantifying the quality of the fit can be employed. Possible objective functions for yield surface fits are summarized in Kolupaev (2018). Using the mean squared error as the objective function

$$f_2 = \frac{1}{N-1}\sum_{i=1}^{N}\left[\sqrt{3I_2'}\,\frac{\Omega(\theta,p_\mathrm{j})}{\Omega(0,p_\mathrm{j})} - \frac{\sigma_\mathrm{eq}-\gamma_1 I_1}{1-\gamma_1}\right]_i^2, \tag{12.48}$$

for the strength criterion (12.12) a rapid convergence of the optimization problem

$$\text{minimize } f_2(\gamma_1, p_\mathrm{j}), \tag{12.49}$$

can be achieved. Here, N measured data points are used. The hydrostatic node γ_1 and the parameters of the deviatoric function p_j are the parameters of the selected strength criterion (12.12).

The objective function

$$f_\infty = \max_{i=1\ldots N} \left| \sqrt{3 I_2'} \frac{\Omega(\theta, p_j)}{\Omega(0, p_j)} - \frac{\sigma_{eq} - \gamma_1 I_1}{1 - \gamma_1} \right|_i . \tag{12.50}$$

with the optimization problem

$$\text{minimize } f_\infty(\gamma_1, p_j), \tag{12.51}$$

can be sometimes preferred for additional solutions and further comparisons.

In order to solve the resulting optimization problem (12.49), the present work employs the `NMinimize` algorithm of the CAS Wolfram Mathematica (Wolfram, 2003). The solution is obtained in the form of parameters of the strength criterion minimizing the objective function. The optimization problem constrains the parameters p_j to ensure convexity of the failure surface in the π-plane. Using `NMinimize` readily allows for implementing different constraints and weights.

Parameters obtained using the objective function f_2 (12.48) approximate measured data well. However, the comparison of different strength criteria against each other is not straightforward. Therefore, we use two "physical" objective functions to compare different approximations with each other:

- Objective function f_{3D}: Evaluates the regression quality in principal stress space. For each measurement the normal distance from data points to the limit surface in principal stress space $(\sigma_I, \sigma_{II}, \sigma_{III})$ is computed and then averaged over all measurements.
- Objective function f_{ray}: The distances between the experimental points and the surface of strength criterion Φ are measured along a line connecting the respective data point with the origin (Wu, 1973) and summarized.

The comparisons are shown in Tables 12.8, 12.10, and 12.12.

12.7.2 Approximation and Restrictions

In the following we will introduce certain restrictions for limit surface approximations (Kolupaev et al, 2016). Using a straight line through the points T and CC on the 0-meridian (12.89) and through the points C and TT on the $\pi/3$-meridian (12.90) in the Burzyński-plane (Figs. 12.24, 12.25, and 12.26), we can estimate the position of the hydrostatic node TTT with its coordinate $1/\gamma_1$ on the axis I_1.

The different estimates are summarized in Table 12.7. The upper limit in

$$\gamma_1 \in [0.2984, 1/3],$$

can be introduced for fitting based on the normal stress hypothesis with $\gamma_1 = 1/3$ (Kolupaev, 2017; Kolupaev et al, 2016).

The approximation of the point K follows based on the points in the Burzyński-plane which are closest to the ordinate with $I_1 = 0$, i.e., T and P_1 (Table 12.4), P_3

(Table 12.5) or P_4 (Table 12.6). Such estimates (Table 12.7) help to control the fitting procedure and are suitable for comparison with known approximations.

In the next step we constrain all approximations with

$$r_0^{\mathrm{CC}} \leq \frac{R^{\mathrm{CC}}}{R^{\mathrm{T}}}, \tag{12.52}$$

such that the data point R^{CC} will not be overestimated. If the approximation contains the point CC, the parameter γ_1 follows directly from r_0^{CC} in Eq. (12.89). Generally, this leads to overdimensioning in the point C. In the next step we try to approximate this point with the restriction

$$r_{60}^{\mathrm{C}} \leq \frac{R^{\mathrm{C}}}{R^{\mathrm{T}}}. \tag{12.53}$$

If the approximation contains the points T, C, and CC, the comparison of the criteria is straightforward (Ottosen, 1977). In this case the value r_{30}^{Cc} is significant for the comparison. The adaptability of the criteria on the data is directly visible.

The value r_{30}^{Cc} can be estimated with the straight line through two adjacent points in the BURZYŃSKI-plane M_1 and M_2 (Table 12.4), M_3 and M_4 (Table 12.5) or M_5 and M_6 (Table 12.6), respectively. The value r_{30}^{Tt} can be estimated from a straight line through two adjacent points in the BURZYŃSKI-plane L_1 and L_2 (Table 12.4), L_3 and L_4 (Table 12.5) or L_5 and L_6 (Table 12.6), respectively. Alternatively, estimates for

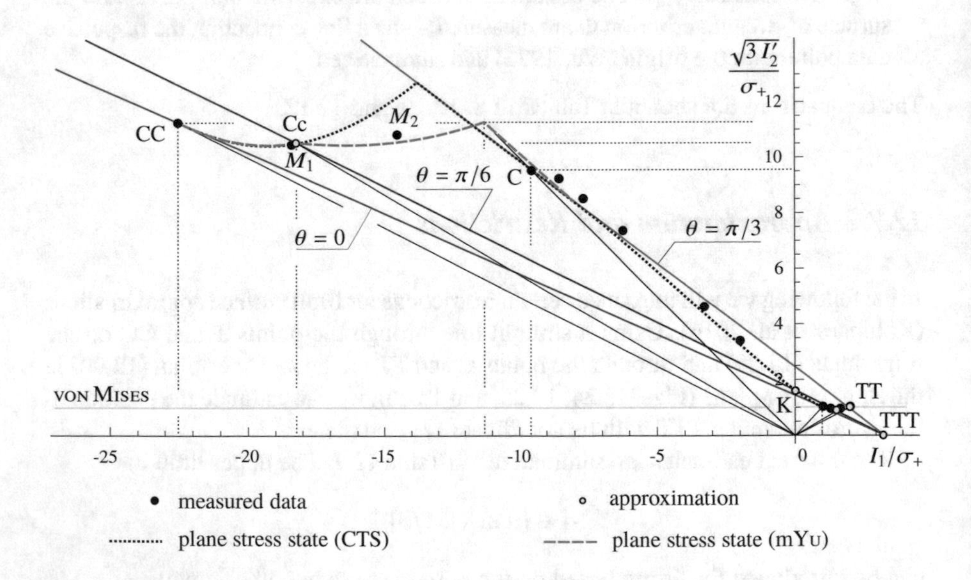

Fig. 12.24 Measured data by KUPFER (Table 12.4) with $R^{\mathrm{C}} = 18.73$ MPa normalized with respect to $R^{\mathrm{T}} = 1.96$ MPa in the BURZYŃSKI-plane approximated using the CTS (12.42). The line of the plane stress state of the modified YU criterion (12.25) with $r^{\mathrm{C}} = 11.25$ and $\chi = 0.2596$ is shown for comparison

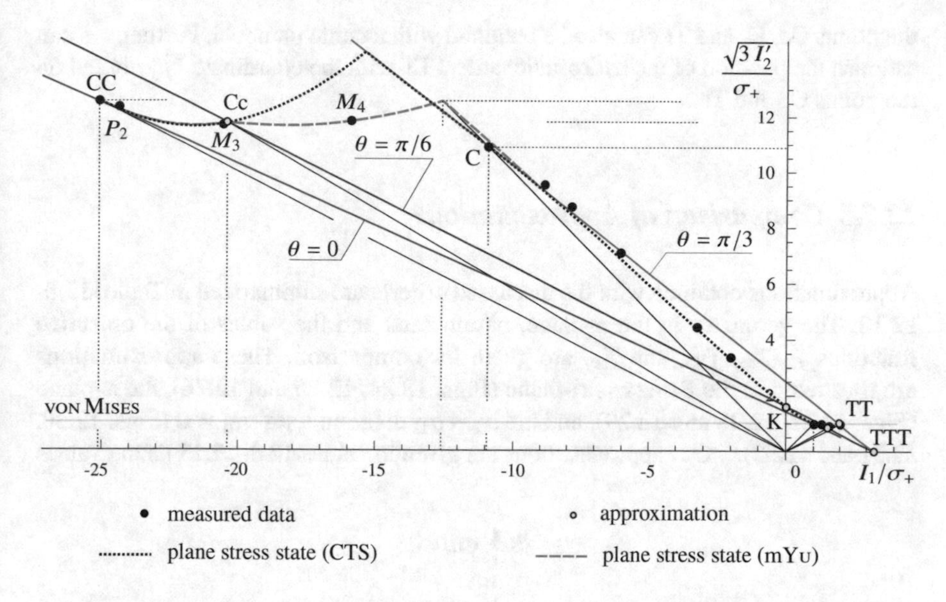

Fig. 12.25 Measured data by KUPFER (Table 12.5) with $R^{\mathrm{C}} = 30.50\,\mathrm{MPa}$ normalized with respect to $R^{\mathrm{T}} = 2.79\,\mathrm{MPa}$ in the BURZYŃSKI-plane approximated using the CTS (12.42). The line of the plane stress state of the modified YU criterion (12.25) with $r^{\mathrm{C}} = 12.57$ and $\chi = 0.2693$ is shown for comparison

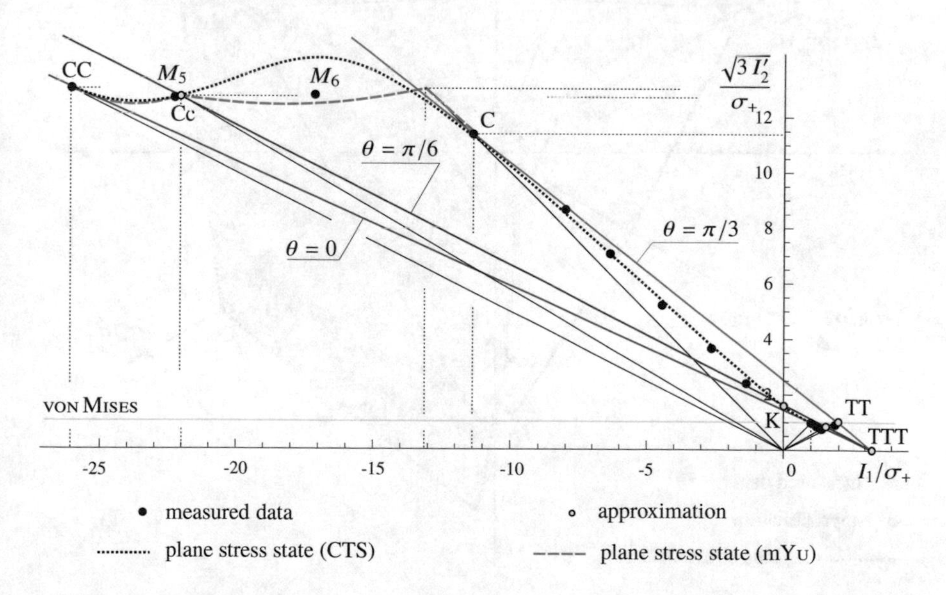

Fig. 12.26 Measured data by KUPFER (Table 12.6) with $R^{\mathrm{C}} = 58.25\,\mathrm{MPa}$ normalized with respect to $R^{\mathrm{T}} = 5.12\,\mathrm{MPa}$ in the BURZYŃSKI-plane approximated using the CTS (12.42). The line of the plane stress state of the modified YU criterion (12.25) with $r^{\mathrm{C}} = 13.05$ and $\chi = 0.4203$ is shown for comparison

the points Cc, K, and Tt can also be obtained with a cubic parabola. Further, we can estimate the position of the hydrostatic node TTT with the coordinate $1/\gamma_1$ based on the points Cc and Tt.

12.7.3 Comparison of Approximations

Approximations obtained with the discussed criteria are summarized in Tables 12.8–12.12. The geometry in the π-plane, parameters, and the values of the objective functions f_2, f_∞, f_{3D}, and f_{ray} are given for comparison. These approximations are illustrated in the BURZYŃSKI-plane (Figs. 12.24, 12.25 and 12.26), the π-plane (Figs. 12.27, 12.28 and 12.29), and the $\sigma_I - \sigma_{II}$ diagram with $\sigma_{III} = 0$ (Figs. 12.30, 12.31 and 12.32). Our approximations are given in Tables 12.8–12.13 listing values

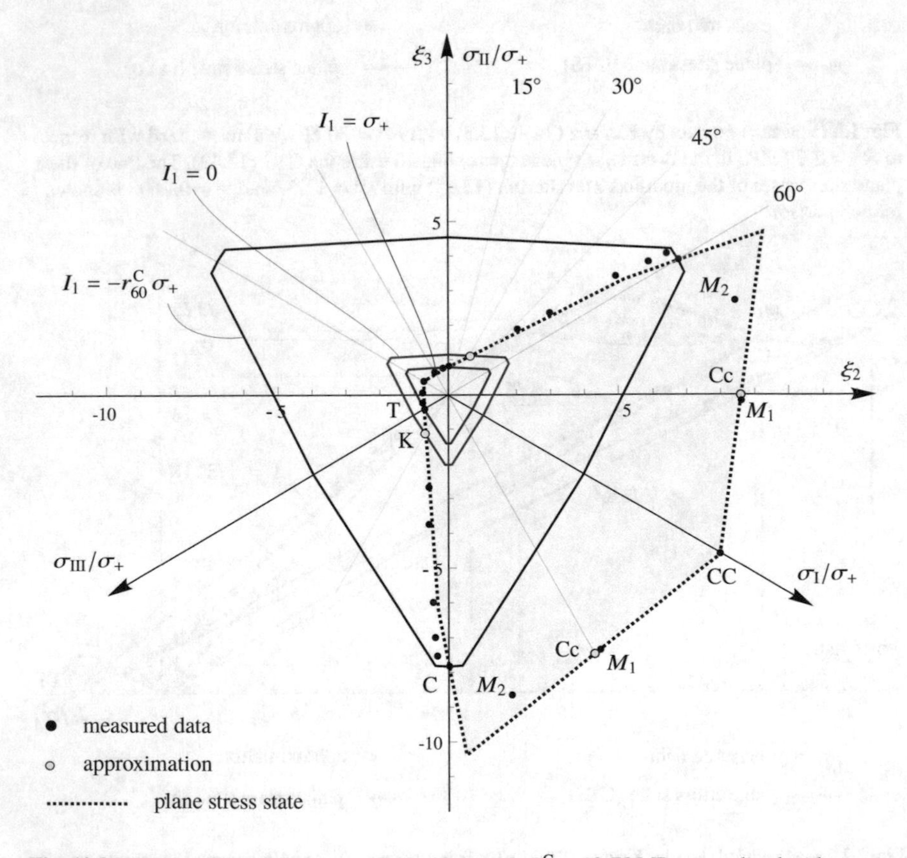

Fig. 12.27 Measured data by KUPFER (Table 12.4) with $R^C = 18.73$ MPa normalized with respect to $R^T = 1.96$ MPa in the π-plane approximated using the CTS (12.42) (Fig. 12.24). The cross sections orthogonal to the hydrostatic axis corresponding to the different values of I_1 are shown

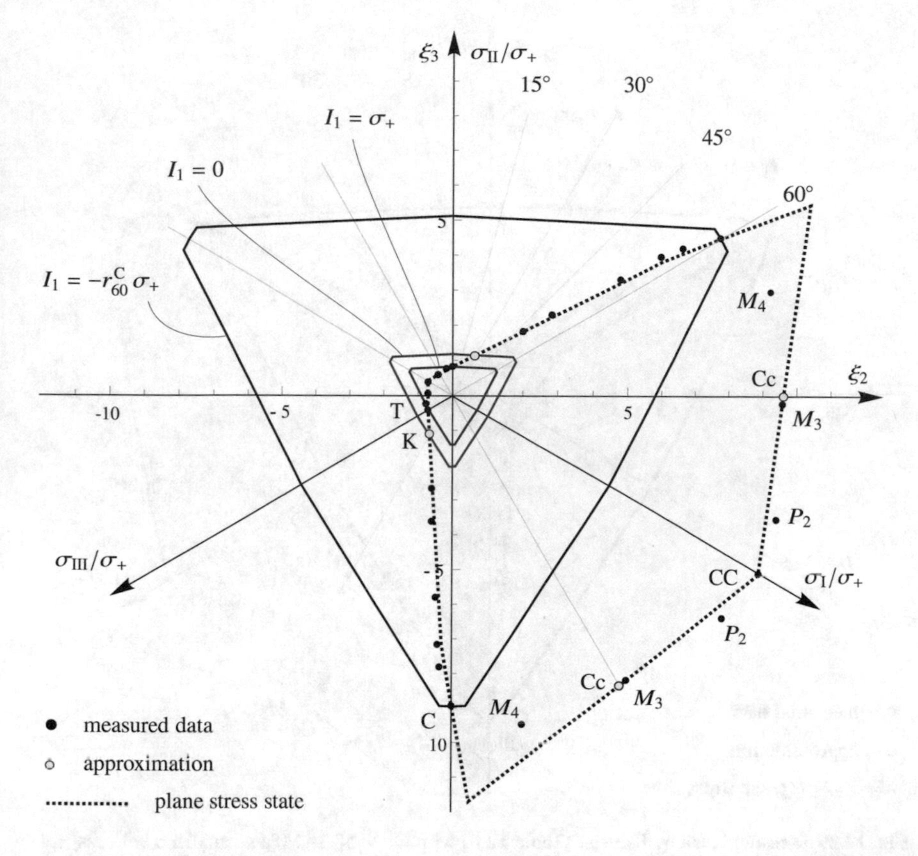

Fig. 12.28 Measured data by Kupfer (Table 12.6) with $R^{\mathrm{C}} = 30.50\,\mathrm{MPa}$ normalized with respect to $R^{\mathrm{T}} = 2.79\,\mathrm{MPa}$ in the π-plane approximated using the CTS (12.42) (Fig. 12.25). The cross sections orthogonal to the hydrostatic axis corresponding to the different values of I_1 are shown

for basic loadings for straightforward comparison. Extrapolations yield the TTT node corresponding to the hydrostatic tensile strength.

The concretes with $R^{\mathrm{C}} = 18.73$ and $R^{\mathrm{C}} = 30.50\,\mathrm{MPa}$ show the "classical" property Eq. (12.24)

$$R^{\mathrm{T}} = R^{\mathrm{TT}} \quad \text{or} \quad r_{60}^{\mathrm{TT}} = 1.$$

The concrete with $R^{\mathrm{C}} = 58.25\,\mathrm{MPa}$ shows

$$R^{\mathrm{T}} > R^{\mathrm{TT}} \quad \text{or} \quad r_{60}^{\mathrm{TT}} < 1.$$

For all test series

$$R^{\mathrm{CC}} > R^{\mathrm{C}} \quad \text{or} \quad r_{0}^{\mathrm{CC}} > r_{60}^{\mathrm{C}},$$

can be observed in the compression region. Thus, the approximations with the Mohr-Coulomb and the modified Yu criteria can be carried out based on either the point C or based on CC. Approximations using the Mohr-Coulomb criterion through

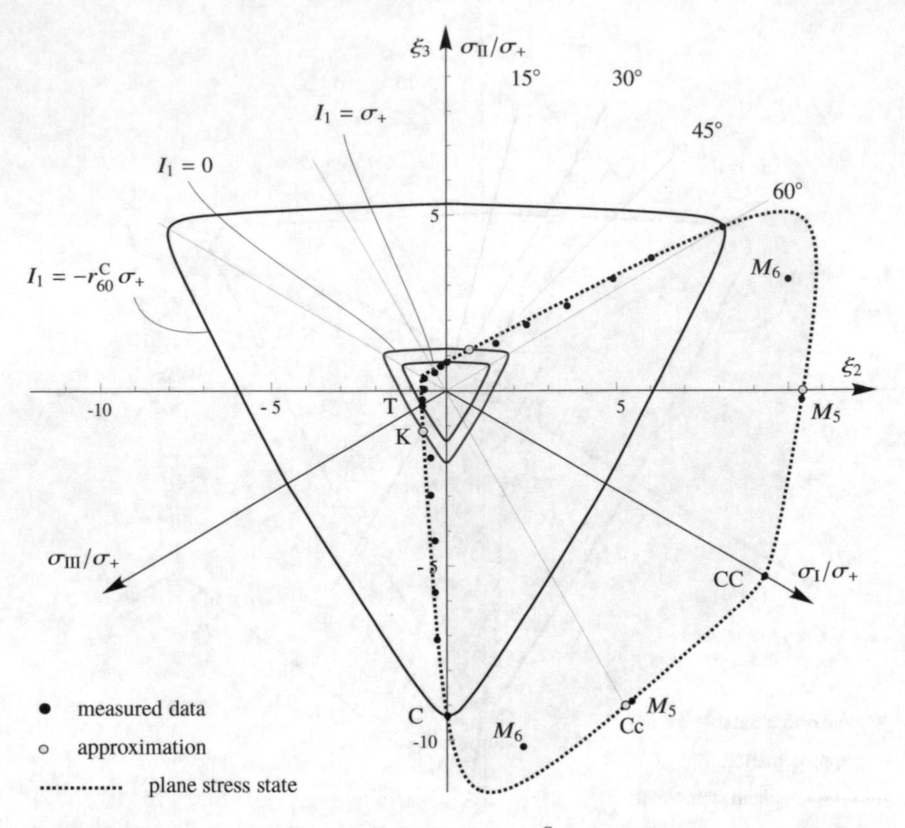

Fig. 12.29 Measured data by Kupfer (Table 12.6) with $R^{\mathrm{C}} = 58.25\,\mathrm{MPa}$ normalized with respect to $R^{\mathrm{T}} = 5.12\,\mathrm{MPa}$ in the π-plane approximated using the CTS (12.42) (Fig. 12.26). The cross sections orthogonal to the hydrostatic axis corresponding to the different values of I_1 are shown

the points C and CC respectively are shown in the $\sigma_{\mathrm{I}} - \sigma_{\mathrm{II}}$ diagram (Figs. 12.30, 12.31, and 12.32) to illustrate the necessity of the generalized criteria. Approximations using the Mohr-Coulomb criterion though the point C are very conservative. Approximations using the modified Yu criterion through the points C or CC show deviations of the measured data from the classical concept. Hence, approximations though either C or CC are not sufficient for concrete.

All approximations except the CTS with the fixed points T, C, and CC show large deviations in the third quadrant of the $\sigma_{\mathrm{I}} - \sigma_{\mathrm{II}}$ diagram (Figs. 12.33, 12.34, and 12.35). In the second (forth) quadrant all approximations practically coincide. Here, deviations of approximations and data lie in a tolerable range. The measured data are approximated with convex criteria whereas the data hint at non-convex cluster in the second quadrant. In the first quadrant all approximations show considerable deviations from the data.

For the present approximations no ideal parameter settings were found. Different weights are realized for different approaches. The most important approximations are

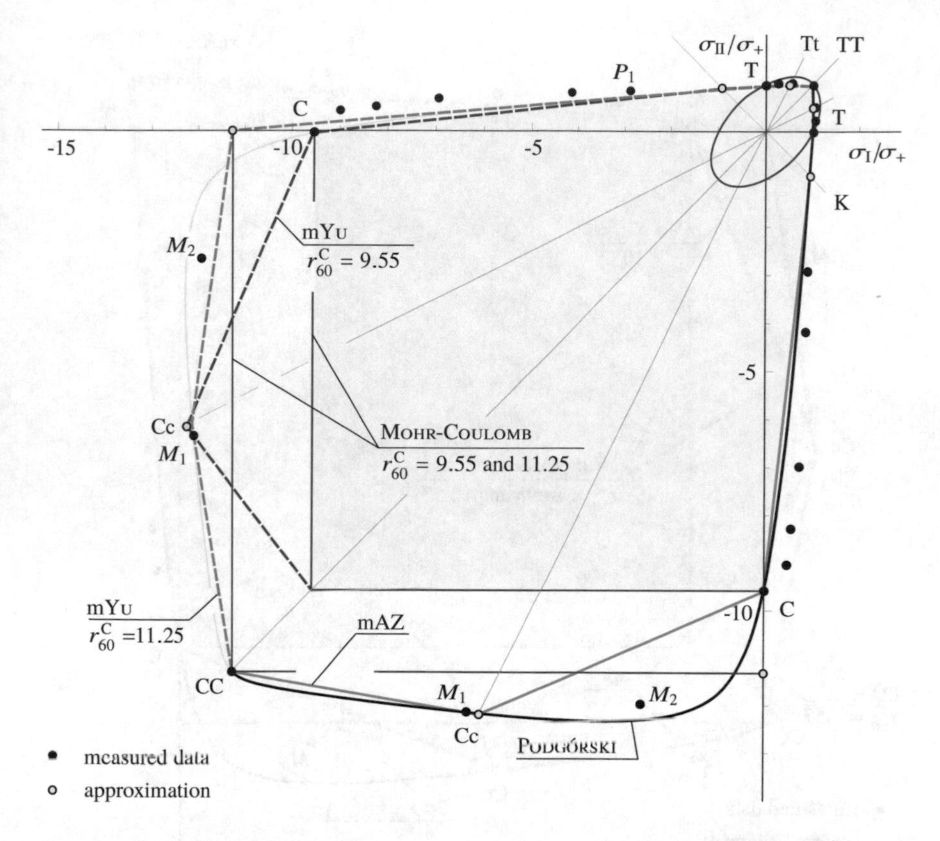

Fig. 12.30 Measured data by Kupfer (Table 12.4) with $R^{\mathrm{C}} = 18.73\,\mathrm{MPa}$ normalized with respect to $R^{\mathrm{T}} = 1.96\,\mathrm{MPa}$. Plane stress state $\sigma_{\mathrm{I}} - \sigma_{\mathrm{II}}$, $\sigma_{\mathrm{III}} = 0$: the Podgórski criterion (12.30), modified Altenbach-Zolochevsky (mAZ) criterion (12.35), and the modified Yu (mYu) criterion (12.25) with two settings $r_{60}^{\mathrm{C}} = 9.55$ and 11.25 (Table 12.8). The von Mises hypothesis and the Mohr-Coulomb criterion with two settings $r_{60}^{\mathrm{C}} = 9.55$ and 11.25 are shown for comparison

summarized in Appendix A.8 in order to illustrate the fitting capabilities of different criteria. Additional to T, C, and CC different points $\mathrm{M_i}$ can be required to be fitted exactly rather than just approximated. Corresponding approximations are shown in Figs. 12.33, 12.34, and 12.35. The objective functions $f_{3\mathrm{D}}$ and f_{ray} do not vary significantly.

Deviations of the the approximations using the CTS (12.42) from measured points can be explained with a straight meridian line of the linear I_1-substitution (12.11) used for simplification and by requiring the points T, C, and CC to fall onto the curve exactly. Due to the strength-differential (SD) effect $r_{60}^{\mathrm{C}} \geq 9.55$, the difference in the approximations in the first quadrant are not directly visible. However, it is to point out that using the CTS these approximations are better than known from literature.

Comparing the values r_{30}, r_{60}, and γ_1 from the approximations Kolupaev et al (2018) for the measured data by Tasuji (Tasuji, 1976; Tasuji et al, 1978, 1979) and

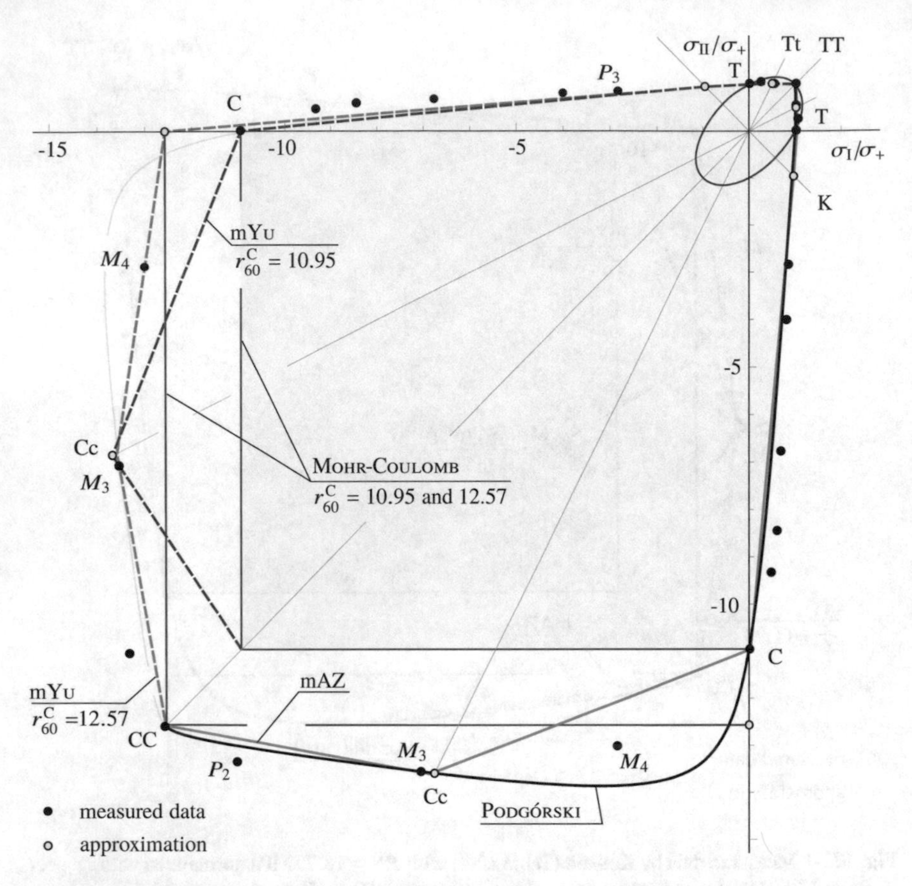

Fig. 12.31 Measured data by Kupfer (Table 12.5) with $R^{\mathrm{C}} = 30.50\,\mathrm{MPa}$ normalized with respect to $R^{\mathrm{T}} = 2.79\,\mathrm{MPa}$. Plane stress state $\sigma_{\mathrm{I}} - \sigma_{\mathrm{II}}$, $\sigma_{\mathrm{III}} = 0$: the Podgórski criterion (12.30) modified Altenbach-Zolochevsky (mAZ) criterion (12.35), and the modified Yu (mYu) criterion (12.25) with two settings $r^{\mathrm{C}}_{60} = 10.95$ and 12.57 (Table 12.10). The von Mises hypothesis and the Mohr-Coulomb criterion with two settings $r^{\mathrm{C}}_{60} = 10.95$ and 12.57 are shown for comparison

by Lee-Song-Han (Lee et al, 2004) for concrete with the values computed for the measured data by Kupfer (Tables 12.8–12.13) shows good agreement. This confirms the applicability of the introduced method.

Data measured by Kupfer are well discussed in the literature (de Borst et al, 2012; Chen, 1975; Chen and Saleeb, 1982; Moradi et al, 2018; Schimmelpfennig, 1971; Speck, 2008; Sun et al, 2011). These data are approximated in Aubertin et al (2000); Boswell and Chen (1987); Brencich and Gambarotta (2001); Brünig and Michalski (2017); Comi (2001); Donida and Mentrasti (1982); Fan and Wang (2002); Folino and Etse (2011); François (2008); Hinchberger (2009); Lade (1982); Link (1976); Menétrey and Willam (1995); Nielsen (1984); Park and Kim (2005); Ren et al (2008); Seow and Swaddiwudhipong (2005); Willam and Warnke (1975), among others.

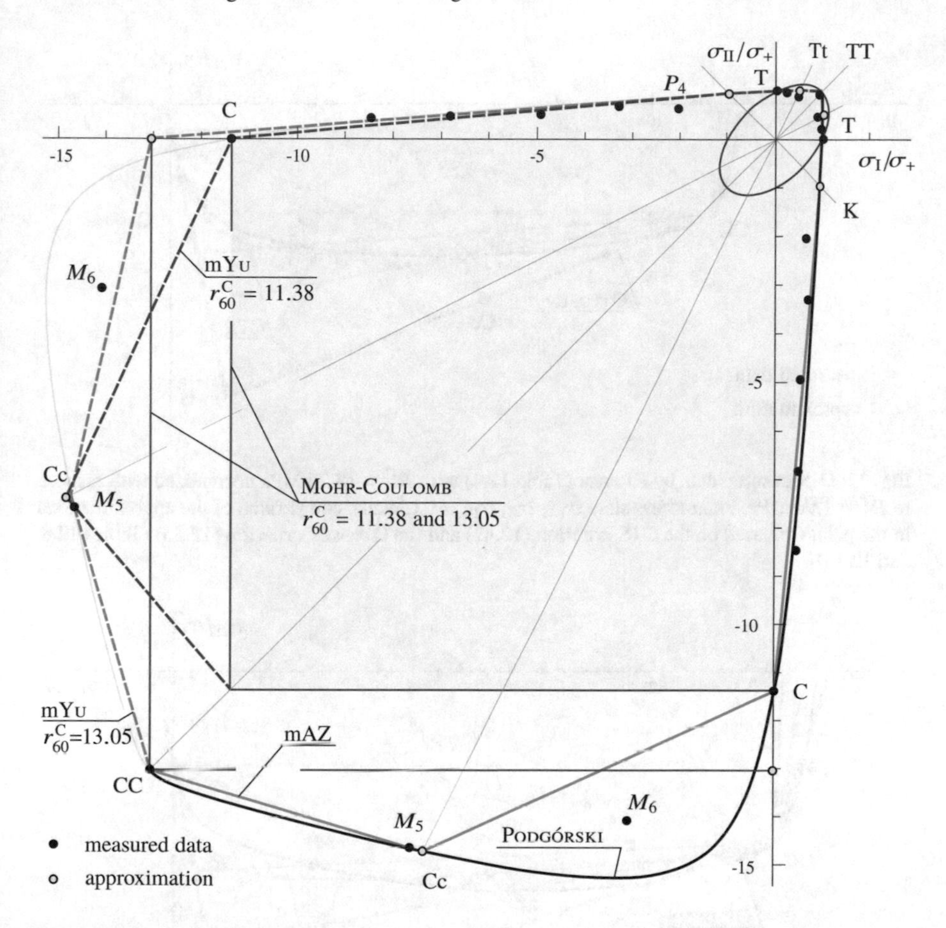

Fig. 12.32 Measured data by Kupfer (Table 12.6) with $R^{\mathrm{C}} = 58.25$ MPa normalized with respect to $R^{\mathrm{T}} = 5.12$ MPa. Plane stress state $\sigma_{\mathrm{I}} - \sigma_{\mathrm{II}}$, $\sigma_{\mathrm{III}} = 0$: the Podgórski criterion (12.30) modified Altenbach-Zolochevsky (mAZ) criterion (12.35), and the modified Yu (mYu) criterion (12.25) with two settings $r_{60}^{\mathrm{C}} = 11.38$ and 13.05 (Table 12.12). The von Mises hypothesis and the Mohr-Coulomb criterion with two settings $r_{60}^{\mathrm{C}} = 11.38$ and 13.05 are shown for comparison

Let us examine the following approximations, which promise good fits of the data, closely:

- Ottosen et al. (Ottosen, 1975, 1977, 1980; Ottosen and Ristinmaa, 2005; Xiaoping et al, 1989) for $R^{\mathrm{C}} = 58.25$ MPa series and
- Chen et al. (Boswell and Chen, 1987; Chen and Han, 2007; Chen, 2007; Chen and Han, 1988; Hsieh et al, 1979, 1980, 1982) for measured series likely either with $R^{\mathrm{C}} = 30.50$ or $R^{\mathrm{C}} = 58.25$ MPa (without exact specification). Further note that data shown by Chen, e.g., Fig. 5.32 in Chen (2007), do not correspond exactly to the measured data published by Kupfer.

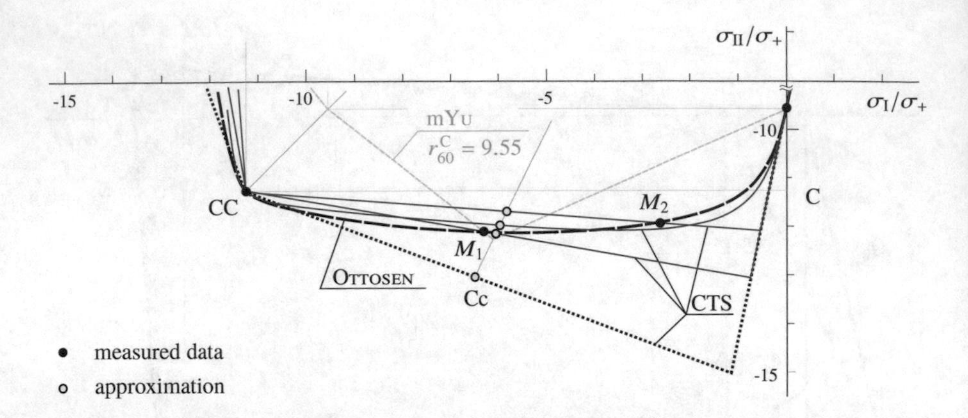

Fig. 12.33 Measured data by Kupfer (Table 12.4) with $R^{\mathrm{C}} = 18.73\,\mathrm{MPa}$ normalized with respect to $R^{\mathrm{T}} = 1.96\,\mathrm{MPa}$. Plane stress state $\sigma_{\mathrm{I}} - \sigma_{\mathrm{II}}$, $\sigma_{\mathrm{III}} = 0$ (detail): comparison of the approximations in the point Cc based on the CTS criterion (12.42) and the Ottosen criterion (12.55) (Tables 12.8 and 12.14)

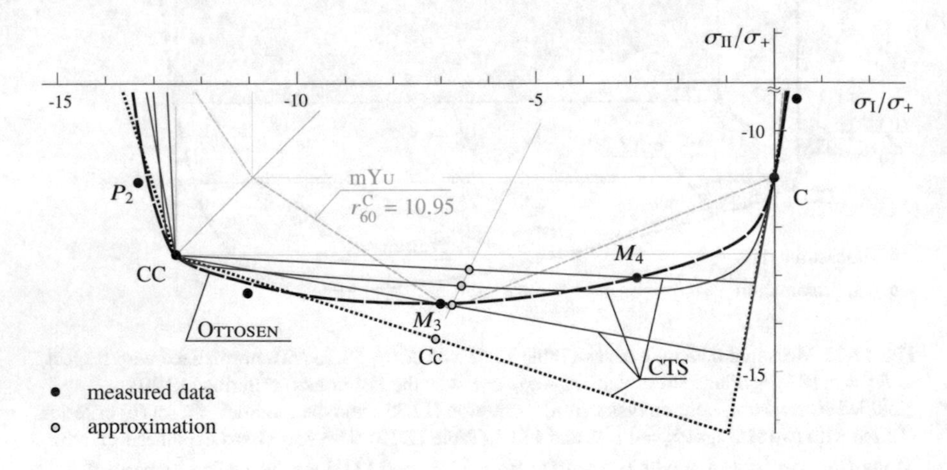

Fig. 12.34 Measured data by Kupfer (Table 12.5) with $R^{\mathrm{C}} = 30.50\,\mathrm{MPa}$ normalized with respect to $R^{\mathrm{T}} = 2.79\,\mathrm{MPa}$. Plane stress state $\sigma_{\mathrm{I}} - \sigma_{\mathrm{II}}$, $\sigma_{\mathrm{III}} = 0$ (detail): comparison of the approximations in the point Cc based on the CTS criterion (12.42) and the Ottosen criterion (12.55) (Tables 12.10 and 12.14)

12.7.3.1 Ottosen Criterion

Instead of generalizing possible shapes in the π-plane, Ottosen allowed for shape variation in the π-plane along the hydrostatic axis. Using the Podgórski shape function Ω (12.28), the Ottosen criterion can be written as

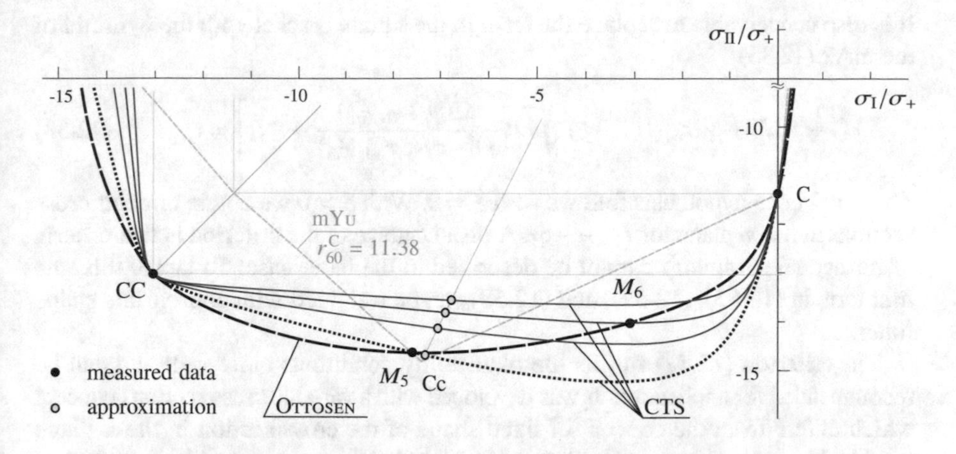

Fig. 12.35 Measured data by KUPFER (Table 12.6) with $R^{\mathrm{C}} = 58.25$ MPa normalized with respect to $R^{\mathrm{T}} = 5.12$ MPa. Plane stress state $\sigma_{\mathrm{I}} - \sigma_{\mathrm{II}}$, $\sigma_{\mathrm{III}} = 0$ (detail): comparison of the approximations in the point Cc based on the CTS criterion (12.42) and the OTTOSEN criterion (12.55) (Tables 12.12 and 12.14)

$$A\,\frac{I_1}{R^{\mathrm{C}}} + B\,\frac{3I_2'}{\left(R^{\mathrm{C}}\right)^2} + C\,\frac{\sqrt{3I_2'}}{R^{\mathrm{C}}}\,\frac{\Omega(\theta,\,0,\,\eta)}{\Omega(\pi/3,\,0,\,\eta)} = 1,\ A \geq 0,\ B \geq 0,\ \text{and } C \geq 0. \quad (12.54)$$

With $B > 0$ we obtain a parable in the meridian cross sections. The geometry in the π-plane "changes from nearly triangular to more circular shape with increasing hydrostatic pressure" (Ottosen, 1977), see also Lubliner et al (1989).

The parameters of the approximations A, B, C, and η are given by OTTOSEN for the values

$$r^{\mathrm{C}}_{60} = 8.33,\ 10,\ 12.5 \qquad \text{and} \qquad \frac{r^{\mathrm{CC}}_0}{r^{\mathrm{C}}_{60}} = 1.16,$$

and do not correspond precisely to the measured data of concrete (Table 12.14). For comparison with our approximations, the criterion (12.54) can be normalized with respect of the uniaxial tension $\sigma_{\mathrm{eq}} = R^{\mathrm{T}}$ (12.9)

$$3\,(1-\chi)\,I_2' + \chi\,\sigma_{\mathrm{eq}}\left[(1-\xi)\,\sqrt{3\,I_2'}\,\frac{\Omega(\theta,\,\beta,\,\eta)}{\Omega(0,\,\beta,\,\eta)} + \xi\,I_1\right] = \sigma_{\mathrm{eq}}^2, \qquad (12.55)$$

with the parameters

$$\chi \in [0,\,1] \qquad \text{and} \qquad \xi \in [0,\,1]. \qquad (12.56)$$

With $\chi = 1$ we obtain the conical criterion (12.30). In general, it can be replaced with the universal conical criterion (12.42)

$$3\,(1-\chi)\,I_2' + \chi\,\sigma_{\mathrm{eq}}\left[(1-\xi)\,\sqrt{3\,I_2'}\,\frac{\Omega(\theta,\,\alpha,\,\beta,\,\gamma)}{\Omega(0,\,\alpha,\,\beta,\,\gamma)} + \xi\,I_1\right] = \sigma_{\mathrm{eq}}^2. \qquad (12.57)$$

It is also conceivable to replace the term in the square bracket with the pyramid of the mAZ (12.35)

$$3\,(1-\chi)\,I_2' + \chi\,\sigma_{\text{eq}}\left[(1-\xi)\sqrt{3\,I_2'}\,\frac{\Omega(\varphi,\,r_{60},\,\xi_{\text{m}})}{\Omega(-\pi/6,\,r_{60},\,\xi_{\text{m}})} + \xi\,I_1\right] = \sigma_{\text{eq}}^2. \qquad (12.58)$$

The von Mises hypothesis follows with $\xi = 0$. With $\xi > 0$ we obtain circular cross sections in the π-plane for $I_1 \to -\infty$. A disadvantage of the criterion is that criteria of hexagonal symmetry cannot be described in the latter case. To tackle this, the first term in (12.55), (12.57), and (12.58) can be weighted with appropriate shape function.

The criterion (12.57) fulfills the plausibility conditions quite well and can be recommended for application. It was developed with a variable cross sections concept which differs from the concept of fixed shape of the cross section in the π-plane used in this work. Compared with (12.42) this effect is controlled with an additional parameter χ. Note, for a reliable determination of failure surfaces with variable cross section available measured data are usually incomplete and insufficient. Some measurements regarding the change of the cross section as function of I_1 are given in Launay and Gachon (1971, 1972); Launay et al (1970). Further approximations are shown in Fahlbusch (2015); Kolupaev (2018), among others. Fitting of such criteria require increased numerical effort.

The criteria (12.55) and (12.57) can be considered as a generalization of

- the Podgórski criterion (12.30) or its corresponding the CTS-formulation (12.42) with $\chi = 1$,
- the formulation in accordance with (12.54) with $\beta = 0$,
- the strain criterion (Kolupaev, 2018) with $\chi = 1$, $\beta = 0$, and $\eta = 1$ (the normal stress hypothesis follows then with $\xi = 1/3$),
- an alternative formulation of the Pisarenko-Lebedev criterion (Kolupaev, 2018; Pisarenko and Lebedev, 1976) with $\chi \in]0,\,1[$, $\beta = 0$, $\eta = 1$, and $\xi = 1/3$,
- the Drucker-Prager (Drucker and Prager, 1952) criterion (rotationally symmetric cone) with $\chi = 1$, $\beta \in \{0,\,1\}$ and $\eta = 1$, and
- the Burzyński-Torre criterion (rotationally symmetric paraboloid) (Balandin, 1937; Burzyński, 1928; Torre, 1947) with $\chi \in]0,\,1[$ and $\xi = 1$.

The variable cross section approach of the Ottosen criterion is fundamentally different from the simple fixed cross section approach followed to this point in the present work. The criterion also allows for parabolic meridian cross sections as opposed to our assumption of conical meridian cross sections to this point. Because of this greater flexibility it provides good approximations of experimental data, as shown in the details of Figs. 12.33, 12.34 and 12.35.

12.7.3.2 Hsieh-Ting-Chen Criterion

The Hsieh-Ting-Chen criterion (Hsieh et al, 1979, 1980, 1982) follows a similar variable crosse section approach as the Ottosen criterion. It reads

$$A\,\frac{I_1}{R^{\mathrm{C}}} + B\,\frac{3I_2'}{(R^{\mathrm{C}})^2} + C\,\frac{\sqrt{3I_2'}}{R^{\mathrm{C}}} + D\,\frac{\max\{\sigma_{\mathrm{I}}, \sigma_{\mathrm{II}}, \sigma_{\mathrm{III}}\}}{R^{\mathrm{C}}} = 1. \tag{12.59}$$

For approximation of the data of concrete with the values

$$r_{60}^{\mathrm{C}} = 10 \qquad \text{and} \qquad \frac{r_0^{\mathrm{CC}}}{r_{60}^{\mathrm{C}}} = 1.15,$$

the four parameters

$$A = 0.2312, \qquad B = 0.6703, \qquad C = 0.5608, \qquad \text{and} \qquad D = 9.1412,$$

were used. Our approximations with the criterion (12.59) are summarized in the Table 12.15 for comparison.

With the normal stress hypothesis as function of the stress angle θ (Chen, 2007; Chen and Zhang, 1991; Kolupaev, 2018) and the normalization $\sigma_{\mathrm{eq}} = R^{\mathrm{T}}$ (12.9), the HSIEH-TING-CHEN criterion (12.59) can be rewritten as

$$3\,(1-\chi)\,I_2' + \chi\,\sigma_{\mathrm{eq}}\left[(1-\xi-\eta)\,\sqrt{3\,I_2'} + \frac{1}{3}\,\xi\left(2\,\sqrt{3\,I_2'}\,\cos\theta + I_1\right) + \eta\,I_1\right] = \sigma_{\mathrm{eq}}^2, \tag{12.60}$$

with the parameter restrictions

$$\xi \in [0, 1], \qquad \eta \in [0, 1], \qquad \xi + \eta \leq 1, \qquad \text{and} \qquad \chi \in [0, 1]. \tag{12.61}$$

The parameters of the criterion (12.60) can be converted into the parameters of the OTTOSEN criterion but not vice versa.

The modified HSIEH-TING-CHEN criterion (12.60) can be considered as a generalization of

- the LECKIE-HAYHURST criterion (Hayhurst, 1972; Leckie and Hayhurst, 1977) with $\chi = 1$,
- the PISARENKO-LEBEDEV criterion (Lebedev, 1965; Pisarenko and Lebedev, 1976) with $\chi = 1$, $\eta = 0$,
- the strain criterion (Kolupaev, 2018) with $\chi = 1$ and $\xi + \eta = 1$ resulted in $\gamma_1 \in [1/3, 1[$,
- alternative formulation of the PISARENKO-LEBEDEV criterion (Kolupaev, 2018; Pisarenko and Lebedev, 1976) with $\chi \in]0, 1[$, $\xi = 1$, and $\eta = 0$,
- the DRUCKER-PRAGER (Drucker and Prager, 1952) criterion (rotationally symmetric cone) with $\chi = 1$, $\xi = 0$, and
- the BURZYŃSKI-TORRE criterion (rotationally symmetric paraboloid) (Balandin, 1937; Burzyński, 1928; Torre, 1947) with $\chi \in]0, 1[$, $\eta = 1$, and $\xi = 0$.

Some approximations with the criterion (12.60) are summarized in Table 12.16. The weighting parameter χ in Eq. (12.60), which describes meridian curvature, takes the values $\chi \in [0.9846, 1]$ for the discussed measured data (Table 12.16).

The modified criterion (12.60) comprises only the geometries M–R_2 in the π-plane (Fig. 12.5) and thus has limited fitting possibilities compared to the OTTOSEN criterion (12.54). Despite good matches, it is pretentious to infer the material behaviour at $I_1 \to -\infty$ from the plane stress state.

12.8 Summary

Enforcing plausibility assumptions the number of applicable yield and strength criteria reduces to three criteria:

- PODGÓRSKI criterion (12.27)–(12.28),
- modified ALTENBACH-ZOLOCHEVSKY criterion (12.31)–(12.32), and
- modified YU (12.25)–(12.26) criterion.

These criteria describe a single surface in principal stress space without any additional outer contours or plane intersections and as such are preferred over criteria formulated as polynomials. The explicit solvability of the equations of the criteria with respect to the equivalent stress σ_{eq} simplifies the implementation of the criteria in FEM codes and reduces the computational effort. A wide range of possible convex shapes in the π-plane allows the application of these criteria to broad classes of materials.

In the present work, the PODGÓRSKI criterion (12.27) and the modified ALTENBACH-ZOLOCHEVSKY criterion (12.31) are extended with the linear I_1-substitution (12.11) in order to describe pressure-sensitive material behavior. The modified YU criterion (12.25) is intrinsically a function of the first invariant I_1 and can be used for modeling of pressure-sensitive and pressure-insensitive material behavior.

Despite the versatility of these criteria, they show restrictions in their application. The PODGÓRSKI criterion (12.27) is C^1-continuously differentiable except at the border of the CAPURSO criterion, but it cannot describe isogonal hexagons in the π-plane (Fig. 12.7). The modified ALTENBACH-ZOLOCHEVSKY criterion (12.31) is capable of representing all points in the $r_{60} - r_{30}$ diagram (Fig. 12.5), but it is only C^0-continuously differentiable. The modified YU criterion has a fixed relation between the geometry in the π-plane and the inclination of the meridians. Owing to these limitation, none of the three above criteria is universally applicable.

In order to resolve the limitations of the above three criteria, the present study proposes a universal strength criterion of trigonal symmetry (CTS) (12.40) which is capable of describing isogonal and isotoxal hexagons as well as isotoxal dodecagons in the π-plane (Table 12.1). The proposed criterion has one additional parameter in comparison to the PODGÓRSKI and the modified ALTENBACH-ZOLOCHEVSKY criteria.

Introducing a second additional parameter allows for further generalizing the proposed criterion to incorporate arbitrary yield surfaces of hexagonal symmetry, as well. To this end we introduce a universal deviatoric function (UDF) capable of describing isogonal and isotoxal dodecagons. The universal criterion of hexagonal symmetry (CHS) formulated using the UDF thus encompasses basic yield criteria (Table 12.1). Its application can be envisaged in theory of plasticity. Further geometries can be obtained with the UDF (12.45) with the setting $n > 2$.

We implicitly presumed that meridians are straight lines in the Burzyński-plane and the geometry of the surface in the π-plane does not change along the hydrostatic axis. Lifting this constraint and allowing for variable cross sections in the π-plane along the hydrostatic axis, even more accurate approximations are possible. Such approximations for concrete are known by Ottosen, Chen, and also by Cuntze (Cuntze, 2017; Mittelstedt and Becker, 2016). The variable geometry of Ottosen, for instance, may be combined with the herein proposed UDF. Of course, variable cross section criteria require more measured data points.

Despite extensive fitting capabilities, the CTS and CHS do not fulfill all plausibility assumptions. Certain shapes are not associated to a unique set of parameters and parameters do not change continuously when describing all possible shapes in the $r_{60} - r_{30}$ and $r_{30} - r_{15}$ diagrams. This may cause problems in the fitting procedure, which can be treated with appropriate parameter restrictions.

Parameter identification follows the method described in Kolupaev (2018); Kolupaev et al (2016, 2018). The fitting procedure and different visualizations are provided in Wolfram Mathematica (Wolfram (2003)). The source codes are freely available. Approximations obtained with criteria discussed in the present work for three types of concrete are visualized in the Burzyński-plane, π-plane, and $\sigma_{\text{I}} - \sigma_{\text{II}}$ plane. The yield function proposed in the present work provides better accuracy than existing criteria from literature. Further I_1-substitutions (parabola, hyperbola, etc. as meridian) (Kolupaev, 2018) can improve accuracy even further.

The search for an appropriate strength criterion continues driven by the influence of current developments in material research and stringent design requirements. As history demonstrates, only simple and elegant criteria with clear geometrical background have chances to find practical application.

Acknowledgements The authors thank Dr. Alexandre Bolchoun, ISG Industrielle Steuerungstechnik GmbH, Stuttgart for his helpful suggestions and comments. We thank Prof. Dr.-Ing. habil. Wilfried Becker, Technische Universität Darmstadt, for his support of our research. The help of Sophia Bremm is gratefully acknowledged.

Appendices

A.1 Invariants of the Stress Tensor

Eigenvalues of the symmetric second-rank stress tensor are principal stresses (or principal invariants) σ_{I}, σ_{II}, and σ_{III} (Altenbach et al, 1995; Życzkowski, 1981). The axiatoric-deviatoric invariants of the stress tensor are used throughout the present work: The three stress invariants – the trace I_1 of the stress tensor

$$I_1 = \sigma_{\text{I}} + \sigma_{\text{II}} + \sigma_{\text{III}} \tag{12.62}$$

and the invariants I_2', I_3' of the stress deviator

$$I_2' = \frac{1}{6}\left[(\sigma_{\mathrm{I}} - \sigma_{\mathrm{II}})^2 + (\sigma_{\mathrm{II}} - \sigma_{\mathrm{III}})^2 + (\sigma_{\mathrm{III}} - \sigma_{\mathrm{I}})^2\right], \tag{12.63}$$

$$I_3' = \left(\sigma_{\mathrm{I}} - \frac{1}{3}I_1\right)\left(\sigma_{\mathrm{II}} - \frac{1}{3}I_1\right)\left(\sigma_{\mathrm{III}} - \frac{1}{3}I_1\right), \tag{12.64}$$

are expressed in principal stresses (Życzkowski, 1981).

Invariants with geometric meaning of the stress state are often preferred: The scaled invariant I_1 of the stress tensor as the loading coordinate on the hydrostatic axis

$$\xi = I_1/\sqrt{3}, \tag{12.65}$$

the scaled root of the second invariant I_2' of the stress deviator as the radius ρ in the π-plane

$$\rho = \sqrt{2I_2'}, \tag{12.66}$$

and the corresponding stress angle θ in the π-plane

$$\cos 3\theta = \frac{3\sqrt{3}}{2}\frac{I_3'}{(I_2')^{3/2}}, \qquad \theta \in \left[0, \frac{\pi}{3}\right], \tag{12.67}$$

or sometimes the stress angle φ

$$\sin 3\varphi = \frac{3\sqrt{3}}{2}\frac{I_3'}{(I_2')^{3/2}}, \qquad \varphi \in \left[-\frac{\pi}{6}, \frac{\pi}{6}\right]. \tag{12.68}$$

Further, the invariant $\tan\psi$ as the stress triaxiality factor or the inclination of the loading in the Burzyński-plane with coordinates (ξ, ρ)

$$\tan\psi = \frac{\rho}{\xi} \qquad \text{with} \qquad \psi \in [0, \pi] \tag{12.69}$$

is useful for the analysis of measured data. Alternatively, the radius ρ (12.66) can be scaled as

$$\rho^* = \sqrt{\frac{3}{2}}\,\rho = \sqrt{3I_2'} \tag{12.70}$$

and the stress triaxiality factor is defined as relation

$$\tan\psi^* = \frac{\sqrt{3I_2'}}{I_1} \qquad \text{with} \qquad \psi \in [0, \pi]. \tag{12.71}$$

This adjustment results from the equivalence

$$I_1^2 = 3I_2' \tag{12.72}$$

for the uniaxial tensile and compression stresses. The axes of the BURZYŃSKI-plane are then $(I_1, \sqrt{3I_2'})$. Using the axes $(I_1, \sqrt{3I_2'})$ often leads to rational values in evaluations and the comparison of approximations simplifies. The comparison of measured data to approximations using the VON MISES hypothesis (12.7) is straightforward.

A.2 Geometric Properties in the π-plane

In the following, details on the computation of geometric properties in the π-plane are given. The radius $\rho(0)$ for the 0-meridian (meridian with the stress angle $\theta = 0$) in the chosen cross section (12.15) is obtained setting $\cos 3\theta = 1$:

$$\sigma_{\text{I}} = I_1 - 2\,\sigma_{\text{III}}, \qquad \sigma_{\text{II}} = \sigma_{\text{III}} = \frac{1}{3}\left[I_1 - \rho(0)\right], \tag{12.73}$$

or

$$\sigma_{\text{I}} = \sigma_{\text{II}} = \frac{1}{2}\,(I_1 - \sigma_{\text{III}}), \qquad \sigma_{\text{III}} = \frac{1}{3}\left[I_1 + 2\,\rho(0)\right]. \tag{12.74}$$

The radius $\rho(\pi/12)$ for the $\pi/12$-meridian is obtained using $\cos 3\theta = \sqrt{2}/2$:

$$\begin{aligned}
\sigma_{\text{I}} &= I_1 + \frac{1}{\sqrt{3}}\sqrt{(I_1 - 3\,\sigma_{\text{III}})^2} - 2\,\sigma_{\text{III}},\\
\sigma_{\text{II}} &= -\frac{1}{\sqrt{3}}\sqrt{(I_1 - 3\,\sigma_{\text{III}})^2} + \sigma_{\text{III}},\\
\sigma_{\text{III}} &= \frac{1}{3}\left[I_1 - \sqrt{2 - \sqrt{3}}\,\rho(\pi/12)\right],
\end{aligned} \tag{12.75}$$

or

$$\begin{aligned}
\sigma_{\text{I}} &= I_1 + \frac{1}{\sqrt{3}}\sqrt{(I_1 - 3\,\sigma_{\text{III}})^2} - 2\,\sigma_{\text{III}},\\
\sigma_{\text{II}} &= -\frac{1}{\sqrt{3}}\sqrt{(I_1 - 3\,\sigma_{\text{III}})^2} + \sigma_{\text{III}},\\
\sigma_{\text{III}} &= \frac{1}{3}\left[I_1 + \sqrt{2 + \sqrt{3}}\,\rho(\pi/12)\right],
\end{aligned} \tag{12.76}$$

or

$$\begin{aligned}
\sigma_{\text{I}} = \sigma_{\text{II}} &= \frac{1}{6}\left[3\,I_1 \pm \sqrt{3}\sqrt{(I_1 - 3\,\sigma_{\text{III}})^2} - 3\,\sigma_{\text{III}}\right],\\
\sigma_{\text{III}} &= \frac{1}{3}\left[I_1 - \sqrt{2}\,\rho(\pi/12)\right].
\end{aligned} \tag{12.77}$$

The radius $\rho(\pi/6)$ for the $\pi/6$-meridian is obtained with $\cos 3\theta = 0$ or $I_3' = 0$:

$$\sigma_{\text{I}} = \frac{1}{3}\,I_1 \qquad \sigma_{\text{II}} = \frac{2}{3}\,I_1 - \sigma_{\text{III}}, \qquad \sigma_{\text{III}} = \frac{1}{3}\left[I_1 - \sqrt{3}\,\rho(\pi/6)\right], \tag{12.78}$$

or

$$\sigma_{\rm I} = \frac{1}{3}\,I_1 \qquad \sigma_{\rm II} = \frac{2}{3}\,I_1 - \sigma_{\rm III}, \qquad \sigma_{\rm III} = \frac{1}{3}\left[I_1 + \sqrt{3}\,\rho(\pi/6)\right]. \tag{12.79}$$

For the $\pi/3$-meridian setting $\cos 3\theta = -1$ yields

$$\sigma_{\rm I} = I_1 - 2\,\sigma_{\rm III}, \qquad \sigma_{\rm II} = \sigma_{\rm III} = \frac{1}{3}\,[I_1 + \rho(\pi/3)], \tag{12.80}$$

or

$$\sigma_{\rm I} = \sigma_{\rm II} = \frac{1}{2}\,(I_1 - \sigma_{\rm III}), \qquad \sigma_{\rm III} = \frac{1}{3}\,[I_1 - 2\,\rho(\pi/3)]. \tag{12.81}$$

Inserting the above stress states into the equation of the criterion Φ (12.3), the values $\rho(0)$, $\rho(\pi/12)$, $\rho(\pi/6)$, and $\rho(\pi/3)$ in the chosen cross section I_1 = const. can be computed as the respective smallest positive solution.

A.3 Identification of Limit Surface for Pressure-sensitive Materials

The relations (12.14) describe certain geometrical properties in the π-plane. In the following, additional points for pressure-sensitive materials are discussed: The loadings C and TT lie on the $\pi/3$-meridian (Fig. 12.2):

$$\sigma_{\rm I} = \sigma_{\rm II} = 0, \qquad \sigma_{\rm III} = -r_{60}^{\rm C}\,\sigma_{\rm eq} \tag{12.82}$$

and

$$\sigma_{\rm I} = \sigma_{\rm II} = r_{60}^{\rm TT}\,\sigma_{\rm eq}, \qquad \sigma_{\rm III} = 0. \tag{12.83}$$

The loadings K, Tt, and Cc lie on the $\pi/6$-meridian:

$$\sigma_{\rm I} = -\sigma_{\rm III} = \frac{1}{\sqrt{3}}\,r_{30}^{\rm K}\,\sigma_{\rm eq}, \qquad \sigma_{\rm II} = 0, \tag{12.84}$$

$$\sigma_{\rm I} = 0, \qquad 2\,\sigma_{\rm II} = \sigma_{\rm III} = -\frac{2}{\sqrt{3}}\,r_{30}^{\rm Cc}\,\sigma_{\rm eq}, \tag{12.85}$$

and

$$\sigma_{\rm I} = 2\,\sigma_{\rm II} = \frac{2}{\sqrt{3}}\,r_{30}^{\rm Tt}\,\sigma_{\rm eq}, \qquad \sigma_{\rm III} = 0. \tag{12.86}$$

The loadings CC and uniaxial tensile loading T lie on the 0-meridian:

$$\sigma_{\rm I} = 0, \qquad \sigma_{\rm II} = \sigma_{\rm III} = -r_{0}^{\rm CC}\,\sigma_{\rm eq} \tag{12.87}$$

and

$$\sigma_{\rm I} = \sigma_{\rm eq}, \qquad \sigma_{\rm II} = \sigma_{\rm III} = 0. \tag{12.88}$$

Table 12.3 Basic loading: points, coordinates in the Burzyński-plane $(I_1, \sqrt{3I_2'})$, stress angle (12.67)-(12.68), and the triaxiality factor $\tan\psi^*$ (12.71)

Loading	Label	Coordinates	θ	φ	$\tan\psi^*$
hydrostatic tension	TTT	$(3\,r^{\mathrm{TTT}}, 0)$	–	–	0
equibiaxial tension	TT	$(2\,r_{60}^{\mathrm{TT}}, r_{60}^{\mathrm{TT}})$	$\pi/3$	$\pi/6$	1/2
biaxial tension with $I_3' = 0$	Tt	$(\sqrt{3}\,r_{30}^{\mathrm{Tt}}, r_{30}^{\mathrm{Tt}})$	$\pi/6$	0	$1/\sqrt{3}$
tension	T	(1, 1)	0	$-\pi/6$	1
torsion	K	$(0, r_{30}^{\mathrm{K}})$	$\pi/6$	0	∞
compression	C	$(-r_{60}^{\mathrm{C}}, r_{60}^{\mathrm{C}})$	$\pi/3$	$\pi/3$	-1
equibiaxial compression	Cc	$(-\sqrt{3}\,r_{30}^{\mathrm{Cc}}, r_{30}^{\mathrm{Cc}})$	$\pi/6$	0	$-1/\sqrt{3}$
biaxial compression with $I_3' = 0$	CC	$(-2\,r_{0}^{\mathrm{CC}}, r_{0}^{\mathrm{CC}})$	0	$-\pi/6$	$-1/2$
hydrostatic compression	CCC	$(-3\,r^{\mathrm{CCC}}, 0)$	–	–	0

The values for the plane stress state at the angle $\theta = \pi/12$ and $\pi/4$ can be introduced equivalently.

The triaxiality factor $\tan\psi^*$ (12.71) defines the elevation of the straight line in the Burzyński-plane $(I_1, \sqrt{3I_2'})$ through the origin (0, 0) which contains corresponding points of the surface Φ (Table 12.3). Some examples are shown in Tables 12.4, 12.5, and 12.6.

For criteria of pressure-insensitive material behavior Eq. (12.4) with the linear I_1-substitution Eq. (12.11) the estimation of the hydrostatic strength (12.21) is obtained from a straight line through the measured points:

- on the 0-meridian: $\mathrm{CC}(-2\,r_0^{\mathrm{CC}}, r_0^{\mathrm{CC}})$ and T(1, 1)

$$\frac{I_1/R^{\mathrm{T}} - 1}{-2r_0^{\mathrm{CC}} - 1} = \frac{\sqrt{3I_2'}/R^{\mathrm{T}} - 1}{r_0^{\mathrm{CC}} - 1} \tag{12.89}$$

- on the $\pi/6$-meridian: $\mathrm{C}(-r_{60}^{\mathrm{C}}, r_{60}^{\mathrm{C}})$ and $\mathrm{TT}(2r_{60}^{\mathrm{TT}}, r_{60}^{\mathrm{TT}})$

$$\frac{I_1/R^{\mathrm{T}} + r_{60}^{\mathrm{C}}}{2r_{60}^{\mathrm{TT}} + r_{60}^{\mathrm{C}}} = \frac{\sqrt{3I_2'}/R^{\mathrm{T}} - r_{60}^{\mathrm{C}}}{r_{60}^{\mathrm{TT}} - r_{60}^{\mathrm{C}}} \tag{12.90}$$

- on the $\pi/3$-meridian: $\mathrm{K}(0, r_{30}^{\mathrm{K}})$ and $\mathrm{Cc}(-\sqrt{3}\,r_{30}^{\mathrm{Cc}}, r_{30}^{\mathrm{Cc}})$

$$\frac{I_1/R^{\mathrm{T}}}{-\sqrt{3}r_{30}^{\mathrm{Cc}}} = \frac{\sqrt{3I_2'}/R^{\mathrm{T}} - r_{30}^{\mathrm{K}}}{r_{30}^{\mathrm{Cc}} - r_{30}^{\mathrm{K}}}, \tag{12.91}$$

$\mathrm{K}(0, r_{30}^{\mathrm{K}})$ and $\mathrm{Tt}(\sqrt{3}\,r_{30}^{\mathrm{Tt}}, r_{30}^{\mathrm{Tt}})$ or $\mathrm{Cc}(-\sqrt{3}\,r_{30}^{\mathrm{Cc}}, r_{30}^{\mathrm{Cc}})$ and $\mathrm{Tt}(\sqrt{3}\,r_{30}^{\mathrm{Tt}}, r_{30}^{\mathrm{Tt}})$ accordingly.

Setting $\sqrt{3I_2'} = 0$ in the criterion we obtain $I_1 = 3\,R^{\mathrm{TTT}}$ for the hydrostatic tensile loading. The reciprocal value of $3\,R^{\mathrm{TTT}}/R^{\mathrm{T}}$ gives the parameter γ_1 (12.21)

$$\gamma_1 = \frac{1}{3\,r^{\mathrm{TTT}}}. \tag{12.92}$$

The geometrical values r_{30} and r_{60} on the corresponding meridians can be estimated with the discussed meridians (12.89)–(12.91) and the chosen cross section (12.15). For the criteria (12.12) as functions of the stress angle θ the geometrical values at the cross section are

$$r_{15} = \frac{\Omega(0)}{\Omega(\pi/4)}, \qquad r_{30} = \frac{\Omega(0)}{\Omega(\pi/6)}, \qquad \text{and} \qquad r_{60} = \frac{\Omega(0)}{\Omega(\pi/3)}. \tag{12.93}$$

The values for the basic tests as functions of the geometrical values are

- on the 0-meridian

$$\frac{1-\gamma_1\cdot(-2\,r_0^{\mathrm{CC}})}{1-\gamma_1} = r_0^{\mathrm{CC}}\,\frac{\Omega(0)}{\Omega(0)} \qquad \text{or} \qquad r_0^{\mathrm{CC}} = \frac{1}{1-3\,\gamma_1}, \tag{12.94}$$

- on the $\pi/6$-meridian

$$\frac{1-\gamma_1\cdot(-\sqrt{3}\,r_{30}^{\mathrm{Cc}})}{1-\gamma_1} = \frac{r_{30}^{\mathrm{Cc}}}{r_{30}}, \qquad \frac{1-\gamma_1\cdot 0}{1-\gamma_1} = \frac{r_{30}^{\mathrm{K}}}{r_{30}}, \tag{12.95}$$

and

$$\frac{1-\gamma_1\cdot(\sqrt{3}\,r_{30}^{\mathrm{Tt}})}{1-\gamma_1} = \frac{r_{30}^{\mathrm{Tt}}}{r_{30}}, \tag{12.96}$$

- on the $\pi/3$-meridian

$$\frac{1-\gamma_1\cdot(-r_{60}^{\mathrm{C}})}{1-\gamma_1} = \frac{r_{60}^{\mathrm{C}}}{r_{60}}, \qquad \frac{1-\gamma_1\cdot(2\,r_{60}^{\mathrm{TT}})}{1-\gamma_1} = \frac{r_{60}^{\mathrm{TT}}}{r_{60}}. \tag{12.97}$$

For the criteria (12.12) as functions of the stress angle φ the values follow in analogy (Table 12.3). With $\gamma_1 = 0$ (pressure-insensitive material behavior) the geometrical values (12.93) and the values for the basic tests (12.95)–(12.97) coincide, see Eq. (12.22).

A.4 Derivation of the Modified Yu Strength Criterion

The Yu strength criterion (YSC) contains three classical criteria (the Tresca and the Schmidt-Ishlinsky hypotheses (Fig. 12.3c) and the normal stress hypothesis) and the Sokolovsky criterion. It was formulated in principal stresses (Yu, 2002, 2018; Yu and Yu, 2019) and as a polynomial function (Kolupaev, 2018). The Yu yield criterion (YYC) is obtained setting $r_{60}^{\rm C} = 1$ in the YSC. It contains the criteria of hexagonal symmetry: the Tresca and the Schmidt-Ishlinsky hypotheses and the Sokolovsky criterion (Fig. 12.6, transition H_1–D_1–H_2). In order to avoid plane intersections, the YYC is formulated as a function of the stress angle θ (Kolupaev, 2017, 2018)

$$\Phi_{\rm YYC} = \sqrt{3 I_2'}\,\frac{\Omega(\theta,\chi)}{\Omega(0,\chi)} - \sigma_{\rm eq} = 0, \qquad r_{60} = r_{60}^{\rm C} = 1, \tag{12.98}$$

with the shape function

$$\Omega(\theta,\chi) = \sin\left(\chi\,\frac{\pi}{6} + \arcsin\left[\cos\left(\frac{1}{3}\arcsin\left[\cos 3\,\theta\right]\right)\right]\right), \qquad \chi \in [0, 1]. \tag{12.99}$$

The values r_{15} and r_{30} of the YYC can be computed with (12.93) as

$$r_{15} = \frac{\Omega(0,\chi)}{\Omega(\pi/12,\chi)} = \csc\left[\frac{\pi}{12}\,(5 + 2\,\chi)\right]\,\sin\left[\frac{\pi}{6}\,(2 + \chi)\right] \tag{12.100}$$

and

$$r_{30} = \frac{\Omega(0,\chi)}{\Omega(\pi/6,\chi)} = \frac{1}{2}\left(\sqrt{3} + \tan\left[\frac{\pi\,\chi}{6}\right]\right). \tag{12.101}$$

The YYC yields with

- $\chi = 0$ the Tresca hypothesis,
- $\chi = 1/2$ the Sokolovsky criterion, and
- $\chi = 1$ the Schmidt-Ishlinsky hypothesis.

The linear combination of the YYC Eqs. (12.98)–(12.99) and the normal stress hypothesis as function of the stress angle θ, cf. Chen and Zhang (1991)

$$\Phi_{\rm NSH} = \sqrt{3 I_2'}\,\cos\theta - \frac{\sigma_{\rm eq} - \gamma_1^*\,I_1}{1 - \gamma_1^*} = 0 \qquad \text{with} \qquad \gamma_1^* = \frac{1}{3}, \tag{12.102}$$

leads to the modified Yu strength criterion (mYu) which is similar to the YSC

$$\frac{3}{1 + 2\,r_{60}^{\rm C}}\,\Phi_{\rm YYC} + \left(1 - \frac{3}{1 + 2\,r_{60}^{\rm C}}\right)\Phi_{\rm NSH} = 0 \qquad \text{for} \qquad r_{60}^{\rm C} \geq 1. \tag{12.103}$$

By solving Eq. (12.103) for the equivalent stress $\sigma_{\rm eq}$ we obtain

$$\sigma_{\rm eq} = \frac{1}{3\,r_{60}^{\rm C}}\left[\sqrt{3 I_2'}\left[3\,\frac{\Omega(\theta,\chi)}{\Omega(0,\chi)} + 2\,(r_{60}^{\rm C} - 1)\,\cos\theta\right] + I_1\,(r_{60}^{\rm C} - 1)\right]. \tag{12.104}$$

The basic test values $r_{30}^{\rm K}$, $r_{30}^{\rm Tt}$, and $r_{30}^{\rm Cc}$ are

$$r_{30}^{\mathrm{K}} = \frac{r_{60}^{\mathrm{C}}}{\dfrac{\sqrt{3}}{3}\,(r_{60}^{\mathrm{C}} - 1) + \dfrac{\Omega(\pi/6,\,\chi)}{\Omega(0,\,\chi)}}, \tag{12.105}$$

$$r_{30}^{\mathrm{Tt}} = \frac{r_{60}^{\mathrm{C}}}{\dfrac{2\sqrt{3}}{3}\,(r_{60}^{\mathrm{C}} - 1) + \dfrac{\Omega(\pi/6,\,\chi)}{\Omega(0,\,\chi)}}, \tag{12.106}$$

$$r_{30}^{\mathrm{Cc}} = r_{60}^{\mathrm{C}}\,\frac{\Omega(0,\,\chi)}{\Omega(\pi/6,\,\chi)}. \tag{12.107}$$

The geometrical values r_{15}, r_{30}, and r_{60} follow with

$$r_{15} = \frac{1 + 2\,r_{60}^{\mathrm{C}}}{\sqrt{2+\sqrt{3}}\,(r_{60}^{\mathrm{C}} - 1) + 3\,\dfrac{\Omega(\pi/12,\,\chi)}{\Omega(0,\,\chi)}}, \tag{12.108}$$

$$r_{30} = \frac{1 + 2\,r_{60}^{\mathrm{C}}}{\sqrt{3}\,(r_{60}^{\mathrm{C}} - 1) + 3\,\dfrac{\Omega(\pi/6,\,\chi)}{\Omega(0,\,\chi)}}, \tag{12.109}$$

$$r_{60} = \frac{1 + 2\,r_{60}^{\mathrm{C}}}{(r_{60}^{\mathrm{C}} - 1) + 3\,\dfrac{\Omega(\pi/3,\,\chi)}{\Omega(0,\,\chi)}} = \frac{1 + 2\,r_{60}^{\mathrm{C}}}{2 + r_{60}^{\mathrm{C}}}. \tag{12.110}$$

Setting $r_{60}^{\mathrm{C}} = 1$ we obtain the geometrical values r_{15} (12.100) and r_{30} (12.101) of the YYC Eqs. (12.98)–(12.99). The value γ_1 follows to

$$\gamma_1 = \frac{1}{3}\left(1 - \frac{1}{r_{60}^{\mathrm{C}}}\right). \tag{12.111}$$

A.5 Properties of the Podgórski Criterion

The geometrical values r_{60}, r_{30}, and r_{15} (12.93) of the Podgórski criterion are

$$r_{60} = \frac{\cos\left[\dfrac{1}{3}(\pi\beta - \arccos\eta)\right]}{\cos\left[\dfrac{1}{3}(\pi\beta - \arccos[-\eta])\right]},$$

$$r_{30} = \frac{\cos\left[\dfrac{1}{3}(\pi\beta - \arccos\eta)\right]}{\sin\left[\dfrac{1}{3}\pi\,(\beta + 1)\right]}, \tag{12.112}$$

$$r_{15} = \frac{\cos\left[\dfrac{1}{3}(\pi\beta - \arccos\eta)\right]}{\cos\left[\dfrac{1}{3}(\pi\beta - \arccos[\eta/\sqrt{2}])\right]}.$$

The basic test values follow with Eq. (12.94)–(12.97). Note that in order to avoid numerical issues the real part function Re can be introduced to the shape function (12.28).

$$\Omega(\theta, \beta, \eta) = \mathrm{Re}\left[\cos\left[\frac{1}{3}(\pi\beta - \arccos[\,\eta\cos 3\,\theta\,])\right]\right]. \tag{12.113}$$

A.6 Properties of the Modified Altenbach-Zolochevsky Criterion

The geometrical value r_{15} is computed as

$$r_{15} = \frac{\sqrt{2+\sqrt{3}}\,r_{30}}{1 + r_{30}}, \tag{12.114}$$

and does not depend on the value r_{60} (Fig. 12.11), cf. Fig. 12.10. The geometrical value r_{30} reads

$$r_{30} = \frac{2\sqrt{3}\,r_{60}}{2 + 2\,r_{60} - r_{60}\,\xi_{\mathrm{m}}}. \tag{12.115}$$

Solving Eq. (12.115) for the parameter ξ_{m} leads to

$$\xi_{\mathrm{m}} = 2\left(1 + \frac{1}{r_{60}} - \frac{\sqrt{3}}{r_{30}}\right). \tag{12.116}$$

The basic test value r_0^{CC} follows with (12.94). Further basic test values can be obtained in analogy to (12.95-12.97) for the corresponding angle φ (Table 12.3). A C^0-continuously differentiable approximation of the Sayir criterion (R_1–M–R_2) is obtained by equating the values r_{30} in (12.115) and r_{30} of the Sayir criterion (Kolupaev, 2017, 2018)

$$r_{30} = \frac{r_{60}}{\sqrt{1 - r_{60} + r_{60}^2}} \tag{12.117}$$

with

$$\xi_{\mathrm{m}} = \frac{2}{r_{60}} \left(1 + r_{60} - \sqrt{3}\sqrt{1 - r_{60} + r_{60}^2}\right). \tag{12.118}$$

A.7 Measured Concrete Data

Data measured by KUPFER for three types of concrete with uniaxial compressive strength $R^{\mathrm{C}} = 18.73, 30.50$, and 58.25 MPa, respectively, are given in his Table 5 in KUPFER (Kupfer, 1973). These data were further studied by KUPFER et al. in several publications (Kupfer et al, 1969; Kupfer and Zelger, 1968; Kupfer and Gerstle, 1973; Kupfer and Zelger, 1973).

The $\sigma_{\mathrm{I}}-\sigma_{\mathrm{II}}$ diagram by KUPFER contains additional measured points (Kupfer, 1973). These points, normalized with respect to the corresponding uniaxial compressive strength R^{C}, are:

- $R^{\mathrm{C}} = 18.73$ MPa (191 kp/cm^2, 2717 psi): point $P_1(-0.30, 0.09)$,
- $R^{\mathrm{C}} = 30.50$ MPa (311 kp/cm^2, 4424 psi): points $P_2(-0.26, 0.08)$ and $P_3(-1.22, -1.01)$, and
- $R^{\mathrm{C}} = 58.25$ MPa (594 kp/cm^2, 8449 psi): point $P_4(-0.18, 0.06)$.

Data given by KUPFER (Kupfer, 1973) are completed with the above points (Tables 12.4, 12.5, and 12.6) for fitting strength criteria. The basic loading points in these tables are labeled in accordance with Table 12.3: T (tension), TT (equibiaxial tension), C (compression), and CC (equibiaxial compression). The points L_{i} are used for the estimation of the value r_{30}^{Tt} (Table 12.7). The measured points additionally weighted by approximations with the discussed criteria are marked with M_{i}.

A.8 Estimates and Parameter Studies

Table 12.7 shows estimates for the position of the hydrostatic node for each test series by KUPFER. Tables 12.8 and 12.9 list identified parameters and corresponding restrictions for different criteria for the concrete with $R^{\mathrm{C}} = 18.73$ MPa. Tables 12.10–12.11 and Tables 12.12–12.13 complement the data for the concretes with $R^{\mathrm{C}} = 30.50$ MPa and $R^{\mathrm{C}} = 58.25$ MPa, respectively. Tables 12.14 and 12.15–12.16 show approximations obtained using the OTTOSEN and HSIEH-TING-CHEN criteria, respectively.

Table 12.4 Data measured by KUPFER (Kupfer, 1973) for concrete with $R^{\mathrm{C}} = 18.73$ MPa normalized with respect to $R^{\mathrm{T}} = 1.96$ MPa. The axiatoric-deviatoric invariants (12.62)–(12.64), the invariant $\sqrt{3\,I_2'}$, the stress angle θ Eq. (12.67), and the inclination ψ Eq. (12.71) in the BURZYŃSKI-plane are given for the points of the plane stress state $\sigma_{\mathrm{I}} - \sigma_{\mathrm{II}}, \sigma_{\mathrm{III}} = 0$

Loading point	σ_{I} [–]	σ_{II} [–]	I_1 [–]	I_2' [–]	I_3' [–]	$\sqrt{3\,I_2'}$ [–]	$\cos 3\theta$ [–]	θ [deg]	ψ [deg]
CC	−11.25	−11.25	−22.50	42.1875	105.468750	11.25	1	0	−26.6
M_1	−12.10	−6.29	−18.39	36.6220	5.904154	10.48	0.0692	28.7	−29.7
M_2	−11.90	−2.62	−14.52	39.1032	−75.900560	10.83	−0.8065	47.9	−36.7
C	−9.55	0	−9.55	30.4008	−64.517324	9.55	−1	60	−45
–	−9.00	0.47	−8.53	28.4770	−57.985383	9.24	−0.9914	57.5	−47.3
–	−8.25	0.58	−7.67	24.3868	−45.641113	8.55	−0.9846	56.7	−48.1
–	−6.95	0.72	−6.23	17.9300	−28.285879	7.33	−0.9679	55.2	−49.6
–	−4.15	0.84	−3.31	7.1347	−6.530822	4.63	−0.8903	51.0	−54.4
P_1	−2.86	0.88	−1.98	3.8155	−2.232968	3.38	−0.7784	47.0	−59.6
T	0	1	1	0.3333	0.074074	1	1	0	45
L_1	0.25	1.07	1.32	0.3140	0.053341	0.97	0.7874	12.7	36.4
L_2	0.58	1.07	1.65	0.2868	−0.008161	0.93	−0.1380	32.7	29.4
TT	1	1	2	0.3333	−0.074074	1	−1	60	26.6

Table 12.5 Data measured by KUPFER (Kupfer, 1973) for concrete with $R^{\mathrm{C}} = 30.50$ MPa normalized with respect to $R^{\mathrm{T}} = 2.79$ MPa

Loading point	σ_{I} [–]	σ_{II} [–]	I_1 [ ]	I_2' [–]	I_3' [–]	$\sqrt{3\,I_2'}$ [–]	$\cos 3\theta$ [–]	θ [deg]	ψ [deg]
CC	−12.57	−12.57	−25.14	52.6718	147.134955	12.57	1	0	−26.6
P_2	−13.34	−11.04	−24.38	50.8390	123.372578	12.35	0.8843	9.3	−26.9
M_3	−13.56	−7.05	−20.61	45.9681	8.302879	11.74	0.0692	28.7	−29.7
M_4	−13.03	−2.87	−15.89	46.8690	−99.598833	11.86	−0.8065	47.9	−36.7
C	−10.95	0	−10.95	39.9726	−97.273015	10.95	−1	60	−45
–	−9.33	0.49	−8.85	30.6101	−64.620990	9.58	−0.9914	57.5	−47.3
–	−8.45	0.59	−7.86	25.5878	−49.053849	8.76	−0.9846	56.6	−48.1
–	−6.76	0.70	−6.06	16.9659	−26.035380	7.13	−0.9679	55.2	−49.6
–	−4.01	0.81	−3.20	6.6750	−5.909940	4.47	−0.8903	51.0	−54.4
P_3	−2.84	0.85	−1.99	3.7355	−2.187486	3.35	−0.7872	47.3	−59.2
T	0	1	1	0.3333	0.074074	1	1	0	45
L_3	0.25	1.07	1.32	0.3164	0.053933	0.97	0.7874	12.7	36.4
L_4	0.56	1.03	1.59	0.2667	−0.007316	0.89	−0.1380	32.6	29.4
TT	1	1	2	0.3333	−0.074074	1	−1	60	26.6

References

Altenbach H, Kolupaev VA (2014) Classical and non-classical failure criteria. In: Altenbach H, Sadowski T (eds) Failure and Damage Analysis of Advanced Materials, Springer, Wien, Int. Centre for Mechanical Sciences CISM, Courses and Lectures Vol. 560, pp 1–66

Altenbach H, Altenbach J, Zolochevsky A (1995) Erweiterte Deformationsmodelle und Versagenskriterien der Werkstoffmechanik. Deutscher Verlag für Grundstoffindustrie, Stuttgart

Table 12.6 Data measured by KUPFER (Kupfer, 1973) for concrete with $R^{\mathrm{C}} = 58.25\,\mathrm{MPa}$ normalized with respect to $R^{\mathrm{T}} = 5.12\,\mathrm{MPa}$

Loading point	σ_{I} [−]	σ_{II} [−]	I_1 [−]	I_2' [−]	I_3' [−]	$\sqrt{3I_2'}$ [−]	$\cos 3\theta$ [−]	θ [deg]	ψ [deg]
CC	−13.05	−13.05	−26.09	56.7325	164.473547	13.05	1	0	−26.6
M_5	−14.66	−7.62	−22.28	53.7222	10.489972	12.70	0.0692	28.7	−29.7
M_6	−14.04	−3.09	−17.13	54.4485	−124.710928	12.78	−0.8065	47.9	−36.7
C	−11.38	0	−11.38	43.1629	−109.147567	11.38	−1	60	−45
–	−8.45	0.44	−8.01	25.0926	−47.961777	8.68	−0.9914	57.5	−47.3
–	−6.82	0.48	−6.34	16.6650	−25.782965	7.07	−0.9846	56.6	−48.1
–	−4.94	0.51	−4.43	9.0680	−10.173324	5.22	−0.9679	55.2	−49.6
–	−3.30	0.67	−2.63	4.4978	−3.268868	3.67	−0.8903	51.0	−54.4
P_4	−2.02	0.65	−1.37	1.9427	−0.790808	2.41	−0.7588	46.5	−60.5
T	0	1	1	0.3333	0.074074	1	1	0	45
L_5	0.22	0.97	1.19	0.2567	0.039425	0.88	0.7874	12.7	36.4
L_6	0.48	0.89	1.37	0.1971	−0.004649	0.77	−0.1380	32.6	29.4
TT	0.93	0.93	1.86	0.2878	−0.059413	0.93	−1	60	26.6

Table 12.7 Position of the hydrostatic nodes $1/\gamma_1$ on the axis I_1 estimated from a straight line through the points T and CC on the 0-meridian and through the points C and TT on the $\pi/3$-meridian. The values r_{30}^{K}, r_{30}^{Cc}, and r_{30}^{Tt} are estimated from a straight line through the closest points

R^{C} [MPa]	γ_1 [−] straight line T–CC	γ_1 [−] straight line C–TT	r_{30}^{Cc} [−] straight line M_1–M_2	r_{30}^{K} [−] straight line P–T	r_{30}^{Tt} [−] straight line L_1–L_2
18.73	0.3037	0.2984	10.50	1.80	0.93
30.50	0.3068	0.3029	11.75	1.78	0.90
58.25	0.3078	0.3295	12.70	1.60	0.78

Altenbach H, Bolchoun A, Kolupaev VA (2014) Phenomenological yield and failure criteria. In: Altenbach H, Öchsner A (eds) Plasticity of Pressure-Sensitive Materials, Springer, Berlin Heidelberg, Engineering Materials, pp 49–152

Aubertin M, Li L, Simon R (2000) A multiaxial stress criterion for short- and long-term strength of isotropic rock media. Int J of Rock Mechanics and Mining Sciences 37(8):1169–1193

Balandin PP (1937) On the strength hypotheses (in Russ.: K voprosu o gipotezakh prochnosti). Vestnik inzhenerov i tekhnikov 1:19–24

Bellman RE, Cooke KL, Lockett JA (1970) Algorithms, Graphs and Computers. Academic Press, New York

Benvenuto E (1991) An Introduction to the History of Structural Mechanics. Springer, New York

Betten J (1979) Über die Konvexität von Fließkörpern isotroper und anisotroper Stoffe. Acta Mech 32:233–247

Betten J (2001) Kontinuumsmechanik, 2nd edn. Springer, Berlin

Bigoni D, Piccolroaz A (2004) Yield criteria for quasibrittle and frictional materials. Int J of Solids and Structures 41(11):2855–2878

Billington EW (1986) Introduction to the Mechanics and Physics of Solids. Adam Hilger Ltd., Bristol

Bolchoun A, Kolupaev VA, Altenbach H (2011) Convex and non-convex yield surfaces (in German: Konvexe und nichtkonvexe Fließflächen). Forschung im Ingenieurwesen 75(2):73–92

Table 12.8 Approximation of the measured data by KUPFER with $R^{\mathrm{C}} = 18.73\,\mathrm{MPa}$ (Table 12.4) with the discussed criteria. The objective functions f_∞ and f_{ray} are used for comparison of approximations

Crit.	r_{15} [–]	r_{30} [–]	r_{60} [–]	γ_1 [–]	Parameters [–]	f_2 [–]	f_∞ [–]	f_{ray} [–]
(12.25)	1.04	1.15	1.74	0.2984	$r_{60}^{\mathrm{C}} = 9.5500, \chi = 1.0000$	0.101975	0.586026	0.386292
	1.03	1.15	1.74	0.2984	$r_{60}^{\mathrm{C}} = 9.5500, \chi = 0.8658$	0.104695	0.591279	0.389918
	1.05	1.20	1.74	0.2984	$r_{60}^{\mathrm{C}} = 9.5500, \chi = 1.8221$	0.098926	0.551753	0.557997
	1.03	1.14	1.77	0.3037	$r_0^{\mathrm{CC}} = 11.2500, \chi = 0.7408$	0.056954	0.444263	0.331007
	1.02	1.12	1.77	0.3037	$r_0^{\mathrm{CC}} = 11.2500, \chi = 0.2596$	0.060221	0.465772	0.307861
	1.03	1.13	1.77	0.3037	$r_0^{\mathrm{CC}} = 11.2500, \chi = 0.4598$	0.058004	0.456457	0.308258
(12.30)	1.03	1.14	1.70	0.3037	$\beta = 0.0173, \gamma = 0.8404$	0.041897	0.359271	0.332844
	1.02	1.12	1.70	0.3037	$\beta = 0.0526, \gamma = 0.9082$	0.051722	0.432348	0.269141
	1.02	1.12	1.70	0.3037	$\beta = 0.0589, \gamma = 0.9210$	0.056594	0.453178	0.277316
	1.02	1.12	1.70	0.3037	$\beta = 0.0560, \gamma = 0.9149$	0.054139	0.443095	0.272733
(12.35)	1.04	1.15	1.70	0.3037	$r_{60} = 1.7049, \xi_{\mathrm{m}} = 0.1731$	0.099343	0.583903	0.411533
	1.02	1.12	1.70	0.3037	$r_{60} = 1.7049, \xi_{\mathrm{m}} = 0.0828$	0.107750	0.609975	0.368762
	1.04	1.17	1.70	0.3037	$r_{60} = 1.7049, \xi_{\mathrm{m}} = 0.2245$	0.101147	0.569055	0.545360
	1.03	1.13	1.70	0.3037	$r_{60} = 1.7049, \xi_{\mathrm{m}} = 0.1200$	0.102506	0.599250	0.365609
(12.42)	1.03	1.13	1.70	0.3037	$\alpha = 0.1731, \beta = 0.0322, \gamma = 0.9913$	0.024633	0.265907	0.276391
	1.02	1.12	1.70	0.3037	$\alpha = 0.1731, \beta = 0.0497, \gamma = 1.0000$	0.031098	0.349299	0.218284
	1.02	1.11	1.70	0.3037	$\alpha = 0.1731, \beta = 0.0604, \gamma = 1.0000$	0.043493	0.417015	0.243727
	1.02	1.12	1.70	0.3037	$\alpha = 0.1731, \beta = 0.0564, \gamma = 0.9159$	0.054527	0.444733	0.273408

Table 12.9 Restrictions and values for the evaluations of the measured data normalized w.r.t. $R^{\mathrm{T}} = 1.96\,\mathrm{MPa}$ (Table 12.4) for the approximations (Table 12.8). The surfaces do not cross the hydrostatic axis in the compression region: $r^{\mathrm{CCC}} \to \infty$

Crit.	Restrictions: Points	Restrictions: Parameters [–]	r_0^{CC} [–]	r_{30}^{Cc} [–]	r_{60}^{C} [–]	r_{30}^{K} [–]	r_{30}^{Tt} [–]	r_{60}^{TT} [–]	r^{TTT} [–]
(12.25)	C	$\chi \in [0, 1]$	9.55	11.03	**9.55**	1.65	0.89	1	1.12
	C M_1	$\chi \in [0, 1]$	9.55	10.60	**9.55**	1.64	0.89	1	1.12
	C M_2	$\chi \in [0, 1]$	restriction for χ is exceeded						
	CC	$\chi \in [0, 1]$	**11.25**	12.04	11.25	1.64	0.88	1	1.10
	CC M_1	$\chi \in [0, 1]$	**11.25**	10.51	11.25	1.61	0.87	1	1.10
	CC M_2	$\chi \in [0, 1]$	**11.25**	11.12	11.25	1.62	0.88	1	1.10
(12.30)	CC C	$\gamma \in [0, 1], \beta \in [0, 1]$	**11.25**	11.84	**9.55**	1.64	0.88	0.98	1.10
	CC C M_1	$\gamma \in [0, 1], \beta \in [0, 1]$	**11.25**	10.50	**9.55**	1.61	0.87	0.98	1.10
	CC C M_2	$\gamma \in [0, 1], \beta \in [0, 1]$	**11.25**	10.27	**9.55**	1.60	0.87	0.98	1.10
	CC C $M_1 - M_2$	$\gamma \in [0, 1], \beta \in [0, 1]$	**11.25**	10.37	**9.55**	1.61	0.87	0.98	1.10
(12.35)	CC C	$r_{60} \in [1, 2], \xi_{\mathrm{m}} \in [0, 2/r_{60} - 1]$	**11.25**	13.00	**9.55**	1.66	0.89	0.98	1.10
	CC C M_1	$r_{60} \in [1, 2], \xi_{\mathrm{m}} \in [0, 2/r_{60} - 1]$	**11.25**	10.51	**9.55**	1.61	0.87	0.98	1.10
	CC C M_2	$r_{60} \in [1, 2], \xi_{\mathrm{m}} \in [0, 2/r_{60} - 1]$	restriction for ξ is exceeded						
	CC C $M_1 - M_2$	$r_{60} \in [1, 2], \xi_{\mathrm{m}} \in [0, 2/r_{60} - 1]$	**11.25**	11.41	**9.55**	1.63	0.88	0.98	1.10
(12.42)	CC C	$\alpha, \beta, \gamma \in [0, 1]$	**11.25**	11.28	**9.55**	1.63	0.88	0.98	1.10
	CC C M_1	$\alpha, \beta, \gamma \in [0, 1]$	**11.25**	10.51	**9.55**	1.61	0.87	0.98	1.10
	CC C M_2	$\alpha, \beta, \gamma \in [0, 1]$	**11.25**	10.10	**9.55**	1.60	0.87	0.98	1.10
	CC C $M_1 - M_2$	$\alpha, \beta, \gamma \in [0, 1]$	**11.25**	10.36	**9.55**	1.61	0.87	0.98	1.10

Table 12.10 Approximation of the measured data by Kupfer with $R^{\mathrm{C}} = 30.50\,\mathrm{MPa}$ (Table 12.5) with the discussed criteria. The objective functions f_∞ and f_{ray} are used for comparison of approximations

Crit.	r_{15} [−]	r_{30} [−]	r_{60} [−]	γ_1 [−]	Parameters [−]	f_2 [−]	f_∞ [−]	f_{ray} [−]
(12.25)	1.04	1.15	1.77	0.3029	$r_{60}^{\mathrm{C}} = 10.9507, \chi = 1.0000$	0.057316	0.482342	0.332630
	1.03	1.14	1.77	0.3029	$r_{60}^{\mathrm{C}} = 10.9507, \chi = 0.7814$	0.060343	0.490058	0.340084
	1.04	1.18	1.77	0.3029	$r_{60}^{\mathrm{C}} = 10.9507, \chi = 1.4851$	0.056122	0.464921	0.406116
	1.03	1.13	1.79	0.3068	$r_0^{\mathrm{CC}} = 12.5704, \chi = 0.5618$	0.032054	0.360376	0.258222
	1.02	1.12	1.79	0.3068	$r_0^{\mathrm{CC}} = 12.5704, \chi = 0.2693$	0.033323	0.370303	0.248285
	1.02	1.13	1.79	0.3068	$r_0^{\mathrm{CC}} = 12.5704, \chi = 0.2849$	0.033185	0.369747	0.247891
(12.30)	1.03	1.14	1.74	0.3068	$\beta = 0.0281, \gamma = 0.8882$	0.020901	0.281872	0.255183
	1.02	1.12	1.74	0.3068	$\beta = 0.0462, \gamma = 0.9240$	0.024478	0.329869	0.205349
	1.02	1.12	1.74	0.3068	$\beta = 0.0573, \gamma = 0.9467$	0.031512	0.372026	0.222628
	1.02	1.12	1.74	0.3068	$\beta = 0.0527, \gamma = 0.9372$	0.028009	0.353001	0.211167
(12.35)	1.03	1.15	1.74	0.3068	$r_{60} = 1.7411, \xi_{\mathrm{m}} = 0.1245$	0.054737	0.487200	0.309511
	1.02	1.12	1.74	0.3068	$r_{60} = 1.7411, \xi_{\mathrm{m}} = 0.0687$	0.058247	0.503709	0.288861
	1.04	1.16	1.74	0.3068	$r_{60} = 1.7411, \xi_{\mathrm{m}} = 0.1570$	0.055925	0.477593	0.381836
	1.03	1.13	1.74	0.3068	$r_{60} = 1.7411, \xi_{\mathrm{m}} = 0.0905$	0.056043	0.497271	0.284960
(12.42)	1.03	1.13	1.74	0.3068	$\alpha = 0.1487, \beta = 0.0326, \gamma = 0.9991$	0.010906	0.229776	0.226458
	1.02	1.12	1.74	0.3068	$\alpha = 0.1487, \beta = 0.0441, \gamma = 1.0000$	0.013683	0.233645	0.170687
	1.02	1.12	1.74	0.3068	$\alpha = 0.1487, \beta = 0.0575, \gamma = 1.0000$	0.026592	0.336903	0.209145
	1.02	1.12	1.74	0.3068	$\alpha = 0.1462, \beta = 0.0533, \gamma = 0.9384$	0.028414	0.355348	0.212328

Table 12.11 Restrictions and values for the evaluations of the measured data normalized w.r.t. $R^{\mathrm{T}} = 2.79\,\mathrm{MPa}$ (Table 12.5) for the approximations (Table 12.10). The surfaces do not cross the hydrostatic axis in the compression region: $r^{\mathrm{CCC}} \to \infty$

	Restrictions		r_0^{CC}	r_{30}^{Cc}	r_{60}^{C}	r_{30}^{K}	r_{30}^{Tt}	r_{60}^{TT}	r^{TTT}
Crit.	Points	Parameters [−]	[−]	[−]	[−]	[−]	[−]	[−]	[−]
(12.25)	C	$\chi \in [0, 1]$	10.95	12.64	**10.95**	1.66	0.89	1	1.10
	C M_3	$\chi \in [0, 1]$	10.95	11.86	**10.95**	1.64	0.88	1	1.10
	C M_4	$\chi \in [0, 1]$	restriction for χ is exceeded						
	CC	$\chi \in [0, 1]$	**12.57**	12.79	12.57	1.64	0.88	1	1.09
	CC M_3	$\chi \in [0, 1]$	**12.57**	11.78	12.57	1.62	0.87	1	1.09
	CC M_4	$\chi \in [0, 1]$	**12.57**	11.83	12.57	1.62	0.87	1	1.09
(12.30)	CC C	$\gamma \in [0, 1], \beta \in [0, 1]$	**12.57**	12.64	**10.95**	1.64	0.88	0.99	1.09
	CC C M_3	$\gamma \in [0, 1], \beta \in [0, 1]$	**12.57**	11.77	**10.95**	1.62	0.87	0.99	1.09
	CC C M_4	$\gamma \in [0, 1], \beta \in [0, 1]$	**12.57**	11.25	**10.95**	1.61	0.87	0.99	1.09
	CC C $M_3 - M_4$	$\gamma \in [0, 1], \beta \in [0, 1]$	**12.57**	11.46	**10.95**	1.62	0.87	0.99	1.09
(12.35)	CC C	$r_{60} \in [1, 2], \xi_{\mathrm{m}} \in [0, 2/r_{60} - 1]$	**12.57**	13.56	**10.95**	1.65	0.88	0.99	1.09
	CC C M_3	$r_{60} \in [1, 2], \xi_{\mathrm{m}} \in [0, 2/r_{60} - 1]$	**12.57**	11.78	**10.95**	1.62	0.87	0.99	1.09
	CC C M_4	$r_{60} \in [1, 2], \xi_{\mathrm{m}} \in [0, 2/r_{60} - 1]$	restriction for ξ is exceeded						
	CC C $M_3 - M_4$	$r_{60} \in [1, 2], \xi_{\mathrm{m}} \in [0, 2/r_{60} - 1]$	**12.57**	12.42	**10.95**	1.63	0.87	0.99	1.09
(12.42)	CC C	$\alpha, \beta, \gamma \in [0, 1]$	**12.57**	12.39	**10.95**	1.63	0.87	0.99	1.09
	CC C M_3	$\alpha, \beta, \gamma \in [0, 1]$	**12.57**	11.78	**10.95**	1.62	0.87	0.99	1.09
	CC C M_4	$\alpha, \beta, \gamma \in [0, 1]$	**12.57**	11.14	**10.95**	1.61	0.87	0.99	1.09
	CC C $M_3 - M_4$	$\alpha, \beta, \gamma \in [0, 1]$	**12.57**	11.43	**10.95**	1.62	0.87	0.99	1.09

Table 12.12 Approximation of the measured data by KUPFER with $R^{\mathrm{C}} = 58.25$ MPa (Table 12.6) with the discussed criteria. The objective functions f_∞ and f_{ray} are used for comparison of approximations

Crit.	r_{15} [–]	r_{30} [–]	r_{60} [–]	γ_1 [–]	Parameters [–]	f_2 [–]	f_∞ [–]	f_{ray} [–]
(12.25)	1.04	1.15	1.78	0.3040	$r_{60}^{\mathrm{C}} = 11.3793, \chi = 0.9879$	0.023397	0.284273	0.245538
	1.03	1.15	1.78	0.3040	$r_{60}^{\mathrm{C}} = 11.3793, \chi = 0.9244$	0.023464	0.281898	0.243994
	1.04	1.15	1.78	0.3040	$r_{60}^{\mathrm{C}} = 11.3793, \chi = 1.0000$	0.023399	0.284727	0.246131
	1.02	1.13	1.80	0.3078	$r_{0}^{\mathrm{CC}} = 13.0460, \chi = 0.3293$	0.018787	0.290799	0.171696
	1.03	1.13	1.80	0.3078	$r_{0}^{\mathrm{CC}} = 13.0460, \chi = 0.4203$	0.018919	0.294128	0.174632
	1.03	1.14	1.80	0.3078	$r_{0}^{\mathrm{CC}} = 13.0460, \chi = 0.6138$	0.020000	0.300903	0.199590
(12.30)	1.02	1.12	1.75	0.3078	$\beta = 0.0534, \gamma = 0.9454$	0.017017	0.291169	0.137864
	1.03	1.13	1.75	0.3078	$\beta = 0.0343, \gamma = 0.9067$	0.021042	0.320538	0.149306
	1.02	1.12	1.75	0.3078	$\beta = 0.0489, \gamma = 0.9359$	0.017288	0.298592	0.120174
	1.02	1.13	1.75	0.3078	$\beta = 0.0429, \gamma = 0.9237$	0.018373	0.307979	0.117224
(12.35)	1.03	1.14	1.75	0.3078	$r_{60} = 1.7495, \xi_{\mathrm{m}} = 0.0920$	0.018788	0.274852	0.176708
	1.03	1.13	1.75	0.3078	$r_{60} = 1.7495, \xi_{\mathrm{m}} = 0.0831$	0.018888	0.271949	0.173781
	1.04	1.17	1.75	0.3078	$r_{60} = 1.7495, \xi_{\mathrm{m}} = 0.1864$	0.030082	0.363400	0.492240
	1.03	1.14	1.75	0.3078	$r_{60} = 1.7495, \xi_{\mathrm{m}} = 0.1084$	0.019127	0.280194	0.196362
(12.42)	1.02	1.12	1.75	0.3078	$\alpha = 0.0711, \beta = 0.0534, \gamma = 0.9454$	0.017017	0.291167	0.137871
	1.03	1.13	1.75	0.3078	$\alpha = 0.1431, \beta = 0.0343, \gamma = 0.9067$	0.021042	0.320538	0.149306
	1.02	1.12	1.75	0.3078	$\alpha = 0.1432, \beta = 0.0491, \gamma = 0.9398$	0.017283	0.298525	0.120510
	1.02	1.13	1.75	0.3078	$\alpha = 0.0101, \beta = 0.0436, \gamma = 0.9252$	0.018190	0.306785	0.116231

Table 12.13 Restrictions and values for the evaluations of the measured data normalized w.r.t. $R^{\mathrm{T}} = 5.12$ MPa (Table 12.6) for the approximations (Table 12.12). The surfaces do not cross the hydrostatic axis in the compression region: $r^{\mathrm{CCC}} \to \infty$

Crit.	Restrictions: Points	Restrictions: Parameters [–]	r_0^{CC} [–]	r_{30}^{Cc} [–]	r_{60}^{C} [–]	r_{30}^{K} [–]	r_{30}^{Tt} [–]	r_{60}^{TT} [–]	r^{TTT} [–]
(12.25)	C	$\chi \in [0, 1]$	11.38	13.09	**11.38**	1.66	0.89	1	1.10
	C M_5	$\chi \in [0, 1]$	11.38	12.85	**11.38**	1.65	0.88	1	1.10
	C M_6	$\chi \in [0, 1]$	11.38	13.14	**11.38**	1.66	0.89	1	1.10
	CC	$\chi \in [0, 1]$	**13.05**	12.43	13.05	1.63	0.87	1	1.08
	CC M_5	$\chi \in [0, 1]$	**13.05**	12.76	13.05	1.64	0.87	1	1.08
	CC M_6	$\chi \in [0, 1]$	**13.05**	13.47	13.05	1.65	0.88	1	1.08
(12.30)	CC C	$\beta \in [0, 1], \gamma \in [0, 1]$	**13.05**	11.76	**11.38**	1.62	0.87	0.99	1.08
	CC C M_5	$\beta \in [0, 1], \gamma \in [0, 1]$	**13.05**	12.74	**11.38**	1.64	0.87	0.99	1.08
	CC C M_6	$\beta \in [0, 1], \gamma \in [0, 1]$	**13.05**	11.99	**11.38**	1.62	0.87	0.99	1.08
	CC C $M_5 - M_6$	$\beta \in [0, 1], \gamma \in [0, 1]$	**13.05**	12.30	**11.38**	1.63	0.87	0.99	1.08
(12.35)	CC C	$r_{60} \in [1, 2], \xi_{\mathrm{m}} \in [0, 2/r_{60} - 1]$	**13.05**	13.05	**11.38**	1.64	0.88	0.99	1.08
	CC C M_5	$r_{60} \in [1, 2], \xi_{\mathrm{m}} \in [0, 2/r_{60} - 1]$	**13.05**	12.76	**11.38**	1.64	0.87	0.99	1.08
	CC C M_6	$r_{60} \in [1, 2], \xi_{\mathrm{m}} \in [0, 2/r_{60} - 1]$	restriction for ξ is exceeded						
	CC C $M_5 - M_6$	$r_{60} \in [1, 2], \xi_{\mathrm{m}} \in [0, 2/r_{60} - 1]$	**13.05**	13.64	**11.38**	1.65	0.88	0.99	1.08
(12.42)	CC C	$\alpha, \beta, \gamma \in [0, 1]$	**13.05**	11.76	**11.38**	1.62	0.87	0.99	1.08
	CC C M_5	$\alpha, \beta, \gamma \in [0, 1]$	**13.05**	12.74	**11.38**	1.64	0.87	0.99	1.08
	CC C M_6	$\alpha, \beta, \gamma \in [0, 1]$	**13.05**	11.98	**11.38**	1.62	0.87	0.99	1.08
	CC C $M_5 - M_6$	$\alpha, \beta, \gamma \in [0, 1]$	**13.05**	12.26	**11.38**	1.63	0.87	0.99	1.08

Table 12.14 Approximation of the measured data of KUPFER containing the points T, C, and CC with the criterion (12.55). With bold the setting of the extremal values are highlighted

R^{C} [MPa]	r^{C}_{60} [–]	r^{CC}_{0}/r^{C}_{60} [–]	χ [–]	β [–]	η [–]	ξ [–]	f_2 [–]	f_∞ [–]	f_{ray} [–]	Points [–]	[–]	r^{Cc} [–]	r^{K} [–]	r^{Tt} [–]	r^{TT} [–]	r^{TTT} [–]
18.73	9.55	1.17	0.9932	0	0.9850	0.3294	0.0161	0.2559	0.1868			11.29	1.70	0.87	0.97	1.02
			0.9845	**0**	**1**	0.3624	0.0334	0.3951	0.1879			10.80	1.75	0.85	0.96	0.93
			1	0.0173	0.9688	0.3037	0.0203	0.2501	0.3328			11.84	1.64	0.88	0.98	1.10
			0.9985	0.0501	0.9923	0.3093	0.0248	0.2986	0.2530	M_1		10.48	1.62	0.87	0.98	1.08
			0.9907	0	0.9922	0.3387	0.0170	0.2751	0.1689		M_2	11.10	1.71	0.86	0.97	0.99
			0.9985	0.0501	0.9923	0.3091	0.0248	0.2985	0.2533	M_1	M_2	10.48	1.62	0.87	0.98	1.08
30.50	10.95	1.15	0.9937	0.0059	0.9953	0.3334	0.0060	0.1541	0.1169			12.39	1.70	0.87	0.98	1.01
			0.9929	**0**	0.9952	0.3364	0.0061	0.1459	0.1153			12.46	1.72	0.87	0.98	1.00
			0.9890	**0**	**1**	0.3533	0.0102	0.2176	0.1152			12.12	1.74	0.86	0.97	0.95
			1	0.0281	0.9846	0.3068	0.0100	0.1954	0.2552			12.65	1.64	0.88	0.99	1.09
			0.9969	0.0397	0.9978	0.3199	0.0102	0.2239	0.1617	M_3		11.73	1.65	0.87	0.98	1.05
			0.9923	0.0081	0.9984	0.3392	0.0066	0.1747	0.1082		M_4	12.21	1.71	0.86	0.98	0.99
			0.9973	0.0407	0.9972	0.3181	0.0102	0.2235	0.1650	M_3	M_4	11.73	1.65	0.87	0.98	1.05
58.25	11.38	1.15	1	0.0534	0.9963	0.3078	0.0082	0.2016	0.1378			11.76	1.62	0.87	0.99	1.08
			1	**0**	0.9695	0.3078	0.0191	0.2516	0.4329			14.61	1.66	0.88	0.99	1.08
			0.9898	**0**	**1**	0.3523	0.0360	0.3581	0.2243			12.58	1.75	0.86	0.97	0.96
			1	0.0343	0.9893	0.3078	0.0100	0.2219	0.1493	M_5		12.74	1.64	0.87	0.99	1.08
			1	0.0489	0.9949	0.3078	0.0083	0.2067	0.1202		M_6	11.99	1.62	0.87	0.99	1.08
			0.9963	0.0207	0.9951	0.3239	0.0169	0.2718	0.1574	M_5	M_6	12.69	1.68	0.87	0.98	1.03

Table 12.15 Approximation of the measured data of Kupfer containing the points T, C, and CC with the criterion (12.59). With bold the setting of the fixed values are highlighted

R^{C} [MPa]	r^{C}_{60} [–]	r^{CC}_{0}/r^{C}_{60} [–]	A [–]	B [–]	C [–]	D [–]	f_2 [–]	f_∞ [–]	f_{ray} [–]	r^{Cc} [–]	r^{K} [–]	r^{Tt} [–]	r^{TT} [–]	r^{TTT} [–]
18.73	9.55	1.17	0.4004	1.4004	**0**	9.0030	0.033391	0.395254	0.1886	10.77	1.76	0.85	0.96	0.94
			0.1511	**0**	1.1511	8.2478	0.046772	0.408334	0.3490	10.74	1.62	0.87	0.98	1.10
30.50	10.95	1.15	0.3248	1.3248	**0**	10.5049	0.010160	0.217591	0.1152	12.12	1.75	0.86	0.97	0.95
			0.1289	**0**	1.1289	9.6930	0.024731	0.334309	0.2707	12.09	1.63	0.87	0.99	1.09
58.25	11.38	1.15	0.3213	1.3213	**0**	10.9419	0.035998	0.358074	0.2286	12.58	1.75	0.86	0.97	0.96
			0.1278	**0**	1.1278	10.1238	0.0086383	0.195168	0.1498	12.55	1.63	0.87	0.98	1.08

Table 12.16 Approximation of the measured data of Kupfer containing the points T, C, and CC with the criterion (12.60). With bold the setting of the extremal values are highlighted

R^{C} [MPa]	r^{C}_{60} [–]	r^{CC}_{0}/r^{C}_{60} [–]	χ [–]	ξ [–]	η [–]	f_2 [–]	f_∞ [–]	f_{ray} [–]	Points [–]	r^{Cc} [–]	r^{K} [–]	r^{Tt} [–]	r^{TT} [–]	r^{TTT} [–]
18.73	9.55	1.17	0.9846	0.9574	0.0426	0.033391	0.395253	0.1886		10.77	1.76	0.85	0.96	0.94
			1	0.8636	0.0158	0.046772	0.408333	0.3490		10.74	1.62	0.87	0.98	1.10
			1.0075	0.8189	0.0031	0.062857	0.483807	0.4784	M_1	restriction for χ exceeded				
			1.0100	0.8043	-0.0011	0.069519	0.510186	0.5883	M_2	restrictions for χ, η exceeded				
30.50	10.95	1.15	0.9890	0.9700	0.0300	0.010160	0.217591	0.1152		12.12	1.74	0.86	0.97	0.95
			1	0.8851	0.0118	0.024731	0.334308	0.2707		12.09	1.63	0.87	0.99	1.09
			1.0063	0.8374	0.0015	0.045929	0.421217	0.4141	M_3	restriction for χ exceeded				
			1.0084	0.8220	-0.0018	0.054846	0.454874	0.6451	M_4	restrictions for χ, η exceeded				
58.25	11.38	1.15	1	0.8897	0.0112	0.008638	0.195168	0.1468		12.55	1.63	0.87	0.99	1.08
			0.9898	0.9715	0.0285	0.035998	0.358074	0.2286		12.58	1.74	0.86	0.97	0.96
			1.0110	0.8037	-0.0070	0.029809	0.371660	1.0699	M_5	restrictions for χ, η exceeded				
			1.0070	0.8347	-0.0004	0.016049	0.294948	0.3792	M_6	restrictions for χ, η exceeded				

de Borst R, Crifield MA, Remmers JJC, Verhoosel CV (2012) Nonlinear finite element analysis of solids and structures. Wiley, Chichester

Boswell LF, Chen Z (1987) A general failure criterion for plain concrete. Int J of Solids and Structures 23(5):621–630

Botkin AI (1940a) Equilibrium of granular and brittle materials (in Russ.: O ravnovesii sypuchikh i khrupkikh materialov). Transactions of the Scientific Research Institute of Hydrotechnics, Izvestija NIIG, Leningrad 28:189–211

Botkin AI (1940b) Theories of elastic failure of gramelar and of brittle materials (in Russ.: O prochnosti sypuchikh i khrupkikh materialov). Transactions of the Scientific Research Institute of Hydrotechnics, Izvestija NIIG, Leningrad 26:205–236

Brencich A, Gambarotta L (2001) Isotropic damage model with different tensile–compressive response for brittle materials. Int J of Solids and Structures 38(34):5865–5892

Brünig M, Michalski A (2017) A stress-state-dependent continuum damage model for concrete based on irreversible thermodynamics. Int J of Plasticity 90:31–43

Burzyński W (1928) Study on Material Effort Hypotheses, (in Polish: Studjum nad Hipotezami Wytężenia). Akademia Nauk Technicznych, Lwów

Capurso M (1967) Yield conditions for incompressible isotropic and orthotropic materials with different yield stress in tension and compression. Meccanica 2(2):118–125

Chen WF (1975) Limit Analysis and Soil Plasticity. Elsevier, Amsterdam

Chen WF (2007) Plasticity in Reinforced Concrete. J. Ross Publishing, Plantation

Chen WF, Han DJ (1988) Plasticity for Structural Engineers. Springer, New York

Chen WF, Han DJ (2007) Plasticity for Structural Engineers. J. Ross Publishing, New York

Chen WF, Saleeb AF (1982) Elasticity and Modeling. Elsevier, Amsterdam

Chen WF, Zhang H (1991) Structural Plasticity - Theory, Problems, and CAE Software. Springer, New York

Cicala P (1961) Presentazione geometrica delle relazioni fondamentali d'elastoplasticità. Giornale del Genio Civile 99:125–137

Comi C (2001) A non-local model with tension and compression damage mechanisms. European J of Mechanics-A/Solids 20(1):1–22

Coulomb CA (1776) Essai sur une application des regles des maximis et minimis a quelques problemes de statique relatifs, a la architecture. Mem Acad Roy Div Sav 7:343–387

Cuntze RG (2017) Fracture failure bodies of porous concrete (foam-like), normal concrete, ultra-high-performance-concrete and of the lamella sheet – generated on basis of Cuntze's Failure-Mode-Concept (FMC). In: NAFEMS World Congress NWC, 11-14 June 2017, Stockholm, pp 1–13

Donida G, Mentrasti L (1982) A linear failure criterion for concrete under multiaxial states of stress. Int J of Fracture 19(1):53–66

Drucker DC, Prager W (1952) Soil mechanics and plastic analysis or limit design. Quarterly of Appl Mathematics 10:157–165

Fahlbusch NC (2015) Entwicklung und Analyse mikromechanischer Modelle zur Beschreibung des Effektivverhaltens von geschlossenzelligen Polymerschäumen. Diss., Fachbereich Maschinenbau der Technischen Universität Darmstadt, Darmstadt

Fan SC, Wang F (2002) A new strength criterion for concrete. ACI Structural Journal 99-S33(3):317–326

Filonenko-Borodich MM (1960) Theory of Elasticity. P. Noordhoff W. N., Groningen

Folino P, Etse G (2011) Validation of performance-dependent failure criterion for concretes, Title no. 108–M28. ACI Materials Journal 108(3):261–269

François M (2008) A new yield criterion for the concrete materials. Comptes Rendus Mécanique 336(5):417–421

Hayhurst DR (1972) Creep rupture under multi-axial states of stress. J of the Mechanics and Physics of Solids 20(6):381–390

Haythornthwaite RM (1961) Range of yield condition in ideal plasticity. Proc ASCE J Eng Mech Division EM6 87:117–133

Haythornthwaite RM (1962) Range of yield condition in ideal plasticity. Transactions ASCE 127(1):1252–1269

Hinchberger SD (2009) Simple single-surface failure criterion for concrete. J of Eng Mechanics 135(7):729–732

Hsieh SS, Ting EC, Chen WF (1979) An elastic-fracture model for concrete. In: Proceedings, Third Engineering Mechanics Division Specialty Conference, September 17-19, 1979, American Society of Civil Engineers, Engineering Mechanics Division, University of Texas at Austin, pp 437–440

Hsieh SS, Ting EC, Chen WF (1980) A plastic-fracture model for concrete. In: Hsieh SS, Ting EC (eds) Fracture in Concrete, Proceedings of a session sponsored by the Committee on Properties of Materials of the ASCE Engineering Mechanics Division at the ASCE National Convention in Hollywood, Florida, October 27-31, 1980, American Society of Civil Engineers, New York, pp 50–64

Hsieh SS, Ting EC, Chen WF (1982) A plastic-fracture model for concrete. Int J of Solids and Structures 18(3):181–197

Huber MT (1904) Specific strain work as a measurment of material effort (in Polish: Właściwa praca odkształcenia jako miara wytężenia materyału). Czasopismo Techniczne 22:34–40, 49–50, 61–62, 80–81

Ishlinsky AY (1940) Hypothesis of strength of shape change (in Russ.: Gipoteza prochnosti formoizmeneniya). Uchebnye Zapiski Moskovskogo Universiteta, Mekhanika 46:104–114

Ivlev DD (1959) The theory of fracture of solids (in Russ.: K teorii razrusheniya tverdykh tel). J of Applied Mathematics and Mechanics 23(3):884–895

Ivlev DD (1960) On extremum properties of plasticity conditions (in Russ.: Ob ekstremal'nykh svoistvakh uslovij plastichnosti). J of Applied Mathematics and Mechanics 24(5):1439–1446

Koca M, Koca NO (2011) Quasi regular polygons and their duals with Coxeter symmetries D_n represented by complex numbers. In: J. of Physics: Conference Series, Group 28: Physical and Mathematical Aspects of Symmetry, IOP Publishing, vol 284, pp 1–10

Kolupaev VA (2017) Generalized strength criteria as functions of the stress angle. J of Eng Mechanics 143(9), doi:10.1061/(ASCE)EM.1943-7889.0001322

Kolupaev VA (2018) Equivalent Stress Concept for Limit State Analysis. Springer, Cham

Kolupaev VA, Becker W, Massow H, Dierkes D (2014) Design of test specimens from hard foams for the investigation of biaxial tensile strength (in German: Auslegung von Probekörpern aus Hartschaum zur Ermittlung der biaxialen Zugfestigkeit). Forschung im Ingenieurwesen 78(3-4):69–86

Kolupaev VA, Yu MH, Altenbach H (2016) Fitting of the strength hypotheses. Acta Mechanica 227(2):1533–1556

Kolupaev VA, Yu MH, Altenbach H, Bolchoun A (2018) Comparison of strength criteria based on the measurements on concrete. J of Eng Mechanics 144(6), doi:10.1061/(ASCE)EM.1943-7889.0001419

Kupfer H (1973) Das Verhalten des Betons unter mehrachsiger Kurzzeitbelastung unter besonderer Berücksichtigung der zweiachsigen Beanspruchung, Ernst & Sohn, Berlin, pp 1–105. Deutscher Ausschuss für Stahlbeton, Vol. 229

Kupfer H, Zelger C (1968) Biaxial Strength of Concrete (in German: Zweiachsige Festigkeit von Beton: Bau und Erprobung der Belastungseinrichtung). Technische Hochschule München, Materialprüfungsamt für das Bauwesen

Kupfer H, Zelger C (1973) Bau und Erprobung einer Versuchseinrichtung für zweiachsige Belastung, Ernst & Sohn, Berlin, pp 108–131. Deutscher Ausschuss für Stahlbeton, Vol. 229

Kupfer H, Hilsdorf HK, Rusch H (1969) Behavior of concrete under biaxial stresses, Title no. 66–52. ACI Journal 66(8):656–666

Kupfer HB, Gerstle KH (1973) Behavior of concrete under biaxial stresses. J of the Eng Mechanics Division 99(4):853–866

Lade PV (1982) Three-parameter failure criterion for concrete. J of the Engineering Mechanics Division 108(5):850–863

Launay P, Gachon H (1971) Strain and ultimate strength of concrete under triaxial stress. In: Proceedings of the First International Conference on Structural Mechanics in Reactor Technology, Berlin, September 20-24, 1971. Commission of the European Communities, Brussels (EUR-4820), Vol. 3, paper Hl/3, pp 23–40

Launay P, Gachon H (1972) Strain and ultimate strength of concrete under triaxial stress, paper SP 34-13. In: Kesler CE (ed) Concrete for Nuclear Reactors, Bundesanstalt für Materialprüfung in Berlin, Oct. 5-9, 1970, ACI Publication SP-34, American Concrete Institute, Detroit, pp 269–282

Launay P, Gachon H, Poitevin P (1970) Déformation et résistance ultime du béton sous étreinte triaxiale. Annales de l'institut Technique du Batiment et de Travaux Publics, Série: Essais et Mesures (123) 269(5):21–48

Lebedev AA (1965) Generalized criterion for the fatigue strength (in Russ.: Obobshchennyi kriterij dlitel'noj prochnosti). In: Thermal Strength of Materials and Structure Elements (in Russ.: Termoprochnost' materialov i konstrukcionnykh elementov), vol 3, Naukova Dumka, Kiev, pp 69–76

Leckie FA, Hayhurst DR (1977) Constitutive equations for creep rupture. Acta Metallurgica 25(9):1059–1070

Lee SK, Song YC, Han SH (2004) Biaxial behavior of plain concrete of nuclear containment building. Nuclear Engineering and Design 227(2):143–153

Link J (1976) Eine Formulierung des zweiaxialen Verformungs- und Bruchverhaltens von Beton und deren Anwendung auf die wirklichkeitsnahe Berechnung von Stahlbetonplatten. Deutscher Ausschuss für Stahlbeton 270:1–119

Lubliner J, Oliver J, Oller S, Onate E (1989) A plastic-damage model for concrete. Int J of Solids and Structures 25(3):299–326

Mariotte E (1718) Traité du Mouvement des Eaux et des Autres Corps Fluides. J. Jambert, Paris

Marti P (1980) Zur plastischen Berechnung von Stahlbeton. Diss., Institut für Baustatik und Konstruktion ETH Zürich, Birkhäuser Verlag, Basel

Menétrey P, Willam KJ (1995) Triaxial failure criterion for concrete and its generalization. ACI Structural Journal 92(3)

Mirolyubov IN (1953) On the generalization of the strengt theory based on the octaedral stresses in the case of brittle materials (in Russ.: K voprosu ob obobshchenii teorii prochnosti oktaedricheskikh kasatel'nykh napryazhenij na khrupkie materialy). Trudy Leningradskogo Technologicheskogo Instituta pp 42–52

von Mises R (1913) Mechanik des festen Körpers im plastischen deformablen Zustand. Nachrichten der Königlichen Gesellschaft der Wissenschaften Göttingen, Mathematisch-physikalische Klasse pp 589–592

Mittelstedt C, Becker W (2016) Strukturmechanik ebener Laminate. Technische Universtät Darmstadt, Darmstadt

Mohr O (1900a) Welche Umstände bedingen die Elastizitätsgrenze und den Bruch eines Materials. Zeitschrift des VDI 45:1524–1530

Mohr O (1900b) Welche Umstände bedingen die Elastizitätsgrenze und den Bruch eines Materials. Zeitschrift des VDI 46:1572–1577

Mohr O (1914) Abhandlungen aus dem Gebiete der technischen Mechanik. Wilhelm & Sohn, Berlin

Moradi M, Bagherieh AR, Esfahani MR (2018) Damage and plasticity constants of conventional and high-strength concrete. Part II: Statistical equation developmentt using genetic programming. Int J Optim Civil Eng 8(1):135–158

Nielsen MP (1984) Limit analysis and concrete plasticity. Prentice-Hall, Englewood Cliffs

Ottosen NS (1975) Failure and elasticity of concrete. Report Risø-M-1801, Danish Atomic Energy Commission, Research Establishment Risö, Engineering Department, Roskilde

Ottosen NS (1977) A failure criterion for concrete. J of the Engineering Mechanics Division 103(4):527–535

Ottosen NS (1980) Nonlinear finite element analysis of concrete structures. Tech. rep., Risø-R-411, Roskilde: Risø National Laboratory

Ottosen NS, Ristinmaa M (2005) The Mechanics of Constitutive Modeling. Elsevier Science, London

Park H, Kim JY (2005) Plasticity model using multiple failure criteria for concrete in compression. Int J of Solids and Structures 42(8):2303–2322

Paul B (1968) Macroscopic plastic flow and brittle fracture. In: Liebowitz H (ed) Fracture: An Advanced Treatise, vol II, Academic Press, New York, pp 313–496

Pisarenko GS, Lebedev AA (1976) Deformation and Strength of Materials under Complex Stress State (in Russ.: Deformirovanie i prochnost' materialov pri slozhnom napryazhennom sostoyanii). Naukowa Dumka, Kiev

Podgórski J (1984) Limit state condition and the dissipation function for isotropic materials. Archives of Mechanics 36(3):323–342

Podgórski J (1985) General failure criterion for isotropic media. J of Engineering Mechanics 111(2):188–201

Ren XD, Yang WZ, Zhou Y, Li J (2008) Behavior of high-performance concrete under uniaxial and biaxial loading. ACI Materials Journal 105(6):548–557

Sandel GD (1919) Über die Festigkeitsbedingungen: Ein Beitrag zur Lösung der Frage der zulässigen Anstrengung der Konstruktionsmaterialen. Diss., TeH, Stuttgart

Sayir M (1970) Zur Fließbedingung der Plastizitätstheorie. Ingenieur-Archiv 39:414–432

Sayir M, Ziegler H (1969) Der Verträglichkeitssatz der Plastizitätstheorie und seine Anwendung auf räumlich unstetige Felder. Zeitschrift für angewandte Mathematik und Physik ZAMP 20(1):78–93

Schimmelpfennig K (1971) Concrete strength under multiaxial stress (in German: Die Festigkeit des Betons bei mehraxialer Belastung). Bericht Nr. 5, Institut für Konstruktiven Ingenieurbau, Forschungsgruppe Reaktordruckbehälter, Ruhr-Universität Bochum

Schmidt R (1932) Über den Zusammenhang von Spannungen und Formänderungen im Verfestigungsgebiet. Ingenieur-Archiv 3(3):215–235

Seow PEC, Swaddiwudhipong S (2005) Failure surface for concrete under multiaxial load - A unified approach. J of Materials in Civil Engineering, ASCE 17(2):219–228

Shesterikov SA (1960) On the theory of ideal plastic solid (in Russ.: K postroeniju teorii ideal'no plastichnogo tela). Prikladnaja Matematika i Mekhanika, Otdelenie Tekhnicheskikh Nauk Akademii Nauk Sojusa SSR 24(3):412–415

Speck K (2008) Beton unter mehraxialer Beanspruchung. Diss., Fakultät Bauingenieurwesen, Technische Universität Dresden

Sun L, Huang WM, Purnawali H (2011) Constitutive model for concrete under multiaxial loading conditions. Advanced Materials Research 163–167:1171–1174

Tasuji ME (1976) The behavior of plan concrete subject to biaxial stress. Research report no. 360, Department of Structural Engineering, Cornell University, Ithaca

Tasuji ME, Slate FO, Nilson AH (1978) Stress-strain response and fracture of concrete in biaxial loading. ACI J Proceedings 75(7):306–312

Tasuji ME, Nilson AH, Slate FO (1979) Biaxial stress-strain relationships for concrete. Magazine of Concrete Research 31(109):217–224

Timoshenko SP (1953) History of Strength of Materials: With a Brief Account of the History of Theory of Elasticity and Theory of Structure. McGraw-Hill, New York

Torre C (1947) Einfluß der mittleren Hauptnormalspannung auf die Fließ- und Bruchgrenze. Österreichisches Ingenieur-Archiv I(4/5):316–342

Tóth LF (1964) Regular Figures. Pergamon Press, Oxford

Tresca H (1868) Mémoire sur l'ecoulement des corps solides. Mémoires Pres par Div Savants 18:733–799

Troost A, Betten J (1974) Zur Frage der Konvexität von Fließbedingungen bei plastischer Inkompressibilität und Kompressibilität. Mechanics Research Communications 1:73–78

Walser H (2018) Isogonal polygons (in German: Isogonale Vielecke). http://www.walser-h-m.ch, Frauenfeld

Willam KJ, Warnke EP (1975) Constitutive model for the triaxial behavior of concrete. In: Colloquium on Concrete Structures Subjected to Triaxial Stresses, vol 19, pp 1–30

Wolfram S (2003) The Mathematica Book: The Definitive Best-Selling Presentation of Mathematica by the Creator of the System. Wolfram Media, Champaign

Wronski AS, Pick M (1977) Pyramidal yield criteria for epoxides. J of Materials Science 12(1):28–34

Wu EM (1973) Phenomenological anisotropic failure criterion. In: Broutman LJ, Krock RH, Sendeckyj GP (eds) Treatise on Composite Materials, Academic Press, New York, vol 2, pp 353–431

Xiaoping V, Ottosen NS, Thelandersson S, Nielsen MP (1989) Review of constructive models for concrete, EUR 12394 EN. Final Report Ispra, Reactor Safety Programme 1985-1987, Commission of the European Communities, Nuclear Science and Technology, Contract No. 3301-87-12 ELISPDK, Luxembourg

Yagn YI (1931) New methods of strength prediction (in Russ.: Novye metody rascheta na prochnost'). Vestnik inzhenerov i tekhnikov 6:237–244

Yu MH (1961) General behaviour of isotropic yield function (in Chinese). Scientific and Technological Research Paper of Xi'an Jiaotong University pp 1–11

Yu MH (2002) Advances in strength theories for materials under complex stress state in the 20th century. Applied Mechanics Reviews 55(3):169–218

Yu MH (2018) Unified Strength Theory and its Applications. Springer, Singapore

Yu MH, Yu SQ (2019) Introduction to Unified Strength Theory. CRC Press/Balkema, London

Życzkowski M (1981) Combined Loadings in the Theory of Plasticity. PWN-Polish Scientific Publ., Warszawa

Chapter 13
On the Derivation and Application of a Finite Strain Thermo-viscoelastic Material Model for Rubber Components

Jonas Schröder, Alexander Lion, and Michael Johlitz

Abstract This contribution deals with a modified material model of the finite thermoviscoelasticity for the efficient calculation of the dissipative self-heating of elastomer components. The occurrence of critical temperatures, which can lead to loss of functionality or component failure, can be identified at an early stage. Here, the focus lies on industrial applicability, which, in addition to calculation time and quality, also includes the experimental effort required to identify the material parameters. This contribution starts with the formulation of a thermomechanically consistent constitutive model. For this purpose, an appropriate description of the kinematics and the derivation of the constitutive relationships is carried out. These are transferred in a suitable way into the form used by the commercial finite element software ABAQUS and implemented as a thermomechanically fully coupled problem. Furthermore, an industrially applied elastomer material is characterised and the model is parameterized in a special method by selecting the potential function. Finally, the validation of the model and its parameterization are carried out by means of experimental component tests.

Keywords: Finite element implementation · Fully coupled · Finite thermoviscoelasticity · Dissipative heating · Thermomechanics · Elastomer

13.1 Introduction

Due to their typical material characteristics, elastomer components are used in almost all areas of engineering and across industries (Elsner et al, 2012). In addition

---

Jonas Schröder · Alexander Lion · Michael Johlitz
Institute of Mechanics, Faculty for Aerospace Engineering, Bundeswehr University Munich, Germany,
e-mail: jonas.schroeder@unibw.de, alexander.lion@unibw.de, michael.johlitz@unibw.de

H. Altenbach and A. Öchsner (eds.), *State of the Art and Future Trends in Material Modeling*, Advanced Structured Materials 100,
https://doi.org/10.1007/978-3-030-30355-6_13

to the chemical properties, the physical behaviour of the material is of primary importance in the selection process. These include the reversible absorption of large deformations at comparatively low loads as well as the vibration and noise decoupling properties (Koltzenburg et al, 2013). In many cases, these components are subject to large cyclic deformations which result in dissipation-induced self-heating. Depending on the application, elastomer components may also be exposed to elevated ambient temperatures. Increased component temperatures can lead to impermissible changes in the material properties, i.e. to loss a function or total failure. Therefore, it is important to identify critical temperatures and loads early in the development process. The aim is to replace cost-intensive prototype tests with FEM[1] simulations. The concept shown in Fig. 13.1 is used for the simulation of component temperatures. The input parameters of this concept are primarily the component geometry and the material properties. By the suitable formulation of the material model, the relevant material behaviour is taken into account and used for the calculation. The parameterization of the model is carried out by experimental material characterization. Subsequently, boundary conditions and material properties are assigned to the discretized component geometry such that the local temperature and the load curves can be calculated.

The second section focuses on the phenomenological analysis of elastomer materials: First the typical material behaviour is explained based on the material structure, then a selection of the modelled relevant phenomena is made by assessing their responsibility for self-heating. The following section contains the continuum mechanical material modelling. A suitable description of the kinematics is introduced by multiplicative decomposition of the deformation gradient in order to represent different types of deformation. In addition, the constitutive relationships are derived by evaluation of the Clausius-Duhem inequality for a general potential function. The

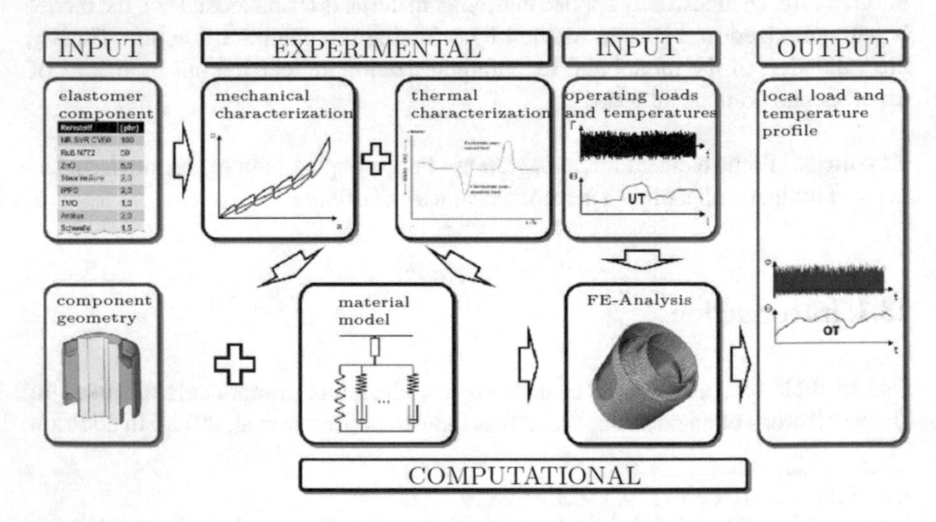

Fig. 13.1 Concept for the derivation of the dissipative heating of elastomer components

[1] FEM is the abbreviation of finite element method

ABAQUS-related heat conduction equation is also specified. The fourth section deals with the FE implementation. From the heat conduction equation in combination with the balance of momentum, a fully coupled functional is obtained by applying Galerkin's method. This is linearised for the iterative calculation with the Newton method. Next, the required constitutive relationships and their associated consistent tangent operators are derived. The selection of the potential functions and the resulting evaluations of the state vectors and tangent operators are similary carried out in the fifth section, as well as the conversion into the formulation required for the UMAT[2] implementation. In section six, the validation of the material model and its parametrization is carried out. This begins with the development of the computational model and concludes with the comparison of simulated and experimentally determined characteristics. Chapter seven concludes with a discussion and evaluation of the model quality and efficiency with regard to industrial applicability. In addition, an outlook on future trends and prospects is given.

13.2 Elastomer Structure and Behaviour

This section provides the basic considerations leading to the selection of a suitable material model. For this purpose, the typical behaviour of elastomer materials is first explained on the basis of the chemical structure and then assessed with regard to the influence on the dissipative self heating process. Elastomers are weakly crosslinked polymers which exhibit the characteristic entropy-elastic behaviour at operating temperatures. The meaning of this term is deduced from the macroscopic view of the material. Detailed descriptions of the molecular structure can be found in the standard literature, e.g. Treloar (1975); Tobolsky et al (1971) or Schwarzl (2013). The chain molecules perform thermally disordered movements and have a large number of moving segments. Thus, they occupy the thermodynamically and statistically most probable arrangement, namely the one of maximum entropy. Therefore, the molecules are strongly entangled in the material. The elasticity can be traced back to the mobility of the molecules above the glass transition, which is limited by cross-links and entanglements. Polymer chains are initially entangled. Under external stress they rearrange themselves such that the chains get stretched and thereby change the state of order. Due to that, the directed chains exhibit less entropy and have the tendency to take up a state of higher disorder or entropy. Therefore, rubber elasticity is also called entropy-induced (Tobolsky, 1967; Treloar, 1975). This enables large reversible deformations with almost incompressible material behaviour, whereby the stress depends non-linearly on the strain (Rivlin and Saunders, 1951), but approximately linearly on the temperature (Anthony et al, 1942). However, under sufficiently small constant deformations, a drop in the engineering stress with increasing temperature can be observed. This phenomenon is generally known as the thermo-elastic inversion and is based on the overlay of thermal expansion and entropy elasticity. The deformation

[2] User-defined material model (UMAT) in ABAQUS can be used to define the mechanical constitutive behaviour of a material.

during an adiabatic process results in a temperature change, also known as Joule-Gough effect (Gough, 1805; Joule, 1859). The reason for this is the compensation of the deformation-related decrease in entropy due to an increase in temperature. Since, time-dependent internal sliding and rearrangement processes of molecules also occur, elastomers are also referred as viscoelastic materials. Infinitely slow deformation processes, denoted as quasi-static, lead to thermodynamic equilibrium states. On this occasion the viscous components play a minor role. In the case of cyclic dynamic loading, a load history and strain rate dependent hysteresis loop occurs. The enclosed area is a measure of the mechanical energy converted into thermal energy and is defined as dissipation. The temperature influence on the viscoelastic material properties, which can be also observed, is based on the fact that sliding and rearrangement processes accelerate with increasing temperature. The relationship between temperature and rate dependence can be described with the time-temperature superposition principle. In order to adapt the material properties to the technical application, fillers in addition to chemical additives are added to improve the mechanical properties. Due to the different types of interaction, the defined properties must be adjusted with regard to a common optimum. Filled elastomers show significant differences in their characteristic behaviour compared to unfilled elastomers. With filled materials, the complex temperature behaviour of the interactions between the fillers and the elastomer matrix superimposes to the entropy elasticity, such that a completely different stiffness characteristic can be observed. The elastomer/filler interaction leads to a characteristic softening within the first loading cycles. The so-called Mullins-effect (Mullins, 1948) results from the breaking and rearrangement of weak polymer chains and the successive breakage of certain sections of the filler network until a more or less "constant" material behaviour is achieved. The viscoelastic behaviour shows a non-linear dependence on the loading. This amplitude dependence is also known as the Payne-effect (Payne, 1962).

In summary, it can be stated that elastomers, due to their molecular structure, can take up elongations of several 100%, show no major volume change and return to their original shape completely when the load is removed. In addition, elastomers exhibit a marked viscoelastic behaviour, such that cyclic mechanical loads at adequate amplitudes and frequencies are leading to significant energy dissipation. Therefore, elastomer components can heat up strongly under insufficient heat removal. The temperature change leads to a change in the viscoelastic material properties and generates thermal strains or thermally induced stresses. In the case of filled elastomers, thermo-elastic effects play only a minor role in self-heating due to the dependencies described above.

13.3 Continuum Mechanical Material Modelling

Material theory is a subsection of continuum mechanics which deals with material models. It provides general principles and systematic methods for the formulation of mathematically and thermodynamically consistent models to describe the individual properties of a material body. The derivation of constitutive relations follows the principles of rational thermomechanics. Here, the second law of thermodynamics acts as a restriction to obtain a thermomechanically consistent constitutive equation. In order to formulate consistent models, the dissipation postulate must be fulfilled. In addition, the axiomatic principles of material theory must be followed. Multiplicative decomposition of the deformation gradient allow to consider different deformation mechanisms. The free energy density is an appropriate thermodynamic potential to model the material properties, whereby its independent variables, or arguments must be defined. At the beginning of this section, the basics and contexts, which are necessary for understanding the following considerations, are explained. This includes the balance relations of thermomechanics and the resulting principle of irreversibility. In addition, a proper description of the kinematics is introduced, the independent variables are defined and the constitutive relationships are derived from the potential.

13.3.1 Balance Equations

This section presents the classical balance equations of thermomechanics. They are independent of the special properties of the continuum, since they describe universally valid laws of nature. They can be formulated globally for the entire material body in integral form, or locally in differential form. Furthermore, the balance equations can be formulated for each configuration. In the following, the equations are described in local form using variables related to the reference configuration.

13.3.1.1 Conservation of Mass

$$\frac{\partial}{\partial t}\rho_0\big(\mathbf{X},t\big) = 0 \qquad \Rightarrow \qquad \rho_0 = \rho_0\big(\mathbf{X}\big) = \text{constant} \tag{13.1}$$

ρ_0 is the density related to the reference configuration. It does not depend on time, thus it depends only on the vector of the material points in the reference configuration $\mathbf{X}$.

13.3.1.2 Balance of Linear Momentum

$$\rho_0\dot{\mathbf{V}}(\mathbf{X},t) = \text{Div}\,\big(\mathbf{P}\big) + \rho_0\mathbf{b} \tag{13.2}$$

The time derivative of the momentum on the left hand side is expressed by the time derivative of the material velocity field $\dot{\mathbf{V}}$ which is weighted with the density. On the right hand side the force density, composed of the divergence in relation to the material coordinates of the first Piola-Kirchhoff stress tensor **P** and body force per unit volume $\rho_0 \mathbf{b}$.

13.3.1.3 Balance of the Angular Momentum

$$\mathbf{S} = \mathbf{S}^{\mathrm{T}} \quad \text{or} \quad \mathbf{P} \cdot \mathbf{F}^{\mathrm{T}} = \mathbf{F} \cdot \mathbf{P}^{\mathrm{T}} \tag{13.3}$$

The quantity **S** is the second Piola-Kirchhoff stress tensor. Its symmetry follows from the local form of the balance of rotational momentum. The first Piola-Kirchhoff stress tensor **P** is generally not symmetric and the characteristic described above holds, where $\mathbf{F} = \mathrm{Grad}\,(\mathbf{x})$ is the deformation gradient. Here $\mathrm{Grad}(\circ)$ is the gradient operator with respect to the material coordinates. The vector **x** is the current position of the material point **X** at time t in the current configuration.

13.3.1.4 Balance of Energy

$$\rho_0 \dot{e} = \mathbf{S} : \dot{\mathbf{E}} - \mathrm{Div}\,(\mathbf{q}_0) + \rho_0 r \tag{13.4}$$

The energy balance provides the temporal change of the specific internal energy $\rho_0 \dot{e}$. It consists of the volume-related stress power $\mathbf{S} : \dot{\mathbf{E}}$ where $\dot{\mathbf{E}}$ is the time derivative of the Green-Lagrange strain tensor and the heat exchange. The vector $\mathbf{q}_0$ denotes the Piola-Kirchhoff heat flux vector and $\rho_0 r$ is the heat source per unit volume. This equation is also known as the first law of thermodynamics.

13.3.1.5 Balance of Entropy

$$\rho_0 \dot{\eta} + \mathrm{Div}\left(\frac{\mathbf{q}_0}{\theta}\right) - \rho_0 \frac{r}{\theta} = \rho_0 \tilde{\eta} \geq 0 \tag{13.5a}$$

$$\Leftrightarrow \tilde{\eta} \geq 0 \tag{13.5b}$$

On the left hand side, the temporal change in the entropy per unit volume is described by the expression $\rho_0 \dot{\eta}$. The heat supply per unit time related to the thermodynamic temperature θ is used to calculate the entropy supply. Here, $\frac{\mathbf{q}_0}{\theta}$ is the entropy flux and $\frac{r}{\theta}$ is the specific entropy source, whereby θ denotes a time-dependent scalar field. On the right hand side the specific entropy production $\tilde{\eta}$ is opposed. For all thermomechanical admissible processes, the entropy production $\tilde{\eta}$ must be greater than or equal to zero. The balance is also known as the second law of thermodynamics.

13.3.1.6 Dissipation Inequality

As mentioned above, the constitutive relations are derived from the Helmholtz free energy density as a function of deformation and temperature. The equations presented previously are valid for all material models of continuum mechanics, such that the

Legendre transform of thermodynamic potentials is used to transform the specific internal energy $\rho_0 e$ into the Helmholtz free energy per unit mass:

$$\Psi = e - \theta\eta \tag{13.6}$$

The insertion of the time derivative of the free energy function (13.6) into the entropy inequality (13.5a) leads to the well known Clausius-Duhem inequality:

$$\rho_0\dot{\Psi} + \mathbf{S} : \dot{\mathbf{E}} - \rho_0\dot{\theta}\eta - \frac{\mathbf{q}_0}{\theta}\,\mathrm{Grad}\,\theta \geq 0 \tag{13.7}$$

Here Grad($\circ$) is the gradient operator with respect to the material coordinates. The entire model must satisfy this inequality, which represents the second law of thermodynamics, to obtain a thermomechanically consistent material model.

13.3.2 Quasi-incompressible Modified Thermoviscoelasticity

Subsequently, this contribution emphasises thermomechanically consistent material modelling. For this purpose, the concept of the model is motivated on the basis of a rheological representation. Based on these findings, the description of suitable kinematics and the definition of independent variables is carried out. Finally, the constitutive relations are derived respecting thermomechanical consistency. The development of thermomechanical material models has been the focus of the following research activities Lion (2000); Johlitz (2015); Dippel et al (2014); Reese (2001). It should be mentioned that this list is not complete. The implementation of thermomechanically coupled material models has been the subject of Miehe (1988); Simo and Miehe (1992); Anand (1985); Arruda et al (1995); Heimes (2004); Bröcker and Matzenmiller (2008); Anand et al (2009); Naumann and Ihlemann (2011); Bröcker (2013); Hamkar (2013) or Lejeunes et al (2018). The solution of thermomechanical coupled processes has been investigated among by Glaser (1992) or Erbts and DüSter (2012). Based on the model of classical viscoelasticity, the deduced material model can be motivated. The usage of rheological elements as a method of representation has not only the advantage of special illustration, it also leads to thermomechanically consistent models. Moreover, these elements can be extended easily to three-dimensional states of stress and strain and nonlinearities. In consequence, the model presented in Fig. 13.2 is introduced. The part of the free energy that depends only on the temperature is described by Ψ_{th}. The total energy stored in the springs can be additively allocated to the respective springs. The equilibrium part of the free energy Ψ_{eq} is assigned to the single spring. In addition to the elastic behaviour, it represents the equilibrium part of the stress $\mathbf{P}_{eq}$. The springs of the Maxwell elements represent the overstresses $\mathbf{P}_{neq}^{(k)}$ and can be linear or non linear. The free energies $\Psi_{neq}^{(k)}$ are related to them. The temperature-dependent viscosities $\breve{\eta}^{(k)}(\theta)$ describe the rate dependence of the damper elements. The structure

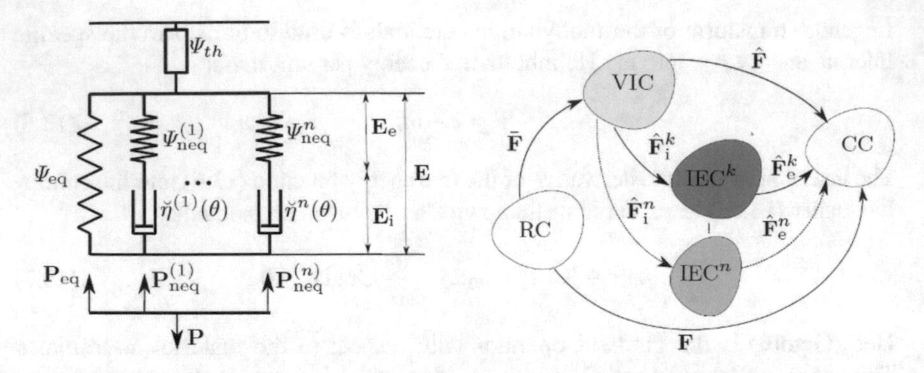

Fig. 13.2 Rheological representation of modified finite thermoviscoelasticity (left), decomposition of the deformation gradient (right): The reference configuration (RC), volumetric-isochoric intermediate configuration (VIC), elastic-inelastic intermediate configurations (EIC) and the current configuration (CC)

of the model mapping of the dissipation implies a separation of elastic and inelastic deformations.

13.3.2.1 Kinematics

First of all, large deformation require a distinction between the reference and the current configuration. In order to distinguish different types of deformation, it is necessary to split the deformation gradient multiplicatively. Some intermediate configurations will be introduced on this occasion. The first multiplicative decomposition of the deformation gradient is carried out to split the local deformation into volumetric and isochoric parts and simplifies the representation of quasi incompressible behaviour. For this purpose the volumetric-isochoric intermediate configuration (Flory, 1961) is introduced

$$\mathbf{F} = \hat{\mathbf{F}} \cdot \bar{\mathbf{F}} \qquad \text{with} \qquad \hat{\mathbf{F}} = J^{-\frac{1}{3}}\mathbf{F} \qquad \text{and} \qquad \bar{\mathbf{F}} = J^{\frac{1}{3}}\mathbf{I} \tag{13.8}$$

with the definition of the volumetric part $\bar{\mathbf{F}}$ of the deformation and the isochoric part $\hat{\mathbf{F}}$ using the determinant $\det(\mathbf{F}) = J$. Finally, the isochoric part is divided multiplicatively into purely elastic components $\hat{\mathbf{F}}_{\mathrm{e}}^{(k)}$ and inelastic components $\hat{\mathbf{F}}_{\mathrm{i}}^{(k)}$. This multiplicative split introduces elastic-inelastic intermediate configurations (Lubliner, 1985):

$$\hat{\mathbf{F}} = \hat{\mathbf{F}}_{\mathrm{e}}^{(k)} \cdot \hat{\mathbf{F}}_{\mathrm{i}}^{(k)}, \tag{13.9}$$

where the index $[\circ]^{(k)}$ indicates the respective Maxwell element. At this point, the right elastic Cauchy-Green deformations tensor $\hat{\mathbf{C}}_e^j$ and the right inelastic Cauchy-Green deformations tensor $\hat{\mathbf{C}}_i^j$ are defined:

$$\hat{\mathbf{C}}_e^{(k)} = \hat{\mathbf{F}}_\mathrm{e}^{\mathrm{T}(k)} \cdot \hat{\mathbf{F}}_\mathrm{e}^{(k)} \tag{13.10}$$

$$\hat{\mathbf{C}}_i^{(k)} = \hat{\mathbf{F}}_\mathrm{i}^{\mathrm{T}(k)} \cdot \hat{\mathbf{F}}_\mathrm{i}^{(k)} \tag{13.11}$$

The tensor $\hat{\mathbf{L}}_\mathrm{i}^j$, known as inelastic spatial velocity gradient of the corresponding Maxwell element and the inelastic rate of deformation tensor $\hat{\mathbf{D}}_\mathrm{i}^{(k)}$ is defined as:

$$\hat{\mathbf{L}}_\mathrm{i}^{(k)} = \dot{\hat{\mathbf{F}}}_\mathrm{i}^{(k)} \cdot \hat{\mathbf{F}}_\mathrm{i}^{-1(k)} \tag{13.12}$$

$$\hat{\mathbf{D}}_\mathrm{i}^{(k)} = \frac{1}{2}\left(\hat{\mathbf{L}}_\mathrm{i}^{(k)} + \hat{\mathbf{L}}_\mathrm{i}^{\mathrm{T}(k)}\right) \tag{13.13}$$

13.3.2.2 Derivation of the Potential Expressions

The free energy density $\rho_0\Psi$ for the considered case depends on the right Cauchy-Green deformation tensor $\mathbf{C}$, the elastic right Cauchy-Green deformation tensors $\hat{\mathbf{C}}_e^j$ and the thermodynamical temperature θ:

$$\rho_0\Psi = \rho_0\Psi(\mathbf{C}, \hat{\mathbf{C}}_\mathrm{e}^1, \ldots, \hat{\mathbf{C}}_\mathrm{e}^n, \theta) \tag{13.14}$$

Using the temporally free energy , the dissipation inequality (13.7) leads to:

$$\mathbf{S} : \frac{1}{2}\dot{\mathbf{C}} - \rho_0\left(2\frac{\partial\Psi}{\partial\mathbf{C}} : \frac{1}{2}\dot{\mathbf{C}} + \sum_{j=1}^{n}\frac{\partial\Psi}{\partial\hat{\mathbf{C}}_\mathrm{e}^j} : \dot{\hat{\mathbf{C}}}_\mathrm{e}^j + \frac{\partial\Psi}{\partial\theta}\dot{\theta}\right) - \rho_0\dot{\theta}\eta - \frac{\mathbf{q}_0}{\theta}\,\mathrm{Grad}\,\theta \geq 0 \tag{13.15}$$

The constitutive relations are obtained by fulfilling the Clausius-Duhem inequality (13.15). If Fourier's law is applied, the last term is non-negative, due to the negative proportionality between heat flux and temperature gradient,

$$\mathbf{q}_0 = -\lambda\mathbf{C}^{-1}\,\mathrm{Grad}\,\theta \qquad \text{with } \lambda \geq 0, \tag{13.16}$$

where λ is the heat conduction coefficient. After some transformations under consideration of the kinematic relations, one obtains the following inequality:

$$\begin{aligned} &\left[\mathbf{S} - 2\rho_0\left(\frac{\partial\Psi}{\partial\mathbf{C}} + \sum_{j=1}^{n}(\det\mathbf{C})^{-\frac{1}{3}}\hat{\mathbf{F}}_\mathrm{i}^{j-1} \cdot \frac{\partial\Psi}{\partial\hat{\mathbf{C}}_\mathrm{e}^j} \cdot \hat{\mathbf{F}}_\mathrm{i}^{j-\mathrm{T}} - \sum_{j=1}^{n}\frac{1}{3}\left(\frac{\partial\Psi}{\partial\hat{\mathbf{C}}_\mathrm{e}^j} : \mathbf{I}\right)\mathbf{C}^{-1}\right)\right] : \frac{1}{2}\dot{\mathbf{C}} \\ &- \rho_0\left[\eta + \frac{\partial\Psi}{\partial\theta}\right]\dot{\theta} + 2\rho_0\sum_{j=1}^{n}\frac{\partial\Psi}{\partial\hat{\mathbf{C}}_\mathrm{e}^j} \cdot \hat{\mathbf{C}}_\mathrm{e}^{j\mathrm{T}} : \frac{1}{2}(\hat{\mathbf{L}}_\mathrm{i}^{j\mathrm{T}} + \hat{\mathbf{L}}_\mathrm{i}^j) \\ &+ \frac{\lambda}{\theta}(\mathrm{Grad}\,\theta) \cdot (\mathbf{C}^{-1}\,\mathrm{Grad}\,\theta) \geq 0 \end{aligned} \tag{13.17}$$

$\hat{\mathbf{L}}_{\rm i}^{j}$ is known as the inelastic spatial velocity gradient of the corresponding Maxwell element. The inequality is evaluated according to Coleman and Noll (1963). This means that each dependent variable is completely characterized by the values of the process variables and thus independent of their temporal changes. In addition, (13.17) has to be satisfied for arbitary values of $\dot{\theta}$ and tensors $\dot{\mathbf{C}}$. In this way, the constitutive equations can be obtained:

$$\mathbf{S} = 2\rho_0 \left(\frac{\partial \Psi}{\partial \mathbf{C}} + \sum_{j=1}^{n} (\det \mathbf{C})^{-\frac{1}{3}} \hat{\mathbf{F}}_{\rm i}^{j-1} \cdot \frac{\partial \Psi}{\partial \hat{\mathbf{C}}_{\rm e}^{j}} \cdot \hat{\mathbf{F}}_{\rm i}^{j-\rm T} - \sum_{j=1}^{n} \frac{1}{3} \left(\frac{\partial \Psi}{\partial \hat{\mathbf{C}}_{\rm e}^{j}} : \mathbf{I} \right) \cdot \mathbf{C}^{-1} \right) \tag{13.18}$$

$$\eta = -\frac{\partial \Psi}{\partial \theta} \tag{13.19}$$

To satisfy the residual inequality, proportionality relations with temperature-dependent functions $\breve{\eta}^{j}(\theta) \geq 0$ are introduced,

$$\hat{\mathbf{D}}_{\rm i}^{j} = \frac{2}{\breve{\eta}^{j}(\theta)} \frac{\partial \Psi}{\partial \hat{\mathbf{C}}_{\rm e}^{j}} \cdot \hat{\mathbf{C}}_{\rm e}^{j\rm T} \tag{13.20}$$

where $\hat{\mathbf{D}}_{\rm i}^{j}$ represents the symmetric part of the inelastic spatial velocity gradient. Furthermore, $\breve{\eta}^{j}(\theta)$ are interpreted as temperature-dependent viscosity functions, which are expressed by the standard Williams-Landel-Ferry equation (Williams et al, 1955):

$$\breve{\eta}^{j}(\theta) = \breve{\eta}_{t}^{j} \exp \left(-\frac{C_1(\theta - \theta_t)}{C_2 + \theta - \theta_t} \right) \tag{13.21}$$

In this context, $\breve{\eta}_{t}^{j}$ is the viscosity that belongs to the reference temperature θ_t and C_1, C_2 are empirical constants adjusted to fit the experimentally observed temperature dependence. Moreover, the deviatoric form of the evolution equation can be derived using the condition of incompressibility $(\det \hat{\mathbf{F}}_{\rm i})^{\cdot} = 0$. Taking into account the kinematic relations, the evolution equation can be reformulated as:

$$\dot{\hat{\mathbf{C}}}_{\rm i}^{j} = \hat{\mathbf{F}}_{\rm i}^{j\rm T} \cdot \left\{ \frac{2}{\breve{\eta}^{j}(\theta)} \frac{\partial \Psi}{\partial \hat{\mathbf{C}}_{\rm e}^{j}} \cdot \left[\hat{\mathbf{C}}_{\rm e}^{j\rm T} - \frac{1}{3} \mathrm{tr}\left(\hat{\mathbf{C}}_{\rm e}^{j\rm T} \right) \mathbf{I} \right] \right\} \cdot \hat{\mathbf{F}}_{\rm i}^{j} \tag{13.22}$$

Here, $\dot{\hat{\mathbf{C}}}_{\rm i}^{j}$ denotes the time derivative of the isochoric right Cauchy-Green deformation tensor related to the respective Maxwell element. The trace of a second order tensor is defined as $\mathrm{tr}(\circ) = (\circ) : \mathbf{I}$.

13.3.3 Heat Conduction Equation

From the first law of thermodynamics (13.4) in combination with the usage of the Legendre transform (13.6), one obtains:

$$\mathbf{S} : \frac{1}{2}\dot{\mathbf{C}} - \rho_0(\dot{\Psi} + \eta\dot{\theta}) - \rho_0\theta\dot{\eta} - \mathrm{Div}\,(\mathbf{q}_0) + \rho_0 r = 0 \tag{13.23}$$

Inserting the time derivative of the free energy density (13.14) leads to:

$$\rho_0\dot{\eta}\theta = \mathrm{Div}\,(\mathbf{q}_0) + \rho_0 r + 2\rho_0 \sum_{j=1}^{n} \frac{\partial\Psi}{\partial\hat{\mathbf{C}}_\mathrm{e}^j} \cdot \hat{\mathbf{C}}_\mathrm{e}^j : \hat{\mathbf{D}}_\mathrm{i}^j \tag{13.24}$$

In addition, the time derivative of the entropy density reads as:

$$\rho_0\dot{\eta} = \rho_0\left(\frac{\partial\eta}{\partial\theta}\dot{\theta} + \frac{\partial\eta}{\partial\mathbf{C}} : \dot{\mathbf{C}} + \sum_{j=1}^{n} \frac{\partial\eta}{\partial\hat{\mathbf{C}}_\mathrm{e}^j} : \dot{\hat{\mathbf{C}}}_\mathrm{e}^j\right) \tag{13.25}$$

Furthermore, the following simplifying assumption postulates a constant specific heat capacity c which is approximately equal to the isobaric specific heat capacity c_p:

$$c_p \approx c \approx -\frac{\partial\Psi}{\partial\theta\partial\theta}\theta \approx \mathrm{const} \tag{13.26}$$

Thus, the heat conduction equation is written in the current configuration as:

$$\rho\, c\, \dot{\theta} = -\,\mathrm{div}\,(\mathbf{q}) + \rho\, r + \rho\,\delta + \rho\,\pi \tag{13.27}$$

Here, $\mathbf{q}$ is the Cauchy heat flux vector with the associated operator div($\circ$) that relates to the spatial coordinates. Furthermore, ρ is the density in the current configuration, $\rho\,\delta$ corresponds to the dissipation term and $\rho\,\pi$ represents the thermoelastic coupling term. Within the UMAT interface in the ABAQUS-software, these terms are added to the term ρr_mat such that the isobaric specific heat capacity remains on the left side (Abaqus, 2002).

$$\rho\, c_p\, \dot{\theta} = -\,\mathrm{div}\,(\mathbf{q}) + \rho\, r + \rho\, r_{mat} \tag{13.28}$$

13.4 Finite Element Implementation

After the formulation of the material model and the determination of the equations for the stress and entropy calculation, this section presents the material independent basic equations and methods for the implementation of the model in the commercial finite element software ABAQUS. Starting with the initial boundary value problem in the local form, the variation formulation is required for the approximate calculation. First, the required weak forms of the local quasi-static momentum balance and the heat conduction equation are presented and linearized for an iterative method. Here, it is focused on the constitutive equations and consistent tangent operators.

The balance equations (13.2) and (13.38) are general field equations for the determination of the displacement field $\mathbf{u}$ and the temperature θ. They are completed by the constitutive relations (13.18), (13.19). However, the determination of initial and boundary conditions is mandatory for the unique description of the initial boundary value problem. From now on, quasi-static processes are considered whereby the initial conditions for the temperature field and the internal variables are required.

$$\theta(\mathbf{X}, t_0) = {}_{t=0}\theta(\mathbf{X}) \quad \text{and} \quad \hat{\mathbf{C}}_i^j(\mathbf{X}, t_0) = {}_{t=0}\hat{\mathbf{C}}_i^j(\mathbf{X}) \quad \text{for} \quad \mathbf{X} \in \Omega_0 \tag{13.29}$$

The specification of boundary conditions requires that the boundary $\partial\Omega_0$ of a body $\mathcal{B}_0$, which occupies the domain Ω_0, is divided into disjoint parts. The second subscript index indicates the type of boundary condition on the partial boundary, such that the following conditions are valid:

$$\partial\Omega_0 = \partial\Omega_{0u} \cup \partial\Omega_{0\sigma} \qquad \text{with} \qquad \partial\Omega_{0u} \cap \partial\Omega_{0\sigma} = \varnothing \tag{13.30}$$

$$\partial\Omega_0 = \partial\Omega_{0\theta} \cup \partial\Omega_{0q} \cup \partial\Omega_{0\theta q} \qquad \text{with} \qquad \partial\Omega_{0\theta} \cap \partial\Omega_{0q} \cap \partial\Omega_{0\theta q} = \varnothing \tag{13.31}$$

The Dirichlet boundary conditions are assigned to the values that a solution needs to take a long the boundary of the domain:

$$\mathbf{u}(\mathbf{X}, t) = \bar{\mathbf{u}}(\mathbf{X}, t) \text{ on } \partial\Omega_{0u} \quad \text{and} \quad \theta(\mathbf{X}, t) = \bar{\theta}(\mathbf{X}, t) \text{ on } \partial\Omega_{0\theta} \tag{13.32}$$

The Neumann boundary conditions are assigned to the values in which the derivative of a solution is applied with

$$\mathbf{q}_0 \cdot \mathbf{n}_0 = \bar{q}_0(\mathbf{X}, t) \text{ on } \partial\Omega_{0q} \quad \text{and} \quad \mathbf{t}_0 = \mathbf{P}\mathbf{n}_0 = \bar{\mathbf{t}}_0(\mathbf{X}, t) \text{ on } \partial\Omega_{0\sigma}. \tag{13.33}$$

The mixed boundary surfaces, where the condition additionally dependents on the surface temperature

$$\mathbf{q}_0 \cdot \mathbf{n}_0 = \bar{q}_0(\mathbf{X}, t, \theta) \text{ on } \partial\Omega_{0\theta q} \tag{13.34}$$

is specified. This results in a well-defined initial boundary value problem. The operator $(\bar{\circ})$ denotes a prescribed function on the boundary where $\mathbf{n}_0$ is the outward normal to the boundary $\partial\Omega_0$ and $\mathbf{t}_0$ depicts the first Piola-Kirchhoff traction vector which is associated to the reference configuration. An analytical solution of the field problem is usually not possible. However, an approximate solution can be calculated using the finite element method exemplary. This requires the formulation of balance equations in the form of variational principles.

The weak formulation of the problem is mandatory for the finite element implementation. Therefore, the balance of linear momentum (13.2) has to be rearranged in terms of quantities which are related to the current configuration first. Secondly, it is assumed that the acceleration is zero for quasi-static processes. This leads to the spatial quasi-static balance of momentum.

$$\operatorname{div} \boldsymbol{\sigma} + \rho \mathbf{b} = \mathbf{0} \tag{13.35}$$

As the next steps to derive the weak formulation. The balance equation is multiplied with the test function $\delta\mathbf{v}$ and integrated over the area Ω where $\delta\mathbf{v}$ is the first variation of the spatial velocity vector. Using the Gaussian integral theorem and the boundary condition (13.33) provides the weak form of mechanical equilibrium as a mechanical functional $\mathcal{M}$, where the gradient operator grad(∘) corresponds to the current configuration.

$$\mathcal{M}(\mathbf{u},\theta,\delta\mathbf{v}) = \int\limits_{\Omega} \boldsymbol{\sigma} : \mathrm{grad}\,\delta\mathbf{v}\,\mathrm{d}V - \int\limits_{\partial\Omega_\sigma} \bar{\mathbf{t}}\cdot\delta\mathbf{v}\,\mathrm{d}A - \int\limits_{\Omega} \rho\mathbf{b}\cdot\delta\mathbf{v}\,\mathrm{d}V = 0 \tag{13.36}$$

To obtain the variation formulation of the heat conduction equation, the heat conduction equation (13.28) is multiplied by the variation of the temperature $\delta\theta$. The thermal functional $\mathcal{T}$ follows from the subsequent reformulation and insertion of the boundary condition (13.34):

$$\mathcal{T}(\mathbf{u},\theta,\delta\theta) = \int\limits_{\Omega} \rho c_p \dot{\theta}\delta\theta\mathrm{d}V - \int\limits_{\partial\Omega_q} \bar{q}_0\delta\theta\mathrm{d}A + \int\limits_{\Omega} \mathbf{q}\cdot\mathrm{grad}\,\delta\theta\mathrm{d}V + \int\limits_{\Omega} \rho(r + r_{mat})\delta\theta\mathrm{d}V = 0 \tag{13.37}$$

In the following, the thermomechanical problem is formulated. Both functionals (13.36) and (13.37) have to be fulfilled. The two requirements are combined with weighting factors to get a fully coupled functional $\mathcal{G}$, such that a closed solution is achieved.

$$\begin{aligned}\mathcal{G} &= \mathcal{M}_u\mathcal{M} + \mathcal{T}_0\mathcal{T} \\ &= \mathcal{M}_u\left\{\int\limits_{\Omega} \boldsymbol{\sigma} : \mathrm{grad}\,\delta\mathbf{v}\,\mathrm{d}V - \int\limits_{\partial\Omega_\sigma} \bar{\mathbf{t}}\cdot\delta\mathbf{v}\,\mathrm{d}A - \int\limits_{\Omega} \rho\mathbf{b}\cdot\delta\mathbf{v}\,\mathrm{d}V\right\} \\ &+ \mathcal{T}_\theta\left\{\int\limits_{\Omega} \rho c_p\dot{\theta}\delta\theta\mathrm{d}V - \int\limits_{\partial\Omega_q} \bar{q}_0\delta\theta\mathrm{d}A + \int\limits_{\Omega} \mathbf{q}\cdot\mathrm{grad}\,\delta\theta\mathrm{d}V + \int\limits_{\Omega} \rho(r + r_{mat})\delta\theta\mathrm{d}V\right\}\end{aligned} \tag{13.38}$$

Since the functionals are non-linear in temperature and displacement, further considerations are necessary. To solve nonlinear functions, they have to be linearized, e.g. with the Gâteux derivative:

$$\mathcal{D}_{\Delta u}(\circ(\mathbf{x})) = \mathcal{D}_u(\circ(\mathbf{x}))\Delta\mathbf{u} = \Delta(\circ) = \frac{\mathrm{d}}{\mathrm{d}\epsilon}(\circ(\mathbf{x} + \epsilon\boldsymbol{\Delta}\mathbf{u}))\Big|_{\epsilon=0} \tag{13.39}$$

The subscript $\mathcal{D}_{(\circ)}$ indicates the direction of the linearization. Thus, the solution at time t_k can be determined iteratively from the solution at the time t_{k-1}, for example with the Newton method. For the i-th iteration step at time $t = t_k$ the notation ${}^{i}_{k}(\circ)$ is introduced. The temperature velocity occurring in the thermal functional (13.37) is approximated from the time discretization with the Euler backward method.

$${}^{i}_{k}\dot{\theta} = \frac{{}^{i}_{k}\theta - {}_{k-1}\theta}{\Delta t} \tag{13.40}$$

Then, one obtains the linearized functional in the form

$$D\mathcal{M}({}^{i+1}_{k}\mathbf{u}, {}^{i+1}_{k}\theta, \delta\mathbf{v}) = \mathcal{M}({}^{i}_{k}\mathbf{u}, {}^{i}_{k}\theta, \delta\mathbf{v}) + D_u\mathcal{M}({}^{i}_{k}\mathbf{u}, {}^{i}_{k}\theta, \delta\mathbf{v})\Delta\mathbf{u} + D_\theta\mathcal{M}({}^{i}_{k}\mathbf{u}, {}^{i}_{k}\theta, \delta\mathbf{v})\Delta\theta = 0 \tag{13.41}$$

$$D\mathcal{T}({}^{i+1}_{k}\mathbf{u}, {}^{i+1}_{k}\theta, \delta\theta) = \mathcal{T}({}^{i}_{k}\mathbf{u}, {}^{i}_{k}\theta, \delta\theta) + D_u\mathcal{T}({}^{i}_{k}\mathbf{u}, {}^{i}_{k}\theta, \delta\theta)\Delta\mathbf{u} + D_\theta\mathcal{T}({}^{i}_{k}\mathbf{u}, {}^{i}_{k}\theta, \delta\theta)\Delta\theta = 0 \tag{13.42}$$

or simplified in matrix notation:

$$\begin{pmatrix} D_u\mathcal{M} & D_\theta\mathcal{M} \\ D_u\mathcal{T} & D_\theta\mathcal{T} \end{pmatrix} \begin{pmatrix} \Delta\mathbf{u} \\ \Delta\theta \end{pmatrix} = \begin{pmatrix} -\mathcal{M} \\ -\mathcal{T} \end{pmatrix} \tag{13.43}$$

The implementation of the material model requires the calculation of the state vector and the definition of the state vector dependent contribution to the tangent stiffness matrix. Therefore, the linearization of the terms δP_{int} and δK_{mat} are of special importance and defined as follows:

$$\delta P_{\text{int}} = \int_\Omega \boldsymbol{\sigma} : \operatorname{grad}\delta\mathbf{v}\,\mathrm{d}V, \qquad \delta K_{\text{mat}} = \int_\Omega \delta\theta\rho r_{mat}\,\mathrm{d}V \tag{13.44}$$

The linearization of the mechanical part (13.36) in the direction of the incremental displacement field $\Delta\mathbf{u}$ leads to the mechanical contribution of the tangent stiffness matrix follows:

$$D_{\Delta u}\delta P_{\text{int}} = \int_\Omega [\delta\mathbf{D} : \overset{\nabla}{\mathbb{c}} : \Delta\mathbf{D} + \delta\mathbf{D} : \Delta\mathbf{W}\boldsymbol{\sigma} - \delta\mathbf{D} : \boldsymbol{\sigma}\Delta\mathbf{W}]\,\mathrm{d}V \tag{13.45}$$

The additive decomposition of the spatial velocity gradient $\mathbf{L}$ contains a symmetric part known as rate of deformation tensor and the antisymmetric part, the spin tensor, denoted as $\mathbf{D}$ and $\mathbf{W}$, follows respectively. The objective spatial tangent operator has to be implemented in the Jaumann formulation $\overset{\nabla}{\mathbb{c}}$. This includes the rotated parts of the stress and the spatial tangent operator $\overset{4}{\mathbb{c}}$, which can be calculated from the material tangent operator $\overset{4}{\mathbb{C}}$. The transposition of the indices is defined as $[(\circ)]^{\overset{ij}{\mathrm{T}}} = (\circ)_{ijkl}\mathbf{e}_j \otimes \mathbf{e}_i \otimes \mathbf{e}_k \otimes \mathbf{e}_l$. The definition of the required tangent operator is shown as follows:

$$\overset{\nabla}{\mathbb{c}} = \frac{1}{J}\overset{4}{\mathbb{c}} + [\mathbf{I} \otimes \boldsymbol{\sigma}]^{\overset{23}{\mathrm{T}}} + [\boldsymbol{\sigma} \otimes \mathbf{I}]^{\overset{23}{\mathrm{T}}} \tag{13.46}$$

$$\overset{4}{\mathbb{c}} = [\mathbf{F} \otimes \mathbf{F}]^{\overset{23}{\mathrm{T}}} : \overset{4}{\mathbb{C}} : [\mathbf{F}^{\mathrm{T}} \otimes \mathbf{F}^{\mathrm{T}}]^{\overset{23}{\mathrm{T}}} \qquad \text{with} \qquad \overset{4}{\mathbb{C}} = 4\rho_0 \frac{\partial^2\Psi}{\partial\mathbf{C}\partial\mathbf{C}} \tag{13.47}$$

After the mechanical part (13.36) is linearized in the direction of $\Delta\theta$, the mechanical-thermal contribution is obtained:

$$D_{\Delta\theta}\delta P_{\text{int}} = \int\limits_{\Omega} \delta\mathbf{D} : \mathbf{t}^{\theta}\Delta\theta \,\mathrm{d}V \tag{13.48}$$

It contains The spatial mechanical-thermo coupling tangent to be implemented in the form of the stress temperature tensor $\mathbf{t}^{\theta}$:

$$\mathbf{t}^{\theta} = \frac{2}{J}\rho_0\mathbf{F} \cdot \frac{\partial^2\Psi}{\partial\theta\partial\mathbf{C}} \cdot \mathbf{F}^{\mathrm{T}} \tag{13.49}$$

The same procedure is applied to the thermal component. The thermo-mechanical contribution is expressed by

$$D_{\Delta\theta}\delta K_{\text{mat}} = \int\limits_{\Omega} \delta\theta\mathbf{d}^{\mathbf{u}} : \Delta\mathbf{D} \,\mathrm{d}V \tag{13.50}$$

with the following definition of the thermo-mechanical coupling tangent $\mathbf{d}^{\mathbf{u}}$ with respect to spatial coordinates

$$\mathbf{d}^{\mathbf{u}} = \frac{2}{J}\rho_0\mathbf{F} \cdot \frac{\partial r_{mat}}{\partial\mathbf{C}} \cdot \mathbf{F}^{\mathrm{T}} \tag{13.51}$$

and finally the thermal contribution is:

$$D_{\Delta\theta}\delta K_{\text{mat}} = \int\limits_{\Omega} \delta\theta d^{\theta}\Delta\theta \,\mathrm{d}V \tag{13.52}$$

Accordingly, the thermal tangent operator is calculated as:

$$d^{\theta} = \frac{1}{J}\rho_0\frac{\partial r_{mat}}{\partial\theta} \tag{13.53}$$

13.5 Material Model

The linearization leads to the definition of the material independent tangent operators (13.46), (13.49), (13.51), (13.53) which are mandatory besides the state vector in order to solve the fully coupled problem. Since elastomers behave almost incompressible under isothermal deformations, a volumetric-isochoric separation is advantageous. Furthermore, the isochoric part of the elastic Cauchy-Green tensor is introduced as a variable in the isochoric part of the free energy density. There are also different approaches for the determination of the thermal part of the free energy. The total free energy is calculated as follows:

$$\rho_0\Psi\big(I_{\hat{\mathbf{C}}}, II_{\hat{\mathbf{C}}}, I_{\hat{\mathbf{C}}_e^j}, J, \theta\big) = \rho_0\Psi_{\text{th}}(\theta) + \rho_0\Psi_{\text{vol}}(J) \tag{13.54}$$

$$+ \rho_0 \Psi_{\mathrm{eq}}\left(I_{\hat{\mathbf{C}}}, II_{\hat{\mathbf{C}}}\right) + \sum_{j=1}^{n} \rho_0 \Psi_{\mathrm{neq}}^{j}\left(I_{\hat{\mathbf{C}}_{\mathrm{e}}^{j}}\right)$$

In various studies, the temperature-dependent part is specified by the requirement of a constant heat capacity at a constant deformation. Here, the following approach is chosen for thus thermal part of the free energy density (Holzapfel, 2000):

$$\rho_0 \Psi_{\mathrm{th}}(\theta) = \rho_0 c\left(\left(\theta - \theta_0\right) - \theta \ln \frac{\theta}{\theta_0}\right) \tag{13.55}$$

The incompressible material behaviour must be taken into account in the stress calculation, which can lead to numerical difficulties. The volumetric approaches usually use the determinant of the deformation gradient as independent variable (Simo and Taylor, 1982) :

$$\rho_0 \Psi_{\mathrm{vol}}(J) = \frac{1}{2}\kappa\left[(J-1)^2 + (\ln J)^2\right] \tag{13.56}$$

where, the material parameter κ has the function of a penalty parameter. In the literature, a number of approaches for the isochoric part of the free energy density can be found e.g. Mooney (1940); Rivlin (1948-1951) or Rivlin (1997). They use the invariants of the Cauchy-Green tensors $I_{\square}$ and $II_{\square}$ as variables.

The equilibrium part of the free energy density reads as follows (Mooney, 1940):

$$\rho_0 \Psi_{\mathrm{eq}}\left(I_{\hat{\mathbf{C}}}, II_{\hat{\mathbf{C}}}\right) = C_{10}\left(I_{\hat{\mathbf{C}}} - 3\right) + C_{20}\left(I_{\hat{\mathbf{C}}} - 3\right)^2 + C_{01}\left(II_{\hat{\mathbf{C}}} - 3\right) \tag{13.57}$$

The non-equilibrium parts of the free energy density (Mooney, 1940) are assumed as:

$$\sum_{j=1}^{n} \rho_0 \Psi_{\mathrm{neq}}^{j}\left(I_{\hat{\mathbf{C}}_{\mathrm{e}}^{j}}\right) = \sum_{j=1}^{n} C_{\mathrm{e}10}^{j}\left(I_{\hat{\mathbf{C}}_{\mathrm{e}}^{j}} - 3\right) \tag{13.58}$$

Where the material parameters $C_{10}, C_{20}, C_{01}, C_{\mathrm{e}10}^{1}, \ldots, C_{\mathrm{e}10}^{n}$ are temperature independent and therefore constant. In the following, the free energy densities are used to describe stresses, heat sources and internal variables. In addition to that, tangent operators for the mechanical, thermal and coupling behaviour, are presented.

The internal variables are described by the evolution equations (13.22). Using (13.58), they can be solved numerically according to a method proposed by Shutov et al (2013) and expressed further by the isochoric elastic left Cauchy-Green tensor $\hat{\mathbf{B}}_e = \hat{\mathbf{F}} \cdot \hat{\mathbf{C}}_{\mathrm{i}}^{-1} \cdot \hat{\mathbf{F}}^{\mathrm{T}}$:

$$\dot{\hat{\mathbf{C}}}_{i}^{j} = \frac{4C_{e10}^{j}}{\breve{\eta}^{j}(\theta)}\left[\hat{\mathbf{C}} - \frac{1}{3}\operatorname{tr}\left(\hat{\mathbf{C}} \cdot \hat{\mathbf{C}}_{i}^{j-1}\right)\hat{\mathbf{C}}_{i}^{j}\right] \tag{13.59}$$

Inserting the free energy densities (13.56)-(13.58) into the stress definition (13.18) and transforming the quantities to the current configuration, the Cauchy stress is obtained:

$$\boldsymbol{\sigma} = \boldsymbol{\sigma}_{vol} + \boldsymbol{\sigma}_{eq} + \sum_{j=1}^{n} \boldsymbol{\sigma}_{neq}^{j} \tag{13.60}$$

The volumetric part of Cauchy stress is:

$$\boldsymbol{\sigma}_{vol} = J^{-1}\kappa\Big(J(J-1) + \ln J\Big)\mathbf{I} \tag{13.61}$$

The equilibrium part of Cauchy stress is:

$$\begin{aligned}\boldsymbol{\sigma}_{eq} = \frac{2}{J}\Big[&\Big(C_{10} + 2C_{20}(I_{\hat{\mathbf{B}}} - 3) + C_{01}I_{\hat{\mathbf{B}}}\Big)\hat{\mathbf{B}} - C_{01}\hat{\mathbf{B}}^2 \\ &-\frac{1}{3}\Big(I_{\hat{\mathbf{B}}}\big(C_{10} + 2C_{20}(I_{\hat{\mathbf{B}}} - 3)\big) + 2C_{01}II_{\hat{\mathbf{B}}}\Big)\mathbf{I}\Big]\end{aligned} \tag{13.62}$$

The non-equilibrium part of the Cauchy stress is:

$$\boldsymbol{\sigma}_{neq} = \sum_{j=1}^{n} C_{e10}^{j}\big(I_{\hat{\mathbf{B}}_e^j} - 3\big) \tag{13.63}$$

The inelastic stress power is represented by variables of the current configuration:

$$r_{mat} = \sum_{j=1}^{n} \frac{1}{\breve{\eta}^j(\theta)}\boldsymbol{\sigma}_{neq}^{j} : \boldsymbol{\sigma}_{neq}^{j} \tag{13.64}$$

Using the equation (13.47) and (13.56)-(13.58) yields to the tangent of the current configuration

$$\overset{4}{\mathbb{C}} = \overset{4}{\mathbb{C}}_{vol} + \overset{4}{\mathbb{C}}_{eq} + \sum_{j=1}^{n} \overset{4}{\mathbb{C}}{}_{neq}^{j} \tag{13.65}$$

The volumetric part of the tangent is:

$$\overset{4}{\mathbb{C}}_{vol} = \frac{1}{J}\Big[\kappa\Big(J(2J-1) + 1\Big)\Big[\mathbf{I}\otimes\mathbf{I}\Big] + 2\kappa\Big(J(1-J) - \ln J\Big)\Big[\mathbf{I}\otimes\mathbf{I}\Big]^{\overset{23}{\mathrm{T}}}\Big] \tag{13.66}$$

The equilibrium part of the tangent is:

$$
\begin{aligned}
\overset{4}{\mathbb{C}}_{eq} = \frac{4}{3J}\Big[&- \Big(C_{10} + 2C_{20}(I_{\hat{\mathbf{B}}} - 3) + 2I_{\hat{\mathbf{B}}}(C_{20} + C_{01})\Big)\Big[\mathbf{I}\otimes\hat{\mathbf{B}} + \hat{\mathbf{B}}\otimes\mathbf{I}\Big] \\
&+ \Big(2C_{20} + C_{01}\Big)\Big[\hat{\mathbf{B}}\otimes\hat{\mathbf{B}}\Big] \\
&+ \frac{2}{3}C_{01}\Big[\mathbf{I}\otimes\hat{\mathbf{B}}^2 + \hat{\mathbf{B}}^2\otimes\mathbf{I}\Big] \\
&+ C_{01}\Big[\hat{\mathbf{B}}\otimes\hat{\mathbf{B}}\Big]^{\overset{23}{T}} \\
&+ \frac{1}{9}\Big(I_{\hat{\mathbf{C}}}(C_{10} + 2C_{20}(I_{\hat{\mathbf{C}}} - 3) + 2C_{20}I_{\hat{\mathbf{C}}}) + 4C_{01}II_{\hat{\mathbf{C}}}\Big)\Big[\mathbf{I}\otimes\mathbf{I}\Big] \\
&+ \frac{1}{3}\Big(I_{\hat{\mathbf{C}}}\Big(C_{10} + 2C_{20}(I_{\hat{\mathbf{C}}} - 3)\Big) + 2C_{01}II_{\hat{\mathbf{C}}}\Big)\Big[\mathbf{I}\otimes\mathbf{I}\Big]^{\overset{23}{T}}\Big]
\end{aligned}
\tag{13.67}
$$

Non-equilibrium part of the tangent:

$$
\overset{4}{\mathbb{C}}_{neq} = \sum_{j=1}^{n}\frac{4}{3J}C^j_{e10}\Big[\operatorname{tr}(\hat{\mathbf{B}}^j_e)\Big[\frac{1}{3}\mathbf{I}\otimes\mathbf{I} + \Big[\mathbf{I}\otimes\mathbf{I}\Big]^{\overset{23}{T}}\Big] - \Big[\hat{\mathbf{B}}^j_e\otimes\mathbf{I} + \mathbf{I}\otimes\hat{\mathbf{B}}^j_e\Big]\Big] \tag{13.68}
$$

The remaining tangents of the current configuration are formulated by definitions (13.49), (13.51), (13.53) derived in the previous section. The stress temperature tensor is zero:

$$
\mathbf{t}^\theta = \mathbf{0} \tag{13.69}
$$

The thermal-mechanical tangent is:

$$
\mathbf{d}^{\mathbf{u}} = \sum_{j=1}^{n}\frac{4}{\breve{\eta}^j(\theta)}\boldsymbol{\sigma}^j_{neq}\cdot\Big(J\boldsymbol{\sigma}^j_{neq} - \frac{1}{3}\operatorname{tr}(J\boldsymbol{\sigma}^j_{neq})\mathbf{I}\Big) \tag{13.70}
$$

The thermal tangent is:

$$
d^\theta = \sum_{j=1}^{n}\frac{\breve{\eta}^j(\theta)^{-2}}{J}\frac{C_1\,C_2\,\breve{\eta}^j_t\exp(-\frac{C_1(\theta-\theta_t)}{C_2+\theta-\theta_t})}{\big(-C_2-(\theta-\theta_t)\big)^2}\boldsymbol{\sigma}^j_{neq}:\boldsymbol{\sigma}^j_{neq} \tag{13.71}
$$

The last step is to formulate the required quantities in the Voigt notation. To this end, it is essential to symmetrize the tensors if necessary. Now, all tangent operators (13.46), (13.69), (13.70), (13.71) and state vectors (13.58) and (13.64) required for the fully coupled calculation are defined correctly.

13.6 Model Validation

The material model is validated as discussed in this section. First, the derivation of the parameter set is explained. Secondly, the structure of the calculation of the model is explained as well as the derivation of the boundary and initial conditions. With these definitions, the simulation results are compared with the experimentally determined data.

13.6.1 Parameter Identification

First, the parameter Set used for the current model is listed in Table 13.1. With the exceptions explained below, these are all independently determined parameters. One characteristic of the parameter set is that the compression modulus κ is chosen at least three orders higher than the shear modulus in order to formulate the quasi-incompressible material behaviour and thus is used as a penalty parameter. Furthermore, the optimized heat capacity c_p^{opt} is of particular importance. The integration of the heat conduction equation with respect to the current configuration over a period of time T in the stationary range shows the independence of the stationary state from the heat capacity c_p. Where $\vartheta(\mathbf{X},t)$ is a scalar temperature field depending on the location coordinate and time. Regarding the left side of the term, one can conclude that:

Table 13.1 Parameter set used for the modified finite strain thermo-viscoelastic material model

parameter	value	unit
Quasi-incompressible hyperelasticity: Mooney-RivlinMooney (1940)		
C_{10}	0.0788	MPa
C_{20}	0.0107	MPa
C_{01}	0.1739	MPa
ρ_0	$1.0748 \cdot 10^{-9}$	kg/m^3
κ	1000	MPa
viscoelasticity: Neo-HookeMooney (1940)		
C_{e10}^1	0.0042	MPa
$\breve{\eta}_t^1$	0.005	MPas
C_{e10}^2	0.1020	MPa
$\breve{\eta}_t^2$	0.1201	MPas
temperature dependence $\breve{\eta}^j(\theta)$, Williams, Landel, FerryWilliams et al (1955)		
C_1	−16	-
C_2	−730	K
θ_t	296	K
thermal material properties		
λ	0.3280	mW/K · mm
c_p	$1.639 \cdot 10^9$	mJ/t · K
c_p^{opt}	$1.639 \cdot 10^7$	mJ/t · K
heat transfer		
$\breve{\alpha}^{ES}$	$22 \cdot 10^{-3}$	mW/mm^2 · K
$\breve{\alpha}^{EA}$	$5 \cdot 10^{-3}$	mW/mm^2 · K

$$\frac{1}{T}\int_{t}^{t+T} \rho c_p \dot{\vartheta}(s)\mathrm{d}s = \rho c_p\Big(\vartheta(t+T) - \vartheta(t)\Big) = 0. \tag{13.72}$$

This motivates to the usage of an optimized heat capacity to reduce the calculation time to reach the stationary state. Finally, the heat transfer coefficients from elastomer to air $\breve{\alpha}^{EA}$ is calculated from Stephan (2002) and that from elastomer to steel $\breve{\alpha}^{ES}$ is taken from Schlanger (1983) or Klauke (2015).

13.6.2 Computational Model

For the computational model (Fig. 13.3), the formulation of the boundary conditions is mandatory. This section describes the boundary condition formulation as shown in Fig. 13.3. For the mixed boundary conditions, the heat transfers from the elastomer to the ambient air and from the elastomer to the steel are defined as heat flows $\mathbf{q}_0^{\mathrm{EA}} = -\breve{\alpha}^{\mathrm{EA}}(\theta_{\partial\Omega} - \theta_{\mathrm{t}})$ and $\mathbf{q}_0^{\mathrm{ES}} = -\breve{\alpha}^{\mathrm{ES}}(\theta_{\partial\Omega} - \theta_{\mathrm{t}})$, respectivley. In addition, the displacement boundary conditions are defined in the form $\bar{\mathbf{u}}_0^{\mathrm{FIX}} = \mathbf{0}$ for the fixed constraint and $\bar{\mathbf{u}}_0^{\mathrm{SYMX}} = [0\, u_2\, u_3]^{\mathrm{T}}$ or $\bar{\mathbf{u}}_0^{\mathrm{SYMY}} = [u_1\, 0\, u_3]^{\mathrm{T}}$ for the symmetry constraint. The force is applied at the reference point P_{REF} by $\mathbf{f} = [F_0 \sin(\omega\, t)\, 0\, 0]$ where $F_0 = 1.4/4$ [kN] and $\omega = 4\pi$ [1/s]. This results in a completely thermomechanically coupled problem.

13.6.3 Analysis

The simulation results are now compared to the experimental data. Multiple experiments were carried out as part of the so-called Elasto-Opt II project at the Fraunhofer LBF in Darmstadt (Schröder and Parra Pelaez, 2019). First, the force-displacement curve at the reference point is considered and shown in Fig. 13.4 (right). A nearly

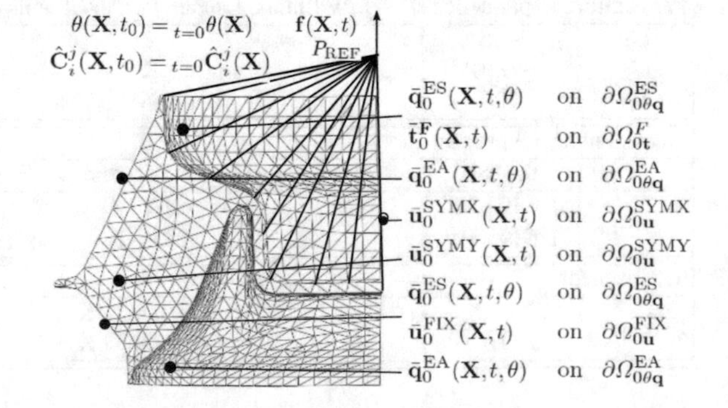

Fig. 13.3 Computational model showing mixed boundary conditions and displacement boundary conditions as well as fixed and symmetry constraints

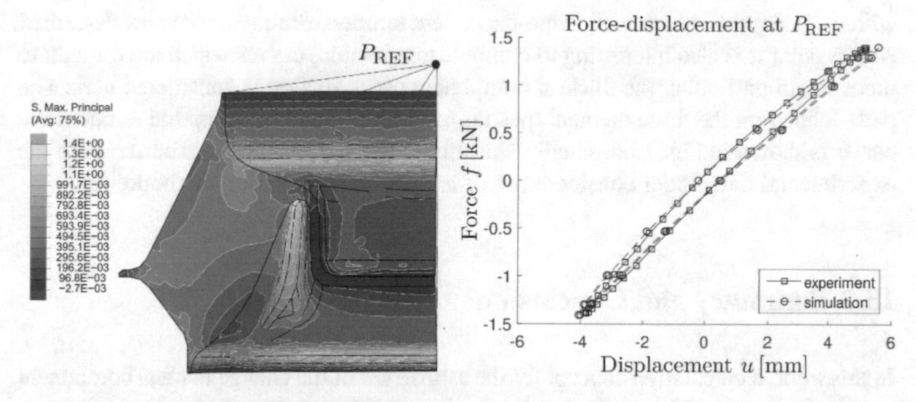

Fig. 13.4 Local stresses on the engine bearing (left). Force-displacement hysteresis of the engine bearing at the reference point (right)

identical hysteresis is observed. This means that in addition to the mechanical i.e. the force-displacement, behaviour, the dissipative behaviour, which is characterized by the hysteresis area, is also mapped. Accordingly, the local loads can be deduced at this point as shown in Figure 13.4.
The self-heating caused by dissipation is shown in Fig. 13.5. The surface temperatures were experimentally determined using an infrared camera technology. For reasons of clarity, the experimental data smoothed. This was performed with the calculated temperature curve. In Fig. 13.5 (right), the agreement of the stationary temperature values can be recognized. However, it is observed that the calculation duration is significantly reduced by the suitable selection of the heat capacity. Furthermore, the local temperature profile in the stationary state can be inferred. In consequence, a temperature rise of approximately 18 K can be observed under the given conditions. The point in time at which the stationary temperature equilibrium is reached is marked

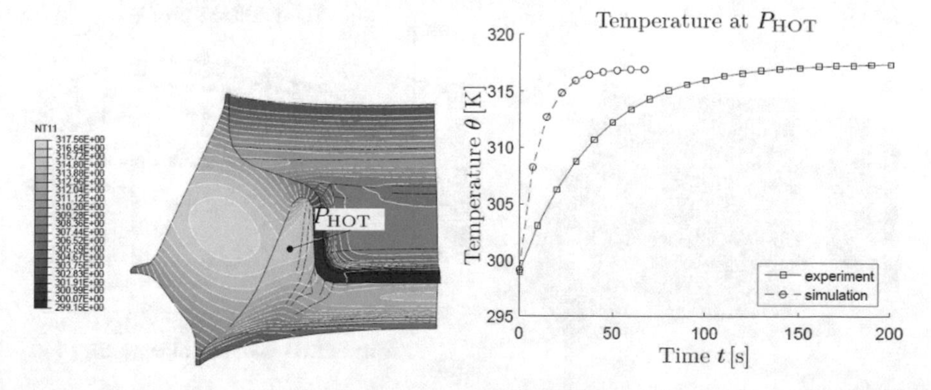

Fig. 13.5 Local temperature distribution in the engine bearing (left). Temperature evolution with respect to the time of the engine bearing at the hotspot (right)

with τ_s. Using this variable, the time-dependent temperature curves are now described. At this point it is also interesting to compute temperature curves which are difficult to measure. In particular, the thickest component cross-section is considered here. The path defined on the finite element mesh is used to evaluate the temporal temperature curve as shown in Fig. 13.6. Finally, the simulation shows a high concordance to the experimental data under consideration of an efficient simulation methodology.

13.7 Summary and Conclusion

In this work, a calculation concept for the estimation of the change in local component temperatures caused by dissipative heating was presented. Based on the phenomenological consideration of elastomer materials, phenomena relevant to self-heating were identified and used as a basis for the constitutive modelling. A modified model of the finite thermoviscoelasticity was continuum mechanically modelled. In addition to the kinematic description, a thermomechanically consistent derivation of the constitutive relations as well as the formulation of the heat conduction equation was performed. Within the framework of finite element implementation, the unique initial and boundary value problem was presented and a fully coupled functional was derived using the variation principle. The tangent operators were subsequently determined by linerisation. A special approach of the total free energy density was defined and used for the analytical calculation of state vectors and consistent tangent operators. Finally, the model was validated, starting with the explanation of the parameterisation and model calculations, followed by comparison of experiments and simulations as well as their discussion. Last but not least, not only the experimental setup but also the calculation effort can be significantly reduced compared to the classical model of finite thermal viscoelasticity by neglecting the thermoelastic effects. In summary, it can be said that the implemented concept is a suitable instrument for the

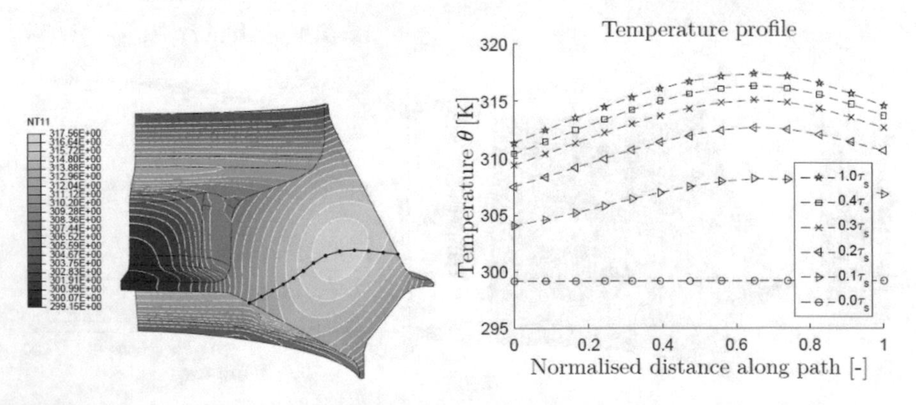

Fig. 13.6 Local temperature distribution in the engine bearing (left). Temperature evolution with respect to the time of the engine bearing at the path (right)

robust, cost-effective and valid estimation of dissipation-related temperature changes in components. In the future, characteristic diagrams of the stationary component temperatures over large amplitude and frequency ranges should be created with this methodology. The necessity of the experimental determination of a parameter or the use of literature data shall be determined by means of a parameter analysis. Furthermore, of different components and elastomer compounds may be validated as part of a follow-up project.

References

Abaqus (2002) ABAQUS/CAE 6.14 User's Manual. Hibbitt, Karlsson & Sorensen, Incorporated

Anand L (1985) Constitutive equations for hot-working of metals. International Journal of Plasticity 1(3):213–231

Anand L, Ames NM, Srivastava V, Chester SA (2009) A thermo-mechanically coupled theory for large deformations of amorphous polymers. part i: Formulation. International Journal of Plasticity 25(8):1474–1494

Anthony RL, Caston RH, Guth E (1942) Equations of state for natural and synthetic rubber-like materials. i. unaccelerated natural soft rubber. The Journal of Physical Chemistry 46(8):826–840

Arruda EM, Boyce MC, Jayachandran R (1995) Effects of strain rate, temperature and thermomechanical coupling on the finite strain deformation of glassy polymers. Mechanics of Materials 19(2-3):193–212

Bröcker C (2013) Materialmodellierung für die simultane Kalt /Warmumformung auf Basis erweiterter rheologischer Modelle. kassel university press GmbH

Bröcker C, Matzenmiller A (2008) Modellierung und simulation thermo-mechanisch gekoppelter umformprozesse. In: Proceedings in Applied Mathematics and Mechanics, Wiley Online Library, vol 8, pp 10,485–10,486

Coleman BD, Noll W (1963) The thermodynamics of elastic materials with heat conduction and viscosity. Archive for Rational Mechanics and Analysis 13:167–178

Dippel B, Johlitz M, Lion A (2014) Thermo-mechanical couplings in elastomers - experiments and modelling. Journal of Applied Mathematics and Mechanics doi:10.1002/zamm.201400110

Elsner P, Eyerer P, Hirth T (2012) Domininghaus - Kunststoffe: Eigenschaften und Anwendungen. VDI-Buch, Springer Berlin Heidelberg

Erbts P, DüSter A (2012) Accelerated staggered coupling schemes for problems of thermoelasticity at finite strains. Computers & Mathematics with Applications 64(8):2408–2430

Flory PJ (1961) Thermodynamic relations for high elastic materials. Transactions of the Faraday Society 57:829–838

Glaser S (1992) Berechnung gekoppelter thermomechanischer Prozesse. Dissertation, Institut für Baumechanik und Numerische Mechanik, Universität Hannover, Bericht-Nr. F88/6

Gough J (1805) A description of a property of caoutchouc, or indian rubber. Memories of the Literacy and Philosophical Society of Manchester 1:288–295

Hamkar AW (2013) Eine iterationsfreie Finite-Elemente Methode im Rahmen der finiten Thermoviskoelastizität. Universitätsbibliothek Clausthal

Heimes T (2004) Finite Thermoviskoelastizität. Forschungs- und Seminarbericht aus dem Gebiet Technische Mechanik und Flächentragwerke, Universität der Bundeswehr München

Holzapfel GA (2000) Nonlinear Solid Mechanics: A Continuum Approach for Engineering Science. Wiley

Johlitz M (2015) Zum Alterungsverhalten von Polymeren: Experimentell gestützte, thermochemomechanische Modellbildung und numerische Simulation. Habilitationsschrift Institut für Mechanik an der Universität der Bundeswehr München

Joule JP (1859) On some thermo-dynamic properties of solids. Philosophical Transactions of the Royal Society of London 149:91–131

Klauke R (2015) Lebensdauervorhersage mehrachsig belasteter Elastomerbauteile unter besonderer Berücksichtigung rotierender Beanspruchungsrichtungen. Fakultät für Maschinenbau der Technischen Universität Chemnitz, Institut für Mechanik und Thermodynamik

Koltzenburg S, Maskos M, Nuyken O (2013) Polymere: Synthese, Eigenschaften und Anwendungen. Springer-Verlag

Lejeunes S, Eyheramendy D, Boukamel A, Delattre A, Méo S, Ahose KD (2018) A constitutive multiphysics modeling for nearly incompressible dissipative materials: application to thermo–chemo-mechanical aging of rubbers. Mechanics of Time-Dependent Materials 22(1):51–66

Lion A (2000) Thermomechanik von Elastomeren. Berichte des Instituts für Mechanik der Universität Kassel (Bericht 1/2000)

Lubliner J (1985) A model of rubber viscoelasticity. Mechanics Research Communications 12:93–99

Miehe C (1988) Zur numerischen Behandlung thermomechanischer Prozesse. Dissertation, Institut für Statik und Dynamik der Luft- und Raumfahrtkonstruktionen Universität Stuttgart

Mooney M (1940) A theory of large elastic deformation. Journal of Applied Physics 11:582–592

Mullins L (1948) Effect of stretching on the properties of rubber. Rubber Chemistry and Technology 21(2):281–300

Naumann C, Ihlemann J (2011) Thermomechanical material behaviour within the concept of representative directions. In: Constitutive Models for Rubber VII, Balkema Leiden, pp 107–112

Payne AR (1962) The dynamic properties of carbon black-loaded natural rubber vulcanizates. part i. Journal of applied polymer science 6(19):57–63

Reese S (2001) Thermomechanische Modellierung gummiartiger Polymerstrukturen. Habilitation, Bericht-Nr. F01/4 des Instituts für Baumechanik und Numerische Mechanik, Universität Hannover

Rivlin RS (1948-1951) Large elastic deformations of isotropic materials part:vii. Philosophical Transactions of the Royal Society of London Series A, Mathematical and Physical Sciences

Rivlin RS (1997) Some applications of elasticity theory to rubber engineering. In: Collected papers of R. S. Rivlin, Springer, pp 9–16

Rivlin RS, Saunders DW (1951) Large elastic deformations of isotropic materials vii. experiments on the deformation of rubber. Philosophical Transactions of the Royal Society of London Series A, Mathematical and Physical Sciences 243(865):251–288

Schlanger HP (1983) A one-dimensional numerical model of heat transfer in the process of tire vulcanization. Rubber Chemistry and Technology 56(2):304–321

Schröder J, Parra Pelaez G (2019) Erfassung, simulation und bewertung der thermomechanischen schädigungsmechanismen von elastomerbauteilen unter dynamischen mechanischen beanspruchungen ii. FKM-Abschlussbericht Heft 334(FKM-Vorhaben Nr.603)

Schwarzl FR (2013) Polymermechanik: Struktur und mechanisches Verhalten von Polymeren. Springer-Verlag

Shutov AV, Landgraf R, Ihlemann J (2013) An explicit solution for implicit time stepping in multiplicative finite strain viscoelasticity. Comp Meth Appl Mech Engrg 265:213–225

Simo JC, Miehe C (1992) Associative coupled thermoplasticity at finite strains: Formulation, numerical analysis and implementation. Computer Methods in Applied Mechanics and Engineering 98(1):41–104

Simo JC, Taylor RL (1982) Penalty function formulations for incompressible nonlinear elastostatics. Computer Methods in Applied Mechanics and Engineering 35:107–118

Stephan P (2002) Vdi wärmeatlas: Berechnungsblätter für den wärmeübergang. Verein Deutscher Ingenieure

Tobolsky AV (1967) Mechanische Eigenschaften und Struktur von Polymeren. Berliner Union

Tobolsky AV, Mark HF, Bondi AA, Deanin RD, DuPre DB, Gent AN, Mark H, Peterlin A, Rebenfeld L, Samulski E, et al (1971) Polymer science and materials. Wiley-Interscience New York

Treloar L (1975) The physics of rubber elasticity. Oxford University Press, USA

Williams ML, Landel RF, Ferry JD (1955) The temperature dependence of relaxation mechanisms in amorphous polymers and other glass-forming liquids. Journal of the American Chemical Society 77(14):3701–3707

Chapter 14
Additive Manufacturing: A Review of the Influence of Building Orientation and Post Heat Treatment on the Mechanical Properties of Aluminium Alloys

Enes Sert, Leonhard Hitzler, Markus Merkel, Ewald Werner, and Andreas Öchsner

Abstract Selective laser melting is one of the powder bed-based processes that allows a layered fabrication of components, which has become the most widely utilized additive manufacturing technique for metal processing. The ability to create futuristic designs and non-standard topology-optimized structures is one of the biggest advantages of additive manufacturing. On the other hand, one of the major challenges is to account for the anisotropic and inhomogeneous material properties. This work presents an overview of the most relevant studies, concerning the influence of building directions and post heat treatments on the mechanical properties of selective laser melted aluminium alloys.

Keywords: Selective laser melting · Homogenization · Anisotropic material behavior · Hardness · Tensile strength

Enes Sert & Andreas Öchsner
Faculty of Mechanical Engineering, Esslingen University of Applied Sciences, Kanalstrasse 33, 73728 Esslingen, Germany,
e-mail: enes.sert@hs-esslingen.de, andreas.oechsner@hs-esslingen.de

Leonhard Hitzler & Ewald Werner
Institute of Materials Science and Mechanics of Materials, Technical University of Munich, Boltzmannstrasse 15, 85748 Garching, Germany,
e-mail: hitzler@wkm.mw.tum.de, werner@wkm.mw.tum.de

Markus Merkel
Institute for Virtual Product Development, Aalen University of Applied Sciences, 73430 Aalen, Germany,
e-mail: markus.merkel@hs-aalen.de

H. Altenbach and A. Öchsner (eds.), *State of the Art and Future Trends in Material Modeling*, Advanced Structured Materials 100,
https://doi.org/10.1007/978-3-030-30355-6_14

14.1 Nomenclatur

SLM	Selective laser melting
AM	Additive manufacturing
3D-CAD	Three dimensional computer aided design
STL	Standard triangulation language
STEP	Standard for the the exchange of model data
BO	Building orientation
OT	Operating temperature
HIP	Hot isostatic pressing
SH	Surface hardness
UE	Upper edge
LE	Lower edge
CH	Core hardness
E	Young's modulus
$R_{p0.2}$	Yield strength
R_m	Ultimate tensile strength
A_t	Elongation at failure
CT	Computed tomography
VED	Volumetric energy density
P	Laser power
d	Layer thickness
x	Hatch spacing
v_{scan}	Scan speed
Φ	Polar angle

14.2 Introduction

The additive manufacturing (AM) of metals is a key technology and an emerging industry for the manufacture of highly specialized components. Recently, extensive research has been performed on achievable material properties, as well as on the correlation between process parameters and material characteristics.

The main focus on titanium-based alloys seen in this area results from the main sectors of medicine, as well as the aerospace industry, in which additive manufacturing technologies have already partially established themselves (Buchbinder et al, 2015; Hitzler et al, 2017c). Aluminium alloys offer a further possibility for additive manufacturing, opening up utilization in the automotive segment. Additive manufacturing using aluminium alloys is an interesting alternative, especially for prototypes and for components with complex geometry, rendering prototype production via die casting very costly (Sert et al, 2019a). Due to the large growth in different areas of application, the structural mechanical properties are becoming increasingly important (Aboulkhair et al, 2015). The specification of anisotropies, preferred directions and weak points in the microstructure vary with the considered material, and completely opposite behavior patterns are not uncommon (Hitzler et al, 2016a). Thus, it is becoming increasingly important to gain precise knowledge about the mechanical properties of additively manufactured components. The aim of this work is to provide an overview on the mechanical properties of additively produced aluminium alloys. In particular, the influence of various heat treatments and the orientation of the components in the built environment is examined. Furthermore, the selective laser melting process (SLM), alternative also called laser beam melting (LBM) , which is the most common AM process for aluminium alloys, will be considered in more detail to prepone a possible digital data chain.

14.3 Additive Manufacturing - Selective Laser Melting

The selective laser melting (SLM) process is closely related to selective laser sintering (SLS) and it is based on the further development of laser-sintering of metal powder. As with all layered powder bed fusion processes, the construction process takes place in individual layers, incrementally lowering the work platform step by step downwards (z-direction). In order to prevent thermal distortion, the working platform for aluminum alloys is usually heated to an operating temperature of 200°C, thereby ensuring that the temperature conditions during the manufacturing process remain constant, thus allowing for a continuous construction process. The layer thicknesses vary between 20 μm and 100 μm, depending on the laser capabilities and the desired accuracy and productivity (Frazier, 2014; Beese and Carroll, 2015). During the manufacturing process, the unexposed powder remains unfused, whereby the unbounded powder serves on the one hand as a protection for the components and on the other hand as support for the creation of thin-walled or overhanging components and the stacking of several components one above the other. By attaching support structures between the work platform and the component, the heat energy generated during the melting process is conducted into the platform and is dissipated by the thermal management system. SLM systems are often equipped with an automatic powder recycling unit in the closed protective gas atmosphere (commonly either argon or nitrogen). The production space is emptied of unused powder by means of vacuum suction. This powder is then sieved using a vibrating screen and is returned into the storage container for the creation of a new layer (see Fig. 14.1).
Given the similarities amongst the various AM procedures, the bonding mechanism is used to distinguish one process from another. Within SLM the raw material is fully molten and the process returns a dense part. The older selective laser sintering process corresponds to liquid phase sintering with different binder and structural material. After the plastic coating of the metal powder particles has melted, the emerging porous green body has to be thermally treated and infiltrated in a multi-step procedure. The disadvantage of multiple steps is avoided in SLM processing due to complete melting of the metallic material. During the binderless bonding process, only small punctiform connections between the articles are created or the parts are

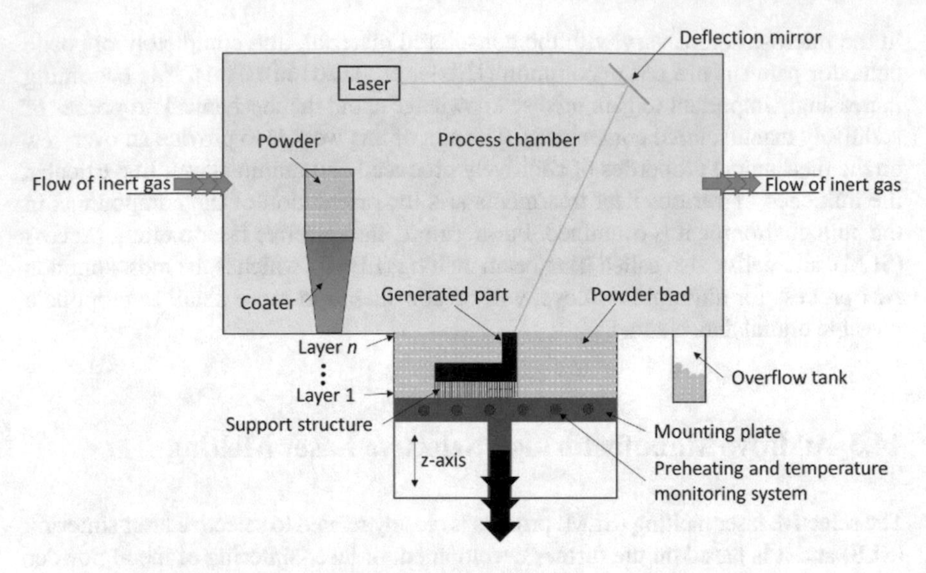

Fig. 14.1 Schematic setup of powder-bed fusion techniques

manufactured by powder metallurgy under high pressure and temperature, dense parts being formed by diffusion processes in solids. The porosity of the components depends on how much heat energy is supplied. The alloyed or unalloyed particles can be melted and bonded to a residual porosity of about 2% (Frazier, 2014; Beese and Carroll, 2015; Bikas et al, 2015; Singh et al, 2017; Lewandowski and Seifi, 2016; Wong and Hernandez, 2012; Gong et al, 2014).

The starting point of additive manufacturing is a complete 3D model of the component to be built, usually obtained via 3D-CAD modeling (newly designed components and prototypes). A second method of capturing the model data, important for spare parts (duplicating) and for medical applications, is the use of reverse engineering. Thereby, the data sets are generated from measurements, surface scans or CT scans. From these data sets surface and/or volumetric models are generated, utilizing polygonization and triangulation algorithms. With the obtained refined models containing the geometric information of the part to be fabricated or duplicated STL data sets can be obtained and are then transferred to the generative production process in STL format. This data conversion is software-aided and automatized as far as possible (VDI-Richtlinie, 2014, see Fig. 14.2).

14.4 Mechanical Properties

Additive manufacturing has the advantages of flexible geometric design, and correspondingly an outstanding potential for mass reduction (Sert et al, 2019a). The full exploitation of these benefits requires predictable mechanical properties. Due to the

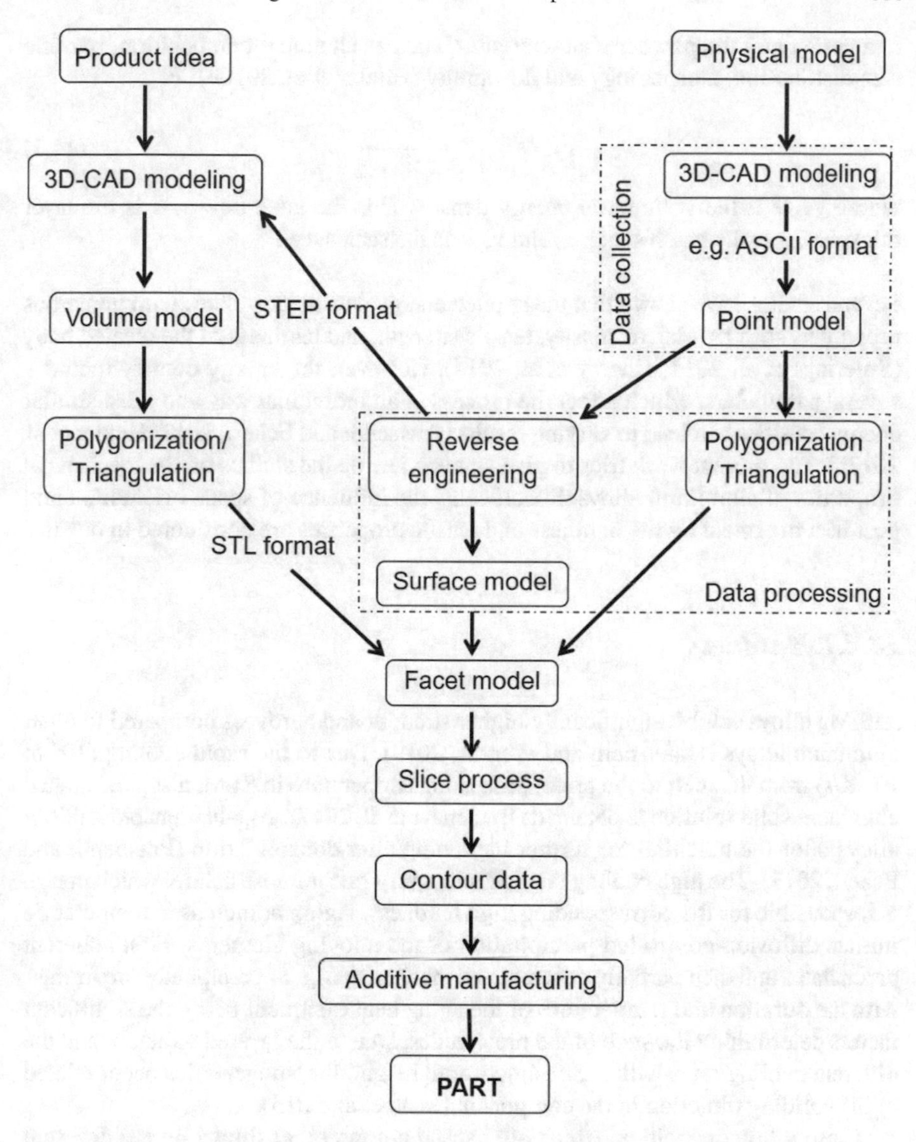

Fig. 14.2 Schematic representation of a possible digital data chain for additive manufacturing

complete melting of the powder in the selective laser melting process, it is possible to produce almost completely dense parts. The relative density of the components produced is an important quality characteristic that is related to the mechanical properties and process parameters. The influence of the latter is mainly attributed to the surface condition and pore formation (Dai and Gu, 2015; Aboulkhair et al, 2014). In contrast, the melt pool itself mainly depends on the energy density of the irradiation and the powder characteristics, such as chemical composition, particle size distribution, morphology and flowability (Hitzler et al, 2016b):

$$VED = \frac{P}{d \times x \times v_{\text{scan}}} \tag{14.1}$$

where VED is the volumetric energy density, P is the laser power, d is the layer thickness, x is the hatch spacing, and v_{scan} is the scan speed.

Several studies have shown that the applied energy density correlates with numerous properties, such as relative density, tensile strength, and hardness of the created body (Spierings et al, 2011; Cherry et al, 2014). However, the energy density includes several parameters, which affect the process in an individual way and thus, similar energy densities can lead to varying results (Prashanth and Eckert, 2017; Hitzler et al, 2017b). The present work tries to give an overview on the studies on the mechanical properties of aluminum alloys. Specifically, the influence of space orientation and post heat treatment on the hardness and tensile properties are considered in detail.

14.4.1 Hardness

AlSiMg alloys exhibit significantly higher strength and hardness compared to other aluminum alloys (Hafenstein and Werner, 2019). Due to the rapid cooling (10^5 to 10^6 K/s) from the melt to the preset preheating temperature in SLM, a supersaturated aluminum solid solution is obtained (Buchmayr et al, 2017). Age-hardenable AlSiMg alloys offer the potential for further hardening after consolidation (Prashanth and Eckert, 2017). The high cooling rate leads to a very fine microstructure, which in turn is responsible for the corresponding high hardness. Aging at increased temperature initiate diffusion-controlled precipitation of the alloying elements. First coherent precipitates and then partially coherent and incoherent Mg_2Si precipitates are formed, with the duration and temperature of the aging heat treatment being the significant factors determining the state of the precipitates. Due to the layered structure and the different cooling rates with increasing overall height, inhomogeneities occur related to the building direction in the component (Read et al, 2015).

These inhomogeneities, which are caused among other things by the different stages of precipitation hardening, can be detected by hardness measurements. Recent studies focus on the correlation between surface and core hardness, as the process parameters for the surface and core differ. However, the studies demonstrate that the hardness values between surface and core are very similar in most cases (Sert et al, 2018; Hitzler et al, 2018).

The inhomogeneous nature, coupled with the dependencies of the manufacturing parameters, evidence a tendency for the hardness to decrease with increasing built height, Fig. 14.3 (Hitzler et al, 2016a; Sert et al, 2018; Aboulkhair et al, 2016a; Buchbinder et al, 2009). This trend was also confirmed by other studies, see Table 14.1.

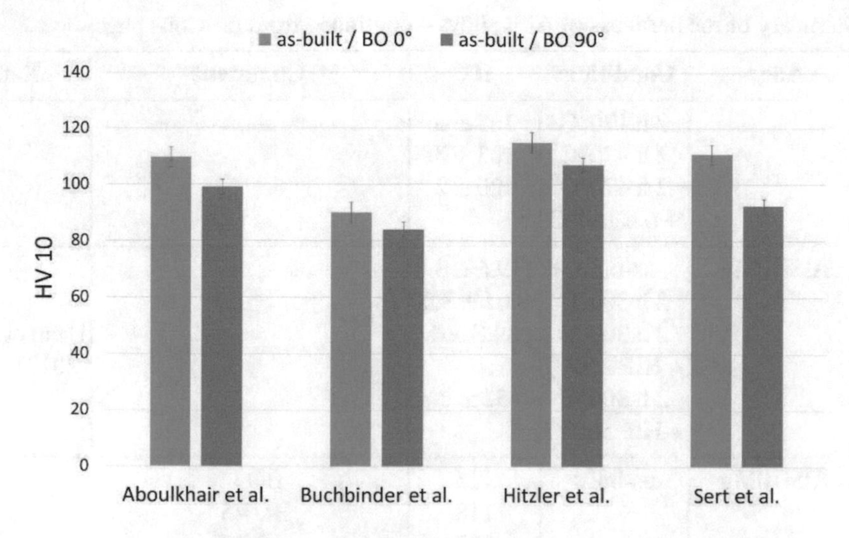

Fig. 14.3 Vickers hardness in terms of as built and post heat treatment from different studies (Aboulkhair et al, 2016a; Buchbinder et al, 2009; Hitzler et al, 2016a; Sert et al, 2018)

Table 14.1 Summary of the hardness of AlSi alloys

Alloy	Conditions	HV	Comment	Ref.
AlSi10Mg	as-built	109.7 ± 0.9 99.07 ± 2	BO 0° BO 90°	Spierings et al (2011)
AlSi10Mg	as-built 1 h 520°C + 6 h 160°C 6 h 520°C + 7 h 160°C	125 ± 1 100 ± 1 103 ± 2		Aboulkhair et al (2016b)
AlSi10Mg	as-built	134 ± 4 90 ± 8 130 ± 7 84 ± 3	BO 90° OT 200°C / BO 90° BO 0° OT 200°C / BO 0°	Buchbinder et al (2015)
Al7075 Al7075 + 1%Si Al7075 + 2%Si Al7075 + 3%Si Al7075 + 4%Si	as-built as-built as-built as-built as-built 10 h 120°C 6 h 150°C	130 ± 10 150 ± 10 147 ± 15 156 ± 9 159 ± 9 163 ± 5 171 ± 4		Montero-Sistiaga et al (2016)

Continued on next page

Summary of the hardness of AlSi alloys – continued from previous page

Alloy	Conditions	HV	Comment	Ref.
	2 h 170°C	165 ± 3		
	2 h 470°C	103 ± 3		
	2 h 470°C + 6 h 150°C	120 ± 2		
AlSi10Mg	as-built	119.6 ± 3.8		Uzan et al (2017)
	2 h 300°C	93.1 ± 3.2		
	2 h 300°C + HIP 250°C	89.2 ± 4		
	2 h 300°C + HIP 500°C	52 ± 2		
AlSi10Mg	as-built	115	BO 0°	Hitzler et al (2016a)
		118	BO 45°	
		107	BO 90°	
	6 h 525°C + 6 h 165°C	56	BO 0°	
		52	BO 45°	
		51	BO 90°	
	6 h 300°C + 6 h 180°C	66	BO 0°	
		70	BO 45°	
		63	BO 90°	
	4 h 500°C + 6 h 180°C	110	BO 0°	
		110	BO 45°	
		109	BO 90°	
AlSi10Mg	as-built	103 ± 1	SH / BO 0° / right	Sert et al (2019a)
		102 ± 2	SH / BO 0° / left	
		107.3 ± 1.2	SH / BO 0°/90° / right	
		95.7 ± 3.1	SH / BO 0°/90° / left	
		115 ± 2.6	SH / BO 30° / UE	
		89.7 ± 2.5	SH / BO 30° / LE	
		114.7 ± 1.2	SH / BO 45° / UE	
		85.7 ± 2.5	SH / BO 45° / LE	
		105.3 ± 2.1	SH / BO 75° / UE	
		96 ± 1	SH / BO 75° / LE	
		101.7 ± 1.5	SH / BO 90° / UE	
		90.3 ± 3.5	SH / BO 90° / LE	
AlSi10Mg	4 h 170°C	103 ± 1	SH / BO 0° / right	
		104.3 ± 2.1	SH / BO 0° / left	
		110.7 ± 1.5	SH / BO 0°/90° / right	
		97.3 ± 1.5	SH / BO 0°/90° / left	

Continued on next page

Summary of the hardness of AlSi alloys – continued from previous page

Alloy	Conditions	HV	Comment	Ref.
		115.7 ± 0.6	SH / BO 30° / UE	Sert et al (2019a)
		87.3 ± 1.2	SH / BO 30° / LE	
		116.3 ± 0.6	SH / BO 45° / UE	
		80 ± 0	SH / BO 45° / LE	
		127.3 ± 1.5	SH / BO 75° / UE	
		79 ± 2	SH / BO 75° / LE	
		124.3 ± 2.5	SH / BO 90° / UE	
		94.3 ± 4.2	SH / BO 90° / LE	
AlSi10Mg	as-built	110.7 ± 2.1	CH / BO 0° / right	Sert et al (2019a)
		107.3 ± 1.5	CH / BO 0° / left	
		105.7 ± 0.6	CH / BO 0°/90° / right	
		111 ± 1.7	CH / BO 0°/90° / left	
		112.3 ± 2.1	CH / BO 30° / UE	
		107 ± 3	SH / CH 30° / LE	
		101.3 ± 1.5	CH / BO 45° / UE	
		105 ± 1	SH / CH 45° / LE	
		93.3 ± 1.5	CH / BO 75° / UE	
		109 ± 2.6	CH / BO 75° / LE	
		92.3 ± 2.5	CH / BO 90° / UE	
		101.3 ± 2.1	CH / BO 90° / LE	
AlSi10Mg	4 h 170°C	115.3 ± 2.5	CH / BO 0° / right	Sert et al (2019a)
		110.7 ± 4	CH / BO 0° / left	
		106 ± 2	CH / BO 0°/90° / right	
		112.7 ± 0.6	CH / BO 0°/90° / left	
		113.3 ± 3.1	CH / BO 30° / UE	
		105.7 ± 1.5	SH / CH 30° / LE	
		101.3 ± 1.5	CH / BO 45° / UE	
		105 ± 1	SH / CH 45° / LE	
		102 ± 2.6	CH / BO 75° / UE	
		108.7 ± 2.1	CH / BO 75° / LE	
		117.3 ± 1.2	CH / BO 90° / UE	
		115 ± 1.7	CH / BO 90° / LE	

This phenomenon is caused by differing stakes of the precipitated resulting from varying dwell times in the built chamber after consolidation. By post heat treatments, the inhomogeneities related to the height can be reduced. Therefore, the choice of temperature and the aging time are of significant importance. The T6 heat treatment or the stress-relief annealing at 300°C for 2 hours results in a homogenization of the hardness profile, but at the expense of a coarsening of the microstructure, Table 14.1. Hitzler et al (2016a); Sert et al (2019a) showed that it is possible to homogenize the hardness profile while maintaining the as-fabricated microstructure by means of a second aging heat treatment.

14.4.2 Tensile Strength

In order to assess the quality of an additive manufactured component destructive material testing is inevitable. The studies summarized in Table 14.2 documents the anisotropic material behavior, which was determined via tensile testing.

Table 14.2 Summery of tensile properties of AlSi alloy

Alloy	Conditions	E [GPa]	$R_{p0.2}$ [MPa]	R_m [MPa]	A_t [%]	Ref.
AlSi10Mg	as-built	77 ± 5	268 ± 2	333 ± 15	1.4 ± 0.3	Aboulkhair et al (2016b)
	2 h 535°C + 10 h 158°C	73 ± 4	239 ± 2	292 ± 4	3.9 ± 0.5	
A357	as-built					Aversa et al (2017)
	OT 100°C		245 ± 4	389 ± 3	5.2 ± 0.2	
	OT 140°C		284 ± 3	408 ± 5	4.9 ± 0.2	
	OT 170°C		288 ± 7	397 ± 9	3.8 ± 0.3	
	OT 190°C		246 ± 4	362 ± 2	4.4 ± 0.3	
AlSi10Mg	as-built					Buchbinder et al (2015)
	OT 220°C / BO 0°		130	300	6.5	
	OT 170°C / BO 90°		150	250	4	
	BO 0°		210	400	8	
	BO 90°		240	450	5.4	
	BO 0°		210	400	6	
	BO 90°		240	450	3.2	
Al7075+ 4%Si	as-built		279 ± 10			Montero-Sistiaga et al (2016)
			338 ± 13			
AlSi12	as-built					
	BO 0°		227.31	261.8	0.87	

Continued on next page

Summery of tensile properties of AlSi alloy – continued from previous page

Alloy	Conditions	E [GPa]	$R_{p0.2}$ [MPa]	R_m [MPa]	A_t [%]	Ref.
			±20.16	±37.01	±0.004	
	BO 45°		262.36	367.33	2.57	
			±10.08	±29.88	±0.008	Rashid et al
	BO 90°		224.78	398.57	3.42	(2018)
			±33.52	±16.48	±0.004	
AlSi10Mg	as-built		319 ± 2.8	477.5 ± 4.9	4 ± 4.9	Zhang et al (2018)
	2 h 300°C		266 ± 4	369 ± 4	7.5 ± 1.5	
	1 h 530°C		151 ± 2.8	253 ± 5.6	10.5 ± 2	
	2 h 535°C + 10 h 158°C		197.5 ± 2	253 ± 5.6	10.5 ± 2	
AlSi10Mg	as-built					Wang et al (2018a)
	BO 0°			325	3	
	BO 90°			375	8	
	3/4 h 380°C					
	BO 0°		105.7	171.3	9.4	
	BO 90°			171.7	16.5	
AlSi10Mg	as-built					Wang et al (2018b)
	BO 0°			334	3.64	
	BO 90°			358		
	2 h 535°C +10 h 158°C					
	BO 0°		174	267.3	9.28	
	BO 90°		160	278		
AlSi10Mg	as-built		241 ± 10	384 ± 16	6 ± 1	Uzan et al (2017)
	2 h 300°C		205 ± 8	253 ± 18	18 ± 3	
	2 h 300°C + HIP 200°C		186 ± 5	233 ± 7	22 ± 2	
	2 h 300°C + HIP 500°C		115 ± 5	141 ± 8	35 ± 3	
AlSi10Mg	as-built	72.322	206.74	366.43	6.12	
		±2.9953	±4.419	±12.506	±1.096	
	BO 0°/5°	72.888	241.15	399.10	6.47	
		±1.1788	±5.697	±7.33	±0.361	
	BO 0°/85°	71.715	222.83	360.27	5.33	
		±1.1462	±9.301	±10.442	±0.457	
	BO 45°/0°	65.64	188.15	330.11	4.47	Hitzler et al
		±3.5145	±7.038	±10.385	±0.152	(2017b)
	BO 45°/5°	69.515	179.71	314.32	3.97	

Continued on next page

Summery of tensile properties of AlSi alloy – continued from previous page

Alloy	Conditions	E [GPa]	$R_{p0.2}$ [MPa]	R_m [MPa]	A_t [%]	Ref.
		±2.3033	±8.313	±7.236	±0.449	
	BO 90°/5°	70.422	208.57	357.49	3.15	
		±2.6857	±16.942	±19.6	±0.080	
	BO 90°/45°	62.560	198.13	344.73	3.2	
		±3.7283	±13.635	±20.564	±0.189	
AlSi10Mg	as-built					Dai et al (2018)
	BO 0°		287.2	476.8	7.33	
	2 h 300°C					
	BO 0°		201.3	320.5	13.3	
Al3.5Cu1.5Mg	as-built		233 ± 4	366 ± 7	5.3 ± 0.3	Wang et al (2018c)
	2 h 535°C + 10 h 158°C		368 ± 6	455 ± 10	6.2 ± 1.8	
AlSi10Mg	4 h 170°C					Sert et al (2019b)
	BO 0°	66.71	226.14	352.05	5.21	
		±1.6449	±8.0951	±8.875	±0.4802	
	BO 45°	68.43	190.07	322.60	4.41	
		±2.1194	±3.9837	±3.5846	±0.33	
	BO 60°	66.50	205.69	348.39	4.29	
		±0.9991	±5.4356	±6.4003	±0.2656	
	BO 70°	68	201.41	356.62	4.18	
		±5.1241	±7.0740	±10.1685	±0.2018	
	BO 80°	67.91	188.11	344.35	4.55	
		±2.1674	±5.1051	±6.6534	±0.2592	
	BO 90°	66.11	185.87	344.78	4.85	
		±1.5225	±3.9003	±5.9526	±0.25	
AlSi10Mg	as-built					VDI-Richtlinie (2017)
	BO 0°	67...72	239...292	372...473	4...7	
	BO 45°	71...76	213...295	370...478	4...6	
	BO 90°	68...78	210...272	353...482	2...5	
	6 h 525°C					
	BO 0°	66...73	132...151	236...257	10...17	
	BO 45°	60...72	134...156	239...260	12...18	
	BO 90°	57...73	126...160	221...254	11...18	
	6 h 525°C + 7 h 165°C					

Continued on next page

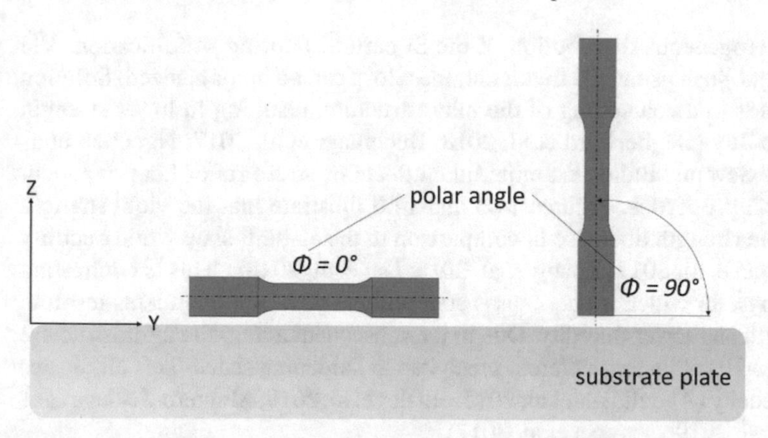

Fig. 14.4 Applied nomenclature for the description of the orientation of tensile specimens

Summery of tensile properties of AlSi alloy – continued from previous page

Alloy	Conditions	E [GPa]	$R_{p0.2}$ [MPa]	R_m [MPa]	A_t [%]	Ref.
	BO 0°	71...76	225...262	287...311	5...10	
	BO 45°	70...80	226...271	289...320	5...9	
	BO 90°	69...77	222...260	281...309	6...10	

In order to clarify the comprehensibility with respect to the building orientation, the polar angle was introduced as shown in Fig. 14.4. The polar angle varies between 0° (horizontal) and 90° (vertical). As tensile specimens are loaded only along their longitudinal direction, the consideration of the alignment of their longitudinal axis in the fabrication / built space is sufficient for a unique designation. The directional dependency is such that the highest material characteristics measured samples with layer oriented parallel to the macroscopic load (Buchbinder, 2013; Sert et al, 2019b). This seems to stem from the increased likelihood that irregularities and voids will occur as the number of layers increases.

However, this is not necessarily synonymous with an increased degree of porosity, but it has been shown that the connections between individual laser tracks and layers predominantly change the importance of anisotropy on the tensile properties (Zhao et al, 2018; Reschetnik et al, 2016; Qiu et al, 2013; Shifeng et al, 2014; Cloots et al, 2016). As Table 2 demonstrates, this wording cannot be applied generally but the qualitative dependence should be similar. The specified values vary depending on the settings and machine conditions. Some studies indicate that removal of the surface and the associated removal of the increased pore density in the transition zone between the surface and the core can lead to an increase in strength (Aboulkhair et al, 2015; Bagherifard et al, 2018; Tang and Pistorius, 2017; Hitzler et al, 2017a; Zhang, 2004). Heat treatment can markedly influence the mechanical properties. AlSi alloys show an inhomogeneous composition and grain size over each individual layer, which lead to a heterogeneous distribution of the Si particles during solidification. Via heat treatments, such as the T6 treatment, the alloys can be homogenized. Solution annealing leads to a coarsening of the microstructure, resulting in lower strength and high ductility (Bagherifard et al, 2018; Buchmayr et al, 2017; Ngnekou et al, 2017, 2018). Several studies examined the effects of stress relief heat treatment on mechanical properties. Figures 14.5 and 14.6 illustrate that the yield strength and the tensile strength decrease in comparison to the as-built state while ductility increases (Uzan et al, 2017; Zhang et al, 2018; Dai et al, 2018). This is contrasting cast AlSi alloys, for which in most cases conventional post heat treatments result in higher strength and lower ductility. Due to the subsequent aging, the microstructure is homogenized by adjusting different precipitation hardening states. This eliminates the inhomogeneity (Aboulkhair et al, 2015; Hitzler et al, 2018; Montero-Sistiaga et al, 2016; Sert et al, 2019b; Aversa et al, 2017).

14.5 Conclusions

AM techniques combine the advantages of geometric design freedom with good mechanical properties. Based on the process-related layered production in the powder bed environment, the components have anisotropic properties that are more or less emphasized depending on the manufacturing conditions and the raw materials. In addition, anisotropy may be superimposed with inhomogeneities resulting from microstructural developments over different dwell times at elevated temperatures in the built environment. These directional and spatial dependencies make it more

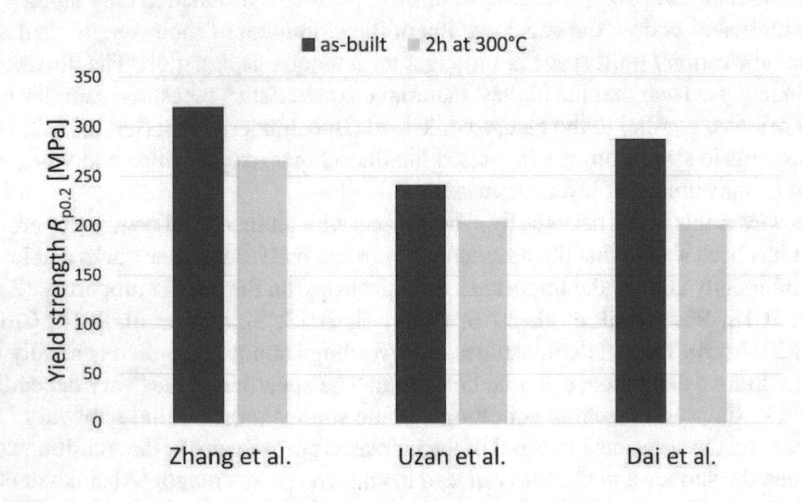

Fig. 14.5 Yield strength in terms of as bulit and post heat treatment from different studies (Zhang et al, 2018; Uzan et al, 2017; Dai et al, 2018)

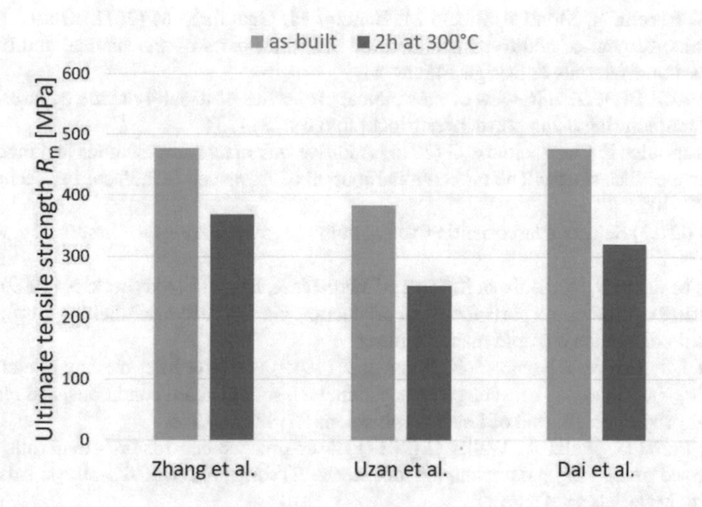

Fig. 14.6 Ultimate tensile strength of as-built and post heat treated taken from different studies (Zhang et al, 2018; Uzan et al, 2017; Dai et al, 2018)

challenging to characterize material properties and render their full description difficult. This, however, is of utmost importance when predicting part properties by employing micromechanical approaches, since components manufactured with identical machines can have different mechanical properties simply by changing their orientation in space during manufacture. Therefore, it is very important to consider the manufacturing conditions and their effects on the resulting properties in the design and dimensioning processes.

References

Aboulkhair NT, Everitt NM, Ashcroft I, Tuck C (2014) Reducing porosity in alsi10mg parts processed by selective laser melting. Additive Manufacturing 1-4:77–86

Aboulkhair NT, Stephens A, Maskery I, Tuck C, Ashcroft I, Everitt NM (2015) Mechanical properties of selective laser melted alsi10mg: Nano, micro, and macro properties. In: Solid Freeform Fabrication Symposium, pp 1026–1035

Aboulkhair NT, Maskery I, Tuck C, Ashcroft I, Everitt NM (2016a) Improving the fatigue behaviour of a selectively laser melted aluminium alloy: Influence of heat treatment and surface quality. Materials & Design 104:174–182

Aboulkhair NT, Maskery I, Tuck C, Ashcroft I, Everitt NM (2016b) The microstructure and mechanical properties of selectively laser melted alsi10mg: The effect of a conventional t6-like heat treatment. Materials Science and Engineering: A 667:139–146

Aversa A, Lorusso M, Trevisan F, Ambrosio E, Calignano F, Manfredi D, Biamino S, Fino P, Lombardi M, Pavese M (2017) Effect of process and post-process conditions on the mechanical properties of an a357 alloy produced via laser powder bed fusion. Metals 7:68

Bagherifard S, Beretta N, Monti S, Riccio M, Bandini M, Guagliano M (2018) On the fatigue strength enhancement of additive manufactured alsi10mg parts by mechanical and thermal post-processing. Materials & Design 145:28–41

Beese AM, Carroll BE (2015) Review of mechanical properties of ti-6al-4v made by laser-based additive manufacturing using powder feedstock. Jom 68:724–734

Bikas H, Stavropoulos P, Chryssolouris G (2015) Additive manufacturing methods and modelling approaches: a critical review. The International Journal of Advanced Manufacturing Technology 83(1-4):389–405

Buchbinder D (2013) Selective laser melting von aluminiumgusslegierungen. Dissertation, RWTH Aachen

Buchbinder D, Mainers W, Wissenbach K, Müller-Lohmeier K, Brandl E, Skrynecki N (2009) Rapid manufacturing of aluminium parts for serial production via selective laser melting (slm), in 4th international conference on rapid manufacturing

Buchbinder D, Meiners W, Wissenbach K, Poprawe R (2015) Selective laser melting of aluminum die-cast alloy—correlations between process parameters, solidification conditions, and resulting mechanical properties. Journal of Laser Applications 27(S2):S29,205

Buchmayr B, Panzl G, Walzl A, Wallis C (2017) Laser powder bed fusion - materials issues and optimized processing parameters for tool steels, AlSiMg- and CuCrZr-alloys. Advanced Engineering Materials 19(4):n/a

Cherry JA, Davies HM, Mehmood S, Lavery NP, Brown SGR, Sienz J (2014) Investigation into the effect of process parameters on microstructural and physical properties of 316l stainless steel parts by selective laser melting. The International Journal of Advanced Manufacturing Technology 76(5-8):869–879

Cloots M, Kunze K, Uggowitzer PJ, Wegener K (2016) Microstructural characteristics of the nickel-based alloy in738lc and the cobalt-based alloy mar-m509 produced by selective laser melting. Materials Science and Engineering: A 658:68–76

Dai D, Gu D (2015) Tailoring surface quality through mass and momentum transfer modeling using a volume of fluid method in selective laser melting of tic/alsi10mg powder. International Journal of Machine Tools and Manufacture 88:95–107

Dai D, Gu D, Zhang H, Zhang J, Du Y, Zhao T, Hong C, Gasser A, Poprawe R (2018) Heat-induced molten pool boundary softening behavior and its effect on tensile properties of laser additive manufactured aluminum alloy. Vacuum 154:341–350

Frazier WE (2014) Metal additive manufacturing: A review. Journal of Materials Engineering and Performance 23(6):1917–1928

Gong X, Anderson T, Chou K (2014) Review on powder-based electron beam additive manufacturing technology. Manufacturing Review 1

Hafenstein S, Werner E (2019) Pressure dependence of age-hardenability of aluminum cast alloys and coarsening of precipitates during hot isostatic pressing. Materials Science and Engineering: A 757:62–69

Hitzler L, Charles A, Öchsner A (2016a) The influence of post-heat-treatments on the tensile strength and surface hardness of selective laser melted alsi10mg. Defect and Diffusion Forum 370:171–176

Hitzler L, Janousch C, Schanz J, Merkel M, Mack F, Öchsner A (2016b) Non-destructive evaluation of alsi10mg prismatic samples generated by selective laser melting: Influence of manufacturing conditions. Materialwissenschaft und Werkstofftechnik 47(5-6):564–581

Hitzler L, Hirsch J, Merkel M, Hall W, Öchsner A (2017a) Position dependent surface quality in selective laser melting. Materialwissenschaft und Werkstofftechnik 48(5):327–334

Hitzler L, Janousch C, Schanz J, Merkel M, Heine B, Mack F, Hall W, Öchsner A (2017b) Direction and location dependency of selective laser melted alsi10mg specimens. Journal of Materials Processing Technology 243:48–61

Hitzler L, Williams P, Merkel M, Hall W, Öchsner A (2017c) Correlation between the energy input and the microstructure of additively manufactured cobalt-chromium. Defect and Diffusion Forum 379:157–165

Hitzler L, Merkel M, Hall W, Öchsner A (2018) A review of metal fabricated with laser- and powder-bed based additive manufacturing techniques: Process, nomenclature, materials, achievable properties, and its utilization in the medical sector. Advanced Engineering Materials 20(5)

Lewandowski JJ, Seifi M (2016) Metal additive manufacturing: A review of mechanical properties. Annual Review of Materials Research 46(1):151–186

Montero-Sistiaga ML, Mertens R, Vrancken B, Wang X, Van Hooreweder B, Kruth JP, Van Humbeeck J (2016) Changing the alloy composition of al7075 for better processability by selective laser melting. Journal of Materials Processing Technology 238:437–445

Ngnekou JND, Nadot Y, Henaff G, Nicolai J, Ridosz L (2017) Influence of defect size on the fatigue resistance of alsi10mg alloy elaborated by selective laser melting (slm). Procedia Structural Integrity 7:75–83

Ngnekou JND, Henaff G, Nadot Y, Nicolai J, Ridosz L (2018) Fatigue resistance of selectively laser melted aluminum alloy under t6 heat treatment. Procedia Engineering 213:79–88

Prashanth KG, Eckert J (2017) Formation of metastable cellular microstructures in selective laser melted alloys. Journal of Alloys and Compounds 707:27–34

Qiu C, Adkins NJE, Attallah MM (2013) Microstructure and tensile properties of selectively laser-melted and of hiped laser-melted ti–6al–4v. Materials Science and Engineering: A 578:230–239

Rashid R, Masood SH, Ruan D, Palanisamy S, Rahman Rashid RA, Elambasseril J, Brandt M (2018) Effect of energy per layer on the anisotropy of selective laser melted alsi12 aluminium alloy. Additive Manufacturing 22:426–439

Read N, Wang W, Essa K, Attallah MM (2015) Selective laser melting of alsi10mg alloy: Process optimisation and mechanical properties development. Materials & Design (1980-2015) 65:417–424

Reschetnik W, Brüggemann JP, Aydinöz ME, Grydin O, Hoyer KP, Kullmer G, Richard HA (2016) Fatigue crack growth behavior and mechanical properties of additively processed en aw-7075 aluminium alloy. Procedia Structural Integrity 2:3040–3048

Sert E, Hitzler L, Merkel M, Öchsner A (2018) Entwicklung von topologieoptimierten adapterelementen für die fertigung mittels additiver verfahren: Vereinigung von reinelektrischem antriebsstrang mit konventionellem chassis. Materialwissenschaft und Werkstofftechnik 49(5):674–682

Sert E, Hitzler L, Heine B, Merkel M, Werner A, Öchsner A (2019a) Influence of heat treatment on the microstructure and hardness of additively manufactured alsi10mg samples. Practical Metallography

Sert E, Schuch E, Öchsner A, Hitzler L, Werner E, Merkel M (2019b) Tensile strength performance with determination of the poisson's ratio of additively manufactured alsi10mg samples. Materialwissenschaft und Werkstofftechnik 50(5):539–545

Shifeng W, Shuai L, Qingsong W, Yan C, Sheng Z, Yusheng S (2014) Effect of molten pool boundaries on the mechanical properties of selective laser melting parts. Journal of Materials Processing Technology 214(11):2660–2667

Singh S, Ramakrishna S, Singh R (2017) Material issues in additive manufacturing: A review. Journal of Manufacturing Processes 25:185–200

Spierings AB, Herres N, Levy G (2011) Influence of the particle size distribution on surface quality and mechanical properties in am steel parts. Rapid Prototyping Journal 17(3):195–202

Tang M, Pistorius PC (2017) Oxides, porosity and fatigue performance of alsi10mg parts produced by selective laser melting. International Journal of Fatigue 94:192–201

Uzan NE, Shneck R, Yeheskel O, Frage N (2017) Fatigue of alsi10mg specimens fabricated by additive manufacturing selective laser melting (am-slm). Materials Science and Engineering: A 704:229–237

VDI-Richtlinie (2014) Additive fertigungsverfahren grundlagen, begriffe, verfahrensbeschreibungen, 3405

VDI-Richtlinie (2017) Additive fertigungsverfahren laser-strahlschmelzen metallischer bauteile materialkenndatenblatt aluminiumlegierung alsi10mg, 3405, blatt 2.1

Wang L, Sun J, Zhu X, Cheng L, Shi Y, Guo L, Yan B (2018a) Effects of t2 heat treatment on microstructure and properties of the selective laser melted aluminum alloy samples. Materials (Basel) 11(1)

Wang LF, Sun J, Yu XL, Shi Y, Zhu XG, Cheng LY, Liang HH, Yan B, Guo LJ (2018b) Enhancement in mechanical properties of selectively laser-melted alsi10mg aluminum alloys by t6-like heat treatment. Materials Science and Engineering: A

Wang P, Gammer C, Brenne F, Prashanth KG, Mendes RG, Rümmeli MH, Gemming T, Eckert J, Scudino S (2018c) Microstructure and mechanical properties of a heat-treatable al-3.5cu-1.5mg-1si alloy produced by selective laser melting. Materials Science and Engineering: A 711:562–570

Wong KV, Hernandez A (2012) A review of additive manufacturing. ISRN Mechanical Engineering 2012:1–10

Zhang C, Zhu H, Liao H, Cheng Y, Hu Z, Zeng X (2018) Effect of heat treatments on fatigue property of selective laser melting alsi10mg. International Journal of Fatigue 116:513–522

Zhang D (2004) Entwicklung des selective laser melting (slm) für aluminumwerkstoffe. Dissertation, RWTH Aachen

Zhao J, Easton M, Qian M, Leary M, Brandt M (2018) Effect of building direction on porosity and fatigue life of selective laser melted alsi12mg alloy. Materials Science and Engineering: A 729:76–85

Chapter 15
Efficient Numerics for the Analysis of Fibre-reinforced Composites Subjected to Large Viscoplastic Strains

Alexey V. Shutov and Igor I. Tagiltsev

Abstract Fibre-reinforced composites which sustain large multi-axial inelastic strains are of great importance for modern engineering. Besides, numerous biological soft tissues like blood vessels and heart valves as well as their artificial substitutes can be idealized as fibre-reinforced composites as well. Therefore, there is a growing demand for sufficiently accurate and numerically efficient modelling approaches which can reproduce the mechanical behaviour of such materials. In the current study we focus on the phenomenological material modelling and the related numerics. The kinematics of inelastic body is based on the well-proven multiplicative decomposition of the deformation gradient in combination with hyperelastic relations between stresses and elastic strains. An efficient numerical algorithm is suggested for the implementation of a phenomenological material model which accounts for the plasticity both in matrix and fibre. The performance of the algorithm is tested and its applicability is exemplified in terms of a demonstration problem.

Keywords: Fibre-reinforced composite · Large strain · Elasto-visco-plasticity · Efficient numerics

15.1 Introduction

Applications involving multi-axial straining of fibre-reinforced composites in the plastic and viscoplastic modes are versatile (Uddin, 2013). Originally, the interest in such composite materials was related to engineering purposes like a design of tires and air springs (Helnwein et al, 1993; Meschke and Helnwein, 1994; Holzapfel and Gasser, 2001; Donner and Ihlemann, 2013). Since the 80s, biomechanical agenda

---

Alexey V. Shutov · Igor I. Tagiltsev
Lavrentyev Institute of Hydrodynamics, pr. Lavrentyeva 15 & Novosibirsk State University, ul. Pirogova 1, 630090, Novosibirsk, Russia,
e-mail: alexey.v.shutov@gmail.com, i.i.tagiltsev@gmail.com

H. Altenbach and A. Öchsner (eds.), *State of the Art and Future Trends in Material Modeling*, Advanced Structured Materials 100,
https://doi.org/10.1007/978-3-030-30355-6_15

motivates an in-depth study of mechanical behaviour of "fibre-reinforced" living soft tissues (Chuong and Fung, 1986; Holzapfel et al, 2000; Gasser et al, 2002; Marino et al, 2018); a closely related issue is the analysis of their artificial substitutes like blood vessel prosthesis and engineered heart valves (Chernonosova et al, 2018).

Although a substantial progress has been achieved within the micromechanical modelling of fibre-reinforced composites (see, for example, Wongsto and Li, 2005; Maligno et al, 2009; Vaughan and McCarthy, 2011), the phenomenological approach is a method of choice when working with large and complex systems as well as multiple time scales. The main advantages of the phenomenological approach are its relative simplicity combined with the efficient numerical implementation, see, among others, Holzapfel et al (2000); Gasser et al (2002); Tagiltsev et al (2018); Liu et al (2019).

In the current study we deal with the multiplicative approach, used both for the matrix and the fibre. Within that approach, the deformation gradient is decomposed into the inelastic (viscoplastic) and elastic (hyperelastic) parts by a multiplicative split. Since both parts can be finite, the setting is geometrically exact. Using the multiplicative split one may create a model which is a-priory objective, thermodynamically consistent, and free from non-physical shear oscillations (Shutov, 2016). The chosen framework also allows for a pure isochoric/volumetric split (Shutov, 2016) and a big number of hyperelastic potentials can be employed for more accurate simulations. Other studies which also employ the multiplicative approach to the inelasticity of fibre and matrix are presented in Nguyen et al (2007); Huang et al (2012); Liu et al (2019).

Unfortunately, in case of viscoelasticity and viscoplasticity, the underlying evolution equations are stiff. For that reason, implicit time discretization methods must be used. Typically, the resulting system of nonlinear algebraic equations is solved by a Newton-Raphson-like iteration, which is time consuming. Within a large scale FEM computation the material law needs to be evaluated for a big number of times, the number of evaluations ranges up to 10^{10} in some applications with globally implicit FEM; even a larger computational effort may arise when using a globally explicit FEM. The very large computational effort may become a serious issue in the analysis of surgical operations which needs to be carried out in real time and in the solution of inverse problems, typical in engineering. In this contribution we present a simple iteration-free algorithm of implicit integration for the fibres and a simple one-equation-integrator for the matrix. The overall performance of the numerical scheme is tested; the applicability of the framework is validated in terms of a simple demonstration problem which involves large elastic and plastic strains of the composite material.

15.2 Material Model of the Fibre-reinforced Composite

Following the conventional iso-strain approach, the material model of the fibre-reinforced composite is a parallel connection of elasto-viscoplastic submodels of two

Fig. 15.1 Rheological model of the fibre-reinforced composite

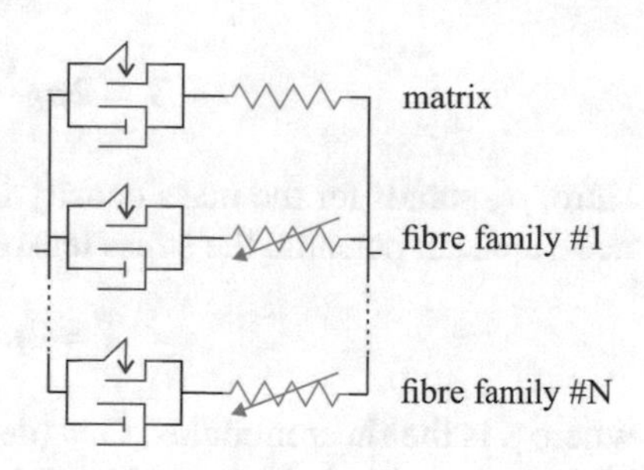

types: for the matrix and for the fibre, see Fig. 15.1. Dealing with strain-controlled loadings, the submodels are completely uncoupled. The coupling appears if a stress-controlled loading is considered, or in a more general setting of a boundary value problem.

15.2.1 Isotropic Viscoplasticity for the Matrix

To capture the behaviour of the matrix we employ the viscoplastic model originally proposed in Simo and Miehe (1992). Later this model was re-written on the reference configuration in Lion (1997). The application of this framework to viscoelasticity was popularized in Reese and Govindjee (1998). The advantages of this approach are discussed in Shutov (2016) in case of viscoelasticity and in Shutov and Ihlemann (2014) in case of elasto-plasticity.

The deformation gradient $\mathbf{F}$ is decomposed into the elastic part $\mathbf{F}_{\mathrm{e}}^{\mathrm{matrix}}$ and the inelastic (viscoplastic) part $\mathbf{F}_{\mathrm{i}}^{\mathrm{matrix}}$; the inelastic part gives rise to the inelastic right Cauchy-Green tensor $\mathbf{C}_{\mathrm{i}}$:

$$\mathbf{F} = \mathbf{F}_{\mathrm{e}}^{\mathrm{matrix}} \mathbf{F}_{\mathrm{i}}^{\mathrm{matrix}}, \quad \mathbf{C}_{\mathrm{i}} = \mathbf{F}_{\mathrm{i}}^{\mathrm{matrix}\,\mathrm{T}} \mathbf{F}_{\mathrm{i}}^{\mathrm{matrix}}. \tag{15.1}$$

The viscoplastic flow is governed by the following initial value problem:

$$\dot{\mathbf{C}}_{\mathrm{i}} = \sqrt{6}\dot{s}\frac{1}{\mathfrak{F}}(\mathbf{C}\tilde{\mathbf{T}})^{\mathrm{D}}\mathbf{C}_{\mathrm{i}}, \quad \mathbf{C}_{\mathrm{i}}\,|_{t=0} = \mathbf{C}_{\mathrm{i}}^{0}. \tag{15.2}$$

Here, $\dot{s}$ is the rate of the accumulated plastic arc-length; $\mathbf{C}$ is the right Cauchy-Green tensor; $\tilde{\mathbf{T}}$ is the second Piola-Kirchhoff stress tensor. The norm of the Kirchhoff stress deviator is denoted by $\mathfrak{F}$. It can be computed in terms of the referential Mandell stress $\mathbf{C}\tilde{\mathbf{T}}$:

$$\mathfrak{F} = \sqrt{\mathrm{tr}((\mathbf{C}\tilde{\mathbf{T}})^{\mathrm{D}})^{2}}. \tag{15.3}$$

Let Ψ_{matrix} be the specific free energy per unit mass; it defines the hyperelastic properties of the matrix. The following expression can be obtained for the stresses in the matrix by the Coleman-Noll procedure

$$\tilde{\mathbf{T}} = 2\rho_R \frac{\partial \Psi_{\text{matrix}}(\mathbf{C}\mathbf{C}_{\text{i}}^{-1})}{\partial \mathbf{C}}. \tag{15.4}$$

Here, ρ_R stands for the mass density in the reference configuration. In case of the neo-Hookean potential the stress tensor equals

$$\tilde{\mathbf{T}} = 2\mu \mathbf{C}^{-1}(\overline{\mathbf{C}}\mathbf{C}_{\text{i}}^{-1})^{\text{D}}, \tag{15.5}$$

where μ is the shear modulus; $\overline{\mathbf{C}} = (\det \mathbf{C})^{-1/3}\mathbf{C}$ is the isochoric part of $\mathbf{C}$. The rate of the accumulated plastic arc-length is related to the rate of the viscoplastic flow; it is computed using the Perzyna rule

$$\dot{s} = \sqrt{\frac{2}{3}} \frac{1}{\eta_{\text{matrix}}} \left\langle \frac{\mathfrak{F} - \sqrt{\frac{2}{3}} K_{\text{matrix}}}{k_0} \right\rangle^{m_{\text{matrix}}}, \quad k_0 = 1\text{KPa}. \tag{15.6}$$

Here, K_{matrix} is the uniaxial yield stress of the matrix; m_{matrix} and η_{matrix} are the viscosity parameters; $\langle x \rangle = \max(0, x)$ is the positive part of a real number. In the Perzyna rule (15.6), the norm of the Kirchhoff stress deviator $\mathfrak{F}$ is seen as the driving force of the viscoplastic flow; k_0 is not a material parameter.

The model of Simo and Miehe (1992) summarized here is objective and thermodynamically consistent. Moreover, it is w-invariant under isochoric changes of the reference configurations, see Shutov and Ihlemann (2014). As shown in Shutov and Kreißig (2010), the exact solution is exponentially stable with respect to perturbations of initial data, which brings advantages with respect to efficient numerics.

15.2.2 Anisotropic Viscoplasticity for the Fibre

In practice, fibre-reinforced materials are usually reinforced by *N families* of fibres exhibiting different orientations and structural properties. Again, following the iso-strain approach, we neglect the interaction between the fibre families by assuming a parallel connection of corresponding submodels (see Fig. 15.1). There are different approaches to the analysis of inelastic properties of individual fibre families. For instance, in Holzapfel and Gasser (2001) and related papers the authors use convolution integrals, which (for general relaxation kernels) requires the integration over the entire deformation history. An alternative approach is advocated in the current contribution. It is based on the multiplicative decomposition in combination with hyperelastic relations and evolution equations governing the changes of internal variables. As in the previous subsection, the fibre kinematics is based on the multiplicative decomposition:

$$\mathbf{F} = \mathbf{F}_{\text{e}}^{\text{fibre}} \mathbf{F}_{\text{i}}^{\text{fibre}}. \tag{15.7}$$

Although the matrix and the fiber experience the same deformation $\mathbf{F}$, the intermediate (stress-free) configuration for the considered fibre family differs from the one of the matrix. Therefore, superscripts “matrix” and “fibre” are used in the current contribution. Let a unit vector $\tilde{\mathbf{a}}$ represent the direction of the chosen fibre family in the reference configuration. The corresponding structure tensor $\tilde{\mathbf{M}}$ is as follows

$$\tilde{\mathbf{M}} = \tilde{\mathbf{a}} \otimes \tilde{\mathbf{a}}, \quad \|\tilde{\mathbf{a}}\| = 1. \tag{15.8}$$

Next, let $\lambda = \|\mathbf{F}\tilde{\mathbf{a}}\|$ be the stretch of the fibre family. Considering the isochoric part of the deformation, the stretch can be computed by the double contraction

$$\lambda = \sqrt{\tilde{\mathbf{C}} : \tilde{\mathbf{M}}}. \tag{15.9}$$

Let $\lambda_\mathrm{i} = \|\mathbf{F}_\mathrm{i}^\mathrm{fibre}\tilde{\mathbf{a}}\|$ be the inelastic stretch of the fibre. We assume that the inelastic deformation does not change the fibre direction: $\mathbf{F}_\mathrm{i}^\mathrm{fibre}\tilde{\mathbf{a}} = \lambda_\mathrm{i}\tilde{\mathbf{a}}$. In that case it is natural to introduce the elastic stretch as $\lambda_\mathrm{e} = \|\mathbf{F}_\mathrm{e}^\mathrm{fibre}\tilde{\mathbf{a}}\|$. Thus, the multiplicative decomposition (15.7) yields a decomposition of the fibre stretch: $\lambda = \lambda_\mathrm{e}\lambda_\mathrm{i}$.

In the simplest case one may assume that the free energy stored in the fiber family is a function of the elastic stretch λ_e:

$$\Psi_\mathrm{fibre} = \Psi_\mathrm{fibre}(\lambda_e^2), \quad f = \frac{d\,\Psi_\mathrm{fibre}(\lambda_e^2)}{d(\lambda_e^2)}. \tag{15.10}$$

Here, the derivative f is introduced for convenience. To be definite, we use the potential proposed by Holzapfel et al (2000)

$$\rho_R f(\lambda_e^2) = k_1^\mathrm{fibre}(\lambda_e^2 - 1)\exp(k_2^\mathrm{fibre}(\lambda_e^2 - 1)^2), \tag{15.11}$$

where $k_1^\mathrm{fibre} \geq 0$ is the stiffness parameter of the fibre and $k_2^\mathrm{fibre} > 0$ is a non-dimensional parameter attributed to the nonlinearity of the stress response under finite tension. The potential proposed by Holzapfel and co-workers is one the most common energy storage functions, extensively used in biomechanical applications. Its modifications are presented in Tagiltsev et al (2018); Marino et al (2018).

As shown in Tagiltsev et al (2018), the Clausius-Duhem inequality for the considered constitutive assumptions takes the following form

$$f \cdot \dot{\lambda}_\mathrm{i} \geq 0. \tag{15.12}$$

This inequality states that the fibre subjected to tensile stress can not contract ($f > 0 \Rightarrow \dot{\lambda}_\mathrm{i} \geq 0$) and compressed fibres can not elongate ($f < 0 \Rightarrow \dot{\lambda}_\mathrm{i} \leq 0$). In this paper we formulate a flow rule which identically satisfies this inequality.

Our intention now is to build a flow rule similar to the Perzyna rule (15.6) of the matrix. The axial component of the Kirchhoff stress is given by

$$\sigma = \frac{4}{3}\rho_R f\left(\lambda_\mathrm{e}^2\right)\lambda_\mathrm{e}^2 = \frac{4}{3}\rho_R f\left(\frac{\lambda^2}{\lambda_\mathrm{i}^2}\right)\frac{\lambda^2}{\lambda_\mathrm{i}^2},$$

the rate of the logarithmic strain equals $\frac{\dot{\lambda}_\mathrm{i}}{\lambda_\mathrm{i}}$. Thus, we naturally obtain the following initial value problem for the inelastic stretch:

$$\frac{\dot{\lambda}_{\rm i}}{\lambda_{\rm i}} = \frac{1}{\eta_{\rm fibre}} \left\langle \frac{\frac{4}{3}\rho_R f(\frac{\lambda^2}{\lambda_{\rm i}^2})\frac{\lambda^2}{\lambda_{\rm i}^2} - K_{\rm fibre}}{k_0} \right\rangle^{m_{\rm fibre}}, \quad \lambda_{\rm i}\mid_{t=0} = \lambda_{\rm i}^0, \tag{15.13}$$

where $\eta_{\rm fibre}$, $K_{\rm fibre}$, and $m_{\rm fibre}$ are the material parameters; again, $k_0 = 1$ KPa. Note that according to this flow rule the fibre yields under tension only. In case of a constant fiber stretch $\lambda = const$, the flow rule takes an equivalent form

$$-\frac{\dot{\lambda}_{\rm e}}{\lambda_{\rm e}} = \frac{1}{\eta_{\rm fibre}} \left\langle \frac{\frac{4}{3}\rho_R f(\lambda_{\rm e}^2)\lambda_{\rm e}^2 - K_{\rm fibre}}{k_0} \right\rangle^{m_{\rm fibre}}. \tag{15.14}$$

After some algebra (see Tagiltsev et al, 2018), the contribution of the fibre family to the second Piola-Kirchhoff stress takes the form

$$\tilde{\mathbf{T}} = 2\rho_R \frac{f(\lambda_{\rm e}^2)}{\lambda_{\rm i}^2} \mathbb{P}_{\mathbf{C}} : \tilde{\mathbf{M}}, \quad \mathbb{P}_{\mathbf{C}} : \tilde{\mathbf{M}} = \tilde{\mathbf{M}} - \frac{1}{3}\mathrm{tr}(\mathbf{C}\tilde{\mathbf{M}})\mathbf{C}^{-1}. \tag{15.15}$$

15.3 Efficient Numerics

15.3.1 Isotropic Viscoplasticity of the Matrix

Consider a typical time-step $t_n \mapsto t_{n+1}$, $\Delta t = t_{n+1} - t_n > 0$. Assume that the previous $\mathbf{C}_{\rm i}$ and the current $\mathbf{C}$ are known. Denote them by ${}^n\mathbf{C}_{\rm i}$ and ${}^{n+1}\mathbf{C}$, respectively. Within the time step we need to update $\mathbf{C}_{\rm i}$ and to compute the current value of the second Piola-Kirchhoff stress ${}^{n+1}\tilde{\mathbf{T}}$. Combining the evolution equation (15.2) with the neo-Hookean relation (15.5) we arrive at the following differential equation

$$\dot{\mathbf{C}}_{\rm i} = \sqrt{6}\dot{s}\frac{2\mu}{\mathfrak{F}}(\overline{\mathbf{C}}\mathbf{C}_{\rm i}^{-1})^{\rm D}\mathbf{C}_{\rm i}. \tag{15.16}$$

This equation is stiff. Therefore, it needs to be integrated using an implicit time-stepping method. If $\dot{s}/\mathfrak{F}$ is assumed to be constant within the time step, the flow rule (15.16) coincides with the evolution equation governing the Maxwell fluid, see Shutov et al (2013). Therefore, an update formula from Shutov et al (2013) can be used:

$${}^{n+1}\mathbf{C}_{\rm i} = \overline{{}^n\mathbf{C}_{\rm i} + 2\mu\frac{\xi}{\mathfrak{F}}\,\overline{{}^{n+1}\mathbf{C}}}, \tag{15.17}$$

where $\xi = \Delta t\sqrt{3/2}\dot{s}$ is the so-called inelastic strain increment. Note that, according to the Perzyna rule (15.6), the norm of the driving force $\mathfrak{F}$ can be obtained as a function of the strain increment ξ. Indeed, multiplying both sides of (15.6) with $\sqrt{3/2}\Delta t$ we obtain

$$\xi = \frac{\Delta t}{\eta_{\rm matrix}} \left\langle \frac{\mathfrak{F} - \sqrt{\frac{2}{3}}K_{\rm matrix}}{k_0} \right\rangle^{m_{\rm matrix}}. \tag{15.18}$$

In case of the inelastic flow we have $\mathfrak{F} > \sqrt{\frac{2}{3}} K_{\text{matrix}}$ and

$$\mathfrak{F} = \tilde{\mathfrak{F}}(\xi) = \sqrt{2/3} K_{\text{matrix}} + k_0 (\xi \eta_{\text{matrix}} / \Delta t)^{1/m_{\text{matrix}}}. \tag{15.19}$$

Using this function and the update formula (15.17), we define

$$^{n+1}\mathbf{C}_{\text{i}}(\xi) = \overline{{}^{n}\mathbf{C}_{\text{i}} + 2\mu \frac{\xi}{\tilde{\mathfrak{F}}(\xi)} \, \overline{{}^{n+1}\mathbf{C}}}. \tag{15.20}$$

Employing the function $^{n+1}\mathbf{C}_{\text{i}}(\xi)$ we define two dependencies:

$$^{n+1}\tilde{\mathbf{T}}(\xi) = 2\mu \, {}^{n+1}\mathbf{C}^{-1} \big(\overline{{}^{n+1}\mathbf{C}} \, ({}^{n+1}\mathbf{C}_{\text{i}}(\xi))^{-1} \big)^{\text{D}}, \tag{15.21}$$

$$\hat{\mathfrak{F}}(\xi) = \sqrt{\operatorname{tr}(({}^{n+1}\mathbf{C} \, {}^{n+1}\tilde{\mathbf{T}}(\xi))^{\text{D}})^2}. \tag{15.22}$$

The unknown strain increment ξ is found from the yield condition (15.18). In the incremental form it can be written as $H(\xi) = 0$ or $D(\xi) = 0$ with

$$H(\xi) = \xi \eta_{\text{matrix}} / \Delta t - \left(\frac{\hat{\mathfrak{F}}(\xi) - \sqrt{2/3} K_{\text{matrix}}}{k_0} \right)^{m_{\text{matrix}}}, \tag{15.23}$$

$$D(\xi) = (\xi \eta_{\text{matrix}} / \Delta t)^{1/m_{\text{matrix}}} - \frac{\hat{\mathfrak{F}}(\xi) - \sqrt{2/3} K_{\text{matrix}}}{k_0}. \tag{15.24}$$

Elastic predictor. First, we compute a trial driving force $\mathfrak{F}^{\text{trial}} = \hat{\mathfrak{F}}(0)$, which corresponds to a purely hyperelastic load increment with a frozen inelastic flow ($\xi = 0$). If $\mathfrak{F}^{\text{trial}} \leq \sqrt{\frac{2}{3}} K_{\text{matrix}}$ then we put $\xi = 0$, ${}^{n+1}\mathbf{C}_{\text{i}} = {}^{n}\mathbf{C}_{\text{i}}$, ${}^{n+1}\tilde{\mathbf{T}} = {}^{n+1}\tilde{\mathbf{T}}(\xi = 0)$; the time step is complete. Otherwise, an inelastic corrector step is needed.

Inelastic corrector. A reasonable correction procedure is as follows (see the discussion in Shutov and Kreißig (2008)). First, a single iteration of the Newton method is carried out for the equation $H(\xi) = 0$ using $\xi = 0$ as the initial approximation. The remaining iterations of the Newton method are performed for $D(\xi) = 0$. After the inelastic strain increment ξ is found, the internal variables and the stresses are updated using (15.20) and (15.21). The corrector step is thus complete.

Since the entire procedure boils down to a single equation, it is the so-called one-equation-integrator. A similar one-equation-integrator can also be obtained if the free energy is given by a more general assumption of the Mooney-Rivlin potential, see Shutov (2018). Another important property of the algorithm is that it preserves the incompressibility of the inelastic flow: $\det(\mathbf{C}_{\text{i}}) = 1$. As shown in Shutov and Kreißig (2010), such an integrator does not accumulate numerical errors even when working with big time spans and large time steps.

15.3.2 *Anisotropic Viscoplasticity of the Fibre*

15.3.2.1 Conventional Iteration-based Approach

For the time-step $t_n \mapsto t_{n+1}$ assume that ${}^{n+1}\lambda$ and ${}^{n}\lambda_{\mathrm{i}}$ are known. To complete the time step, we need to compute the current inelastic stretch ${}^{n+1}\lambda_{\mathrm{i}}$ and the current stress ${}^{n+1}\tilde{\mathbf{T}}$. Just as in the isotropic case, we subdivide the time step into the elastic predictor and the inelastic corrector.

Elastic predictor. First, we assume that the fibre stretch λ changes instantly from ${}^{n}\lambda$ to ${}^{n+1}\lambda$. The corresponding trial elastic stretch equals ${}^{n+1}\lambda_e^{\mathrm{trial}} = {}^{n+1}\lambda / {}^{n}\lambda_{\mathrm{i}}$. The trial stress ${}^{n+1}\sigma^{\mathrm{trial}}$ is then obtained using ${}^{n+1}\lambda_e^{\mathrm{trial}}$:

$$
{}^{n+1}\sigma^{\mathrm{trial}} = \frac{4}{3}\rho_R f\Big(({}^{n+1}\lambda_e^{\mathrm{trial}})^2\Big)({}^{n+1}\lambda_e^{\mathrm{trial}})^2. \tag{15.25}
$$

If ${}^{n+1}\sigma^{\mathrm{trial}} < K$ then we set ${}^{n+1}\lambda_{\mathrm{e}} = {}^{n+1}\lambda_e^{\mathrm{trial}}$, ${}^{n+1}\sigma = {}^{n+1}\sigma^{\mathrm{trial}}$; the evaluation of the time step is thus complete. Otherwise we proceed to the inelastic corrector step.

Inelastic corrector. Within the inelastic corrector, the total fibre stretch remains constant: $\lambda \equiv {}^{n+1}\lambda$. The elastic stretch is computed using the evolution equation (15.14) and assuming the trial elastic stretch as the initial condition: $\lambda_{\mathrm{e}} |_{t=t_n} = {}^{n+1}\lambda_e^{\mathrm{trial}}$. The classical Euler-backward method for (15.14) is used here to obtain an unconditionally stable procedure:

$$
\frac{{}^{n+1}\lambda_{\mathrm{e}} - {}^{n}\lambda_{\mathrm{e}}}{\Delta t} = -\frac{{}^{n+1}\lambda_{\mathrm{e}}}{\eta}\left\langle \frac{\frac{4}{3}\rho_R f({}^{n+1}\lambda_{\mathrm{e}}^2){}^{n+1}\lambda_{\mathrm{e}}^2 - K_{\mathrm{fibre}}}{k_0} \right\rangle^{m_{\mathrm{fibre}}}. \tag{15.26}
$$

This nonlinear algebraic equation is equivalent to the equation $H({}^{n+1}\lambda_{\mathrm{e}}) = 0$ or $D({}^{n+1}\lambda_{\mathrm{e}}) = 0$, where

$$
H(x) = (\lambda_e^{\mathrm{trial}} - x)\frac{\eta}{\Delta t} - x\left(\frac{\frac{4}{3}\rho_R f(x^2)x^2 - K_{\mathrm{fibre}}}{k_0}\right)^{m_{\mathrm{fibre}}}; \tag{15.27}
$$

$$
D(x) = \left(\frac{\lambda_e^{\mathrm{trial}} - x}{x\Delta t}\eta\right)^{1/m_{\mathrm{fibre}}} - \left(\frac{\frac{4}{3}\rho_R f(x^2)x^2 - K_{\mathrm{fibre}}}{k_0}\right). \tag{15.28}
$$

Just as in the previous section, an efficient Newton-like iteration process can be organised as follows. First Newton iteration is carried out for the equation $H({}^{n+1}\lambda_{\mathrm{e}}) = 0$ using ${}^{n+1}\lambda_e^{\mathrm{trial}}$ as the initial approximation for the unknown ${}^{n+1}\lambda_{\mathrm{e}}$. All the subsequent Newton iterations are carried out for the equation $D({}^{n+1}\lambda_{\mathrm{e}}) = 0$. After ${}^{n+1}\lambda_{\mathrm{e}}$ is found, we make the update: ${}^{n+1}\lambda_{\mathrm{i}} = {}^{n+1}\lambda / {}^{n+1}\lambda_{\mathrm{e}}$.

15.3.2.2 Iteration-free Approach

The iterative approach described above can lead to a very large computational effort, especially when a big number of fibre families is involved. An efficient iteration-free stress update algorithm will be presented here. Note that for fixed time step size Δt and fixed material parameters, the current elastic stretch ${}^{n+1}\lambda_e$ is a unique function of the trial elastic stretch ${}^{n+1}\lambda_e^{\mathrm{trial}}$. A typical transition curve ${}^{n+1}\lambda_e^{\mathrm{trial}} \mapsto {}^{n+1}\lambda_e$ is presented in Fig. 15.2.

Note that it is smooth and it is given by a straight line for ${}^{n+1}\lambda_e^{\mathrm{trial}} \leq \lambda_e^{cr}$, where λ_e^{cr} is the critical elastic stretch corresponding to the initial plastification of the fibre. Therefore, the transition curve can be approximated by a cubic spline with a sufficient accuracy. For instance, we consider the key-points of the spline interpolation listed in Table 15.2. In order to build the interpolation, the transition curve is pre-computed at these key-points using the conventional iteration-based approach described above; the interpolation result is shown in Fig. 15.2.

The new efficient approach is as follows. First, the elastic predictor is carried out: ${}^{n+1}\lambda_e^{\mathrm{trial}} = {}^{n+1}\lambda / {}^{n}\lambda_{\mathrm{i}}$. Next, an approximated elastic stretch is estimated using the cubic spline: ${}^{n+1}\lambda_e^{\mathrm{est}} = \mathrm{Spline}({}^{n+1}\lambda_e^{\mathrm{trial}})$. Finally, ${}^{n+1}\lambda_e$ is computed using a single Newton iteration for the equation $D({}^{n+1}\lambda_{\mathrm{e}}) = 0$ starting from ${}^{n+1}\lambda_e^{\mathrm{est}}$: ${}^{n+1}\lambda_e = {}^{n+1}\lambda_e^{\mathrm{est}} - D({}^{n+1}\lambda_e^{\mathrm{est}})/D'({}^{n+1}\lambda_e^{\mathrm{est}})$. The inelastic stretch is updated as in the previous scheme: ${}^{n+1}\lambda_{\mathrm{i}} = {}^{n+1}\lambda / {}^{n+1}\lambda_{\mathrm{e}}$.

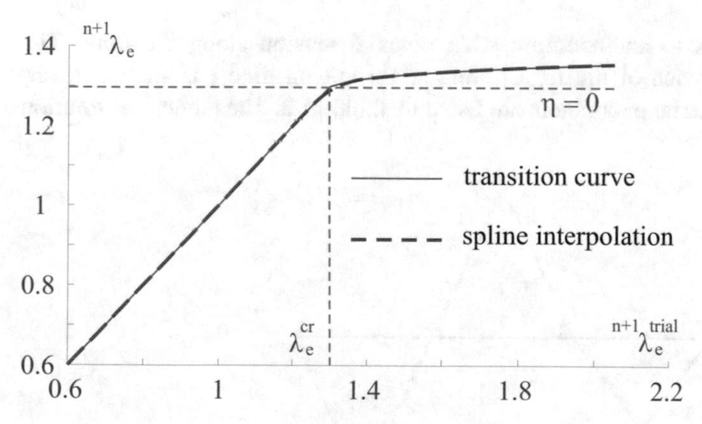

Fig. 15.2 Transition curve with parameters from Table 15.1 (solid line) and its interpolation by cubic splines using key-points from Table 15.2 (dashed line). The transition curve is a straight line for ${}^{n+1}\lambda_{\mathrm{e}}^{\mathrm{trial}} \leq \lambda_e^{cr}$

Table 15.1 Simulation parameters for transition curve

k_1^{fibre}	k_2^{fibre}	η_{fibre}	K_{fibre}	m_{fibre}	Δt
130 KPa	0.5	4000 s	500 KPa	2	2^{-5} s

Table 15.2 Key points for the spline-approximation of the transition curve

X_1	X_2	X_3	X_4	X_5
0.6	1.25	1.4	1.55	2.05

15.4 Tests and Applications

15.4.1 Single Fibre

It is instructive to test the stress update algorithm proposed in Subsubect. 15.3.2.2 using a single fibre family. In a suitable Cartesian coordinate system the deformation gradient tensor and the orientation vector of the fibre family are given by

$$\mathbf{F} = \begin{pmatrix} \lambda & 0 & 0 \\ 0 & \lambda^{-\frac{1}{2}} & 0 \\ 0 & 0 & \lambda^{-\frac{1}{2}} \end{pmatrix}, \quad \tilde{\mathbf{a}} = \begin{pmatrix} 1 \\ 0 \\ 0 \end{pmatrix}, \tag{15.29}$$

which corresponds to an incompressible uniaxial tension along the fibre. The prescribed dependence of the logarithmic strain $\ln\lambda$ on time t is shown in Fig. 15.3. The used material parameters are listed in Table 15.3. The numerical solution

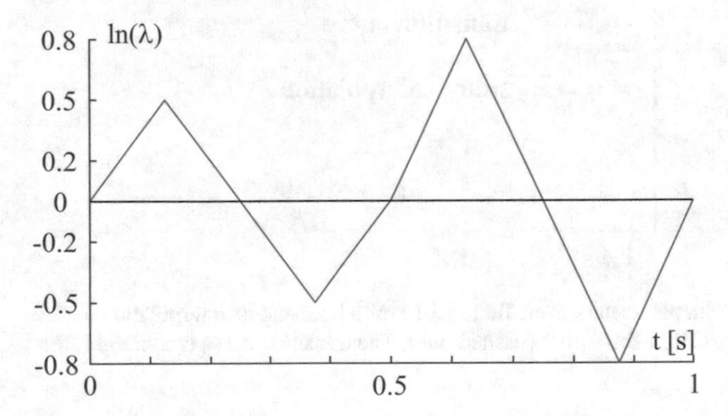

Fig. 15.3 Prescribed dependence of the logarithmic strain $\ln\lambda$ on the time t used for testing

Table 15.3 Numerical parameters for the test with a single fibre family

k_1^{fibre}	k_2^{fibre}	η_{fibre}	K_{fibre}	m_{fibre}	Δt
130 KPa	0.5	4000 s	500	2	2^{-7} s

obtained with a very small time step size using the conventional iteration-based

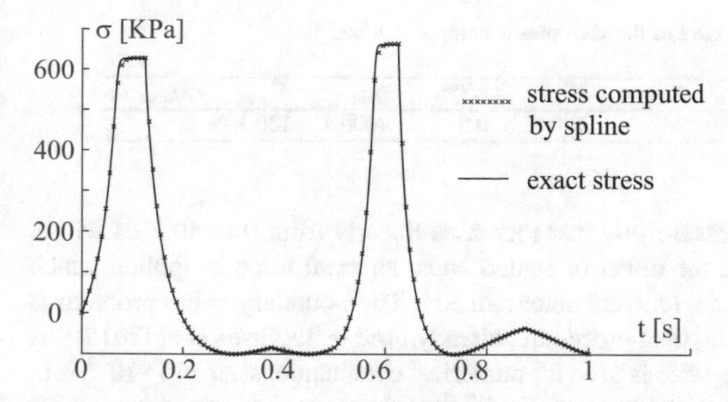

Fig. 15.4 Computed axial component of the Cauchy stress: the exact solution and the solution with $\Delta t = 2^{-7}$ s

method is referred to as the exact solution. The numerically computed Cauchy stresses with a finite time step size and the corresponding exact solution are plotted against time in Figure 15.4. As is seen from the figure, the efficient algorithm is sufficiently accurate for engineering and biomechanical applications.

15.4.2 Inflation of a Viscoplastic Composite Tube

Now we proceed to the demonstration problem which is related to the inflation of a thick-walled tube. In the initial state, the inner and outer radii equal 7 mm and 10 mm, respectively. The tube is made of a single layer of the composite material, the material is reinforced by two families of fibres, each of them is inclined at the angle $\beta = 64^o$ to the hoop (circumferential) direction, see Fig. 15.5. The material parameters are summarized in Table 15.4. The loading is controlled by a

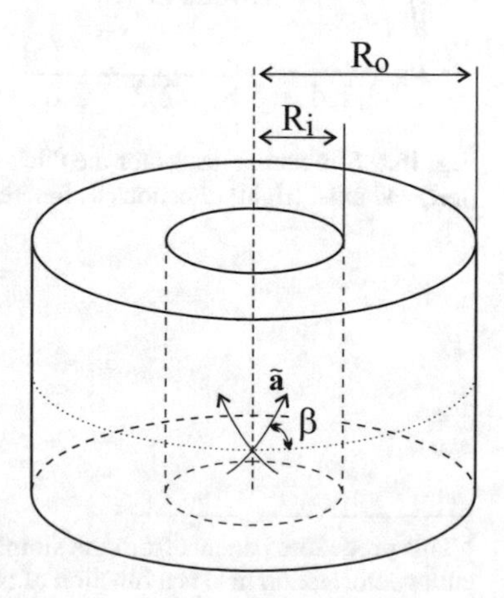

Fig. 15.5 Composite tube reinforced by two families of fibres

Table 15.4 Material parameters of the viscoplastic composite tube

η_{matrix}	μ_{matrix}	K_{matrix}	m_{matrix}	k_1^{fibre}	k_2^{fibre}	η_{fibre}	K_{fibre}	m_{fibre}
1600 s	80 KPa	20 KPa	2	130 KPa	0.5	4000 s	130 KPa	2

prescribed internal pressure $p(t)$ that increases linearly from 0 to 40 KPa within one second. To mimic the effect of sealed ends, an axial force is applied which equals $F(t) = p(t) \times \pi \times (\text{current inner radius})^2$. The boundary-value problem is solved using a semi-analytical procedure, already used in Tagiltsev et al (2018)[1]. A relatively large time step size is used for numerical computations: $\Delta t = 5 \cdot 10^{-3}$ s. In Fig. 15.6, the stretches in the hoop and axial directions are plotted as functions of the applied pressure. The hoop-stretch-curve (Fig. 15.6, left) indicates two distinct yielding events which correspond to the plastification of the matrix and the fibre. Before the fibre yields, the tube shortens under action of the internal pressure (see the axial-stretch-curve in Fig. 15.6, right). This shortening effect is attributed to the fibre-reorientation along the hoop direction, which is the dominant mechanism of the volume increase. After the plastification of fibres, the tube wall stretches both in hoop and axial directions. Ultimately, this leads to the loss of stability.

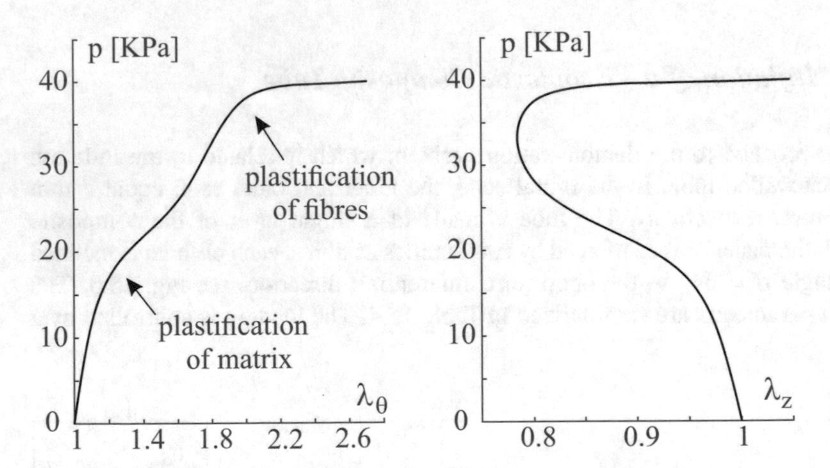

Fig. 15.6 Simulation results for the thick-walled pressurized composite tube. Stretches in the hoop (left) and axial (right) directions as functions of the applied pressure

[1] This procedure benefits from the simplified kinematics. In case of incompressible material the entire deformation field is a function of two scalars: the internal radius and the length of the tube.

15.5 Discussion and Conclusion

In the current contribution, a phenomenological approach to material modelling and the corresponding numerics are developed. It is shown that the model kinematics based on the multiplicative split of the deformation gradient allows one to capture complex geometrically nonlinear mechanical phenomena. Within the advocated multiplicative approach it is possible to incorporate differen kinds of hyperelastic potentials with various flow rules of viscous type. The resulting models are objective, thermodynamically consistent and they respect the material symmetries. Another advantage of the multiplicative approach lies in the possibility to model residual (initial) stresses: Different constituents of the composite can posses different stress-free configurations, which brings additional freedom into the modelling framework. Such a flexibility is especially important when dealing with natural grown biological tissues. The presented numerical integration approach based on pre-computed splines allows one to carry out highly efficient numerical computations, which is demonstrated by the solution of a boundary value problem.

In opinion of the authors, the future of composite modelling will be an increased use of physics-based multi-scale material models combined with development of refined phenomenological approaches. Computationally heavy micromechanical models will be used for development, calibration, and validation of computationally efficient phenomenological models. Another promising direction of research are phenomenological models which provide an entry point for physically sound constitutive assumptions.

Acknowledgements The financial support provided by RFBR (grant number 17-08-01020) and by the integration project of SB RAS (project 0308-2018-0018) is acknowledged.

References

Chernonosova VS, Gostev AA, Gao Y, Chesalov YA, Shutov AV, Pokushalov EA, Karpenko AA, Laktionov PP (2018) Mechanical properties and biological behavior of 3d matrices produced by electrospinning from protein-enriched polyurethane. BioMed Research International ID 1380606:10, doi:10.1155/2018/1380606

Chuong CJ, Fung YC (1986) Residual stress in arteries. In: Schmid-Schönbein GW, Woo SLY, Zweifach BW (eds) Frontiers in Biomechanics, Springer, New York

Donner H, Ihlemann J (2013) On the efficient finite element modelling of cord-rubber composites. In: Constitutive Models for Rubber, Taylor & Francis Group, vol VIII, pp 149–155

Gasser TC, Schulze-Bauer CAJ, Holzapfel GA (2002) A three-dimensional finite element model for arterial clamping. Journal of Biomechanical Engineering 124:355–363

Helnwein P, Liu CH, Meschke G, Mang HA (1993) A new 3-D finite element model for cord-reinforced rubber composites - application to analysis of automobile tires. Finite Elements in Analysis and Design 14:1–16

Holzapfel GA, Gasser TC (2001) A viscoelastic model for fiber-reinforced composites at finite strains: Continuum basis, computational aspects and applications. Computer Methods in Applied Mechanics and Engineering 190(34):4379–4403

Holzapfel GA, Gasser TC, Ogden RW (2000) A new constitutive framework for arterial wall mechanics and a comparative study of material models. Journal of Elasticity and the Physical Science of Solids 61(1-3):1–48

Huang R, Becker AA, Jones IA (2012) Modelling cell wall growth using a fibre-reinforced hyperelastic–viscoplastic constitutive law. Journal of the Mechanics and Physics of Solids 60:750–783

Lion A (1997) A physically based method to represent the thermo-mechanical behaviour of elastomers. Acta Mechanica 123(1-4):1–25

Liu H, Holzapfel GA, Skallerud BH, Prot V (2019) Anisotropic finite strain viscoelasticity: Constitutive modeling and finite element implementation. Journal of the Mechanics and Physics of Solids 124:172–188

Maligno AR, Warrior NA, Long AC (2009) Effects of inter-fibre spacing on damage evolution in unidirectional (UD) fibre-reinforced composites. European Journal of Mechanics - A/Solids 28(4):768–776

Marino M, von Hoegen M, Schröder J, Wriggers P (2018) Direct and inverse identification of constitutive parameters from the structure of soft tissues. part 1: micro- and nanostructure of collagen fibers. Biomechanics and Modeling in Mechanobiology 17(4):1011–1036

Meschke G, Helnwein P (1994) Large-strain 3D-analysis of fibre-reinforced composites using rebar elements: hyperelastic formulations for cords. Computational Mechanics 13:241–254

Nguyen TD, Jones RE, Boyce BL (2007) Modeling the anisotropic finite-deformation viscoelastic behavior of soft fiber-reinforced composites. International Journal of Solids and Structures 44:8366–8389

Reese S, Govindjee S (1998) A theory of finite viscoelasticity and numerical aspects. International Journal of Solids and Structures 35(26-27):3455–3482

Shutov AV (2016) Seven different ways to model viscoelasticity in a geometrically exact setting. Proceedings of the 7th European Congress on Computational Methods in Applied Sciences and Engineering 1:1959–1970

Shutov AV (2018) Efficient time stepping for the multiplicative Maxwell fluid including the Mooney-Rivlin hyperelasticity. International Journal for Numerical Methods in Engineering 113(12):1851–1869

Shutov AV, Ihlemann J (2014) Analysis of some basic approaches to finite strain elasto-plasticity in view of reference change. International Journal of Plasticity 63:183–197

Shutov AV, Kreißig R (2008) Finite strain viscoplasticity with nonlinear kinematic hardening: Phenomenological modeling and time integration. Computer Methods in Applied Mechanics and Engineering 197:2015–2029

Shutov AV, Kreißig R (2010) Geometric integrators for multiplicative viscoplasticity: analysis of error accumulation. Computer Methods in Applied Mechanics and Engineering 199:700–711

Shutov AV, Landgraf R, Ihlemann J (2013) An explicit solution for implicit time stepping in multiplicative finite strain viscoelasticity. Computer Methods in Applied Mechanics and Engineering 256:213–225

Simo JC, Miehe C (1992) Associative coupled thermo-plasticity at finite strains: formulation, numerical analysis and implementation. Computer Methods in Applied Mechanics and Engineering 98:41–104

Tagiltsev II, Laktionov PP, Shutov AV (2018) Simulation of fiber-reinforced viscoelastic structures subjected to finite strains: multiplicative approach. Meccanica 53(15):3779–3794

Uddin N (ed) (2013) Developments in Fiber-Reinforced Polymer (FRP) Composites for Civil Engineering. No. 45 in Woodhead Publishing Series in Civil and Structural Engineering, Woodhead Publishing

Vaughan TJ, McCarthy CT (2011) Micromechanical modelling of the transverse damage behaviour in fibre reinforced composites. Composites Science and Technology 71(3):388–396

Wongsto A, Li S (2005) Micromechanical FE analysis of UD fibre-reinforced composites with fibres distributed at random over the transverse cross-section. Composites Part A: Applied Science and Manufacturing 36(9):1246–1266

Chapter 16
An Artificial Intelligence-based Hybrid Method for Multi-layered Armour Systems

Filipe Teixeira-Dias, Samuel Thompson, and Mariana Paulino

Abstract The design of protective structures is a complex task mostly due to threat-related unknowns, such as the exact kinetic energy of the impactor and the dominant energy dissipation mechanisms. The design process is often costly and inefficient due to the number of these unknowns and to the cost of necessary steps such as laboratory testing and numerical modelling. In this chapter the authors propose a hybrid method that significantly increases the efficiency of the design process, and consequently decreasing its cost. The method combines an energy-based analytical approach with a set of deep learning (DL) models. Finite Element Analysis (FEA) and experimental results are used to train the artificial intelligence (AI) models and verify and validate the design process. The energy-based analytical method generates solutions for the DL algorithms, which can then be used to find optimal configurations for the protective structure. The proposed deep learning model is a neural network which is trained using experimental results and analytical data, to understand the ballistic response of a specific material, and predict the residual velocity for a given impact velocity, layer thickness and material properties. Networks trained for individual layers of the armour system are then interconnected in order to predict the residual velocity of blunt projectiles perforating multi-layered composite structures. Validation tests are done on systems including single and multi-layered targets.

Keywords: Multi-layered protective structures · Ballistic impact · Perforation · Armour systems · Artificial intelligence · Neural networks · Analytical modelling · Numerical analysis

---

Filipe Teixeira-Dias · Samuel Thompson
School of Engineering, The University of Edinburgh, Edinburgh, United Kingdom,
e-mail: F.Teixeira-Dias@ed.ac.uk, Samuel.Thompson@ed.ac.uk

Mariana Paulino
School of Engineering, Faculty of Science, Engineering and Built Environment, Deakin University, Geelong, Australia,
e-mail: Mariana.Paulino@deakin.edu.au

H. Altenbach and A. Öchsner (eds.), *State of the Art and Future Trends in Material Modeling*, Advanced Structured Materials 100,
https://doi.org/10.1007/978-3-030-30355-6_16

16.1 Introduction

Protective structures are used for a number of different purposes, ranging from protection from the environment to blast and ballistic impact. The design of protective barriers, structures and armour systems is often complex due to the number of unknowns associated with the threat, which often include the kinetic energy of the impactor (velocity and mass) and the mechanisms of energy dissipation within the protective structure or armour system. These mechanisms have been thoroughly studied and can include, for example, dissipation through plastic deformation, ductile hole growth, petalling or plugging. Multi-layered structures are known to potentially increase the protection capability without significant increase in weight (Liu et al, 2018; Ali et al, 2017; Elek et al, 2005). The design of said structures is thus very closely associated with known factors (e.g. the specific application) and unknown parameters such as those associated to the threat. The design process is often expensive and inefficient due to the number of unknowns and to the cost of involved steps such as laboratory testing and numerical modelling.

A hybrid method is proposed that increases the efficiency of the design process while at the same time significantly decreasing its cost. The method relies on a combination of a sound analytical method, which is inherently cost efficient, and artificial intelligence (AI) deep learning (DL) models. Experimental results are used not only to inform and train the AI models but also to validate the whole design process, together with Finite Element Analysis. The energy-based analytical method is developed to generate a set of solutions for the DL algorithm in order to find an optimal configuration for the protective structure, considering the most relevant energy dissipation mechanisms, and to determine perforation and residual velocity.

The deep learning model is a neural network trained using experimental results and analytical data, with the aim of understanding the ballistic response of a specific material or set of materials, and predicting the residual velocity for given impact conditions and layer thicknesses.

16.1.1 The Hybrid Methodology

The proposed method relies on experimental data for the training of the AI. Finite element analysis is used as a second validation stage, albeit not strictly necessary. Verification and validation tests are done on multiple systems, including single and multi-layered, in-contact target plates. This chapter describes the methods and presents the advantages of the proposed hybrid method over conventional FEA and experimental testing-based methodologies.

The impact of a projectile on a target can result in penetration or perforation. The former is defined as a projectiles entrance into a target without fully completing its passage through the body (Backman and Goldsmith, 1978). This means that the striker leaves an indentation on the target, without completely perforating it. The latter describes a ballistic impact which completely pierces the target (Zukas, 1980).

In this scope, the Ballistic Limit Velocity (BLV) is the minimum projectile velocity that ensures perforation (Børvik, 200; Zhang and Stronge, 1996). This velocity is a property of the armour system and is determined by a number of parameters, such as the projectile and target material properties, projectile mass and target configuration (e.g. thickness). The residual velocity is the projectile velocity after it has perforated the target. The definition of the BLV implies that if $v_0 = v^{\star}$ then $v_r = 0$, where v_0 is the projectile velocity just before impact, $v^{\star}$ is the BLV of the target and v_r is the residual velocity of the projectile. The residual velocity is zero if the target is struck by a projectile at its BLV (Sikarwar et al, 2014).

The following sections detail the analytical models and AI methods used and how they are integrated in a tool that can be used to predict post-perforation residual velocities from ballistic impacts on metallic layered targets.

16.2 Plugging of Ductile Plates: Analytical Modelling

Protective structures and plates can be perforated in a number of different ways, which are often grouped in six main distinct perforation mechanisms (Jia et al, 2014; Woodward, 1987; Taylor, 1948; Thomson, 1955; Atkins et al, 1998; Landkof and Goldsmith, 1985). The most common in ductile plates are ductile hole growth and plugging, shown schematically in Figs. 16.1(a) and (b) (Teixeira-Dias et al, 2018). This chapter focuses on perforation by orthogonal plugging, which occurs in finite thickness targets impacted at right angles by blunt cylindrical projectiles travelling close to or above the target's ballistic limit velocity (BLV). The impactor forms a *plug* of target material of similar diameter to the projectile by adiabatic shearing. Plugging is initiated by plastic strains caused by high stress concentrations in a small area (and thus high stress and strain gradients), leaving the remainder of the target unaffected. Plastic strain energy is converted into heat, increasing the temperature in the shearing zone, which leads to further localisation of the plastic strain, as shown in Fig. 16.1(c). In plugging this process continues until the plug completely exits the target (Krauthammer, 2008; Børvik et al, 2001).

This section describes the main principles and derivations involved in analysing ballistic perforation of ductile plates using energy-based principles. Conservation of energy, which is the basic principle behind these approaches, can be stated as $\sum E_{\text{in}} = \sum E_{\text{out}}$ where E_{in} and E_{out} are the input (initial) and output (final) energies of the system. The basic assumption is that all energy is transformed during the impact (Greszczuk et al, 1982; Horne, 2014; Goldsmith, 2001).

These energy-based approaches are often simplistic and thus based on a large number of geometrical, mechanical and physical assumptions and simplifications. In such models it is often assumed, for example, that thermal effects can be neglected. In the case of plugging, where adiabatic shearing is the predominant deformation mechanism, this is potentially too big an assumption. The analytical model proposed and described in this chapter tries to compensate for this by considering an additional friction term between the projectile and the target.

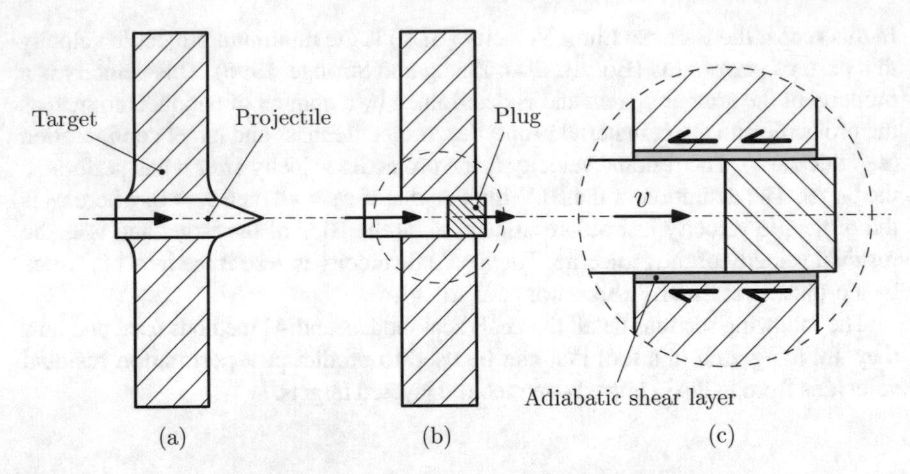

Fig. 16.1 Schematic representation of the most common mechanisms of ductile plate perforation: (a) ductile hole growth, (b) plugging and (c) detail of the adiabatic shear layer, where v is the current projectile and plug velocity (adapted from Teixeira-Dias et al, 2018)

The model assumes a rigid (non-deformable) projectile, is based on the relationship between stiffness and impact velocity and on the conservation of momentum. The elastic wave velocities of the projectile and target, c_p and c_t are, respectively,

$$c_p = \sqrt{\frac{E_p}{\rho_p}} \quad \text{and} \quad c_t = \sqrt{\frac{(1-\nu_t)E}{\rho_t(1+\nu_t)(1-2\nu_t)}} \tag{16.1}$$

where E_p and ρ_p are the projectile's Young's modulus and density, and E_t and ρ_t are the target's Young's modulus and density, respectively. Based on the above and on the compatibility relation between the projectile and target, the contact compressive stress σ_c, dependent on the relative velocity V, is (Teixeira-Dias et al, 2018; Sikarwar et al, 2014)

$$\sigma_c = \varphi V = \left(\frac{\rho_t c_t \rho_p c_p}{\rho_t c_t + \rho_p c_p}\right) V \tag{16.2}$$

The conservation of momentum condition applied to the whole system is

$$M_p v_0 = v_f M_p + v_f M_g \tag{16.3}$$

where v_f is the free impact final velocity, M_p is the mass of the projectile, M_g is the mass of the plug (the material from the target) and v_0 is the projectile impact velocity. On a purely inelastic collision, the kinetic energy of the projectile is converted into deformation and heat during the impact (E_{fn}) and loss of work due to adiabatic shearing (W_n). When the projectile perforates the target there are two additional kinetic energy terms that need to be accounted for: (i) the kinetic energy of the projectile after impact (E_{kp}) and (ii) the kinetic energy of the plug after impact (E_{kg}). The energy balance equation can be written as (Recht and Ipson, 1963)

$$\frac{1}{2}M_{\rm p}v_0^2 = E_{\rm fn} + W_{\rm n}\frac{1}{2}M_{\rm p}v_{\rm r}^2 + \frac{1}{2}M_{\rm g}v_{\rm r}^2 \tag{16.4}$$

where $v_{\rm r}$ is the residual velocity of the projectile, assumed to be the same for the plug. The total energy fraction lost to deformation and heat during free impact, $E_{\rm fn}$, must equal the difference between initial and final kinetic energies, that is,

$$E_{\rm fn} = \frac{1}{2}\left(\frac{M_{\rm g}}{M_{\rm p} + M_{\rm g}}\right) M_{\rm p}v_0^2 \tag{16.5}$$

The work due to adiabatic shearing, $W_{\rm n}$, is

$$W_{\rm n}^\star = \frac{1}{2}\left(\frac{M_{\rm p}}{M_{\rm p} + M_{\rm g}}\right) M_{\rm p}\left(v^\star\right)^2 \tag{16.6}$$

which is derived for an initial velocity equal to the ballistic limit velocity (BLV) of the target, that is $v_0 = v^\star$. $W_{\rm n}$ is insensitive to changes in velocity as long as the dynamic shear stress of the target material remains constant (Sikarwar et al, 2014).

For targets with multiple layers, the projectile is subjected to increasing compression contact stresses due to the formation of multiple plugs. Figure 16.2 shows the plug formation sequence for a multi-layered target (Teixeira-Dias et al, 2018). By considering this incremental contact stress, which can be determined by considering the mechanical impedance resistance caused by the peripheral shear area, the energy fraction lost to deformation and heat for layer i becomes

$$E_{\rm fn}^i = \frac{1}{2}\left(\frac{M_{\rm g}^i\Omega^j}{M_{\rm g}^i + \Omega^j}\right)\frac{\left(\sigma_{\rm c}^i\right)^2 + \sigma_\tau^2}{\left(\varphi^i\right)^2} \tag{16.7}$$

where

$$\Omega^j = M_{\rm p}^{i-1} + \sum_{j=1}^{i-1} M_{\rm g}^j \qquad \text{and} \qquad \sigma_\tau = \frac{4h^i\tau^i}{d} \tag{16.8}$$

The additional energy dissipated into the peripheral shear area $W_{\rm n}^i$ can be defined by considering the average work done by the projectile in order to displace the plug of the i-th layer

$$W_{\rm n}^i = \frac{1}{2}\pi d\left(h^i\right)^2\tau^i \tag{16.9}$$

Assuming the residual velocity is zero ($v_{\rm r} = 0$) and substituting Eqs. (16.7) and (16.9) into Eq. (16.4), rearranging for the ballistic limit velocity $v^\star$ and simplifying yields

$$v^{\star i} = \frac{4h^i\tau^i\varphi^i M_{\rm g}^i}{d\Omega^j}\left[1 + \sqrt{\frac{\Omega^j}{M_{\rm g}^i}\left(1 + \frac{\pi d^3}{16\tau^i\left(\varphi^i\right)^2 M_{\rm g}^i}\right)}\right] \tag{16.10}$$

This however, does not account for the velocity loss due to friction between the projectile and the hole, for each layer of the target. Based on geometrical and kinematic

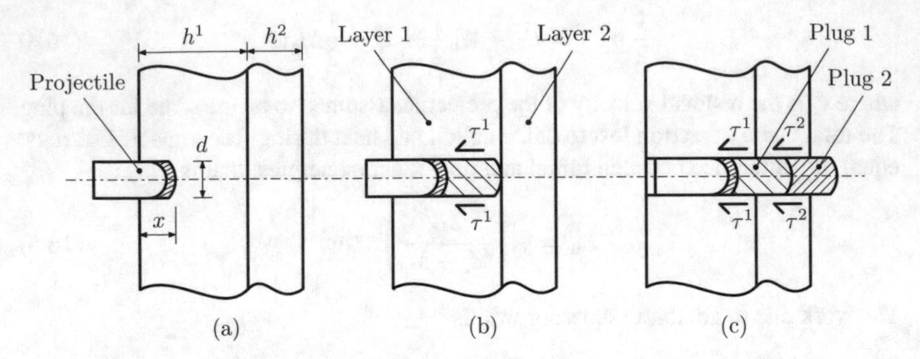

Fig. 16.2 Plugging in an in-contact multi-layered target: (a) initial penetration stage, (b) formation of first plug and (c) formation of second plug (Teixeira-Dias et al, 2018)

considerations, this velocity loss can be described by the relation

$$v_{\text{fi}} = \left(\frac{\sigma_c^i \pi d \bar{L} \mu_k^i}{M_p} \right) \left[\frac{v^{i-1} \pm \sqrt{\left(v^{i-1}\right)^2 - 2a^i h^i}}{a^i} \right] \tag{16.11}$$

where $\bar{L}$ is the friction length — the total thickness of the target or the length of the projectile, whichever is smaller. The coefficients of kinetic friction are μ_k^i, the deceleration of the projectile going through layer i is $a^i = (v_r^i - v_r^{i-1})/t^i$. The projectile contact time with each layer is t^i.

A generalised expression for the residual velocity can now be derived by rewriting Eq. (16.4) for multi-layered targets as

$$\frac{1}{2} M_p^{i-1} \left(v_r^{i-1}\right)^2 = E_{\text{fn}}^i + W_n^i + \frac{1}{2} M_p^{i-1} \left(v_r^i\right)^2 + \frac{1}{2} M_g^i \left(v_r^i\right)^2 \tag{16.12}$$

Rearranging the previous equation for the residual velocity of the i-th layer v_r^i and including the friction term yields

$$v_r^i = \frac{M_p^{i-1}}{M_p^{i-1} + M_g^i} \sqrt{\left(v_r^{i-1}\right)^2 - \left(v^{\star i}\right)^2 - v_{\text{fi}}^2} \tag{16.13}$$

16.3 Plugging of Ductile Plates: Neural Network Model

Neural networks have been applied successfully to a wide variety of application cases, from autonomous vehicles to voice-controlled home assistants. They are entirely data-driven and the quality of the output from the trained network is dependent on the quality of the data used and the available computational resources. As such, in recent years, deep learning techniques have become much more commonplace amongst researchers due to the increasing availability of data and advances in hardware and parallel computing. This section details the use of a Multi-Layer Perceptron (MLP) network to predict the residual velocity of blunt, cylindrical projectiles perforating metallic plates. A MLP network consists of at least three layers of nodes; an input layer, a hidden layer and an output layer. MLP networks utilise a supervised learning technique called back propagation for training and can distinguish between linear and non-linear data. They are particularly suitable for predicting solutions to problems where a series of numeric inputs correspond to a single target output.

16.3.1 Training Process

The layers within a MLP network are made of nodes (perceptrons). A node can be visualised as a place where a computation occurs. A node combines data from the input with a set of coefficients, or weights, that either amplify or dampen that input, thereby assigning significance to inputs with regard to the task the algorithm is trying to learn. These input-weight products are summed and passed through the nodes activation function to determine whether, and to what extent, the signal should progress further through the network and influence the final predictions. For clarity, a node within a MLP network performs a function that takes in multiple inputs and produces a single output. This function is made up of two parts: a weighted sum of all the inputs plus a constant (bias), and an activation function. The operation at a node can be expressed mathematically as

$$y = f\left(b + \sum_{i=1}^{n} w_i x_i\right) \tag{16.14}$$

where y is the output, w_i is the vector of weights, x_i is the vector of inputs, b is the bias constant and f is the activation function. Adjusting the weights and bias at the node makes it possible to change y to more closely match the desired output, hence training the network. Figure 16.3 shows a schematic of the operations that occur at a single node within a neural network. On its own, however, the active node is trivial — complex operations can be performed when these nodes are combined and arranged into layers to create a mesh-like network. The term deep learning is given to networks composed of multiple hidden layers.

A hidden layer is a vector of these nodes that switch on or off as the input is fed through the network. Each layer's output becomes the subsequent layer's input, stemming from the initial input layer that receives the data that was fed into the network. Pairing the model's adjustable weights with input features makes it possible to assign significance to those features with regard to how the network classifies and clusters input. Deep learning networks are distinguished from the more commonplace single hidden-layer neural networks by their depth (see Fig. 16.4), that is, the number of node layers through which data must pass through before reaching the output. Once

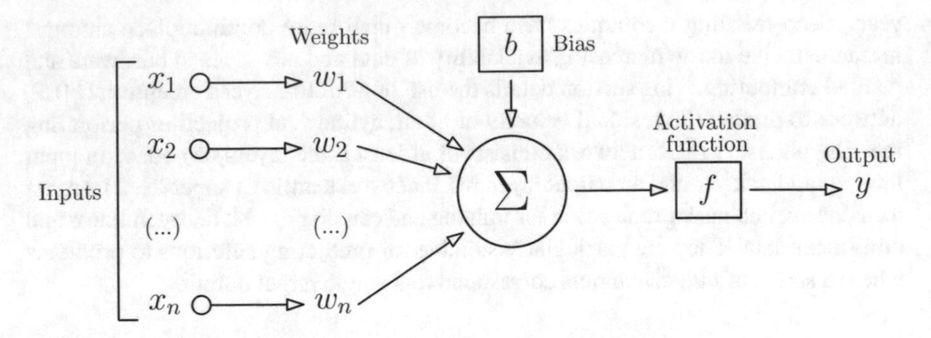

Fig. 16.3 Schematic representation of an active node in a MLP neural network

again, the output for each layer can be expressed mathematically as in Eq. (16.14), where *y* is now an output vector, *b* is the bias vector and *f* is now a vector operation. The mathematical reasoning behind why layering is useful is denoted by Taylor's theorem where a function can be represented as an infinite linear combination of polynomials. This is analogous to a layered network and thus with an infinitely large network it is possible to model any function precisely.

The MLP network is structured such that key parameters including the number of inputs, outputs, number of hidden layers and the type of activation functions that are selected are best suited to model the problem case. MLP are well suited to regression-based prediction problems where a real quantity is desired given a set of inputs. The network is trained on a tabular data-set arranged in samples, where a single sample consists of a series of inputs and a corresponding target output.

In the present case, the target output is the residual velocity of the projectile v_r. A training algorithm cycles the data-set through the network such that the model *learns* the relationship between its inputs and outputs. Once the network has been trained, it is possible to provide new inputs that the model has not been trained on to create new predictions.

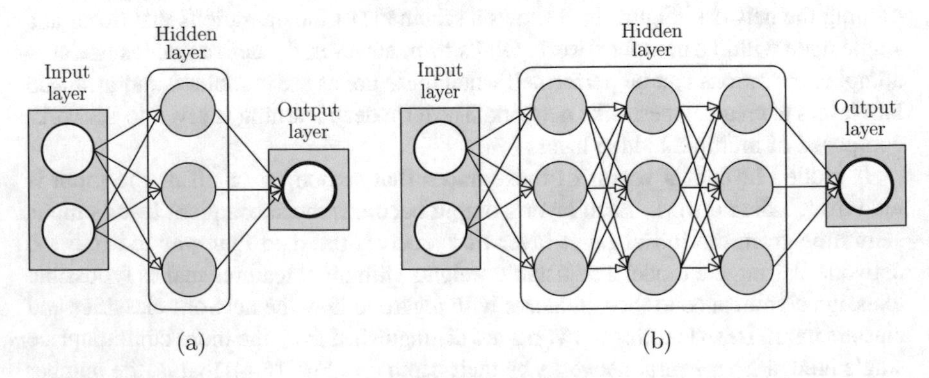

Fig. 16.4 Schematic diagram of (a) a neural network and (b) a deep neural network

The Levenberg-Marquadt algorithm is an efficient optimisation algorithm and has been selected to train the MLP network. More detailed information regarding the algorithm can be found in Ahmadian (2016). Put simply, the Levenberg-Marquadt algorithm is a combination of a loss function and an optimiser that assigns weights and biases to each node in order to best represent the function. The loss function, $E(y,t)$ depends on two parameters: the values predicted by the model y and the target values t. However, y depends on the previous layer's outputs and the current neuron's weights and activation function. Therefore it is possible to use the chain rule to differentiate $E(y,t)$ with respect to the current neurons weights,

$$\frac{\partial E}{\partial w_{nm}} = \frac{\partial E}{\partial o_m}\frac{\partial o_m}{\partial i_m}\frac{\partial i_m}{\partial w_{nm}} \tag{16.15}$$

where w_{nm} is the weight from neuron n in the previous layer to the current neuron m. The output of and input to m are o_m and i_m, respectively. The error is then fed back through the network in a process referred to as back-propagation. Using this information, the algorithm adjusts the weights of each connection in order to reduce the value of the error function. This process is repeated for a sufficiently large number of training cycles as the network converges to a state where the error in the calculations meets a target criteria. There are multiple ways to do this, but general methods for non-linear optimisation are improvements on the stochastic gradient descent. This is where a constant learning rate, η, is defined across all the weights and adjusts weights such that

$$\Delta w_{ij} = -\eta\frac{\partial E}{\partial w_{ij}} \tag{16.16}$$

where Δw_{ij} is the change in the weight at position (i,j). The negative sign ensures that Δw is always reducing the error. The back-propagation process is the method used to calculate the gradient needed in the calculation of the weights to be used in the network. The term back-propagation is used since the error is computed at the output and distributed backwards through the networks layers. The larger the gradient, the larger the adjustments and vice-versa. This is done so that on the next iteration, the prediction from the forward pass is closer to the ground truth, ultimately creating a model that is 'trained' to give a particular response given certain inputs.

16.3.2 Problem Setting

The main aim of this work is to develop a robust hybrid method to predict residual velocities from plugging metallic layered armour plates. It combines a number of experimental results by Børvik et al (2003), an analytical model developed by the authors, numerical simulation using the finite element method and two separate neural networks. One of these is trained on a data-set generated by the analytical model and the other entirely on experimental data collected from the literature.

All experiments use a blunt-nosed cylindrical projectile with geometry and material properties listed in Table 16.1. The target plates were manufactured from Weldox 460 E steel and the corresponding material properties of the plate can be found in Table 16.2. The target is assumed to be fully clamped at the supports, which is a reasonable assumption as far boundary conditions are of minor importance in ballistic penetration by small mass projectiles in the ordnance velocity range when the target diameter is greater than just a few projectile diameters.

16.3.3 Artificial Intelligence Setup

16.3.3.1 Data Collection

In this study, two different MLP neural networks have been trained and the results compared. One network was trained only on experimental data and the other on data generated using the analytical model described in Sect. 16.2. Experimental data was collected from a series of publications regarding the perforation of metal plates by blunt cylindrical projectiles (Xiao et al, 2019b,a; Rosenberg et al, 2016; Wei et al, 2012a; Børvik et al, 1999; Huang et al, 2018; Zhou and Stronge, 2008; Rodriguez-Millan et al, 2018; Yunfei et al, 2014a,b; Holmen et al, 2016; Børvik et al, 2003; Wei et al, 2012b; Awerbuch and Bodner, 1973). The data includes key experimental parameters such as the diameter, impact velocity and residual velocity of the projectile and the thickness, modulus of elasticity, yield stress and density of the metal target plate. Data from various aluminium alloys and steel were collected to form a data-set of 232 samples. Evidently, the experimental data used to train the model was limited by what experiments have been performed, what materials and projectile type the researchers selected and finally what was made accessible in the literature. As a result, the collected experimental data-set is not optimal. A perfect data-set to train the neural network would include the residual velocities associated with a wider range of impact parameters, across a range of plate thicknesses and for a number of different materials. This would give a neural network the best opportunity

Table 16.1 Geometry and material properties of the blunt, cylindrical projectile (Børvik et al, 2003)

Diameter [mm]	Length [mm]	Density [kgm^{-3}]	Elastic modulus [GPa]	Yield stress [MPa]
20	80	7850	204	490

Table 16.2 Mechanical properties of the metal plate (Børvik et al, 2003)

Test #	Thickness [mm]	Density [kg/m^3]	Elastic modulus [GPa]	Yield stress [MPa]
1	10	7850	290	300
2	16	7850	290	300

to understand how the input parameters affect the residual velocities of the projectiles as they perforate metal plates. It is for this reason that a separate MLP network was trained on data generated by the analytical model. This represents a best case scenario where the *ideal* data-set can be replicated and presents an opportunity to assess the suitability of using neural networks to make predictions in this domain. It should be noted that the experimental data-set collected from publications is subject to instrumental error, which is frequently defined in the region of 10% when measuring the velocity of the projectile (Børvik et al, 2003). Random noise in the range of 0 to 10% was added to impact and residual velocities in the analytical training data-set to simulate measurement error.

16.3.3.2 Setup and Training Parameters

This section details the parameters and criteria defined for the training of the neural network, which was done on both the experimental and analytical data-sets. Each network has six nodes in the input layer and one node in the output layer. The six input nodes allow each sample containing information regarding the diameter and impact velocity of the projectile, and the thickness, modulus of elasticity, yield stress and density of the metallic target plate to be introduced into the network. The single output node is reserved for the residual velocity of the projectile, as can be seen in Fig. 16.5. The MLP network is a deep network consisting of 15 hidden layers and utilises a Rectified Linear Unit (ReLu) activation function. The experimental data set was split such that 70% of the samples were used for training, 15% for validation and the remaining 15% to test the performance of the model. The data set allocated for training is used to fit the model and determine the weights between connections and biases associated with each node.

The goal of neural networks is to be able to make accurate predictions on new data, which the network has not been trained on; the validation data-set is used here to expose the trained network to a new data-set and measure its performance. This allows for a second opportunity to modify the network parameters defined before training (hyper-parameters), such as the initial weights, biases and number of hidden layers, to improve its performance on new data. Without this step, the network may be susceptible to over-fitting. This is where the error on the training set is driven down to a very small value, but when new data is presented to the network the error

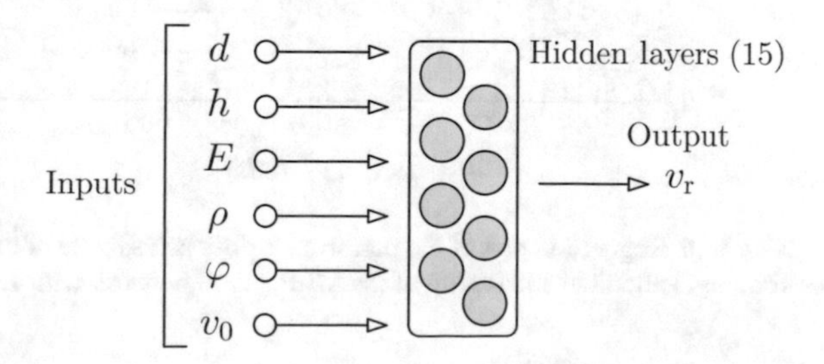

Fig. 16.5 Schematic diagram of the architecture of the deep neural network with 15 hidden layers, highlighting the six input nodes (projectile diameter d, plate thickness h, elastic modulus E, density ρ, yield constant φ and impact velocity v_0) and single output node (residual velocity v_r)

is large. This occurs when the network has become extremely tailored to the training examples, but has not learned to generalise to new situations and input combinations. The validation data-set is therefore useful to moderate this phenomenon. Finally, the test set is another independent data-set used to test the performance of the network.

16.4 Results and Discussion

The results in Fig. 16.6 illustrate a regression plot during the training of the neural network on the experimental data-set. They show the performance of each sample during the training, validation and testing phases of training the network, and the distribution of residual velocities in the data-set, where it can be seen that the majority lie in the range of [0 – 400] m/s. The line $v_r^O = v_r^T$ is plotted to show how much the predicted output from the neural network deviates from the actual target. The R-square statistical measure was used to measure how successful the fitting model was in explaining the variation of data and achieved a score of $R = 0.988$.

$$R = 1 - \frac{\sum_{i=1}^{n}(y_i - f_i)^2}{\sum_{i=1}^{n}(y_i - \overline{y})^2} \tag{16.17}$$

where y refers to values from the data-set, f refers to fitted values predicted by the network and $\overline{y}$ is the mean of the data-set values such that $\overline{y} = \frac{1}{n}\sum_{i=1}^{n} y_i$. The

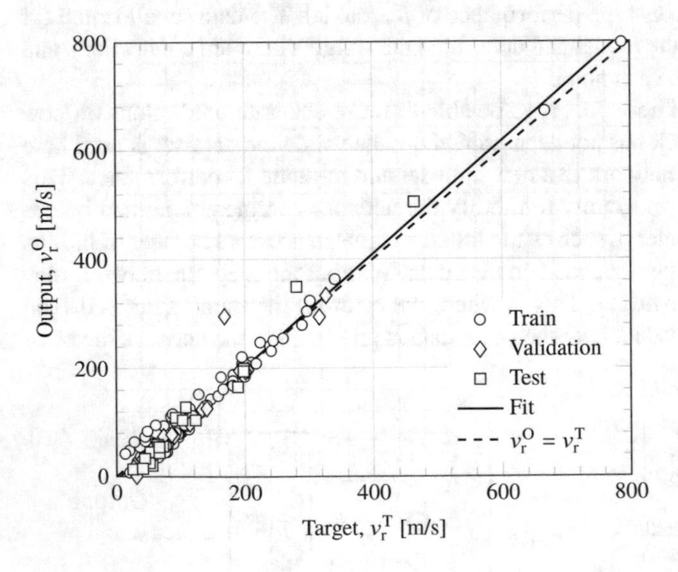

Fig. 16.6 Regression plot of the performance of each sample of experimental data used during the training, validation and testing of the MLP neural network with 15 hidden layers

R-square value of 0.988 means that the fit explains 98.8% of the total variation in the data about the average.

Figures 16.7(a) and 16.7(b) present the results from the analytical model and the predictions from each MLP network with the experimental results published by Børvik et al (2003) for the perforation of a blunt, cylindrical projectile perforating 10 mm and 16 mm Weldox 460 E plates.

The predictions from the analytical model and each neural network on Figs. 16.7(a) and 16.7(b) show good agreement with the experimental data published by Børvik et al (2003). The analytical model and MLP network trained on the analytical data-set, shown in Fig. 16.7(a), predict a BLV lower than that found by Børvik et al. The $\text{MLP}_{\text{a}}^{\text{N}}$ predictions match very closely to that predicted by the analytical model, this relationship is expected as the network was trained on data produced by the model, albeit with added noise to compensate for the 10% measurement error stated in Børvik et al.'s experiments (Børvik et al, 2003). The predictions made by the $\text{MLP}_{\text{e}}^{\text{N}}$, i.e. the predictions made by the MLP network trained on the experimental data-set, predicted a projectile response that matches very closely with the experimental results with respect to both the BLV and the shape of the ballistic curve. It should be noted that the results produced by the $\text{MLP}_{\text{e}}^{\text{N}}$ for higher velocities, i.e. greater than 400 m/s, are less reliable than the predictions made at relatively lower impact velocities. This is because the predictions from the AI model are trained solely on the experimental data-set. Referring back to Fig. 16.6, the majority of training samples used to train the network are within the range of [0 – 400] m/s. The network has therefore been exposed to more experimental data in that range and as such the respective weights and bias at each node have been optimised to best match these values. As a result, the predictions for higher velocities are less reliable and it would not be advised to use

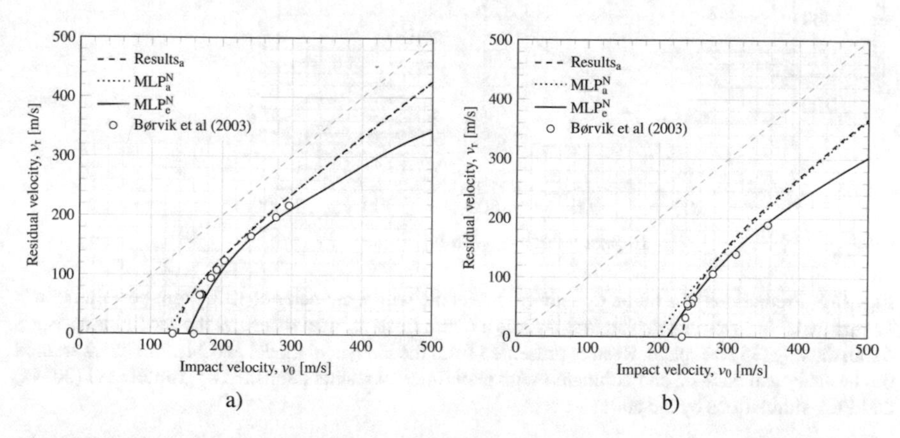

Fig. 16.7 Analytical, neural network and experimental residual versus impact velocity curves of a blunt cylindrical projectile with a diameter of 20 mm perforating (a) a 10 mm and (b) a 16 mm Weldox 460 E plate, where Results$_a$ refers to the analytical model, $\text{MLP}_{\text{a}}^{\text{N}}$ are the predictions from the MLP network trained on the analytical data-set and $\text{MLP}_{\text{e}}^{\text{N}}$ are the predictions of the network trained on the experimental data-set

this network to study the ballistic response of materials at impact velocities above 400 m/s.

The network was additionally tested with a set of multi-layered targets. Information in literature regarding experimental tests on multi-layered targets with blunt rigid projectiles is, however, almost nonexistent. As such, the authors used a set of experimental tests by Yunfei et al (2014a) to validate the proposed AI network and method. These authors impacted 12 mm bi-layer steel targets with deformable blunt cylindrical projectiles with a diameter of 12.67 mm. The first layer is 45 steel with 6 mm thickness and the second layer is Q235 steel, also with 6 mm thickness, and the results are shown in Fig. 16.8. There is a clear and significant discrepancy between the experimental and the MLP results. This is, however, justified, because the multi-layered target MLP was trained with analytical results, which assume the projectile is rigid. A deformable projectile, such as the one used by Yunfei et al (2014a), needs significantly more energy to perforate the target, as is evident in Fig. 16.8. To account for this, the authors developed FEA models with both deformable and rigid projectiles.

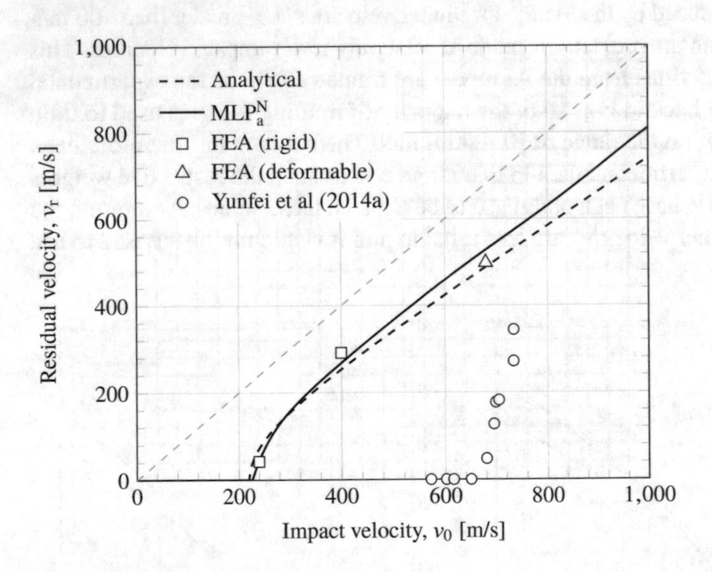

Fig. 16.8 Predictions of a blunt, cylindrical projectile with a diameter of 12.67 mm perforating a 12 mm metal bi-layer target. The first layer is a 6 mm thick 45 steel plate and the second layer is a 6 mm thick Q235 steel plate. Results presented from the analytical model and MLP network trained on the analytical data-set and compared with experimental results published by Yunfei et al (2014a) and FEA simulations by the authors

16.4.1 Finite Element Modelling

The efficiency of the AI described above, namely when predicting residual velocities of blunt projectiles impacting layered metallic armour plates, was tested with a set of numerical examples. The authors used finite element analysis (FEA) and the hydrocode LS-Dyna to replicate a number of ballistic impact tests. On a first instance the FEA models were calibrated for the ballistic limit velocity using experimental results obtained by Børvik et al (2003). Validation models were developed for plates with the characteristics listed in Table 16.3. Figure 16.9 shows a snapshot of finite element analysis Tests 2, for a ballistic limit velocity of 240 m/s, and the corresponding projectile velocity profile. The formation of the plug can be seen in Fig. 16.9(a) and a residual velocity of approximately 18 m/s was obtained for this model, indicating that the corresponding BLV will be slightly lower than 240 m/s.

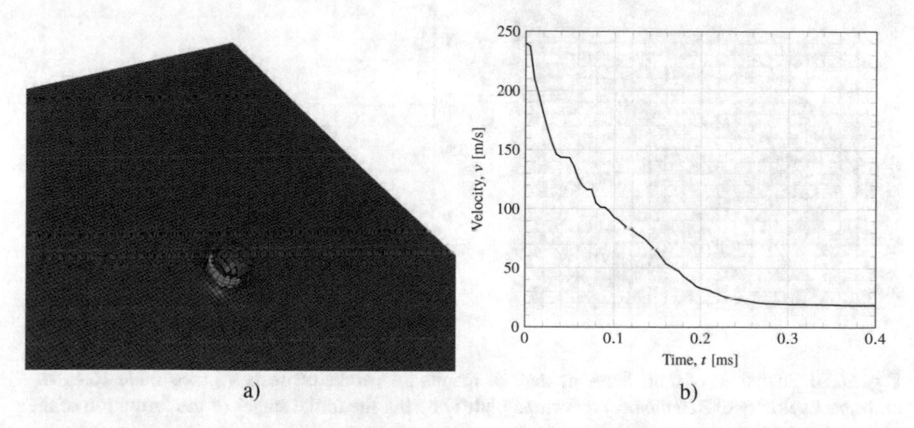

Fig. 16.9 Snapshots of finite element analysis results for Test 2 (see Table 16.3) for a ballistic limit velocity of 240 m/s: (a) deformed plate showing the formation of the plug and (b) velocity profile for the projectile

A number of additional numerical tests were ran to further validate the hybrid-method here proposed, which combines analytical and experimental results to train a predictive neural network. The models, which were defined with specifications outside the set of results used for training the AI, are listed in Table 16.4. All these

Table 16.3 Characteristics of the tests used to validate the finite element models and corresponding ballistic limit velocities (BLV); numerical BLV shown in brackets. Material properties can be found in Table 16.2

Test number	Plate material	Thickness [mm]	Ballistic limit velocity (FEA) [m/s]	Number of elements
1	Weldox 460 E	10	165.6 (166)	135,000
2	Weldox 460 E	16	236.6 (240)	225,000

tests were done at impact velocities above the ballistic limit velocity and the residual velocity was used as the validation parameter. As can be seen from these results, there is very good agreement with the experimental results with the AI model always lower than 5.1%. Discrepancies become significant (as high as 35.2% for Test V4) when comparing with the FEA results, however, this is believed to be related to the method used to model adiabatic shearing and the formation of the plug. The finite element analyses use element deletion for this, which is known to be inaccurate and potentially deviating from mass conservation of the system. Figure 16.10 shows a snapshot of validation Test V3, for an impact velocity of 320 m/s and the corresponding projectile velocity profile.

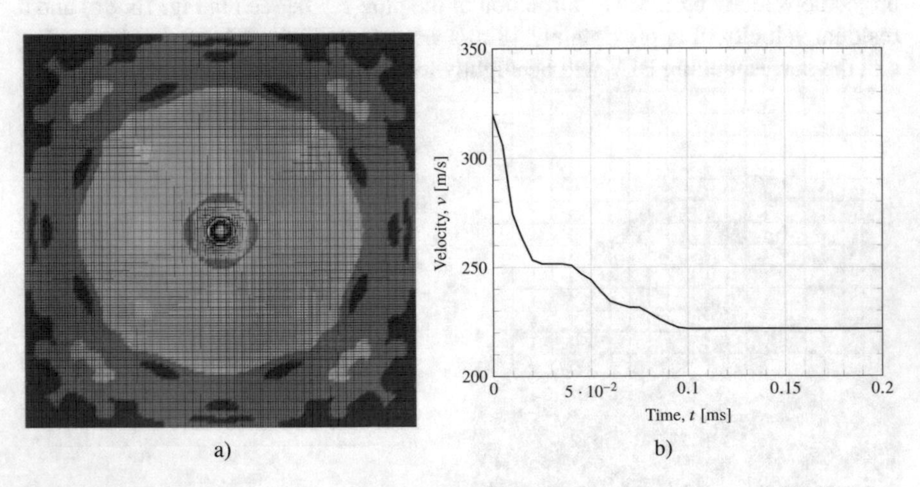

Fig. 16.10 Snapshots of finite element analysis results for validation Tests V3 (see Table 16.4) for an impact velocity of 320 m/s: (a) deformed plate showing the initial stages of the formation of the plug and von Mises stresses, and (b) velocity profile for the projectile

Table 16.4 Specifications, results and comparison of the tests used to validate the AI algorithm. Tests V1 to V6 are monolithic plates and V7 to V9 are multi-layered targets. FEA models of Tests V7/V8 are with rigid projectile, and Test V9 is with a deformable projectile

Test ID	h [mm]	v_0 [m/s]	Experimental v_r [m/s]	FEA v_r [m/s]	AI v_r [m/s]	FEA/AI [%]	Exp/AI [%]
V1	10	220.0	143.1	126.6	136.2	7.0 (−)	5.1 (+)
V2	10	280.0	201.7	184.1	196.6	6.4 (−)	2.6 (+)
V3	10	320.0	234.6	221.7	226.4	2.1 (−)	3.6 (+)
V4	16	260.0	83.5	108.7	79.9	35.2 (+)	4.5 (+)
V5	16	280.0	111.9	143.3	108.7	31.8 (+)	2.9 (+)
V6	16	320.0	153.3	189.4	157.6	20.2 (+)	2.7 (−)
V7	2×6	240.0	–	42.3	39.9	6.0 (+)	–
V8	2×6	400.0	–	291.5	268.0	8.8 (+)	–
V9	2×6	680.0	45.4	498.5	–	–	–

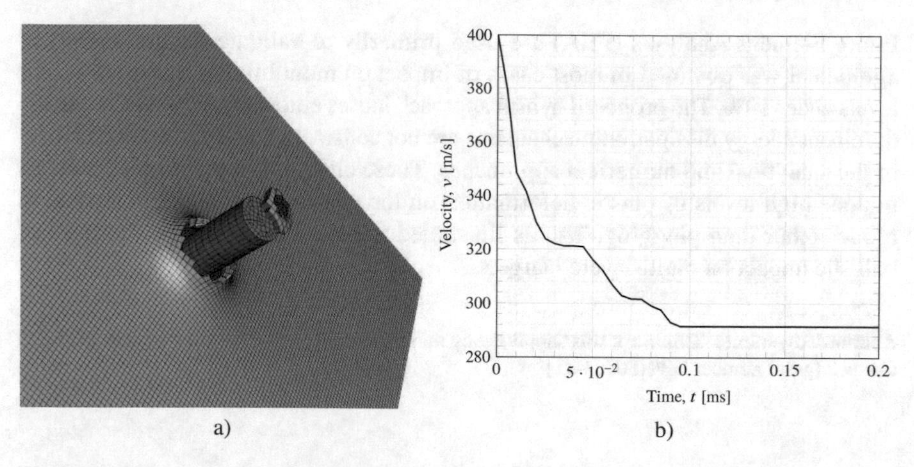

Fig. 16.11 Impact on a multi-layered target: (a) detail of the interaction between the projectile and target for Test V8 (rigid projectile) showing the formation of the double plug, and (b) the corresponding projectile velocity profile

To further validate the AI method, finite element analyses were run on models of multi-layered targets with rigid and deformable projectiles, and the results are shown in Fig. 16.8 and listed in Table 16.4 (Tests V7 to V9). As expected, the FEA results match very well the AI predictions, with differences not exceeding 8.8%, as these were trained assuming a rigid projectile. To support the justification on the discrepancy between the AI and experimental results by Yunfei et al (2014a), an FEA model was developed with a deformable projectile. The corresponding results, however, do not match the experimental observations within reasonable differences. This is most probably due to a combination of factors, namely, the FEA model not being able to capture correctly the plastic deformation levels observed on the real projectile and the extremely high strain levels experienced by the projectile, leading to excessive and unnatural element deformation and distortion. This also explains why the results from the FEA model with a deformable projectile are closer to the AI predictions. Figures 16.11(a) and 16.11(b) illustrate the interaction between the projectile and bi-layer target for Test V8 (rigid projectile), and the corresponding velocity profile, respectively.

16.5 Conclusions and Final Remarks

This chapter presents a new approach to efficiently predict residual velocities from impacts on monolithic and multi-layered metallic ductile targets. The proposed method is based on a combination of experimental, analytical and numerical methods, and a set of Deep Learning (DL) models. The main aim is to increase design efficiency by significantly reducing testing and computational costs. The experimental and analytical results are used to train the Artificial Intelligence (AI) models, while the

Finite Element Analyses (FEA) are used primarily to validate results. Excellent agreement was obtained in most cases of impact on monolithic targets, with error levels under 10%. The proposed hybrid approach looses efficiency, however, when the dominant energy dissipation mechanisms are not considered or are dificult to model in the analytical and numerical approaches. These energy dissipation mechanisms include high levels of plastic deformation on the projectile or target deformation modes other than plugging. This is illustrated in this chapter in the examples of ballistic impact on multi-layered targets.

Acknowledgements This work was supported by the Engineering and Physical Sciences Research Council [grant number EP/N509644/1].

References

Ahmadian AS (2016) Numerical models for submerged breakwaters - coastal hydrodynamics and morphodynamics. In: Ahmadian AS (ed) Numerical Models for Submerged Breakwaters, Butterworth-Heinemann, Boston, pp 93 – 108, doi:10.1016/B978-0-12-802413-3.00006-7

Ali MW, Mubashar A, Uddin E, Haq SWU, Khan M (2017) An experimental and numerical investigation of the ballistic response of multi-level armour against armour piercing projectiles. International Journal of Impact Engineering 110:47 – 56, doi:10.1016/j.ijimpeng.2017.04.028, special Issue in honor of Seventy Fifth Birthday of Professor N. K. Gupta

Atkins A, khan MA, Liu J (1998) Necking and radial cracking around perforations in thin sheets at normal incidence. International Journal of Impact Engineering 21(7):521 – 539, doi:10.1016/S0734-743X(98)00010-4

Awerbuch J, Bodner SR (1973) Experimental investigation of normal perforation of projectiles in metallic plates. Technical report, National Technical Information service, Springfield

Backman ME, Goldsmith W (1978) The mechanics of penetration of projectiles into targets. International Journal of Engineering Science 16(1):1 – 99, doi:doi.org/10.1016/0020-7225(78)90002-2

Børvik T (200) Ballistic penetration and perforation of steel plates. Technical report, Norwegian University of Science and Technology – NUST, Trondheim

Børvik T, Langseth M, Hopperstad OS, Malo KA (1999) Ballistic penetration of steel plates. International Journal of Impact Engineering 22(9):855 – 886, doi:10.1016/S0734-743X(99)00011-1

Børvik T, Leinum JR, Solberg JK, Hopperstad OS, Langseth M (2001) Observations on shear plug formation in Weldox 460 E steel plates impacted by blunt-nosed projectiles. International Journal of Impact Engineering 25(6):553 – 572, doi:10.1016/S0734-743X(00)00069-5

Børvik T, Hopperstad OS, Langseth M, Malo KA (2003) Effect of target thickness in blunt projectile penetration of Weldox 460 E steel plates. International Journal of Impact Engineering 28(4):413 – 464, doi:10.1016/S0734-743X(02)00072-6

Elek P, Jaramaz S, Micković D (2005) Modeling of perforation of plates and multi-layered metallic targets. International Journal of Solids and Structures 42(3):1209 – 1224, doi:10.1016/j.ijsolstr.2004.06.053

Goldsmith W (2001) Impact: The Theory and Physical Behaviour of Colliding Solids. Dover Publications Inc., New York

Greszczuk LB, Zukas JA, Nicholas T, Swift HF, Curran DR (1982) Impact Dynamics. John Wiley & Sons

Holmen JK, Johnsen J, Hopperstad OS, Børvik T (2016) Influence of fragmentation on the capacity of aluminum alloy plates subjected to ballistic impact. European Journal of Mechanics - A/Solids 55:221 – 233, doi:10.1016/j.euromechsol.2015.09.009

Horne MR (2014) Plastic Theory of Structures. Elsevier

Huang X, Zhang W, Deng Y, Jiang X (2018) Experimental investigation on the ballistic resistance of polymer-aluminum laminated plates. International Journal of Impact Engineering 113:212 – 221, doi:10.1016/j.ijimpeng.2017.12.002

Jia X, xiang Huang Z, dong Zu X, hui Gu X, qiang Xiao Q (2014) Theoretical analysis of the disturbance of shaped charge jet penetrating a woven fabric rubber composite armor. International Journal of Impact Engineering 65:69 – 78, doi:10.1016/j.ijimpeng.2013.11.005

Krauthammer T (2008) Modern Protective Structures. CRC Press

Landkof B, Goldsmith W (1985) Petalling of thin, metallic plates during penetration by cylindro-conical projectiles. International Journal of Solids and Structures 21(3):245 – 266, doi:10.1016/0020-7683(85)90021-6

Liu J, Long Y, Ji C, Liu Q, Zhong M, Zhou Y (2018) Influence of layer number and air gap on the ballistic performance of multi-layered targets subjected to high velocity impact by copper efp. International Journal of Impact Engineering 112:52 – 65, doi:10.1016/j.ijimpeng.2017.10.001

Recht RF, Ipson TW (1963) Ballistic perforation dynamics. Trans ASME Journal of Applied Mechanics 30(3):384 – 390, doi:10.1115/1.3636566

Rodriguez-Millan M, Garcia-Gonzalez D, Rusinek A, Abed F, Arias A (2018) Perforation mechanics of 2024 aluminium protective plates subjected to impact by different nose shapes of projectiles. Thin-Walled Structures 123:1 – 10, doi:10.1016/j.tws.2017.11.004

Rosenberg Z, Kositski R, Dekel E (2016) On the perforation of aluminum plates by 7.62 mm APM2 projectiles. International Journal of Impact Engineering 97:79 – 86, doi:10.1016/j.ijimpeng.2016.06.003

Sikarwar RS, Velmurugan R, Gupta NK (2014) Influence of fiber orientation and thickness on the response of glass/epoxy composites subjected to impact loading. Composites Part B: Engineering 60:627 – 636, doi:10.1016/j.compositesb.2013.12.023

Taylor GI (1948) The formation and enlargement of a circular hole in a thin plastic shee. The Quarterly Journal of Mechanics and Applied Mathematics 1(1):103–124, doi:10.1093/qjmam/1.1.103

Teixeira-Dias F, Smith N, Galiounas F (2018) Analytical and energy based methods for penetration mechanics. In: Altenbach H, Öchsner A (eds) Encyclopedia of Continuum Mechanics, Springer Nature, doi:10.1007/978-3-662-53605-6_207-1

Thomson WT (1955) An approximate theory of armor penetration. Journal of Applied Physics 26(1):80–82, doi:10.1063/1.1721868

Wei Z, Yunfei D, Sheng CZ, Gang W (2012a) Experimental investigation on the ballistic performance of monolithic and layered metal plates subjected to impact by blunt rigid projectiles. International Journal of Impact Engineering 49:115 – 129, doi:10.1016/j.ijimpeng.2012.06.001

Wei Z, Yunfei D, Sheng CZ, Gang W (2012b) Experimental investigation on the ballistic performance of monolithic and layered metal plates subjected to impact by blunt rigid projectiles. International Journal of Impact Engineering 49:115 – 129, doi:10.1016/j.ijimpeng.2012.06.001

Woodward RL (1987) A structural model for thin plate perforation by normal impact of blunt projectiles. International Journal of Impact Engineering 6(2):129 – 140, doi:10.1016/0734-743X(87)90015-7

Xiao X, Pan H, Bai Y, Lou Y, Chen L (2019a) Application of the modified Mohr–Coulomb fracture criterion in predicting the ballistic resistance of 2024-T351 aluminum alloy plates impacted by blunt projectiles. International Journal of Impact Engineering 123:26 – 37, doi:10.1016/j.ijimpeng.2018.09.015

Xiao X, Wang Y, Vershinin VV, Chen L, Lou Y (2019b) Effect of Lode angle in predicting the ballistic resistance of Weldox 700 E steel plates struck by blunt projectiles. International Journal of Impact Engineering 128:46 – 71, doi:10.1016/j.ijimpeng.2019.02.004

Yunfei D, Wei Z, Guanghui Q, Gang W, Yonggang Y, Peng H (2014a) The ballistic performance of metal plates subjected to impact by blunt-nosed projectiles of different strength. Materials & Design (1980-2015) 54:1056 – 1067, doi:10.1016/j.matdes.2013.09.023

Yunfei D, Wei Z, Yonggang Y, Lizhong S, gang W (2014b) Experimental investigation on the ballistic performance of double-layered plates subjected to impact by projectile of high strength. International Journal of Impact Engineering 70:38 – 49, doi:10.1016/j.ijimpeng.2014.03.003

Zhang TG, Stronge WJ (1996) Theory for ballistic limit of thin ductile tubes hit by blunt missiles. International Journal of Impact Engineering 18(7):735 – 752, doi:10.1016/S0734-743X(96)00033-4
Zhou DW, Stronge WJ (2008) Ballistic limit for oblique impact of thin sandwich panels and spaced plates. International Journal of Impact Engineering 35(11):1339 – 1354, doi:10.1016/j.ijimpeng.2007.08.004
Zukas JA (1980) Impact dynamics: Theory and experiments. Technical report, US Army Armament Research and Development Command, Maryland

Chapter 17
A Review on Numerical Analyses of Martensitic Phase Transition in Mono and Polycrystal Transformation-induced Plasticity Steel by Crystal Plasticity Finite Element Method with Length Scales

Truong Duc Trinh and Takeshi Iwamoto

Abstract Strain-induced martensitic transformation (SIMT) plays an essential role for enhancing the mechanical properties of TRIP steel such as high strength, ductility and toughness. Thus, the mechanical responses including the SIMT are significantly important for an engineering design of the materials with microstructural predictions. It is proven that length scales such as a grain size of the parent phase influence the deformation behavior of TRIP steel, and it is necessary to understand deeply the SIMT behavior considering with specific length scales. This review focuses on computational analyses of SIMT in both single crystal and polycrystal TRIP steel by the crystal plasticity finite element method (CPFEM) for predicting the SIMT behavior with the appropriate length scales such as grain size. Then, in order to discuss the size-dependency, examples of computational results under an assumption of plane strain tension with two planar slip systems are shown for both single crystal and polycrystal TRIP steel by the proposed framework of CPFEM in the past without any length scales.

Keywords: Transformation-induced plasticity · Strain-induced martensitic transformation · Crystal plasticity · Finite element method · Length scales

Truong Duc Trinh
Graduate School of Engineering, Hiroshima University, 1-4-1 Kagamiyama, Higashi-Hiroshima, Hiroshima, 739 - 8527 Japan,
e-mail: d184860@hiroshima-u.ac.jp

Takeshi Iwamoto
Academy of Science and Technology, Hiroshima University, 1-4-1 Kagamiyama, Higashi-Hiroshima, Hiroshima, 739 - 8527 Japan,
e-mail: iwamoto@mec.hiroshima-u.ac.jp

H. Altenbach and A. Öchsner (eds.), *State of the Art and Future Trends in Material Modeling*, Advanced Structured Materials 100,
https://doi.org/10.1007/978-3-030-30355-6_17

17.1 Introduction

Steel containing metastable austenitic phase has significantly important roles in structural materials because it owns outstanding combinations of mechanical properties such as strength, toughness and ductility. These mechanical properties are due to an effect of the transformation-induced plasticity (TRIP) (Zackay et al, 1967; Tamura, 1982), and the steels with this kind of phenomenon are called TRIP steels in a wide meaning. In engineering applications, TRIP steels are considered as one of suitable alloys for automotive industries because of the greatly-improved formability without degradation of the mechanical properties. There are two different types of TRIP steels based on the chemical composition of TRIP steels. In particular, the high-alloyed steel has only metastable austenitic constituent in the microstructure at room temperature, for instance, type 304 austenitic stainless steel could be one of the high-alloyed TRIP steel. On the other hand, the low-alloyed TRIP steel, originally it is called TRIP-assisted (Bhadeshia, 2002) or TRIP-aided (Barbe et al, 2001) steel, contains not only metastable austenite but also ferrite and bainite at room temperature. Thus, the low-alloyed TRIP steel has naturally a hyponym “multiphase”. It is important to notice that martensitic-transformed phase exists inside austenitic phase with some geometrical and crystallographic patterns at a scale of crystal lattices in both types of TRIP steels during the process of plastic deformation. From the viewpoint of solid mechanics, the TRIP effect can be realized by an appropriate combination between the plastic deformation behavior in the both phases and the strain-induced martensitic transformation (SIMT) process at the moment in which the retained or single metastable austenitic phase (Iwamoto et al, 1998; Fischer et al, 2000; Iwamoto and Pham, 2015).

In order to predict and control accurately the superior mechanical properties of TRIP steel, it is indispensable to understand deeply the SIMT with different crystallographic patterns at the microstructural level. For such purpose, a computational simulation becomes a quite effective tool. With the rapid development of the tools to solve micromechanically-based problems, a lot of physical aspects at the microstructures can be included and explained by an appropriate modeling such as a multi-scale modeling. As the great majority of the microstructures in materials from the nature is in forms of crystalline, crystalline-based numerical models have got many concentrations from researchers. Among the models, continuum crystal plasticity (CCP) (Asaro, 1983; Peirce et al, 1983) showed its great benefits when it is combined with finite element method (FEM). This framework is so-called crystal plasticity finite element method (CPFEM). As a consequence, the CCP is continuously developed to be an independent group of constitutive models which is able to deal with huge number of problems in mechanics of crystals (Roters et al, 2010). One of the most important advantages of the CCP is its high flexibility to include and extend to constitutive equations based on various frameworks for a flow theory and hardening mechanisms in plasticity. Also, the CPFEM is capable effectively when dealing with complicated boundary conditions to express the nature of mechanics in crystallites as the interaction at a micro scale between grains or inside the grain itself, for example.

The importance and positive effects of the CCP in solving the problems on any elasto-viscoplastic deformation have begun to be recognized (Barbe et al, 2001; Staroselsky and Anand, 2003; Diard et al, 2005; Graff et al, 2007; Kratochvil, 2012; Forest and Aifantis, 2010). The capability of CPFEM is demonstrated in a texture prediction of pure aluminum during the equal channel angular extrusion (Li et al, 2005) to investigate the mechanism of microstructural deformation due to an crystallographic orientation effect. In Ardeljan et al (2014), a hardening model based on dislocation density, which can include a specific length scale, is employed successfully for polycrystalline of two-phase metallic composite into the framework of CPFEM. The CCP keeps being enhanced in a recent study by a new constitutive model based on the elasto-viscoplastic self-consistent theory subjected to complicated loading histories in the austenitic stainless steel (Wang et al, 2017). Noticeably, though the CCP is not applied directly, the theoretical concept of the CCP is also applicable and continuously improved in various frameworks of numerical studies. Originally, constitutive formulations for the shape memory alloy with the stress-induced martensitic transformation, and TRIP steel are proposed based on the analogy of the CCP in some greatly influent research works (Tokuda et al, 1998; Cherkaoui et al, 1998; Diani and Parks, 1998). Among the improvements of the constitutive model, a constitutive model for single crystal TRIP steel is fulfilled as an exceptional mark to explore the martensitic transformation behaviors by coupling the effects between slip deformation on a slip plane and variants in the SIMT on a habit plane using automata cellular approach (Iwamoto and Tsuta, 2004). The numerical studies based on CCP for martensitic transformation keep receiving great contributions of researchers (Tjahjanto et al, 2008; Sidhoum et al, 2018; Taejoon et al, 2019). Recently, CPFEM in solving the martensitic transformation problems is continuously developing by coupling with a phase-field approach (Schmitt et al, 2017; Levitas, 2014; Levitas and Javanbakht, 2015). Remarkably, Levitas and Javanbakht (2015) have proposed a phase field model based on a framework of nonlinear continuum thermodynamics with large strains to investigate the phase interface propagation during martensitic transformation, especially, the transformation shear strain is considered as one of order parameters and coupled successfully with plastic shear strain. Nonetheless, studies on martensitic phase transition in polycrystal TRIP steel including specific length scales as well as a good approach to describe the morphology of polycrystal material are still very little.

The present review focuses on the significant effects of length scales as the size of austenitic grains and the nuclei of martensite on the strain-induced martensitic transformation (SIMT) in TRIP steel. To complete the review, the computational results of SIMT and deformation behavior in not only mono but also polycrystal TRIP steel under the plane strain condition with two planar slip systems are shown by the numerical simulation based on the platform of CPFEM (Iwamoto and Tsuta, 2004; Trinh and Iwamoto, 2019). For analyses here, the CPFEM coupling with the crystallography of martensitic transformation (Bowles and Mackenzie, 1954) is employed to derive the constitutive formulation of single crystal TRIP steel. In order to express the important effect of the length scales on the SIMT and mechanical behavior in polycrystal TRIP steel, two different patterns of the initial crystallographic orientations with two different numbers of grains by the Voronoi tessellation approach are assigned randomly for each grain.

17.2 Literature Survey of Problems on Length Scales Regarding with Martensitic Phase Transformation

17.2.1 Effects of Length Scales in the Parent Phase

At the microstructural scale, huge shear and dilatation which generate unavoidable dislocations in crystal structure are induced by the martensitic transformation. The increasing number of dislocations is able to affect plastic deformation behaviors, as a result, the dislocations provide more hardening for the materials. As a critical point of view, the length scale effect is significantly important and complicated since it is changed due to the deformation of microstructures. Thus, some length scales including Burgers vector are dependent on the grain size (Voyiadjis and Abu Al-Rub, 2005; Cheong et al, 2005; Ehrler et al, 2008; Faghihi and Voyiadjis, 2012; Dunstan and Bushby, 2014; Yeddu, 2018) which is the one of length scales. As evidences, by introducing the geometrical effect of austenitic grain into a generalized macroscopic model for the kinetics of SIMT, the size-dependent deformation and SIMT behavior in TRIP steel can be expressed by combining with the Hall-Petch relation (Iwamoto and Tsuta, 2000). Furthermore, the dependence of austenitic grain size on the martensitic phase transition is studied by a newly-developed experimental technique which is an accurate heat treatment combined with a dilatometry measurement, so it is able to obtain the precise starting temperature and finishing temperature of the martensitic transformation as well as the size-dependent microstructure of martensite (Hanamura et al, 2013).

On the other hand, the martensite is transformed through a formation process of microstructures due to the energy minimization, while the appropriate combination between plastic deformation and the microstructure by SIMT is established from inherent inhomogeneity in parent phase, as a consequence, the combination makes the microstructure inside martensite become size-dependent. There are various concepts of length scales at microstructure relevant to plastic deformation. For instance, at the smallest scale, in order to measure the dimension of a crystal, the magnitude of the Burgers vector is used representing for the lattice spacing. Additionally, a diameter in a dislocation core or stacking fault width of a partial dislocation can be used as other length scale (Eisenmann et al, 2005; Hunter and Beyerlein, 2013). The length scale effect is confirmed in several research works. For instance, the length scale effect is included as a function of plastic strain and grain size to express the physical phenomenon by introducing a gradient plasticity theory (Fleck and Hutchinson, 1993) for metal which is mainly relied on the framework of the hardening model of crystallographic dislocation mechanics through the Taylor model (Abu Al-Rub and Voyiadjis, 2006; Liu and Dunstan, 2017).

Among the methods to describe the grain size effect, another approach by geometrically necessary dislocations (GNDs) representing the length scale in plasticity is discussed (Gao and Huang, 2003; Gurtin, 2008; Littlewood et al, 2011). It is pointed out that the length scales where GNDs occur play an important role in the deformation behavior of material. To extend the theory of plasticity and give a clearer physical understanding, the hardening by GNDs is taken into account with a specific material length scale based on the framework of higher-order governing equations (Fleck and Hutchinson, 1993). Next, in order to understand the grain size effect to initial yield stress of polycrystal material (Ohno and Okumura, 2007), the self-energy of GND is considered to derive the higher-order stress work-conjugate incorporating with strain gradient plasticity theory of Gurtin (2008). In Gurtin (2008), the microstress is defined as the work-conjugate to the slip gradient for a slip system (Gurtin, 2002). Though it is not necessary to introduce the higher order stress in their analyses as Aifantis and Ngan (2007), the grain size effect is successfully expressed by using an internal length scale parameter to measure the effect of the GND. The size dependence of the mechanical properties of materials can be resulted from an evolution in strain gradient at a very small area. The strain gradient is induced by interactions of dislocations, as a result, an additional hardening is provided. Therefore, GNDs to Taylor's hardening law Voyiadjis and Abu Al-Rub (2005), stacking fault width created by partial dislocation emission from grain boundary (Hunter and Beyerlein, 2013), or Burgers vector related to mobile dislocation density (Richeton et al, 2018) are applied to the same concept expressing the length scale via the mechanism of dislocation motions. Importantly, those research works emphasizes that plasticity is naturally a multi-scale problem in which the related size considerations can vary from atom scale to millimeter scale, and depend on the dislocation motion. In another word, the length scale changes with the microstructural deformation due to the evolution of dislocation and strongly depended on the grain size.

Next, an analysis on SIMT based-deformation gradient crystallographic theory with the influence of the austenitic grain size on the mechanical behavior of multiphase carbon steel is done by 3D CPFEM model (Turteltaub and Suiker, 2006). In this work, the grain size is considered via a surface energy parameter in the Helmholtz free energy associated with regions near the habit planes as an indirect length scale (Turteltaub and Suiker, 2006). Recently, the direct dependence of kinetics of athermal martensite on the austenitic grain structure and prior austenitic grain size in martensitic transformation process is also studied theoretically by applying the Koistinen-Marburger kinetic equation, while the fcc lattice expansion process is taken into account by dilatometry is applied to obtain the volume fraction of martensite and describe the kinetics of phase transformation (Carola et al, 2019). In another research work, the TRIP effect with different austenitic grain size on the properties of austenitic stainless steel during the bending process of a single crystal beam is studied by using ferrite-scope and thermographic camera to measure the martensitic volume fraction and temperature during phase transformation process, after that, a numerical simulation is implemented to confirm experimental work by a commercial software (Gupta et al, 2015). In their simulation (Gupta et al, 2015), a higher order non-local crystal plasticity model is applied in which the different hardening effects caused by statistically stored dislocation (SSD) and geometrically necessary dislocation can be taken into account simultaneously.

17.2.2 Effects of Length Scales in the Product Phase

As reviewed above, there are numbers of research works investigating the effect of length scales in the plasticity of the parent phase into SIMT. Nevertheless, there are very limited publications mentioning about the influence of length scale in martensitic phase to the phase transformation process regardless its incompliant understanding. Since the morphology and kinetics of the martensitic transformation depend directly on the austenitic grain size (Iwamoto and Tsuta, 2000), it is clear that the grain size of austenite could induce the expected mechanical properties of TRIP steel. From the above discussions, the plasticity in martensitic phase with GNDs can be considered as one of the mechanisms of the length scale effect.

On the other hand, by applying the Ginzburg-Landau equation, the size-dependent microstructure by the interface motion between the parent austenite and martensitic product phase is studied (Levitas, 2014; Levitas and Javanbakht, 2015; Militzer, 2011; Tuma and Stupkiewicz, 2016). Although the interface motion can be introduced successfully by the evolution of an order parameter with the length scale as an interface thickness, there are still limitations to describe clearly the effect of grain size as a length scale as well as the martensitic nucleation. At the athermal transformation, martensite nucleates due to the dislocation-initiating slip in the austenitic parent phase for the accommodated process through plastic deformation. The nucleation of martensite is a significantly complicated and rapid process which is very difficult to understand well by in situ. It is demonstrated theoretically or computationally that there is a specific size of the martensitic embryo which plays a dominant role in the evolution of martensitic microstructure (Suezawa and Cook, 1980; Ghosh and Olson, 1994). Therefore, a knowledge on the critical size in the martensitic embryos through simulating martensitic transformation is able to obey better understanding of aspects on martensitic nucleation as well as the effect of this length scale on SIMT behaviors. It is noticeable that the most direct mechanism of such heterogeneous nucleation is the size dependence of martensitic transformation.

Additionally, the volume fraction of martensite is able to influence greatly the mechanical properties of multiphase steel at a macroscopic scale. Using the law of energy balance, a macroscopic model is proposed to study the martensitic transformation of Fe-Ni alloy without external stress and SIMT as a first order phase transition (Yu, 1997). Hueper et al (1999) examines the influence of the volume fraction of the harder phase by FEM to indicate the macroscopic stress-strain relationship of dual phase material in three stages of elastoplastic deformation. The role of martensitic volume fraction is continuously studied more significant in an another micromechanical model based an axisymmetric unit cell to capture the macroscopic elastoplastic response of dual phase steel model (Lai et al, 2016). Nevertheless, a use of volume fraction of martensite as an internal state variable in the continuum theory of plasticity for a macroscopic analysis eliminates the length scale effect in martensitic product phase which is much smaller than the macroscopic length scale. Generally, based on the framework with the CPFEM, the kinetics model with the volume fraction should not be employed. However, it is necessary to have a good approach to capture the length scale effect in martensitic phase as well as the complex of microstructural heterogeneity of martensite. For the purpose, several research works have focused on martensitic volume fraction in order to understand quantitatively the kinetics of a phase transition process by different approaches such as the mean-field analysis, or the self-consistent method (Wang et al, 2017; Davies, 1978; Taillard et al, 2008; Perdahcioglu and Geijselaers, 2012). As well as these approaches, an asymptotic homogenization method (Iwamoto, 2004) might be a solution by direct coupling with the CPFEM.

17.3 Computational Aspects

The easiest recognizable feature of martensitic transformation is the microstructure that it produces under the effect of martensitic transformation. The morphology with various kinds of complicated geometrical and crystallographic patterns in the material is formed in the microstructure at the length scales which varies from nano to micrometers, and it depends on the grain size. The complicated microstructure is affected by the crystallographic orientation as well as its rotation in the parent and product phases during the plastic deformation (Iwamoto and Tsuta, 2004). Besides that, polycrystal models also play an important role in the numerical analyses since almost actual steels including TRIP steel is in the form of polycrystalline. In the polycrystal materials, there is a number of grains aggregating and the amount of martensite varies among the grains. Thus, the variation of grain size is able to express a great change in the SIMT behavior. In addition, it is necessary to have an appropriate method to describe the structural geometry of grains for the polycrystal materials. Recently, the Voronoi polyhedral is an ideal and useful approach to describe the geometrical shape of grains in the polycrystalline materials since the produced crystals would have easy geometrical descriptions respecting to faces, edges and vertices (Kobayashi and Sugihara, 2002; Kirubel and Lori, 2015). In this section, the constutitve model for monocrystal TRIP steel proposed by Iwamoto and Tsuta (2004) and computational polycrystal model employed by Trinh and Iwamoto (2019) are briefly reviewed.

17.3.1 A Model of Single Crystal Transformation-induced Plasticity Steel Based on Continuum Crystal Plasticity Suggested by Iwamoto and Tsuta (2004)

Let us consider the mono-crystal TRIP steel which undergoes a thermo-elasto-plastic deformation with the martensitic transformation. The total deformation gradient $\mathbf{F}$ can be decomposed multiplicatively into three parts as stretching and rotation of crystal lattice $\mathbf{F}^e$, slip deformation on slip planes $\mathbf{F}^p$ and martensitic transformation on a habit plane $\mathbf{F}^{tr}$ as follows (Levitas, 1998)

$$\mathbf{F} = \mathbf{F}^e \mathbf{F}^{tr} \mathbf{F}^p . \tag{17.1}$$

Similar to the conventional derivation process of CCP (Asaro, 1983) and when the tangent modulus method (Peirce et al, 1983) is applied to enhance the stability of FE computation, the following constitutive equation can be obtained

$$\begin{aligned} \overset{\triangledown}{S}_{ij} &= D^v_{ijkl} d_{kl} - \sum_a R^{(a)}_{ij} \dot{b}^{(a)} - \sum_I R^{tr(I)}_{ij} \dot{b}^{tr(I)} - B^e_{ij} \dot{T}, \\ D^v_{ijkl} &= D^e_{ijkl} - \sum_a R^{(a)}_{ij} C^{(a)}_{kl} - \sum_I R^{tr(I)}_{ij} C^{tr(I)}_{kl}, \\ R^{(a)}_{ij} &= D^e_{ijkl} P^{(a)}_{kl} + \beta^{(a)}_{ij} \text{ and } R^{tr(I)}_{ij} = D^e_{ijkl} Q^{(I)}_{kl} + \beta^{tr(I)}_{ij} . \end{aligned} \tag{17.2}$$

Here, $\overset{\triangledown}{S}_{ij}$ is the Jaumann rate of Kirchhoff stress and d_{ij} is the stretching tensor. $\dot{b}^{(a)}$, $\dot{b}^{tr(I)}$, $C^{(a)}_{ij}$ and $C^{tr(I)}_{ij}$ are derived from a procedure related to the tangent modulus method (Peirce et al, 1983). $P^{(a)}_{ij}$ and $Q^{(I)}_{ij}$ are Schmid tensors for slip and variant systems, respectively. D^e_{ijkl} is the elastic modulus tensor and B^e_{ij} is the tensor related to the thermal expansion. The other variables in Eq. (17.2) can be referred in Iwamoto and Tsuta (2004) and Trinh and Iwamoto (2019).

Here, the transformation strain rate is dependent on the resolved shear stress (Tokuda et al, 1998; Stringfellow et al, 1992). In order to obtain the accurate amount of transformation strain on the active variant system as similar to slip systems, the rate-dependent constitutive equation for transformation strain rate $\dot{\gamma}^{tr(I)}$ is assumed following a power law

$$\dot{\gamma}^{tr(I)} = \dot{a}^{(I)} \frac{\tau^{tr(I)}}{g^{tr(I)}} \left| \frac{\tau^{tr(I)}}{g^{tr(I)}} \right|^{\frac{1}{m}-1} . \tag{17.3}$$

where m is strain rate sensitivity exponent and $\dot{a}^{(I)}$ is the reference strain rate and can be expressed by Dirac's δ function since the martensitic transformation occurs explosively.

$$\dot{a}^{(I)} = \gamma^{tr(I)} \delta \left[G_{(I)} - G_0 \right] \dot{G}_{(I)} \tag{17.4}$$

where $G_{(I)}$ is the transformation driving force on the I th variant and G_0 is the critical driving force for transformation. The Dirac's δ function is derived from the time derivative of Heaviside step function with respect to the transformation condition shown in the above equation. The concept to employ the Heaviside function is following the sharp interface theory. The resistance $g^{tr(I)}$ against the martensitic transformation and its evolution equation for the whole variant systems can be formulated by the cross-variant hardening through the analogy with slip deformation (Iwamoto and Tsuta, 2004). All the details on the model can be found in Iwamoto and Tsuta (2004), and Trinh and Iwamoto (2019).

17.3.2 Computational Models and Conditions for Single and Polycrystal Transformation-induced Plasticity Steel

Figures 17.1 and 17.2 show rectangular blocks made of mono and polycrystalline TRIP steel with two planar slip systems. Two planar slip systems is considered for both mono and polycrystal models in which the initial crystallographic orientation ϕ are given as shown in Fig. 17.1(d). The height L and width W of the block are 1.0

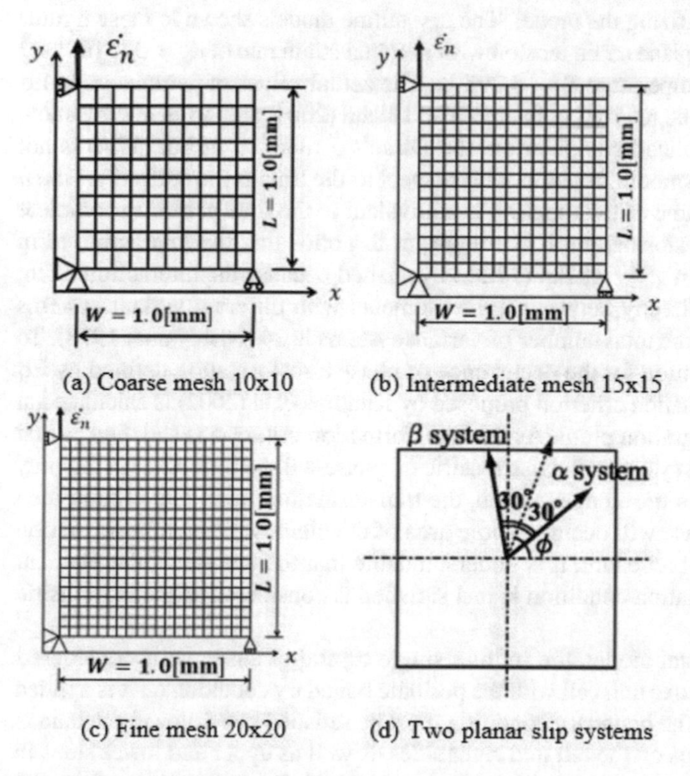

Fig. 17.1 Finite element models for single crystal TRIP steel and two planar slip systems with different finite element discretizations

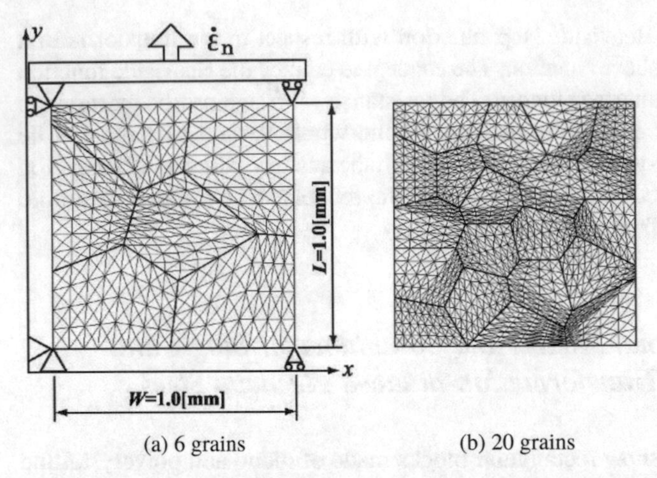

Fig. 17.2 Finite element models for polycrystal TRIP steel with different numbers of grains

mm, respectively. The each quadrilateral indicates a four crossed-triangular plane strain element discritizing the model. The crystalline models shown in these figures are subjected to the plane strain tension with nominal strain rate of $\dot{\varepsilon}_n = 5 \times 10^{-4}$ s^{-1} at environmental temperature T_{env} = 298 K. The actual values of parameters in Eq. (17.2) to (17.4) can be referred in Iwamoto and Tsuta (2004) and Trinh and Iwamoto (2019). For the stablized computation, the Dirac's δ function in Eq. (17.4) is not realistic. Thus, the smooth function with respect to the time is introduced for the δ function. The rise time of the function is equivalent to the time period to complete the martensitic transformation. It is important to notice that the interaction term of variant systems in $g^{tr(I)}$ of Eq. (17.3) is vanished because the interaction effect can be expressed directly between finite elements with different variant systems transformed. Here, the total number of variant systems is 24 (Nishiyama, 1978). To determine the condition for the occurrence of phase transformation defined by Eq. (17.4), the transformation criterion proposed by Kitajima et al (2002) is calculated at each Gaussian integration point. As the transformation criterion is satisfied at first in one of 24 variant systems, the martensitic α' phase will be formed with the only the variant system in the element. Then, the transformation strain rate is computed and martensitic phase will occupy whole area of the element when transformation finishes. From this procedure, it is understandable that the one triangular element when the transformation condition is just satisfied is considered as the martensitic embryo.

In the monocrystal model, the infinite single crystal of austenite is considered. Thus, the representative unit cell with the periodic boundary condition on it is applied (Smit et al, 1998). The boundary condition must be satisfied two following demands. Firstly, the two edges on the left and right sides as well as upper and lower sides in the unit cell must remain the identical shapes during deformation. Secondary, the traction vectors acting on the boundaries must have an opposite sign. According to these demands, the stress and strain will be continuous though the boundaries.

As shown in Fig. 17.1(a) to (c), the unit cell is discretized regularly by 10 x 10, 15 x 15, and 20 x 20 in horizontal and vertical directions with the element. Hereafter, the mesh divisions are called a coarse mesh, intermediate mesh, and finer mesh, respectively. ϕ is constantly set to 60° for all the mesh divisions. In order to realize tensile deformation, the additional boundary conditions at three nodes are imposed as shown in this figure.

In the polycrystal model, each Voronoi polygon is assumed to correspond to the crystal grain in order to express the various crystal grains with different crystal orientation. A set of Voronoi tessellation with the numbers of crystal grains of 6 and 20 is chosen as shown in Fig. 17.2 (a) and (b). The ϕ is randomly given from 0° to 90° in the each grain. The Voronoi polygons are discretized by 32 quadrilaterals as the crossed-triangular plane strain elements. One triangular element is considered as each crystal lattice. Furthermore, the polycrystal TRIP steel with an ideal texture which all the grains have a preferred orientation of $\phi = 60°$ is also simulated in the case of 20 grains. It must be emphasized that there are no concepts of the length scale effect in the current simulation except for the martensitic nuclei.

17.4 Computational Results and Discussions

17.4.1 Effect of Mesh Discritization for Single Crystal Transformation-induced Plasticity Steel

Figure 17.3 describes (a) nominal stress versus nominal strain ε_n and (b) volume fraction of martensitic phase versus ε_n in the case of monocrystal TRIP steel when ϕ is set as 60° in comparison of coarse, intermediate and finer meshes. Though the nominal stress of TRIP steel in all three cases of mesh discretization is almost the

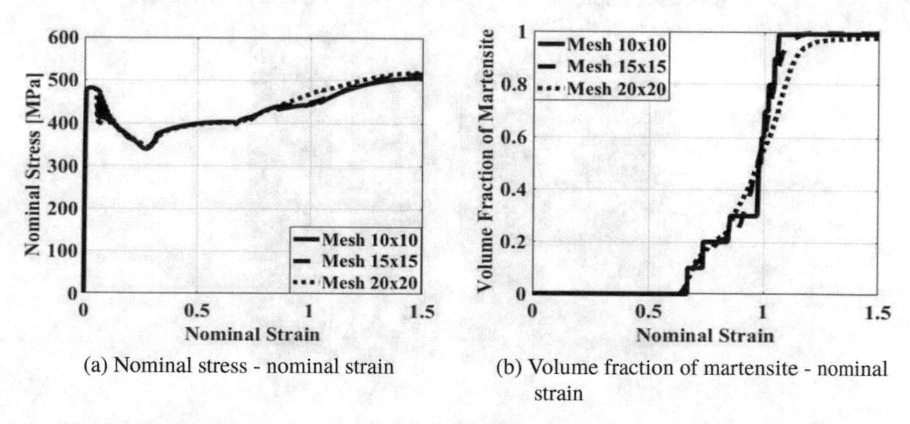

(a) Nominal stress - nominal strain

(b) Volume fraction of martensite - nominal strain

Fig. 17.3 Dependence of mesh refinement on macroscopic behavior of single crystal TRIP steel

same as shown in Fig. 17.3 (a), there is a clear difference in volume fraction of martensite.

In the case of the rough mesh, the discretization behavior in volume fraction of martensite is appeared, while it is increased eventually linearly in the case of intermediate and finer meshes. Thus, the mesh dependence induces a significant effect to the numerical results. In order to eliminate the dependence on mesh size, the use of strain gradient plasticity theory as above mentioned is recently recognized and considered as an effective tool to overcome effects related to the mesh dependency. Here, the present review focuses on the dependence of length scale in SIMT behavior.

Figure 17.4 presents the distribution of the martensitic α' phase as $\varepsilon_n = 0.8 \sim 1.2$ in the comparison among three meshes. In this figure, the regions with white and black color denote γ and α' phases, respectively. As shown in this figure, the band-like structures are observed as the evolution of martensitic phase transition. Importantly, it can be observed that at the same step for nominal strain, the evolutions of α' phase between the rough and the intermediate meshes are different. Even though the stepwise changes are not appeared in the evolution of martensitic volume fraction as

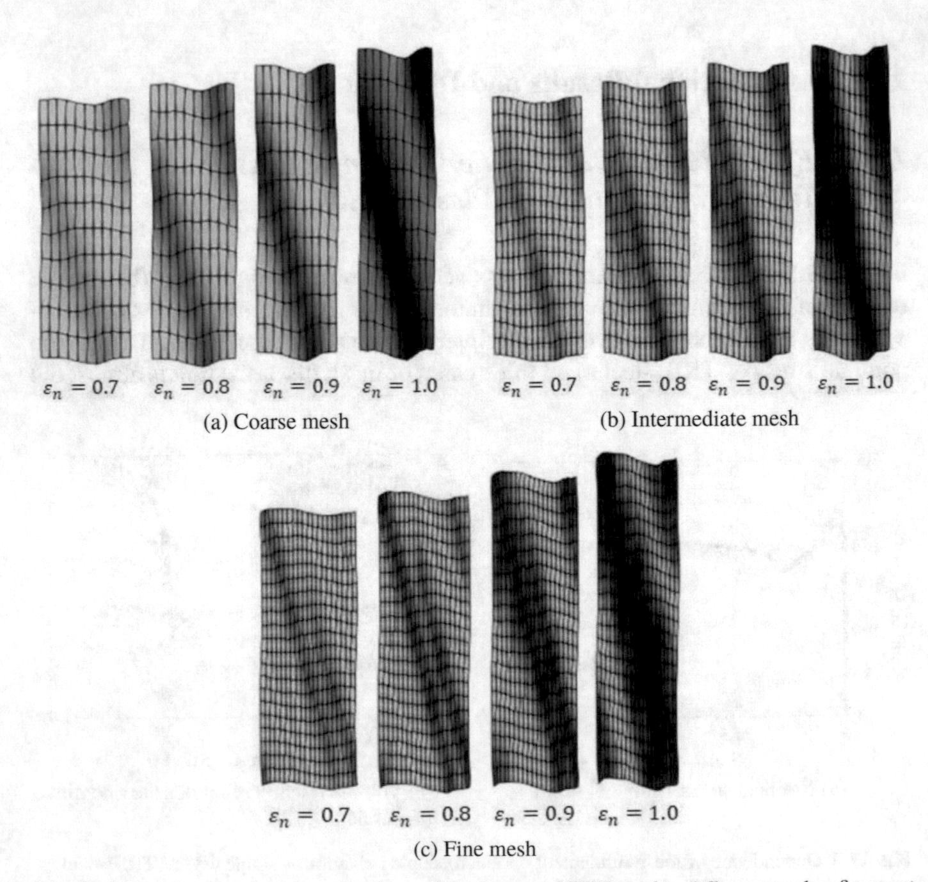

Fig. 17.4 Distribution of martensitic phase of single crystal TRIP steel in different mesh refinement

shown in Fig 17.3 for two cases of intermediate and finer meshes as shown in Fig. 17.4, the distributions of martensitic phases have a clear difference in the band-like structure. As a consequence, the size of martensitic nuclei is influenced.

17.4.2 Polycrystal Transformation-induced Plasticity Steel

In this part, the numerical results of polycrystalline TRIP steel are presented. Figure 17.5 shows (a) nominal stress versus ε_n and (b) volume fraction of martensitic phase versus ε_n in the consideration of 6 and 20 grains. In addition, the result of the material with ideal texture is also included. As shown in this figure, the nominal stress and the volume fraction of martensite at the saturation level in the case of 20 grains is higher than those in the case of 6 coarse grains. It is reasonable since the hardness of austenitic grain is relatively increased as the size of the austenitic grain reduced.

On the other hand, it is obviously that ϕ is able to induce great effects to SIMT and deformation process. When polycrystal material has a preferred orientation, it cannot be estimated that the material with the texture is weak, moderate or strong before the simulation. In this case, the imposed ideal texture provides a weaker material property. Thus, the nominal stress is greatly small. The volume fraction of martensite will be intermediate between 6 and 20 grains with random patterns of ϕ.

Figure 17.6 represents the distribution of martensitic phase from $\varepsilon_n = 0 \sim 1.0$ as a comparison between 6 and 20 grains. As similar to Fig. 17.4, the regions with white and black color denote γ and α' phases, respectively. It is clear that the α' phase in 6 grain starts generating in the center and left upper regions on the side surface while it appears around the center and right bottom regions in the case of 20 grains at the same ε_n. It is noticed that these regions is developed with the promotion of deformation and near necking-like regions for both grains. Finally, not only the

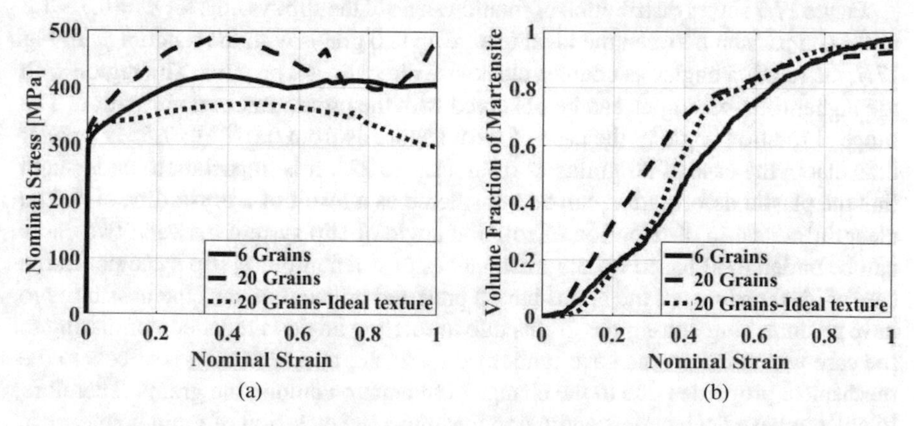

Fig. 17.5 (a) Nominal stress versus nominal strain and (b) the volume fraction of martensite versus nominal strain under various initial crystallographic orientations in parent phase

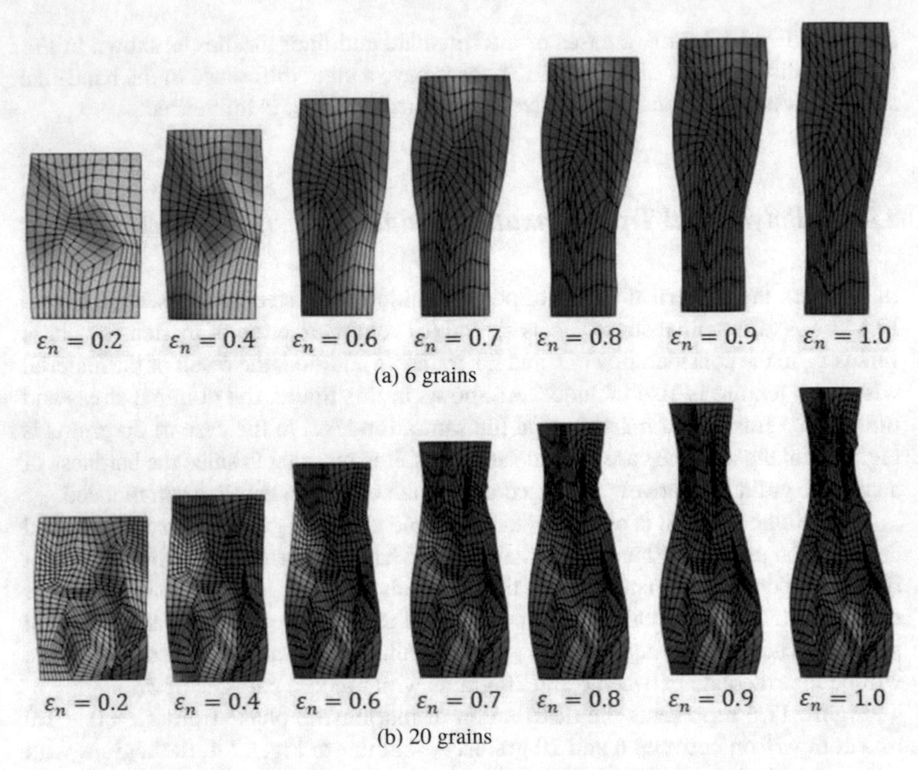

Fig. 17.6 Distribution of martensitic phase of polycrystal TRIP steel in the case of different number of grains

distribution of α' but also the growth of deformation profiles are different compared between two numbers of grains.

Figure 17.7 shows distribution of rotation angle of the slip systems for $\varepsilon_n = 0.7 \sim 1.2$ in the comparison between the ideal texture and 20 grains with the random ϕ. In Fig. 17.7, the rotation angles as counter-clockwise direction are positive. The region with the higher rotation angles can be observed with the promotion of deformation. The range of rotation angle in the case of ideal texture as from 0.078° to 7.5° is smaller than that in the case of 20 grains as from −12° to 22°. It is important to understand that the plastic deformation can be considered as a result of a crystalline slip. The clear difference in distribution of rotation angle of slip system between two cases can be understood based on the mismatches of orientation of slip systems. In the case of ideal texture, all the grains have a preferred orientation, thus the possibility to have mismatching among the grains due to rotation angle is reduced comparing to the case which the grains have random ϕ for each grain. Partially, it affects to the mechanical properties due to the change of interaction among the grains. Therefore, in order to have deep understanding on this topic, the inclusion of grain boundary in grain size effect is necessary to analyze the phenomenon of grain mismatching in polycrystalline materials.

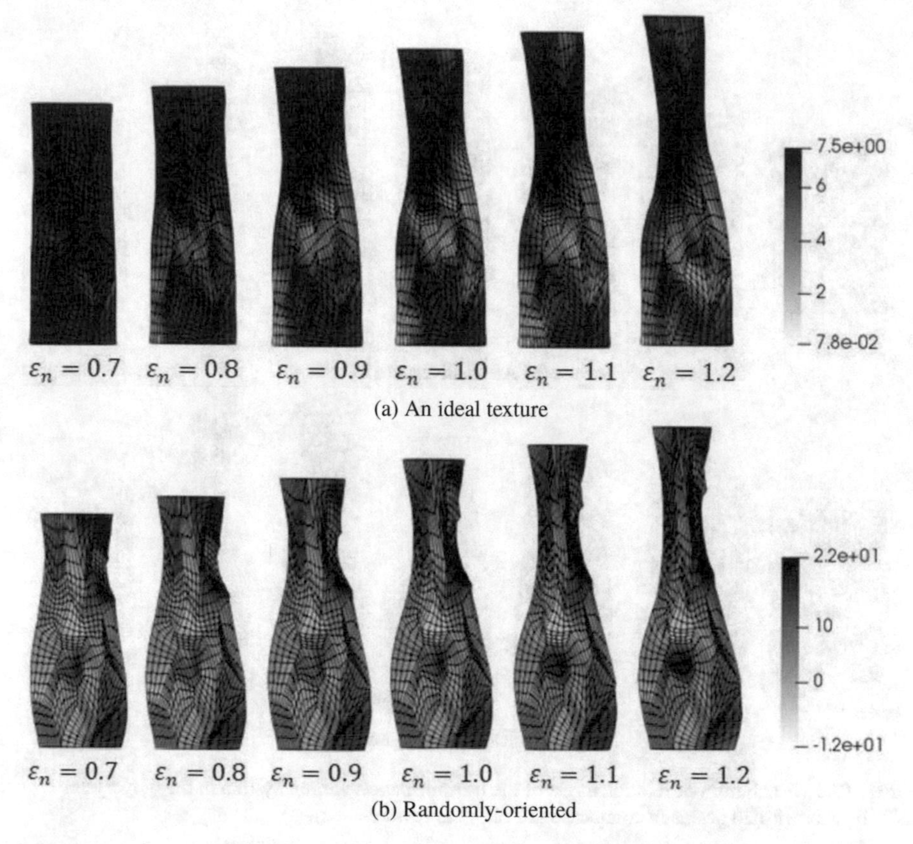

Fig. 17.7 Distribution of rotation angle of the slip system in the polycrystal TRIP steel with 20 grains in comparison of different textures

Next, the distribution of rotation angle of transformation variant in the case of 20-grain polycrystal TRIP steel from $\varepsilon_n = 0.2 \sim 1.2$ as a comparison between the ideal texture and the randomly initial oriented angle ϕ for each grain is shown in Fig. 17.8. As shown in the figure, the counter-clockwise direction for the rotation angle is also positive as similar to Fig. 17.7. The range of rotation angle in the case of the ideal texture as $-51°$ to $69°$ is greater than that in the case of randomly-oriented 20 grains as from $-48°$ to $63°$. In the case of the ideal texture, the rotation angle distributes quite homogeneously even though its deformation is heterogeneous. When the material has a preferred orientation, the nucleation will occur at this preferred site, and it is much easier to consume the available free energy for starting the martensitic transformation. The rotation angle increases with a promotion of deformation. However, some huge distorted areas with very low rotation angle appears suddenly at $\varepsilon_n = 1.2$. In the case of randomly-oriented 20 grains, it is noticed that the region with higher rotation angle corresponds to the region with the α' phase as shown in Fig 17.6 (b). The formation of the α' phase and the transformation strain induce a great effect to the change of

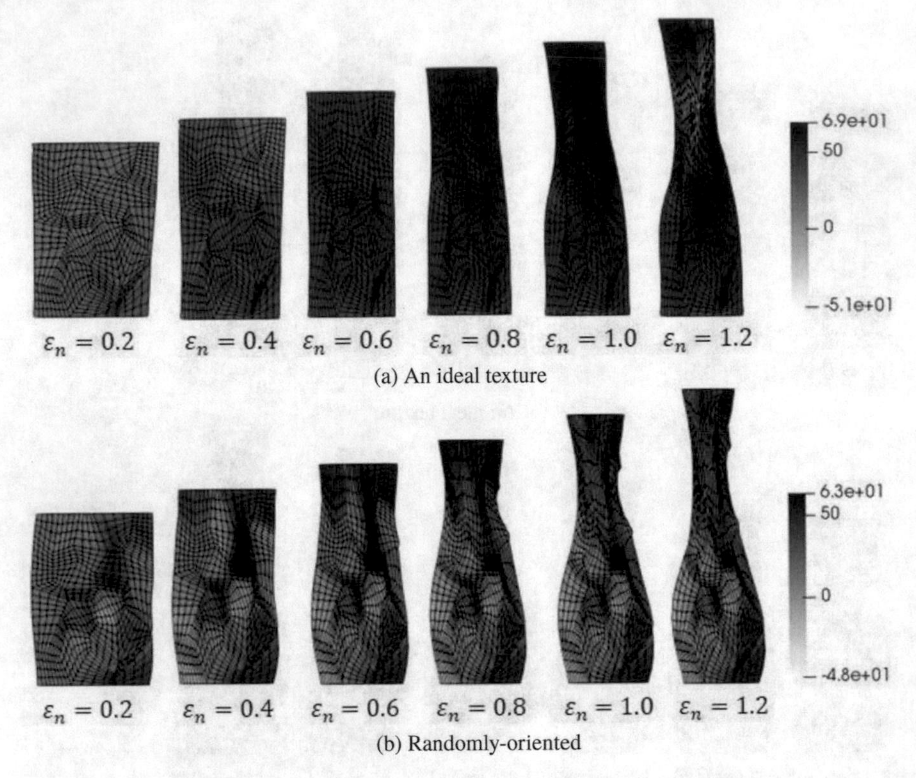

Fig. 17.8 Distribution of rotation angle of the transformation variant system in the polycrystal TRIP steel with 20 grains in comparison of different textures

the distribution of rotation angle. As a consequence, the distribution of rotation angle becomes inhomogeneous and the variant system rotates during deformation process.

Nevertheless, in the analysis of the polycrystal model, there are still limitations on boundary condition and a use of relatively coarse mesh. It is necessary to understand more deeply the size effect at this level of the specific length scale to SIMT behavior. Hence, approaches to describe the hardening during phase transformation such as non-local model, or generalized stresses balance are considered in the future work. Furthermore, to express the infinite medium in the polycrystal model, the unit cell with periodic boundary condition must be employed to fulfill the polycrystal TRIP steel.

Lastly, just one pattern of random distribution on ϕ is imposed for the polycrystal models with different number of grains. It has already been understood that the many trials of simulation with different pattern and statistical processing of the results are necessary to find the true phenomena.

17.5 Summary

This review focused on computational analyses of the martensitic transformation in mono and polycrystal TRIP steel including the appropriate length scales implemented to the crystal plasticity finite element method. The importance of martensitic phase transition in materials and advances in continuum crystal plasticity theory to include the length scale effect at microstructure were reviewed.

In order to discuss the length scale effect, the computational simulation of single and polycrystal TRIP steel based on the CPFEM with the cellular automata and the Voronoi tessellation approaches proposed by Iwamoto and Tsuta (2004) and Trinh and Iwamoto (2019) were performed.

From the numerical results, in the model of single crystal TRIP steel, the mesh affects significantly to the volume fraction of martensite, then the morphology during the phase transition is also different with the promotion of deformation. By a comparison between two numbers of grains which the initial orientations in each grain are set randomly, the strength, the evolution of volume fraction of martensite, and the distribution of martensitic phase are affected greatly.

It is clearly that it is still far to fulfill the study on the strain-induced martensitic transformation. Additionally, the influence of austenitic grain and martensitic nuclei related to the SIMT is needed to examine. Thus, a non-local model could be one of the most suitable approaches since it is able to capture the influence of grain boundaries. Morcover, in order to have the overall observation, it is indispensable to explore thc 2D model into 3D with an appropriate approach to describe the complex morphology of polycrystal materials.

References

Abu Al-Rub RK, Voyiadjis GZ (2006) A physically based gradient plasticity theory. Int J Plast 22:654–684

Aifantis EC, Ngan AHW (2007) Modeling dislocation-grain boundary interactions through gradient plasticity and nanoindentation. Mater Sci Eng A 459:251–261

Ardeljan M, Beyerlein IJ, Knezevic M (2014) A dislocation density based crystal plasticity finite element model: Application to a two-phase polycrystalline HCP/BCC composites. J Mech Phys Solids 66:16–31

Asaro R (1983) Crystal plasticity. Trans ASME J Appl Mech 50:921–934

Barbe F, Decker L, Jeulin D, Cailletaud G (2001) Intergranular and intragranular behavior of polycrystalline aggregates. Part 1: F.E. model. Int J Plast 17:513–536

Bhadeshia HKDH (2002) TRIP-Assisted Steels. ISIJ Int 42:1059–1060

Bowles JS, Mackenzie JK (1954) The crystallography of martensite transformations I. Acta Metal 2:129–137

Carola CC, Jilt S, Maria JS (2019) The role of the austenite grain size in the martensitic transformation in low carbon steels. Mater Des 167:107,625

Cheong K, Busso E, Arsenlis A (2005) A study of microstructural length scale effects on the behaviour of FCC polycrystals using strain gradient concepts. Int J Plast 21:1797–1814

Cherkaoui M, Berveiller M, Saber H (1998) Micromechanical modeling of martensitic transformation induced plasticity (TRIP) in austenitic single crystals. Int J Plast 14:597–626

Davies RG (1978) Influence of martensite composition and content on properties of dual phase steels. Metall Trans A 9:671–679
Diani JM, Parks D (1998) Effect of strain state on the kinetics of strain-induced martensite in steels. J Mech Phys Solids 46:1613–1635
Diard O, Leclercq S, Rousselier G, Cailletaud G (2005) Evaluation of finite element based analysis of 3D multicrystalline aggregates plasticity: application to crystal plasticity model identification and the study of stress and strain fields near grain boundaries. Int J Plast 21:691–722
Dunstan DJ, Bushby AJ (2014) Grain size dependence of the strength of metals: the Hall-Petch effect does not scale as the inverse square root of grain size. Int J Plast 53:56–65
Ehrler B, Hou XD, P'Ng KMY, Walker CJ, Bushby A, Dunstan DJ (2008) Grain size and sample size interact to determine strength in a soft metal. Phil Mag 88:3043–3050
Eisenmann C, Gasser U, Keim P, Maret G, von Grünberg HH (2005) Pair interaction of dislocations in two-dimensional crystals. Phys Rev Let 95:185,502
Faghihi D, Voyiadjis GZ (2012) Determination of nanoindentation size effects and variable material intrinsic length scale for body-centered cubic metals. Mech Mater 44:189–211
Fischer FD, Reisner G, Werner E, Tanaka K, Cailletaud G, Antretter T (2000) A new view on transformation induced plasticity (TRIP). Int J Plast 16:723–748
Fleck NA, Hutchinson JW (1993) A phenomenological theory for strain gradient effects in plasticity. J Mech Phys Solids 41:1825–1857
Forest S, Aifantis E (2010) Some links between recent gradient thermo-elasto-plasticity theories and the thermos-mechanics of generalized continua. Int J Solids Struct 47:3367–3376
Gao H, Huang Y (2003) Geometrically necessary dislocation and size-dependent plasticity. Scr Mater 48:113–118
Ghosh G, Olson GB (1994) Kinetics of F.C.C. → B.C.C. heterogeneous martensitic nucleation-I. The critical driving force for athermal nucleation. Acta Metall Mater 42:3361–3370
Graff S, Brocks W, Steglich D (2007) Yielding of magnesium: From single crystal to polycrystalline aggregates. Int J Plast 23:1957–1978
Gupta S, Ma A, Hartmaier A (2015) Investigating the influence of crystal orientation on bending size effect of single crystal beams. Comp Mater Sci 101:201–210
Gurtin ME (2002) A gradient theory of single-crystal viscoplasticity that accounts for geometrically necessary dislocations. J Mech Phys Solids 50:5–32
Gurtin ME (2008) A finite-deformation gradient theory of single-crystal plasticity with free energy dependent on densities of geometrically necessary dislocations. Int J Plast 24:702–725
Hanamura T, Torizuka S, Tamura S, Enokida S, Takechi H (2013) Effect of austenite grain size on transformation behavior microstructure and mechanical properties of 0.1C - 5Mn Martensitic steel. ISIJ Int 53:2218–2225
Hueper T, Endo S, Ishikawa N, Osawa K (1999) Effect of volume fraction of constituent phases on the stress-strain relationship of dual phase steels. ISIJ Int 39:288–294
Hunter A, Beyerlein IJ (2013) Unprecedented grain size effect on stacking fault width. APL Mater 1:4820,427
Iwamoto T (2004) Multiscale computational simulation of deformation behavior of TRIP steel with growth of martensitic particles in unit cell by asymptotic homogenization method. Int J Plast 20:841–869
Iwamoto T, Pham HT (2015) Review on spatio-temporal multiscale phenomena in TRIP steels and enhancement of its energy absorption. In: Altenbach H, Okumura D, Matsuda T (eds) From Creep Damage Mechanics to Homogenization Methods, Adv Struct Mater., vol 64, Springer-Verlag, Switzerland, pp 143–161
Iwamoto T, Tsuta T (2000) Computational simulation of the dependence of the austenitic grain size on the deformation behavior of TRIP steels. Int J Plast 16:791–804
Iwamoto T, Tsuta T (2004) Finite element simulation of martensitic transformation in single-crystal TRIP steel based on crystal plasticity with cellular automata approach. Key Eng Mater 274-276:679–684

Iwamoto T, Tsuta T, Tomita Y (1998) Investigation on deformation mode dependence of strain-induced martensitic transformation in TRIP steels and modeling of transformation kinetics. Int J Mech Sci 40:173–182

Kirubel T, Lori GB (2015) Tessellation growth models for polycrystalline microstructures. Comp Mater Sci 102:57–67

Kitajima Y, Sato N, Tanaka K, Nagaki S (2002) Crystal-based simulations in transformation thermomechanics of shape memory alloys. Int J Plast 18:1527–1559

Kobayashi K, Sugihara K (2002) Crystal Voronoi diagram and its applications. Future Gene Comp Sys 18:681–692

Kratochvil J (2012) Crystal plasticity treated as a quasi-static material flow through adjustable crystal lattice. Acta Phys Polon A 122:482–484

Lai Q, Brassart L, Bouaziz O, Goune M, Verdier M, Parry G, Perlade A, Brechet Y, Pardoen T (2016) Influence of martensite volume fraction and hardness on the plastic behavior of dual-phase steels: Experiments and micromechanical modeling. Int J Plast 80:187–203

Levitas VI (1998) Thermomechanical theory of martensitic phase transformations in inelastic materials. Int J Solids Struct 35:889–940

Levitas VI (2014) Phase field approach to martensitic phase transformations with large strains and interface stresses. J Mech Phys Solids 70:154–189

Levitas VI, Javanbakht M (2015) Interaction between phase transformations and dislocations at the nanoscale. Part1. General phase field approach. J Mech Phys Solids 82:287–319

Li S, Kalidindi SR, Beyerlein IJ (2005) A crystal plasticity finite element analysis of texture evolution in equal channel angular extrusion. Mater Sci Eng A 410-411:207–212

Littlewood PD, Britton TB, Wilkinson AJ (2011) Geometrically necessary dislocations density distributions in Ti-6Al-4V deformed in tension. Acta Mater 59:6489–6500

Liu D, Dunstan DJ (2017) Material length scale of strain gradient plasticity: a physical interpretation. Int J Plast 98:156–174

Militzer M (2011) Phase field modeling of microstructure evolution in steels. Cur Opinion Solid State Mater Sci 15:106–115

Nishiyama Z (1978) Martensitic Transformation. Academic Press, New York

Ohno N, Okumura D (2007) Higher-order stress and grain size effects due to self-energy of geometrically necessary dislocations. J Mech Phys Solids 55:1879–1898

Peirce D, Asaro R, Needleman A (1983) Material rate dependence and localized deformation in crystalline solids. Acta Metal 31:1951–1976

Perdahcioglu ES, Geijselaers HJM (2012) A macroscopic model to simulate the mechanically induced martensitic transformation in metastable austenitic stainless steels. Acta Mater 60:4409–4419

Richeton T, Wagner F, Chen C, Toth LS (2018) Combined effects of texture and grain size distribution on the tensile behavior of α-titanium. Materials 11:1088

Roters F, Eisenlohr P, Hantcherli L, Tjahjanto D, Bieler T, Raabe D (2010) Overview of constitutive laws, kinematics, homogenization and multiscale methods in crystal plasticity finite-element modeling: Theory, experiments, applications. Acta Mater 58:1152–1211

Schmitt R, Kuhn C, Mueller R (2017) On a phase field approach for martensitic transformations in a crystal plastic material at a loaded surface. Cont Mech Thermodyn 29:957–968

Sidhoum Z, Ferhoum R, Almansba M, Bensaada R, Habak M, Aberkane M (2018) Experimental and numerical study of the mechanical behavior and kinetics of the martensitic transformation in 304L TRIP steel: applied to folding. Int J Adv Manuf Tech 97:2757–2765

Smit RJM, Brekelmans WAM, Meijer HEH (1998) Prediction of the mechanical behavior of nonlinear heterogeneous systems by multi-level finite element modeling. Computer Methods in Applied Mechanics and Engineering 155(1):181 – 192

Staroselsky A, Anand L (2003) A constitutive model for hcp materials deforming by slip and twinning: Application to magnesium alloy AZ31B. Int J Plast 19:1843–1864

Stringfellow RG, Parks DM, Olson GB (1992) A constitutive model for transformation plasticity accompanying strain-induced martensitic transformations in metastable austenitic steels. Acta Metall 40:1703–1716

Suezawa M, Cook HE (1980) On the nucleation of martensite. Acta Metall 28:423–432

Taejoon P, Louis GH, Xiaohua H, Fadi A, Michael RF, Hyunki K, Rasoul E, Farhang P (2019) Crystal plasticity modeling of 3rd generation multi-phase AHSS with martensitic transformation. Int J Plast p in press

Taillard K, Chirani SA, Calloch S, Lexcellent C (2008) Equivalent transformation strain and its relation with martensite volume fraction for isotropic and anisotropic shape memory alloys. Mech Mater 40:151–170

Tamura I (1982) Deformation-induced martensitic transformation and transformation-induced plasticity in steels. Met Sci 16:245–253

Tjahjanto DD, Turteltaub S, Suiker ASJ (2008) Crystallographically-based model for transformation-induced plasticity in multiphase carbon steels. Cont Mech Thermodyn 19:399–422

Tokuda M, Ye M, Takakura M, Sittner P (1998) Calculation of mechanical behaviors of shape memory alloy under multi-axial loading conditions. Int J Mech Sci 40:227–235

Trinh TD, Iwamoto T (2019) A computational simulation of martensitic transformation in polycrystal TRIP steel by crystal plasticity FEM with Voronoi tessellation. Key Eng Mater 794:71–77

Tuma K, Stupkiewicz S (2016) Phase-field study of size-dependent morphology of austenite-twinned martensite interface in CuAlNi. Int J Solids Struct 97-98:89–100

Turteltaub S, Suiker ASJ (2006) Grain size effects in multiphase steels assisted by transformation-induced plasticity. Int J Solids Struct 43:7322–7336

Voyiadjis GZ, Abu Al-Rub RK (2005) Gradient plasticity theory with a variable length scale parameter. Int J Solids Struct 42:3998–4029

Wang H, Capolungo L, Clausen B, Tome C (2017) A crystal plasticity model based on transition state theory. Int J Plast 93:251–268

Yeddu HK (2018) Phase-field modeling of austenite grain size effect on martensitic transformation in stainless steels. Comp Mater Sci 154:75–83

Yu HY (1997) A new model for the volume fraction of martensitic transformations. Metall Mater Trans A 28:2499–2506

Zackay VF, Parker ER, Fahr D, Busch RA (1967) The enhancement of ductility in high strength steels. ASTM Trans Quart 60:252–259

Chapter 18
On Micropolar Theory with Inertia Production

Elena Vilchevskaya

Abstract This paper presents a new aspect in generalized continuum theory, namely micropolar media showing structural change. Initially the necessary theoretical framework for a micropolar continuum is presented. To this end the standard macroscopic equations for mass and linear and angular momentum are complemented by a recently proposed balance equation for the moment of inertia tensor containing a production term. The new balance and, in particular, the production is interpreted mesoscopically by taking the inner structure of micropolar media into account. Various of examples for the term are presented.

Keywords: Micropolar continua · Structural transformations · Inertia production

18.1 Introduction

Mechanics of Micropolar Continua is a theory with independent force and moment (couple) actions. That theory incorporates local rotations of points as well as translations assumed in classical elasticity. The idea of a couple stress can be traced to Voigt (1887) where the effects of couple stresses were investigated and a generalization of the classical theory of symmetric elasticity to a non-symmetric theory was made. The approach was further elaborated by the Cosserat brothers (Cosserat and Cosserat, 1909) who suggested to consider the rotational degrees of freedom of material particles as independent variables and so every particle contains six degrees of freedom: three displacements assigned to the macro-element, plus three rotations

---

Elena Vilchevskaya
Peter the Great St. Petersburg Polytechnic University, Department of Theoretical Mechanics, Institute of Applied Mathematics and Mechanics, Politechnicheskaya 29, 195251 St. Petersburg & Laboratory of Mathematical Methods in Mechanics of Materials, Institute for Problems in Mechanical Engineering of Russian Academy of Sciences, Bolshoy pr. V.O. 61, 199178 St. Petersburg, Russia,
e-mail: vilchevska@gmail.com

H. Altenbach and A. Öchsner (eds.), *State of the Art and Future Trends in Material Modeling*, Advanced Structured Materials 100,
https://doi.org/10.1007/978-3-030-30355-6_18

referring to the micro-structure. Thus, force and moment actions in the continuum were introduced independently, as it was done by Euler, and the angular momentum equation was explicitly used instead of being reduced to a symmetry statement of the stress tensor. This material model referred to as the Cosserat continuum provides the mathematical characterization of solid bodies with microstructure in which couple stresses, body couples, and local motions are included. These peculiarities of the Cosserat continuum model give a possibility to describe more complex media, for example, micro-inhomogeneous materials, particle assemblies, viscous fluids, fiber suspensions, liquid crystals, etc. Later Eringen and associates started to use the term micropolar to describe Cosserat media.

Although Hellinger (1914) paid tribute to the potential of the theory and obtained the general constitutive relations for stress and couple-stress, the ideas of the Cosserat continuum were not widely accepted and it was not until the 1960's that fully developed microstructure theories evolved. In fact, it was only after a paper by Ericksen and Truesdell (1957) that the ideas of the Cosserat brothers were revived. In this paper, the purely kinematical description of Cosserat continua emphasizing the one- and two-dimensional cases of rods and shells was developed. The original concept was modified in two ways. Firstly, the concept of directors defining the orientation of the material particle was introduced and secondly, these directors were also allowed to deform to describe a deformation of the material particle at the microscale. These are micromorphic continua in Eringen's classification (Eringen, 1999, 2001), which have nine degrees of freedom (three for microrotation and six for microdeformation). A particular case is that of continua with microstretch (Eringen, 1969), where the directors are orthogonal, but permit isotropic expansion or contraction in addition to rotation. This means, particles of microstretch continua have four additional degrees of freedom more than classical continua.

Later Günther (1958) developed a linear theory of the Cosserat continuum with an application to the continuum theory of dislocations, Grioli (1960) elaborated a theory of elasticity with a non-symmetric stress tensor, and Ericksen (1960c,a,b, 1961) developed a theory of anisotropic fluids and liquid crystal assuming that a fluid is a three-dimensional point continuum with one director at each point. Since the mid of 20th century a lot of publications devoted to the Cosserat continuum have appeared. Not even trying to give detailed information about various contributions we just refer to Mindlin and Tiersten (1962), Aero and Kuvshinskii (1961), Kröner (1963), Palmov (1964), Toupin (1962, 1964) as pioneers in the field. Later the micropolar elasticity was considered for example in Maugin (1998); Neff and Jeong (2009); Jeong and Neff (2010); Dyszłewicz (2004); Pietraszkiewicz and Eremeyev (2009); Ramezani and Naghdabadi (2007). There are also many publications on the plastic and visco-elastic Cosserat continuum, among them Lippmann (1969); de Borst (1993); Steinmann (1994); Forest and Sievert (2003, 2006); Grammenoudis and Tsakmakis (2005); Neff (2006). Variational problems in the micropolar continuum are investigated in Steinmann and Stein (1997); Nistor (2002). It is also worth mentioning recent collections (Maugin and Metrikine, 2010; Altenbach et al, 2011; Altenbach and Eremeyev, 2012; Sansour and Skatulla, 2012; Altenbach and Forest, 2016) and references therein where modern views on the micropolar media are presented.

The essential developments in the field of micropolar theory were made by Eringen and Suhubi (1964a,b); Eringen (1999, 2001); Eringen and Kafadar (1976); Eringen (1997). Unlike Ericksen and Truesdell (1957) and other early contributions, where the orientation of the material particle was defined by directors, they considered a field of orthogonal transformations (rotations) and not the directors themselves. In analogy to rigid body dynamics Eringen extended the Cosserat theory to include body microinertia effects and used the microinertia tensor $\boldsymbol{J}$ as the orientational descriptor. A truly new notion in his approach is an establishment of existence of a conservation law of micro-inertia. It is based on the concept of an indestructible "material particle" (polar particle) that is phenomenologically equivalent to a rigid body, see for example Eringen (1997); Truesdell and Toupin (1960); Mindlin (1964); Eringen and Kafadar (1976), where it is supposed that the inertia tensor changes only due to rotation of the material particle as a rigid body. So the inertia tensor in the current configuration can be written as follows:

$$\boldsymbol{J} = \boldsymbol{Q}\cdot\boldsymbol{J}_0\cdot\boldsymbol{Q}^T, \tag{18.1}$$

where the inertia tensor in the reference configuration, $\boldsymbol{J}_0$, is a priory known constant tensor, $\boldsymbol{Q}$ is the tensor of microrotation.

Equation (18.1) can be rewritten in differential form:

$$\frac{\delta \boldsymbol{J}}{\delta t} = \boldsymbol{\omega}\times\boldsymbol{J} - \boldsymbol{J}\times\boldsymbol{\omega}, \tag{18.2}$$

where the Poisson equation

$$\frac{\delta \boldsymbol{Q}}{\delta t} = \boldsymbol{\omega}\times\boldsymbol{Q} \tag{18.3}$$

is taken into account. Here $\boldsymbol{\omega}$ is the angular velocity. We denote by

$$\frac{\delta(\cdot)}{\delta t} = \frac{d(\cdot)}{dt} + (\mathbf{v}-\mathbf{w})\cdot\boldsymbol{\nabla}(\cdot) \tag{18.4}$$

the substantial derivative of a field quantity that characterizes the rate of change a property of the material point that was in the observation point at the certain moment of time, $\frac{d(\cdot)}{dt}$ is the total derivative that determines the rate of change of property in an observation point, $\mathbf{v}$ is the velocity of the material point and $\mathbf{w}$ is the mapping velocity of the observational point (see Ivanova et al (2016)).

Even if a micromorphic structure is considered, following Eringen (1976); Eringen and Kafadar (1976); Eringen (1999), many papers use the balance law for the conservation of inertia (e.g., see Oevel and Schröter, 1981; Chen, 2007). A different approach was suggested in Dłuzewski (1993), where it was assumed that the inertia of polar particles may change as the continuum deforms. Furthermore, in order to take the interaction of suspensions with viscous fluids surrounding the suspensions into account, Eringen (1984, 1985, 1991) proposed a modified balance law for microinertia:

$$\frac{\delta \boldsymbol{J}}{\delta t} = \boldsymbol{\omega}\times\boldsymbol{J} - \boldsymbol{J}\times\boldsymbol{\omega} - \boldsymbol{F}, \tag{18.5}$$

where the additional term $\boldsymbol{F}$ describes changes of the microinertia of rigid suspensions due to the fluid sticking to the suspensions.

This idea was further elaborated by Ivanova and Vilchevskaya (2016) who clearly stated that the tensor of inertia should be treated as an independent field. They considered the micropolar theory based on the spatial description. Within the spatial description, it is customarily to refer thermodynamic state quantities to an elementary volume V, fixed in space, containing an ensemble of micro-particles. In their approach, the tensor of inertia associated with the elementary volume was obtained as a result of averaging of the inertia tensors of micro-particles that constitute V. Because the elementary volume is an open system, its inertia tensor can change due to the inertia flux as micro-particles travel across the bounding surface. Moreover, to take into account internal structural transformations, such as combination or fragmentation of micro-particles, chemical reactions, or changes of anisotropy of the material, the authors in Ivanova and Vilchevskaya (2016) assumed that the inertia tensor in the reference configuration is no longer a constant tensor but an additional independent variable characterizing structural transformations in the media. As a result, they proposed a governing equation for the inertia tensor, which in contrast to former theories contains a production term. On the macroscopic continuum level, the production term must be considered as a new constitutive quantity for which an additional constitutive equation has to be formulated. The form of the constitutive equation depends on the problem under consideration and can be a function of many physical quantities, such as temperature, pressure, flow rate, etc.

For a better understanding of these new concepts, the authors in Ivanova and Vilchevskaya (2016) also presented a mesoscopic theory. The main idea was to connect information on a mesoscale by taking the intrinsic microstructure within the elementary volume into account with the balances of micropolar continua on the continuum level in combination with suitable constitutive equations. A similar approach for the case of material description was presented in the series of papers by Stojanović et al (1964) and Rivlin (1968), where the discrete structure of macro-particles constituting the medium was taken into account. It was assumed that each macro-particle consists of a number of micro-particles and characterizes by a position vector to the center of gravity of micro-particles and a number of directors. Later a transition from the dynamics of single particles to micropolar continuum was done by many researchers, for example, the homogenization approaches in Ehlers et al (2003) was based on the volume averages and in Mandadapu et al (2018) it was derived by means of the Irving-Kirkwood procedure. Various homogenization procedures also were used for the determination of the micropoloar moduli, see for example, van der Sluis et al (1999); Larsson and Diebels (2007); Larsson and Zhang (2007). However, the production term in the kinetic equation for the inertia tensor has never been considered from this point of view, to the best of our knowledge.

In this paper different examples of the production term introduction will be discussed and a potential of this approach for modeling materials with higher inner degrees of freedom by various example problems will be illustrated. It will be shown that the new term in the balance of inertia allows to model additional features of materials, namely processes inherent of considerable structural changes.

In what follows, we firstly consider the theoretical aspects of micropolar theory with the inertia production from the mesoscopic point of view, which results in an answer to the question of how to determine the inertial and kinematic characteristics of the polar particle within the spatial description. Then, because the balance for the inertia tensor field is extended by the production term, we also discuss possible forms of the production term on the continuum level in relation with properties of micro-particles which are located in the elementary volume.

18.2 Outline of the Theory

Within the spatial description, it is customary to refer thermodynamic state quantities to an elementary volume, fixed in space. If the length scale difference between an elementary or micro-particle (microscale) and the whole medium under consideration (macroscale) is sufficiently large (e.g., a sand grain in a sand heap), then the elementary volume containing sufficiently many micro-particles can be introduced in the sense of a representative volume element (RVE). It means that within that approach a continuum is understood as a manifold of RVEs (Fig. 18.1). The RVE, by turn, is constructed as a manifold of micro-particles and links the micro- and mesoscales. Note that the presence of a very large number of micro-particles in the RVE is required since establishing a continuous field theory would not be possible otherwise, and fluctuations would become dominant.

Let us consider a volume element containing $N(\mathbf{r}, t)$ micro-particles. The position vector $\mathbf{r}$ corresponds to the geometrical center of the volume and is independent of the motion of the medium. Generally the volume may be considered as moving but we will suppose that it is fixed in space. It means that the position vector is

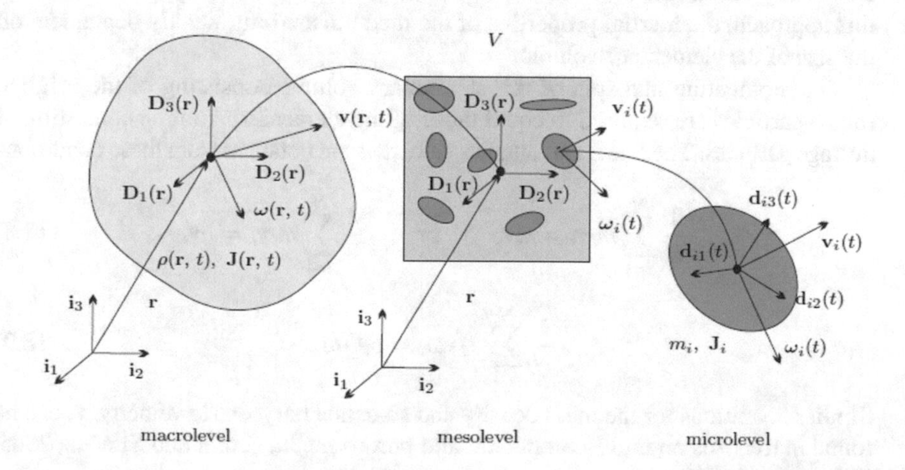

Fig. 18.1 Continuum as a manifold of elementary volumes

independent on time, t, and the velocity of the observational point in (18.4) equals to zero. Note that the volume is an open system since, as the medium moves, different micro-particles each having their own mass, inertia tensor, translational and angular velocities pass through the volume.

Since at different moments the volume consists of different micro-particles each having their own mass and inertia tensor, one has to introduce corresponding fields at the macrolevel as effective characteristics. The micro-particles within the RVE are assumed to be replaceable by an ensemble of identical average particles each having an average mass and an average tensor of inertia.

$$m(\mathbf{r}, t) = \frac{1}{N}\sum_{i=1}^{N} m_i, \qquad \hat{\boldsymbol{J}}(\mathbf{r}, t) = \frac{1}{N}\sum_{i=1}^{N} \boldsymbol{J}_i. \tag{18.6}$$

The second equation calls for a short explanation. The field $\hat{\boldsymbol{J}}(\mathbf{r}, t)$ characterizes the size, shape and orientation of the average particle rather than the mass distribution over the RVE. In fact, it is nothing more than an effective characteristic of rotational inertia that should not be associated with a real material particle. Note that if the micro-particles are randomly oriented within the RVE, then, due to symmetry consideration, the averaged tensor $\hat{\boldsymbol{J}}(\mathbf{r}, t)$ must be a spherical tensor.

If $n(\mathbf{r}, t) = N/V$ denotes the number of particles per unit volume, then the mass density, ρ, and the volumetric density of the moment of inertia are expressed as:

$$\rho = nm, \qquad \rho\boldsymbol{J} = n\hat{\boldsymbol{J}}, \tag{18.7}$$

where $\boldsymbol{J} = \hat{\boldsymbol{J}}/m$ is the average geometrical moment of inertia of a single particle. However, from the continuum point of view, $\boldsymbol{J}$ stands for the specific density of the moment of inertia of the elementary volume and $\rho\boldsymbol{J}$ refers to the volumetric density of the inertia tensor of the elementary volume. Thus, the inertial characteristics of the elementary volume are assumed to coincide with those of the average particle. Within this approach, the inertial properties of the medium are only weakly dependent on the size of the elementary volume.

The momentum and spin of the elementary volume consisting of the original micro-particles are required to equal those of the elementary volume consisting of average particles. The linear and angular velocities are obtained from these conditions

$$\frac{1}{N}\sum_{i=1}^{N} m_i\mathbf{v}_i = m\mathbf{v}, \qquad \text{or} \qquad \frac{1}{V}\sum_{i=1}^{N} m_i\mathbf{v}_i = \rho\mathbf{v}, \tag{18.8}$$

$$\frac{1}{N}\sum_{i=1}^{N} \boldsymbol{J}_i\cdot\boldsymbol{\omega}_i = \rho\boldsymbol{J}\cdot\boldsymbol{\omega}. \tag{18.9}$$

Similar definitions for the mass density and so-called barycentric velocity, $\mathbf{v}$, can be found in treatises on multi-component and porous media (Loret and Simões, 2005; Wilmanski, 2008).

It should be noted that with respect to the translational degrees of freedom, the spatial description only considers the current configuration. Nevertheless, the concept of a reference configuration for the rotational degrees of freedom should be introduced. To describe the average particle rotation, we choose some fixed state of the medium that may be taken at $t = 0$ or another fixed instant and call this state the reference configuration. In order to determine the orientations of particles, reference directors $\boldsymbol{D}_k(\mathbf{r})$, $(k = 1, 2, 3)$ must be locally introduced at each point of the space (Fig. 18.1a). These directors may coincide with the base vectors of the reference coordinate system or can be chosen independently and, say, coincide with the primary axes of $\boldsymbol{J}(\mathbf{r}, 0)$.

Note also that within the spatial description the translational and angular velocities are the primary quantities. The displacement, $\mathbf{u}$, and the microrotation tensor, $\boldsymbol{Q}$, have to be found as solutions from the corresponding differential equations:

$$\mathbf{v} = \frac{\delta \mathbf{u}}{\delta t}, \quad \frac{\delta \boldsymbol{Q}}{\delta t} = \boldsymbol{\omega} \times \boldsymbol{Q}, \tag{18.10}$$

provided $\mathbf{v}$ and $\boldsymbol{\omega}$ are known. Furthermore, the microrotational tensor is different from the rotation tensor of the elementary volume as a rigid body as well as from the rotation tensor obtained by averaging over all micro-particles found in a given volume element at a given time. In fact, it describes the change of directors from the reference to the current position. To this end, it suffices to postulate that the tensors of rotation of all macroparticles in the reference configuration are identity tensors $\boldsymbol{Q}(\mathbf{r}, 0) = \boldsymbol{I}$.

Now we assume that the tensor of inertia in spatial description has a representation similar to (18.1):

$$\boldsymbol{J}(\mathbf{r}, t) = \boldsymbol{Q}(\mathbf{r}, t) \cdot \boldsymbol{J}_0(\mathbf{r}, t) \cdot \boldsymbol{Q}^T(\mathbf{r}, t), \quad \boldsymbol{J}_0(\mathbf{r}, 0) = \tilde{\mathbf{J}}_0(\mathbf{r}). \tag{18.11}$$

The key point within this approach is that the inertia tensor in the reference configuration is no longer a known characteristic of the medium. Indeed, let us suppose that $\boldsymbol{\omega}(\mathbf{r}, t) = \mathbf{0}$, then $\boldsymbol{Q}(\mathbf{r}, t) = \boldsymbol{I}$ and Eqns. (18.6), (18.7) determine the reference inertia tensor $\boldsymbol{J}_0(\mathbf{r}, t)$. Being the averaged characteristic of the micro-particles within the elementary volume the inertia tensor in the reference configuration may change due to different reasons.

- In the case of an inhomogeneous distribution of micro-particles within the medium, the tensor of inertia associated with the elementary volume changes in a certain way as micro-particles travel across its boundary surface S. Mathematically, the balance of $\rho \boldsymbol{J}$ can be expressed as

$$\frac{d}{dt} \int_V \rho \mathbf{J}_0 \, dV = - \int_S (\mathbf{n} \cdot \mathbf{v}) \rho \mathbf{J}_0 \, dS, \tag{18.12}$$

  where $\mathbf{n}$ is the outward unit normal to S.
  Equation (18.12), after application of Gauss's theorem and pulling the differentiation under the integral, leads to the following local statement for the balance of inertia:

$$\frac{\delta \boldsymbol{J}_0}{\delta t} = \boldsymbol{0}, \quad \Longleftrightarrow \quad \frac{\partial \boldsymbol{J}_0}{\partial t} + \mathbf{v} \cdot \nabla \boldsymbol{J}_0 = \boldsymbol{0}. \tag{18.13}$$

Here the local conservation of mass is taken into the account. The solution of equation (18.13) determines the inertia tensor change due to inertia flux, however, there is no inertia production.

- The size and shape of the micro-particles within the elementary volume may change due to phase transitions and chemical reactions or due to fragmentation or combination of the micro-particles. Thus, the size and shape of the average particle also change and leads to a change of $\boldsymbol{J}_0$.
- Nonspherical micro-particles may have a tendency to align with an external applied field or conversely to realign due to thermal motion. This describes a transition from the isotropic state to nonisotropic one or vice versa. It will reflect a change in the average tensor of inertia. Note that if the micro-particles within the elementary volume remain the same and only change their orientation in space, from (18.6), (18.7) follows that the spherical part of the inertia tensor is constant and the changes of anisotropy of the material are characterized only by the deviatoric part of $\boldsymbol{J}_0$.

In view of the above remarks, we may conclude that the inertia tensor in the reference configuration should be treated as a variable rather then a parameter. As a result the reference inertia tensor has to satisfy the following balance equation:

$$\frac{\delta \boldsymbol{J}_0}{\delta t} = \boldsymbol{\chi}_0, \tag{18.14}$$

or in explicit form

$$\frac{\partial \boldsymbol{J}_0}{\partial t} = -\mathbf{v} \cdot \nabla \boldsymbol{J}_0 + \boldsymbol{\chi}_0, \tag{18.15}$$

where the first term on the right side describes the inertia flux and the production term $\boldsymbol{\chi}_0$ reflects structural transformations of the media. The form of the production term depends on the physical interpretation of microstructural changes. It can, for instance, depend on $\boldsymbol{J}_0$ as well as other characteristics of the medium, such as density, temperature, stresses, etc. It can also depend on external stimuli such as external electric or magnetic fields.

Then from (18.11), $(18.10)_2$ and (18.14) it follows that the tensor of inertia in the current configuration has the form:

$$\frac{\delta \boldsymbol{J}}{\delta t} = \boldsymbol{\omega} \times \boldsymbol{J} - \boldsymbol{J} \times \boldsymbol{\omega} + \boldsymbol{\chi}, \qquad \boldsymbol{\chi} = \boldsymbol{Q} \cdot \boldsymbol{\chi}_0 \cdot \boldsymbol{Q}^T, \tag{18.16}$$

where the first two terms describe the inertia tensor change due to rotation of the average particle and the last term is responsible for the inertia production due to internal structural transformations.

The inertia tensor production leads to the production of spin. Then the balance of moment of momentum is formulated as follows

$$\rho \frac{\delta(\boldsymbol{J} \cdot \boldsymbol{\omega})}{\delta t} = \nabla \cdot \boldsymbol{\mu} + \boldsymbol{\sigma}_\times + \rho \mathbf{m} + \rho \boldsymbol{\chi} \cdot \boldsymbol{\omega}. \tag{18.17}$$

Here $\boldsymbol{\sigma}$ and $\boldsymbol{\mu}$ are the non-symmetric Cauchy and couple stress tensors, $(\mathbf{a}\otimes\mathbf{b})_\times = \mathbf{a}\times\mathbf{b}$ is the Gibbsian cross, and $\mathbf{m}$ is the specific couple density. The extra term in the balance equation, with $\boldsymbol{\chi}$, describes the moment of momentum production due to structural transformations.

However, by taking into account the balance equation for the inertia tensor in the current configuration (18.16) we obtain the spin balance equation in classical form:

$$\rho \boldsymbol{J} \cdot \frac{\delta \boldsymbol{\omega}}{\delta t} = -\boldsymbol{\omega} \times \boldsymbol{J} \cdot \boldsymbol{\omega} + \boldsymbol{\nabla} \cdot \boldsymbol{\mu} + \boldsymbol{\sigma}_\times + \rho \mathbf{m}. \tag{18.18}$$

Thus in the suggested model, the basic equations are essentially the same as in the classical approach except for replacement of the classical conservation law of micro-inertia by the inertia tensor balance equation containing the production term.

It should be also noted that traditionally the tensor of inertia of a continuum particle plays a role only in context with rotations. However, within this approach the balance equation for $\boldsymbol{J}$ and hence the production term in Eqn.(18.14) are physical meaningful by themselves, independent of the angular velocity and may serve as an indicator of the internal structural changes.

18.3 Special Cases for the Production Term

We shall now proceed and illustrate the theory by several examples. By the first example, it is intended to show what happens to the tensor of inertia if the number of micro-particles and their size within the elementary volume change, for example due to the presence of a crusher. By the second example, the impact of a changing moment of inertia onto rotational motion will be demonstrated. In particular, the change of the state of rotation of a isotropic thermoelastic continuum will be studied. The average particle will undergo a nonuniform change of external temperature affecting its moment of inertia. Note that within the classical framework of micropolar theory a change of temperature would not influence rotation. However, within the to-be-presented theory, changes in temperature will influence the inertia tensor and hence couple to angular velocity. The last example will describe changes of anisotropy of a material under an external electrical field.

18.3.1 Milling Matter in a Crusher

As the first very simple example let us consider a continuous flow of granular matter of height H moving on a conveyor belt in the x-direction at a constant, prescribed speed, $\mathbf{v}_0 = v_0\mathbf{e}_x$. On its way it enters a region $0 \le x \le L$, where it is continuously crushed by an external distributed force, $\mathbf{p}_0 = p_0\mathbf{e}_y$, applied at the top so that smaller and smaller particles will form. Note that in spite of the fact that the micro-particles have a very irregular shape the homogenized tensor of inertia on a continuum level is isotropic due to the statistically random distribution of micro-particles of different size and shape, as illustrated in Fig. 18.2 in the left inset. During the milling process the mass and characteristic size of a micro-particle will decrease over time (right inset on the top of the figure), which, under the assumption that the distribution stays isotropic, leads to a decreasing moment of inertia on the macro-level. At the same time the mass density of the elementary volume remains the same. The isotropy means that the production term also has to be a spherical tensor, $\chi = \chi \boldsymbol{I}$, and Eq. (18.16) turns into a scalar one:

$$\frac{\delta J}{\delta t} = \chi. \tag{18.19}$$

Here the identity $\mathbf{a} \times \boldsymbol{I} = \boldsymbol{I} \times \mathbf{a}$ is taken into account.

We assume that the production term is given by the following expression:

$$\chi = -\alpha_0 \operatorname{tr} \boldsymbol{\sigma} \, (J - J_*)\big(H(x) - H(x - L)\big)\boldsymbol{I}, \tag{18.20}$$

where $\boldsymbol{\sigma}$ is the stress tensor, $H(x)$ is the Heaviside step function, J_* and α_0 are positive constants, which can be interpreted intuitively as being related to the minimum grain size the particles can be crushed to, and to the inverse of the particle toughness. Thus, being the characteristics of the material and not of the crusher, they may be considered as constitutive properties. At the same time $\operatorname{tr} \boldsymbol{\sigma}$ describes a conversion of the crusher action to a material response. In other words, it is related to the effectiveness of the crusher and transmission of its external forces into the material. Hence, in this case, the production term depends on the material properties, external action, and space.

Since the material in the crusher is under a significant pressure we will model it as a linear-elastic material. For linear elasticity the Cauchy stresses $\boldsymbol{\sigma}$ is related to the strain $\boldsymbol{\varepsilon}$ by:

$$\boldsymbol{\sigma} = \boldsymbol{C} : \boldsymbol{\varepsilon}, \qquad \boldsymbol{C} = k(J)\boldsymbol{I}\boldsymbol{I} + 2\mu(J)\left({}^4\boldsymbol{I} - \frac{1}{3}\boldsymbol{I}\boldsymbol{I}\right), \tag{18.21}$$

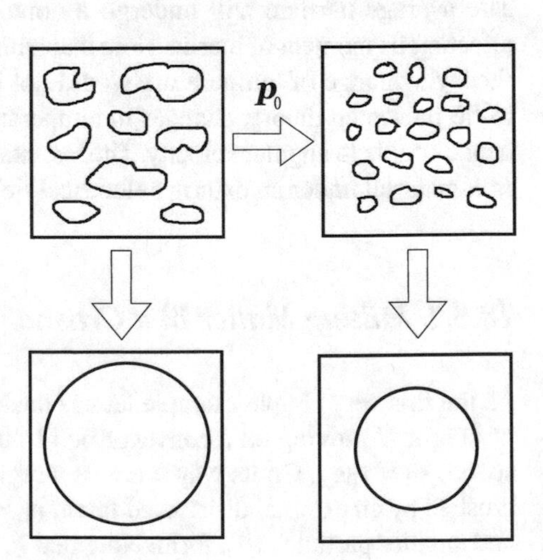

Fig. 18.2 Structural shape change and corresponding homogenization

where ${}^4\boldsymbol{I}_{ijkl} = (\delta_{ik}\delta_{lj} + \delta_{il}\delta_{kj})/2$ denotes the fourth rank identity tensor, δ_{ij} is the Kronecker symbol. It is assumed that the bulk and shear, moduli, k and μ, depend on the particle size:

$$k(J) = k_* f_k\,(J/J_*), \quad \mu(J) = \mu_* f_\mu\,(J/J_*), \quad f_k(1) = f_\mu(1) = 1. \tag{18.22}$$

Here k_* and μ_* are the bulk and shear moduli of the material consisting of the particles of the minimal size. The functions f_k and f_μ depend on the material and have to be obtained from experiments. For example, in Hamilton (1971) it was shown that the bulk modulus decreases and the shear modulus increases with decreasing grain size. Thus we assume that the elastic modules depend on the moment of inertia as follows:

$$k(J) = k_*\,(J/J_*)^2\,, \quad \mu(J) = \mu_*\,(J/J_*)^{-2}\,. \tag{18.23}$$

Furthermore it is assumed that the material moves freely along the x-axis while its movement in the y-direction is limited by the walls. Then in the absence of a body force the equilibrium condition $\boldsymbol{\nabla}\cdot\boldsymbol{\sigma} = \boldsymbol{0}$, (18.19) and (18.20) yield:

$$\frac{\partial \bar{J}}{\partial \bar{t}} + \frac{\partial \bar{J}}{\partial \bar{x}} = -\xi\,(\bar{J} - \bar{J}_*)\big(H(\bar{x}) - H(\bar{x}-1)\big), \quad \xi = \frac{9\bar{\alpha}}{3 + \dfrac{\mu_*}{k_*}\left(\dfrac{\bar{J}_*}{\bar{J}}\right)^4}. \tag{18.24}$$

The bar on symbols refers to dimensionless quantities, namely,

$$\bar{x} = \frac{x}{L}, \qquad \bar{t} = \frac{v_0}{L}t, \qquad \bar{J} = \frac{J}{J^0}, \qquad \bar{J}_* = \frac{J_*}{J^0}, \qquad \bar{\alpha} = \frac{L p_0 \alpha_0}{v_0}, \tag{18.25}$$

where J^0 is the maximal moment of inertia.

Eqn. (18.24) has to be supplemented with initial and boundary conditions. In order to obtain a non-trivial solution, the initial distribution of the particles along the vertical axis at the left side of the crusher at $x = 0$ has to be inhomogeneous. For simplicity we will consider a linear distribution:

$$\bar{J}(0, \bar{z}, \bar{t}) = \left(\bar{J}_* - \bar{J}^0\right)\bar{z} + \bar{J}^0. \tag{18.26}$$

The numerical solutions of (18.24), (18.26) based on an explicit method of discrete integration are presented in Fig. 18.3, where the distributions of the moment of inertia for a vertical and a horizontal cut within the crusher area (stationary case) are visualized in dimensionless form. It is clearly visible that the moment of inertia decreases linearly from the bottom to the top of the crusher. At the same time, the distributions of the moment of inertia along the x-axis have an exponential shape.

During the computation the following parameters were used

$$\bar{\alpha} = 1.5, \qquad \frac{\mu_*}{k_*} = 0.1.$$

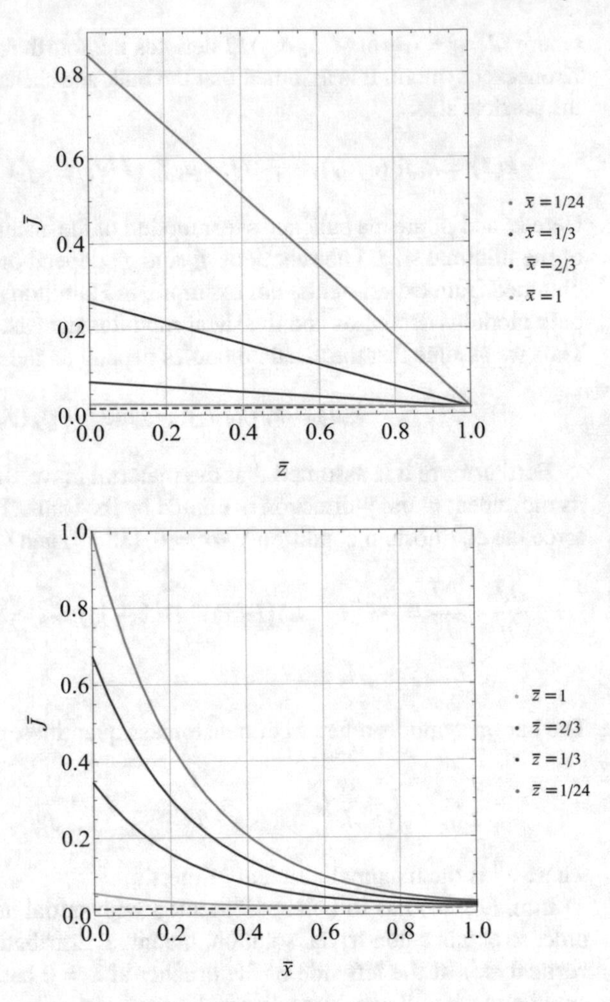

Fig. 18.3 Distribution of the moment of inertia (top inset – vertical cut, bottom inset – horizontal cut through crusher region)

The influence of the last parameter on the crushing process is shown in Fig. 18.4, where the stationary distributions of the moment of inertia along the crusher region for constant and for variable elastic moduli are presented. As can be expected the difference is more pronounced for large values of the moment of inertia. In Müller et al (2017) an analytical solution to a very simple one dimensional initial-boundary value problem for non-homogeneous crushing of particles was found based on the method of characteristics. The similar problem for viscous material was considered in Fomicheva et al (2019).

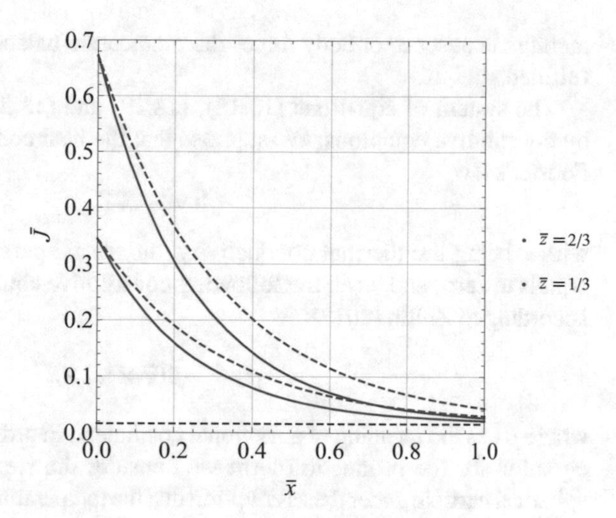

Fig. 18.4 Profiles of the moment of inertia for variable and constant elastic moduli (solid and dashed lines, respectively)

18.3.2 Turning Heat Conduction into Space-varying Rotational Motion

We next consider a medium consisting of thermoelastic micro-particles homogeneously distributed over a rectangular plate: $x \in [0, L]$, $y \in [-l_y, l_y]$, $z \in [-L_z, L_z]$. The medium represents the behavior of a homogeneous mix of micro-particles of arbitrary size and shape on the mesoscale, that results in the isotropic tensor of inertia on a continuum level, $\boldsymbol{J} = J\boldsymbol{I}$.

Initially the temperature of the media is also homogeneous and equal to T_0. By positioning the medium in between two reservoirs kept at temperatures T_0 and T_L and attached at positions $x = 0$ and $x = L$ of the region, respectively, the temperature of this medium will gradually change. The temperature development is described by the heat conduction equation after (Zhilin, 2012):

$$\rho\, c_v \frac{\delta T}{\delta t} = \boldsymbol{\sigma}_d : (\boldsymbol{\nabla} \otimes \mathbf{v} + \boldsymbol{I} \times \boldsymbol{\omega}) + \boldsymbol{\mu}_d : (\boldsymbol{\nabla} \otimes \boldsymbol{\omega}) + \rho q - \boldsymbol{\nabla} \cdot \mathbf{h}\,. \tag{18.27}$$

Here T is the absolute temperature, c_v is the specific heat capacity at constant volume, double convolution means $(\mathbf{a} \otimes \mathbf{b}) : (\mathbf{c} \otimes \mathbf{d}) = (\mathbf{a} \cdot \mathbf{c})(\mathbf{b} \cdot \mathbf{d})$, q is the specific heat source, $\mathbf{h}$ is the heat flux, and $\boldsymbol{\sigma}_d$ and $\boldsymbol{\mu}_d$ are the inelastic (dissipative) parts of the stress tensor and couple stress tensor:

$$\boldsymbol{\sigma} = \boldsymbol{\sigma}_e + \boldsymbol{\sigma}_d, \quad \boldsymbol{\mu} = \boldsymbol{\mu}_e + \boldsymbol{\mu}_d, \tag{18.28}$$

where $\boldsymbol{\sigma}_e$ and $\boldsymbol{\mu}_e$ are the elastic (velocity independent) parts of the stress tensor and couple stress tensor.

For simplicity we suppose that the macro-particles have only rotational degrees of freedom and their translational velocities are equal to zero. Then, for an unconstrained medium in absence of body forces the momentum balance equation is automatically fulfilled.

The system of equations (18.18), (18.19) and (18.27) has to be supplemented by constitutive equations. We suppose that the heat conduction is governed by the Fourier's law

$$\mathbf{h} = -\kappa \nabla T,$$

with κ being the thermal conductivity, the elastic part of the couple stress tensor equals to zero, and write the following constitutive equation for its dissipative part according to Zhilin (2012):

$$\boldsymbol{\mu}_d = -\beta(\nabla \times \boldsymbol{\omega}) \times \boldsymbol{I}, \tag{18.29}$$

where β has the meaning of a frictional coefficient. In order to formulate a constitutive equation for the production term we consider the free thermal expansion of the spherical particle under the assumption that the temperature increase is instantaneously assumed by the particle. Then the moment of inertia changes in accordance with the temperature field:

$$J(x,t) = J_0 \left[1 + \alpha(T(x,t) - T_0)\right]^2, \tag{18.30}$$

with J_0 being the initial moment of inertia, and α being the linear coefficient of thermal expansion. Therefore the production term can be written as:

$$\chi = \frac{\partial J}{\partial t} = 2J_0\alpha\left(1 + \alpha(T - T_0)\right)\frac{\partial T}{\partial t}. \tag{18.31}$$

As one can see the production depends on the material properties and vanishes at the constant temperature field.

To keep the problem one-dimensional we also assume that:

$$\boldsymbol{\omega}(x,t) = \omega(x,t)\mathbf{e}_z, \quad \mathbf{m}(x,t) = m_0\mathbf{e}_z. \tag{18.32}$$

Thus the development of temperature, the moment of inertia and angular velocity can be obtained as a result of solution of a coupled system of partial differential equations in dimensionless form:

$$\begin{aligned} \frac{\partial \bar{T}}{\partial \bar{t}} &= \delta\left(\frac{\partial \bar{\omega}}{\partial \bar{x}}\right)^2 + \frac{\partial^2 \bar{T}}{\partial \bar{x}^2}, \\ \frac{\partial \bar{J}}{\partial \bar{t}} &= 2\bar{\alpha}[1 + \bar{\alpha}(\bar{T} - 1)]\frac{\partial \bar{T}}{\partial \bar{t}}, \\ \bar{J}\frac{\partial \bar{\omega}}{\partial \bar{t}} &= \eta\frac{\partial^2 \bar{\omega}}{\partial \bar{x}^2} + \bar{m}, \end{aligned} \tag{18.33}$$

$$\bar{\alpha} = \alpha T_0, \quad \delta = \frac{\beta m_0}{\kappa T_0 J_0}, \quad \eta = \frac{\beta c_v}{\kappa J_0}, \quad \bar{m} = \omega_0\frac{L^2}{D}, \quad \omega_0 = \sqrt{\frac{m_0}{J_0}}, \quad D = \frac{\kappa}{\rho c_v},$$

$$\bar{x} = \frac{x}{L}, \qquad \bar{t} = \frac{D}{L^2}t, \qquad \bar{T} = \frac{T}{T_0}, \qquad \bar{J} = \frac{J}{J_0}, \qquad \bar{\omega} = \frac{\omega}{\omega_0}.$$

with the following initial and boundary conditions:

$$\bar{T}(\bar{x}, \bar{t} = 0) = 1, \quad \bar{J}(\bar{x}, \bar{t} = 0) = 1, \quad \bar{\omega}(\bar{x}, \bar{t} = 0) = 0,$$
$$\bar{T}(\bar{x} = 0, \bar{t}) = 1, \quad \bar{T}(\bar{x} = 1, \bar{t}) = \frac{T_L}{T_0}, \quad \left.\frac{\partial \bar{\omega}}{\partial \bar{x}}\right|_{\bar{x}=0;1} = 0.$$

Note that the angular velocity related boundary conditions are necessary only if the viscosity is taken into the account. The proper choice of boundary conditions for the angular velocity is a complex issue. Generally, two types of boundary conditions are considered. The first type is so-called “strict adhesion”, see, for example, Eringen (2001) ($\bar{\omega}(\bar{x} = 0; 1, \bar{t}) = 0$). The second one used here corresponds to an absent couple stress on the boundary.

The developments of angular velocity at three dimensionless times, $\bar{t} = 0.005$ (green), $\bar{t} = 0.01$ (blue), and $\bar{t} = 0.03$ (red) are depicted in Fig. 18.5. It is seen that the obtained profile of angular velocity is nonlinear in contrast to the classical approach where the angular velocity does not change over the sample length. The nonlinear behavior reflects the fact that distribution of the inertia moment over the sample mimics the temperature profile and as a result it follows from Eq. (18.33)

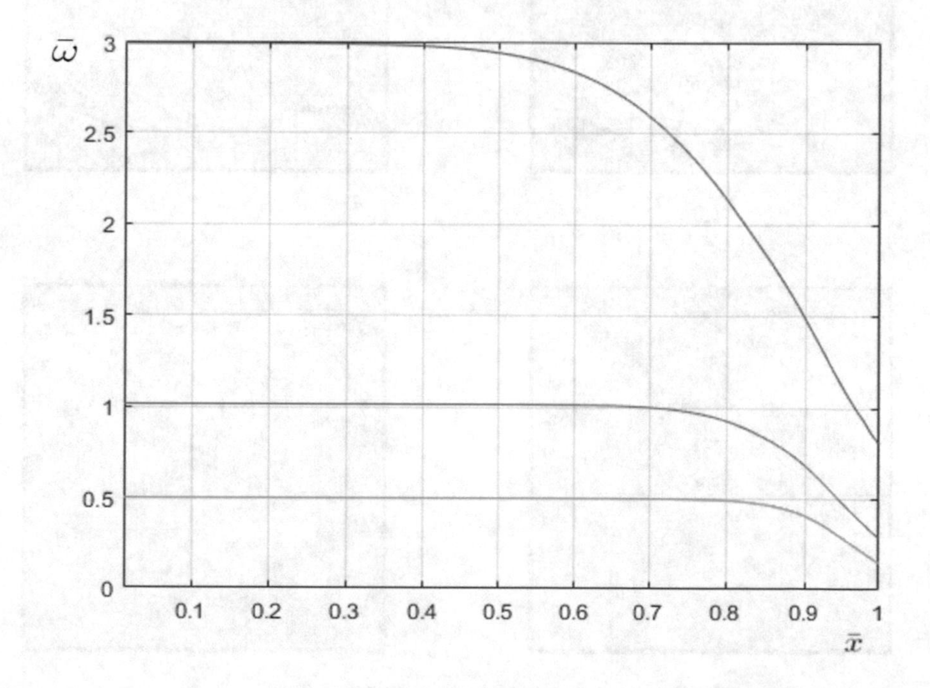

Fig. 18.5 Angular velocity distribution over the sample at different moments of time ($\bar{m} = 100$, $\bar{T}(\bar{x} = 1, \bar{t}) = 2$, $\eta = 1$, $\delta = 1$)

that the angular acceleration varies for particles with different temperature. Different boundary conditions and time-dependent the volume moment couple density was considered in Morozova et al (2019).

18.3.3 Dipolar Polarization

In order to describe anisotropic changes let us consider a material, which, on a mesoscale, consists of an assembly of dipoles. Due to thermal motion the dipoles are randomly oriented in the substance so that there is no overall charge in the material (Fig. 18.6, top left). The initial, homogenized macro-inertia tensor is then the isotropic spherical tensor.

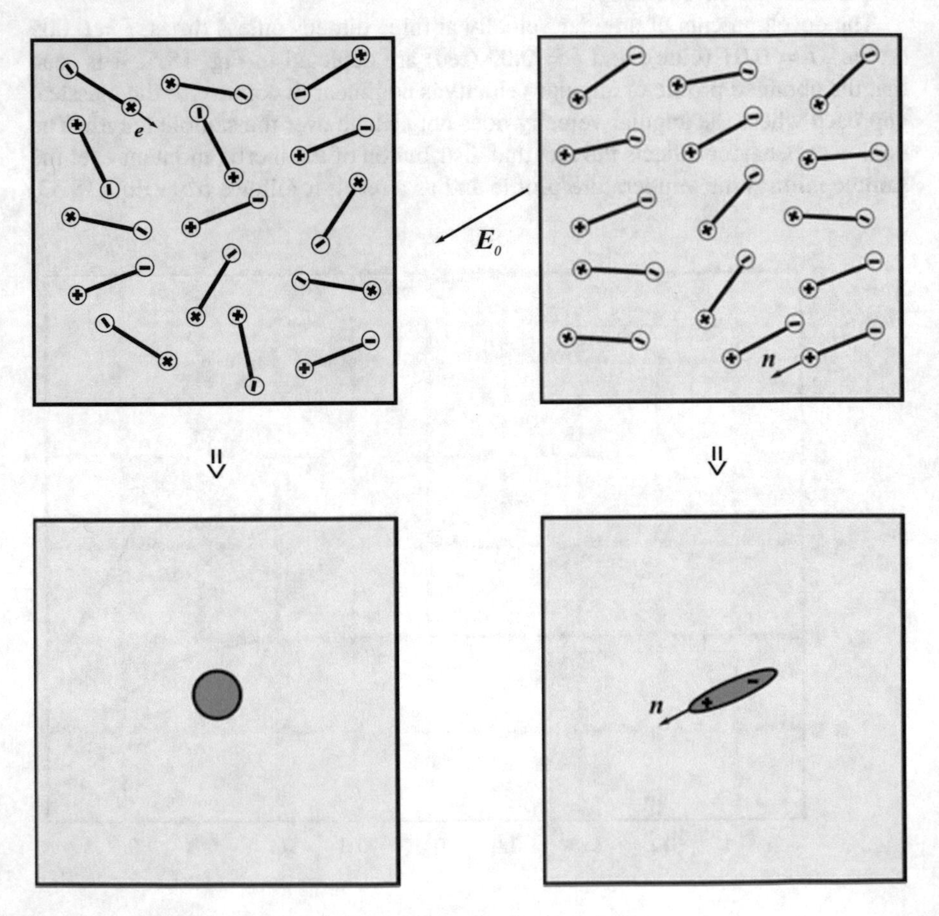

Fig. 18.6 Structural shape change and corresponding homogenization

When an external electric field $\boldsymbol{E} = E_0\mathbf{n}$ is applied, the dipoles tend to align with the applied field to lower their electrostatic energy, basically the positive end of the dipole would like to join the negative end of the applied field. Such behavior can be observed in dipolar polarization, for materials with build-in dipoles that are independent of each other, i.e. they don't interact and they can be rotated freely by an applied field (Kestelman et al, 2013). However, even in case of liquids or gases, where molecules are free to rotate, a complete alignment is impossible because the tendency of dipoles to orient along into the field direction will be counteracted by thermal motion. Thus, as a result of the combined action of the external and internal actions, a dominating orientation of the molecules along the direction of the electric field occurs and the transversally-isotropic state is achieved (Fig. 18.6, right).

Upon the applied electric field removal the thermal motion randomizes the alignment of the dipoles and returns the material into the initial isotropic state. The objective is now to describe the transition processes.

First note that switching back and forth between the isotropic and transversally-isotropic states is characterized only by the deviatoric part of the inertia tensor since the micro-particles within the elementary volume do not change. Thus, we have a purely deviatoric production. Second, since the micro-particles rotate in all possible directions during the orientation and disorientation processes, the macroscopic spin and, therefore, the macroscopic angular velocity is zero. Hence, the macrorotation tensor is an identity tensor. Under an assumption that the process occurs in the same manner at all points of space the balance of the inertia tensor simplifies and reads:

$$\boldsymbol{J}^{sh}(t) = J_0\boldsymbol{I} = \text{const}, \qquad \frac{\mathrm{d}\boldsymbol{J}^{\mathrm{dev}}(\mathrm{t})}{\mathrm{dt}} = \boldsymbol{\chi}, \tag{18.34}$$

where

$$\boldsymbol{J}^{dev}(t) = \frac{1}{15}\left(c(t)^2 - a(t)^2\right)(\boldsymbol{I} - 3\mathbf{nn}). \tag{18.35}$$

$a(t)$ is the semi-axis of the plane of isotropy and $c(t)$ is the semi-axis in the direction of $\mathbf{n}$.

The production term describing the microparticle alignment has to depend on the direction of the external field and on the time of particle orientation under the field action, τ_p, which defines the polarization setting time. The higher the magnitude of the electric field and lower the temperature the shorter is the time of dipolar polarization setting. For instance, $\tau_p = \alpha(T)/E_0$ can be taken as a simple example. Here the parameter α in units of $\mathrm{V\cdot s/m}$ is an increasing function of temperature. Thus, we can postulate the following form of the production term:

$$\boldsymbol{\chi}(t) = \frac{J_\infty}{\tau_p}\exp(-t/\tau_p)(\boldsymbol{I} - 3\mathbf{nn}), \tag{18.36}$$

where $J_\infty = (c_\infty^2 - a_\infty^2)/15$ is the value reached at $t \to \infty$. Note that in that case, the production is explicitly time-dependent. However, since the exponent is a rapidly decreasing function one can assume that the production stops when $t > 5\tau_p$. It also worth mentioning that τ_p characterizes the combined effect of the electric field magnitude and thermal motion. Nevertheless the direction of the electric field appears explicitly in the production term.

Integration of the Eq. (18.34) with zero initial condition gives a development of the inertia tensor in time:

$$\boldsymbol{J}(t) = J_0\boldsymbol{I} + \frac{1}{15}\left(c_\infty^2 - a_\infty^2\right)\left(1 - \exp(-t/\tau_p)\right)(\boldsymbol{I} - 3\mathbf{nn})\,. \tag{18.37}$$

A few comments are now in order. First, note that the inertia tensor (18.37) corresponds to a particle that is oriented in the direction of the electric fields. Then the moment couple on the macro level will be zero and therefore zero angular velocity satisfies the spin balance in the absence of stress tensors. Second, the parameter J_∞ cannot be defined easily. If all dipoles would be uniformly aligned in **n**-direction then the homogenized tensor of inertia would coincide with the inertia tensor of the dipole. But the perfect alignment is not reachable because of randomizing effect of the thermal motion. Hence J_∞ characterizes the "equilibrium" distribution of dipoles over orientations. It is tempting to interrelated J_∞ to the maximal polarization density **P** that can be reached in the material at the given electric field and temperature, but we leave it with this remark.

Now let us consider the reverse process. The production term associated with the thermal motion of the dipoles depends on the temperature and has to disappear as soon as the isotropic case is reached. Having that in mind, we choose the production of microinertia in the most simple form:

$$\boldsymbol{\chi}_{tm}(t) = -\frac{1}{\tau_r(T)}\boldsymbol{J}^{dev}(t), \tag{18.38}$$

where τ_r defines the relaxation time. The smaller it is, the faster the transition from order to disorder will be achieved. Being a quantity associated with the thermal motion the relaxation time has to be a decreasing function of the temperature.

Then it follows that the deviatoric part of the inertia tensor decreases exponentially in time and the inertia tensor turns eventually into a spherical tensor:

$$\boldsymbol{J}(t) = J_0\boldsymbol{I} + \frac{1}{15}\left(c_1^2 - a_1^2\right)\exp(-t/\tau_r)\,(\boldsymbol{I} - 3\mathbf{nn})\,. \tag{18.39}$$

Here c_1 and a_1 are the spheroid axes at the moment t_1, when the external field was removed.

18.4 Conclusions and Outlook

The intention of this paper is to draw attention on some recent activities in the field of micropolar media capable of structural change. One of its main feature is a new balance equation for the tensor of inertia containing a production term. The new balance and in particular the production are interpreted mesoscopically by taking the inner structure of micropolar matter into account. In fact, it is an attempt to generalize the classical approach based on the concept of an indestructible material particle consisting of a statistically significant number of subunits on a mesoscopic scale. Within the classical approach, there should be no exchange of subunits between the material particles. Furthermore, the polar continuum particle assumed to be equivalent to a rigid body and can be neither destroyed nor generated. However, this means that within this framework certain processes and effects in materials can simply not be modeled.

The new approach emphasizes the idea that it may become necessary to abandon the concept of the rigid material particle if one wishes to describe micropolar matter in which structural changes or chemical reactions occur. The approach is based on the spatial description where a representative volume element is treated as a continuum polar particle. It does not impose strict constraints on the motion of micro-particles, rather it embraces the idea of an open system, allowing a priori for exchange of mass, momentum, energy, tensor of inertia, etc., between and within the representative volume elements. For a better understanding of this new concept an underlying mesoscopic theory is presented. The main idea is to connect information on a mesoscale by taking the intrinsic microstructure within RVE into account with the macroscopic world, i.e., with the balances of micropolar continua in combination with suitable constitutive equations. This new approach enables us to study the temporal development of rotational inertial characteristics. In this context, the tensor of inertia is an additional internal variable characterizing structural transformations of the media. Moreover, in contrast to the material description where all neighboring material particles have to remain in the neighborhood during their motion, the spatial description does not impose strict constraints on the motion of material points. As a result, the neighboring particles can separate and travel significant distances from one another as happens in soils, granular and powder-like materials.

The extended theory seems particularly promising in context with the description of materials with complex or variable structure, such as suspensions or liquid crystals where the structural state of the fluid matter is actively controlled by applying external electromagnetic fields and temperature changes. Also, the approach has a potential for modeling processes going on under an influence of different physical and thermo-mechanical factors or taking into account their mutual influence. For example, a dielectric polarization in an alternating electric and temperature fields or electret production where a dielectric is placed in strong electric field and subjected to additional physical action or mechanoelectrets where the electret state can occur from mechanical deformation without external electrical field.

Acknowledgements The author is deeply grateful to E.A. Ivanova and W.H. Müller for useful discussions on the subject.

References

Aero EL, Kuvshinskii EV (1961) Fundamental equations of the theory of elastic media with rotationally interacting particles. Soviet Physics - Solid State 2(7):1272–1281

Altenbach H, Eremeyev VA (eds) (2012) Generalized Continua - From the Theory to Engineering Applications, CISM International Centre for Mechanical Sciences, vol 541. Springer, Wien

Altenbach H, Forest S (eds) (2016) Generalized Continua as Models for Classical and Advanced Materials, Advanced Structured Materials, vol 42. Springer, Switzerland

Altenbach H, Maugin GA, Erofeev V (eds) (2011) Mechanics of Generalized Continua, Advanced Structured Materials, vol 7. Springer, Berlin

de Borst R (1993) A generalization of J2-flow theory for polar continua. Computer Methods in Applied Mechanics and Engineering 103(3):347–362

Chen K (2007) Microcontinuum balance equations revisited: The mesoscopic approach. Journal of Non-Equilibrium Thermodynamics 32:435–458

Cosserat F, Cosserat E (1909) Théorie des corps déformables. A. Hermann et fils

Dłuzewski PH (1993) Finite deformations of polar elastic media. International Journal of Solids and Structures 30(16):2277–2285

Dyszłewicz J (2004) Micropolar Theory of Elasticity. Springer Science, Berlin

Ehlers W, Ramm E, Diebels S, D'Addetta G (2003) From particle ensembles to Cosserat continua: homogenization of contact forces towards stresses and couple stresses. International Journal of Solids and Structures 40:6681–6702

Ericksen JL (1960a) Anisotropic fluids. Archive for Rational Mechanics and Analysis 4:231–237

Ericksen JL (1960b) Theory of anisotropic fluids. Transactions of the Society of Rheology 4:29–39

Ericksen JL (1960c) Transversely isotropic fluids. Kolloid-Zeitschrift 173:117–122

Ericksen JL (1961) Conservation laws for liquid crystals. Transactions of the Society of Rheology 5:23–34

Ericksen JL, Truesdell C (1957) Exact theory of stress and strain in rods and shells. Archive for Rational Mechanics and Analysis 1:295–323

Eringen AC (1969) Micropolar fluids with stretch. International Journal of Engineering Science 7:115–127

Eringen AC (1976) Continuum Physics, vol IV. Academic Press, New York

Eringen AC (1984) A continuum theory of rigid suspensions. International Journal of Engineering Science 22:1373–1388

Eringen AC (1985) Rigid suspensions in viscous fluid. International Journal of Engineering Science 23:491–495

Eringen AC (1991) Continuum theory of dense rigid suspensions. Rheologica Acta 30:23–32

Eringen AC (1997) A unified continuum theory of electrodynamics of liquid crystals. International Journal of Engineering Science 35(12/13):1137–1157

Eringen AC (1999) Microcontinuum Field Theory, vol I: Foundations and Solids. Springer, New York

Eringen AC (2001) Microcontinuum Field Theories, vol II: Fluent Media. Springer, New York

Eringen AC, Kafadar CB (1976) Polar field theories. In: Continuum physics IV. Academic Press, London

Eringen AC, Suhubi ES (1964a) Nonlinear theory of simple microelastic solids. International Journal of Engineering Science 2:189–203

Eringen AC, Suhubi ES (1964b) Nonlinear theory of simple microelastic solids. International Journal of Engineering Science 2:389–404

Fomicheva M, Vilchevskaya E, Müller W, Bessonov N (2019) Milling matter in a crusher: modeling based on extended micropolar theory. Continuum Mechanics and Thermodynamics doi:10.1007/s00161-019-00772-4

Forest S, Sievert R (2003) Elastoviscoplastic constitutive frameworks for generalized continua. Acta Mechanica 160(1-2):71–111

Forest S, Sievert R (2006) Nonlinear microstrain theories. International Journal of Solids and Structures 43(24):7224–7245

Grammenoudis P, Tsakmakis C (2005) Predictions of microtorsional experiments by micropolar plasticity. Proceedings of the Royal Society A Mathematical Physical and Engineering Sciences 461(2053):189–205

Grioli G (1960) Elasticita asimmetrica. Annali di Matematica Pura ed Applicata 50(1):389–417

Günther W (1958) Zur Statik und Kinematik des Cosseratschen Kontinuums. Abhandlungen der Königlichen Gesellschaft der Wissenschaften in Göttingen 10:196–213

Hamilton EL (1971) Elastic properties of marine sediments. Journal of Geophysical Research 76(2):579–604

Hellinger E (1914) Die allgemeinen Ansätze der Mechanik der Kontinua. In: Klein F, Wagner K (eds) Encykl. Math. Wiss., Springer, Berlin, pp 602–694

Ivanova E, Vilchevskaya E, Müller WH (2016) Time derivatives in material and spatial description – What are the differences and why do they concern us? In: Naumenko K, Aßmus M (eds) Advanced Methods of Mechanics for Materials and Structures, Springer, Advanced Structured Materials, vol 60, pp 3–28

Ivanova EA, Vilchevskaya EN (2016) Micropolar continuum in spatial description. Continuum Mechanics and Thermodynamics 28(6):1759–1780

Jeong J, Neff P (2010) Existence, uniqueness and stability in linear cosserat elasticity for weakest curvature conditions. Mathematics and Mechanics of Solids 15(1):78–95

Kestelman VN, Pinchuk LS, Goldade VA (2013) Electrets in engineering: fundamentals and applications. Springer Science & Business Media

Kröner E (1963) On the physical reality of torque stresses in continuum mechanics. International Journal of Engineering Science 1(2):261–278

Larsson R, Diebels S (2007) A second-order homogenization procedure for multi-scale analysis based on micropolar kinematics. International Journal for Numerical Methods in Engineering 69(12):2485–2512

Larsson R, Zhang Y (2007) Homogenization of microsystem interconnects based on micropolar theory and discontinuous kinematics. Journal of the Mechanics and Physics of Solids 55(4):819–841

Lippmann H (1969) Eine Cosserat-Theorie des plastischen Fließens. Acta Mechanica 8(3-4):93–113

Loret B, Simões FMF (2005) A framework for deformation, generalized diffusion, mass transfer and growth in multi-species multi-phase biological tissues. European Journal of Mechanics A/Solids 24:757–781

Mandadapu KK, Abali BE, Papadopoulos P (2018) On the polar nature and invariance properties of a thermomechanical theory for continuum-on-continuum homogenization. arXiv preprint arXiv:180802540

Maugin GA (1998) On the structure of the theory of polar elasticity. Philosophical Transactions of the Royal Society of London Series A: Mathematical, Physical and Engineering Sciences 356(1741):1367–1395

Maugin GA, Metrikine AV (eds) (2010) Mechanics of Generalized Continua: One Hundred Years After the Cosserats, Advances in Mechanics and Mathematics, vol 21. Springer, New York

Mindlin RD (1964) Micro-structure in linear elasticity. Archive for Rational Mechanics and Analysis 16(1):51–78

Mindlin RD, Tiersten HF (1962) Effects of couple–stresses in linear elasticity. Archive for Rational Mechanics and Analysis 11:415–448

Morozova AN, Vilchevskaya EN, Müller WH, Bessonov NM (2019) Interrelation of heat propagation and angular velocity in micropolar media. In: Altenbach H, Belyaev A, Eremeyev V, Krivtsov A, A P (eds) Dynamical Processes in Generalized Continua and Structures, Springer, Cham, Advanced Structured Materials, vol 103, pp 413 –425

Müller WH, Vilchevskaya EN, Weiss W (2017) A meso-mechanics approach to micropolar theory: A farewell to material description. Physical Mesomechanics 20(3):13–24

Neff P (2006) A finite-strain elastic-plastic Cosserat theory for polycrystals with grain rotations. International Journal of Engineering Science 44:574–594

Neff P, Jeong J (2009) A new paradigm: the linear isotropic cosserat model with conformally invariant curvature energy. ZAMM-Journal of Applied Mathematics and Mechanics/Zeitschrift für Angewandte Mathematik und Mechanik 89(2):107–122
Nistor I (2002) Variational principles for cosserat bodies. International Journal of Non-Linear Mechanics 37:565–569
Oevel W, Schröter J (1981) Balance equation for micromorphic materials. Journal of Statistical Physics 25(4):645–662
Palmov VA (1964) Fundamental equations of the theory of asymmetric elasticity. Journal of Applied Mathematics and Mechanics 28(3):496–505
Pietraszkiewicz W, Eremeyev VA (2009) On natural strain measures of the non-linear micropolar continuum. International journal of solids and structures 49(3-4):774–787
Ramezani S, Naghdabadi R (2007) Energy pairs in the micropolar continuum. International Journal of Solids and Structures 44(14-15):4810–4818
Rivlin RS (1968) The formulation of theories in generalized continuum mechanics and their physical significance. Symposia Mathematica 1:357–373
Sansour C, Skatulla S (eds) (2012) Generalized Continua and Dislocation Theory. Springer
van der Sluis O, Vosbeek PHJ, Schreurs PJG, Meijer HEH (1999) Homogenization of heterogeneous polymers. International Journal of Solids and Structures 36(21):3193–3214
Steinmann P (1994) A micropolar theory of finite deformation and finite rotation multiplicative elastoplasticity. International Journal of Solids and Structures 31(8):1063–1084
Steinmann P, Stein E (1997) A uniform treatment of variational principles for two types of micropolar continua. Acta Mechanica 121:215–232
Stojanović R, Djuriić S, Vujoševiić L (1964) On finite thermal deformations. Archiwum Mechaniki Stosowanej 16:102–108
Toupin RA (1962) Elastic materials with couple–stresses. Archives for Rational Mechanics and Analysis 11:385–414
Toupin RA (1964) Theories of elasticity with couple-stress. Archives for Rational Mechanics and Analysis 17:85–112
Truesdell C, Toupin RA (1960) The classical field theories. In: Flügge S (ed) Encyclopedia of Physics / Handbuch der Physik, Springer, Berlin, Heidelberg, vol III/1: Principles of Classical Mechanics and Field Theory / Prinzipien der Klassischen Mechanik und Feldtheorie, pp 226–858
Voigt W (1887) Theoretische Studien über die Elasticitätsverhältnisse der Krystalle. I. Abhandlungen der Königlichen Gesellschaft der Wissenschaften in Göttingen 34:3–52
Wilmanski K (2008) Continuum Thermodynamics. Part I: Foundations. WorldScientific, Singapore
Zhilin PA (2012) Rational Continuum Mechanics (in Russ.). St. Petersburg Polytechnical University, St. Petersburg

Chapter 19
Hencky Strain and Logarithmic Rate for Unified Approach to Constitutive Modeling of Continua

Si-Yu Wang, Lin Zhan, Hui-Feng Xi, Otto T. Bruhns, and Heng Xiao

Abstract A survey is presented for new approaches and main results in developing finite deformation elastic and inelastic constitutive models of continua based on Hencky's logarithmic strain and the co-rotational logarithmic rate. Emphasis is placed on four aspects, including (i) a new set of Hencky invariants by means of which new and explicit approaches are established to obtain multi-axial elastic potentials for rubber-like solids; (ii) log-rate-based self-consistent elastoplastic constitutive models for finite deformation behaviors of usual metals, shape memory alloys and soft solids; (iii) innovative elastoplastic J_2-flow models automatically incorporating cyclic and non-cyclic failure effects as inherent constitutive features; as well as (iv) the latest discovery of the deformable micro-continua that display all known quantum effects exactly as do quantum entities at atomic scale, such as electrons, etc. These suggest that both Hencky strain and the co-rotational logarithmic rate play a unified role in modeling large elastic and inelastic deformation behaviors of a wide variety of continua covering usual metals and alloys, shape memory alloys, polymeric solids and, perhaps unexpectedly, quantum entities at atomic scale. In particular, complete responses over the entire strain range up to failure are also covered in a broad, unified sense. In passing, most recent issues raised concerning the appropriateness of the logarithmic rate with reference to the elastoplastic J_2-flow model are clarified by

Si-Yu Wang · Lin Zhan · Hui-Feng Xi · Heng Xiao
School of Mechanics and Construction Engineering and MOE Key Lab of Disaster Forecast and Control in Engineering, Jinan University, 510632 Guangzhou, China,
e-mail: siyu0904@foxmail.com, notzhanlin@foxmail.com, xihuifeng@jnu.edu.cn

Otto T. Bruhns
Institute of Mechanics, Ruhr-University Bochum,D-44780 Bochum, Germany,
e-mail: otto.bruhns@rub.de

Corresponding author: Heng Xiao
Shanghai Institute of Applied Mathematics and Mechanics & School of Mechanics and Engineering Science, Shanghai University, 200072 Shanghai, China,
e-mail: xiaoheng@shu.edu.cn

H. Altenbach and A. Öchsner (eds.), *State of the Art and Future Trends in Material Modeling*, Advanced Structured Materials 100,
https://doi.org/10.1007/978-3-030-30355-6_19

examining the applicability ranges of both the Hencky elastic potential and the von Mises yield function, etc.

Keywords: Metals · Elastomers · SMAs · SMPs · Quantum entities · Finite deformation · Logarithmic strain · Logarithmic rate · Elasticity · Plasticity · Constitutive Modeling · Unified approach

19.1 Introduction

Large deformation elastic and inelastic constitutive models of metallic solids and rubber-like solids play essential roles in effective design of components and parts in engineering structures and, in particular, in assessing their in-service safety and reliability. Such models for the purpose of engineering design are required to establish direct relationships between stress, deformation and temperature as well as their histories. Here, attention is directed to these direct models for large elastic and inelastic deformation behaviors of continua.

On one hand, it is customary to model large elastic deformation behaviors of rubber-like solids with the scalar function of a strain tensor, known as the elastic strain-energy function or the elastic potential. For this purpose, work-conjugate pairs of strain and stress tensors are usually taken into consideration (Ogden, 1984) and, in particular, the choice of a suitable strain measure proves to be essential. On the other hand, the Eulerian rate-type formulation[1] of finite elastoplastic deformation behaviors establishes objective Eulerian rate constitutive equations for the elastic and the plastic part of the natural deformation rate, i.e. the stretching tensor. This formulation presents a direct, natural extension of the well established Prandtl-Reuss equations to finite deformations (Bruhns, 2014a). Use of objective tensor rates is characteristic of such formulation.

In the past two decades, Hencky's logarithmic strain and the co-rotational logarithmic rate have been found to play substantial roles both in resolving fundamental issues and in leading to new and explicit approaches toward modeling large elastic and inelastic deformation behaviors of continua. Here, a survey[2] will be presented for new approaches and main results in six topics, including (i) a new set of Hencky invariants by means of which a new, explicit approach is established to obtain hyper-elastic potentials for rubber-like solids; (ii) log-rate-based self-consistent elastoplastic constitutive models for usual metals and alloys; (iii) log-rate-based elastoplastic J_2−flow models for shape memory effects and pseudo-elastic effects of shape memory alloys; (iv) log-rate-based elastoplastic constitutive models for rate effects and the Mullins effect of polymeric solids; (v) innovative elastoplastic J_2−flow models automatically incorporating cyclic and non-cyclic failure effects as inherent constitutive features; as well as (vi) the latest discovery of the deformable micro-continua that display

[1] Other formulations, such as the Lagrangian formulation and the multiplicative formulation etc., will be merely touched on in subsequent account.

[2] Cf. an early survey (Xiao, 2005) for the origin of Hencky strain and its applications.

all known quantum effects exactly as do quantum entities at atomic scale, such as electrons, etc. In passing, most recent issues raised concerning the reasonableness of the logarithmic rate with reference to the elastoplastic J_2−flow model will be clarified by examining the applicability ranges of both the Hencky elastic potential and the von Mises yield function.

19.2 Hencky Invariants and Rubber-like Elasticity

19.2.1 Modeling of Rubber-like Elasticity

Large elastic deformation behaviors of rubber-like solids are characterized by the elastic potential as the scalar function of a strain tensor. The central issue in rubber-like elasticity is finding out forms of this potential for various kinds of highly elastic solids. This central issue has attracted much attention in the past decades. Indeed, "the area of rubber-like elasticity has had one of the longest and most outstanding histories in all of polymeric science", as pointed out in Erman and Mark (1989). After 80 years' continuing efforts, however, it appears that a complete solution is still to be worked out and much remains to be done.

The essential complexity lies in the fact that the nonlinear elastic behavior of a rubber-like solid is strongly dependent on deformation modes, namely, distinct nonlinear response features will be expected for various deformation modes. Accordingly, toward comprehensive characterization of the nonlinear elasticity of a rubbery solid, nonlinear response features for a few benchmark deformation modes, including uniaxial extension, equi-biaxial extension and plane-strain extension, should be known from testing (cf. Jones and Treloar, 1975; Treloar, 1958, for early results) and, then, a multi-axial elastic potential should be presented to reproduce as closely as possible such nonlinear response features for these benchmark modes and, in the meanwhile, should provide reasonable predictions for response features for other deformation modes.

Both statistical approach and phenomenological approach have been developed in obtaining forms of the elastic potential for various kinds of elastomeric materials. On the one hand, the statistical approach derives various forms of elastic potentials in terms of certain micro-structural parameters via averaging procedures based on certain approximations and idealizations concerning the network structures of long chainlike macromolecules. Numerous results in this respect are available, see, e.g. the survey articles Boyce et al (2000); Ogden et al (2006) and the references therein for some representative works. On the other hand, the phenomenological approach directly presents various nonlinear forms of the elastic potential in terms of either strain invariants or principal stretches with a number of unknown parameters. Then, predictions of the presented elastic potentials are compared with benchmark test data in identifying the unknown parameters introduced. Many results have also been obtained in this respect. For details, refer to, e.g. Ogden (1984); Treloar (1958) for early results and some reviews (Beatty and Millard, 1987; Horgan and Saccomandi, 2006) for certain recent results.

Of the existing results, those with fewer parameters (e.g., two parameters) display excellent agreement with uniaxial extension data up to a very large stretch ratio of 800%, but merely achieve approximate agreement with biaxial extension data. On the other hand, those with more parameters (e.g., six parameters) can achieve good agreement with both uniaxial and biaxial data, but have to cope with the cumbersome task of identifying more parameters. Details may be found in the foregoing references.

Moreover, it appears that all the existing results are still at the stage with no reference to the bounded stress and the softening effect up to failure. On the contrary, unbounded strain energies would be predicted with unbounded stresses.

Recently, new approaches have been established based on a new set of invariants of the Hencky strain. These new approaches are explicit and straightforward in the sense that benchmark test data for uniaxial and biaxial extension as well as plane-strain extension can be automatically reproduced by means of single-variable stress-strain functions for the three benchmark deformation modes. Furthermore, realistic effects over the entire strain range can be simulated, including bounded stresses and softening effects up to eventual failure. Main results in these recent developments will be explained below.

19.2.2 Direct Potential with Hencky Strain

Infinitely many strain tensors may be introduced to prescribe multi-axial strained states at finite deformations. Details may be found in Ogden (1984). Of them, Hencky's logarithmic strain or Hencky strain (Hencky, 1928) was given prominence (cf., e.g. Hill, 1968, 1970, 1979). Let $\boldsymbol{F}$ be the deformation gradient. The three eigenvalues of the left Cauchy-Green deformation tensor $\boldsymbol{G} = \boldsymbol{F} \cdot \boldsymbol{F}^T$ are designated by λ_1^2, λ_2^2, λ_3^2 and the three corresponding orthonormal eigenvectors by $\boldsymbol{n}_1$, $\boldsymbol{n}_2$, $\boldsymbol{n}_3$. Then, the Hencky strain, denoted $\boldsymbol{h}$, is given by

$$\boldsymbol{h} = \frac{1}{2} \ln \boldsymbol{G} = \sum_{s=1}^{3} (\ln \lambda_s)\, \boldsymbol{n}_s \otimes \boldsymbol{n}_s\,. \tag{19.1}$$

For an isotropic hyper-elastic solid, the Kirchhoff stress $\boldsymbol{\tau}$, i.e. the Cauchy stress $\boldsymbol{\sigma}$ weighted by the volumetric ratio $J = \det\boldsymbol{F}$, is directly derivable from an elastic potential of the Hencky strain $\boldsymbol{h}$, denoted $W = W(\boldsymbol{h})$, namely,

$$\boldsymbol{\tau} = J\boldsymbol{\sigma} = \frac{\partial W}{\partial \boldsymbol{h}}\,, \tag{19.2}$$

where the Hencky-strain-based potential[3] $W = W(\boldsymbol{h})$ is an isotropic scalar function of the Hencky strain $\boldsymbol{h}$. Details may be found in Hill (1968, 1970); Fitzgerald (1980); Xiao and Chen (2003).

For an incompressible, isotropic hyper-elastic solid, the direct potential relation Eq. (19.2) is reducible to (cf., Hill, 1968, 1970; Xiao et al, 2004)

$$\sigma = \frac{\partial W}{\partial \boldsymbol{h}} + q\boldsymbol{I}\,, \tag{19.3}$$

with the incompressibility constraint below:

$$\mathrm{tr}\boldsymbol{h} = 0\,. \tag{19.4}$$

Here, q is the all-around stress, $\boldsymbol{I}$ is the identity tensor and $\mathrm{tr}\boldsymbol{A} = A_{11} + A_{22} + A_{33}$.

19.2.3 Bridging Invariants and Mode Invariant

Since the potential $W = W(\boldsymbol{h})$ is isotropic, as indicated before, it is reducible to a function of three basic invariants of the Hencky strain $\boldsymbol{h}$. The latter are given by $\mathrm{tr}\boldsymbol{h}^r$, $r = 1, 2, 3$. For our purpose, new Hencky invariants are needed, as given below.

The first Hencky invariant is

$$i_1 = \mathrm{tr}\boldsymbol{h}\,. \tag{19.5}$$

This invariant provides the simple, direct condition for the incompressibility constraint as given in Eq. (19.4).

Two bridging invariants are given as follows:

$$\varphi = \sqrt{\frac{2}{3} j_2}\,, \tag{19.6}$$

$$\phi = \sqrt{\frac{1}{2} j_2}\,, \tag{19.7}$$

with

$$j_2 = \mathrm{tr}\tilde{\boldsymbol{h}}^2\,, \tag{19.8}$$

where $\tilde{\boldsymbol{h}}$ is the deviatoric part of the Hencky strain $\boldsymbol{h}$, namely,

[3] The elastic potential is expressible as the scalar function of any chosen strain tensor other than the Hencky strain, but the direct potential relationships as in Eqs. (19.2)-(19.3) would no longer hold true.

$$\tilde{\boldsymbol{h}} = \boldsymbol{h} - \frac{1}{3}(\mathrm{tr}\boldsymbol{h})\boldsymbol{I}\,. \tag{19.9}$$

For incompressible solids, Eq. (19.4) holds true and hence $\boldsymbol{h} = \tilde{\boldsymbol{h}}$. The two invariants φ and ϕ exactly supply the axial Hencky strains for the uniaxial extension mode and the plane-strain extension mode, respectively, and they are accordingly named bridging invariants for these two modes.

Moreover, the following Hencky invariant is introduced:

$$\gamma = \sqrt{6}\frac{j_3}{j_2^{1.5}}\,, \tag{19.10}$$

where j_2 is given by Eq. (19.8) and

$$j_3 = \mathrm{tr}\tilde{\boldsymbol{h}}^3\,. \tag{19.11}$$

The invariant γ ranges from -1 to 1, viz., $-1 \leq \gamma \leq 1$, and can exactly distinguish between all different deformation modes and hence it is named the mode invariant. In particular,

$$\gamma = \begin{cases} +1 & \text{for uniaxial extension mode}\,, \\ -1 & \text{for uniaxial compression mode}\,, \\ 0 & \text{for plane-strain extension mode}\,. \end{cases} \tag{19.12}$$

Details may be found in Xiao (2015b).

19.2.4 Elastic Potentials Automatically Reproducing Uniaxial and Biaxial Responses

Let σ and h denote the axial Cauchy stress and the axial Hencky strain in the uniaxial extension and compression modes, respectively. For any given uniaxial stress-strain relation, namely,

$$\sigma = f(h)\,, \tag{19.13}$$

the strain energy below is available for the uniaxial extension and compression modes:

$$P(h) = \int_0^h \sigma dh = \int_0^h f(h)dh\,. \tag{19.14}$$

With the bridging invariant and the mode invariant given in Eqs. (19.6) and (19.10) as well as the single-variable potential given above, a multi-axial elastic potential is obtainable for incompressible, isotropic hyper-elastic solids, as given below (Xiao, 2012):

$$W = \frac{1}{2}(1+\gamma)P(\varphi) + \frac{1}{2}(1-\gamma)P(-\varphi). \tag{19.15}$$

It may be evident that the above potential can automatically reproduce the uniaxial tensile and compressive response given in Eq. (19.13) for $h \geq 0$ and $h \leq 0$, respectively. As a consequence, whenever the function $f(h)$ can fit test data given for the uniaxial extension and compression modes, the multi-axial potential Eq. (19.15) can automatically fit such data.

It is indicated that the uniaxial function Eq. (19.13) may be given by either a polynomial interpolating function (Wang et al, 2014) or piecewise spline functions (Li et al, 2014) in a sense of achieving accurate agreement with any given data without errors. In particular, it is found (Xiao, 2012; Zhang et al, 2014a) that a simple rational function may be used to fit Treloar's large strain data (Treloar, 1958) with strain-stiffening effects at both extension and compression. Such a function is of the following form:

$$\sigma = f(h) = \frac{2Eh}{\left(1 - \frac{h}{h_e}\right)\left(1 + \frac{h}{h_c}\right)} - Eh, \tag{19.16}$$

where h_e and h_c are referred to as the extension limit and the compression limit. Whenever the strain is approaching either of these two limits, the strain-stiffening effects is displayed with rapidly increasing stress. Results in fitting Treloar's data are shown in Fig. 19.1.

As contrasted with other forms of the elastic potential with a number of adjustable parameters, the two strain limits h_e and h_c in the new potential Eq. (19.15) with Eq. (19.16) are of direct physical meanings and obtainable directly from sufficient test data in arriving at good agreement with Treloar's data (Treloar, 1958) for all three benchmark modes.

The new potential Eq. (19.15) is further developed to treat general biaxial data (Zhang et al, 2014b) and, moreover, further results are given to cover bounded strain energies (Jin et al, 2015; Yu et al, 2015b) and failure effects (Yu et al, 2015a).

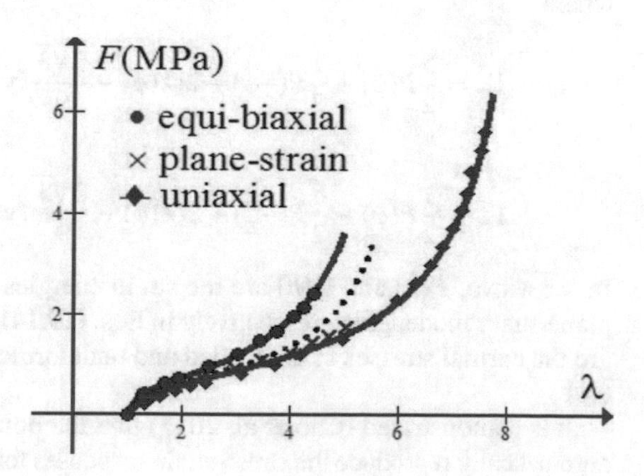

Fig. 19.1 Model predictions compared with Treloar's data (Treloar, 1958): the upper, middle, lower predicted curves for biaxial, plane-strain, uniaxial extension; F and λ for the nominal stress and the stretch in the loading direction

19.2.5 Elastic Potentials Automatically Reproducing both Uniaxial and Plane-strain Responses

Although the new potential Eq. (19.15) can automatically reproduce the uniaxial tensile and compressive responses, for the plane-strain mode it can not simultaneously achieve good agreement with test data for the two normal stresses in the loaded and undeformed directions. Most recently, new multi-axial potentials (Cao et al, 2017) have been explicitly constructed based on Hermitian interpolating procedures and the three Hencky invariants given in Eqs. (19.6)-(19.7) and (19.10). For the plane-strain mode, let h be the Hencky strain in the loaded direction and let σ_1 and σ_2 be the two normal stresses in the loaded and undeformed directions. The stress responses in the plane-strain mode are given by two functions below:

$$\begin{cases}\sigma_1 = y(h)\,,\\ \sigma_2 = s(h)\,.\end{cases} \tag{19.17}$$

Then, the strain energy for the plane-strain mode is as follows:

$$Q(h) = \int_0^h \sigma_1 dh = \int_0^h y(h)dh\,. \tag{19.18}$$

A new multi-axial potential is obtained by extending Hermitian interpolating procedures to the three benchmark deformation modes, i.e. the uniaxial extension and compression and the plane-strain extension. The essential idea is to use the mode invariant γ (cf., Eq. (19.10)) in representing the three benchmark modes (cf., Eq. (19.12)). The new potential is of the following form (Cao et al, 2017):

$$W = (\gamma+1)^2\left[\frac{2-\gamma}{4}P(\varphi) + \frac{\gamma-1}{4}\Gamma_+\right] + (\gamma-1)^2\left[\frac{2+\gamma}{4}P(-\varphi) + \frac{\gamma+1}{4}\Gamma_-\right], \tag{19.19}$$

where

$$\Gamma_+ = \frac{5}{2}P(\varphi) - \frac{1}{2}P(-\varphi) - 2Q(\phi) - \frac{2\sqrt{3}}{9}\left(y(\phi) - 2s(\phi)\right)\,, \tag{19.20}$$

$$\Gamma_- = \frac{1}{2}P(\varphi) - \frac{5}{2}P(-\varphi) + 2Q(\phi) - \frac{2\sqrt{3}}{9}\left(y(\phi) - 2s(\phi)\right)\,. \tag{19.21}$$

In the above, $P(h)$ and $Q(h)$ are the strain energies for the uniaxial mode and the plane-strain mode, given respectively in Eqs. (19.14) and (19.18), and $y(h)$ and $s(h)$ are the normal stresses in the loaded and undeformed directions in the plane-strain case.

It is demonstrated (Cao et al, 2017) that the potential Eqs. (19.19)-(19.21) can automatically reproduce the stress-strain responses for all the three benchmark modes.

With the three response functions $f(h)$, $y(h)$ and $s(h)$ (cf., Eqs. (19.13) and (19.17)) given by rational functions of the simple form as in Eq. (19.16), results are presented in fitting Jones and Treloar's data (Jones and Treloar, 1975), as shown in Figs. 19.2 and 19.3. Results are also given in fitting data for gels up to breaking. Details may be found in Cao et al (2017).

Further results are also obtained in a few respects, including the best approximation with error estimation (Xiao, 2015b; Gu et al, 2015) and extensions to compressibility effects (Xiao, 2013a; Yuan et al, 2015; Xiao et al, 2017), as well as treatments of breaking effects (Cao et al, 2017; Xiao et al, 2017) for gellan gels.

19.3 Self-consistent Prandtl-Reuss Equations with Log-rate

We now direct attention to finite elastoplastic deformations for metallic solids. In the past decades, numerous constitutive formulations of finite elastoplastic deformations have been proposed from various standpoints. Representatives of them are the Prandtl-Reuss formulation, the Lagrangian formulation and the multiplicative formulation.

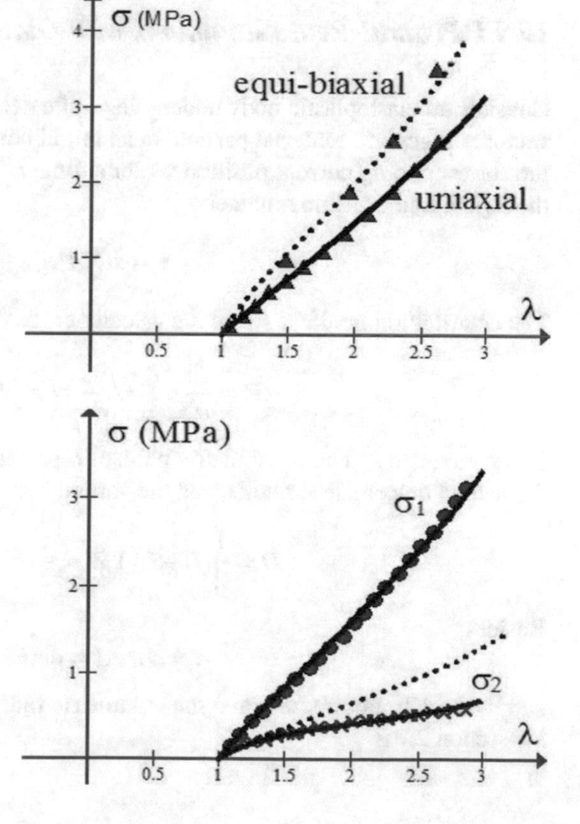

Fig. 19.2 Model predictions compared with Jones and Treloar's data (Jones and Treloar, 1975): Uniaxial and equi-biaxial extension (solid triangles for test data)

Fig. 19.3 Model predictions compared with Treloar's data (Jones and Treloar, 1975): Plane-strain extension (solid dots and crosses for test data; dotted line for the stress σ_2 derived from the potential given in Eq. (19.15)

As indicated in the introduction section, the Prandtl-Reuss formulation is of Eulerian rate type and establishes elastoplastic constitutive equations between the stretching and objective stress rates etc. As a natural, direct extension of the classical Prandtl-Reuss formulation for small deformations, the just mentioned formulation for finite deformations is deeply rooted in the long tradition (Bruhns, 2014a) of the classical theory of elastoplasticity. After certain inherent inconsistency issues have been disclosed, however, it appears that this formulation has been discredited in a broad sense and relegated to a status that it could merely be used in an approximate sense with very small elastic strain. As such, essentially different formulations have been established, such as the Lagrangian formulation and the multiplicative formulation, etc. For details in this respect, refer to the early critical review by Naghdi (1990) and the recent survey by Xiao et al (2006a).

In a series of recent developments, new self-consistent Prandtl-Reuss equations have been established by eliminating the inconsistency issues inherent in the Prandtl-Reuss formulation with objective stress rates. The essential idea is to use the newly discovered co-rotational logarithmic rate that is based on the unique kinematic feature of the Hencky strain Eq. (19.1). Results in this respect and most recent advances will be explained in this and the subsequent sections.

19.3.1 Prandtl-Reuss Equations with Objective Rates

Consider an elastoplastic body undergoing finite deformation. Let $\mathbf{X}$ be the position vector of a generic material particle in an initial configuration at $t = 0$ and let $\mathbf{x}$ be the corresponding current position vector at time t. Then, the latter is a mapping of the former and the time t, namely,

$$\mathbf{x} = \mathbf{x}(\mathbf{X}, t).$$

The deformation gradient $\boldsymbol{F}$ and the velocity gradient $\boldsymbol{L}$ are given by

$$\boldsymbol{F} = \frac{\partial \mathbf{x}}{\partial \mathbf{X}}, \boldsymbol{L} = \frac{\partial \dot{\mathbf{x}}}{\partial \mathbf{x}} = \dot{\boldsymbol{F}} \cdot \boldsymbol{F}^{-1} . \tag{19.22}$$

The symmetric and anti-symmetric parts of $\boldsymbol{L}$ provide the stretching tensor, namely, the natural deformation rate $\boldsymbol{D}$, and the vorticity tensor $\boldsymbol{W}$, viz.,

$$\boldsymbol{D} = \frac{1}{2}(\boldsymbol{L} + \boldsymbol{L}^T), \boldsymbol{W} = \frac{1}{2}(\boldsymbol{L} - \boldsymbol{L}^T). \tag{19.23}$$

Besides,

$$\boldsymbol{\tau} = J\boldsymbol{\sigma},\ J = \det\boldsymbol{F}, \tag{19.24}$$

supply the Kirchhoff stress and the volumetric ratio (the Jacobian), as indicated in subsection 2.2.

The Eulerian rate formulation of finite elastoplasticity is based on the additive separation of the stretching $\boldsymbol{D}$ below (Xiao et al, 2006a):

$$\boldsymbol{D} = \boldsymbol{D}^e + \boldsymbol{D}^p, \tag{19.25}$$

With this separation, objective Eulerian rate constitutive equations need be established to prescribe the elastic part $\boldsymbol{D}^e$ and the plastic part $\boldsymbol{D}^p$, as will be done below.

First, a hypo-elastic equation of Eulerian rate type should be given to relate the elastic part $\boldsymbol{D}^e$ in the separation Eq. (19.25) to a stress rate. For this purpose, an objective stress rate, denoted $\mathring{\boldsymbol{\tau}}$, should be used in order to fulfill the objectivity requirement. Such an elastic rate equation is of the form (cf., e.g. Xiao et al, 2006a; Bruhns et al, 1999, 2003, 2005)

$$\boldsymbol{D}^e = \frac{\partial^2 \overline{W}}{\partial \boldsymbol{\tau}^2} : \mathring{\boldsymbol{\tau}}, \tag{19.26}$$

where $\overline{W}$ is a complementary elastic potential characterizing finite hyper-elastic behavior. Equation (19.26) is for initial isotropy. Refer to Xiao et al (2007b) for treatment of initial anisotropy.

The next step is to formulate a flow rule for the plastic part $\boldsymbol{D}^p$ in the separation Eq. (19.25). It has been demonstrated Bruhns et al (2005) that a normality flow rule of the following form (cf., e.g., Xiao et al, 2006a; Bruhns et al, 1999, 2003, 2005) is derivable from a weakened form of Ilyushin's postulate:

$$\boldsymbol{D}^p = \frac{1}{2}\frac{\zeta}{\breve{h}}\left(\breve{f} + |\breve{f}|\right)\frac{\partial f}{\partial \boldsymbol{\tau}}. \tag{19.27}$$

In the above, ζ and f are the plastic indicator taking values 1 and 0 in the loading and unloading cases (Bruhns et al, 1999, 2003) and the yield function, while $\breve{f}$ and $\breve{h}$ are the strain-rate-based loading factor and the normalized plastic modulus. They are given below, separately.

First, the yield function f is the von Mises function with combined hardening effects, namely,

$$f = \frac{1}{2}\mathrm{tr}\,(\tilde{\boldsymbol{\tau}} - \boldsymbol{\chi})^2 - \frac{1}{3}r^2. \tag{19.28}$$

In the above, the yield strength r is of the form:

$$r = r(\vartheta) \tag{19.29}$$

with the dissipated work ϑ specified by

$$\dot{\vartheta} = (\tilde{\boldsymbol{\tau}} - \boldsymbol{\chi}) : \boldsymbol{D}^p. \tag{19.30}$$

The ϑ-dependent yield strength r characterizes the isotropic hardening effect. Moreover, the kinematic or anisotropic hardening effect is characterized by the back stress $\boldsymbol{\chi}$ governed by the evolution equation below:

$$\mathring{\chi} = c\boldsymbol{D}^p - \xi\dot{\vartheta}\chi\,, \tag{19.31}$$

where c and ξ are referred to as the Prager modulus and the hysteresis modulus, respectively.

Next, the strain-rate-based loading factor $\check{f}$ and the normalized plastic modulus $\check{h}$ are of the following forms:

$$\check{f} = \frac{\partial f}{\partial \tau} : \mathbb{S} : \boldsymbol{D}\,, \tag{19.32}$$

$$\check{h} = \frac{\partial f}{\partial \tau} : \mathbb{S} : \frac{\partial f}{\partial \tau} - \frac{\partial f}{\partial \vartheta}\left((\tilde{\tau} - \chi) : \frac{\partial f}{\partial \tau}\right) - \frac{\partial f}{\partial \chi} : \mathbb{H} : \frac{\partial f}{\partial \tau}\,, \tag{19.33}$$

where $\mathbb{S}$ is the elastic stiffness tensor given by

$$\mathbb{S} = \left(\frac{\partial^2 \overline{W}}{\partial \tau^2}\right)^{-1} \tag{19.34}$$

and $\mathbb{H}$ is the hardening tensor by

$$\mathbb{H} = c\mathbb{I} - \xi\chi \otimes (\tilde{\tau} - \chi) \tag{19.35}$$

with the 4th-order identity tensor $\mathbb{I}$.

Finally, the plastic indicator ζ is given by (cf., Xiao et al, 2007b):

$$\zeta = \begin{cases} 1 \text{ for } f = 0,\ \check{f} > 0, \\ 0 \text{ for } f < 0 \text{ or } (f = 0, \check{f} \leq 0). \end{cases} \tag{19.36}$$

The choice of the objective rates in Eq. (19.26) and Eq. (19.31) will be discussed later on. Moreover, the elastoplastic deformation features are characterized by the three constitutive quantities including the complementary elastic potential $\overline{W}$, the yield strength r, the Prager modulus c and the hysteresis modulus ξ. Their forms will be presented in the sequel for different cases.

19.3.2 Inconsistency Issues with Zaremba-Jaumann Rate

During the years from 1950 to 1960, attention was directed to the definition of an appropriate objective stress rate. Of the four objective rates then known, the Zaremba-Jaumann rate was selected earlier by Prager (1960) based on the yielding stationarity criterion and had been widely used since then. However, the foundation of the Prandtl-Reuss formulation was shaken by the discoveries of the spurious shear oscillating responses (Lehmann, 1964, 1972; Dienes, 1979; Nagtegaal and Jong, 2010). According to such responses, the shear stress would not be increasing but oscillating with increasing shear strain, indicating that there would be something wrong with the Prandtl-Reuss formulation based on the Zaremba-Jaumann rate.

The inherent inconsistency was further substantiated by the non-integrability issue of the elastic rate equation. For metallic solids with small elastic strain, the complementary elastic potential $\overline{W}$ in the elastic rate equation Eq. (19.26) is given as follows:

$$\overline{W} = \frac{1}{4G}\mathrm{tr}\boldsymbol{\tau}^2 - \frac{\nu}{2E}(\mathrm{tr}\boldsymbol{\tau})^2. \tag{19.37}$$

In the above, E, ν and G are used to denote the Young's modulus, the Poisson ratio and the shear modulus evaluated at infinitesimal strain, respectively, and $E = 2G(1+\nu)$. The elastic rate equation (19.26) with the above potential reduces to

$$\boldsymbol{D}^e = \frac{1}{2G}\mathring{\boldsymbol{\tau}} - \frac{\nu}{E}(\mathrm{tr}\mathring{\boldsymbol{\tau}})\boldsymbol{I}. \tag{19.38}$$

For the pure elastic deformation case with $\boldsymbol{D}^e = \boldsymbol{D}$, it was found (Simo and Pister, 1984) that the above elastic rate equation with the Zaremba-Jaumann rate below:

$$\mathring{\boldsymbol{\tau}} = \dot{\boldsymbol{\tau}} + \boldsymbol{\tau}\cdot\boldsymbol{W} - \boldsymbol{W}\cdot\boldsymbol{\tau} \tag{19.39}$$

would not be self-consistent, in the sense that it would fail to be integrable to deliver an elastic stress-strain relation. Furthermore, it was known (Simo and Pister, 1984) that that would be case for a few objective rates then known.

The above finding means that none of the rates then known is compatible with the definition of elasticity. Further issues are studied, e.g., in Xiao et al (2005, 2006b). Since then, there had been a general tendency to believe (Lubliner, 1992; Khan and Huang, 1995; Simo and Hughes, 2008) that the disclosed inconsistency issue might be inherent in the Prandtl-Reuss formation.

19.3.3 *Self-consistent Formulation with Log-rate*

Toward clarifying the above situation, it has been indicated (Xiao et al, 2000a) that, in addition to the fundamental criterion from the objectivity (frame-indifference) requirement, a consistent formulation of Eulerian elastoplasticity should fulfill certain further consistency criteria, including the self-consistency criterion for the elastic rate equation and Prager's yielding stationarity criterion for the composite structure of Eulerian elastoplastic rate equations. In this sense, self-consistent Eulerian elastoplastic J_2−flow models have been established in a previous study (Bruhns et al, 1999) based on the newly discovered logarithmic rate (cf., e.g., Xiao et al, 1997). These models have been shown (Xiao et al, 1999, 2000a) to be unique among all Eulerian models based on all possible co-rotational stress rates and other known stress rates by virtue of the consistency criteria in the foregoing.

The above development arises from the discovery of the unique kinematic feature of the Hencky strain $\boldsymbol{h}$ in correlation with the stretching $\boldsymbol{D}$. It is found (Xiao et al,

1997) that, in a spinning frame with the spin $\boldsymbol{\Omega}^{\log}$, the changing rate of the Hencky strain $\boldsymbol{h}$ is identical to the stretching, viz.,

$$\dot{\boldsymbol{h}} + \boldsymbol{h} \cdot \boldsymbol{\Omega}^{\log} - \boldsymbol{\Omega}^{\log} \cdot \boldsymbol{h} = \boldsymbol{D} . \tag{19.40}$$

It is known (Xiao et al, 1997) that any given strain tensor other than the Hencky strain $\boldsymbol{h}$ could not enjoy the above property, namely,

$$\dot{\boldsymbol{S}} + \boldsymbol{S} \cdot \boldsymbol{\Omega} - \boldsymbol{\Omega} \cdot \boldsymbol{S} \neq \boldsymbol{D} \tag{19.41}$$

for any given strain tensor $\boldsymbol{S} \neq \boldsymbol{h}$ and for every possible spin (skew-symmetric tensor) $\boldsymbol{\Omega}$.

The spin $\boldsymbol{\Omega}^{\log}$ is uniquely determined from the kinematic identity Eq. (19.40) and hence referred to as the logarithmic spin. Its expression is given in Xiao et al (1997, 1999, 2000a) (see also Xiao et al, 1998a,b).

With the logarithmic spin $\boldsymbol{\Omega}^{\log}$, the logarithmic stress rate, denoted $\mathring{\boldsymbol{\tau}}^{\log}$, is defined as follows (Xiao et al, 1997):

$$\mathring{\boldsymbol{\tau}}^{\log} \equiv \dot{\boldsymbol{\tau}} + \boldsymbol{\tau} \cdot \boldsymbol{\Omega}^{\log} - \boldsymbol{\Omega}^{\log} \cdot \boldsymbol{\tau} . \tag{19.42}$$

It is demonstrated in Xiao et al (1997, 1999, 2000a) that the inconsistency issue in the foregoing can be eliminated simply by replacing the rate $\mathring{\boldsymbol{\tau}}$ with the logarithmic stress rate given above. As such, the elastic rate equation Eq. (19.26), in particular, Eq. (19.38), is exactly integrable to deliver a hyper-elastic equation based on the Hencky strain, namely,

$$\boldsymbol{D} = \frac{\partial^2 \overline{W}}{\partial \boldsymbol{\tau}^2} : \mathring{\boldsymbol{\tau}}^{\log} \quad \Longleftrightarrow \quad \boldsymbol{h} = \frac{\partial \overline{W}}{\partial \boldsymbol{\tau}} . \tag{19.43}$$

Note that the elastic equation in the above is just the dual formulation of Eq. (19.2) via the Legendre transformation (Xiao and Chen, 2003) below:

$$W + \overline{W} = \boldsymbol{\tau} : \boldsymbol{h} . \tag{19.44}$$

With Prager's yielding stationarity criterion (Prager, 1960), it is further demonstrated in Xiao et al (2000a) that the rate of the back stress $\boldsymbol{\chi}$ in the evolution equation (19.31) should also be given by the logarithmic rate.

Further results may be found in Bruhns et al (2004) for non-corotational rates and in Xiao et al (2007b) both for thermal effects and for thermodynamic consistency, and simple shear and torsion problems with the Swift effect are studied in Bruhns et al (2001b); Xiao et al (2001); Bruhns et al (2001c).

19.3.4 Remarks on Recently Raised Issues

In a most recent article (Jiao and Fish, 2017), the Prandtl-Reuss formulation based on the separation of the natural deformation rate has been given prominence by highlighting certain favorable constitutive features and, however, the applicability of the logarithmic rate in this formulation has been questioned by presenting numerical experiments on presumably unreasonable stress ratcheting responses for a certain type of strain cycles starting at an induced plastic state, among other things.

In response to the above concern, constitutive implications of elastoplastic J_2−flow models based on the logarithmic rate will further be explained by examining the applicability ranges of the Hencky elastic potential and the von Mises yield function incorporated. Response features of this model will be studied for strain cycles starting at a plastic state.

Toward modeling different elastoplastic features of various kinds of realistic materials, suitable forms of the elastic potential W in the elastic rate equation (19.26) and the yield function in the flow rule Eq. (19.27) should be presented in a sense consistent with experimental facts. It may be essential to note that a given form of either W or f is merely applicable for a certain range of deformation and would have no relevance to any realistic material behavior in an extreme case far beyond such a range, irrespective of the fact that from a formal mathematical standpoint it may be well defined over the whole range of deformation. Below, details will be explained for the widely used quadratic elastic potential as given in Eq. (19.37) and the von Mises yield function as given in Eq. (19.28) and, moreover, remarks will be given for other cases. For the sake of simplicity, the anisotropic hardening effect is neglected with $\boldsymbol{\chi} = \boldsymbol{O}$.

19.3.4.1 The Hencky Potential for Moderate Elastic Strain

A simple form of the complementary elastic potential W as given in Eq. (19.37) is quadratic and referred to as Hencky potential. The elastic rate equation (19.26) with the Hencky potential above is given by Eq. (19.38). Prior to the initial yielding, from Eq. (19.43) it follows that the integration of Eq. (19.38) with $\boldsymbol{D}^e = \boldsymbol{D}$ produces a finite strain hyper-elastic equation of Hookean type below:

$$\boldsymbol{h} = \frac{1}{2G}\boldsymbol{\tau} - \frac{\nu}{E}(\mathrm{tr}\boldsymbol{\tau})\boldsymbol{I}\,. \tag{19.45}$$

Equation (19.45) is known as Hencky elastic equation (Hencky, 1928). It is known (Anand, 1979, 1986) that this equation can well represent moderate elastic deformations with each principal stretch falling within the range [0.7,1.3]. Beyond this range, however, the Hencky elastic equation (19.45) would be no longer applicable (cf., Anand, 1979, 1986; Bruhns et al, 2001a) and merely of formal sense.

19.3.4.2 Small Volumetric Strain: von Mises Yield Function

On the other side, a simple form of the yield function f is also quadratic and in the absence of anisotropic hardening it is given by the von Mises function below:

$$\begin{cases} f = \frac{1}{2}J_2 - \frac{1}{3}r^2, \\ J_2 = \frac{1}{2}\mathrm{tr}\tilde{\boldsymbol{\tau}}^2, \\ r = r(\kappa), \end{cases} \tag{19.46}$$

where κ is the plastic work specified by

$$\dot{\kappa} = \boldsymbol{\tau} : \boldsymbol{D}^p. \tag{19.47}$$

It is well known that the von Mises yield function above is applicable for metals with small volumetric strain, e.g., $|J-1| < 0.001$. In this case, the Kirchhoff stress $\boldsymbol{\tau} = J\boldsymbol{\sigma}$ agrees nearly with the Cauchy stress $\boldsymbol{\sigma}$. Note that the latter is used in the original form of von Mises function.

The von Mises function (19.46) would not be applicable for cases beyond small volumetric strain. Here, a relevant point is that, for a volumetric ratio J not close to 1 the Kirchhoff stress $\boldsymbol{\tau}$ deviates considerably from the Cauchy stress $\boldsymbol{\sigma}$.

19.3.4.3 The J_2–flow Model with Isotropic Hardening

The Hencky potential Eq. (19.37) and von Mises yield function Eq. (19.46) jointly lead to the following elastoplastic J_2–flow model with isotropic hardening:

$$\boldsymbol{D} = \frac{1}{2G}\overset{\circ}{\boldsymbol{\tau}}{}^{\log} - \frac{\nu}{E}\left(\mathrm{tr}\overset{\circ}{\boldsymbol{\tau}}{}^{\log}\right)\boldsymbol{I} + \frac{1}{2}\frac{\zeta}{\breve{h}}\left(\breve{f} + |\breve{f}|\right)\tilde{\boldsymbol{\tau}}, \tag{19.48}$$

$$\dot{\kappa} = \frac{1}{3}\frac{\zeta}{\breve{h}}\left(\breve{f} + |\breve{f}|\right)r^2, \tag{19.49}$$

with the plastic indicator ζ given by Eq. (19.36) and the loading factor $\breve{f}$ and the normalized plastic modulus $\breve{h}$ by

$$\breve{f} = 2G\tilde{\boldsymbol{\tau}} : \boldsymbol{D}, \tag{19.50}$$

$$\breve{h} = \frac{4}{9}r^2\left(3G + rr'\right). \tag{19.51}$$

Here, the notation r' is used to denote the derivative of r with respect to κ.

It follows from the last two subsections that the J_2–flow model is applicable for elastoplastic behavior with moderate elastic strain and small volumetric strain. This model could not be used for elastoplastic behavior either with large elastic strain or with large volumetric strain, as will be explained in the next subsection.

19.3.4.4 On Large Elastic Strain and Large Volumetric Strain

Large volumetric strain is involved in elastoplastic behavior of porous materials such as porous metals (Gurson, 1977) and geo-materials (Nemat-Nasser, 1983), etc., while large elastic strain incorporated in elastoplastic deformations is related to elastoplastic behavior of rubber-like materials (cf., e.g., Xiao, 2015a) such as elastomers etc. Both cases would go beyond the applicability range of either the Hencky potential Eq. (19.37) or the von Mises yield function Eq. (19.46) and, accordingly, results derived from the J_2–flow model with large elastic strain and large volumetric strain would be merely of formal sense, irrespective of the fact that, from a mathematical standpoint, both the Hencky potential Eq. (19.37) and the von Mises yield function Eq. (19.46) are well defined over the whole deformation range, as indicated before.

With the above facts in mind, we are going to address the main concern raised in Jiao and Fish (2017).

19.3.4.5 On the Unloading Stress Ratcheting Obstacle Test

A numerical test for the J_2–flow model as described in Subsubsect. 19.3.4.4, called the unloading ratcheting obstacle test, was performed in Jiao and Fish (2017) by calculating the stress response of a rectangular block to a certain type of two-dimensional strain cycles starting at a plastic state. With two directions held fixed, the block is stretched under tensile load along the other direction to reach a plastic state and then brought back to its initial shape by removing the tensile load and applying compressive load. Whenever the block reaches its initial shape, the block is subjected to undergo two-dimensional strain cycles. Specifically, the current position vector $\boldsymbol{x} = x_i \boldsymbol{e}_i$ of a generic particle relies on the initial coordinate variables X_i of this particle and the time t and is given by

$$x_1 = X_1,\ x_2 = \lambda(t) X_2,\ x_3 = X_3\,, \tag{19.52}$$

where

$$\lambda(t) = \begin{cases} 1+t,\ 0 \le t \le 2, \\ 5-t,\ 2 \le t \le 4, \end{cases} \tag{19.53}$$

for the first stage, and

$$\begin{cases} x_1 = X_1 + 0.7(\sin(t-4)\pi) X_2\,, \\ x_2 = (1 + 0.7(1 - \cos(t-4)\pi)) X_2\,, \\ x_3 = X_3\,, \end{cases} \tag{19.54}$$

for strain cycles starting at $t = 4$.

The parameter values below were assumed in Jiao and Fish (2017):

$$E = 1.95 \times 10^{11}\text{MPa},\ \nu = 0.3,\ r_0 = 9 \times 10^{10}\text{MPa}, \tag{19.55}$$

where r_0 is the initial yield strength[4]. It was found in Jiao and Fish (2017) that, for the strain cycles at issue, the log-rate-based J_2-flow model with the parameter values assumed above could not produce the expected cyclic stress responses in the unloading cases but, instead, produce unloading stress ratcheting responses.

However, the above unreasonable responses reported would be merely of formal mathematical sense. In fact, by utilizing the Hencky elastic equation (19.45) and the yield condition

$$\frac{1}{2}J_2 - \frac{1}{3}r_0^2 = 0$$

at the initial yielding, it may be inferred that, at the first stage of the deformation path as given in Eqs. (19.52)-(19.53), the elastic stretch at the initial yielding is given by

$$e^{r_0/2G} = 2.0927$$

for the parameter values assumed and, in particular, it follows from Eq. (19.54) that, at $t = 5$, the principal stretch in the $\boldsymbol{e}_2-$direction is given by 2.4.

On the other side, the volumetric ratio J is given by

$$J = \frac{\partial x_2}{\partial X_2},$$

and, therefore,

$$J = \begin{cases} 3.0,\ t = 2, \\ 2.4,\ t = 5, \end{cases}$$

for the assumed parameter values.

Comparisons with the bounding values 1.3 for elastic stretch and 0.001 for volumetric strain, as indicated in Subsubsects 19.3.4.3 and 19.3.4.4, suggest that the above values for the elastic stretch and the volumetric ratio go drastically beyond the applicability ranges of the Hencky elastic potential and the von Mises yield function. That may particularly be the case for the volumetric strain. As indicated in Subsubsect. 19.3.4.5, for large elastic strain and large volumetric strain, the Hencky elastic potential Eq. (37) and the von Mises yield function Eq. (19.46) would be of no physical relevance and should be replaced by other suitable forms, as has been known in developing elastoplasticity models for elastomers and porous media, etc.

[4] The parameter values for hardening effects, also listed in Jiao and Fish (2017), are not needed here.

19.3.4.6 Remarks

From the foregoing results it follows that the obstacle test reported in Jiao and Fish (2017) would be merely of formal mathematical sense, albeit results therein are sound from a mathematical standpoint.

With the requirement that the elastic property should be kept intact in each process of repeatedly loading and unloading, a modified form of the logarithmic stress rate is proposed in Jiao and Fish (2017). As such, it is assumed that, at each current elastoplastic strained state, there exist an unstressed intermediate state via an elastic loading process, so that the constitutive formulation be unchanged relative to such unstressed intermediate states. It appears that the requirement just mentioned would merely be an idealized assumption concerning elastoplastic deformation behaviors, which could not be consistent with realistic elastoplastic deformation behaviors in two respects. On the one hand, it would be at variance with fatigue effects under cyclic loadings, as will be explained later on in Sect. 19.6. In fact, after experiencing an elastoplastic deformation process, an elastoplastic material should inevitably undergo changes in its initial mechanical behavior with irreversible changes in internal micro-structures such as dislocations and micro-defects, etc., as commonly known from micro-mechanisms responsible for the elastoplastic behavior.

Furthermore, it has been demonstrated in Xiao et al (2000b, 2007a) that the multiplicative elastoplastic formulation should be consistently incorporated as particular cases in the log-rate-based self-consistent elastoplastic formulation, in a mathematical sense of eliminating the non-uniqueness inherent in the latter. Perhaps more essentially, even the unstressed intermediate state via elastic unloading might be inaccessible, as has been indicated in Xiao et al (2006a) and will be exemplified in the next section. Indeed, even in the uniaxial tension case of a bar sample the unstressed intermediate configuration could not be attained by performing an unloading elastic process, as shown in Fig. 19.4, since the inverse yielding would be induced just at a certain tensile stress.

In passing, it should be pointed out that, just like the Zaremba-Jaumann rate (cf., Eq. (19.39)), an objective stress rate as kinematic quantity should be well-defined with no reference to any particular material behavior. However, the modified rate in Jiao and Fish (2017) would be different for different kinds of elastic behaviors.

Finally, short remarks are presented on the so-called weak invariance requirement imposed on finite deformation behaviors of solids, which was cited in Jiao and Fish (2017) and postulated in Shutov and Ihlemann (2014). According to this newly-postulated requirement, the elastoplastic behavior of a solid should be invariant under every volume-preserving change superimposed on a given reference state. It is noted that this assumption would be inconsistent with realistic material behaviors indicated in the foregoing. Also, it may be quite puzzling just with the basic feature of any solid behavior. In fact, change in stress would always be induced under any change relative to an undistorted reference state, except for certain rotations representing the material symmetry. In particular, it may readily be demonstrated that the weak invariance requirement would imply that every elastic solid should be an elastic fluid.

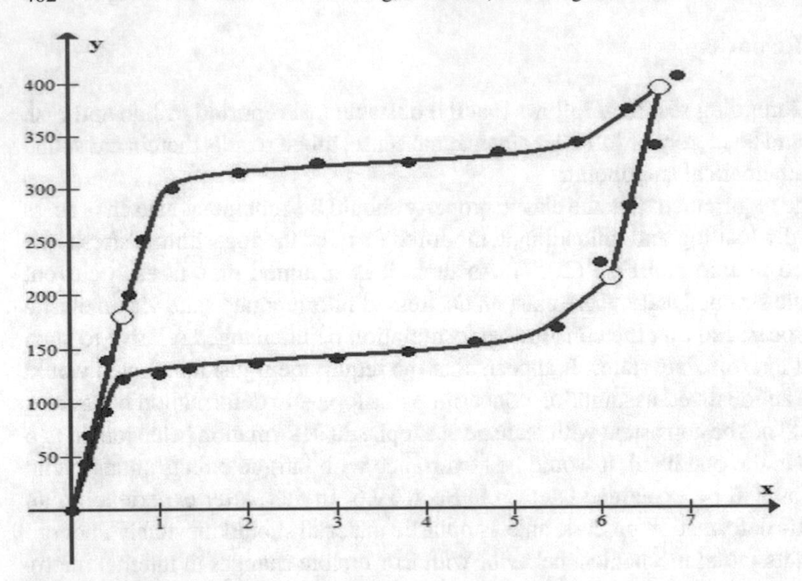

Fig. 19.4 Simulation results compared with test data in Saburi et al (1982) for the pseudo-elastic hysteresis loop of a Ni-Ti alloy (test data in solid dots, the axial Hencky strain x in percent and the axial stress y in MPa)

19.4 Log-rate-based Elastoplastic J_2–flow Equations for Shape Memory Alloy Pseudo-elasticity

A great number of constitutive models have been established to simulate shape memory effects and pseudo-elastic effects of shape memory alloys (SMAs) from various standpoints. Such models are based on either micro-structural mechanisms or phase variables related to solid-solid phase transitions. Details may be found, e.g., in the survey articles (Lagoudas et al, 2006; Patoor et al, 2006).

In most recent developments, it is found that deformation effects of SMAs may be directly simulated based on elastoplastic J_2–flow equations with nonlinear combined hardening, in a sense with involving no phase variables. Early results may be found in Xiao et al (2010a,b) for treating pseudo-elastic loops of simple shapes and in Xiao et al (2011) for the shape memory effect.

A straightforward, explicit approach (Xiao, 2013b) is developed to treat pseudo-elastic loops in which the stress-strain curve at the loading case may be of any given shape. Toward this objective, the elastoplastic J_2–flow equations (19.25)-(19.31) with the Hencky potential in Eq. (19.37) are used. Let

$$h = p(\tau), \tag{19.56}$$

where h and τ are the axial strain and the axial Kirchhoff stress, represent the uniaxial stress-strain curve generated at loading (cf., the upper curve in Fig. 19.4). The yield strength r, the Prager modulus c and the hysteresis modulus ξ are given below in explicit forms:

$$r = \frac{1}{2} r_0 \left(1 + e^{-\beta\vartheta/r_0}\right), \tag{19.57}$$

$$c = \frac{2}{3}\left(K(r) - r r'\right), \tag{19.58}$$

$$r^{-1}\xi = r' - \frac{K(r + |\Lambda|) - K(r)}{\Lambda} \tag{19.59}$$

with

$$\Lambda = 1.5 r^{-1} (\tilde{\tau} - \chi) : \chi, \tag{19.60}$$

$$K(\tau) = p(\tau) p'(\tau). \tag{19.61}$$

With the three quantities given above, the elastoplastic J_2−flow equations (19.25)-(19.31) with Eq. (19.37) automatically reproduce a pseudo-elastic hysteresis loop in each loading-unloading cycle, in which the upper stress-strain curve at loading is just given by the curve represented by the function $h = p(\tau)$.

With the above results, each pseudo-elastic hysteresis loop can be automatically simulated simply by choosing a suitable form of the function $h = p(\tau)$ in fitting test data. A simple form of the function $h = p(\tau)$ is given in Xiao (2013b) with a few parameters. Comparison of simulation results with test data are shown in Fig. 19.4.

The approach proposed in Xiao (2013b) applies merely to pseudo-elastic loops with the upper and lower stress-strain curves parallel with each other. A new approach is developed in Xiao (2014a) for pseudo-elastic loops of any given shape. In this case, the three quantities r, c and ξ can also be given in explicit forms. Details may be found in Xiao (2014a).

Most recently, further results have been obtained in treating a few issues, including tension-compression asymmetry (Wang et al, 2015) of pseudo-elastic loops, the plastic effect (Xiao et al, 2016) in a loading-unloading cycle, as well as the plastic-to-pseudoelastic transition (Zhan et al, 2019a) under multiple loading-unloading cycles.

With the results reported in the foregoing references, a perhaps unexpected yet noticeable finding is as follows: In a loading-unloading cycle in the uniaxial tension, *both the usual plastic flow at the loading stage and the unusual plastic flow at the subsequent unloading stage are induced, as indicated in Fig.* 19.4, *and such two plastic flows jointly lead to the strain recovery effect, i.e. the pseudo-elastic effect, and it turns out that the latter represents such an exceptional kind of anisotropic hardening that the reverse yielding be induced not at a compressive but a tensile stress.* In contrast with this, the reverse yielding for a usual metal is invariably induced at a compressive stress.

19.5 Log-rate-based Elastoplastic Equations for Shape Memory Effects

It has been demonstrated (Xiao, 2015a; Xiao et al, 2011; Li et al, 2017) that the elastoplastic J_2-flow equations (19.25)-(19.31) may be extended to cover thermal effects and such thermo-coupled equations can exactly simulate shape memory effect of both shape memory alloys and polymers.

The main finding is a perhaps noticeable phenomenon, namely, *plastic flow may also be induced in a process of pure heating in the absence of stress and such thermo-induced plastic flow may be responsible for recovery of a pre-strain prior to heating*. This represents the very feature of shape memory materials at heating, including both shape memory alloys (SMAs) and shape memory polymers (SMPs). As will be shown below, simple, exact results may be derived for general multiaxial cases.

19.5.1 Log-rate-based Elastoplastic Equations with Thermal Effects

For our purpose, a complementary thermo-elastic potential is introduced as follows:

$$\hat{W} = \overline{W}(\boldsymbol{\tau}, T), \tag{19.62}$$

where the T is the absolute temperature.

Now, the elastic part $\boldsymbol{D}^e$ in the separation Eq. (19.25) is determined by a log-rate-based thermo-elastic rate equation below (Xiao et al, 2011; Li et al, 2017):

$$\boldsymbol{D}^e = \frac{\partial^2 \overline{W}}{\partial \boldsymbol{\tau}^2} : \overset{\circ}{\boldsymbol{\tau}}{}^{\log} + \frac{\partial^2 \overline{W}}{\partial \boldsymbol{\tau} \partial T} \dot{T}, \tag{19.63}$$

Next, the plastic part $\boldsymbol{D}^p$ in the decomposition Eq. (19.25) is prescribed by the following normality rule Xiao et al (2007b):

$$\boldsymbol{D}^p = \frac{\zeta}{\hat{h}} \left(\frac{\partial f}{\partial \boldsymbol{\tau}} : \overset{\circ}{\boldsymbol{\tau}}{}^{\log} + \frac{\partial f}{\partial T} \dot{T} \right) \frac{\partial f}{\partial \boldsymbol{\tau}}, \tag{19.64}$$

where the yield function f is the von Mises function given in Eq. (19.28). The yield strength r now relies on the temperature T, viz.,

$$r = r(T). \tag{19.65}$$

In addition, the plastic indicator ζ is specified by Xiao et al (2011):

$$\zeta = \begin{cases} 1 \quad \text{for} \quad \left[f = 0, \ \frac{1}{\hat{h}} \left(\frac{\partial f}{\partial \boldsymbol{\tau}} : \overset{\circ}{\boldsymbol{\tau}}{}^{\log} + \frac{\partial f}{\partial T} \dot{T} \right) \geq 0 \right], \\ 0 \quad \text{for} \quad f < 0 \quad \text{or} \quad \left[f = 0, \ \frac{\partial f}{\partial \boldsymbol{\tau}} : \overset{\circ}{\boldsymbol{\tau}}{}^{\log} + \frac{\partial f}{\partial T} \dot{T} \leq 0 \right]. \end{cases} \tag{19.66}$$

The plastic modulus $\hat{h}$ is given in Xiao et al (2011) (cf., Eq. (19.82) given later on).

The back stress $\boldsymbol{\chi}$ is governed by the evolution equation below:

$$\begin{cases} \overset{\circ}{\boldsymbol{\chi}}{}^{\log} = c\boldsymbol{D}^p, \\ c = c(T). \end{cases} \tag{19.67}$$

19.5.2 Plastic Flow Induced at Pure Heating

Now we take into account a heating process with the temperature T changing from T_0 to T_*, in the absence of stress, i.e., $\boldsymbol{\tau} = \boldsymbol{O}$. In this case, the yield strength r is assumed to rely merely on the temperature T. Usually, the yield strength r should decrease with increasing temperature. Namely,

$$\frac{dr}{dT} < 0. \tag{19.68}$$

At an initial temperature T_0, a pre-strain or a plastic strain is induced in a material sample. Let r_0, ϑ_0, $\boldsymbol{F}_0$ and $\boldsymbol{\chi}_0$ be the yield strength, the dissipated work, the deformation gradient and the back stress at $T = T_0$, respectively. At $T = T_0$ following the unloading, the value of the yield function f is negative, i.e.,

$$f_0 = \frac{1}{2}|\boldsymbol{\chi}_0|^2 - \frac{1}{3}(r(T_0))^2 < 0. \tag{19.69}$$

In a heating process with $\dot{T} > 0$, we are going to known whether plastic flow will be induced as from a certain temperature $T_0^* \geq T_0$. To this end, we first study the changing of the yield function with increasing temperature, namely,

$$f = \frac{1}{2}|\boldsymbol{\chi}_0|^2 - \frac{1}{3}(r(T))^2. \tag{19.70}$$

Since the yield strength $r(T)$ is decreasing with increasing temperature (cf., Eq. (19.68)), we deduce that the foregoing value is increasing with increasing temperature. Then, it is possible that f may increase from the initial negative value $f_0 < 0$ (cf., Eq. (19.69)) at T_0 to zero at a certain temperature T_0^*. In this case, we have

$$\frac{1}{2}|\boldsymbol{\chi}_0|^2 - \frac{1}{3}(r(T_0^*))^2 = 0. \tag{19.71}$$

This means that the yield condition is met. Moreover, with vanishing stress $\boldsymbol{\tau} = \boldsymbol{O}$ and increasing temperature $\dot{T} > 0$ as well as Eq. (19.66) we deduce that the following holds true:

$$\frac{1}{\hat{h}}\left(\frac{\partial f}{\partial \boldsymbol{\tau}}:\overset{\circ}{\boldsymbol{\tau}}{}^{\log} + \frac{\partial f}{\partial T}\dot{T}\right) = -(cr)^{-1}\frac{dr}{dT}\dot{T} > 0\,. \tag{19.72}$$

For details, refer to Xiao et al (2011). From Eqs. (19.71)-(19.72) it may be deduced that the loading condition given by Eq. $(19.66)_1$ may be fulfilled. As such, plastic flow may indeed be induced in a pure heating process and emerges as from $T \geq T_0^*$.

A closed-form solution for the response of the above plastic flow at pure heating is derivable from the constitutive equations in Subsubsect. 19.5.1. Results are as follows (Xiao et al, 2011):

$$\boldsymbol{\chi} = \sqrt{\frac{2}{3}}\frac{\boldsymbol{\chi}_0}{|\boldsymbol{\chi}_0|}r(T), \quad T \geq T_0^* \tag{19.73}$$

for the back stress, and

$$\boldsymbol{h} = \frac{1}{2}\ln\left(\boldsymbol{F}_0 \cdot \boldsymbol{F}_0^T\right) + \sqrt{\frac{2}{3}}\frac{\boldsymbol{\chi}_0}{|\boldsymbol{\chi}_0|}\int_{T_0^*}^{T}\frac{1}{c}\frac{dr}{dT}dT \tag{19.74}$$

for the Hencky strain.

19.5.3 Recovery Effect

As indicated before, an initial plastic deformation, $\boldsymbol{F}_0$, and an initial back stress $\boldsymbol{\chi}_0$ are produced in a loading-unloading process at a certain temperature T_0. Then, a process of pure heating follows as from T_0 in the absence of stress. Thermo-induced plastic flow emerges as from a temperature $T_0^* > T_0$ (cf., Eq. (19.71)). The last term in Eq. (19.74) gives the contribution from the plastic flow for $T \geq T_0^*$. It turns out that the minimum of the magnitude of this thermo-induced plastic strain $\boldsymbol{h}$ is always smaller than the initial value $|\boldsymbol{h}_0|$ at T_0. This implies that *the thermo-induced plastic flow in a process of pure heating always leads to recovery of a pre-strain.* As a result, the thermal recovery at pure heating is derived as a direct, natural consequence of the thermo-coupled elastoplastic J_2–flow equations established in Subsubsect. 19.5.1. Furthermore, it may be inferred that the complete recovery corresponds to the following case:

$$\boldsymbol{\chi}_0 = c_0\boldsymbol{h}_0 \tag{19.75}$$

Details may be found in Xiao et al (2011).

As an example, consider a Prager modulus of the form below:

$$c = \frac{1}{2}(c_0 + c_0^*) - \frac{1}{2}(c_0 - c_0^*)\tanh\left[3\frac{T - T_0}{T_0^* - T_0}\right]\,.$$

Then, Eq. (19.74) results in

$$\begin{cases} \boldsymbol{h} = \left(\frac{c_0}{c_0^*} \frac{r}{r_0^*} - \frac{c_0}{c_0^*} + 1 \right) \boldsymbol{h}_0 , \\ r_0^* = r|_{T=T_0^*} = \sqrt{1.5 c_0 |\boldsymbol{h}_0|} . \end{cases} \tag{19.76}$$

his applies to general multi-axial cases. For the uniaxial case, it may be evident that, given the uniaxial temperature-strain relation of any given form, the temperature-dependent yield strength r is accordingly determined. This implies that any given uniaxial data for the shape memory effect may be fitted in an accurate, explicit sense.

An example for the yield strength r with the shape memory effect at heating is given as follows:

$$r = r_0^* \left[1 - \frac{c_0^*}{c_0} \tanh 3 \frac{T - T_0^*}{T_m - T_0^*} \right] , \tag{19.77}$$

where the temperature $T_0^* > T_0$ at the initial yielding (cf. Eq. (19.62)) and the temperature $T_m > T_0^*$ may be of any given values. Comparisons with test data are shown in Fig. 19.5. At $T = 348$K, the strain drops from the pre-strain $h_0 = 0.18$ to 0.008 and the strain recovery is actually completed.

19.5.4 Further Results

Two-way memory effects for SMAs are studied in Xiao et al (2011). In a broad sense (Xiao, 2015a), the elastoplastic J_2–flow equations in Subsubsect. 19.5.1 are extended

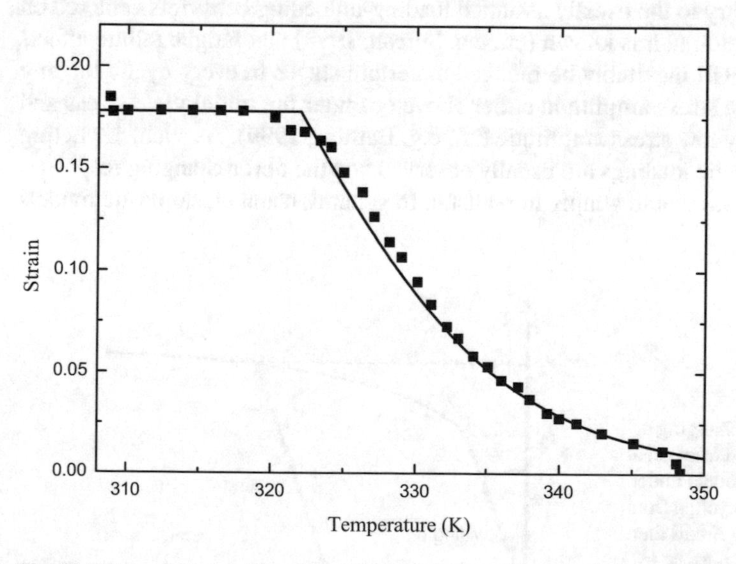

Fig. 19.5 Comparisons with test data in Tobushi et al (2001) for the strain recovery effect of a SMP sample

to cover both rate effects and thermal effects, and new rate-dependent elastoplastic equations with thermal effects are established with a smooth, natural transition to the rate-independent case. Such equations apply to both hard and soft solids, such as metals and polymers, etc. In particular, it is shown (Xiao et al, 2011) that the Mullins effect may be simulated by finding out a complementary elastic potential $\overline{W} = \overline{W}(\boldsymbol{\tau}, \vartheta)$ changing with both the stress $\boldsymbol{\tau}$ and the dissipated work ϑ. Explicit expressions are presented for the yield strength r and such a potential, in a sense of automatically reproducing the unloading curves related to the Mullins effect. For details, refer to Xiao et al (2011); Li et al (2017).

19.6 Innovative Elastoplastic Equations Automatically Incorporating Failure Effects

The plastic indicator ζ either in Eq. (19.36) or in Eq. (19.66) demarcates the loading and the unloading case centered on the notion of yielding and represents the very feature of plastic behavior in an idealized sense. In fact, it is assumed that no plastic deformation would be induced at unloading, namely, $\zeta = 0$ at unloading, whereas plastic deformation would be induced as the yielding is reached and maintained, namely, $\zeta = 1$ at loading. There arise troublesome issues with such a plastic indicator; refer to Xiao et al (2014); Xiao (2014b) for details. One of them, viz., the never-changing response under every cyclic loading process, as schematically shown in Fig. 19.6, would be plainly at variance with observed fatigue effects of metals (cf., e.g. Suresh, 1998).

On the contrary to the usually assumed loading-unloading behaviors centered on the notion of yielding, it is known (cf., e.g. Suresh, 1998) that fatigue failure effects for metals etc. will inevitably be induced at certain stages in every cyclic loading process with the stress amplitude either above or under the initial yield stress and even with a very low stress amplitude (cf., e.g. Bathias, 1999). As such, ratcheting effects under cyclic loadings are usually observed and the never-changing responses shown in Fig. 19.6 would simply unrealistic. In general, usual elastoplastic models

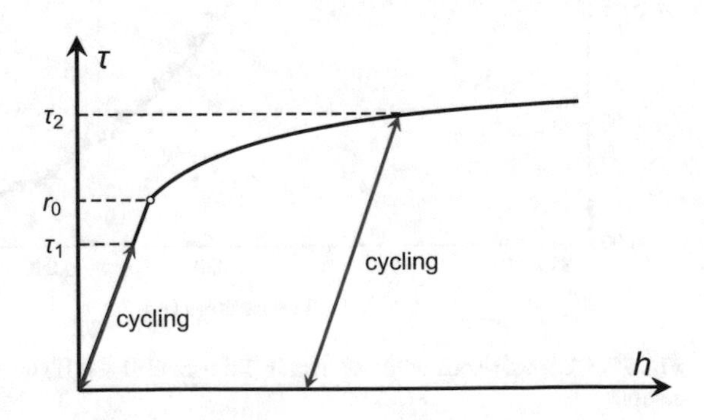

Fig. 19.6 Never-changing responses predicted from usual elastoplasticity models under the uniaxial stress cycling from 0 to the amplitude S and then back to 0 with $S = \tau_1 < r_0$ or $S = \tau_2 > r_0$ (here r_0 is the initial yield stress)

based on the notion of yielding would fail to simulate any fatigue failure effects under cyclic loadings. Details will be explained slightly later.

Very recently, innovative elastoplastic constitutive equations have been established with a smooth transition from the elastic to the plastic state. It has been demonstrated that such new equations can automatically incorporate failure effects as inherent constitutive features. Developments in this respect will be surveyed in this section.

19.6.1 New Elastoplastic Constitutive Equations

The separation Eq. (19.25) still serves as the starting point and elastic rate equation is still given as in Eq. (26). As mentioned before, the notion of yielding is central to usual formulations of elastoplasticity and leads to non-smooth transitions from the elastic to the plastic state. However, such non-smooth transitions would be foreign to realistic materials, as reported, e.g., in Bell and Truesdell (1984). The occurrence of a non-smooth transition from the elastic to the plastic state means that not only the constitutive equations have to be presented by two distinct forms applicable separately to two cases, but also these two forms manifest themselves in a non-smooth manner with strong discontinuities in tangent moduli. As such, a consequence is that, without assuming additional variables and failure criteria, no fatigue failure effects under cyclic loadings, such as high and low cycle fatigue failure effects, could be simulated by any usual elastoplasticity model centered on the notion of yielding.

Unlike the usual flow rule Eq. (19.27) with the discontinuous plastic indicator ζ as given in Eq. (19.36), a new plastic indicator ζ smoothly taking values between 0 and 1 may be introduced. The value of this smooth indicator ζ grows up from 0 to 1 whenever the stress point is going from the center onto the yield surface $f = 0$ in the classical sense, whereas the value of this ζ becomes vanishingly small whenever the stress point stays far away from the surface $f = 0$. With such a new, smooth plastic indicator, possible issues involved in a usual flow rule may be rendered irrelevant and, in particular, both high and low cycle fatigue failure may be incorporated into inherent constitutive features. For details, reference may be made to the general framework in Xiao (2014b). Here, a new normality flow rule in an innovative sense with a smooth plastic indicator ζ is given by (cf., Xiao et al, 2014; Xiao, 2014b):

$$\boldsymbol{D}^p = \frac{1}{2}\frac{\zeta}{\breve{h}}\left(\breve{f} + |\breve{f}|\right)\frac{\partial f}{\partial \boldsymbol{\tau}}, \tag{19.78}$$

with

$$\begin{cases} \zeta = \dfrac{g}{\eta}e^{-m(1-g/\eta)}, \\ g = \dfrac{1}{2}\mathrm{tr}\,(\tilde{\boldsymbol{\tau}} - \boldsymbol{\chi})^2, \\ \eta = \dfrac{1}{3}r^2, \end{cases} \tag{19.79}$$

where both $\check{f}$ and $\check{h}$ are given in Eqs. (19.32)-(19.33) and $m > 0$ is a dimensionless parameter.

The r in the von Mises function f in Eq. (19.28) and Eq. (19.79) is rephrased as the stress limit. The isotropic hardening-softening effects are characterized by the stress limit r changing with the dissipated work ϑ (cf., Eq. (19.30)), as shown in Eq. (19.29), while the anisotropic hardening-softening effects are prescribed by Eq. (19.31).

19.6.2 A Criterion for Critical Failure States

With the new flow rule in Eqs. (19.78)-(19.79), several issues involved in the usual flow rule (cf., Eq. (19.27) with $\zeta = 0$ or 1) based on the notion of yielding may be rendered irrelevant, as explained in Xiao et al (2014); Xiao (2014b). In particular, it has been demonstrated (Xiao, 2014b) that failure effects under cyclic and non-cyclic loadings may be incorporated as inherent constitutive features of the new model, namely, failure effects under cyclic and non-cyclic conditions may be automatically derived as direct consequences of the simple asymptotic properties below (Xiao, 2014b):

$$\begin{cases} \lim_{\vartheta\to\infty} r = 0\,, \\ \lim_{\vartheta\to\infty} c = 0\,, \\ \lim_{\vartheta\to\infty} \xi > 0\,. \end{cases} \tag{19.80}$$

It is noted that the stress limit r in Eq. (19.79) and the two moduli c and ω in the evolution equation (19.31) is just the constitutive quantities representing the mechanical strength withstanding plastic flow. Hence, the above asymptotic properties just mean a natural, basic fact, namely, failure effects would eventually be induced with asymptotic loss of the material strength in each process of accumulation of the dissipated work. Since the dissipated work ϑ is constantly growing with development of any plastic flow, it follows that failure effects would inevitably be induced at a certain stage in each cyclic or non-cyclic loading process.

In a newest development (Wang and Xiao, 2017a), it has been shown that there exist critical states heralding eventual failure. A unified criterion for prescribing such states is derivable from the new elastoplastic equations and of the following form (Wang and Xiao, 2017a):

$$4G(1-\rho)g + \hat{h} = 0\,, \tag{19.81}$$

where

$$\hat{h} = -\frac{\partial f}{\partial \vartheta}(\tilde{\tau} - \chi) : \frac{\partial f}{\partial \tau} - \frac{\partial f}{\partial \chi} : \mathbb{H} : \frac{\partial f}{\partial \tau}\,. \tag{19.82}$$

Here, perhaps an essential point is as follows: the new flow rule Eq. (19.78) with a smooth plastic indicator ζ as given by Eq. (19.79) ensures continuing accumulation of the dissipated work ϑ in every cyclic process. From this it follows that the strength property becomes degraded in each cyclic process, since each ϑ-dependent strength quantity decreases as the ϑ is increasing due to its cumulative effect. Thus, with the strength quantities of the asymptotic properties specified by Eq. (19.80), the new model ensures (cf., Wang and Xiao, 2017a) that the criterion Eq. (19.81) will be met at a certain stage of each cyclic process, namely, the fatigue failure will be inevitably induced as the dissipated work is constantly growing in every cyclic process, as will be exemplified from the full-range responses up to failure shown in Fig. 19.7 below.

19.6.3 Full-strain-range Response up to Failure

As load cycling is constantly progressing, a material would eventually fail with loss of its load-bearing capacity, as schematically shown in Fig. 19.7.

As sharply contrasted with the never-changing response shown in Fig. 19.6 under cyclic loadings, now the full-strain-range response predicted from the new elastoplastic equations indeed displays the ratcheting effect prior to the critical failure state. After the latter is exceeded, eventual failure will inevitably be induced.

19.6.4 Failure Effects Under Various Stress Amplitudes

In a series of most recent developments (Wang and Xiao, 2017a; Wang et al, 2017b; Wang and Xiao, 2017b; Wang et al, 2017a; Zhan et al, 2018; Wang et al, 2018; Zhan et al, 2019b), new elastoplasticity models at finite deformations are proposed for the purpose of simulating the fatigue failure effects for metals etc. With such

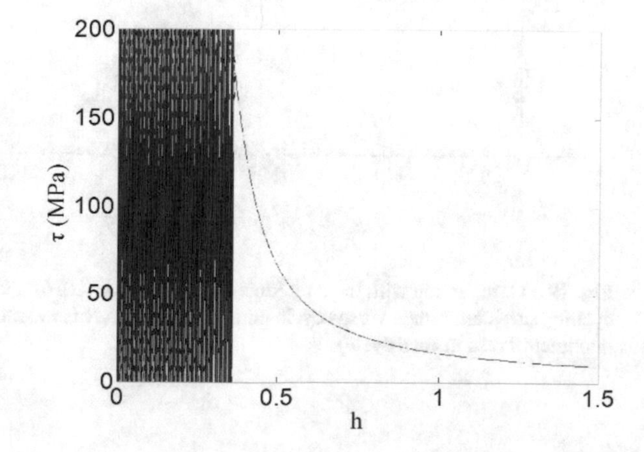

Fig. 19.7 Full-strain-range response up to failure predicted from the new elastoplastic equations under uniaxial stress cycling from 0 to 200MPa and then back to 0, in which a critical failure state is reached at the cycle number of 3089

new models, both the yield condition and the loading-unloading conditions in usual sense need not be involved but may be automatically incorporated into inherent constitutive features and, as such, both low and high cycle fatigue failure effects may be straightforwardly represented by simple asymptotic conditions as given in Eq. (19.80), without involving any additional damage-like variables and any failure criteria of ad hoc nature. Ratcheting effects up to eventual failure may be automatically, reasonably simulated based on these new models. To illustrate these, consider the simple case in the absence of the back stress, namely, $\chi = \boldsymbol{O}$. In this case, only the stress limit r is involved and the dissipated work ϑ is reduced to the plastic work κ given in Eq. (19.47). A simple example of the stress limit r meeting the asymptotic property Eq. $(19.80)_1$ is as follows:

$$r = \frac{1}{2} r_0 \left[1 - \tanh \beta \left(\frac{\kappa}{\kappa_c} - 1 \right) \right], \tag{19.83}$$

where $\beta > 0$ is a dimensionless parameter and κ_c is the critical value of the plastic work. Ratcheting responses up to failure are predicted from the new model under cycling uniaxial stresses. A numerical example is depicted in Fig. 19.7. Moreover, the $S - N$ curve of the stress amplitude versus the cycle number to failure is also obtainable and may be compared with test data. Numerical examples are shown in Fig. 19.8.

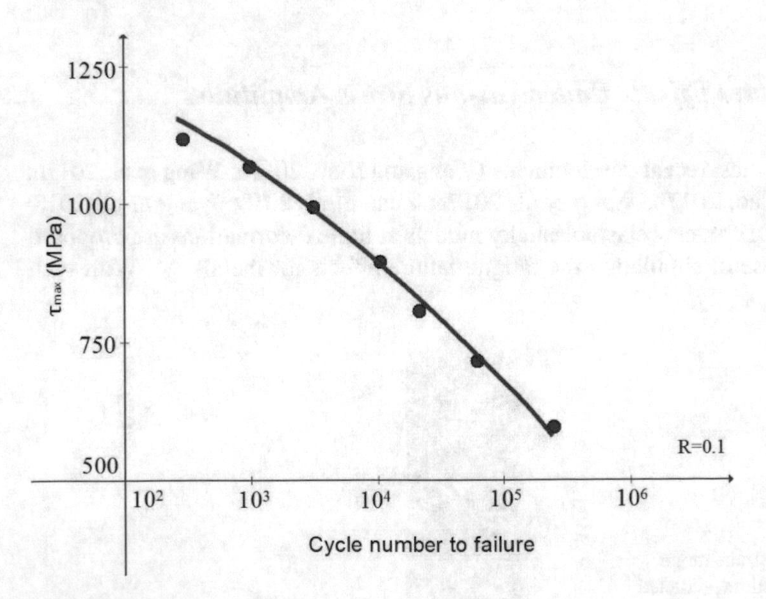

Fig. 19.8 Comparison with the data (dots) in Koster et al (2016) under non-symmetric stress cycling: stress amplitude versus cycle number to failure (R is the ratio of the minimum to the maximum stress in each cycle)

With the new J_2−flow model in the absence of the back stress effect, it has been demonstrated that simulation results for various rate-independent cases (Wang and Xiao, 2017a; Wang et al, 2017b; Wang and Xiao, 2017b) and rate-dependent cases (Wang et al, 2017a) of uniaxial stress cycling and uniaxial strain cycling (Wang et al, 2017a) as well as thermo-mechanical cycling (Zhan et al, 2018) compare well with test data for fatigue failure effects. Moreover, comprehensive simulation results (Wang et al, 2018; Zhan et al, 2019b) for monotonic and cyclic failure effects of twisted cylindrical tubes with either free-ends or fixed-ends are derived from new models with combined hardening and good agreement with test data are achieved.

19.7 Deformable Micro-continua for Quantum Entities at Atomic Scale

Now we come to the last topic, namely, the Hencky strain leads to the newest but perhaps unexpected discovery of the quantum-continua at atomic scale, referred to as the quantum-continua. The latter are deformable micro-continua that display all known quantum effects exactly as do quantum entities at atomic scale, such as electrons etc. The main results will be summarized below. For details, refer to the latest references (Xiao, 2017b,a, 2019).

19.7.1 The Quantum-continua

At each instant a quantum-continuum occupies the entire space but assemblies an overwhelming part of its whole mass into a tiny region at atomic scale. How it can behave this way depends on its own deformability nature prescribed slightly later as well as external actions. Let $\boldsymbol{x}$ denote the current position vector of a generic point in such a continuum at instant t. It is moving and deforming in a force field and a torque field, designated respectively by $\boldsymbol{b}(\boldsymbol{x},t)$ and $\boldsymbol{\beta}(\boldsymbol{x},t)$ with values measured per unit current volume. At each instant t it displays itself with a spatial configuration prescribed and hence represented by four basic field variables including the velocity field $\boldsymbol{u}(\boldsymbol{x},t)$, the mass density field $\theta(\boldsymbol{x},t)$ and the all-around stress field $q(\boldsymbol{x},t)$, as well as the field for the intrinsic angular momentum per unit current volume, called the intrinsic angular momentum density field and denoted $\boldsymbol{\alpha}(\boldsymbol{x},t)$.

During a process of deformation, a quantum-continuum presents itself with a continuous family of time-varying configurations. As indicated in the foregoing, each current configuration at time t is prescribed by the four field variables $\theta(\boldsymbol{x},t)$, $\boldsymbol{u}(\boldsymbol{x},t)$, $q(\boldsymbol{x},t)$ and $\alpha(\boldsymbol{x},t)$. Of them, the mass density field $\theta(\boldsymbol{x},t)$, the velocity field $\boldsymbol{u}(\boldsymbol{x},t)$ and the all-around stress field $q(\boldsymbol{x},t)$ are well known, whereas the intrinsic angular momentum density field $\boldsymbol{\alpha}(\boldsymbol{x},t)$ is in response to the applied torque field $\boldsymbol{\beta}(\boldsymbol{x},t)$ and not involved in usual deformable continua. The field $\alpha(\boldsymbol{x},t)$ is inherent in the dynamic effects of quantum entities at atomic scale. In fact, each infinitesimal mass element with an infinitesimal volume dv is constantly spinning around its mass center in response to the applied torque $\boldsymbol{\beta} dv$ and, then, an intrinsic angular momentum, i.e., $\boldsymbol{\alpha} dv$, is induced, in addition to the usual orbital angular momentum. As such, the field $\boldsymbol{\alpha}(\boldsymbol{x}, t)$ naturally gives rise to the total intrinsic angular momentum of a quantum-continuum.

Let m be the total mass, i.e., $m = \int \theta dv$. Here and henceforth, $\int (\cdots) dv$ is used to designate the volume integration over the whole space. In the sequel, the usual mass density field θ will no longer be treated but, instead, will be replaced by the normalized density field $\rho = \theta / m$ with

$$\begin{cases} \lim_{|\boldsymbol{x}| \to \infty} \rho = 0 \,, \\ \displaystyle\int \rho dv = 1 \,, \end{cases} \tag{19.84}$$

for the purpose of conforming to certain standard forms.

19.7.2 Continuity Equation and Balance Equations

In the non-relativistic case, the mass conservation and balances of the linear and angular momenta produce (Xiao, 2017b):

$$\frac{\partial \rho}{\partial t} + \nabla \cdot (\rho \boldsymbol{u}) = 0 \,, \tag{19.85}$$

$$\nabla q - \rho \nabla V = m\rho \left[\frac{\partial \boldsymbol{u}}{\partial t} + \boldsymbol{u} \cdot \nabla \boldsymbol{u} \right] , \tag{19.86}$$

$$\frac{d}{dt} \left(\frac{\boldsymbol{\alpha}}{\rho} \right) = g_s \frac{\boldsymbol{\alpha}}{\rho} \times \boldsymbol{B} \,. \tag{19.87}$$

Here, the notation ∇ is used to denote the differential vector, i.e.

$$\nabla = \frac{\partial}{\partial x_i} \boldsymbol{e}_i \,,$$

and the force field $\boldsymbol{b}$ and the torque field $\boldsymbol{\beta}$ are given by

$$\boldsymbol{b} = -\rho \nabla V \,,$$

$$\boldsymbol{\beta} = g_s \boldsymbol{\alpha} \times \boldsymbol{B} \,,$$

where V is the potential for a conservative force field and $\boldsymbol{B}$ is the external magnetic field and g_s is a gyro-magnetic ratio.

19.7.3 Constitutive Equation for Deformability Nature

The deformability nature of the quantum-continua is prescribed or defined by the constitutive equation below (Xiao, 2017b,a, 2019):

$$\frac{q}{\rho} = \frac{\hbar^2}{4m}\,\nabla^2\,(\ln\rho) \tag{19.88}$$

where the $\hbar$ is Planck's constant, i.e., $\hbar = 1.05457 \times 10^{-34}$Js, and m is the mass of the quantum-continuum at issue.

In Eq. (19.88), the $\ln\rho$ is referred to as the Hencky density and directly related to the logarithmic volumetric strain $\ln J$. The strain-energy density per unit mass is given by H/m with

$$H = \frac{\hbar^2}{8m}\,|\nabla(\ln\rho)|^2 = \frac{\hbar^2}{8m}\left[\left(\frac{\partial\ln\rho}{\partial x_1}\right)^2 + \left(\frac{\partial\ln\rho}{\partial x_2}\right)^2 + \left(\frac{\partial\ln\rho}{\partial x_3}\right)^2\right]. \tag{19.89}$$

This strain energy density is directly derivable from a reduced form of the Hencky potential given in Eq. (19.37) by considering the localized property of the quantum-continua at issue. Details may be found in Xiao (2019).

19.7.4 Inherent Response Features of the Quantum-continua

The responses of a quantum-continuum in external fields are governed by the coupled system formed by the nonlinear field equations (19.85)-(19.88) governing the four field variables ρ, $\boldsymbol{u}$, q and α. It is perhaps surprising that this coupled system is exactly reducible to a single linear dynamic field equation governing a complex field variable, denoted $\Psi = \Psi(\boldsymbol{x}, t)$, from which exact closed-form solutions for the foregoing four field variables are obtainable in terms of the complex field variable $\Psi = \Psi(\boldsymbol{x}, t)$. Namely, Eqs. (19.84)-(19.88) are exactly reducible to

$$i\,\hbar\,\frac{\partial\Psi}{\partial t} = -\frac{\hbar^2}{2m}\,\nabla^2\Psi + V\Psi, \tag{19.90}$$

$$\lim_{|\boldsymbol{x}|\to\infty}\Psi = 0, \tag{19.91}$$

$$\int \overline{\Psi}\Psi dv = 1, \tag{19.92}$$

with the exact closed-form solutions below:

$$\begin{cases} \rho = \overline{\Psi}\Psi, \\ \boldsymbol{u} = \dfrac{\hbar}{m}\nabla\left[\arctan\left(i\,\dfrac{\overline{\Psi}-\Psi}{\overline{\Psi}+\Psi}\right)\right], \\ q = \dfrac{\hbar^2}{4m}\left(\overline{\Psi}\Psi\right)\nabla^2\left[\ln\left(\overline{\Psi}\Psi\right)\right], \\ \boldsymbol{\alpha} = \left(\overline{\Psi}\Psi\right)\exp\left[\boldsymbol{E}\cdot\left(\int_0^t g_s\boldsymbol{B}dt\right)\right]\cdot\boldsymbol{S}_0\,. \end{cases} \tag{19.93}$$

In the above, i is used to denote the imaginary unit, i.e., $i = \sqrt{-1}$, and $\overline{\Psi}$ is the complex conjugate of Ψ. In the last expression above, the $\boldsymbol{E}$ is the 3rd-order Eddington tensor and the $\boldsymbol{S}_0$ is the total initial intrinsic angular momentum of the quantum-continuum at issue.

It turns out that Eq. (19.90) is just the Schrödinger equation in quantum mechanics and, moreover, that the complex field variable Ψ is just the wave function.

It appears that the above findings disclose for the first time the physical meanings and the dynamic features of both the Schrödinger equation and the wave function. It is noted that it has long been mysterious concerning what the Schrödinger equation and the wave function really mean. Now it is known that both emerge out of the inherent response features of the quantum-continua, as summarized in Fig. 19.9.

With the discovery of the quantum-continua, various puzzling patterns of probabilistic nature for quantum entities may be rendered irrelevant and, accordingly, certain long-standing issues may be clarified. Details may be found in Xiao (2017b,a, 2019).

19.7.5 New Patterns for Hydrogen Atom as Quantum-continuum

As an illustrative example, we are going to find out how the quantum-continuum representing an electron responds to the centripetal force field generated by the charge e of the proton in a hydrogen atom, viz., the Coulomb potential (in SI units) below is taken into consideration:

$$V = -\frac{e^2}{4\pi\epsilon_0}\frac{1}{r}\,. \tag{19.94}$$

Here, $r = |\boldsymbol{x}|$. In the time-independent force field prescribed by the Coulomb potential above, the stationary density distribution of the electron as a deformable micro-continuum, i.e. $\rho = \rho(\boldsymbol{x})$, is determined by

$$\frac{\hbar^2}{4m}\nabla\left(\rho\nabla^2(\ln\rho)\right) + \frac{e^2}{4\pi\epsilon_0}\frac{\boldsymbol{x}}{r^3}\rho = \boldsymbol{0} \tag{19.95}$$

with the boundary condition $\lim_{|\boldsymbol{x}|\to\infty}\rho = 0$ at infinity.

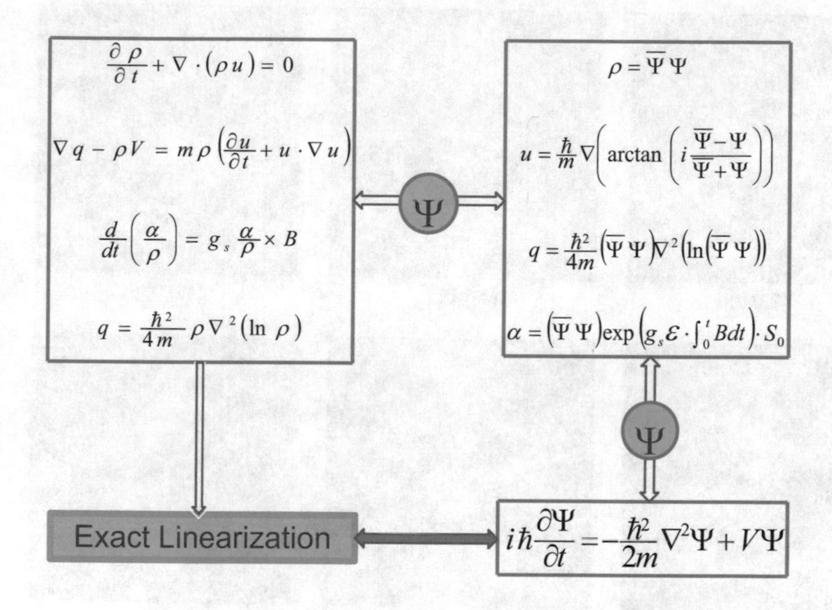

Fig. 19.9 The Schrödinger equation (SE) and the complex wave function Ψ emerge out of the inherent nonlinear response features of the deformable micro-continua governed by Eqs. (19.84)-(19.88): The clockwise arrowed direction shows that the SE is just the exact linearization (EL) of the coupled nonlinear governing equations (19.84)-(19.88) for the quantum-continua via Ψ, while the anti-clockwise arrowed direction indicates that as the EL of Eqs. (19.84)-(19.88) the SE generates exact closed-form solutions of Eqs. (19.84)-(19.88) via Ψ

It may come as a surprise that, for the Coulomb potential as given in Eq. (19.94), infinitely many solutions for the density distribution $\rho(\boldsymbol{x})$ may be derived from the highly nonlinear differential equation (19.95). These solutions give rise to infinitely many stationary configurations for the hydrogen atom. Then, the hydrogen atom presents itself with infinitely many stationary configurations of discretized nature in response to the Coulomb force field generated by the charge of its heavy nucleus. Each such configuration gives rise to different spatial distributions of the mass density, the internal stress and the intrinsic angular momentum density, as given in Eq. (19.93). Examples for these stationary configurations are depicted in Fig. 19.10 for six cases of the three quantum numbers (n, l, k). Sharply localized nature at a scale of the Bohr radius a is evidenced from these examples.

Both the allowed energies and the intrinsic angular momentum of the hydrogen atom emerge as natural consequences of the above new pattern. Now, long-standing issues with the usual particle-based pattern of probabilistic nature may be rendered irrelevant. Below is the new insight into the most outstanding issue, namely, the uncertainty principle. Details for this and other issues may be found in Xiao (2017b,a, 2019).

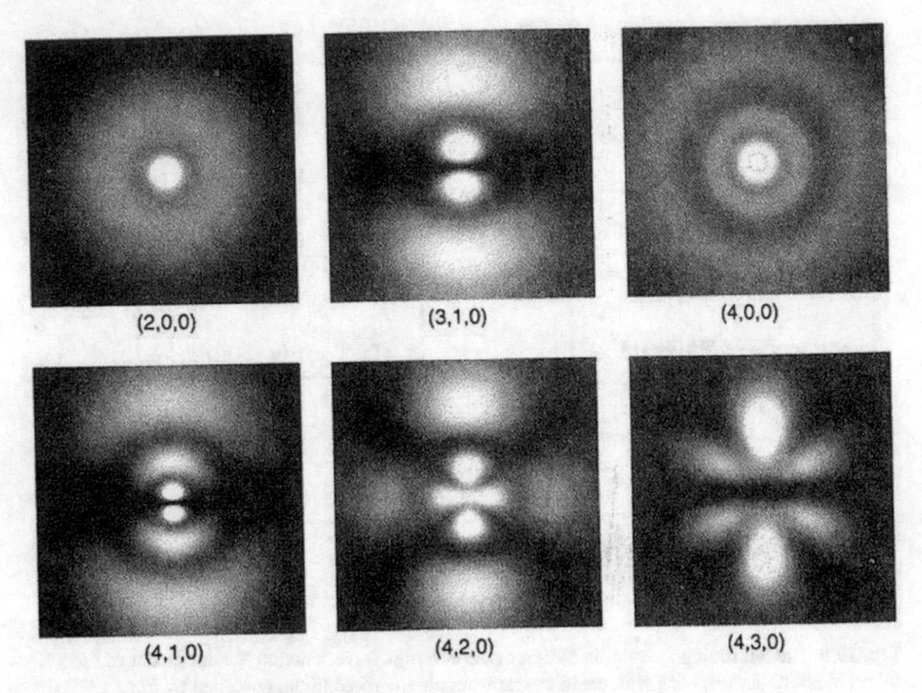

Fig. 19.10 Localized configurations of the hydrogen atom with six combinations of the quantum numbers (n, k, l): the mass density growing from dark to bright and size of each bright region is around $2 \sim 10a$ with the Bohr radius $a = 5.3 \times 10^{-11}$m

19.7.6 New Insight into the Uncertainty Principle

With the response patterns of the quantum-continua, new insights may be gained into such quantities as intended for quantifying uncertainties, namely, the standard deviations of observables. According to the current quantum theory, these quantities have to be evaluated as expectation values from the statistical standpoint based simply on formal mathematical procedures via certain ad hoc Hermitian operators and, therefore, they might possibly deviate from the intended meanings assigned to them. Here, as illustrative examples the standard deviations of the position and the momentum, are taken into consideration.

First, the standard deviation of the a-th coordinate of the position, i.e. σ_{x_a}, is specified by

$$\sigma_{x_a}^2 = \int \overline{\Psi}(x_a - [\hat{x}_a])^2 \Psi dv = \int \rho(x_a - [\hat{x}_a])^2 dv, \tag{19.96}$$

where

$$[\hat{x}_a] = \int \overline{\Psi} x_a \Psi dv = \int \rho x_a dv$$

is the expectation value of the a-th coordinate of the position, which is just the a-th coordinate of the mass center of the quantum-continuum.

Next, via the Hermitian operator $\hat{p}_a = -i\hbar\frac{\partial}{\partial x_a}$ for the a-th component of the linear momentum, the quantity below may be evaluated:

$$\sigma_{p_a}^2 = \int \overline{\Psi}(\hat{p}_a - [\hat{p}_a])^2 \Psi dv\,, \tag{19.97}$$

where $[\hat{p}_a]$ is the expectation value of the a-th component of the linear momentum given by

$$[\hat{p}_a] = \int \overline{\Psi}\hat{p}_a \Psi dv = -i\hbar \int \overline{\Psi}\frac{\partial \Psi}{\partial x_a} dv\,, \tag{19.98}$$

which is just the a-th component of the total momentum of the quantum-continuum, namely (cf., Xiao, 2017b, 2019),

$$[\hat{p}_a] = \int m\rho u_a dv\,.$$

Now Eq. (19.97) may be converted to (cf., Xiao, 2019)

$$\sigma_{p_a}^2 = \int \rho(mu_a - [\hat{p}_a])^2 dv + \frac{\hbar^2}{4}\int \rho\left(\frac{\partial \ln \rho}{\partial x_a}\right)^2 dv\,. \tag{19.99}$$

Note in the above that the first term on the right-hand side is in a direct sense the standard deviation of the a-th momentum component.

According to the current quantum theory, the quantity σ_{p_a} given by Eq. (19.97) is intended for the standard deviation of the a-th momentum component. However, this quantity itself is of ad hoc nature and could not justify the physical relevance intended for the standard deviation of the a-th momentum component. In fact, from Eq. (19.99) it turns out that not only the squared standard deviation of the a-th momentum component, but also an additional term, i.e., $2m$ times the a-th component of the strain energy (cf., Eq. (19.89)), are included in Eq. (19.97). This suggests that the quantity σ_{p_a} in Eq. (19.97) would not represent what it is intended for, i.e., the uncertainty for the a-th momentum component.

Now the commonly-known inequality $\sigma_{x_a}\sigma_{p_a} \geq \frac{\hbar}{2}$ can still be derived, but the two conjugate quantities σ_{x_a} and σ_{p_a} need be reinterpreted based on the response patterns of the quantum-continua, as described below:

(i) As σ_{p_a} tends to vanish, a quantum-continuum tends to the spreading extreme with a uniform density field over the whole space and, accordingly, the position deviation σ_{x_a} tends to infinity; and

(ii) As σ_{x_a} tends to vanish, a quantum-continuum tends to the localization extreme with a concentrated mass and, accordingly, the momentum deviation tends to vanish, viz.,

$$\int \rho(mu_a - [\hat{p}_a])^2 dv \to 0\,,$$

and, in particular, the total strain energy tends to infinity.

Since the energy could not be unbounded, the inequality Eq. (19.72) suggests that the localization extreme would be inaccessible. Namely, the inequality Eq. (19.72) would just set a limit for the particle pattern.

19.7.7 Remarks

The newest discovery summarized above suggests that each quantum entity such as electron would not be a zero-dimensional particle but turns out to be a deformable micro-continuum at atomic scale, as exemplified in Fig. 19.10 for a hydrogen atom. Such micro-continua are the quantum-continua with the deformability feature prescribed by the constitutive equation (19.88). It is found that the Hencky density $\ln\rho$ plays a central role in characterizing both the deformability feature and the strain energy of the quantum-continua, as evidenced in Eqs. (19.88)-(19.89). Now the origins and the meanings of the Schrödinger equation (19.90) and the complex wave function Ψ are emergent exactly from the inherent nonlinear dynamic features of the quantum-continua, as shown in Fig. 19.9. With new response patters of the quantum-continua, long-standing issues concerning quantum entities would become irrelevant. Details may be found in Xiao (2017b,a, 2019).

19.8 Concluding Remarks

In the previous sections, new approaches and main results have been surveyed in modeling finite elastic and inelastic deformation behaviors of continua based on the Hencky strain and the newly-discovered logarithmic rate. The Hencky strain as a coherent thread has played an essential, unified role in these recent developments. With the Hencky strain and the logarithmic rate, seemingly unrelated finite deformation effects may be simulated in a unified constitutive framework, including rubber-like elasticity, metal elastoplasticity, pseudo-elasticity, shape memory effects, cyclic and non-cyclic failure effects, as well as quantum effects at atomic scale. Such developments may be unexpected in the respects of pseudo-elastic and shape memory effects and perhaps quite surprising in the respect of quantum effects.

Impressed probably by the above developments, one likes to know more about the creator of the Hencky strain, that is Heinrich Hencky. Two excellent articles Tanner and Tanner (2003); Bruhns (2014b) present accounts of his life and achievements. Also, certain profound results in geometry of the Hencky strain have been disclosed and expounded very recently in the reference Neff et al (2016), in which a short biography of Heinrich Hencky is also included in an appendix.

References

Anand L (1979) On H. Hencky's approximate strain-energy function for moderate deformations. J Appl Mech 46:78–82

Anand L (1986) Moderate deformations in extension-torsion of incompressible isotropic elastic materials. J Mech Phys Solids 34:293–304

Bathias C (1999) There is no infinite fatigue life in metallic materials. Fatigue Fract Eng Mater Struct 22:559–565

Beatty, Millard F (1987) Topics in finite elasticity: hyperelasticity of rubber, elastomers, and biological tissues–with examples. Appl Mech Rev 40:1699–1734

Bell JF, Truesdell C (1984) The experimental foundations of solid mechanics. Springer, Berlin

Boyce, Mary C, Arruda EM (2000) Constitutive models of rubber elasticity: a review. Rubber Chem Techn 73:504–523

Bruhns OT (2014a) The Prandtl-Reuss equations revisited. ZAMM - J Appl Math Mech 94:187–202

Bruhns OT (2014b) Some remarks on the history of plasticity-Heinrich Hencky, a pioneer of the early years. In: E. Stein (ed.), The History of Theoretical, Material and Computational Mechanics - Mathematics Meets Mechanics and Engineering, Springer, Berlin, pp 133–152

Bruhns OT, Xiao H, Meyers A (1999) Self-consistent eulerian rate type elastoplasticity models based upon the logarithmic stress rate. Int J Plasticity 15:479 – 520

Bruhns OT, Xiao H, Meyers A (2001a) Constitutive inequalities for an isotropic elastic strain-energy function based on Hencky's logarithmic strain tensor. Proc R Soc London A 457:2207–2226

Bruhns OT, Xiao H, Meyers A (2001b) Large simple shear and torsion problems in kinematic hardening elastoplasticity with logarithmic rate. Int J Solids Struct 38:8701–8722

Bruhns OT, Xiao H, Meyers A (2001c) Large strain response of isotropic hardening elastoplasticity with logarithmic rate: Swift effect in torsion. Arch Appl Mech 71:389–404

Bruhns OT, Xiao H, Meyers A (2003) Some basic issues in traditional Eulerian formulations of finite elastoplasticity. Int J Plasticity 19:2007–2026

Bruhns OT, Meyers A, Xiao H (2004) On non-corotational rates of Oldroyd's type and relevant issues in rate constitutive formulations. Proc R Soc London A 460:909–928

Bruhns OT, Xiao H, Meyers A (2005) A weakened form of Ilyushin's postulate and the structure of self-consistent Eulerian finite elastoplasticity. Int J Plasticity 21:199–219

Cao J, Ding XF, Yin ZN, Xiao H (2017) Large elastic deformations of soft solids up to failure: new hyperelastic models with error estimation. Acta Mech 228:1165–1175

Dienes JK (1979) On the analysis of rotation and stress rate in deforming bodies. Acta Mech 32:217–232

Erman B, Mark JE (1989) Rubber-like elasticity. Annu Rev Phys Chem 40:351–374

Fitzgerald J (1980) A tensorial Hencky measure of strain and strain rate for finite deformations. J Appl Phys 51:5111–5115

Gu ZX, Yuan L, Yin ZN, Xiao H (2015) A multiaxial elastic potential with error-minimizing approximation to rubberlike elasticity. Acta Mech Sinica 31:637–646

Gurson AL (1977) Continuum theory of ductile rupture by void nucleation and growth: Part i—yield criteria and flow rules for porous ductile media. J Engrg Mater Tech 99:2–15

Hencky H (1928) Über die Form des Elastizitätsgesetzes bei ideal elastischen Stoffen. Zeit Tech Phys 9:215–220

Hill R (1968) On constitutive inequalities for simple materials-I. J Mech Phys Solids 16:229–242

Hill R (1970) Constitutive inequalities for isotropic elastic solids under finite strain. Proc R Soc London A 314:457–472

Hill R (1979) Aspects of invariance in solid mechanics. Adv Appl Mech 18:1–75

Horgan CO, Saccomandi G (2006) Phenomenological hyperelastic strain-stiffening constitutive models for rubber. Rubber Chem Techn 79:152–169

Jiao Y, Fish J (2017) Is an additive decomposition of a rate of deformation and objective stress rates passé? Comput Meth Appl Mech Engrg 327:196 – 225

Jin TF, Yu LD, Yin ZN, Xiao H (2015) Bounded elastic potentials for rubberlike materials with strain-stiffening effects. ZAMM - J Appl Math Mech 95:1230–1242
Jones DF, Treloar LRG (1975) The properties of rubber in pure homogeneous strain. J Phys D 8:1285–1304
Khan AS, Huang SJ (1995) Continuum Theory of Plasticity. John Wiley, New York
Koster M, Lis A, Lee WJ, Kenel C, Leinenbach C (2016) Influence of elastic-plastic base material properties on the fatigue and cyclic deformation behavior of brazed steel joints. Int J Fatigue 82:49–59
Lagoudas DC, Entchev PB, Popov P, Patoor E, Brinson CL, Gao XJ (2006) Shape memory alloys, part II: Modeling of polycrystals. Mech Mater 38:430–462
Lehmann T (1964) Anisotrope plastische Formänderungen. Rheologica Acta 3:281–285
Lehmann T (1972) Einige Bemerkungen zu einer allgemeinen Klasse von Stoffgesetzen für große elasto-plastische Formänderungen. Ing Arch 41:297–310
Li H, Zhang YY, Wang XM, Yin ZN, Xiao H (2014) Obtaining multi-axial elastic potentials for rubber-like materials via an explicit, exact approach based on spline interpolation. Acta Mech Solida Sinica 27:441–453
Li H, Ding XF, Yin ZN, Xiao H (2017) An explicit approach toward modeling thermo-coupled deformation behaviors of SMPs. Applied Sciences 7
Lubliner J (1992) Plasticity Theory. Macmillan, New York
Naghdi PM (1990) A critical review of the state of finite plasticity. Z Angew Math Phys 41:315–394
Nagtegaal JC, Jong JED (2010) Some computational aspects of elastic-plastic large strain analysis. Int J Numer Meth Engrg 17:15–41
Neff P, Eidel B, Martin RJ (2016) Geometry of logarithmic strain measures in solid mechanics. Arch Rat Mech Anal 222:507–572
Nemat-Nasser S (1983) On finite plastic flow of crystalline solids and geomaterials. J Appl Mech 50:15
Ogden RW (1984) Nonlinear Elastic Deformations. Ellis Horwood, Chichester
Ogden RW, Saccomandi G, Sgura I (2006) On worm-like chain models within the three-dimensional continuum mechanics framework. Proc R Soc Lond A 462:749–768
Patoor E, Lagoudas C Dimitris, Entchev B Pavlin, Brinson LC, Gao XJ (2006) Shape memory alloys, part I: General properties and modeling of single crystals. Mech Mater 38:391–429
Prager W (1960) An elementary discussion of definitions of stress rate. Quart Appl Math 18:403–407
Saburi T, Tatsumi T, Nenno S (1982) Effects of heat treatment on mechanical behavior of TiNi alloys. J de Phys 43:261–266
Shutov AV, Ihlemann J (2014) Analysis of some basic approaches to finite strain elastoplasticity in view of reference change. Int J Plasticity 63:183–197
Simo JC, Hughes TJR (2008) Computational Inelasticity. Springer, Berlin
Simo JC, Pister KS (1984) Remarks on rate constitutive equations for finite deformation problems: computational implications. Comput Meth Appl Mech Engrg 46:201–215
Suresh S (1998) Fatigue of Materials. Cambridge University Press, Cambridge
Tanner RI, Tanner E (2003) Heinrich Hencky: a rheological pioneer. Rheologica Acta 42:93–101
Tobushi H, Okumura K, Hayashi S, Norimitsu (2001) Thermomechanical constitutive model of shape memory polymer. Mech Mater 33:545–554
Treloar LRG (1958) The Physics of Rubber Elasticity. Oxford University Press, Oxford
Wang SY, Zhan L, Wang ZL, Yin ZN, Xiao H (2017a) A direct approach toward simulating cyclic and non-cyclic fatigue failure of metals. Acta Mech 228:4325–4339
Wang XM, Li H, Yin ZN, Xiao H (2014) Multiaxial strain energy functions of rubberlike materials: an explicit approach based on polynomial interpolation. Rubber Chem Technol 87:168–183
Wang XM, Wang ZL, Xiao H (2015) SMA pseudo-elastic hysteresis with tension-compression asymmetry: explicit simulation based on elastoplasticity models. Continuum Mech Thermodyn 27:959–970
Wang YS, Zhan L, Xi HF, Xiao H (2018) Coupling effects of finite rotation and strain-induced anisotropy on monotonic and cyclic failure of metals. Acta Mech 229:4963–4975

Wang ZL, Xiao H (2017a) Direct modeling of multi-axial fatigue failure for metals. Int J Solids Struct 125:216–231

Wang ZL, Xiao H (2017b) A simulation of low and high cycle fatigue failure effects for metal matrix composites based on innovative J_2-flow elastoplasticity model. Materials 10:11–26

Wang ZL, Li H, Yin ZN, Xiao H (2017b) A new, direct approach toward modeling thermo-coupled fatigue failure behavior of metals and alloys. Acta Mech Solida Sinica 30:1–9

Xiao H (2005) Hencky strain and Hencky model: extending history and ongoing tradition. Multidiscipline Modeling in Materials and Structures 1:1–52

Xiao H (2012) An explicit, direct approach to obtaining multiaxial elastic potentials that exactly match data of four benchmark tests for rubbery materials part 1: incompressible deformations. Acta Mech 223:2039–2063

Xiao H (2013a) An explicit, direct approach to obtain multi-axial elastic potentials which accurately match data of four benchmark tests for rubbery materials part 2: general deformations. Acta Mech 224:479–498

Xiao H (2013b) Pseudoelastic hysteresis out of recoverable finite elastoplastic flows. Int J Plasticity 41:82–96

Xiao H (2014a) An explicit, straightforward approach to modeling SMA pseudoelastic hysteresis. Int J Plasticity 53:228–240

Xiao H (2014b) Thermo-coupled elastoplasticity models with asymptotic loss of the material strength. Int J Plasticity 63:211–228

Xiao H (2015a) A direct, explicit simulation of finite strain multiaxial inelastic behavior of polymeric solids. Int J Plasticity 71:146–169

Xiao H (2015b) Elastic potentials with best approximation to rubberlike elasticity. Acta Mech 226:331–350

Xiao H (2017a) Deformable media with quantized effects. J Astrophys Aerospace Tech 5:87

Xiao H (2017b) Quantum enigma hidden in continuum mechanics. Appl Math Mech-Engl Ed 38:39–56

Xiao H (2019) Deformable micro-continua in which quantum mysteries reside (in press). Appl Math Mech-EnglEd 40

Xiao H, Chen LS (2003) Henckys logarithmic strain and dual stress-strain and strain-stress relations in isotropic finite hyperelasticity. Int J Solids Struct 40:1455–1463

Xiao H, Bruhns OT, Meyers A (1997) Logarithmic strain, logarithmic spin and logarithmic rate. Acta Mech 124:89–105

Xiao H, Bruhns OT, Meyers A (1998a) On objective corotational rates and their defining spin tensors. Int J Solids Struct 35:4001–4014

Xiao H, Bruhns OT, Meyers A (1998b) Strain rates and material spins. J Elasticity 52:1–41

Xiao H, Bruhns OT, Meyers A (1999) Existence and uniqueness of the integrable-exactly hypoelastic equation and its significance to finite inelasticity. Acta Mech 138:31–50

Xiao H, Bruhns OT, Meyers A (2000a) The choice of objective rates in finite elastoplasticity: general results on the uniqueness of the logarithmic rate. Proc R Soc London A 456:1865–1882

Xiao H, Bruhns OT, Meyers A (2000b) A consistent finite elastoplasticity theory combining additive and multiplicative decomposition of the stretching and the deformation gradient. Int J Plasticity 16:143–177

Xiao H, Bruhns OT, Meyers A (2001) Large strain responses of elastic-perfect plasticity and kinematic hardening plasticity with the logarithmic rate: Swift effect in torsion. Int J Plasticity 17:211–235

Xiao H, Bruhns OT, Meyers A (2004) Explicit dual stress-strain and strain-stress relations of incompressible isotropic hyperelastic solids via deviatoric Hencky strain and Cauchy stress. Acta Mech 168:21–33

Xiao H, Bruhns OT, Meyers A (2005) Objective stress rates, path-dependence properties and non-integrability problems. Acta Mech 176:135–151

Xiao H, Bruhns OT, Meyers A (2006a) Elastoplasticity beyond small deformations. Acta Mech 182:31–111

Xiao H, Bruhns OT, Meyers A (2006b) Objective stress rates, cyclic deformation paths, and residual stress accumulation. ZAMM-J Appl Math Mech 86:843–855

Xiao H, Bruhns OT, Meyers A (2007a) The integrability criterion in finite elastoplasticity and its constitutive implications. Acta Mech 188:227

Xiao H, Bruhns OT, Meyers A (2007b) Thermodynamic laws and consistent Eulerian formulation of finite elastoplasticity with thermal effects. J Mech Phys Solids 55:338–365

Xiao H, Bruhns OT, Meyers A (2010a) Finite elastoplastic J_2-flow models with strain recovery effects. Acta Mech 210:13–25

Xiao H, Bruhns OT, Meyers A (2010b) Phenomenological elastoplasticity view on strain recovery loops characterizing shape memory material. ZAMM-J Appl Math Mech 90:544–564

Xiao H, Bruhns OT, Meyers A (2011) Thermoinduced plastic flow and shape memory effects. Theor Appl Mech 38:155–207

Xiao H, Bruhns OT, Meyers A (2014) Free rate-independent elastoplastic equations. ZAMM-J Appl Math Mech 94:461–476

Xiao H, Wang XM, Wang ZL, Yin ZN (2016) Explicit, comprehensive modeling of multi-axial finite strain pseudo-elastic SMAs up to failure. Int J Solids Struct 88:215–226

Xiao H, Ding XF, Cao J, Yin ZN (2017) New multi-axial constitutive models for large elastic deformation behaviors of soft solids up to breaking. Int J Solids Struct 109:123–130

Yu LD, Jin TF, Yin ZN, Xiao H (2015a) A model for rubber-like elasticity up to failure. Acta Mech 226:1445–1456

Yu LD, Jin TF, Yin ZN, Xiao H (2015b) Multi-axial strain-stiffening elastic potentials with energy bounds: explicit approach based on uniaxial data. Appl Math Mech-Engl Ed 36:883–894

Yuan L, Gu ZX, Yin ZN, Xiao H (2015) New compressible hyper-elastic models for rubber-like materials. Acta Mech 226:4059–4072

Zhan L, Wang SY, Xi HF, Xiao H (2018) Direct simulation of thermo-coupled fatigue failure for metals. ZAMM-J Appl Math Mech 98:856–869

Zhan L, Wang SY, Xi HF, Xiao H (2019a) An explicit and accurate approach toward simulating plastic-to-pseudoelastic transitions of SMAs under multiple loading-unloading cycles (in revision). Int J Solids Struct

Zhan L, Wang SY, Xi HF, Xiao H (2019b) Innovative elastoplastic J_2-flow equations incorporating failure effects of metals into inherent constitutive features (in revision). ZAMM-J Appl Math Mech

Zhang YY, Li H, Wang XM, Yin ZN, Xiao H (2014a) Direct determination of multi-axial elastic potentials for incompressible elastomeric solids: an accurate, explicit approach based on rational interpolation. Continuum Mech Thermodyn 26:207–220

Zhang YY, Li H, Xiao H (2014b) Further study of rubber-like elasticity: elastic potentials matching biaxial data. Appl Math Mech-Engl Ed 35:13–24

Chapter 20
A Multi-disciplinary Approach for Mechanical Metamaterial Synthesis: A Hierarchical Modular Multiscale Cellular Structure Paradigm

Mustafa Erden Yildizdag, Chuong Anthony Tran, Mario Spagnuolo, Emilio Barchiesi, Francesco dell'Isola, and François Hild

Abstract Recent advanced manufacturing techniques such as 3D printing have prompted the need for designing new multiscale architectured materials for various industrial applications. These multiscale architectures are designed to obtain the desired macroscale behavior by activating interactions between different length scales and coupling different physical mechanisms. Although promising results have been recently obtained, the design of such systems still represents a challenge in terms of mathematical modeling, experimentation, and manufacturing. In this paper, some research perspectives are discussed aiming to determine the most efficient methodology needed to design novel metamaterials. A multidisciplinary approach based on Digital Image Correlation (DIC) techniques may be very effective. The main feature of the described DIC-based approach consists of the integration of different methodologies to create a synergistic relationship among the different steps

Mustafa Erden Yildizdag
Department of Naval Architecture and Ocean Engineering, Istanbul Technical University, 34469, Maslak, Istanbul, Turkey & International Research Center for the Mathematics and Mechanics of Complex Systems, University of L'Aquila, Italy,
e-mail: yildizdag@itu.edu.tr

Chuong Anthony Tran · Emilio Barchiesi · Mario Spagnuolo
International Research Center for the Mathematics and Mechanics of Complex Systems, University of L'Aquila, Italy,
e-mail: tcanth@outlook.com, barchiesiemilio@gmail.com, mario.spagnuolo.memocs@gmail.com

Francesco dell'Isola
International Research Center for the Mathematics and Mechanics of Complex Systems, University of L'Aquila & Dipartimento di Ingegneria Civile, Edile-Architettura e Ambientale, Università degli Studi dell'Aquila, L'Aquila, Italy,
e-mail: francesco.dellisola.me@gmail.com

François Hild
Laboratoire de Mécanique et Technologie (LMT), ENS Paris-Saclay, CNRS, Université Paris-Saclay, 94235 Cachan Cedex, France,
e-mail: francois.hild@ens-paris-saclay.fr

H. Altenbach and A. Öchsner (eds.), *State of the Art and Future Trends in Material Modeling*, Advanced Structured Materials 100,
https://doi.org/10.1007/978-3-030-30355-6_20

from design to fabrication and validation. Experimental techniques and modeling approaches are envisioned to be combined in feedback loops whose objective is to determine the required multiscale architectures of newly designed metamaterials. Moreover, it is necessary to develop appropriate mathematical models to estimate the behavior of such metamaterials. Within this new design approach, the manufacturing process can be effectively guided by a precise theoretical and experimental framework. In order to show the applicability of the proposed approach, some preliminary results are provided for a particular type of mechanical metamaterial, namely, pantographic metamaterials. Lastly, the most relevant challenges are highlighted among those that must be addressed for future applications.

Keywords: Synthesis of metamaterials · Generalized models · Analog circuits · Pantographic structures · Digital image correlation · Homogenization

20.1 Introduction

It is possible to find natural materials that exhibit very exotic and unusual behavior due to their microstructures organized with complex hierarchies (Lakes, 1993). These hierarchical architectures consist of a combination of numerous structural patterns at different length scales, and each pattern is made of architectured microstructures characterized by lower length scales. Here, the overall response generated at the macroscale is related not only to each of the lower-scale microstructures but also to their interactions. The most common example of such natural materials is bone tissues (Maggi et al, 2017; Giorgio et al, 2017; Chia and Wu, 2015; Cima et al, 1994). In Fig. 20.1, their structural hierarchy is illustrated from macro- to nano-scales. The overall response of bone is obtained by the interactions of various features at different length scales. As can be seen from Fig. 20.1, the microstructure of bone also gives very inspirational ideas to design new metamaterials, namely, different parts of a material may have various microstructural patterns depending on the desired macroscale response. In this particular example, osseous tissues (i.e. cancellous and cortical bones) have different structural patterns at the microscale, lamellae are arranged in different manners to form trabeculae and ostea. Consequently, different responses are obtained at particular locations.

Plant stems are another example of natural multiscale materials. They need to resist both axial load from their own mass and bending moment from the wind. Fig. 20.2 shows an example of an internal microstructure enabling for such a strength. A scanning electron micrograph of a hawthorn stem reveals its foam-like interior structure. Gibson et al (1995) showed that this foam-like architecture improves the buckling resistance of the plant.

It can be noted that multiscale natural materials have been inherently optimized by natural selection through a very long process. For instance, bone tissues living now on Earth are the result of a very long (many million year) selection and adaptation process. During the so-called Cambrian explosion, the diversification of living species

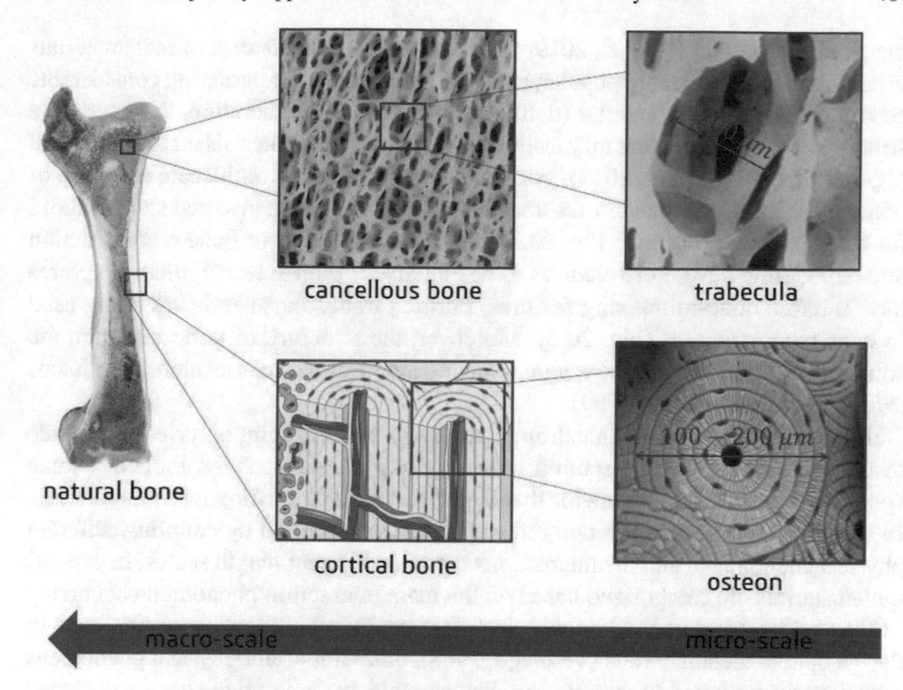

Fig. 20.1 Structural elements of bone at different length scales

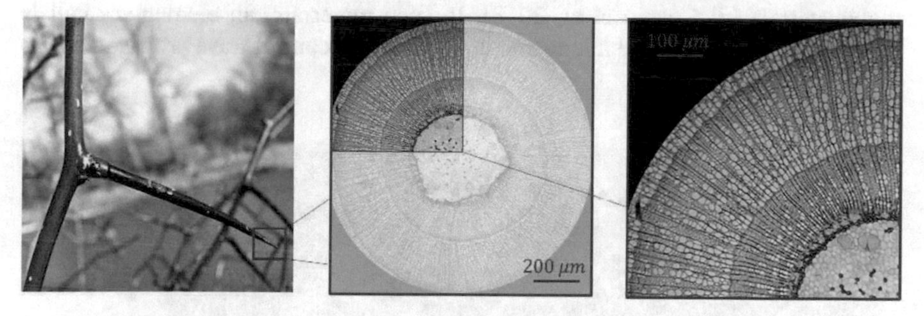

Fig. 20.2 Microstructure (right and center) of hawthorn stem (left)

experienced an exponential growth, and in the most recent taxonomy list, it is possible to find at least 69,276 different species. Therefore, many adaptations occurred in the evolution of bone tissues, and different structures at various length scales are observed nowadays. If enough time were given to natural selection, one would still discover new multiscale materials!

With the newest manufacturing techniques, in particular with 3D printing, many researchers are trying to design novel materials whose exotic macroscopic properties are obtained with suitably designed multiscale microstructures (Liu et al, 2013; Geers et al, 2003). Materials that do not exist in nature, and whose design is based on multiscale modeling to exhibit desired performances, are sometimes called metamaterials (Barchiesi et al, 2019; Gatt et al, 2015). The concept of metamaterials is becoming more and more popular, and their applications are garnering considerable academic and industrial interest (dell'Isola et al, 2019a,b). Therefore, the multiscale structures observed in nature may inspire the design of such materials for technological applications (Wegst et al, 2015). For instance, based on the multiscale structure of bones (Fig. 20.1), artificial bio-resorbable materials have been invented and produced for bone grafting processes (Fig. 20.3). Scaffolds used to favor bone reconstruction and remodeling have more chances to be effective if their internal microstructures have suitable bone-mimicking features. Further, trabecular metals are being used in bone reconstruction (Fig. 20.3). Moreover, the structure of bone has been the source of inspiration for light-weight structure applications such as aluminum foams (Fig. 20.3, Andrews et al, 1999).

In addition to biomechanical applications, a lot of attempts have been made to design multiscale architectured materials (e.g. metamaterials) inducing some specific types of overall behavior that is not observed in existing natural materials. In such designs, application-tailored responses are obtained by coupling different physical phenomena, and the interactions between different length scales. In general, metamaterials are categorized based on the main interaction phenomena occurring in their microstructures. Although electromagnetic interactions were first used to design optical metamaterials (Veselago, 1968), other important physical phenomena are currently exploited in their design. For example, metamaterials that are designed to control the propagation of acoustic (elastic) waves are referred to as acoustic metamaterials (di Cosmo et al, 2018). In such materials, an elementary cell is periodically repeated in the microstructure. In order to control wave propagation, the

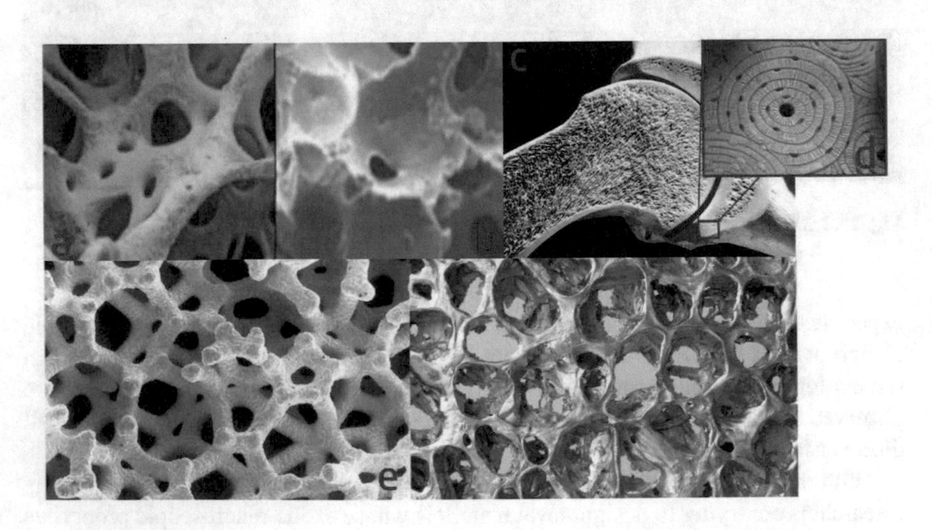

Fig. 20.3 Some multiscale materials. Example of bone tissue (a) and a bio-resorbable artificial graft (b) Giorgio et al (2016b). In (c) and (d) the multi-scale structure of bone is evident: from trabeculae to osteons. In (e) trabecular metal Andreykiv et al (2005) and in (f) aluminum foam are shown

elementary cell is designed with a smaller length scale compared to that involved in the targeted application. In the field of optical and acoustic metamaterials, many novel products have been designed. Typical examples are materials with negative index of refraction (Veselago, 1968, 1967), and those behaving like a low-density plasma with an effective dielectric constant that becomes negative below the effective plasma frequency (Pendry et al, 1996). Other important trends in metamaterial design are images focusing below the diffraction limit (Deng et al, 2009; Zhang et al, 2009; Ambati et al, 2007; Ao and Chan, 2008; Jia et al, 2010; Liu et al, 2007) (e.g. hyperlenses are able to transform evanescent waves into propagating waves, which can be detected at large distance, and superlenses amplify these evanescent waves) and metafluids (Norris, 2009).

This paper focuses on mechanical metamaterials, namely multiscale materials whose behavior is only determined in terms of mechanical interactions among different structures at different scales. Mechanical metamaterials have been investigated in a large number of different studies (e.g. see Kadic et al, 2012; Lee et al, 2012; dell'Isola et al, 2015c; Vangelatos et al, 2018, 2019; Barchiesi et al, 2018; Misra et al, 2018; Laudato et al, 2018; Del Vescovo and Giorgio, 2014a; Carcaterra et al, 2015; Turco et al, 2017a; Barchiesi and Placidi, 2017; Placidi et al, 2017b)). Auxetic structures (Lakes, 1987) (i.e. materials which have negative Poisson's ratio) and locally resonant microstructured materials (Liu et al, 2000) with negative refraction index are typical examples. There are many interesting mathematical problems to be solved in the design of such metamaterials. In reality, the long natural selection process that did manage to optimize the functionality of many natural materials has to be sped up because some applications cannot wait so long.

A main change in research paradigm is needed for the design of new metamaterials. Usually, in mathematical physics, a model is built by conjecturing some postulates assumed to be satisfied to model some specific aspects of the physical reality. For instance, if one wants to model a deformable body in the elastic regime, a time-dependent field of placement and an action functional (e.g. see Germain, 1973; Auffray et al, 2014)) in the set of admissible motions is introduced to describe its evolving shape. Once the postulated action functional is conjectured, the motions predicted by means of the Principle of Least Action can be compared with experimental evidence. If the material parameters appearing in the action functional are usually determined with a small set of measurements, and allow for the description of many more experiments, then one can say that the experimental evidence supports the validity of conjectured models. In this way, the mathematical model for a given class of phenomena is tailored to predict the overall performance of the given material under different design conditions.

Conversely, in the design of metamaterials, an approach that reverses the above-described conceptual order is followed (dell'Isola et al, 2016a). A mathematical model that a priori describes the desired overall behavior is first proposed. Then, the corresponding synthesis problem is solved, namely, finding a (possibly multiscale and/or multiphysics) micro-architecture whose overall behavior is modeled with the selected mathematical model. The synthesized multiscale structure is then fabricated, and its behavior experimentally tested. The final steps of the described "reversed order" process have been made possible with the recent developments in 3D printing technologies. A specific example of such novel metamaterials is given by the so-called pantographic sheets. In Alibert et al (2003); Seppecher et al (2011), a synthesis problem was solved to model this type of structures. In these studies, to find microstructures for one-dimensional and two-dimensional continua, the governing equations for the described microstructures were obtained by a Lagrangian whose potential energy depends on the second gradient of displacement fields at the macroscale in the case of plane motions.

The current increasing interest in metamaterials is mainly due to the availability of new advanced manufacturing techniques such as 3D printing (Rumpf et al, 2013), optical lithography (Madou, 2011), roll-to-roll processing (Ok et al, 2012), electrospinning (Teo and Ramakrishna, 2006), dry and wet etching(Pearton et al, 1993), micro-molding (Heckele and Schomburg, 2003), and micro-machining (Masuzawa et al, 1985). With this spectacular progress obtained in advanced manufacturing techniques in the past ten years, it is much easier to design and manufacture multiscale architectures performing desired overall responses in different industrial applications (Engheta and Ziolkowski, 2006). All these new manufacturing techniques are seen as solutions in a more and more complex manufacturing environment, specifically in terms of customization, multifunctionality, innovative design, and geometry. These new manufacturing technologies not only enable for accurate fabrication with characteristic lengths of the order of micrometers and even less but they are also getting less expensive and more reliable. Thus, with these new techniques, it is possible to manufacture multiscale materials obtained as a result of the solution to the synthesis problem mentioned above. As an example, the relevant length scales of a pantographic sheet that was designed and 3D printed for light-weight structural applications are shown in Fig. 20.4.

Furthermore, the reassessment of the existing mathematical models for the description of deformable bodies is unavoidable from theoretical points of view as another consequence of this progress in material technology. Since materials may have complex hierarchical architectures, the classical description of continuum mechanics is no longer applicable to model exotic responses (dell'Isola et al, 2017). Therefore, researchers have to develop and reformulate many well-known classical concepts such as stress, strain, strain energy, constitutive laws, and balance equations (Eugster and Glocker, 2017). The improvement of existing theoretical frameworks can be achieved with variational approaches and suitable homogenization techniques, which provide efficient micro-to-macro identification (Francfort and Murat, 1986; Abdoul-Anziz and Seppecher, 2018; dell'Isola et al, 2016b).

Although many results have been presented in the literature (and they are really promising), the design of new metamaterials still remains a formidable challenge. The main issue to overcome corresponds to the "complexity" that is involved at every stage of the process in terms of modeling, experimentation, and manufacturing. The sought description requires a robust design approach that creates a synergistic interplay among all involved highly complex design stages to provide an efficient feedback loop in data analysis. This kind of approach may provide the expected progress in the field of the design of novel metamaterials. In order to mitigate such

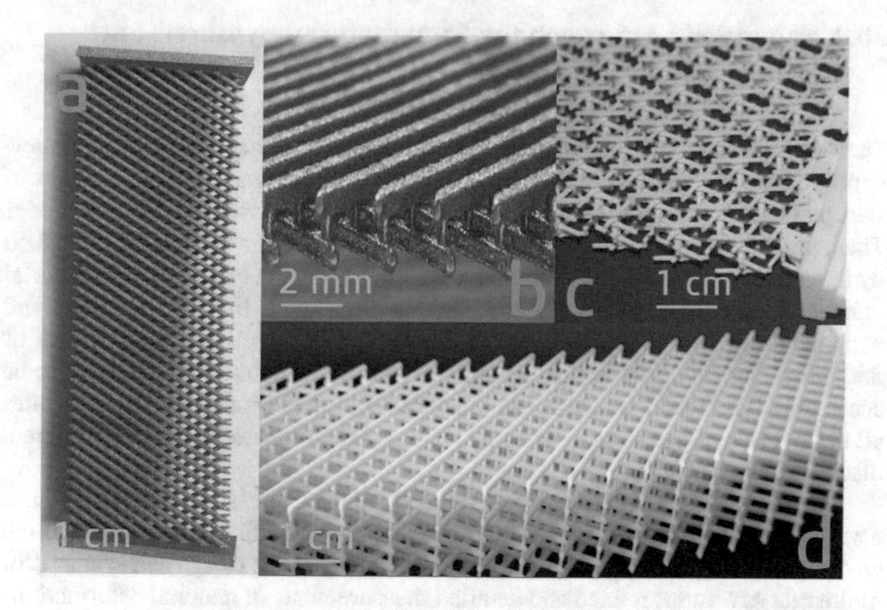

Fig. 20.4 Relevant scales from macro to micro (a-d) for pantographic sheets

challenge, it is suggested to focus, as a first stage, on the design of mechanical metamaterials. For such materials, the behavior at the macroscale is achieved with mechanical interactions of structural networks arranged at different length scales.

The design strategies need to be improved. A systematic methodology, which combines modeling, experimental and manufacturing points of view simultaneously, is called for. Instead of conjecturing metamaterial microstructures without any theoretical guidance, and then trying to experimentally investigate their mechanical properties, an *a priori* synthesis can precede any 3D printing activity, while the feedback from experiments allows for verifying the quality of the theoretical elaboration and, possibly, guide new theoretical investigations. Furthermore, using Lagrangian variational formalisms, one can carefully and efficiently study both static and dynamic responses of every type of materials and design new metamaterials for different industrial applications (Del Vescovo and Giorgio, 2014b; Placidi et al, 2014; Rosi et al, 2013).

The organization of the remainder of the paper is as follows. In Sect. 20.2, more details are given on the proposed synergistic design approach. Some conjectures about the steps required in the design approach are detailed in Sect. 20.3, and promising preliminary results are presented for an additively manufactured mechanical metamaterial in Sect. 20.4.

20.2 Synergistic Approach for Metamaterial Synthesis and Fabrication

The design and fabrication of new metamaterials is a challenging task. A new approach can be followed as the usual logical order is reversed. First, one has to start by characterizing, with suitable Lagrangians, the desired constitutive model. Then, the microstructure of the material whose macroscopic behavior is described by the *a priori* chosen Lagrangian is identified. Lastly, the designed metamaterial is manufactured for the targeted applications. From modeling, experimental and manufacturing standpoints, this procedure is challenging, namely, it consists of designing, fabricating and validating suitable multiscale architectures. In order to be successful, it is crucial to develop an efficient conceptual framework that integrates all the involved design steps by creating a synergistic feedback loop among different disciplines and techniques.

Among many different techniques, Digital Image Correlation (DIC) may have a very prominent role (Sutton et al, 2009; Grédiac and Hild, 2012) to create the envisioned synergistic approach. To check the validity of the design and synthesis of multiscale structures, refined and detailed measurements of material deformations is essential to guide the synthesis process and to validate its results. DIC is an (automatic) image analysis method that measures the deformation of tested specimens and generates displacement and strain fields at prescribed resolution. DIC is very popular in experimental mechanics (Sutton, 2013). This non-contact technique is carried out by using mathematical/numerical registration procedures to process digital images of specimens recorded during the experiment. Sophisticated DIC methods have been recently developed (Hild and Roux, 2012b; Sutton et al, 2009; Hild and Roux, 2012a; Tomičević et al, 2013), which are applicable to many mechanical situations, in particular, in the case of large deformations (Chevalier et al, 2001; Hild et al, 2002). The DIC techniques can efficiently enable the comparison between experimental evidence and theoretical models (Leclerc et al, 2009).

To transform digital images into data, experimental and numerical tools have to be used. The surface of the specimen has generally to be first prepared to make the motion of material points distinguishable for the DIC process. During the experiment, digital images are to be recorded with possibly high-definition cameras. At the beginning of the experiment, a reference digital image is recorded, to represent the reference configuration, and then the displacement field is calculated with a correlation between the reference image and subsequent images of the deformed configuration. The concepts at the basis of continuum mechanics, in particular, its kinematics and “deformatics”, play a central role in DIC (Sutton et al, 2009; Hild and Roux, 2012b).

The DIC techniques have proven to be effective in analyzing experimental results, and they can provide a rapid feedback to guide numerical and theoretical applications in metamaterial design (dell'Isola et al, 2019a,c). Due to the complexity of the considered mechanical systems, no closed-form solutions for their deformation problems are generally available. Hence, numerical simulations must be performed to predict deformation patterns. In general, these simulations must consider large deformation phenomena, and therefore sophisticated algorithms. By comparing the DIC results with numerical results (e.g. finite element simulations, see for example Niiranen et al (2017); Khakalo and Niiranen (2017); Niiranen et al (2016); Khakalo and Niiranen (2018); Eugster et al (2014); Cazzani et al (2016d); Turco et al (2016b); Cazzani et al (2016a,c,b); Grillanda et al (2019); Cazzani et al (2018b,a)), it is possible to validate the results of theoretical and practical syntheses. The detailed analysis of deformations made possible by DIC will point toward weak points in the process. DIC can also be used to analyze the image sets by using displacement fields generated via numerical simulations (i.e. via integrated frameworks (Leclerc et al, 2009; Mathieu et al, 2015)). Let us note that in some previous studies (Quiligotti et al, 2002), the calibration of material parameters was based on the choice of few geometrical properties of the specimen used in the experiments and in the analysis of the difference between these measured quantities and their predicted values. DIC analyses allow for more thorough and systematic comparisons between predicted and measured displacement fields.

Further, DIC techniques are also capable of measuring displacement fields at different length scales (Turco et al, 2018; dell'Isola et al, 2019a,b). This is another essential feature of DIC that will have to be exploited in a more extensive way in the present context as multiscale models are developed in the description and design of metamaterials. By using multiscale DIC analyses, and considering both the desired overall behavior and its microscopic features, the mathematical synthesis process and its transformation into 3D printed specimens can be modified or developed again and again based on the data provided by the DIC-based synergistic procedure.

The theoretical synthesis process of a specific metamaterial produces an architectured microstructure that is represented by the CAD modeler (e.g. standard tessellation language or STL file), and then used in the fabrication step. Within the described design framework, this file can be used in both numerical simulations and 3D printing processes. Among all the advanced manufacturing techniques, 3D printing is one of the most promising technologies for the fabrication of complex materials and geometries. It can be easily optimized to produce specimens made of multiscale architectured materials. Among its main features, 3D printing has a very significant advantage in comparison with conventional manufacturing techniques, namely, it can easily make complex 3D objects with its layer-wise approach, by eliminating the dependence on additional design constraints. One can easily deal with geometric complexity and control the microstructure of fabricated parts in detail. The process of 3D printing also enables for effective multimaterial fabrication. This feature will increase its range of applications in the design of metamaterials.

A further step in the development of the proposed methodology will consist in the formulation of a corrective algorithm, which must automatically modify the STL file once the results of some experiments are analyzed with DIC techniques. A DIC-based system, which couples DIC registration algorithms with synthesis and finite element procedures, may expedite and make effective the feedback redesigning action. In this way, one can automatically unify theoretical and experimental studies by integrating the whole design and verification processes.

20.3 Digital Image Correlation-based Metamaterial Design Process

In this section, the main steps of the DIC-based metamaterial design process is discussed and its main features are delineated. *In the first step of the process*, the required macroscopic behavior has to be identified carefully. At this step, it would be ideal to find out possible design constraints due to the applied manufacturing technology. One cannot 3D print any kind of designed microstructure because of the limits related to geometry, material and resolution of the printing device.

Many interesting macroscopic responses may be sought for different applications. For instance, one can

- require that the designed material remain elastic in large deformation regimes;
- demand the design of a material to exhibit wide frequency band gaps;
- look for an optimized bone scaffold, favoring the reconstruction and remodeling of bone tissues.

A clear understanding of all involved phenomena is an unavoidable prerequisite for this type of design processes, and a precise mathematical formulation is needed for the description of designed metamaterials. For example, in the design of bio-resorbable grafts (Madeo et al, 2011) for bone healing applications (Fig. 20.1), the resorption mechanism must be understood as the material is expected to have a successful and effective integration with the bone structure, biological activities and healthy tissues (Giorgio et al, 2016a; Eugster and Glocker, 2013). Therefore, it is important to understand the driving features of newly designed metamaterials and their compatibility with existing systems. All these phenomenological aspects of the designed metamaterial must be specified by means of Lagrangian action functionals (and possibly Rayleigh dissipation functionals), which are assumed *a priori* to govern the behavior of the designed metamaterials (dell'Isola and Placidi, 2011).

In the second step, the hierarchical architecture of the metamaterial is synthesized. From the theoretical standpoint, a mathematical model, which describes the desired behavior, has been already proposed at the previous step. Due to the hierarchical complexity of the material, a multiscale modeling procedure must be followed in the synthesis scheme. It has to be noted that only few materials (i.e. very restricted classes of Lagrangian and Rayleigh functionals) can be synthesized by using a single scale architecture. Instead of trying to implement ineffective trial and error computations between microscopic and macroscopic scales, an extra intermediate step may be included in the synthesis scheme. Different discrete mechanical systems at several intermediate scales are to be introduced.

The proposed process is very similar to that used in the theory of synthesis of analogue circuits (Giorgio et al, 2015). As every passive linear n-port circuit can be synthesized by using an algorithmically produced graph and by linking any pair of points of this graph with four specific circuital elements (i.e. resistances, inductances, capacitors and transformers), it is expected that the most general microstructures for mechanical metamaterials can be built by reproducing some basic microstructures at different length scales. Moreover, by introducing only discrete mesoscopic models, the numerical algorithms are implemented efficiently, and the micro-to-macro transition process can be performed more easily (Turco et al, 2016a). As a further perspective, deduced from the analogue circuits field, it would be interesting and useful to produce piezoelectromechanical microstructures to be controlled by means of piezoelectric actuators. Some relevant results already present in the literature about this perspective can be found in Casadei et al (2012); Bergamini et al (2006, 2015).

In the third step, the synthesis scheme previously obtained must be transformed into real-world specimens, for instance by means of 3D printing techniques. Every basic microstructure must be built by supplying a suitable STL file to the selected 3D printer. These files can efficiently be used as a basis for *a posteriori* finite elements analyses and, at the same time, as guide for DIC data collection. This step requires the development of innovative engineering solutions (Golaszewski et al, 2018; Turco et al, 2017b; Gunenthiram et al, 2017; Haboudou et al, 2003; Andreau et al, 2019).

In the fourth step, the validation of the synthesis and construction steps must be performed. This step requires the systematic use of DIC-based techniques. Due to the multiscale nature of the considered microstructures, some computational meshes must be generated at different length scales by using, for instance, the gray level images of the tested specimens.

We give here an example already available in dell'Isola et al (2019b). In Fig. 20.5, meshes at macroscopic and mesoscopic levels are shown in the case of a pantographic sheet. These meshes overlap with the gray images of the test specimen. In the analysis, a coarse discretization of the region of interest is first created with triangular elements independent from its mesostructure or microstructure (Fig. 20.5 a). Then, the mesh is successively refined to increase the accuracy of the results, and in this way, the convergence of the analysis is expected. From the multiscale standpoint, this step looks like the transitions from continuum to discrete models.

The final steps consist of going back, thanks to the experimental results as elaborated by DIC, to the synthesis step and/or to the construction step. The discrepancies between the desired and the measured responses, as revealed by multiscale DIC analyses, redirect both the synthesis process and the scheme of specimens production. This feedback loop is made easier by the fact that the DIC meshes are tailored for micro-to-macro model identifications (Grédiac and Hild, 2012).

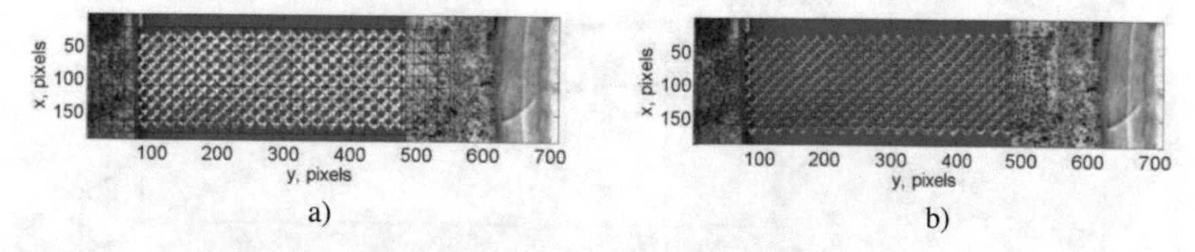

Fig. 20.5 Example of multiscale mesh applied to a pantographic structure

20.4 Preliminary Results

In this section, some preliminary results are presented to show the applicability of the proposed approach for the design of mechanical metamaterials. For this purpose, pantographic structures are considered. This is an example of a theoretical problem formulated as the result of experimental observations via DIC analyses. This multiscale design approach can be further utilized to develop more sophisticated DIC techniques to design and fabricate metamaterials (dell'Isola et al, 2019a,b).

The studies related to the design of higher gradient continua (Mindlin, 1965; dell'Isola et al, 2015a) would be addressed to show the potential of such a synergistic approach. For classical continuum media, the Cauchy theory is applied in terms of balance equations. This theory assumes that the strain energy is only a function of the first gradient of the displacement field. However, with the design of new advanced materials, it was shown that the strain energy can be a function of higher gradients of displacement fields (Seppecher et al, 2011; Alibert et al, 2003). Thus, higher gradient theories are developed to derive the macroscopic behavior of multiscale materials. In Fig. 20.6, the design of a beam whose strain energy depends on higher gradients of the displacement field in the axial direction is presented. A unit cell of the beam is arranged as shown in Fig. 20.6a), and the different levels of the structure in Fig. 20.6c). Then, by using appropriate homogenization techniques, it was proven that the strain energy depends on the second gradient of the displacement field in both vertical and horizontal directions (Seppecher et al, 2011). To simplify the micro-to-macro upscaling, a discrete model consisting of a network of mass particles connected with rotational and extensional springs was introduced (dell'Isola et al, 2016b). For the fabrication of the designed metamaterials, the theoretical data are transformed into a manufacturing process. For this particular structure, the following design has been proposed. Two layers of beams are oriented orthogonally and connected with a set of cylinders or joints allowing for the relative displacement of the beams (dell'Isola et al, 2015b). The alternation of empty and filled spaces enables DIC analyses for the resulting specimen (Turco et al, 2018). As seen from this example, one can extend

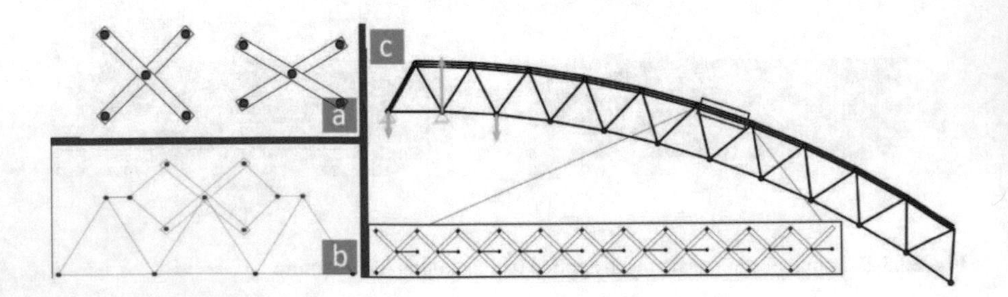

Fig. 20.6 Example of a multiscale scheme for higher gradient one-dimensional material (Seppecher et al, 2011)

this model and fabricate higher order gradient systems by exploiting this multiscale design approach (see Fig. 20.6b) for third gradient model).

The equilibrium shapes of the pantographic structures are shown in Fig. 20.7 for different cases. In Fig. 20.7a)-b), the experimental results are presented for shear and torsion loadings of a pantographic sheet. These pantographic sheets are made of aluminum alloy, and they are 3D printed and designed based on composition of elementary blocks. In Fig. 20.7c), the shear deformation is tracked by performing local DIC registrations.

The displacement fields of the macro- and meso-scale meshes (Fig. 20.5) are reported in Fig. 20.8. In this particular example, the results are presented for longitudinal displacement fields measured during a tensile (i.e. bias) test.

Regarding the experimental study, three major challenges were observed. First, the extension of this application to three dimensional problems might be difficult as the fabrication of beam lattices deforming in 3D is a more complex procedure. It is clearly more complicated to design a material exhibiting the desired overall behavior in three dimensional applications. Using ball joint links would be helpful to make this design possible. In Fig. 20.9, the design of a pivot/hinge link is illustrated. They can be fabricated with 3D printing technologies. Second, a multiscale architecture with nonlinear macroscale responses might be synthesized. This can be avoided by exploiting the synergistic nature of the design framework. Third, possible instability and buckling at the microscale may create a dramatic change in the macroscopic

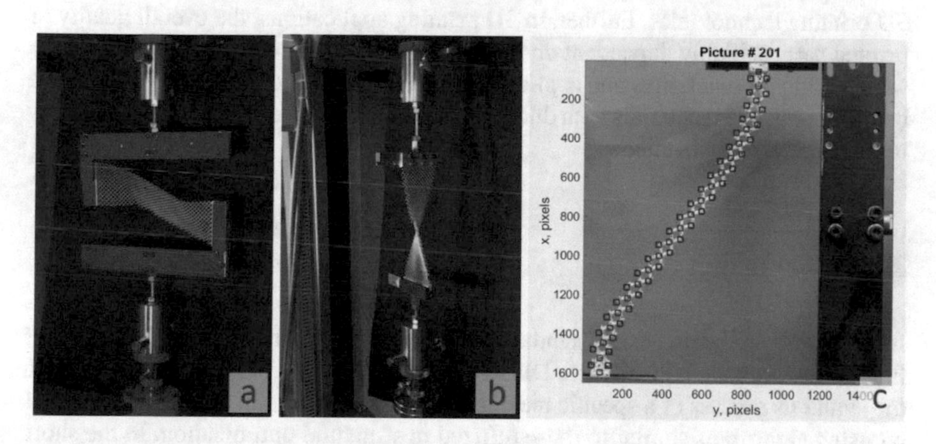

Fig. 20.7 Experiments on 3D printed pantographic sheet

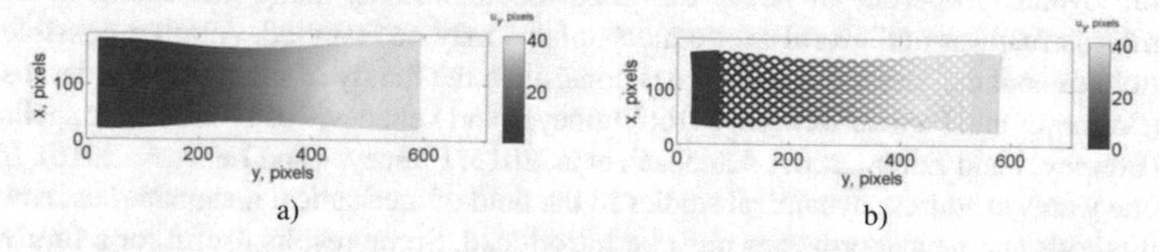

Fig. 20.8 Deformation of different DIC meshes in longitudinal direction

Fig. 20.9 Designed and printed pivot/hinge link dell'Isola et al (2019c)

response of the material. Hence, different critical phenomena must be taken into account to increase the reliability of the approach.

Moreover, from the manufacturing point of view, the structural pattern of the material must be arranged regarding the technological limits of the selected 3D printing technique. Some design rules must be standardized for 3D printing applications. Although it seems that 3D printing can easily deal with any geometric complexity, some important criteria have to be considered before fabricating the designed metamaterials. In general, these rules are applied for the design of supported/unsupported walls, overhangs, holes, connecting/moving parts, and engravings, and may vary for different 3D printing technologies. Further, in 3D printing applications, the overall quality of printed parts is highly dependent on the processing parameters. The latter ones may vary for different materials and applications. Therefore, it is crucial to investigate the behavior of printed materials with different processing parameters and their feasibility in metamaterial applications.

20.5 Conclusion

In this paper, in the process of synthesis and construction of novel metamaterials, it is proposed to systematically use DIC-based methodologies. Based on DIC output, the synthesis process of a specific metamaterial may be partially or totally automated by using algorithms similar to those utilized in structural optimization. In the short term, it is expected that by using DIC techniques to design, characterize and validate the overall properties of newly designed metamaterials, many interesting novel microstructures and useful exotic mechanisms may be invented. Another possible field of application for this techniques consists in the family of micropolar materials (Eremeyev and Pietraszkiewicz, 2016; Eremeyev and Lebedev, 2011) and elastic shells (Eremeyev and Zubov, 2007; Altenbach et al, 2015; Eremeyev and Lebedev, 2016). If one wants to address dynamical studies in the field of mechanical metamaterials, new methods and new approaches must be introduced. Some results useful for a future characterization of the dynamics in memamaterials can be found in Cazzani and

Ruge (2016, 2013); Piccardo et al (2014); Ferretti and Piccardo (2013); Luongo and Zulli (2012); Luongo et al (2008).

To show the applicability of the introduced approach, some preliminary results were presented, namely, those concerning so-called pantographic structures (see for example Placidi et al, 2017a; Scerrato et al, 2016; Boutin et al, 2017). Possible issues related to the design and manufacturing phases have been discussed and highlighted for the future applications.

It is envisioned that the proposed synergistic approach can be extended to the design of the following solutions:

1. metamaterials remaining in their elastic regime for large deformations,
2. metamaterials maintaining their mechanical properties under large temperature changes and experiencing only very limited creep phenomena,
3. metamaterials for bone scaffolds that are optimized for being bio-resorbable and bio-compatible with the host tissues.

Concerning this last class of metamaterials (e.g. see Madeo et al, 2012; Lekszycki and dell'Isola, 2012), the DIC-based framework may design bone scaffolds with adaptive optimal behavior. The latter is obtained when the metamaterial exhibits a proper response to a vast variety of external stimuli. Further, the desired overall response of biomechanical metamaterials can be achieved by enriching their microstructure with other exotic materials such as *shape memory alloys*.

References

Abdoul-Anziz H, Seppecher P (2018) Strain gradient and generalized continua obtained by homogenizing frame lattices. Mathematics and mechanics of complex systems 6(3):213–250

Alibert JJ, Seppecher P, dell'Isola F (2003) Truss modular beams with deformation energy depending on higher displacement gradients. Mathematics and Mechanics of Solids 8(1):51–73

Altenbach H, Eremeyev VA, Naumenko K (2015) On the use of the first order shear deformation plate theory for the analysis of three-layer plates with thin soft core layer. ZAMM-Journal of Applied Mathematics and Mechanics/Zeitschrift für Angewandte Mathematik und Mechanik 95(10):1004–1011

Ambati M, Fang N, Sun C, Zhang X (2007) Surface resonant states and superlensing in acoustic metamaterials. Physical Review B 75(19):195,447

Andreau O, Koutiri I, Peyre P, Penot JD, Saintier N, Pessard E, De Terris T, Dupuy C, Baudin T (2019) Texture control of 316l parts by modulation of the melt pool morphology in selective laser melting. Journal of Materials Processing Technology 264:21–31

Andrews E, Sanders W, Gibson LJ (1999) Compressive and tensile behaviour of aluminum foams. Materials Science and Engineering: A 270(2):113–124

Andreykiv A, Prendergast P, Van Keulen F, Swieszkowski W, Rozing P (2005) Bone ingrowth simulation for a concept glenoid component design. Journal of biomechanics 38(5):1023–1033

Ao X, Chan C (2008) Far-field image magnification for acoustic waves using anisotropic acoustic metamaterials. Physical Review E 77(2):025,601

Auffray N, dell'Isola F, Eremeyev V, Madeo A, Placidi L, Rosi G (2014) Least action principle for second gradient continua and capillary fluids: A lagrangian approach following piola's point of view. In: dell'Isola F, Maier G, Perego U, Andreaus U, Esposito R, Forest S (eds) The Complete

Works of Gabrio Piola: Volume I, Springer, Cham, Advanced Structured Materials, vol 38, pp 606–694

Barchiesi E, Placidi L (2017) A review on models for the 3d statics and 2d dynamics of pantographic fabrics. In: Wave Dynamics and Composite Mechanics for Microstructured Materials and Metamaterials, Springer, pp 239–258

Barchiesi E, Spagnuolo M, Placidi L (2018) Mechanical metamaterials: a state of the art. Mathematics and Mechanics of Solids 24(1):212–234, doi:10.1177/1081286517735695

Barchiesi E, Spagnuolo M, Placidi L (2019) Mechanical metamaterials: a state of the art. Mathematics and Mechanics of Solids 24(1):212–234, doi:10.1177/1081286517735695

Bergamini A, Christen R, Maag B, Motavalli M (2006) A sandwich beam with electrostatically tunable bending stiffness. Smart materials and structures 15(3):678

Bergamini AE, Zündel M, Flores Parra EA, Delpero T, Ruzzene M, Ermanni P (2015) Hybrid dispersive media with controllable wave propagation: A new take on smart materials. Journal of Applied Physics 118(15):154,310

Boutin C, Giorgio I, Placidi L, et al (2017) Linear pantographic sheets: Asymptotic micro-macro models identification. Mathematics and Mechanics of Complex Systems 5(2):127–162

Carcaterra A, dell'Isola F, Esposito R, Pulvirenti M (2015) Macroscopic description of microscopically strongly inhomogeneous systems: A mathematical basis for the synthesis of higher gradients metamaterials. Archive for Rational Mechanics and Analysis 218(3):1239–1262

Casadei F, Delpero T, Bergamini A, Ermanni P, Ruzzene M (2012) Piezoelectric resonator arrays for tunable acoustic waveguides and metamaterials. Journal of Applied Physics 112(6):064,902

Cazzani A, Ruge P (2013) Rotor platforms on pile-groups running through resonance: A comparison between unbounded soil and soil-layers resting on a rigid bedrock. Soil Dynamics and Earthquake Engineering 50:151–161, doi:10.1016/j.soildyn.2013.02.022

Cazzani A, Ruge P (2016) Stabilization by deflation for sparse dynamical systems without loss of sparsity. Mechanical Systems and Signal Processing 70–71:664–681, doi:10.1016/j.ymssp.2015.09.027

Cazzani A, Malagù M, Turco E (2016a) Isogeometric analysis: a powerful numerical tool for the elastic analysis of historical masonry arches. Continuum Mechanics and thermodynamics 28(1-2):139–156

Cazzani A, Malagù M, Turco E (2016b) Isogeometric analysis of plane-curved beams. Mathematics and Mechanics of Solids 21(5):562–577

Cazzani A, Malagù M, Turco E, Stochino F (2016c) Constitutive models for strongly curved beams in the frame of isogeometric analysis. Mathematics and Mechanics of Solids 21(2):182–209

Cazzani A, Stochino F, Turco E (2016d) An analytical assessment of finite element and isogeometric analyses of the whole spectrum of timoshenko beams. ZAMM-Journal of Applied Mathematics and Mechanics/Zeitschrift für Angewandte Mathematik und Mechanik 96(10):1220–1244

Cazzani A, Rizzi N, Stochino F, Turco E (2018a) Modal analysis of laminates by a mixed assumed-strain finite element model. Mathematics and Mechanics of Solids 23(1):99–119, doi:10.1177/1081286516666405

Cazzani A, Serra M, Stochino F, Turco E (2018b) A refined assumed strain finite element model for statics and dynamics of laminated plates. Continuum Mechanics and Thermodynamics pp 1–28, doi:10.1007/s00161-018-0707-x

Chevalier L, Calloch S, Hild F, Marco Y (2001) Digital image correlation used to analyze the multiaxial behavior of rubber-like materials. Eur J Mech A/Solids 20:169–187

Chia HN, Wu BM (2015) Recent advances in 3d printing of biomaterials. Journal of Biological Bngineering 9(1):4

Cima M, Sachs E, Cima L, Yoo J, Khanuja S, Borland S, Wu B, Giordano R (1994) Computer-derived microstructures by 3d printing: bio-and structural materials. In: Solid Freeform Fabr Symp Proc: DTIC Document, pp 181–90

di Cosmo F, Laudato M, Spagnuolo M (2018) Acoustic metamaterials based on local resonances: homogenization, optimization and applications. In: Generalized Models and Non-classical Approaches in Complex Materials 1, Springer, pp 247–274

Del Vescovo D, Giorgio I (2014a) Dynamic problems for metamaterials: review of existing models and ideas for further research. International Journal of Engineering Science 80:153–172

Del Vescovo D, Giorgio I (2014b) Dynamic problems for metamaterials: review of existing models and ideas for further research. International Journal of Engineering Science 80:153–172

dell'Isola F, Placidi L (2011) Variational principles are a powerful tool also for formulating field theories. In: Variational models and methods in solid and fluid mechanics, Springer, pp 1–15

dell'Isola F, Andreaus U, Placidi L (2015a) At the origins and in the vanguard of peridynamics, non-local and higher-gradient continuum mechanics: An underestimated and still topical contribution of gabrio piola. Mathematics and Mechanics of Solids 20(8):887–928

dell'Isola F, Lekszycki T, Pawlikowski M, Grygoruk R, Greco L (2015b) Designing a light fabric metamaterial being highly macroscopically tough under directional extension: first experimental evidence. Zeitschrift für angewandte Mathematik und Physik 66(6):3473–3498

dell'Isola F, Steigmann D, Della Corte A (2015c) Synthesis of fibrous complex structures: designing microstructure to deliver targeted macroscale response. Applied Mechanics Reviews 67(6):060,804

dell'Isola F, Bucci S, Battista A (2016a) Against the fragmentation of knowledge: The power of multidisciplinary research for the design of metamaterials. In: Advanced Methods of Continuum Mechanics for Materials and Structures, Springer, pp 523–545

dell'Isola F, Giorgio I, Pawlikowski M, Rizzi N (2016b) Large deformations of planar extensible beams and pantographic lattices: heuristic homogenization, experimental and numerical examples of equilibrium. Proc R Soc A 472(2185):20150,790

dell'Isola F, Della Corte A, Giorgio I (2017) Higher-gradient continua: The legacy of piola, mindlin, sedov and toupin and some future research perspectives. Mathematics and Mechanics of Solids 22(4):852–872

dell'Isola F, Seppecher P, Alibert JJ, Lekszycki T, Grygoruk R, Pawlikowski M, Steigmann D, Giorgio I, Andreaus U, Turco E, Golaszewski M, Rizzi N, Boutin C, Eremeyev VA, Misra A, Placidi L, Barchiesi E, Greco L, Cuomo M, Cazzani A, Corte AD, Battista A, Scerrato D, Zurba Eremeeva I, Rahali Y, Ganghoffer JF, Müller W, Ganzosch G, Spagnuolo M, Pfaff A, Barcz K, Hoschke K, Neggers J, Hild F (2019a) Pantographic metamaterials: an example of mathematically driven design and of its technological challenges. Continuum Mechanics and Thermodynamics doi:10.1007/s00161-018-0689-8

dell'Isola F, Seppecher P, Spagnuolo M, Barchiesi E, Hild F, Lekszycki T, Giorgio I, Placidi L, Andreaus U, Cuomo M, Eugster SR, Pfaff A, Hoschke K, Langkemper R, Turco E, Sarikaya R, MISRA A, DE ANGELO M, D'Annibale F, Bouterf A, Pinelli X, Misra A, Desmorat B, Pawlikowski M, Dupuy C, Scerrato D, Peyre P, Laudato M, Manzari L, Göransson P, Hesch C, Hesch S, Franciosi P, Dirrenberger J, Maurin F, Vangelatos Z, Grigoropoulos C, Melissinaki V, Farsari M, Muller W, Abali E, Liebold C, Ganzosch G, Harrison P, Drobnicki R, Igumnov LA, Alzahrani F, Hayat T (2019b) Advances in Pantographic Structures: Design, Manufacturing, Models, Experiments and Image Analyses. Continuum Mechanics and Thermodynamics doi:10.1007/s00161-019-00806-x

dell'Isola F, Seppecher P, Spagnuolo M, Barchiesi E, Hild F, Lekszycki T, Giorgio I, Placidi L, Andreaus U, Cuomo M, et al (2019c) Advances in pantographic structures: design, manufacturing, models, experiments and image analyses. Continuum Mechanics and Thermodynamics pp 1–52, doi:10.1007/s00161-019-00806-x

Deng K, Ding Y, He Z, Zhao H, Shi J, Liu Z (2009) Theoretical study of subwavelength imaging by acoustic metamaterial slabs. Journal of Applied Physics 105(12):124,909

Engheta N, Ziolkowski RW (2006) Metamaterials: physics and engineering explorations. John Wiley & Sons

Eremeyev V, Zubov L (2007) On constitutive inequalities in nonlinear theory of elastic shells. ZAMM-Journal of Applied Mathematics and Mechanics/Zeitschrift für Angewandte Mathematik und Mechanik: Applied Mathematics and Mechanics 87(2):94–101

Eremeyev VA, Lebedev LP (2011) Existence theorems in the linear theory of micropolar shells. ZAMM-Journal of Applied Mathematics and Mechanics/Zeitschrift für Angewandte Mathematik und Mechanik 91(6):468–476

Eremeyev VA, Lebedev LP (2016) Mathematical study of boundary-value problems within the framework of steigmann–ogden model of surface elasticity. Continuum Mechanics and Thermodynamics 28(1-2):407–422

Eremeyev VA, Pietraszkiewicz W (2016) Material symmetry group and constitutive equations of micropolar anisotropic elastic solids. Mathematics and Mechanics of Solids 21(2):210–221

Eugster S, Hesch C, Betsch P, Glocker C (2014) Director-based beam finite elements relying on the geometrically exact beam theory formulated in skew coordinates. International Journal for Numerical Methods in Engineering 97(2):111–129

Eugster SR, Glocker C (2013) Constraints in structural and rigid body mechanics: a frictional contact problem. Annals of solid and structural mechanics 5(1-2):1–13

Eugster SR, Glocker C (2017) On the notion of stress in classical continuum mechanics. Mathematics and Mechanics p 299

Ferretti M, Piccardo G (2013) Dynamic modeling of taut strings carrying a traveling mass. Continuum Mechanics and Thermodynamics 25(2-4):469–488

Francfort GA, Murat F (1986) Homogenization and optimal bounds in linear elasticity. Archive for Rational mechanics and Analysis 94(4):307–334

Gatt R, Mizzi L, Azzopardi JI, Azzopardi KM, Attard D, Casha A, Briffa J, Grima JN (2015) Hierarchical auxetic mechanical metamaterials. Scientific reports 5:8395

Geers M, Kouznetsova VG, Brekelmans W (2003) Multiscale first-order and second-order computational homogenization of microstructures towards continua. International Journal for Multiscale Computational Engineering 1(4)

Germain P (1973) The method of virtual power in continuum mechanics. part 2: Microstructure. SIAM Journal on Applied Mathematics 25:556–575, doi:10.1137/0125053, URL doi.org/10.1137/0125053

Gibson LJ, Ashby MF, Karam G, Wegst U, Shercliff H (1995) The mechanical properties of natural materials. ii. microstructures for mechanical efficiency. Proceedings of the Royal Society of London Series A: Mathematical and Physical Sciences 450(1938):141–162

Giorgio I, Galantucci L, Della Corte A, Del Vescovo D (2015) Piezo-electromechanical smart materials with distributed arrays of piezoelectric transducers: current and upcoming applications. International Journal of Applied Electromagnetics and Mechanics 47(4):1051–1084

Giorgio I, Andreaus U, Madeo A (2016a) The influence of different loads on the remodeling process of a bone and bioresorbable material mixture with voids. Continuum Mechanics and Thermodynamics 28(1-2):21–40

Giorgio I, Andreaus U, Scerrato D, dell'Isola F (2016b) A visco-poroelastic model of functional adaptation in bones reconstructed with bio-resorbable materials. Biomechanics and modeling in mechanobiology 15(5):1325–1343

Giorgio I, Andreaus U, Lekszycki T, Corte AD (2017) The influence of different geometries of matrix/scaffold on the remodeling process of a bone and bioresorbable material mixture with voids. Mathematics and Mechanics of Solids 22(5):969–987

Golaszewski M, Grygoruk R, Giorgio I, Laudato M, Di Cosmo F (2018) Metamaterials with relative displacements in their microstructure: technological challenges in 3d printing, experiments and numerical predictions. Continuum Mechanics and Thermodynamics pp 1–20, doi:10.1007/s00161-018-0692-0

Grédiac M, Hild F (2012) Full-field measurements and identification in solid mechanics. John Wiley & Sons

Grillanda N, Chiozzi A, Bondi F, Tralli A, Manconi F, Stochino F, Cazzani A (2019) Numerical insights on the structural assessment of historical masonry stellar vaults: the case of santa maria del monte in cagliari. Continuum Mechanics and Thermodynamics pp 1–24, doi:10.1007/s00161-019-00752-8

Gunenthiram V, Peyre P, Schneider M, Dal M, Coste F, Fabbro R (2017) Analysis of laser–melt pool–powder bed interaction during the selective laser melting of a stainless steel. Journal of Laser Applications 29(2):022,303

Haboudou A, Peyre P, Vannes A (2003) Study of keyhole and melt pool oscillations in dual beam welding of aluminum alloys: effect on porosity formation. In: First International Symposium on

High-Power Laser Macroprocessing, International Society for Optics and Photonics, vol 4831, pp 295–301

Heckele M, Schomburg W (2003) Review on micro molding of thermoplastic polymers. Journal of Micromechanics and Microengineering 14(3):R1

Hild F, Roux S (2012a) Comparison of local and global approaches to digital image correlation. Experimental Mechanics 52(9):1503–1519

Hild F, Roux S (2012b) Digital image correlation. In: Rastogi P, Hack E (eds) Optical Methods for Solid Mechanics. A Full-Field Approach, Wiley-VCH, Weinheim (Germany), pp 183–228

Hild F, Raka B, Baudequin M, Roux S, Cantelaube F (2002) Multi-scale displacement field measurements of compressed mineral wool samples by digital image correlation. Appl Optics IP 41(32):6815–6828

Jia H, Ke M, Hao R, Ye Y, Liu F, Liu Z (2010) Subwavelength imaging by a simple planar acoustic superlens. Applied Physics Letters 97(17):173,507

Kadic M, Bückmann T, Stenger N, Thiel M, Wegener M (2012) On the practicability of pentamode mechanical metamaterials. Applied Physics Letters 100(19):191,901

Khakalo S, Niiranen J (2017) Isogeometric analysis of higher-order gradient elasticity by user elements of a commercial finite element software. Computer-Aided Design 82:154–169

Khakalo S, Niiranen J (2018) Form ii of mindlin's second strain gradient theory of elasticity with a simplification: For materials and structures from nano-to macro-scales. European Journal of Mechanics-A/Solids 71:292–319

Lakes R (1987) Foam structures with a negative poisson's ratio. Science 235:1038–1041

Lakes R (1993) Materials with structural hierarchy. Nature 361(6412):511

Laudato M, Manzari L, Barchiesi E, Di Cosmo F, Göransson P (2018) First experimental observation of the dynamical behavior of a pantographic metamaterial. Mechanics Research Communications 94:125–127, doi:10.1016/j.mechrescom.2018.11.003

Leclerc H, Périé J, Roux S, Hild F (2009) Integrated digital image correlation for the identification of mechanical properties. In: Gagalowicz A, Philips W (eds) Computer Vision/Computer Graphics CollaborationTechniques. MIRAGE 2009, Springer, Berlin, Heidelberg, Lecture Notes in Computer Science, vol 5496, pp 161–171

Lee JH, Singer JP, Thomas EL (2012) Micro-/nanostructured mechanical metamaterials. Advanced materials 24(36):4782–4810

Lekszycki T, dell'Isola F (2012) A mixture model with evolving mass densities for describing synthesis and resorption phenomena in bones reconstructed with bio-resorbable materials. ZAMM-Journal of Applied Mathematics and Mechanics/Zeitschrift für Angewandte Mathematik und Mechanik 92(6):426–444

Liu Y, Greene MS, Chen W, Dikin DA, Liu WK (2013) Computational microstructure characterization and reconstruction for stochastic multiscale material design. Computer-Aided Design 45(1):65–76

Liu Z, Zhang X, Mao Y, Zhu Y, Yang Z, Chan CT, Sheng P (2000) Locally resonant sonic materials. science 289(5485):1734–1736

Liu Z, Lee H, Xiong Y, Sun C, Zhang X (2007) Far-field optical hyperlens magnifying sub-diffraction-limited objects. science 315(5819):1686–1686

Luongo A, Zulli D (2012) Dynamic instability of inclined cables under combined wind flow and support motion. Nonlinear Dynamics 67(1):71–87

Luongo A, Zulli D, Piccardo G (2008) Analytical and numerical approaches to nonlinear galloping of internally resonant suspended cables. Journal of Sound and Vibration 315(3):375–393

Madeo A, Lekszycki T, dell'Isola F (2011) A continuum model for the bio-mechanical interactions between living tissue and bio-resorbable graft after bone reconstructive surgery. Comptes Rendus Mécanique 339(10):625–640

Madeo A, George D, Lekszycki T, Nierenberger M, Rémond Y (2012) A second gradient continuum model accounting for some effects of micro-structure on reconstructed bone remodelling. Comptes Rendus Mécanique 340(8):575–589

Madou MJ (2011) Manufacturing techniques for microfabrication and nanotechnology, vol 2. CRC press

Maggi A, Li H, Greer JR (2017) Three-dimensional nano-architected scaffolds with tunable stiffness for efficient bone tissue growth. Acta Biomaterialia 63:294–305

Masuzawa T, Fujino M, Kobayashi K, Suzuki T, Kinoshita N (1985) Wire electro-discharge grinding for micro-machining. CIRP Annals 34(1):431–434

Mathieu F, Leclerc H, Hild F, Roux S (2015) Estimation of elastoplastic parameters via weighted FEMU and integrated-DIC. Exp Mech 55(1):105–119

Mindlin RD (1965) Second gradient of strain and surface-tension in linear elasticity. International Journal of Solids and Structures 1(4):417–438

Misra A, Lekszycki T, Giorgio I, Ganzosch G, Müller WH, dell'Isola F (2018) Pantographic metamaterials show atypical Poynting effect reversal. Mechanics Research Communications 89:6–10

Niiranen J, Khakalo S, Balobanov V, Niemi AH (2016) Variational formulation and isogeometric analysis for fourth-order boundary value problems of gradient-elastic bar and plane strain/stress problems. Computer Methods in Applied Mechanics and Engineering 308:182–211

Niiranen J, Kiendl J, Niemi AH, Reali A (2017) Isogeometric analysis for sixth-order boundary value problems of gradient-elastic kirchhoff plates. Computer Methods in Applied Mechanics and Engineering 316:328–348

Norris AN (2009) Acoustic metafluids. The Journal of the Acoustical Society of America 125(2):839–849

Ok JG, Seok Youn H, Kyu Kwak M, Lee KT, Jae Shin Y, Jay Guo L, Greenwald A, Liu Y (2012) Continuous and scalable fabrication of flexible metamaterial films via roll-to-roll nanoimprint process for broadband plasmonic infrared filters. Applied Physics Letters 101(22):223,102

Pearton S, Abernathy C, Ren F, Lothian J, Wisk P, Katz A (1993) Dry and wet etching characteristics of inn, aln, and gan deposited by electron cyclotron resonance metalorganic molecular beam epitaxy. Journal of Vacuum Science & Technology A: Vacuum, Surfaces, and Films 11(4):1772–1775

Pendry JB, Holden A, Stewart W, Youngs I (1996) Extremely low frequency plasmons in metallic mesostructures. Physical review letters 76(25):4773

Piccardo G, Ranzi G, Luongo A (2014) A complete dynamic approach to the generalized beam theory cross-section analysis including extension and shear modes. Mathematics and Mechanics of Solids 19(8):900–924

Placidi L, Rosi G, Giorgio I, Madeo A (2014) Reflection and transmission of plane waves at surfaces carrying material properties and embedded in second-gradient materials. Mathematics and Mechanics of Solids 19(5):555–578

Placidi L, Andreaus U, Giorgio I (2017a) Identification of two-dimensional pantographic structure via a linear D4 orthotropic second gradient elastic model. Journal of Engineering Mathematics 103(1):1–21

Placidi L, Barchiesi E, Battista A (2017b) An inverse method to get further analytical solutions for a class of metamaterials aimed to validate numerical integrations. In: Mathematical Modelling in Solid Mechanics, Springer, pp 193–210

Quiligotti S, Maugin GA, dell'Isola F (2002) Wave motions in unbounded poroelastic solids infused with compressible fluids. Zeitschrift für angewandte Mathematik und Physik ZAMP 53(6):1110–1138

Rosi G, Giorgio I, Eremeyev VA (2013) Propagation of linear compression waves through plane interfacial layers and mass adsorption in second gradient fluids. ZAMM-Journal of Applied Mathematics and Mechanics/Zeitschrift für Angewandte Mathematik und Mechanik 93(12):914–927

Rumpf RC, Pazos J, Garcia CR, Ochoa L, Wicker R (2013) 3d printed lattices with spatially variant self-collimation. Progress In Electromagnetics Research 139:1–14

Scerrato D, Giorgio I, Rizzi NL (2016) Three-dimensional instabilities of pantographic sheets with parabolic lattices: numerical investigations. Zeitschrift für angewandte Mathematik und Physik 67(3):53

Seppecher P, Alibert JJ, dell'Isola F (2011) Linear elastic trusses leading to continua with exotic mechanical interactions. Journal of Physics: Conference Series 319(1):012,018

Sutton M (2013) Computer vision-based, noncontacting deformation measurements in mechanics: A generational transformation. Appl Mech Rev 65(AMR-13-1009):050,802

Sutton M, Orteu J, Schreier H (2009) Image correlation for shape, motion and deformation measurements: basic concepts, theory and applications. Springer Science & Business Media

Teo WE, Ramakrishna S (2006) A review on electrospinning design and nanofibre assemblies. Nanotechnology 17(14):R89

Tomičević Z, Hild F, Roux S (2013) Mechanics-aided digital image correlation. The Journal of Strain Analysis for Engineering Design 48(5):330–343

Turco E, dell'Isola F, Cazzani A, Rizzi NL (2016a) Hencky-type discrete model for pantographic structures: numerical comparison with second gradient continuum models. Zeitschrift für angewandte Mathematik und Physik 67(4):85

Turco E, Golaszewski M, Cazzani A, Rizzi NL (2016b) Large deformations induced in planar pantographic sheets by loads applied on fibers: experimental validation of a discrete lagrangian model. Mechanics Research Communications 76:51–56

Turco E, Giorgio I, Misra A, dell'Isola F (2017a) King post truss as a motif for internal structure of (meta) material with controlled elastic properties. Royal Society open science 4(10):171,153

Turco E, Golaszewski M, Giorgio I, D'Annibale F (2017b) Pantographic lattices with non-orthogonal fibres: Experiments and their numerical simulations. Composites Part B: Engineering 118:1–14

Turco E, Misra A, Pawlikowski M, dell'Isola F, Hild F (2018) Enhanced piola–hencky discrete models for pantographic sheets with pivots without deformation energy: Numerics and experiments. International Journal of Solids and Structures

Vangelatos Z, Komvopoulos K, Grigoropoulos C (2018) Vacancies for controlling the behavior of microstructured three-dimensional mechanical metamaterials. Mathematics and Mechanics of Solids p 1081286518810739

Vangelatos Z, Melissinaki V, Farsari M, Komvopoulos K, Grigoropoulos C (2019) Intertwined microlattices greatly enhance the performance of mechanical metamaterials. Mathematics and Mechanics of Solids p 1081286519848041

Veselago V (1967) Properties of materials having simultaneously negative values of the dielectric and magnetic susceptibilities. Soviet Physics Solid State USSR 8:2854–2856

Veselago VG (1968) The electrodynamics of substances with simultaneously negative values of and μ. Soviet physics uspekhi 10(4):509

Wegst UG, Bai H, Saiz E, Tomsia AP, Ritchie RO (2015) Bioinspired structural materials. Nature materials 14(1):23

Zhang S, Yin L, Fang N (2009) Focusing ultrasound with an acoustic metamaterial network. Physical review letters 102(19):194,301